Wetland Techniques

James T. Anderson • Craig A. Davis
Editors

Wetland Techniques

Volume 1: Foundations

Editors
James T. Anderson
Forestry and Natural Resources
 and Environmental Research Center
West Virginia University
Morgantown, WV, USA

Craig A. Davis
Department of Natural Resource Ecology
 and Management
Oklahoma State University
Stillwater, OK, USA

ISBN 978-94-007-6859-8   ISBN 978-94-007-6860-4 (eBook)
DOI 10.1007/978-94-007-6860-4
Springer Dordrecht Heidelberg New York London

Library of Congress Control Number: 2013950172

© Springer Science+Business Media Dordrecht 2013
Chapters 1, 3, 4: © Springer Science+Business Media Dordrecht (outside the USA) 2013
This work is subject to copyright. All rights are reserved by the Publisher, whether the whole or part of the material is concerned, specifically the rights of translation, reprinting, reuse of illustrations, recitation, broadcasting, reproduction on microfilms or in any other physical way, and transmission or information storage and retrieval, electronic adaptation, computer software, or by similar or dissimilar methodology now known or hereafter developed. Exempted from this legal reservation are brief excerpts in connection with reviews or scholarly analysis or material supplied specifically for the purpose of being entered and executed on a computer system, for exclusive use by the purchaser of the work. Duplication of this publication or parts thereof is permitted only under the provisions of the Copyright Law of the Publisher's location, in its current version, and permission for use must always be obtained from Springer. Permissions for use may be obtained through RightsLink at the Copyright Clearance Center. Violations are liable to prosecution under the respective Copyright Law.
The use of general descriptive names, registered names, trademarks, service marks, etc. in this publication does not imply, even in the absence of a specific statement, that such names are exempt from the relevant protective laws and regulations and therefore free for general use.
While the advice and information in this book are believed to be true and accurate at the date of publication, neither the authors nor the editors nor the publisher can accept any legal responsibility for any errors or omissions that may be made. The publisher makes no warranty, express or implied, with respect to the material contained herein.

Printed on acid-free paper

Springer is part of Springer Science+Business Media (www.springer.com)

Preface

Wetlands are generically defined as lentic systems that take on characteristics of both terrestrial and aquatic systems where vegetation capable of growing in shallow water proliferates. However, there are many definitions of wetlands in use around the world, including a number that have ecological and legal significance. Even among these definitions, there are numerous subtle nuances that blur the lines between wetlands and either terrestrial or aquatic systems. Despite the confusion and oftentimes contradictory nature of wetland definitions, wetlands are increasingly being recognized as critical ecosystems throughout the world. In particular, we are seeing an increased awareness about the values and benefits derived from the world's wetlands. As this awareness has grown, we have also seen a greater focus on efforts to better manage, conserve, and protect wetlands. Wetland-related research has been and will continue to be critically important in providing guidance to all the efforts to better manage, conserve, and protect wetlands. In fact, there is a plethora of wetland-related literature available to wetland scientists, regulators, and managers, many of which can be found in at least two journals that are dedicated exclusively to wetlands. However, for most wetland professionals, it may be a daunting task to access much of this literature. Additionally, wetland professionals have not had a book available that covers techniques associated with wetland research, management, and regulation.

The lack of such a book has been a major void in the wetland field. In fact, wetland professionals have discussed for some time the need for a book that focused on wetland research and management techniques. We believe the development of a techniques book for a profession is a sign that the profession, in this case wetland science, is maturing. Scientific progress in a field is often advanced by the development of a techniques book because almost all studies and management actions boil down to choosing appropriate techniques, and a book focused on the topic of wetland techniques will provide fledgling scientists and managers a solid foundation for initiating research and management efforts. We have designed this three volume set for students and professionals interested in wetlands ecology, management, and creation. We are pleased to be a part of the development and progression of our discipline through our involvement with the development of *Wetland Techniques Volume 1: Foundations*, *Volume 2: Organisms*, and *Volume 3: Applications and Management*.

West Virginia University   James T. Anderson
Morgantown, WV, USA
Oklahoma State University   Craig A. Davis
Stillwater, OK, USA

Acknowledgments

Wetland Techniques is our first attempt at a major book project and it was a wonderful learning opportunity as well as an eye-opening experience in regards to all the effort that goes into creating a series of books of this magnitude. We have new-found admiration for all those before us that have successfully tackled book projects for the benefit of science.

We thank the chapter authors for providing freely of their time and expertise. It has been a pleasure working with the authors and we have learned a lot more about wetlands because of them. We thank all of the chapter referees for giving their time and expertise to improve the quality of this three volume *Wetland Techniques* set through constructive reviews that greatly improved the chapters. We especially thank Rachel Hager, undergraduate student in Wildlife and Fisheries Resources at West Virginia University, for all of her help in formatting and verifying literature citations and performing numerous other tasks to improve the book. We also thank Roseanne Kuzmic, research associate in the Natural Resource Ecology and Management Department at Oklahoma State University, for assistance with verifying literature citations.

The following individuals lent their time and expertise to improving these three volumes by serving as expert reviewers and commenting on one or more chapters: Andrew Burgess, Ann Anderson, John Brooks, Crissa Cooey, Diane DeSteven, Adam Duerr, Walter G. Duffy, Andy Dzialowski, Gordon Goldsborough, Mark Gregory, Kim Haag, Patricia Heglund, Wade Hurt, Paul Koenig, James W. LaBaugh, Ted LaGrange, Richard L. Naff, Chris Noble, Aaron Pearse, James Rentch, Wayne Rosing, Stephen Selego, Ken Sheehan, Lora Smith, Gabe Strain, Jered Studinski, Charles H. Theiling, Walter Veselka, Susan Walls, Lisa Webb, and Nicolas Zegre.

Contents

1 **Study Design and Logistics** 1
David A. Haukos

2 **Wetland Bathymetry and Mapping** 49
Marc Los Huertos and Douglas Smith

3 **Assessing and Measuring Wetland Hydrology** 87
Donald O. Rosenberry and Masaki Hayashi

4 **Hydric Soil Identification Techniques** 227
Lenore M. Vasilas and Bruce L. Vasilas

5 **Sampling and Analyzing Wetland Vegetation** 273
Amanda Little

6 **Physical and Chemical Monitoring of Wetland Water** 325
Joseph R. Bidwell

7 **Wetland Biogeochemistry Techniques** 355
Bruce L. Vasilas, Martin Rabenhorst, Jeffry Fuhrmann,
Anastasia Chirnside, and Shreeam Inamdar

Index 443

Contributors

Joseph R. Bidwell Discipline of Environmental Science and Management, School of Environmental and Life Sciences, University of Newcastle, Callaghan, NSW, Australia

Anastasia Chirnside Department of Entomology and Wildlife Ecology, University of Delaware, Newark, DE, USA

Jeffry Fuhrmann Department of Plant and Soil Sciences, University of Delaware, Newark, DE, USA

David A. Haukos Department of Natural Resources Management, Texas Tech University, Lubbock, TX, USA

Kansas Cooperative Fish and Wildlife Research Unit, Division of Biology, Kansas State University, Manhattan, KS, USA

Masaki Hayashi Department of Geoscience, University of Calgary, Calgary, AB, Canada

Marc Los Huertos Science and Environmental Policy Chapman Science Academic Center, California State University Monterey, Seaside, CA, USA

Shreeam Inamdar Department of Plant and Soil Sciences, University of Delaware, Newark, DE, USA

Amanda Little Department of Biology, University of Wisconsin-Stout, Menomonie, WI, USA

Martin Rabenhorst Department of Environmental Science and Technology, University of Maryland, College Park, MD, USA

Donald O. Rosenberry U.S. Geological Survey, Denver Federal Center, Lakewood, CO, USA

Douglas Smith Science and Environmental Policy Chapman Science Academic Center, California State University Monterey, Seaside, CA, USA

Bruce L. Vasilas Department of Plant and Soil Sciences, University of Delaware, Newark, DE, USA

Lenore M. Vasilas Soil Survey Division, U.S. Department of Agriculture-Natural Resources Conservation Service, Beltsville, MD, USA

Chapter 1
Study Design and Logistics

David A. Haukos

Abstract Reliable knowledge is critical for management and conservation of wetlands. Essential to the scientific method and achieving reliable knowledge is study design. The primary purpose of study design is the collection of data in an unbiased and precise manner for an accurate representation of a population. Proper study design includes formulation of study questions and objectives, hypotheses to explain an observed pattern or process, conceptual models, appropriate methodology, and a data management plan. Inference of study results and conclusions can be explicitly bounded by defining an appropriate target population. Deductive, Inductive, and Retroductive reasoning are used to infer study results to target populations. Development of multiple competing hypotheses capable of being tested is at the core of the hypothetico-deductive approach that maximizes potential knowledge from a study. Selection of independent and dependent variables to test hypotheses should be done with cost, efficiency, and understanding of the wetland system being studied. Study type (e.g., experimental, observational, and assessment) influences the certainty of results. Randomization and replication are the foundation of any study type. In wetlands, impact studies (e.g., BACI [before-after/control-impact] design) are common and usually follow unforeseen events (e.g., hurricanes, wild fire, floods). Sampling design is dictated by study objectives, target population, and defined study area. A robust sampling effort is essential for accurate data. Reduction in statistical and mechanical errors and data management protocols are overlooked features of study design. In addition to statistical tests, estimation of the magnitude (i.e., effect size) of an effect is crucial to interpretation of study results. When judging the merits of results from a study, investigators should independently assess the hypothesis,

D.A. Haukos (✉)
Department of Natural Resources Management, Texas Tech University,
Lubbock, TX 79409, USA

Kansas Cooperative Fish and Wildlife Research Unit, Division of Biology,
Kansas State University, Manhattan, KS 66506, USA
e-mail: dhaukos@ksu.edu

methodology, study design, statistical approach, and conclusions without regard to how they would have conducted the study. Doing so will facilitate the scientific process.

1.1 Introduction

Unfortunately, the history of wetland science is relatively brief. In the United States, most scientific effort prior to the early 1970s was devoted to justifying draining and filling of wetlands. Information from such study results contributed to the greater than 50 % decline of wetlands in the conterminous United States since European settlement (Lewis 2001; Dahl 2011). It is unlikely that any other ecosystem has suffered such organized willful efforts of alteration, destruction, and obliteration based primarily on misinformation and spurious "facts" than wetlands. Slowly during the past century, acceptance of wetlands as critical components of the natural world has resulted in a multitude of conservation and education efforts to protect wetland ecosystems. One of the rare historical exceptions was the creation of National Wildlife Refuges to protect wetlands vital to migratory birds, primarily waterfowl because of their value to hunters. Since the passage of the Clean Water Act and other legislation since the early 1970s, the ecological values of wetlands have increasingly been recognized by conservation organizations, policy makers, governmental agencies, and society at large. The foundation for these changes in societal values and policies from those factions advocating wetland destruction to a predominance of activities proposed for restoration, enhancement, and protection of wetlands is reliable knowledge of the ecological structure, function, and provision of services by these systems.

Reliable knowledge is the result of accumulation of credible results from wetland investigations conducted using a logical framework – **study design** (Table 1.1). Study design involves more than **experimental design**, which can be defined as a plan for assigning experimental conditions to subjects and the statistical analyses appropriate for the plan (Kirk 1982). Study design is also more than **statistics**, which is a body of knowledge that allows one to make sense of collected data and generalize results from a sample to a population. Both experimental design and statistics are beyond the scope of this chapter; there are a considerable number of available texts on both subjects (e.g., Quinn and Keough 2002; Box et al. 2005; Montgomery 2012). Proper use of study design allows for the development of research goals, objectives, and hypotheses based on observations, previous studies, and ecological theory. Study design includes a declaration of variables to be measured, techniques to be applied, and approaches to analyses of collected data. Furthermore, use of the appropriate study design allows for inference beyond the immediate subjects being studied. Most importantly, this framework allows for acceptance of study results into the overall knowledge of wetland ecosystems for use in conservation efforts, generation of additional research questions, and accumulation of defensible reliable knowledge regarding wetland ecosystems.

1 Study Design and Logistics

Table 1.1 Glossary of common terms used in study design

Accurate	Having low bias and variance; where resulting estimates are repeatable and close to the true value of a population
Alternative hypotheses	Alternative explanations for an observed pattern or process that are usually represented in competing statistical or predictive models
Bias	Difference between long-term average of a sample estimate from true population value
Conceptual models	Abstractions of reality based on observation of an ecological pattern or process envisioned by an investigator but not typically formalized graphically or mathematically
Control	Group of experimental units for which the factor of interest is excluded or otherwise accounted for in study design
Descriptive inference	Using observations from a study to learn about or predict other unobserved facts
Deterministic	Completely predictable, not involving any random components
Effect size	Magnitude of a measurable effect due to a treatment of interest
Empirical models	Models in which data are used to estimate parameters or test predictions
Experiment	A process that imposes a **treatment** on a group of elements or subjects (experimental units) to measure a response and quantify an effect
Experimental design	A plan for assigning experimental conditions to subjects and experimental units
Experimental error	The inherent variation among experimental units treated alike or variation not explained by treatments or other variables
Experimental unit	Subjects (elements) to which individual experimental treatments are applied
Fixed effect	A variable in which levels are not subject to random variation under repetition of the experiment
Fundamental objective	What a decision maker wants to accomplish
Hypothesis	Specific statement of reality that is frequently testable by comparing predictions to data
Independence	Organisms, samples, experimental units, or other objects that can be represented by a statistical distribution one at a time, without dependence on the values of other objects
Independent variables	Those variables hypothesized (including treatments) to contribute to variation in **dependent** or **response variables**, the values of which depend on levels or types of independent variables
Means objective	Intermediate objectives that must be accomplished to achieve or address a fundamental objective
Metapopulation	When the target population of interest is subdivided into discrete patches across a landscape but movement among patches continues
Model	An abstraction or perceived representation of nature
Objectives	Statements of desired achievements by investigators and decision makers
Random effect	Where repetition of the experiment will result in different levels within the analyses unless the same experimental units are used
Randomization	Assignment of treatments to or section of experimental or sampling units at random
Replication	Assignment or selection of multiple experimental units to an individual treatment

(continued)

Table 1.1 (continued)

Sampling	Selection of a subset of potential experimental units from the target population for measurement of variables of interest
Sampling error	The variation among samples (or observations) of a given experimental unit
Sampled population	Population from which samples are taken
Statistical inference	The process of drawing sound and appropriate conclusions from data subject to random variation
Target population	Population for which inference can be made
Theory	A broad, general conjecture about a process that can be tested using study design
Treatment	Something that an investigator imposes on experimental units in some deliberate manner
Unbiased	Long-term average of sample estimates equals population value

A principal goal of study design is to minimize personal bias, values, beliefs, and subjectivity of the scientist so that conclusions can be supported beyond a reasonable doubt. Basic tenets of modern study design are grounded in statistical theory and have been applied for >75 years (Fisher 1935). Exponential increases in computational ability during the past 25 years have allowed for increasingly complex approaches to study design and data analyses. However, failure to adhere to basic components of study design cannot be overcome even with the most complex analytical tools. A well-designed study will allow investigators to focus on current knowledge gaps, provide rigorous tests of information, and enhance efficient use of resources.

1.2 Role of Study Design in the Scientific Method

Study design is a critical part of the scientific method (Fig. 1.1). There are several variations of the scientific method for studying ecological systems, but all include steps of (1) construction of question(s) that address uncertainties in the ecosystem of interest, (2) formation of theories to explain observations or questions based on observation, which leads to multiple hypotheses that have predictions suitable for testing data, (3) design of a study to test primary and alternative hypotheses and their associated predictions, (4) collection and analyses of data, (5) report conclusions and make inference from results, and (6) communicate results through the peer-review process that adds credibility to the findings (Gauch 2003). Typically, the scientific method is referred to as a process because all studies and resultant conclusions are subject to being repeated, typically as a feed-back loop restarting with step (2) by other scientists. However, it would be a mistake for anyone to perceive the scientific method as a defined sequence of steps leading to knowledge, but rather as a framework for creative and productive processes that can be used to accumulate knowledge that leads to truths corresponding with reality of

1 Study Design and Logistics

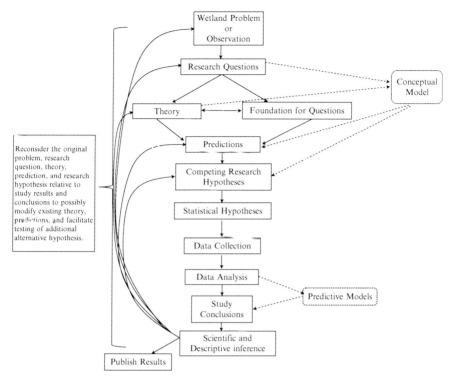

Fig. 1.1 The process of the scientific method to achieve reliable ecological knowledge (Modified from Brown and Guy (2007). Published with kind permission of © American Fisheries Society 2007. All Rights Reserved)

natural systems. Because science is a process, knowledge of natural systems, including wetlands, evolves over time as previous conclusions are subjected to further study, and occasionally rejected due to lack of continued empirical support.

Study design expands on the scientific method and has been presented in an outline or checklist format (Morrison et al. 2001). Although there are multiple texts with step-by-step guides for developing study designs (e.g., Cochran 1983; Cook and Stubbendieck 1986; Martin and Bateson 1993; Lehner 1996), Morrison et al. (2001) characterized the process of study design by 12 discrete steps – (1) question development relative to a natural system, (2) development of >1 hypothesis that may answer the question of interest (testing of competing hypothesis) and associated predictions should the hypothesis be true, (3) determination of a conceptual study design, (4) selection of independent and dependent variables, (5) choose appropriate methods to measure variables of interest, (6) establish acceptable level of precision and accuracy, (7) pilot study or preliminary data collection to test limitations of original design and estimate sample size requirements, (8) establish quality/quantity assurance protocols, (9) conduct data collection, (10) perform data analyses and partition evidence in support of competing hypotheses, (11) interpret and provide context of results and make appropriate inference of results and conclusions, and (12) publication of results in a peer-reviewed outlet. Individual investigators can insert the details specific to their study within these broad steps to initially design a study.

Without study design, the scientific method would be ineffective as a means to attain reliable knowledge. Investigators must be aware that study design, as well as experimental design, statistical analyses, and data modeling, should not be considered a "cookbook" process devoid of critical thought. Once a wetland-related question has been formulated and competing hypotheses developed, there are usually a number of different approaches to appropriately design a study that can be used to accumulate evidence to test one (or more) hypotheses. Furthermore, development of a study design must be considered in the context of accessibility of study sites, equipment and labor costs, time to collect samples, other time-sensitive constraints (e.g., lab availability, sample storage, and occurrence of measurable dependent variables), and a multitude of other potential considerations. In addition, it is important to realize that it common for changes in the design to occur after a study has been implemented. Hopefully, if one is unfamiliar with the wetland type or study question, limitations of or changes to a study design can be addressed using prior knowledge (e.g., experience of the investigator, literature or previously collected data). Without such insight, the initial study effort is typically defined by determining the limitations of the proposed design and, at times, resulting in a potentially unreliable data set. As these issues may compromise acquisition of reliable knowledge, it is recommended that any proposed study design be reviewed by a biometrician or statistician familiar with the inherent quantitative hurdles associated with natural systems and an investigator familiar with the wetland type or issue proposed for study.

1.3 Development of a Study Hypothesis

Reliance on ecological theory and generation of hypotheses are basic to study design. A **theory** as defined for the purpose of study design is a "broad, general conjecture about a process" (Romesburg 1981: 295) or "a conceptual framework through which the world is observed and facts about the world are discerned" (Williams 1997: 1009). A **hypothesis** is a proposed explanation for a process or phenomenon that creates data. Once an intangible theory has been developed relative to an observed phenomenon, hypotheses can be formulated to describe the natural processes that produced the phenomenon that lead to predictions of outcomes should the hypotheses be true. Development of hypotheses is the key step for a successful study design. Without sound, well-developed, hypotheses, it is difficult to conduct an investigation that will result in clear conclusions and advance our understanding of ecological systems.

Hypotheses follow the question being asked and must be explicitly stated so as to provide predictive power or specific conditions under which the hypothesis is true.

1 Study Design and Logistics

Typically, development of a study question and associated hypotheses is considered a basic and simple task that is hurriedly stated prior to designing a study. In practice, a successful study depends on a thoroughly developed question and sound, conceptual hypotheses that can be represented by competing models and tested with data (e.g., Anderson 2008). However, lacking competing quantitative models, hypotheses can be formulated as existential statements, which include an expression that existence of a phenomenon has identifiable characteristics and causal explanations that exist for each occurrence of the phenomenon. All subsequent steps in a study must refer back to the question and its associated hypotheses, which require prioritization and agreement by everyone involved in the study (including funding sources). Questions can be developed from personal experience, expert opinion, literature, intuition, and guesswork but are usually driven by stated goals.

For complex natural resource issues, Structured Decision Making and Adaptive Resource Management have recently been adopted by many governmental agencies; the approach is focused on developing and prioritizing questions related to a natural resource issue (e.g., Martin et al. 2009). Questions or study objectives under this approach are categorized as fundamental (explicit declarations of core concerns or questions), means (typically methodological and represent an intermediary step in reaching the fundamental objective), process (ground rules for decision processes related to the study), and strategic (fundamental to a broader set of decisions than the one in question) objectives and require considerable effort to distinguish among these (e.g., Keeney 2007; Williams 2012). For example, there may be a controversial issue relative to mitigation of a wetland that is slated to be lost to development. There are likely a number of stakeholders (i.e., developers, local government, natural resources agencies, non-governmental organizations) that have competing views on the type, location, and magnitude of mitigation. Use of the Structured Decision Making Approach facilitates the decision-making process by involving all stakeholders in (1) identifying the problem to be addressed, (2) specifying objectives and tradeoffs, (3) identifying the range of potential decisions, (4) specifying assumptions about resource structures and functions, (5) projecting the consequences of alternative actions, and (6) identifying key uncertainties among other steps (Williams and Brown 2012). This approach explicitly addresses decision uncertainty, which is the typical roadblock preventing sound decisions based on scientific knowledge. One should recognize that uncertainty represents incomplete knowledge, not doubt, when addressing specific wetland issues. It is more desirable to design studies to address fundamental objectives (questions) rather than the other objective types. No single study can address every question, and, if attempted, usually results in mediocrity and an inefficient use of resources.

There are several types of reasoning that can be used to obtain knowledge through generation of theories and hypotheses (Morrison et al. 2001). One can use **induction** to create general conclusions based on a collection of individual facts (i.e., conclusions are drawn based on an association between individual facts); frequently an extrapolation of results from a study to a general situation. As an example of inductive reasoning, an investigator has concluded that the development of terrestrial buffer strips of 100 m in width filters 90 % of suspended material present in precipitation run-off; therefore, creating buffer strips around wetlands will minimize or eliminate the potential of a wetland filling due to accumulated terrestrial sediments. **Deduction** is reasoning from the general (premise) to a specific event and includes the development of testable explicit predictions under a specific hypothesis to explain observations from a natural system. An example of deductive reasoning would be if avian body mass varied in relation to environmental conditions, then an investigator may predict that average body mass would decline as temperature declines or number of days below freezing increases. The use of **retroduction** involves the relatively subjective attribution of an underlying cause to an observed pattern and is a common occurrence in the discussion section of scientific papers. Conclusions based on retroduction should be considered hypotheses that require additional testing. Most conclusions from scientific studies are the result of retroduction and induction, which Romesburg (1981) lamented does not result in gaining reliable knowledge as one would when using deduction. This **hypothetico-deductive** approach was advocated by Romesburg (1981) as the preferred approach for study of natural systems. Guthery (2008) described the approach as the classical method to test a hypothesis by deducing events or relationships that should be observed under experimentation if the hypothesis was true. Specific, sound hypotheses result in efficient, organized, and goal-oriented studies that minimize uncertainty in results.

Most commonly, hypotheses are stated in terms of treatment effects. Historically, treatment effects were based on potential outcomes of statistical tests. Classical statistical, "null" hypotheses are usually depicted in shorthand as H_o: and generally specify that "no difference" exists among treatments. Whereas the "alternative" hypothesis H_1: can be more specific and takes several forms related to the stated hypotheses but usually is a statement that a difference exists among treatments. Alternative hypotheses may also define a magnitude and direction of a difference. It is important to remember that studies can reject or disprove a null hypothesis, but a hypothesis cannot be considered proven when a null hypothesis is not rejected. For example, a wetland manager can hypothesize that cattails (*Typha* spp.) will be completely eradicated from a wetland due to herbicide treatment, but the presence of a single cattail would cause the hypothesis to be rejected. For the hypothesis to be "accepted," every plant would have to be identified to prove the null hypothesis and it is impossible to be completely certain that all plants have been correctly identified. Therefore, basic to scientific endeavor is the use of study design to formalize the effort to disprove hypotheses rather than prove them (Peirce 1958).

Recently, use of the null statistical hypotheses has been dismissed as uninformative and nonproductive in scientific endeavors because use of a statistical test that only declares whether a difference exists or not exists between treatment means is relatively uninformative (e.g., Johnson 1999; Cherry 1999; Anderson 2008; Guthery 2008). Instead, multiple alternative hypotheses should be developed that are specific and depict cause and effect, measurable predictions, or explicit outcomes that can be tested using data (Platt 1964; Romesburg 1981; Morrison et al. 2001). In a study design, this can be relative to study objectives as long as one formalizes the question that is being addressed. Popper (1959) and Platt (1964) indicated that science progresses best when hypotheses of natural systems are evaluated empirically by comparing to predicted results to reject hypotheses that are inconsistent with predictions. Further, Anderson (2008) advocated that all plausible alternative hypotheses should be translatable into mathematical models that are subjected to empirical methods to test the relative strength of the evidence for each hypothesis. Chamberlain (1890) urged scientists to conduct studies using the strategy of "multiple working hypotheses". That is, study design should be capable of simultaneously testing multiple plausible hypotheses, eliminating poor hypotheses, and quantifying the relative strength of one hypothesis over the alternatives (Royall 1997). Ultimately, conclusive evidence for a hypothesis (i.e., science answer) can only be possible after all other hypotheses are rejected through study design or additional studies (Williams 1997). Development of hypotheses is a time consuming, challenging process that is critical to overall rigor of a study. Sound hypotheses are based on (1) familiarity of the system being studied, (2) detailed formulation of the question or observation being studied, and (3) working knowledge of the established literature related to the subject being studied.

In wetland science, development of hypotheses is usually study specific. Hypotheses can range among general statements that explain an observation, specific measureable predictions should a hypothesis be true, directionality of a treatment effect, or support for an ecological theory. Turner (1997) proposed four hypotheses to explain the observation of a high rate of coastal wetland loss in the northern Gulf of Mexico (-0.86 %/year): (1) an extensive dredge canal and spoil bank network; (2) decline in sediments in the Mississippi River during the 1950s; (3) Mississippi River navigation and flood protection levees; and (4) salinity changes. The hypotheses were developed following extensive consideration of all potential factors influencing wetland loss and familiarity with the wetland system being studied. A study was designed to address predictions from each hypothesis. Turner (1997) concluded that, based on his study, dredging man-made channels and forming dredge spoil banks had the greatest impact on wetland hydrology and had the most influence in explaining wetland loss.

Testing ecological theory among ecosystems requires testable predictions for each competing hypotheses. Megonigal et al. (1997) tested two competing hypotheses under the subsidy-stress hypothesis for rate of aboveground net primary production in southeastern floodplain forest. Under the subsidy-stress hypothesis, they hypothesized that periodically flooded forests have higher rates of net primary productivity than upland or continuously flooded forests. As a competing hypothesis, they proposed that effects of periodic inputs of nutrients and water on net primary productivity are diminished or offset by stresses associated with anaerobic soils or drought. Using an experimental field study design, they measured aboveground net primary productivity under three categories of mean growing-season water depth. Megonigal et al. (1997) concluded that extensive flooding caused significant stress on forest productivity, but there was insufficient support for the subsidy-stress hypothesis in the description of patterns of net primary productivity in flooded forests and suggested testing a more complex interaction between subsidy and stress factors.

Use of predictive hypotheses led Collins and Storfer (2003) to categorize six hypotheses potentially explaining global amphibian declines into two classes. Class I hypotheses were those in which underlying ecological mechanisms affecting amphibians were well known, but the relative magnitude of the effects was uncertain; these were presence of alien species, over-exploitation, and land use change. For Class II hypotheses, there was a poor, but increasing understanding of the relative effects on amphibian populations; these were global changes in UV radiation and climate, contaminants, and emerging infectious diseases. They concluded that additional research using integrated approaches was necessary to understanding all of the complex interacting predictions of the hypotheses.

One can retrospectively test competing hypotheses by compiling results from previous studies. van der Valk (2012) reviewed theories and multiple hypotheses relative to invasive plant species in wetland systems. There are two principal theories for why so many invasive plants are found in wetlands: (1) wetlands are more vulnerable to invasion because they are landscape sinks and susceptible to disturbance and (2) invasive species are superior competitors. Numerous hypotheses and associated predictions have been advanced within the two theories (e.g., enemy release, hybrid vigor, empty niche). Following a review of the evidence, van der Valk (2012) concluded that while there is some support for the superior competitor theory, hypotheses based on landscape sink/disturbance theory had the most support for explaining the presence of invasive plant species in wetlands.

The key to development of a study design is the ability to conclusively reject a hypothesis (or several hypotheses) such that the scientific process can progress. To test the efficient-community hypothesis (all plant species that can become established and survive under the environmental conditions found at a site will eventually be found growing there and/or will be in its seed bank) for restored wetlands, Galatowitsch and van der Valk (1996) compared the floristic composition of natural and restored wetlands in northern Iowa. Although a few similarities were found between natural and restored wetlands, they rejected the efficient-community hypothesis with a conclusion that dispersal ability of plants had a greater influence on recolonization of plants in restored wetlands than site-specific presence.

In wetland science, theories and hypotheses are not restricted to ecological concepts. For example from an economic perspective, Whitten and Bennett (2005: 45) proposed a theoretical concept that "the production of wetland protection outputs is unlikely to be at the level desired by the community," which essentially means that society values wetlands at a level greater than that being provided by conservation efforts. They formulated two basic hypotheses that could be tested by an appropriate study: (1) "an increase in the production of wetland protection outputs would generate a net benefit to the community" and (2) "policies in alternative to those currently in place would reduce the extent of market or government failure in the protection of wetland production outputs" (Whitten and Bennett 2005: 46–47). From an archaeological viewpoint, Kelly and Thomas (2012) outlined competing hypotheses for human presence in the Carson Desert,

Nevada, USA, based on use of a wetland. If humans associated with the wetland were sedentary (i.e., lived year-round in same place), then the human archaeological evidence should be concentrated in and around the wetland with little evidence, except for hunting parties, in surrounding mountains. If humans were nomadic and they used the wetland as a stop-over point, then one would expect to find transient evidence at the wetland and extensive evidence in surrounding mountains as people roamed throughout the region. They then proposed a study, using proper design to test the hypotheses; the final conclusion was that both hypotheses lacked support and thus, additional hypotheses were generated based on the information generated during the study.

While development and testing of competing hypotheses should be the goal of wetland investigations, it should be noted that descriptive research (i.e., natural history), long-term monitoring of ecological systems and their components, estimation of magnitudes of effects, and documentation of changes in status of ecological systems are valid and informative provided that the methodology is not flawed. Results from these types of studies can be used to generate hypotheses for additional testing, document historical or baseline conditions for future comparison, provide input for policy and economic decision makers, and document ecological conditions and responses for future use. Indeed, these types of studies are quite common, but one must realize that conclusions based on these efforts can be considered premature pending a rigorous test of competing hypotheses intended to explain observed patterns, trends, and relationships.

1.4 Study Population

After hypotheses or objectives are explicitly stated, the focus of study design shifts to the process of devising a study to test the hypotheses. Inherent to proper study design is the identification of a **population** to define the biological entity of interest and placement of bounds on the scope of the experiment. A **biological population** is defined as a "group of organisms of the same species occupying a particular space at a particular time" (Krebs 1985: 157). A **metapopulation** is formed when the population of interest is subdivided into discrete patches across a landscape but movement among patches remains (Levins 1969). Often, in wetland science, one is also interested in **communities**, which is "any assemblage of populations of living organisms in a prescribed area" (Krebs 1985: 435).

However, from a statistical and study design perspective, population has a broader meaning than just an organismal definition. It is the total set of elements or membership of a defined class of organisms, objects, or events. For wetland studies, the population of interest may consist of the wetland type, ecological condition of wetlands, organisms depending on the wetland, or a variety of other elements of the system. The statistical or target population is the foundation of study design and subsequent application of results. The **target population** is statistically (i.e., has measurable parameters with true, but usually unknown, values or distributions that can be estimated) and biologically defined, occupies specific units of time and space, contains measurable attributes, and represents the collection of subjects such as individuals, habitats, communities, or natural systems available to be studied in which results and conclusions would be applicable. Frequently, not every element of the target population is available to be selected for study. Those elements accessible for study form the **sampled population**. For example, if one is conducting aerial waterfowl surveys of wetlands in a defined area, then the target population would be waterfowl on all wetlands in the area; however, if flights are restricted for some reason (e.g., military operations, wind turbines, powerlines), then the sampled population would only be the wetlands available to be surveyed. Under proper study design, one could assume that the sampled population was representative of the target population.

Kentula et al. (1992) indicated that the process of defining the population of wetland elements within a study influences the techniques used in the study, timing of the study, and most aspects of data collection. They highlighted the need to define and record characteristics of the target population early in the planning process. Knowledge of the target element (e.g., wetland, biota) is critical to setting boundaries of the target population. When setting the spatial boundaries of target population, Kentula et al. (1992) emphasized the need that boundaries should be set to include similar hydrologic, climatic, geologic, and other relevant geographic conditions that influence the ecology of wetlands of interest. For example, one would have a much different target population of wetlands if the element of interest is waterfowl compared to amphibians.

All studies have the implicit goal of making **descriptive** or **explanatory inferences** based on empirical data about the natural world (Platt 1964). Both of these inferences have the primary role in study design of allowing the researcher to infer results beyond the immediate data to something broader that is not directly observed (King et al. 1994). Descriptive inference is using observations from a study to learn about other unobserved facts. Statistical inference is the process of drawing sound and appropriate conclusions from data subject to random variation. This includes a formal understanding of the limitations of application of the inference in time and space. That is, one cannot explicitly apply the knowledge to any population outside of the study population or under conditions not experienced during the study. Such limitations must be described during any presentation (verbal or written) of study results.

1.5 Determination of Experimental Unit Variables

Once the target population and all elements within the target population have been defined and identified, then the process of selection of elements for study and variables to measure is initiated. Subjects (elements) to which individual experimental treatments are applied are termed **experimental units**. Experimental units can be individual animals, defined populations of animals, unique ecosystems or

1   Study Design and Logistics

habitats (e.g., isolated wetlands, islands), subdivided units within larger ecosystems or habitats (e.g., management units of a contiguous system, pastures, watersheds), or some measure of time.

Experimental units can be natural features (e.g., wetland, bird, plant) or man-made (e.g., mesocosm, microcosm, greenhouse flat). Balcombe et al. (2005) tested hypotheses that invertebrate family richness, diversity, density, and biomass were similar between mitigation and reference wetlands. As experimental units, they selected 11 mitigation and four reference wetlands across three physiographic regions of West Virginia. Maurer and Zedler (2002) tested hypotheses contributing to the invasion of reed canary grass (*Phalaris arundinacea*) using a parent plant transplanted into a cone-tainer and attached to aluminum troughs to measure tiller growth over time in response to shade and nutrient treatments; each cone-tainer was an experimental unit.

All study designs involve identification, measurements, or estimation of variables considered to affect the hypothesis being tested. There are several classes of variables to consider during study design. Basic to statistical models are independent and dependent variables. **Independent variables** are those hypothesized (including treatments) to contribute to variation in **dependent** or **response variables**, the values of which depend on levels or types of independent variables. In most study designs for wetlands, there is one dependent variable of interest; for example, density of waterbirds, species richness of invertebrates, levels of nutrients in water runoff, and soil moisture. However, there can be several associated independent variables that may be categorical or continuous variables that are hypothesized to influence the variance of the measured dependent variables; for example, wetland type, wetland area, watershed condition, vertebrate sex and age, and time.

Analyses related to a single dependent variable are termed **univariate**, and there is a long history of established methods to test hypotheses involving a single dependent variable for both discrete (i.e., categorical [e.g., chi-square analyses] or factor variables [e.g., analysis of variance]) and continuous (e.g., regression) independent variables. However, simultaneous analyses of greater than one dependent variable are often of interest and use of **multivariate** statistics (e.g., ordination, principal components analysis, path analysis) has greatly increased during the past three decades with advances in computing power necessary to conduct these analyses. Regardless of the approach, the focus of an established study design is to quantify the relationships among dependent and independent variables through some form of data analysis. The goals of data analysis include evaluation of hypotheses, predicting or forecasting an event or response, development of structure of future models, determination of important variables relative to variation of the dependent variable(s), and detection or describing patterns and trends.

Each independent and dependent variable needs to be determined as a fixed or random effect prior to determining the appropriate study design and analyses. A **fixed effect** is a variable in which levels are not subject to random variation under repetition of the experiment (e.g., wetland type, animal age and sex, levels of nutrients applied, number of seedlings planted). A **random effect** is one where repetition of the experiment will result in different levels within the analyses (e.g., time [day, month, year] or environmental condition, wetland area, water depth), unless the same experimental units are used. Statistical inference only can be applied to the target population under the actual (fixed) treatment levels that are within the range of random variables being tested. When a study includes both fixed and random variables, it is considered a mixture of effects and requires some additional consideration during analyses (i.e., mixed models). It is important to define variables as fixed or random when describing the methods used in the study. When more than one independent variable is being assessed in a study, the interaction between effects is of great interest. A significant interaction indicates that the magnitude of differences between levels of one effect depends on the level of the other effect. Many times the interaction between effects is more interesting than individual main effects in explaining data observations and results, albeit this is frequently considered more cumbersome to explain than results for simple main effects. However, an investigator must use their knowledge of the system to ensure that significant interactions have biological meaning and are not a spurious result. Spurious (an apparent relationship between noncausal events or variables) results typically result from the presence of a confounding or nuisance variable. At times, further investigation of interactions is necessary to develop confidence that the interaction is meaningful.

In addition to the proper identification of dependent and independent variables, there are many other types of variables that can impact results and should be considered during development of a study design. It is important to categorize all variables that contribute to the variation of dependent variables into those that are of interest and related to the hypotheses being tested and those that are **nuisance** variables, which are assumed to be of little interest but may affect study results. Extraneous or nuisance variables can have disproportionate impacts on results from a study unless accounted for in the study design. Indeed, the failure to control for nuisance variables frequently results in spurious conclusions. Through study design, nuisance variables can be controlled to account for the potential bias associated with the variable. Examples of methods of controlling nuisance variables in study designs using analysis of variance include grouping experimental subjects into **blocks** (use of some common characteristic to group homogeneous subunits of the sampled population) or use of a **covariate** (a random variable of little interest associated with but varies among experimental units) if the variable is categorical or continuous, respectively.

There are a variety of approaches to account for nuisance variables in a study design. For example, in a study of avian response to prescribed fire in wetlands, Brennen et al. (2005) acknowledged that migration timing (changing avian densities over time) and wetland size (species-area relationship) could influence results, yet these variables were not of primary interest in assessing the effect of spring burning of wetlands. Furthermore, the investigators recognized that conducting a wetland study over a large geographic region could be influenced by varying environmental conditions (e.g., differing precipitation patterns) across the target population of wetland. Therefore, because the primary interest was in avian response to a burning treatment, they paired adjacent burned and unburned (control) wetlands across the

1 Study Design and Logistics

geographic range of the study to remove the nuisance effect of varying environmental conditions to ensure that the effect of burning was determined.

Although it is important to control for as many variables as possible in a study design, there is a limit to the number of controllable or measurable variables for most ecological studies. This is true for both field and laboratory studies. Frequently, it is not possible to identify or recognize all of the potential variables affecting a dependent variable. At times, it is difficult, economically unfeasible, or impossible to measure certain variables. Therefore, it is often appropriate to identify and measure proxy (i.e., correlated) variables that can serve as an index to the variable of interest. Finally, there is a statistical limit to the number of variables that can be addressed through study design as well, primarily because sample size relative to the number of variables dictates the potential analyses. Such uncontrollable variables are typically assumed to be random with the same effect across all samples and controlled variables and thus accounted for in the appropriate experimental design.

Proxy variables can take many forms and do not have to be directly related to the variable of interest. For example, Mackay et al. (2007) measured soil moisture, which affects plant root water availability, to use as a proxy variable for detecting water stress. Pfeiffer (2007) used wetland and other surface water area as a proxy variable for the location of spatially clustered wild bird infection in a study of the influence of wild birds and risk from H5N1 highly-pathogenic avian influenza. In a review study, Elser et al. (2007) identified several variables that were correlated (i.e., proxy) with standing biomass of autotrophs including chlorophyll concentration, ash-free dry mass, carbon mass, biovolume, percent cover, and primary production. They then used a meta-analysis combining results from nitrogen and phosphorus enrichment studies measuring standing biomass and proxies to evaluate nutrient limitation in freshwater, marine, and terrestrial ecosystems. Kantrud and Newton (1996) used the amount of cropland in a wetland watershed as a proxy for their quality. Use of proxy variables is common in wetland studies due to the difficultly in measurements of many ecological characteristics; however, one must be somewhat reserved in stating conclusions using proxy variables unless certainty exists relative to the strength of the relationship between a proxy variable and the variable of interest.

1.6 Conceptual Model and Variable Selection

A common approach to determine which variables to measure for testing competing hypotheses is to transform a conceptual hypothesis to a conceptual model that describes the response of a dependent variable to a set of independent variables. The conceptual model forms the basis of an appropriate statistical model with defined components. The statistical model can be formulated as Y (dependent variable) being a function (F) of some fixed and random independent variables or factors. For example, if one is measuring the total nitrogen load in a wetland, Y could be defined as the amount of total nitrogen in ppt and independent variables of interest could be watershed type (WS), month (MO), precipitation measure (PR), water depth (WD), and vegetation cover (VEG). Such a model could initially be defined conceptually as

$$Y = F[\text{WS}, \text{MO}, \text{PR}, \text{WD}, \text{VEG}].$$

Such a model is considered indeterminate at this stage because the list of explanatory factors is likely not complete and the response is not predictive (i.e., nature of the relationship of each independent variable with the response variable is not defined). However, such a model is useful in designing a study. Fixed effects (if control of experimental units is possible) could be watershed type, water depth, and vegetation cover. Random effects could be month and precipitation. However, final determination of random and fixed effects depends on the amount of control an investigator has on the system and variables used to measure the effects.

An important principle in conducting experiments is to hold all factors, except the one of interest, constant so that any response to treatment can reliably be attributed to the treatment. Unfortunately, this is rarely possible in wetland studies and thus, investigators must design studies to limit the variation within all variables that are not of interest. In this example, one may be only interested in the relationship between watershed type and total nitrogen while recognizing that factors other than watershed type influence nitrogen levels in wetlands. The proper study design can remove or partition the influence of independent variables in such a way that priority treatment effects can be estimated and evaluated. Ideally, the model will be constructed based on the hypothesis, available data, assumptions, and potential unknown parameters.

Conceptual models can be specific, similar to the above example, or encompass an entire wetland type (Fig. 1.1). A critical step in constructing a conceptual model is identification of the problem or question. Brooks et al. (2005) used existing data to develop a conceptual model of wetland degradation and restoration in an effort to improve scenarios for the use of mitigation wetlands to replace lost wetland area and ecological function. They hypothesized that increasing influence of stressors homogenizes wetland diversity and variability. Devito and Hill (1998) developed a conceptual model of wetland sulfate (SO_4) retention and export based on watershed hydrogeology. An investigator should use conceptual models to develop objectives and competing hypotheses for experimental testing. Ogden (2005) developed a conceptual ecological model for anthropogenic stressors on an Everglades ridge and slough system. He identified five major ecosystem stressors (reduced spatial extent, degraded water quality, reduced water storage capacity, compartmentalization, and exotic species) and made predictions on stressor effects. In addition, he identified a series of biological indicators of wetland restoration success that can be incorporated into future studies. In addition, conceptual models can be informed by or produce predictions from ecological theory. Euliss et al. (2004) developed a conceptual model for prairie pothole wetlands – the wetland continuum. The model allows for simultaneous consideration of climate and hydrologic setting on wetland treatment. Because successive measurements are relative to and correlated with the initial measurement, distinctive statistical analyses are necessary to account for the correlation among successive measurements of experimental units due to violation of the assumption for most analyses of independence among experimental units.

The use of randomization and replication in a study design determines the target population and extent of inference from conclusions. These tools also maintain study integrity and scientific objectivity (Morrison et al. 2001). Study designs can be distinguished by the rules used to govern randomization and replication. In addition, the application and extent of randomization and replication are critical factors in judging the reliability of conclusions from studies. **Randomization**, according to Fisher (1935), is at the heart of experimental design. Randomization refers to both (1) random selection of representative study units for sampling and (2) random assignment of treatments to experimental units in an experiment. Most traditional experimental designs are based solely on the rules for randomization (Kirk 1982). The extent of inference is directly related to the degree of random sampling, which ensures that the study units are representative of the target population. Failure to randomly assign treatments to experimental units increases the potential impacts of nuisance variables leading to spurious results and potential bias. Underlying statistical theory demonstrates that randomization ensures that estimates of treatment effects and experimental error are unbiased estimates of their respective population parameters. Random assignment of treatments to experimental units is necessary to satisfy the assumption that experimental errors are independent by minimizing the effects of correlation between experimental units on statistical results. Cox (1980: 313) summarized the importance of randomization to studies, in that "randomization provides a physical basis for the view that the experimental outcome in a given study is simply one of a set of many possible outcomes. The uniqueness of the outcome, its significance, is judged against the reference set of all possible outcomes under an assumption about treatment effects, as such effects are negligible. For the logic of this view to prevail, all outcomes must be equally likely, and this is achieved only by randomization." In wetland studies, true randomization within a target population can be difficult. The reasons for this complicatedness are numerous, but include denial of access to study sites, environmental conditions (e.g., drying of wetland when studying aquatic invertebrates), equipment placement requirements (e.g., insufficient water depth to measure water quality, flooding potential of deployed equipment), lack of defined boundaries identifying experimental units, sampling logistics (e.g., travel, time to collect samples), and presence, or suspected presence, of organisms of interest (e.g., certain amphibian species, habitats used by certain avian species).

Replication is the necessary practice of using more than one experimental unit for each treatment. Replication is required to minimize the uncertainty of concluding that differences among treatments are due to treatment effects rather than inherent differences among experimental units or due to random chance. Replication is required to measure the variation among and within treatments to make conclusions regarding treatment effects, which is the basis for drawing inference about treatment effects using traditional, univariate statistical techniques such as analysis of variance. If only one experimental unit is assigned per treatment, then no statistical inference beyond the experimental units sampled is possible because experimental error cannot be estimated. Much too frequently in wetland science, a study is conducted once with weak evidence for conclusions that results in uncertain knowledge; repeating the study would either strengthen evidence for conclusions or show that the initial conclusions are not supported allowing for the development of alternative hypotheses.

Another form of replication is the practice of repeating a study to strengthen conclusions. Results based on studying a few wetlands in a limited area or constrained environmental conditions (e.g., only wet or dry years) could be confirmed by a similar study or succession of studies conducted over larger temporal and spatial scales to determine if conclusions hold under more general conditions. For example, Luo et al. (1997) concluded that unsustainable accumulation of upland sediment was filling playa wetlands and represented the greatest impact to this unique wetland system. Their results were based on data from 40 playas (20 cropland watersheds and 20 grassland watersheds) in a limited spatial distribution. However, since the initial study, a number of subsequent studies have confirmed that the original conclusion is valid at larger spatial scales and under a variety of environmental conditions with steadily increasing evidence of negative impacts of sediment accumulation on playa wetlands (e.g., Tsai et al. 2007, 2010; Johnson et al. 2011, 2012; Smith et al. 2011; Burgess and Skagen 2012; O'Connell et al. 2012).

The concept of pseudoreplication is frequently confused with true replication (Hurlbert 1984). Replication is based on the number of experimental units whereas pseudoreplication usually refers to multiple measurements from a single experimental unit that are treated as independent experimental units during analyses. For example, if one was interested in biomass production between grazed and ungrazed wetlands (i.e., grazed/ungrazed are treatments) but only applied each treatment to a single wetland of each treatment and clipped and weighed aboveground biomass in 20 plots/wetland, then analyses using plots as experimental units would be pseudoreplicated. Furthermore, because the treatments were applied to the entire wetland, the plots are samples and not individual experimental units randomly assigned to a grazing treatment. Because there is only one experimental unit per treatment, estimation of experimental error is impossible because the variation among samples within each experimental unit would be considered sampling error (i.e., variation among samples of a given experimental unit; see Sources of Error below). Pseudoreplication can easily be avoided when it is understood what the unit is to which the randomization rule applies when assigning treatments to experimental units. The results from studies that include multiple samples from single experimental units should not be considered invalid because it is often difficult or impossible to replicate certain experimental units (e.g., oil spills in a coastal marsh) in applied ecological research (Wester 1992). However, it is critical to realize that inference of results can be strictly extended only to the experimental unit(s) sampled. In most ecological fields, use of pseudoreplication is considered a fatal flaw for studies, but many times this approach is appropriate in wetland studies as there may exist

1 Study Design and Logistics   21

only one experimental unit or environmental condition to be studied (i.e., the Everglades, Great Salt Lake marshes), interest is only in the experimental units being studied (see BACI study below), or the scale of the study is so large that replication is impractical. As an extreme example, it is not possible to use replication to test hypotheses related to global climate change because Earth cannot be replicated.

1.8 Impact Studies

In wetland ecology and management, biologists are frequently interested in the effects of impacts to wetlands. Impacts can be natural or human-made, planned or unplanned, but cause a change in the system state of the wetland. Impacts can be natural disasters such as hurricanes, extensive prolonged drought, or designed for management to improve ecological conditions (e.g., removal of invasive species, restoration of historical hydrology). An **impact assessment study** includes a design common to wetland studies known as a BACI (before-after/control-impact) design (Green 1979). Principally, these types of study designs are the result of some sort of natural or anthropogenic disturbance. The majority of wetland studies include some sort of measurement or modeling of the effects of disturbance on the abiotic and biotic components of the ecosystem. In wetland ecosystems, disturbance is common and, for many wetland types essential to ecological function, differing primarily in degree of disturbance (i.e., short-term flood, multi-year drought, hurricane effects that last decades). Furthermore, included in definitions of wetland, both ecological and legal, are references to disturbance that must occur prior to the system being declared a wetland. For example, coastal marshes are affected daily by the predictable disturbance of tides that raise and lower water depths and adjust salinity levels in the marshes. In many inland, geographically isolated wetlands such as prairie potholes and playa wetlands, fluctuations between wet and dry states are fundamental to the function of these systems.

When conducting an impact study, the type of disturbance will greatly influence the development of a study design. There are three primary categories of disturbance – pulse, press, and those affecting temporal variation (Bender et al. 1984; Underwood 1994). A **pulse disturbance** is not sustained beyond initial disturbance, but effects persist beyond cessation of the disturbance (e.g., fire, hurricane). A **press disturbance** persists beyond the initial event (e.g., flood, drought, invasive species). A **temporal variance disturbance** results in increasing or decreasing amplitudes (i.e., variance) around a constant mean on some sort of meaningful temporal scale. Documenting a temporal variance disturbance is difficult and requires long-term investigation or system monitoring. For example, some wetlands require precipitation runoff events to flood; however, future climate change may increase variation of precipitation between years or years represented by extreme precipitation events can change over time while average annual precipitation remains relatively constant. Therefore, species adapted to the historical precipitation patterns (e.g., timing, intensity, and amount) may not persist as they are replaced by species better adapted to the changing precipitation patterns. From a wetland perspective, this is, as yet, an understudied potential consequence of global climate change. Recognition of a human-defined temporal scale relative to disturbance makes it difficult to fully understand the role of natural disturbance in wetlands beyond a few decades.

Because BACI studies are considered pseudoreplicated due to lack of true replication, the inferential scope of these studies is limited to those wetlands studied and not to a larger target population. Such designs are very common in wetland studies to assess effects of proposed anthropogenic activities, especially when mitigation is involved. The basic approach to a BACI-type study design is the collection of a sample prior to the disturbance and another taken after the disturbance at both the disturbed site and representative "undisturbed" control site. A measurable effect due to the disturbance would be represented as a statistical difference in the average value of the dependent variable between the control and disturbed sites prior to and after the disturbance. A wetland that is proposed to be impacted by some activity (e.g., dredging, filling, change in hydrology) has not been chosen randomly, but the impacts of the activity on wetland function must be known to mitigate any negative effects of the action. A BACI study can be used to quantify these impacts. Following identification of the impact site, a particular wetland that is a geographic neighbor and similar enough to the impact site such that both wetlands would be subjected to the same nuisance variables is subjectively identified as the control site (i.e., reference site not experiencing the impact of interest). None of the experimental units were randomly chosen nor were treatments randomly assigned; the goal of this study design is to make inference for only the impacted wetland by measuring the effect size of the impact. Data are collected simultaneously in both wetlands in the same manner prior to the impact and after the impact. Statistical analyses and resulting inference are on the comparison of the magnitude of the differences in recorded data between impact and control sites prior to and after the impact (Stewart-Oaten et al. 1986). Underwood (1991, 1994) details a variety of approaches for statistical analyses of impact studies.

Assessment of impacts not defined prior to impact occurring can be studied using an impact assessment approach, but the lack of pre-impact data results in weaker inference. Unpredicted environmental events such as hurricanes, wildfire, floods, drought, and wind blowdowns are frequently studied for impacts to wetlands without any pre-impact impact data. Skalski and Robson (1992) described an approach to these types of studies defined as an accident assessment study. They suggested using a control site and creating a time series of measurements for the control and impacted site. Inference regarding impact would be based on comparisons of trajectories of the data as the impacted site recovers. In situations where a reference area is not available, an impact gradient study may be appropriate (Skalski and Robson 1992). This approach assumes that impacts related to a disturbance are greatest at the core of the disturbance and declines as one moves away from the core. Therefore, sampling is conducted on the spatial linear distance radiating from the core location of disturbance. The resulting data are used to model the decline of impact effect with linear distance.

Hannaford and Resh (1999) used a BACI study design to estimate the impact of all-terrain vehicles (ATV) on vegetation in a San Francisco Bay wetland. They found that ATV use caused immediate impact to vegetation but limited use allows for recovery within a year without continuing traffic. Zimmer et al. (2001) used a BACI approach to assess the ecological response to colonization and extinction of minnows in a prairie pothole in Minnesota. The impacted wetland was paired with a fishless site and comparisons were made when both were fishless, following introduction of fish into impact wetland, and after eradication of fish in impacted wetland. They found that introduction of fish into a prairie pothole resulted in increased turbidity, total phosphorus and chlorophyll a in water, and decreased abundance of aquatic insects. Removal of fish reversed these effects. Suren et al. (2011) followed a BACI protocol to evaluate the effect of hydrologic restoration of drains within a wetland in New Zealand. Results indicated that restoration of drains was beneficial as invertebrate communities were similar to natural wetlands and cover of exotic pasture grasses declined. In addition, connectivity was improved for recolonization of native wetland plant and aquatic invertebrate communities.

1.9 Sampling

A consistent and common criticism of scientific studies is the scale to which study results are applied beyond the target population (i.e., inference). It is rare to measure every member of a target population (i.e., a census), which is why experimental design and statistical analyses are crucial for study design. Therefore, a subset of potential experimental units from the target population is usually selected to measure the variables of interest, which is termed **sampling**. The selected sampling design, as detailed later, is a contributing limiting factor of the extent of inference from a study. In order for results to be extended to the target population, the sampled experimental units must be representative of the target population (i.e., random selection and replication). Ultimately, statistical analyses of data collected from an appropriate study design enable scientists to make inferences about a target population from its sample.

Representative sampling of a target population allows for the description of spatial and temporal patterns in nature and, through testing competing hypotheses, linking an ecological process to the observed pattern. Ultimately, proper study design should elucidate the linkages between described patterns and the ecological processes that created the pattern. Unfortunately, few studies go beyond description of a pattern with a conclusion based on retroductive speculation on the processes that created the pattern. However, all investigators must realize that wetland data are created by two classes of processes. The first is the ecological process that generated the true pattern. The second is the process inherent in the sampling effort that resulted in the data of interest. The assumption is that the sampling process

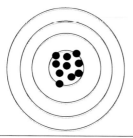

A. Collected data are unbiased and precise, with the sample mean representing the population mean with low variability, which represents an accurate sample

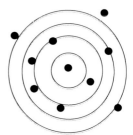

B. Collected data are unbiased, but not precise with the sample mean representing the population mean with high variability, which represents an inaccurate sample.

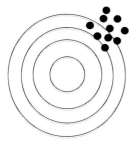

C. Collected data are biased and precise with the sample mean not representative of the population mean but with low variability, which represents an inaccurate sample.

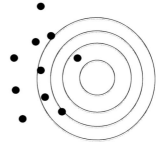

D. Collected data are biased and not precise where the sample mean is not representative of the population mean and high variability is present, which represents an inaccurate sample.

Fig. 1.2 A graphical representation of the concepts of bias, precision, and accuracy where the *center ring* represents the true mean of the target population and *black dots* represents data generated by sampling with the intent to estimate the true mean

does not mask the ecological process and data collected through sampling characterizes the ecological process of interest albeit usually in a much more simplified manner.

Basic to any study design is the goal of sampling randomly-selected experimental units to measure variables of interest. This ensures unbiased inference about some set of population parameters based on a statistic (e.g., mean, variance, standard error) that describes some attribute of interest. The purpose of sampling is to estimate the variable of interest and describe its variation in space and time. When not all members of a target population can be measured (i.e., a census), a sampling design is used to estimate the population value of the variables; statistical methods are used to describe the data and make comparisons regarding tendencies of the data. If one is able to do a census by measuring all subjects of interest (i.e., entire target population), then statistical tests are not necessary. Sampling can be used to both select experimental units for study and control of nuisance variables through a prescribed strategy. Numerous textbooks are available to assist in designing sampling strategies beyond what is described in this chapter (e.g., Cochran 1977; Scheaffer et al. 1979; Thompson 1992). Points, plots, transects, and marking captured animals are among the techniques used to sample experimental units.

The goal of sampling is to achieve an unbiased (closeness of observed values to true value) and precise (close proximity of repeated measurements of the quantity) estimate of a population parameter value (Fig. 1.2). Ideally, sample measurements for an estimator should have a narrow range of variation (i.e., precision) centered on the population value (i.e., unbiased), which represents an accurate estimate. For example, an objective of sampling must be to produce a sample mean $\approx$ population parameter with low variance around the sample mean. Groups of sample measurements that are centered on the population value but yet have a wide range are considered unbiased, but the presence of a high variance will decrease the reliability of detecting treatment effects. Biased samples generate a mean or other statistic that is not representative of the population parameter, but can have a narrow (precise) or wide range of values. A sample that is both unbiased and imprecise yields little information relative to the target population. Unfortunately, it is rarely possible to determine if one has an unbiased and precise sample because rarely are population means and variances known. Finally, wetland systems are exceptionally complex such that strict adherence to sampling schemes may be difficult, even for laboratory studies. However, it is critical that protocols associated with sampling designs be followed as explicitly as possible. Inference from samples to a target population is conditional on the protocol for selection of study sites and subsequent sampling. Thus, information from any sampling design is subject to interpretation based on the context in which the samples were collected.

Sampling protocols can be categorized as (1) haphazard sampling, (2) judgment sampling, (3) search sampling, and (4) probability sampling (Gilbert 1987). There are many variations of the sampling process within these categories (Gilbert 1987; Gilbert and Simpson 1992) that are beyond the scope of this chapter. The sampling designs described below are not meant to be inclusive of all possible approaches, but rather a description of those that would be commonly used in wetland studies. However, as a caveat, a minimal goal for reliable inference of results is some form of probability sampling where all potential experimental units have the same probability of being selected as a sample. This strategy produces unbiased estimates of the population mean, variance, and other attributes. Frequently, the phrase **sampling frame** is used to describe a list of all members of a target population (i.e., elements) that potentially can be sampled (Jessen 1978). In field studies, the sampling frame is usually spatially (study area) or temporally (study period) defined. In laboratory and human dimension studies, the sampling frame is usually defined by a list of all potential elements that could be selected for study. Finally, it is highly recommended that any proposed sampling design be reviewed by a statistician or quantitative biologist to ensure that all possible contingencies have been addressed and the proposed sampling strategy will allow for the desired inference of results.

Haphazard sampling, also known as **convenience sampling**, is frequently justified due to cost, time, and logistics, or, on occasion, historical merit. This sampling approach greatly limits the number of experimental units within a target population available to be sampled because the strategy employs a protocol that limits sampling only to a limited number of potential experimental units, which can have substantial influence on subsequent inference. Results from haphazard sampling must be placed in the context that the data were recorded and are valid only if the target population is homogeneously distributed (Gilbert 1987). Examples of haphazard sampling would be to only sample wetlands on public land or adjacent to field stations, conduct roadside surveys of wetlands, rely on volunteer reporting of flora and fauna outside a defined study, or use an inconsistent temporal sampling schedule. Similar to pseudoreplication, inference from haphazard studies is limited to the sampled experimental units. For example, if one only sampled wetlands with public access, then reliable inference can only be made to similar wetlands with public access. However, at times, investigators may be interested only in the wetland represented by haphazard sampling and thus, inference can be considered valid if the remainder of the study design is appropriate. Haphazard sampling can be used for initial assessments of an area or hypothesis development (Morrison et al. 2001).

Because of the sheer number and small size of available wetlands, Babbitt (2005) used haphazard sampling across a microhabitat gradient to relate wetland size and hydroperiod on the occurrence of amphibians rather than random sampling. She justified the efficiency of the sampling approach by noting that no new species were found in subsequent sampling efforts. In wetland ecology and management, frequently one is interested in the effects of impacts to wetlands. Hornung and Rice (2003) haphazardly selected grazed wetland treatment locations to evaluate the relationship between the presence of Odonata species and wetland quality in Alberta, Canada. They also used haphazard methods to sample invertebrates. Unfortunately, one conclusion from the study was that the haphazard sampling was insufficient to detect a trend for aquatic macroinvertebrate abundance, diversity, and composition along a gradient of grazing intensity. Due to wind conditions, Pierce et al. (2001) haphazardly sampled fixed sampling stations to document the littoral fish community in Spirit Lake, Iowa. Their results indicate a native species decline of 25 % during a 70 year period, which was attributed to a decline in littoral vegetation and other habitat changes. In wetland science, haphazard sampling is relatively common primarily due to access restrictions precluding random selection of sampling units. It is incumbent upon the researchers to declare the context of the sampling effort, which will appropriately restrict inference of results.

Judgment sampling is based on the presumption that prior experience allows for representative selection of a study area or target population (Gilbert 1987). Deming (1990) stated that judgment sampling was a type of nonrandom sampling based on the opinion of an expert. This approach can be considered subjective and representativeness of results relative to the target population difficult to assess (i.e., uncertainty regarding what population is being sampled). As with haphazard sampling, judgment sampling can be used to assess an area, generate questions

1 Study Design and Logistics

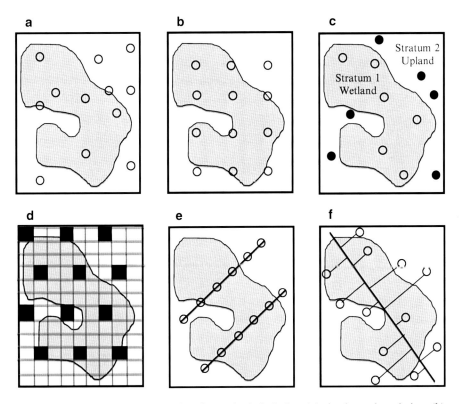

Fig. 1.3 Examples of sampling design for wetlands including (**a**) simple random design, (**b**) systematic sampling, (**c**) stratified random sampling, and (**d**) systematic use of a grid to select sample area. Examples of sampling method are (**e**) plots along transects and (**f**) line transect

that lead to hypotheses, and generate data to be used in a modeling study (Morrison et al. 2001). The most common type of judgment sampling is restricting data collection only to those wetlands that are known to contain the variable of interest (e.g., certain amphibian species, distinctive watershed conditions). Once again, inference is strictly limited to the experimental units sampled, with extension to other subjects of the target population to be considered tentative at best.

Dobbie et al. (2008) stated that professional judgment and opinion were critical in designing monitoring programs of aquatic systems. Cohen et al. (2005) used scientific judgment to rank relative impairment of wetlands and then sampled wetlands based on categories of ecological condition. Hopfensperger et al. (2006) made a case for the use of profession judgment for situations where no prior information is available when evaluating the feasibility of restoring a wetland.

Search sampling is based on historical information. Frequently, this information is available from results of long-term inventory and monitoring programs. Typically, this sampling involves using *a priori* knowledge to select areas to sample. This type of sampling differs from judgment sampling in that sampling locations are selected based on the known occurrence of variables of interest (e.g., certain plant species, known nesting locations of birds) rather than an informed opinion that the variables of interest would be found at a certain location.

The strongest inference comes from data collected using a form of probability sampling. **Probability sampling** is when all elements within a defined population have an equal probability of being selected to be sampled and that probability is known. There are a number of probabilistic sampling schemes including simple random sampling, stratified random sampling, and systematic sampling (see below, Fig. 1.3). By selecting experimental units at random, statistical properties are unbiased and represent the target population. Furthermore, such samples allow for evaluation of the magnitude of treatment effect size. These sampling strategies can range from rather straight forward to increasingly complex depending on restrictions (e.g., subsets of experimental units and nuisance variables that may need to be addressed). In most wetland studies, elements or experimental units are selected without replacement as each element appears only once in a sample (Levy and Lemeshow 1991). Sampling with replacement is when elements are returned to the target population following measurement and have the potential to be sampled again. Sampling without replacement increases precision of the results (Caughley 1977). There are several types of probability sampling strategies.

One type of probability sampling is **simple random sampling**, which occurs when each element of a sampling frame or target population has an equal probability of being selected. The elements or experimental units are not subdivided or stratified in any manner. Random selection of each element is independent of all other elements. Morrison et al. (2001) outlined five basic steps to achieve a simple random sample. First, the investigator assumes that the target population consists of a finite number of elements (i.e., experimental units) available to be selected. All selected elements can be located, accessed, and the variable(s) of interest can be measured without error. The elements must occur throughout the sampling frame and cannot overlap in any manner. Elements do not need to be identical, but as differences among elements increase in magnitude or subsets occur, then a more complex design may be necessary to avoid biasing a sample with overrepresentation by certain element types. All elements are normally sampled (i.e., consist with all other elements) without replacement.

Use of simple random sampling can be problematic if the target population is comprised of groups or subsets of similar elements. In wetland studies, this occurs when elements are clumped and patchy, such that a relatively small sample size (typical for field studies) may result in an overrepresentation of certain groups or elements with distinctive characteristics that can skew results to properties of subgroups rather than the entire target population. Dividing the elements of a target population into independent subsets or groups (i.e., strata) and then applying a random sampling approach within each stratum can increase the likelihood that results are representative of the target population in addition to increasing knowledge for elements of distinct strata that could be missing using a simple random sample.

Stratified random sampling can be used to increase sampling efficiency and statistical estimation. The key for successful stratified sampling is that the basis for stratification is correlated with the measured dependent variable. For example, if an investigator is interested in the effect of watershed condition on water quality of a wetland then sampling should be stratified using identified watershed conditions (e.g., grassland, cultivation, forested) to ensure that each condition is properly represented in the sample. By stratifying, one ensures that a single watershed condition does not dominate the sample and consequently the final results.

Drawbacks to stratified sampling are (1) spatial and temporal scale of relevant stratification variables can be difficult to determine; (2) increased complications for analyses when homogeneous strata do not exist; and (3) sampling costs are increased. Samples can be distributed among strata either by proportion of strata size or an optimal allocation process. An example of this type of sampling would be to stratify an area of coastal marsh by a salinity gradient (i.e., fresh, intermediate, brackish, saline marsh), estimate the proportion of each strata (e.g., fresh = 0.10, intermediate = 0.30, brackish = 0.50, and saline = 0.10), then determine sample size within each strata by dividing the total number of samples to be taken proportionally among the strata (e.g., if 250 total samples are needed to detect a difference between treatment levels, then 25 would be taken in fresh and saline marsh; 125 in brackish marsh; and 75 in intermediate marsh).

Strata can be defined within the study area (e.g., wetland and upland), study period (e.g., seasons), and target population (e.g., small and large wetlands). Strata cannot overlap, and elements cannot be available for selection in greater than one stratum. For stratification to be useful, elements (experimental units) should be more homogeneous within strata than among strata. If this is the case, by stratifying, sampling standard error of the overall population mean should be reduced to the standard error estimated by simple random sampling. Further, estimates of dependent variables for each strata allows for comparisons among strata, which are frequently of interest. However, it is critical to delineate strata based on knowledge that the identified strata influence variables of interest. For example, one would not test effects of herbicide treatments using strata of wetland size, but rather stratification based on wetland hydrology, soil type, or vegetation would be appropriate.

Systematic sampling represents an interesting approach that is rarely used in wetland studies, but has a role in a variety of settings. Such a sampling approach is possible when a population can be ranked in ascending or descending order of some characteristic (e.g., wetland area, watershed area, salinity gradient). Here, one would rank the population of interest relative to the characteristic and then sample based on some rule (e.g., every 10th ranked object). In addition, systematic sampling is often done on a spatial scale whereby a systematic grid of points or units is established and those to be sampled are chosen by randomly selecting a starting point and then establishing a rule to sample the remaining points or units in reference to the starting point. For example, in a large coastal marsh where one is interested in the distribution of a contaminant, use of an appropriately sized grid overlaid on a map of the marsh provides unique sampling units. Upon randomly choosing the initial grid cell to sample, the investigator can then systematically assign the remaining cells to be sampled. The usual assumption for systematic sampling is that the study area is relatively homogeneous and thus, the variable of interest is uniformly distributed across the study area. Occupancy modeling (MacKenzie et al. 2006) frequently utilizes this sampling approach. Advantages to systematic sampling include being easier to establish sampling units than random sampling, and it may be more representative (i.e., more precise) because of the uniform coverage of the entire population (Scheaffer et al. 1990; Morrison et al. 2001).

1.10 Errors to Consider in Study Design

A thread linking all aspects of study design is the minimization of errors that impact results, conclusions, and inference of a study. Because sampling is at the core of any study design and the primary goal of any study is to produce reliable data, one must be aware of the potential biases associated with sampling and strive to eliminate or minimize sources of bias or error. Failure to do so confounds subsequent data analyses and results, obscuring the true inference and, frequently, contributes to incorrect conclusions. There are several types of errors that one should be cognizant of throughout the study design process. Such errors can be categorized as theoretical, statistical, mechanical or procedural. While investigators need to be aware of how each type of error affects their study, the best defense against errors disproportionally affecting one's study is strict adherence to a sound design, sampling protocols, and data collection.

An example of theoretical error is in the interpretation of statistical results. Statistical results should be used to support a conclusion or inference based on totality of evidence from a study, rather than an investigator responding exclusively to each statistical result. However, errors associated with statistical results can be found in the inherent uncertainty of statistical tests and expressed in probabilistic terms. In classical null hypothesis testing, the possibility of conclusion errors should be considered in the study design. There are two predominate theoretical decision conclusion errors that can occur in a study. A Type I Error occurs when the null hypothesis is rejected when it is true. The probability of a Type I Error occurring is α, which is set by the investigator prior to conducting statistical tests of the data (conventionally $\alpha = 0.05$) and commonly referred to as the significance level of a statistical test (i.e., the probability level at which a test results in a significant difference between treatments or levels of a treatment). A Type II Error is more serious than a Type I Error and is defined as the probability (β) of failing to reject the null hypothesis when it is indeed false. Determination of β depends on the defined α-level and the sampling distribution of the estimated variable. More importantly, one can derive the value of $1 - \beta$, which is defined as power of the test and defined as the probability of correctly rejecting a false null hypothesis. Power should only be calculated prior to conducting a study when computing the required sample size; it should not be used following a study to evaluate confidence when failing to reject a null hypothesis (i.e., retrospective power; Gerard et al. 1998).

Statistically, although referred to as error, variation within the target population is important to correctly estimate as it is the foundation of many statistical techniques used for testing differences among levels of dependent variables. Estimation of **experimental error** is the inherent variation among experimental units treated alike or variation not explained by treatments or other variables. Accurate estimation of experimental error is critical for testing treatment effects on response variables. Experimental error differs from **sampling error**, which is the variation among samples (or observations) of a given experimental unit. Sampling error can be due to natural variability among units under study and can result from chance or sampling bias in selecting subjects for sampling (Cochran 1977). Any time that more than one sample or observation is recorded per experimental unit (e.g., multiple plots or water samples/wetland), accounting for sampling error needs to be considered as the study design is developed. An example of experimental error would be variation of above-ground biomass among wetlands; this could be the result of a single sample collected in each wetland or the variation among wetlands of the average of multiple samples taken within a wetland. Sampling error would be the variation among samples within a single experimental unit; that is, the variation of multiple samples of biomass collected within a wetland designated as an experimental unit.

Mechanically, during the course of data collection for a study, a number of errors are possible. Cochran (1977) outlined these and other sources of error in ecological studies for which investigators must be prepared and vigilant. Proper methodology is the primary protection from a study suffering from investigator bias, personal values, and preconceived results. However, if the observations or measurements are made incorrectly or with the inappropriate equipment, then **measurement error** is a likely outcome. For example, species can be misidentified, counts incomplete, flow meters improperly calibrated, and measurements taken at the improper scale (e.g., meters recorded instead of millimeters) are among the countless potential other sources of measurement error. Each observer tasked with data collection must be trained, occasionally assessed, and dedicated to consistent effort to reduce effects of measurement error on final results.

Another source of mechanical error is **missing data** due either to the failure to record the proper measurements or loss of recorded data (e.g., nonfunctional equipment, weather, electronic storage failure, loss of paper copies). Missing data can cause serious issues with subsequent data analysis unless accounted for by an appropriate analysis. Investigators should take steps to avoid missing data by securing data, checking equipment functionality, and ensuring that procedures are understood by all. At times, individuals fail to record an appropriately measured zero in the data, choosing to leave the data cell blank or empty creates the impression of missing data when, in reality, the results may be biased due to lack of a zero. For example, when inventorying plant species in multiple wetlands, one must be careful to ensure that when a species is not detected in a wetland that a zero or absent is recorded and properly transcribed rather than leaving the results for the species/wetland combination as a blank entry. It is important to realize that missing data and data containing zeros represent vastly different representations of the data.

Observer bias is a mechanical error and constant factor to consider in studies and represent variation among observers. Such bias can be represented in differences in skill of ocular or aural estimate of a variable (e.g., number of birds in flock, percent vertical cover of vegetation, soil moisture relative to field capacity, species of calling amphibians), ability in using a technique (i.e., proficiency with an instrument, ability to distinguish the appropriate scale of measurement) to measure a variable, and human error in recording and transcribing data. If one can measure the magnitude and direction of inter-observer variation, then the data can be adjusted for the bias (Morrison et al. 2001). However, it is quite rare to be able to adjust for observer bias. Therefore, it is important that all observers are trained and tested relative to the data being collected prior to sampling. In most instances, it would be appropriate to consider minimizing the number of observers that record noninstrumented data to reduce observer bias (e.g., same person should conduct bird counts, listen for amphibian calls, estimate percent cover of vegetation types). However, even with limited observers, one must be able to determine if a systematic bias resulting from observer bias where a variable is consistently under- or overestimated due to the selection of sampling points or unit of data measurement (Thompson et al. 1998).

A procedural type of error is what Cochran (1977) termed **gross error** where mistakes are made in transcribing, entering, typing, and editing data and results from analyses. Therefore, all study designs must have a well-defined, unambiguous observation/measurement methodology prior to collection of data. In addition, a protocol for data management is necessary prior to initiating any study. For example, it should be mandated that all paper data sheets be copied at the end of each data collection period and copies placed in safe locations. All electronic data should be backed up in at least two locations and paper copies of electronic data should be printed and stored in a safe location. Loss of complete records of historical data, while rare, can occur due to natural disaster (e.g., hurricane), human error (e.g., inadvertently discarded), loss of electronic data (e.g., hard drive failure), misfiling, or mislabeling. Finally, all data sets and preliminary data analyses results should be reviewed and copy-edited prior to conducting final analyses to ensure observations were accurately transcribed. This is critical if one is using voice-recognition software to enter data. All data should be linked to specific observers. In addition, it is preferable for data review to occur shortly after collection or transcription so that (1) technicians collecting the data remain available to answer any questions, (2) illegible handwriting can be deciphered using unsullied memories, and (3) there is an increased likelihood for recollection of data should issues be identified.

Procedural errors occur when design and sampling protocols are not correctly followed for recording measurements, transcribing and storing data, and conducting data analyses. Development of a structured Quality Assurance and Quality Control (QA/QC) program prior to initiating a project will minimize this bias. A QA/QC program is the foundation for risk management in a study. In addition, any ethical questions that may arise during a study should be alleviated with an approved QA/QC program. Quality Assurance is a set of activities designed to ensure that the development and/or maintenance process is adequate to ensure a system will meet its objectives. EPA (2001: Appendix B-3) defined Quality Assurance as "an integrated system of management activities involving planning, implementation, documentation, assessment, reporting, and quality improvement to ensure that a process, item, or service is of the type and quality needed and expected by the client." Quality Control is the application of procedures to minimize errors during data collection and analysis. Furthermore, EPA (2001: Appendix B-3) defined Quality Control as "the overall system of technical activities that measures the attributes and performance of a process, item, or service against defined standards to verify that they meet the stated requirements established by the customer; operational techniques and activities that are used to fulfill requirements for quality."

Quality Assurance requires development of a study plan that includes the objectives, design, and implementation of the study with a stated protocol for data recording, storage, analysis, and reporting. The study plan should be reviewed by peers and a biometrician prior to initiation of data collection. The subsequent study plan becomes a dynamic document that should be updated to account for any changes throughout the duration of the study. Examples of Quality Control are stated calibration and maintenance of equipment, training requirements of personnel, methodology procedures and protocols, and use of any data generated during quality control procedures.

Many agencies, private industry, and other organizations require a detailed QA/QC prior to funding an approved study. However, even if a QA/QC plan is not a formal requirement, adherence to and occasional review of an informal QA/QC protocol preserves data integrity. There are countless examples of acceptable QA/QC plans for the U.S. Federal Government (e.g., EPA 1998, 2001, 2008; U.S. Fish and Wildlife Service (http://www.fws.gov/aah/PDF/QI-FWS%20AAHP%20QA%20Program.pdf)), state agencies (e.g., Minnesota and Connecticut http://files.dnr.state.mn.us/eco/wetlands/nwi_comprehensive_project_plan_021012.pdf, http://www.ct.gov/dep/cwp/view.asp?a=2715&q=324958&depNav_GID=1626), and private industry (e.g., Integrated Ocean Drilling Program http://www.iodp.org/qaqc-taskforce/). The U.S. Environmental Protection Agency (EPA) has developed a Wetlands Quality Assurance Project Plan Guidance (QAPP) to assist wetland grant recipients documenting the procedural and data requirements for projects involving environmental measurements (http://www.epa.gov/region9/qa/pdfs/WetlandsQAPPGuidance.pdf).

The EPA Wetlands QAPP guidance is comprehensive and provides a starting point for any wetland project. The guidance includes nine sections with a variety of subjects within each section for consideration prior to embarking on wetland-related studies. Many of these sections can be useful to most investigators of wetland ecology, management, and conservation. The *Project Description* section includes background information and a justification for the study and includes the following items: (1) Project Purpose and Problem Definition; (2) Project Area

Description; (3) Responsible Agency and Participating Organizations; (4) Project Organization Roles and Responsibilities; (5) Permits for Collection of Environmental Measures; and (6) History, Previous Studies, and Regulatory Involvement. The *Project Data Quality Objectives* section ensures that data quality and data management are sufficient to achieve the objectives of the study. The section *Field Study Design/Measurement Protocols* details how data are to be collected (i.e., variables measured) for a suite of abiotic and biotic features. Included in the section *Field Preparation and Documentation* are details related to data management such as (1) Field Preparation; (2) Field Notes (e.g., logbooks, data sheets and forms, and photographs); (3) Documentation of Sample Collections; (4) Labeling of Sample Collections; and (5) Field Variances. The section *Quality Control for Samples Collected for Off-Site Analysis* details handling of samples to prevent contamination and confirmation of lab analyses (i.e., collection of field samples and transport to a laboratory for analyses). Details related to field samples are provided in the section *Field Sample Collection Protocols for Off-Site Analyses*, which can be the most important section for wetland studies. Details related to laboratories are found in the sections *Laboratory Analyses and Section* and *Sample Shipment of Off-Site Laboratory*.

Quality Control is practiced by any entity producing a product. Industry has quality control guidelines and practices to ensure products are functional and within a margin of acceptable variation. That is, identification of defects in products after development but before release. Quality Control is a system of routine technical activities that measures and controls the quality of the inventory as it is being developed. Most Quality Control systems are designed to: (1) provide routine and consistent checks to ensure data integrity, correctness, and completeness; (2) identify and address errors and omissions; and (3) document and archive inventory material and record all QC activities (Penman et al. 2006).

In scientific investigations, Quality Control is project- and method-specific such that development of a Quality Control plan is difficult to generalize. Examples of items to include in a Quality Control plan are (1) equipment monitoring and recalibration, (2) periodic checks for data errors and transcription accuracy, (3) ensuring software and hardware are working correctly, (4) checking integrity of stored data, (5) retraining of technicians and anyone handling or analyzing samples, and (6) confirming that safety protocols are being followed. Therefore, one must identify all fundamental components of a study design and produce a Quality Control plan that addresses each and maintains the highest possible standard of data integrity and accuracy while maintaining a safe environment.

As a final check of the accuracy of the data prior to analyses, one should calculate descriptive statistics (e.g., mean, range, minimum value, maximum value, and variance) or conduct outlier analysis (Barnett and Lewis 1994) to identify extreme values that are inconsistent with the other data and likely to be a result of an error in transcribing data. However, one must have a prepared approach to statistical analyses prior to checking the data to ensure that perceived patterns in the descriptive analyses do not influence subsequent analyses, which can produce spurious conclusions. There is a simple web-based application for the Grubbs' test for outliers (http://www.graphpad.com/quickcalcs/Grubbs1.cfm) that is available to identify extreme values. If the value came from a contaminated sample, then it is appropriate to recollect the sample if possible or discard the initial observation. Outliers cause considerable problems with most statistical analyses that are based on a particular sampling distribution (e.g., normal) and associated assumptions (e.g., constant variance). If the extreme values were actually recorded and represent a legitimate data point, then the investigator must decide how to handle the value by either removal from the data set, consider data transformation to meet statistical assumptions, or conduct the analyses with techniques robust to outliers (i.e., nonparametric and multivariate methods, or use of generalized linear models linked to a non-normal sampling distribution).

1.11 Sample Size and Effect Size

Although usually an afterthought during study formulation, one must consider the magnitude of a treatment difference or effect size that is biologically meaningful in addition to statistically significant results. **Biological significance** is defined by the investigator but based on a firm understanding of the system being studied and associated literature relative to the system. There is no replacement for sound, extensive biological knowledge of the system that generated the data. If an investigator or reviewer of proposed study design lacks this knowledge, discussions with experienced biologists/ecologists regarding the system and interpretation of results is just as important as use of proper statistical techniques. Not all statistically significant results have biological meaning and, at times, biologically significant differences may not be found to be statistically different. Frequently, the latter is attributed to lack of sample size as an explanation, a situation that would be avoided with proper design prior to collecting the first sample in a study.

Central to a scientific study is the ability to detect a biologically meaningful effect and measure the size of the effect. The primary controlling element for detecting an effect of interest is sample size, where the general rule is "more is better." Increasing sample size decreases the overall variability of the data around a mean for a given treatment, which increases the power of statistical tests (i.e., the probability of finding a difference due to treatment when one truly exists). Therefore, one of the most important aspects of study design is the determination of the appropriate sample size necessary to detect a specified effect. Under classical hypothesis testing (i.e., true experiments), the required sample size to realize a level of power to detect a treatment effect of desired magnitude should be estimated. Calculation of the appropriate sample size primarily depends on the underlying distribution (i.e., variation) of the sample values for the dependent variable, significance level (i.e., α), and minimum effect size to be detected. The concept of effect size is part of study design considerations prior to sampling and after analyzing the collected data. In the effort to determine appropriate sample size prior to conducting a study, investigators need to determine the minimum effect size of scientific interest. This determination is not a statistical decision but one that must be made by the investigator. Such a decision is important in the context of other constraints to sampling effort and should be made with a realistic expectation that one would expect to find at the conclusion of the study.

Appropriate sample size can be computed explicitly via formulas or through simulations. Use of previously collected data for the variables of interest in the system being investigated can be used to explicitly calculate the necessary sample size. Use of data from preliminary (i.e., pilot study – a preliminary period of reduced data collection using the proposed study design) studies and literature values can be used to estimate necessary sample size. Tragically, many published and most unpublished studies with nonstatistically significant findings contain statements apologizing for such findings and blaming it on the lack of a sufficient sample size. Investigators should strive to avoid such situations to the extent possible because concluding that results from a wetland study are essentially meaningless due to insufficient sample size adds little to scientific process and squanders precious resources and time.

Assessment of the sample size and statistical power to measure an effect is critical to study design. Both of these aspects require an acceptable measure of precision. Therefore, one needs to measure or estimate the level of variation associated with each dependent variable to evaluate the ability of the proposed study design to produce meaningful results. One can accomplish this either through use of values in the literature or conducting a pilot study. There are numerous formulae and approaches for sample size determination and determination of power; many of which are available as calculators on a variety of websites or in statistical software packages. There are a number of on-line and software sample size calculators (see http://www.epibiostat.ucsf.edu/biostat/sampsize.html?iframe=true&width=100%&height=100% for a comprehensive list of available programs). The investigator needs to apply the formula appropriate for their particular study design. There are a minimum of three categories of data that need to be determined or estimated for most sample size formulae – effect size (i.e., the biological effect that one desires to detect, usually represented as probability), a measure of variation related to the dependent variable, and alpha level.

The initial step in determining a necessary sample size is to use the appropriate sample size equation. There are equations for nearly every use of sampling scheme to estimate a population parameter or detect a difference. In addition, there are variations for many equations depending on whether the estimate of variation of the dependent variable is from a pilot study, literature, or known population value (very rare in wetland field studies). Examples of situations where sample size calculations are available include (1) estimation of a population mean, (2) estimation of a population proportion, (3) testing of hypotheses concerning a population mean, (4) testing of hypotheses concerning a population proportion, (5) testing mean differences between two or more populations, (6) testing difference in proportions between two or more populations, (7) testing main and interactive effects in traditional experimental designs, and (8) conducting human dimension survey studies.

Estimation of **effect size** following data collection and statistical analyses is a relatively recent addition to reporting of results from wetland and other natural resource studies, but estimation of treatment effect provides additional evidence and weight for conclusions developed during the study. It is a relatively simple concept that should not be made any more complicated than necessary regarding the magnitude of any found effect. Any significant statistical test reported in the literature should also include the magnitude of the effect to assess biological significance of the results. For example, one can achieve a statistically significant difference between means of a treatment and control population but, depending on the variation within each population, a biologically significant effect may not be an appropriate conclusion. Such an occurrence is more likely in a laboratory setting, but can also be found in field studies. Effect size can be as simple as reporting the percent mean change due to an effect (i.e., increases or decreases by X % due to the application of the treatment). In addition, a number of indices have been developed to quantify the strength of the difference between groups (e.g., levels of independent variables). The most common effect size index is Cohen's d (Cohen 1988) or standardized mean difference whereby calculated effect size index values are categorized as 0.20 = small, 0.50 = medium, 0.80 = large. These indices can be calculated for a wide variety of study designs (see for example: http://www.bwgriffin.com/gsu/courses/edur9131/content/Effect_Sizes_pdf5.pdf; http://www.campbellcollaboration.org/resources/effect_size_input.php)

1.12 Other Logistical Considerations of Wetland Study

Wetlands are complex and diverse ecosystems; therefore, it is quite difficult to generalize a logistical approach that can be applied to all wetlands under all study situations. However, there are a number of common information needs to access and become familiar with prior to conducting a wetland study. It is imperative to become well-versed in the system being proposed for study beyond the immediate question being addressed. All biotic and abiotic elements of a wetland ecosystem are potential variables in a wetland study no matter whether one is investigating water quality, hydric soils, plant associations, invertebrates, or animal communities because of the ecological linkages among all elements in the ecosystem. To fully document the effects found in any study, one must consider the totality of effects on all elements of the wetland, which can only be accomplished via a thorough ecological understanding of the system being studied – including the potential ecological states of the system under the environmental variation potentially affecting the wetland that may differ from the state measured during the study.

To define the study population of wetlands and the potential scope of inference relative to research results, one needs to have knowledge of the spatial scale of occurrence of the wetlands of interest. There are a number of sources of wetland occurrence, but quality of locations and associated information varies greatly. Nearly all available mapped locations of wetlands are provided as electronic data files that can be used and manipulated using software associated with Geographic Information Systems (GIS). Historically, most wetlands (e.g., prairie potholes, coastal marsh) have been identified and mapped by the U.S. Department of Interior, Fish and Wildlife Service, National Wetland Inventory (NWI) (http://www.fws.gov/wetlands/). The data from NWI are available electronically (http://www.fws.gov/wetlands/Data/index.html) and used to produce periodic status and trends reports of wetlands in the United States (e.g., Dahl 2011). Other potential sources of wetland occurrence include individual joint ventures associated with the North American Waterfowl Management Plan that focus on conservation of wetlands for migratory birds, state-specific land cover data bases, U.S. Geological Survey topographic maps (http://nationalmap.gov/ustopo/index.html), state highway departments, and U.S. Department Agriculture, Natural Resources Conservation Service (primarily at state and county levels) wetland determination and soils mapping data. Most states and some nongovernmental organizations have layers of GIS data available on regional location of wetlands; however, at times it requires some searching to find the storage locations of these data.

Because of the variation among wetland types and, to some extent, within a wetland type, it is important to fully describe the wetland(s) under study. The two primary wetland classification/description approaches are the Cowardin et al. (1979) and hydrogeomorphic methods (Brinson 1993; Chap. 2 of Vol. 3). The Cowardin method was developed to serve as a "classification, to be used in a new inventory of wetlands and deepwater habitats of the United States, is intended to describe ecological taxa, arrange them in a system useful to resource managers, furnish units for mapping, and provide uniformity of concepts and terms. Wetlands are defined by plants (hydrophytes), soils (hydric soils), and frequency of flooding. Ecologically related areas of deep water, traditionally not considered wetlands, are included in the classification as deepwater habitats" (Cowardin et al. 1979: 1). This classification approach is used by NWI and all users of these data need to be familiar with the Cowardin et al. system. The hierarchical approach uses System, Subsystem, Class, Dominance Types, and Modifiers, and understanding these is important to maximize the use of NWI data and describe the study wetland using a common language. The hydrogeomorphic classification approach emphasizes hydrologic and geomorphic (i.e., abiotic) controls for wetlands using the three components of (1) geomorphic setting, (2) water source and its transport, and (3) hydrodynamics (Brinson 1993). The geomorphic setting refers to the topographic location of the wetland within the surrounding landscape. The types of water sources are precipitation, surface/near surface flow, and groundwater discharge. Hydrodynamics is the direction of flow and strength of water movement within the wetland. A variety of descriptive terms are available to classify each wetland using this approach. There are many other classification approaches that have been developed, and local wetlands experts should be consulted to determine what approach might work best for the wetlands being studied.

Hydrology is the dominant force driving ecological mechanisms and patterns within wetlands. Abiotic factors represent indices to the hydrology of the wetland and biotic elements represent the response to wetland hydrology. All ecological functions are ultimately influenced by wetland hydrology; therefore, it is critical for investigators to fully describe the relative source and fates of water for the study wetland type. It is not necessary to provide a detailed water budget but a general depiction of water dynamics assists in understanding the context of wetland studies. In a similar fashion, one must describe as completely as possible the watershed or drainage area associated with study wetlands in terms of size, soil types, land use, anthropogenic features, and any proposed future changes in these characteristics should it be pertinent to the study.

Once a study question and associated hypotheses have been developed, an investigator must define several characteristics of the study. The temporal period for sampling must be appropriate for the question. For example, one would not attempt to test habitat selection for breeding birds during a nonbreeding season. Therefore, it is crucial to understand the life cycle of any species of interest and responses of each species to changing wetland conditions to avoid sampling during unsuitable periods or environmental conditions. Investigators must carefully list potential dependent and independent variables that will provide the most parsimonious information relative to the proposed hypotheses and study objectives. Typically, but not always, dependent variables are defined by the hypotheses and objectives; however, potential independent variables are not as obvious and require a great deal of thought prior to finalizing a study design. Concurrently, confounding and covariate variables must be identified and addressed to avoid any unwanted influence by nuisance variables on the study results.

Finally, in any study design, the project budget must be known with all of the associated restrictions and time sensitive requirements. No study is possible without funding and continuous accounting of project expenses is necessary to avoid situations that would jeopardize the study. It is usually a benefit to keep all investigators and observers informed regarding the budget status to assist in future planning for efforts related to the study. It may be prudent early in the study to develop a number of contingency study plans in the event of unexpected conditions such as loss of funding, natural disasters, destruction of equipment, or greater than anticipated costs. This would allow for the salvage of some information should the study go awry rather than being a complete loss.

1.13 Summary and Additional Considerations on the Application of Study Design

Designing a research project requires a thorough understanding of the wetland system being studied, the question or issue of interest, and those variables that must be measured to address the biological question of interest and test competing hypotheses. Following establishment of the study question (i.e., objective[s]) and associated hypotheses, investigators should then develop the methodology of the study. It is the methods of a study design that attempts to remove any investigator bias relative to the investigation. Clear, concise, and definite methodology must be produced to not only guide one's study but allow future investigators to replicate and reproduce the original study. During this step, the investigator determines the study population, limits to inference, available resources (e.g., funding, personnel), sampling approach, and initiates an evaluation of the literature relative to the system to be studied. The succeeding step is to state the independent and dependent variables of interest. Each must be expected to have a measureable response or linkage to the treatments, disturbance, impacts, developed hypotheses, or other elements of interest. Again, careful consideration must be made to not attempt to measure all possible variables but only those that are meaningful and unrelated (i.e., not correlated and thus redundant). Pertinent variables can be considered based on literature, prior experience, and results from a pilot study. In addition, one should identify pertinent covariates at this stage so that the influence of typically nuisance or potentially confounding variables can be minimized through appropriate design.

Usually there are a number of potential methods or approaches available for recording data relative to a specific variable. It is most appropriate to choose a measurement method guided by the questions and objectives of the study. Other aspects to consider include techniques used in comparable studies to which collected data will eventually be compared, availability and cost of equipment, type of data being recorded, precision of the measurements, and identification of any identified biases relative to proposed methods. Development of clear, structured, and reliable data recording forms cannot be overstated. Such forms are typically the foundation for data recording, storage, and transfer. A considerable amount of data and information loss occurs with the use of poor data forms. Basic to all data forms are: (1) information to be collected, (2) data collection strategy, (3) order of data recording, and (4) structure of data recording (Levy and Lemeshow 1991). All data must be recorded in a meaningful and legible format that minimizes the probability of recording error. Order of recording data is important for efficiency of data collection and subsequent transfer to electronic format (i.e., data bases). Order of data recording simplifies data collection and minimizes observer effects. The recording structure includes a condensed explanation for sampling protocol that can be referenced in the field; use of "check" boxes or other approaches to minimize mistakes in recording data; and development of variable "keys" to define any shorthand notation or acronyms that can be used on the data sheet. Observers must be trained to consistently complete data forms. Furthermore, during the publication process, a common reason to reject manuscripts is due to poor or inappropriate methodology; therefore, careful consideration of variables to be measured and how to measure and record the variables is necessary for a successful investigation.

Use of pilot studies or some sort of preliminary data collection is recommended, especially for studies where the investigator has little or no experience. As previously mentioned, a pilot study can be used to estimate the variation within data recorded for dependent variables. In addition, a pilot study should be used to develop a suitable sampling protocol allowing all observers to be trained and become familiar with methodology and recording data. Observer bias can be recognized and accounted for by using the results of pilot study usually which leads to increased training sessions for data collection and recording. Identification of potential nuisance variables may be a result of a pilot study. Finally, data from a pilot study should be subjected to the proposed statistical analyses to identify potential issues such as deviation from statistical assumptions, probability distribution of independent variables, and identification of correlated independent variables.

A primary approach to testing data quality is the resampling of a subset of each data set and comparing results. Other proposed actions to ensure data quality include redundant measurements by two or more observers or analysis of duplicate laboratory samples. Recorded data should be proofed shortly after being measured (e.g., completion of sampling period) by an independent observer to eliminate recording and transcription errors. Researchers should assign unique study responsibilities to each observer so that mistakes or errors can be linked to unique individuals and thus, provides an opportunity to correct any incorrect data.

During data collection and at the conclusion of the study (for long-term studies or monitoring efforts at frequent intervals), a final proofing of collected data is essential. In addition, numerous copies of the data forms and electronic data bases need to be made and stored in secured areas for future reference. Data collection points, plots, or units need to be uniquely identified and locations recorded to assist in interpretation of results, recollection of lost or erroneous data, and for ease of relocation for future investigators or return for a comparison study by current investigators. All equipment must be removed from the field, maintained/serviced, and again tested for accuracy. During the data analysis and hypothesis testing stage, a number of additional steps are necessary. Investigators should use statistics to describe the data (e.g., means, measures of variation, missing data, distribution form, and range of values) prior to primary statistical analyses. Graphical representation of the data prior to analyses is appropriate as long as the subsequent planned analyses are not altered due to perceived patterns observed in the data. Researchers should also realistically assess the sample size relative to planned analyses. For example, some multivariate and modeling approaches require relatively large sample sizes that may not be present. All statistical tests have underlying assumptions (e.g., normality, constant variance) and, although most approaches are relatively robust to at least minor violations of assumptions, it is desirable to test the assumptions for each analysis. It is recommended to use a statistical test that is appropriate for the data rather than alter the data through transformation or some other technique just to use a particular statistical test.

Interpretation and eventual publication of the study results represent the conclusion of a study design. Interpretation of results includes not only describing the data and subsequent analyses but also discussing the relevance and context of the results relative to previously published information (i.e., literature). Care must be taken not to inappropriately extend the inference of the results beyond the study population. Here, one must guard against letting the data and analyses determine the study conclusions rather than using the data and results from analyses to support a conclusion based on the accumulation of evidence. Fixation on statistical results is a poor substitute for critical thinking of the results in an ecological context. Reliable conclusions must be supported by data and be capable of withstanding future study results using similar methodology to address a common question. No study is complete until publication of results in an accessible source – preferably in a peer-reviewed scientific journal. Conducting research without publication hinders the scientific process and can be considered an inefficient use of resources requiring future investigators to unknowingly recreate a study that delays the scientific process.

As a final point, there is more than one way to conduct a study, test a hypothesis, measure variables, and generate results. Therefore, when judging the merits of results from a study, investigators should independently assess the hypothesis, methodology, study design, statistical approach, and conclusions reached based on results without regard to how they would have conducted the study. One must consider the evidence in its entirety, not just those bits and pieces that support a preconceived conclusion. At all times, the scientific responsibility is to advance our understanding of natural systems, including wetlands, based on the accumulation of evidence from all reliable sources.

References

Anderson DR (2008) Model based inference in the life science: a primer on evidence. Springer, New York

Babbitt KJ (2005) The relative importance of wetland size and hydroperiod for amphibians in southern New Hampshire, USA. Wetl Ecol Manag 13:269–279

Balcombe CK, Anderson JT, Fortney RH, Kordek WS (2005) Aquatic macroinvertebrates assemblages in mitigated and natural wetlands. Hydrobiologia 541:175–188

Barnett V, Lewis T (1994) Outliers in statistical data, 3rd edn. Wiley, New York

Bender EA, Case TJ, Gilpin ME (1984) Perturbation experiments in community ecology: theory and practice. Ecology 65:1–13

Box GEP Jr, Hunter JS, Hunter WG (2005) Statistics for experimenters: design, innovation, and discovery. Wiley, Hoboken

Brennen EK, Smith LM, Haukos DA, LaGrange TG (2005) Short-term response of wetland birds to prescribed burning in Rainwater Basin wetlands. Wetlands 25:667–674

Brinson MM (1993) A hydrogeomorphic classification for wetlands. Technical report WRP-DE-4, U.S. Army Engineer Waterways Experiment Station, Vicksburg, MS. NTIS No. AD A270 053

Brooks RP, Wardrop DH, Cole CA, Campbell DA (2005) Are we purveyors of wetland homogeneity? A model of degradation and restoration to improve wetland mitigation performance. Ecol Eng 24:331–340

Brown ML, Guy CS (2007) Science and statistics in fisheries research. In: Guy CS, Brown ML (eds) Analysis and interpretation of freshwater fisheries data. American Fisheries Society, Bethesda, pp 1–29

Burgess L, Skagen SK (2012) Modeling sediment accumulation in North American playa wetlands in response to climate change, 1940–2100. Clim Change. doi:10.1007/s10584-012-0557-7

Caughley G (1977) Analysis of vertebrate populations. Wiley, London

Chamberlain TC (1890) The method of multiple working hypotheses. Science 15:92–96

Cherry S (1999) Statistical tests in publications of The Wildlife Society. Wildl Soc Bull 26:947–953

Cochran WG (1977) Sampling techniques, 3rd edn. Wiley, New York

Cochran WG (1983) Planning and analysis of observation studies. Wiley, New York

Cohen J (1988) Statistical power analysis for the behavioral sciences, 3rd edn. Lawrence Erlbaum Associates, Hillsdale

Cohen MJ, Lane CR, Reiss KC, Surdick JA, Bardi E, Brown MT (2005) Vegetation based classification trees for rapid assessment of isolated wetland condition. Ecol Indic 5:189–206

Collins JP, Storfer A (2003) Global amphibian declines: sorting the hypotheses. Divers Distrib 9:89–98

Cook CW, Stubbendieck J (1986) Range research basic problems and techniques. Society of Range Management, Denver

Cowardin LM, Carter V, Golet FC, LaRoe ET (1979) Classification of wetlands and deepwater habitats of the United States. U.S. Department of the Interior, Fish and Wildlife Service, Washington, DC. Northern Prairie Wildlife Research Center Online, Jamestown. http://www.npwrc.usgs.gov/resource/wetlands/classwet/index.htm (Version 04DEC1998)

Cox DR (1980) Design and analysis of nutritional and physiological experimentation. J Dairy Sci 63:313–321

Dahl TE (2011) Status and trends of wetlands in the conterminous United States 2004 to 2009. U.S. Department of the Interior; Fish and Wildlife Service, Washington, DC

Deming WE (1990) Sample design in business research. Wiley, New York

Devito KJ, Hill AR (1998) Sulphate dynamics in relation to groundwater-surface water interactions in headwater wetlands of the southern Canadian Shield. Hydrol Process 11:485–500

Dobbie MJ, Henderson BL, Stevens DL Jr (2008) Sparse sampling: spatial design for monitoring stream networks. Stat Surv 2:113–153

Elser JJ, Bracken MES, Cleland EE, Gruner DS, Harpole WS, Hillebrand H, Ngai JT, Seabloom EW, Shurin JB, Smith JE (2007) Global analysis of nitrogen and phosphorus limitation of primary producers in freshwater, marine, and terrestrial ecosystems. Ecol Lett 10:1–8

EPA (1998) Guidance for quality assurance project plans EPA QA/G-5. EPA/600/R-98/018, Feb 1998. U.S. EPA, Washington, DC. http://www.epa.gov/quality1/qs-docs/g5-final.pdf

EPA (2001) Guidance for quality assurance project plans EPA 240/B-01/003, Mar 2001. U.S. EPA, Washington, DC

EPA (2008) Quality Assurance Project Plan for the mid-Atlantic states regional wetlands assessment, Mar 2008. U.S. EPA, Washington, DC. http://mawwg.psu.edu/resources/MARWA.pdf

Euliss NH Jr, LaBaugh JW, Fredrickson LH, Mushet DM, Laubhan MK, Swanson GA, Winter TC, Rosenberry DO, Nelson RD (2004) The wetland continuum: a conceptual framework for interpreting biological studies. Wetlands 24:448–458

Fisher RA (1935) The design of experiments. Reprinted 1971 by Hafner, New York

Galatowitsch SM, van der Valk AG (1996) The vegetation of restored and natural prairie wetlands. Ecol Appl 6:102–112

Gauch HG (2003) Scientific method in practice. Cambridge University Press, Cambridge

Gerard PD, Smith DR, Weerakkody G (1998) Limits of retrospective power analysis. J Wildl Manag 62:801–807

Gilbert RO (1987) Statistical methods for environmental pollution monitoring. Van Nostrand Reinhold, New York

Gilbert RO, Simpson JC (1992) Statistical methods for evaluating the attainment of cleanup standards, vol 3. Reference-based standards for soils and solid media. Pacific Northwest Laboratory, Battelle Memorial Institute, Richland, Washington for U.S. Environmental Protection Agency and U.S. Department of Energy, Washington, DC. PNI-7409 Vol. 3, Rev. 1/UC-600

Green RH (1979) Sampling design and statistical methods for environmental biologists. Wiley, New York

Guthery FS (2008) A primer on natural resource science. Texas A&M Press, College Station

Hannaford MJ, Resh VH (1999) Impact of all-terrain vehicles (ATVs) on pickleweed (Salicornia virginica L.) in a San Francisco Bay wetland. Wetl Ecol Manag 7:225–233

Hopfensperger KN, Engelhardt KAM, Seagle SW (2006) The use of case studies in establishing feasibility for wetland restoration. Restor Ecol 14:578–586

Hornung JP, Rice CL (2003) Odonata and wetland quality in southern Alberta, Canada: a preliminary study. Odonatologica 32:119–129

Hurlbert SH (1984) Pseudoreplication and the design of ecological field experiments. Ecol Monogr 54:187–211
Jessen RJ (1978) Statistical survey techniques. Wiley, New York
Johnson DH (1999) The insignificance of statistical significance testing. J Wildl Manag 63:763–772
Johnson WP, Rice MB, Haukos DA, Thorpe P (2011) Factors influencing the occurrence of inundated playa wetlands during winter on the Texas High Plains. Wetlands 31:1287–1296
Johnson LA, Haukos DA, Smith LM, McMurry ST (2012) Loss and modification of Southern Great Plains playas. J Environ Manage 112:275–283
Kantrud HA, Newton WE (1996) A test of vegetation-related indicators of wetland quality in the prairie pothole region. J Aquat Ecosyst Stress Recover 5:177–191
Keeney RL (2007) Developing objectives and attributes. In: Edwards W, Miles RFJ, Von Winterfeldt D (eds) Advances in decision analysis: from foundations to applications. Cambridge University Press, Cambridge, pp 104–128
Kelly RL, Thomas DH (2012) Archaeology, 6th edn. Wadsworth, Belmont, California
Kempthorne O (1966) Design and analysis of experiments. Wiley, New York
Kentula ME, Brook RP, Gwin SE, Holland CC, Sherman AD, Sifneos JC (1992) An approach to improving decision making in wetland restoration and creation. Environmental Research Laboratory, U.S. Environmental Protection Agency, Corvallis. EPA/600/R-92/150
King G, Keohane RO, Verba S (1994) Designing social inquiry. Princeton University Press, Princeton
Kirk RE (1982) Experimental design, 2nd edn. Brooks/Cole Publishing Company, Wadsworth, Belmont
Krebs CJ (1985) Ecology: the experimental analysis of distribution and abundance, 3rd edn. Harper & Row, New York
Lehner PN (1996) Handbook of ethological methods, 2nd edn. Cambridge University Press, Cambridge
Levins R (1969) Some demographic and genetic consequences of environmental heterogeneity for biological control. Bull Entomol Soc Am 15:237–240
Levy PS, Lemeshow S (1991) Sampling of populations: methods and applications. Wiley, New York
Lewis WM (2001) Wetlands explained: wetland science, policy, and politics in America. Oxford University Press, New York
Luo HR, Smith LM, Allen BL, Haukos DA (1997) Effects of sedimentation on playa wetland volume. Ecol Appl 7:247–252
Mackay DS, Ewers BE, Cook BD, Davis KJ (2007) Environmental drivers of evapotranspiration in a wetland and an upland forest in northern Wisconsin. Water Resour Res 43:W03442. doi:10.1029/2006WR005149
MacKenzie DI, Nicols JD, Royle JA, Pollock KH, Bailey LL, Hines JE (2006) Occupancy estimation and modeling: inferring patterns and dynamics of species occurrence. Academic, San Diego
Martin P, Bateson P (1993) Measuring behavior: an introductory guide, 2nd edn. Cambridge University Press, Cambridge
Martin J, Runge MC, Nichols JD, Lubow BC, Kendall WL (2009) Structured decision making as a conceptual framework to identify thresholds for conservation and management. Ecol Appl 19:1079–1090
Maurer DA, Zedler JB (2002) Differential invasion of a wetland grass explained by tests of nutrients and light availability on establishment and clonal growth. Oecologia 131:279–288
Megonigal JP, Conner WH, Kroeger S, Sharitz RR (1997) Aboveground production in southeastern floodplain forests: a test of the subsidy-stress hypothesis. Ecology 78:370–384
Montgomery DC (2012) Design and analysis of experiments, 8th edn. Wiley, Hoboken
Morrison ML, Block WM, Strickland MD, Kendall WL (2001) Wildlife study design. Springer, New York

O'Connell J, Johnson L, Smith L, McMurry S, Haukos D (2012) Influence of land-use and conservation programs on wetland plant communities of the semi-arid United States Great Plains. Biol Conserv 146:108–115

Ogden JC (2005) Everglades ridge and slough conceptual ecological model. Wetlands 25:810–820

Peirce CS (1958) Collected papers of Charles Sanders Peirce, vol 7. Science and philosophy, Burks AW (ed). Harvard University Press, Cambridge, MA

Penman J, Kruger D, Galbally I, Hiraishi T, Nyenzi B, Emmanuel S, Buendia L, Hoppaus R, Martinsen T, Meijer J, Miwa K, Tanabe K (2006) Good practice guidance and uncertainty management in national greenhouse gas inventories, vol 1. General guidance and reporting. Intergovernmental Panel on Climate Change, IPCC Secretariat, World Metrological Organization, Geneva

Pfeiffer DU (2007) Assessment of H5N1 HPAI risk and the importance of wild birds. J Wildl Dis 43:547–550

Pierce CL, Sexon MD, Pelham ME (2001) Short-term variability and long-term change in the composition of the littoral zone fish community in Spirit Lake, Iowa. Am Midl Nat 146:290–299

Platt JR (1964) Strong inference. Science 146:347–353

Popper KP (1959) The logic of scientific discovery. Basic Books, New York

Quinn GP, Keough MJ (2002) Experimental design and data analysis for biologists. Cambridge University Press, Cambridge

Romesburg HC (1981) Wildlife science: gaining reliable knowledge. J Wildl Manag 45:293–313

Royall RM (1997) Statistical evidence: a likelihood paradigm. Chapman & Hall, London

Scheaffer RI, Mendenhall W, Ott L (1979) Elementary survey sampling, 2nd edn. Duxbury Press, North Scituate

Scheaffer RL, Mendenhall W, Ott L (1990) Elementary survey sampling. PWS-Kent Publishing, Boston

Skalski JR, Robson DS (1992) Techniques for wildlife investigations: design and analysis of capture data. Academic, San Diego

Smith, LM, Haukos DA, McMurry ST, LaGrange T, Willis D (2011) Ecosystem services provided by playa wetlands in the High Plains: potential influences of USDA conservation programs and practices. Ecol Appl 21:582–592

Stewart-Oaten A, Murdoch WW, Parker KR (1986) Environmental impact assessment: "pseudoreplication" in time? Ecology 67:929–940

Suren AM, Lambert P, Sorrell BK (2011) The impact of hydrological restoration on benthic aquatic invertebrate communities in a New Zealand wetland. Restor Ecol 19:747–757

Thompson SK (1992) Sampling. Wiley, New York

Thompson WL, White GC, Gowan C (1998) Monitoring vertebrate populations. Academic, San Diego

Tsai J-S, Venne LS, McMurry ST, Smith LM (2007) Influences of land use and wetland characteristics onwater loss rates and hydroperiods of playas in the Southern High Plains. Wetlands 27:683–692

Tsai J-S, Venne LS, McMurry ST, Smith LM (2010) Vegetation and land use impact on water loss rate in playas of the Southern High Plains. Wetlands 30:1107–1116

Turner RE (1997) Wetland loss in the northern Gulf of Mexico: multiple working hypotheses. Estuaries 20:1–13

Underwood AJ (1991) Beyond BACI: experimental designs for detecting human impacts on temporal variation in natural populations. Aust J Marsh Freshw Res 42:569–587

Underwood AJ (1994) On beyond BACI: sampling designs that might reliably detect environmental disturbances. Ecol Appl 4:3–15

van der Valk AG (2012) The biology of freshwater wetlands, 2nd edn. Oxford University Press, New York

Wester DB (1992) Viewpoint: replication, randomization, and statistics in range research. J Range Manage 45:285–290

Whitten SM, Bennett J (2005) Managing wetlands for private and social good: theory, policy, and cases from Australia. Edward Elgar Publishing, Northampton

Williams BK (1997) Logic and science in wildlife biology. J Wildl Manag 61:1007–1015
Williams BK (2012) Reducing uncertainty about objective functions in adaptive management. Ecol Model 225:61–65
Williams BK, Brown ED (2012) Adaptive management: the U.S. Department of the Interior Applications Guide. Adaptive Management Working Group, U.S. Department of the Interior, Washington, DC
Zimmer KD, Hanson MA, Butler MG (2001) Effects of fathead minnow colonization and removal on a prairie wetland ecosystem. Ecosystems 4:346–357

Student Exercises

Classroom Exercise

In wetland studies, there are usually a number of acceptable study designs to generate knowledge regarding an observed ecological pattern or process, effects of management or anthropogenic impacts, or approximation to a desirable condition or state. The key is use of a defensible study design that allows an investigator to make reliable conclusions and inference from the results of data collection and statistical analysis. Use of critical thought through the study design process prior to data collection will ensure dependable results that can be used to advance understanding of the wetland system being studied and hypotheses being tested.

Many wetland systems are actively managed for certain ecological responses through application of specific environmental conditions; for example, water-level manipulation. These ecological responses are typically production of food resources (e.g., seeds, tubers, invertebrates) for wetland-dependent wildlife. Development of management prescriptions to maximize food production typically requires a set of manipulative experiments to test wetland response to a variety of different environmental conditions. However, measurements of food resources in wetlands can occur without manipulated experiments by relating (e.g., correlated) resource production to observed environmental conditions. Such an approach does provide some evidence of influential variables relative to production of food resources, but lacks rigor to produce a complete understanding of causal relationships. Therefore, it is crucial for investigators to properly design studies of appropriate rigor to generate knowledge of sufficient scientific quality to meet the study objectives.

When managing wetlands for wildlife-forage resources, characteristic environmental conditions that are frequently tested include frequency and timing of wetland drawdowns (dewater to expose soils and sediments) and flooding that affects soil moisture and temperature; oxygen content in soil and water (i.e., aerobic vs. anaerobic conditions); and nutrient availability (e.g., nitrogen, phosphorus). Typically, investigators collect and measure invertebrate and plant response to (1) determine species composition in response to treatments and (2) estimate available biomass of forage resources. In addition, relative composition, distribution, and variation among studied wetlands of source populations (i.e., seed and egg banks) for food resources are characteristically considered influential on results but

1 Study Design and Logistics

not a primary interest in a study. Finally, the wildlife species of interest are enumerated in some manner to evaluate the response to available food resources. Much of this volume is devoted to descriptions and recommendations for collecting ecological field and laboratory data for wetlands. The purpose of this exercise is to develop a hypothetical field study of wetlands including development of experimental treatments, objectives, and testable hypotheses.

A public land manager has developed 16, 10-ha wetland units on the floodplain of major river in the southwestern United States. Each unit has been laser-leveled to (1) allow ease in flooding and draining each unit using water-control structures and (2) create a relatively uniform elevation across each unit. Each unit can be manipulated independently, but up to four adjacent units can be manipulated simultaneously. The goal of the land manager is to maximize annual production of natural foods for migratory birds, which use the units for migration and wintering.

The four treatments of interest that coincided with availability of water for flooding include a (1) control, (2) early growing-season drawdown, (3) late growing-season drawdown, and (4) early growing-season drawdown with a late growing-season flood to achieve soil field capacity. All wetland units can be flooded at any time during the migratory and wintering period.

Working in small groups, design a study to test the effect of treatments on forage production and wildlife use of the wetland units. Methodology to measure variables does not necessarily need to be included. In your study design include a description or response to the following questions or statements:

1. List 2–4 detailed study objectives
2. Provide at least two testable research hypotheses or predictions
3. Define and describe a study control
4. Provide a minimum of three dependent variables and three independent variables and the units of measurements for each
5. Describe a strategy for allocation of treatments among wetland units
6. Define the sample frame, study population, and extent of inference from the generated results.
7. Describe a potential sampling strategy for each objective
8. Include a statement on data management and storage

Chapter 2
Wetland Bathymetry and Mapping

Marc Los Huertos and Douglas Smith

Abstract Bathymetry is the measurement of underwater topography. In wetlands, development of bathymetric maps can have many applications, including determining water storage capacity and hydroperiod (depth and timing of flooding) of a wetland, assisting with wetland design and restoration and land use planning, and facilitating legal boundary determination. This chapter provides practical steps for mapping and modeling wadeable wetland bathymetry. By characterizing the bathymetry of wetlands, investigators can better understand key hydrologic, geomorphologic, and ecological processes of wetlands. Using standard survey equipment, investigators can plan and implement a relatively simple survey of wetlands. These data can be used to model and quantitatively analyze the surface area, volume, and bottom topography (bathymetry) of wetlands using standard geographic information system software.

2.1 Introduction

Bathymetry is the measurement of underwater topography. The word is a combination of two Greek words: *Bathus*, which means "deep", and *metron* or measure. Bathymetric maps may be charts with various depths printed at specific locations,

---

Disclaimer: Statements regarding the suitability of products (brand names or trademarks) for certain types of applications are based on the authors' knowledge of typical requirements that are often placed in generic applications. Such statements are not binding statements about the suitability of products for a particular application. It is the reader's responsibility to validate a particular product with the properties described by the product's specification.

M. Los Huertos (✉) • D. Smith
Science and Environmental Policy, Chapman Science Academic Center, California State University Monterey Bay, 100 Campus Center, Seaside, CA 93955-8001, USA
e-mail: mloshuertos@csumb.edu

contour lines of equal depth (depth contours or isobaths), or digital elevation models showing bathymetry in shaded relief. Historically, bathymetric maps were used for navigation (i.e., to prevent ships from running aground), but as field biology and environmental sciences have developed, bathymetric mapping has been applied to address a range of hydrologic and ecological questions in wetlands. Wetland bathymetric maps have many applications, including determining water storage capacity and hydroperiod (depth and timing of flooding), assisting with wetland design and restoration and land use planning, and facilitating legal boundary determination.

Hydrologic conditions in wetlands were typically monitored by determining wetland water level at a fixed point near the deepest part of a wetland. However, water level alone tells us very little about the distribution or evolution of hydrologic conditions in a wetland, and how these conditions influence physical, chemical, and biological characteristics. The usefulness of long-term data sets of wetland water levels would greatly increase if the data described not only the depth of water at a point in the wetland, but also the amount of total wetland areas that was inundated at a specific time (Haag et al. 2010). A survey of wetland bathymetry and the surrounding topography can help us understand how water moves through the landscape, and more specifically, how water influences the hydrologic budget of the wetland. The water budget of a wetland depends on the input and output of water where the storage capacity of the wetland and bathymetry determines storage. In addition, wetland bathymetry will influence residence time, flood retention, sediment trapping (Gallardo 2003; Takekawa et al. 2010), and regional surface and ground water interactions (Poole et al. 2006). Bathymetry plays a key role in plant and animal community dynamics (van der Valk 1981; Ripley et al. 2004) and wetland biogeochemistry (Faulkner and Patrick 1992). For example, the depth of the water and hydroperiod can control the presence-absence of taxa (van der Valk 1981; Bliss and Zedler 1997) and their interactions (Pechmann et al. 1989; Corti et al. 1997; Karraker and Gibbs 2009). In particular, the depth of the water may control vegetation dynamics, such as the establishment and growth of various emergent or floating plant species (van der Valk 1981; Keeley and Sandquist 1992). In summary, with adequate bathymetric maps, we can develop a description of the dynamic changes in wetland conditions instead of a simple snapshot (Takekawa et al. 2010). Moreover, we can translate periodic and widely distributed water-level measurements into a regional view of wetland hydrologic status (Lee et al. 2009).

This chapter introduces several survey strategies and methods for measuring wetland bathymetry, and discusses their attributes and limitations. An overview of the use of geographic information system (GIS) is also provided to assist students with an understanding of how to analyze typical bathymetric measurements (Fig. 2.1). Finally, we include an exercise at the end of the chapter that uses a pre-existing survey data set to provide students with experience in using GIS to analyze bathymetry measurements of a wetland.

2 Wetland Bathymetry and Mapping

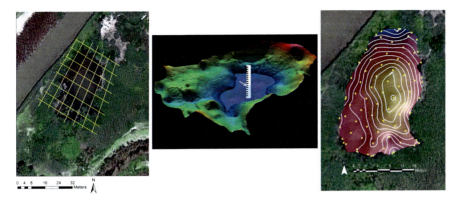

Fig. 2.1 Physical surveys of wetlands and wetland ponds can be used to develop digital models that have the advantage of being visually information-rich and rigorously quantifiable. Popular geographic information system (GIS) software was used to visualize the wetland bathymetry

2.2 Planning for Measuring Wetland Bathymetry

There are three criteria that must be considered when selecting an appropriate method to measure the bottom depths of wetlands: (1) desired accuracy, (2) wetland type, and (3) available resources (e.g., field gear and technology). The level of accuracy of the bathymetry measurements will determine the types of resources needed (e.g., field gear, instruments, and software) as well as the amount of time invested in collecting the measurements. The goals of the survey and mapping project determine the relative accuracy needed to complete the bathymetric analysis. For example, if the goal of the bathymetric analysis is to determine the water storage capacity of a wetland, the relative accuracy for collecting these measurements would be considered low. In contrast, the level of accuracy for collecting the measurements to determine sedimentation rates into a wetland would be considered high. Other goals such as determining the water budget and hydroperiod of the wetland would require medium accuracy.

In general, the type of equipment and time required to collect bathymetry measurements will be limited by whether the wetland is wadeable or non-wadeable. Wadeable wetlands are shallow enough to safely traverse, while non-wadeable wetlands are too deep for wading or may contain a substrate (e.g., muck soil) that is too difficult for walking. In the case of wadeable wetlands, the bottom topography can be measured using a meter tape, hip chain, rotating laser, total station, handheld global positioning system (GPS), or survey-grade GPS. Large wadeable wetlands (~ 0.5 ha) are usually treated similarly to non-wadeable wetlands, and the tools used are determined based on labor and efficiency.

Non-wadeable wetlands are typically large in aerial extent and the tools used to measure their bathymetry include boats that deploy a lead line or a sonar system for depth coupled with either an optical survey or a GPS system for positioning

Fig. 2.2 Graduate student operates a vessel-based terrestrial LiDAR unit as she creates a precise digital elevation model of the Los Padres Reservoir in the Carmel River watershed (Published with kind permission of © Rikk Kvitec 2014. All Rights Reserved)

(Fig. 2.2). Of course, some wetlands have special constraints. For example, quaking bogs are challenging because a portion of the water column is inaccessible from the surface, and might require the use of sonar or SCUBA.

Each survey involves collecting the position (X, Y) and elevation (Z) of a number of points on the landscape. The surveying equipment available to collect these data includes measuring tapes, meter sticks, lead lines, stadia rods, survey levels, laser levels, handheld GPS, total stations, survey grade GPS, and both ground-based and aerial LiDAR. A description of each type of equipment is provided in Table 2.1. The instrument selected will be determined by the factors listed above. We present the survey techniques that are applicable to most wadeable wetlands in order from simplest and least expensive to complex and most expensive. We also provide a description of LiDAR technology, which is most suitable for broad wetland environments such as estuarine tidal flats and large scale hummocky environments without vegetation.

Once the data are collected, there are several software packages available for creating maps from the raw data as well as conducting geomorphic analyses. Two of the common professionally used packages are ESRI ArcGIS and Fledermaus. Both are relatively easy to use. Fledermaus has more flexibility for rendering digital hillshade models, which can export models that can be viewed and rotated in a free viewer.

Defining the range of questions to be addressed in a survey of wetlands may help determine the type of methods to be used. Although the relative accuracy is important, other parameters should be defined. For example, will the field survey work be repeated over time? If so, then it may be important to set up permanent markers or benchmarks so the same transects can be used at a later date. Will the

2 Wetland Bathymetry and Mapping

Table 2.1 Description of commonly used survey instruments for collecting bathymetry data

Survey instruments	Description	Data collected	Notes
Meter stick	Layout orthogonal grid and measure the water depth at grid nodes with a meter stick	Depth, using ambient water surface as the datum. Depth precision of 1 cm is typical	Horizontal positions obtained with 50 or 100 m tapes and cross section stakes. Technique is limited by ambient water depth, so surveying at high water will capture more data. If water is flowing, the water surface (datum) will not be constant through space
Survey level	Lay out orthogonal grid for measurement. Horizontal telescope pointed at a calibrated leveling rod. "Autolevels" are able to fine-level themselves after initial coarse leveling	Elevation precision of 1 cm is typical	Horizontal positions obtained with 50 or 100 m tapes and cross section stakes. Requires two surveyors
Rotating laser level	Same as above, but rotating laser level emits a horizontal plane of laser light that strikes a receiver on a calibrated leveling rod	Same as above	Same as above. Can be operated by one person
Handheld GPS	Wide range of models available. Trimble Explorer is a common model	Coarse horizontal position and very poor vertical positions. Post-processing typically achieves 2 m horizontal precision	Used for horizontal mapping. Should not be used for vertical positions. Data are georeferenced
Total station	Spherical 3D points are obtained through laser-based distance measurements and ultra-precise horizontal and vertical angle measurements. Laser is shot at a pole-mounted prism	3D coordinates with 1 cm precision. With experience and research-grade equipment, sub-centimeter precision is possible	Total stations can obtain a precise 3D fix in a few seconds, so hundreds of positions can be shot in a day
Survey grade GPS	3D points are obtained by a tripod-mounted GPS antenna, with reference to satellites and either a local base station or regional set of base stations	3D position with precision limited by local conditions. One centimeter precision is possible under ideal conditions	Surveys are commonly achieved with a "rover" antenna that corrects positional errors by radio communication with a "base" antenna. This provides "precise" positions. "Accurate" georeferencing is achieved by correcting the base station data using fixed GPS stations in the region
Ground-based LiDAR	Tripod-mounted, vessel-mounted, or vehicle-mounted LiDAR scanner	Millions of 3D points are collected in a LiDAR scan. The precision is limited by vegetation, water, and GPS positioning (if it is a mobile system)	Excellent for expansive wetlands, such as large tidal marsh environments. Very complex data collection and post-processing system. Very expensive to purchase and operate

Table 2.2 Various products that might result from a bathymetric analysis

Mapping dimensions	Products
Two dimensions (X and Y)	Wetland perimeter and area (derived from ground surveys, aerial imagery, topographic maps, vegetation, soils)
	Total area
Three dimensions (X, Y, and Z)	Reference elevation and elevation datum (benchmark/ground control)
	Wetland bottom elevation (Z) at various X, Y locations
	Wetland water level (stage) and water depth
	Outflow elevation and potential surface connections with other wetlands (outflow/inflow)
	Wetland drainage basin boundary

wetlands map need to be placed in a larger geographical context with real world coordinates (e.g., longitude and latitude)? If this is the case, then GPS technology will play a role in the field, and you will use GIS tools and map projections to ensure the data are accurately georeferenced to real world coordinates. If the questions about the wetlands include biophysical features (e.g., vegetation patterns, geologic features, or evidence of animal activity) associated with the bathymetry, the survey work might need to map those features too. Successful characterization of the bathymetry will be guided by clearly defining the products that will be needed for conducting the bathymetric analysis (Table 2.2). The products desired will dictate which mapping dimensions will be needed and what type of analyses will be conducted. Finally, the available resources and budget will ultimately constrain what can be accomplished. In general, a well-defined question will result in efficient use of field and analysis time.

The goals of the wetland survey will dictate the boundaries, number of survey points, and needed resolution. The boundaries of interest may be defined legally (i.e., a jurisdictional wetland) or may include a larger context (i.e., the watershed contribution). In either case, the mapped wetland should include enough area outside the wetted area to avoid interpolation inaccuracies near the defined boundary of interest. "Resolution" is a broad term, generally indicating the smallest physical feature visible in the data set. Resolution is a function of survey point spacing, with higher resolution achieved by closer spacing of elevation data. Your choice of resolution will depend in part on the sources of variation in the wetland itself and how much of that variation needs to be captured, the number of sampling points you can afford to survey, and the precision of individual survey shots. For example, Haag et al. (2005) found that bathymetric data containing a high density of data points provided the most useful stage-area and stage volume-relations characterizing isolated marsh and cypress wetlands in Florida. Moreover, bathymetric maps generated from a low density of data points underestimated by 50–100 % the wetland area and volume over certain ranges of stages compared to maps generated by a high density of data points. This emphasizes the importance of collecting data from an appropriate number of data points when determining wetland bathymetry. From a pragmatic perspective, the size of the wetland feature,

2 Wetland Bathymetry and Mapping

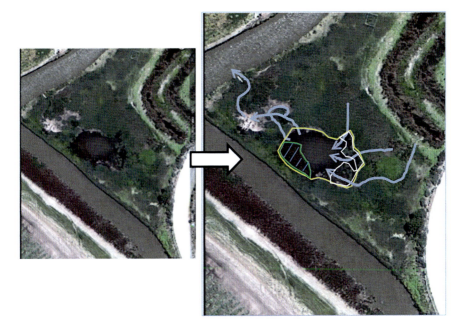

Fig. 2.3 An annotated aerial photograph in the project file indicates that Molera Wetland is a riverine wetland fed by a high water table and upland sources. *Outlines* show the areal extent of vegetative ecosystems present in May 2011, when the photo was taken. Surveys can delimit the true flow pathways and other details

number of sampling points, and resolution will all influence the resources needed to complete the work and ultimately, the quality of the results.

Planning, and obtaining the appropriate resources will increase mapping success. Prior to conducting a survey, we suggest you gather contextual data sets such as U.S. Geological Survey (USGS) topographic maps (http://topomaps.usgs.gov/), historical and recent aerial photographs, and regional digital elevation models. Many of these are available through public databases often overseen by state agencies in the U.S. This overview analysis can help determine the general surface flow patterns that fill and drain the wetland. Review of historic aerial photos can highlight temporal changes, such as gradual infilling of wetlands (sedimentation or land use changes), ecological shifts (such as vegetation changes), and seasonality. The broad view can help constrain the environmental questions and hypotheses, and will serve to plan the survey collection. If resources are limited for initial data collection, a quick tour through Google Earth (http://www.google.com/earth/index.html) can be a useful starting point.

During the planning and preparation process, the following parameters can be explored using the contextual data sets described above or in Google Earth. The general center of the wetland can be described in latitude and longitude or in some other coordinate system. The general setting can be described in terms of access, land use, probable disturbances, position in the watershed, vegetative types, and topography (see Fig. 2.3 as an example). The approximate elevation of the wetland can be derived from digital elevation models, topographic maps, or Google Earth. The general perimeter and area can be estimated, which will help in planning the number of survey points and the time required for surveying. While this broad overview can also provide initial insight on the types of equipment and gear that will be required for conducting a survey, a pre-survey site visit is essential.

In most cases, it is essential to contact property owners or public lands managers for permission to access a wetland. State or county permits may be required if biological sampling is part of the plan. Advanced planning ensures that the field work will not be interrupted or postponed, potentially leading to missed opportunities related to seasonal water levels.

Finally, before beginning field work, being aware of the appropriate safety measures is important. Safety measures may include bringing a first aid kit, adequate communications devices (walky-talkies, cell phones, or satellite phone), and personal protection gear (bug spray, sun screen, hardhat, boots, personal floatation devices, and safety vests). The necessary equipment will vary with local conditions (weather, proximity to infrastructure, etc.). In general, you want to be as prepared as possible to reduce the risks of accidents or injury.

2.3 Wetland Survey Techniques

2.3.1 Recording Field Data

There are many established techniques for bathymetric and topographic mapping. They all have one thing in common: their data are only as valuable as the quality of the comments and notes that describe the methods and features being surveyed. Without clear field notes, survey data are just numbers with no context. Standard survey notes should be adhered to if the goal is long-term monitoring. Professional-quality notes should be unambiguous, and understood by anyone who tries to re-survey the site. Because some long-term monitoring projects can span generations of students and professors, consistency and reproducibility are key features of data collection. Experience shows that a weatherproof notebook (such as "Rite in the Rain") and a #3H mechanical pencil lead will create a long-term archive of survey data and field notes. Electronic data records are in common use, but storage media change through the years, so maintaining a hard copy of original survey data and notes in the office is essential.

Information recorded in standard field notes should include date and time, site description (field sketches of key elements such as bench mark locations are valuable), purpose of the survey, type and serial numbers of survey instruments, names and roles of team members, general weather conditions, and wetland characteristics. The notes that record a specific survey should fully describe the benchmarks and datums that were used. Specific survey shots are recorded in a series of data columns. Depending upon the survey technique, these columns might

2 Wetland Bathymetry and Mapping

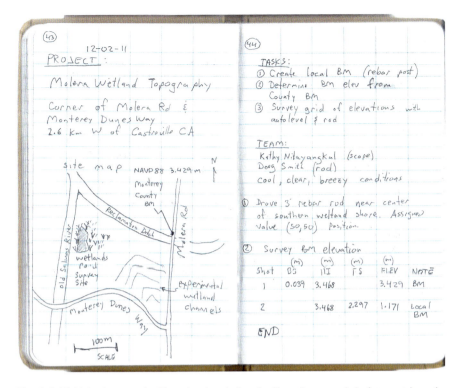

Fig. 2.4 Field book example. Note the descriptive details and map to help interpret how the survey was completed and where the benchmark (BM) is located

include, shot number (a number 1 through n), X coordinate (easting), Y coordinate (northing), Z coordinate (elevation), and a comment column where notes about each shot can be recorded. See Figs. 2.4 and 2.5 for field book examples.

The units of measurement used in each column should be explicitly noted in the column heading, and the units should not change within a column. The numerical values in the columns should reflect the precision of the measurement. For elevations, we commonly read to the millimeter, so an elevation entry might be 3.235 or 4.210, using the zero as a placeholder to show that we are still reading to the millimeter. Harrelson et al. (1994) is an excellent reference for standard environmental survey notes and abbreviations.

2.3.2 Surveying in the Field

Bathymetric survey design is driven by the goals and the available survey tools. Based on the available equipment (Table 2.1) and desired products (Table 2.2), the appropriate survey methods can be selected. For example, if the site is large, but

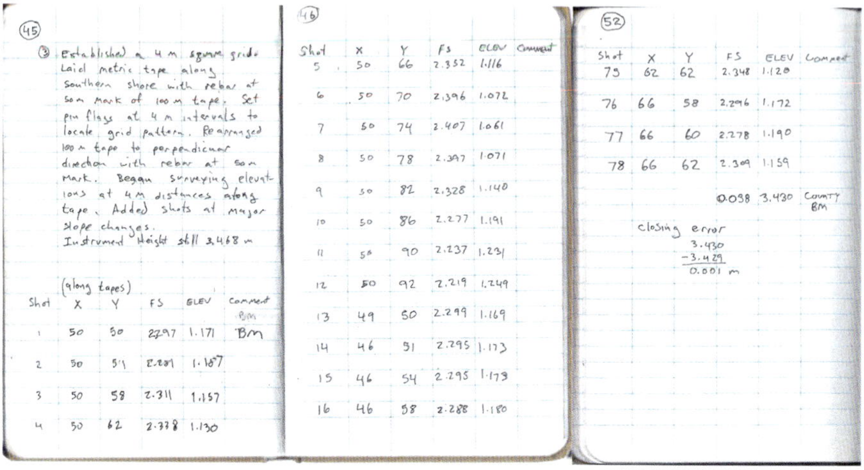

Fig. 2.5 Three pages from the field book showing clear notations for each shot taken during a survey

unvegetated, then LiDAR might be appropriate. On the other hand, if there is a dense of ground cover and the site is small, then an autolevel might suffice. Although automated survey equipment may appear simple, we have found that data processing can be time consuming, so there is a trade-off when the technology exceeds the products needed from the project. The following survey techniques can be implemented using a combination of tools and in conjunction with one another.

The goal of surveying is to collect horizontal coordinates (X, Y), and vertical ordinates (Z). These three-dimensional survey positions may be arbitrarily located in space (e.g., relative to a local benchmark, center of the wetland, or other feature), or they can be georeferenced, which means each point can be positioned on the globe relative to other features. If the data are georeferenced, the horizontal coordinates are often latitude and longitude, or northing and easting in a projected coordinate system. In the georeferenced data set, Z is elevation above sea level, referenced to a standard vertical datum. In some software packages, Z is considered "depth" below a datum such as sea level.

In all mapping projects, you must first establish a horizontal and vertical datum. The datum serves as the reference point from which all measurements will be referenced. Datums can be local (arbitrary) and based on points set in the field, such as rebar stakes, or they can be referenced to published locations. For example, the National Geodetic Service (NGS) has survey data available on the internet (http://www.ngs.noaa.gov/cgi-bin/datasheet.prl). The NGS data system provides a report on each benchmark.

A field survey establishes horizontal and/or vertical locations in relation to a starting point, which is called a benchmark. The selection of permanent horizontal and vertical reference frames is not critical if the wetland is to be surveyed and

Fig. 2.6 National Geodetic Survey Benchmark (*left*) (http://www.ngs.noaa.gov/) and local county benchmark (*right*) used for the Molera Wetland, California, USA

analyzed once. If the object of surveying is to map physical change through time, it is most advantageous to establish at least one long-term, stable benchmark near the survey site. The USGS usually uses brass monuments set in rock, a concrete pylon, or a pipe driven into the ground (Fig. 2.6). Ideally, there will be an NGS or USGS benchmark near your study area and we recommend that you use it. If not, you will need to establish a new local benchmark. This local benchmark can be a wide range of objects such as a chisel mark in exposed bedrock, nails and tin washers driven into a road, concrete pads used for street signs, fencing, or other public infrastructure. You must be sure to select or construct a benchmark that is vertically and horizontally stable, preferably for many years. All bathymetric surveys will follow these basic steps for data collection: (1) establish vertical datum, (2) establish horizontal datum, (3) record wetland water stage, and (4) measure the relative position of wetland features in the X, Y, and Z space. Developing the vertical and horizontal datums allow every survey shot to be referenced to a stationary and stable point—thus, allowing reproducibility and the capacity to measure precision.

The vertical benchmark is the starting point of any topographic and bathymetric survey. We will assume that the ultimate goal of your survey is to develop a "topographic" wetland model, which has elevation values rising in the uplands. "Bathymetric" surveying is analogous, but values rise as you descend into the wetland. Regardless of technique, all surveys will start from a benchmark. There are published benchmarks and local benchmarks that can be used (Fig. 2.6). Local (arbitrary) vertical datums are assigned elevations that are not based on published benchmarks. Assumed elevations can be based on water surfaces, elevations of fixed structures (such as outfalls or crossings), pool points, or staff plate elevations. We recommend you visit the NGS website (http://www.ngs.noaa.gov/) to access published benchmarks. The elevation datum on published benchmarks will either be referenced to the National Geodetic Vertical Datum (NGVD) 1929 or North American Vertical Datum (NAVD) 1988, which are merely vertical scales with a 0 m mark that corresponds to an estimate of mean sea level. You must recognize that even those scales are somewhat arbitrary, given the dynamic nature of sea level at all time scales. Nevertheless, the vertical datum places the wetland site in a vertical framework so that vertical positions can be compared to one another through time, and in the case of an NGS benchmark, referenced to sea level. Using a published benchmark to determine wetland elevations on an established datum is useful for relating the bathymetric data to other data sources such as Federal Emergency Management Agency flood data, tidal ranges, USGS stream gage records, and USGS topographic information.

Local benchmarks can be established in the field for a specific wetland, and referenced to an arbitrary datum. It is common practice to drive a 1 m long, 1.5 cm diameter rebar vertically into the ground within 1.5 cm of the ground surface for use as a local benchmark. Other local benchmarks can be established with spikes, chiseled "x" on concrete or boulders, and nails in pavement. If a high order of vertical accuracy is desired, the survey should use at least two benchmarks on a common datum and check elevations between the benchmarks regularly. Using more than one benchmark is good practice when establishing control for long-term bathymetric monitoring because it allows recovery of the benchmark even if one is lost. A rebar benchmark can be found (recovered) in future surveys even if it is buried by sediment in intervening years. Carefully sketched maps, GPS locations, a shovel, and a metal detector are standard tools for locating benchmarks.

A "staff plate" (sometimes referred to as a staff gauge) is an acceptable supplemental vertical benchmark (Fig. 2.7). A staff plate can be installed using a graduated meter stick extending vertically from the bottom of the wetland that allows for determination of the water surface elevation. The staff plate can be mounted to a piece of lumber or a metal stake that is driven into the substrate for stability. The 0 m mark on the staff plate is another arbitrary vertical reference for surveys and recording data. The water surface elevation can be converted to NAVD 88, or other external references if the 0 m mark (or any other mark) is related to the external reference by surveying to a nearby-published benchmark or via GPS survey.

If a local benchmark will not be referenced to NAVD 88, you can assign a convenient starting elevation to the benchmark. Standard practice is to select a round elevation value (e.g., 10 m), with the constraint being that it is high enough to keep all the elevations in your survey positive. This practice reduces the common math errors stemming from the use of negative numbers. Otherwise, it is strictly a matter of convenience. Any other arbitrary value will do, as long as you record it in your survey book as reference for future surveyors.

The next step will be to establish the horizontal datum and axial framework. The horizontal position of each elevation point in the survey must be recorded. The position can be considered a point in a Cartesian coordinate system (X, Y). The coordinate system requires defining the physical position of at least one reference point, and the direction of one of the axes (X-axis or Y-axis). The direction of the other axis is taken to be orthogonal from the first. If you are working in latitude and longitude (or UTM coordinates), the Y-axis is defined as the direction to true north.

2 Wetland Bathymetry and Mapping

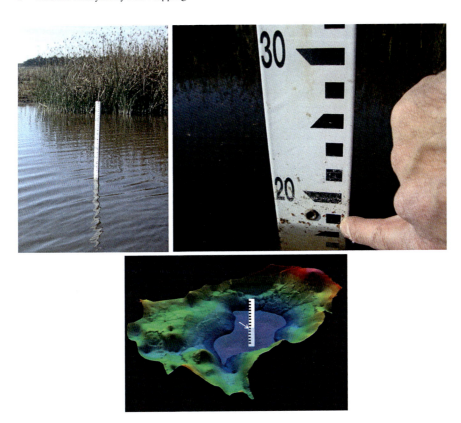

Fig. 2.7 Staff plates indicate the water level elevation with respect to the 0 m mark on the staff plate. The color blocks are 0.01 m tall, and each decimeter is numbered. By interpolating to the millimeter, the water level in the figure is 0.165 m. The staff plate can also be used to relate models with field measurements

Local (arbitrary) horizontal datums are not georeferenced to a standard map projection or coordinate system. Local horizontal datums are assigned arbitrary values for horizontal position relative to a local benchmark to which you have assigned a convenient position such as (0,0), (50,50), or (100,100). As with local elevation datums, it is standard practice to make the coordinates sufficiently large so that the positions will be positive values. Local horizontal datums can be based on fixed structures, such as described for vertical datum, or other objects such as rebar, bridge corners, culverts, or trees. A local horizontal datum allows all measured points to be placed in horizontal space relative to one another, but not necessarily referenced to other datasets that may be available. The other requirement for local horizontal datum to work is to define the direction of one of the axes. It is most convenient to define one of the axes to be the long dimension of the wetland. The axial directions can be recorded as compass bearings from a precise compass, such as a pocket transit.

Georeferencing places the survey shots into a geographic framework, such as latitude and longitude. Georeferencing can be achieved by surveying from a published benchmark, or by placing a GPS antenna above the local benchmark. Published benchmark datasheets will list horizontal coordinates for NGS benchmarks. These published coordinates are either scaled from a USGS topographic map or referenced to State Plane Coordinates based on a horizontal datum such as North American Datum (NAD) of 1983 (NAD83) or of 1927 (NAD27). Although many features are still referenced to NAD27, NAD83 is the official North American datum. It is important to note which datum is used if you plan to make maps that are spatially referenced. The horizontal datum places the wetland site in X, Y space and serves as the initial point from which all measured features will be referenced. Similar to the vertical datum, using a published horizontal datum allows the bathymetric data to be related to other sets of spatial data available.

Once the datum have been established, measurement of wetland features that define the wetland topography can be initiated. The process of measuring wetland feature locations will vary based on the technique employed. The basic bathymetric survey establishes the horizontal and vertical location of points throughout the basin relative to the benchmark. To measure the wetland features, one can establish points along a number of transects traversing the wetland or establish a grid of points. In general, if the wetland is a simple depression, a few transects or a simple grid might be enough to capture the bathymetric variation. But, if the wetland is geometrically complex and large, it might require more numerous measurements. For example, for such a wetland, you may need to establish survey points at a closer spacing where topographic variation is high compared to other parts of the wetland. Determining the density or number of points comes with surveying experience. One strategy is to shoot survey points at major breaks in the slope, but never farther apart than some predetermined value (e.g., 1 or 2 m). Additionally, more points should be used to define key hydrologic features such as the wetland boundary and outlets. If your subsequent analysis shows that more detail is needed, an additional field day can be used to fill in the missing information.

Depending on your objective, you may also be interested in water storage of the wetland. As such, it will also be important to record the wetland water stage while conducting the bathymetry survey because it will be the basis for determining a volume-stage relationship. Stage can be measured by either reading a staff plate (Fig. 2.7) if one is installed or surveying the elevation of the water with respect to your vertical benchmark, which can easily be accomplished at the water's edge.

2.3.2.1 Surveys Using a Taped Grid

The simplest technology for obtaining bathymetric data involves the use of a grid of points determined by long metric tapes placed at set intervals (e.g., every meter) or an equivalent method using a grid pattern (Fig. 2.8). Nodes (points) are created where the tapes cross. The nodes are sampled for elevation data. If the tape positions are referenced with rebar or by some other means, the site can be

2 Wetland Bathymetry and Mapping

Fig. 2.8 Grid layout and photograph of baseline tape with one crossing tape

resurveyed at precisely the same X, Y points through time. If the elevations are not resurveyed at the same place, elevation differences might be related to space rather than time. The following description provides a means of locating your survey points for time series of bathymetric change (e.g., monitoring siltation levels). You can place two pieces of rebar in the ground, and stretch a long metric tape between them. One of the rebar pieces should be held as the horizontal benchmark. That point will be considered the horizontal origin for the survey and assigned X, Y, and Z coordinates. The tape stretched between rebar stakes forms the baseline, and another tape (or tapes) can be set at right angles to the baseline tape along certain horizontal offsets. For example, if the crossing tapes are set at 4 m intervals along the baseline tape, and the elevations are recorded at every 4 m along the crossing tapes, the result is a 4 m square grid of points that can be reconstructed reliably at future times (Fig. 2.8). The choice of spacing will be based upon the time limits and precision requirements of the particular project. An analysis later in this chapter illustrates the benefits of closer spacing. Also, this grid pattern can be established with just two tapes. One is the baseline tape that does not move during the survey, and the other is the "crossing" tape that can be moved along the baseline tape for each subsequent transect. If many persons are involved, you should bring more crossing tapes which will speed up the process considerably.

Once the sampling points are established, the Z value of the wetland bottom can be obtained by determining depth from a vertical reference point, such as the water surface (Fig. 2.9) or by using an autolevel or rotating laser level which will be discussed in the next sections (Table 2.1). The easiest way to measure the Z value for each point in the grid is a depth measurement using a meter stick or survey rod. In this case, the water surface is the vertical datum from which the measurement is

Fig. 2.9 An example of a researcher recording a water depth measurement at a grid node near the staff plate. Note the falling field book. By using waterproof books, this is not a problem

made, but comparisons with future surveys will be problematic unless the water surface elevation present during the survey is somehow linked to the local benchmark using a staff plate, autolevel, rotating laser level, or more sophisticated gear. Features, such as grade breaks, water surface, and vegetation changes, can be recorded along each transect line. In addition to the grid point measurements, taped locations of the shoreline (0 depth) should be recorded. As with any survey, if the grid spacing is too coarse to capture major breaks in slope or details of interest, more Z values can be collected later, along with their corresponding X, Y positions. This technique used alone will only yield topographic information below the current water level; however, the other survey instruments we describe do not have that limitation.

2.3.2.2 Using an Autolevel

When the wetland vegetation is lower than eye-level, autolevel scopes can be used to survey land-surface elevations. While autolevel scopes are suitable for obtaining precise elevations, they are very poor for measuring horizontal positions, and we

2 Wetland Bathymetry and Mapping

Fig. 2.10 Autolevel surveys are a series of "shots" in which a rod reading is recorded. The colored blocks are 0.01 m tall, and every 0.01 m has a number. This shot indicates that the ground elevation is 3.215 m below the optical center or instrument height (HI) of the scope

recommend using a tape grid for positioning. Once a horizontal grid is established (see previous section), the scope and tripod can be set up as close to the wetland as possible, but with a clear view of all points to be surveyed. The tripod might need to be set up multiple times in different locations if the wetland is large, or if the line of sight is limited. Moving the scope requires the use of a "turning point" in the survey to keep the autolevel scope in the original vertical reference frame. In some wetlands, the canopy may be so dense that the use of a scope is impossible. As a substitute, a compass can be used to obtain direction, and a measuring tape can be used to determine the horizontal distance from a known location.

The autolevel is convenient not only because it can measure ground surface elevations beyond the water surface, but also because it can be used under any wadeable condition. The autolevel is an optical telescope with crosshairs that is mounted on a tripod and provides a level view no matter where it is pointing (Fig. 2.10). The scope person views a leveling rod held vertically by the rod person, and records the elevation value indicated by the intersection of the horizontal, center crosshair, and an elevation value marked on the rod (Fig. 2.10). Each reading of the rod is called a survey "shot." Shots are simply measurements between the ground where the rod is placed and the optical center of the scope, indicated by the horizontal cross hair (Fig. 2.10). The standard notes for an auto level survey are shown in Fig. 2.5.

The basic autolevel set up includes the following steps. First, a surveyor must find a location where the instrument has a clear view of the wetland to be surveyed as well as any benchmarks that will be used to vertically control the survey. Next, the tripod feet must be firmly set so that the tripod top height is at a comfortable viewing elevation and the mounting bracket is approximately level. Then, the autolevel is mounted to the bracket, without over-tightening the mounting screw. Precise instrument leveling is accomplished with reference to a bubble level as you adjust three leveling wheels on the autolevel base. When the instrument is leveled in this way, yet more precise leveling occurs within the instrument, which is the basis for the instrument name. Once the instrument is level, the horizontal cross hairs in the scope are focused for an individual's eyesight. The cross hairs delineate a horizontal plane as the instrument is rotated about a vertical axis. In other words, everything that the horizontal cross hair hits is at the same elevation. This elevation is called "instrument height," and the value is denoted "HI" in survey notes. The value of the instrument height is determined by the first shot of the survey when the rod is placed on a benchmark (BM) of known, or arbitrarily assigned, elevation. The shot used to determine the HI is called a "backshot" (BS). The instrument height is determined by summing the BM elevation and the BS reading. The HI value is assumed to remain constant, unless the instrument is moved during the survey.

Once the HI is determined, you can survey the ground elevations of any rod positions where the rod can be seen by the scope person. The shots used to determine unknown ground elevations are called "foreshots" (FS). The ground elevation for each FS is determined by subtracting the FS from HI. All that remains is to place the rod on the ground at the grid points determined by the tape grid described above, record the FS, and calculate the elevation. Then, each X and Y position will have a corresponding elevation Z. As with the water depth method, and any other method, additional shots should be taken at many places along the current shoreline and at any points required to capture the details dictated by your survey goals. The last shot of the survey is the "closing" shot. It is a FS taken with the rod on the BM on which you opened the survey. The resulting ground elevation (HI-FS) should match the real (or assigned) BM elevation. The mathematical difference between the elevation from the closing and the real elevation is the "closing error." Closing error is a measure of the precision of the survey. The source of survey error is typically due to physical changes in the tripod or tripod feet positions that change the HI or scope leveling. Given your calculated closing error, you must decide if the precision is acceptable. Based on our experience, autolevel surveys that last 1–2 h typically may have a closing error of less than 1 cm. Errors greater than 1.5 cm are uncommon. Resurveying may be necessary if higher precision is required. If low precision is acceptable, then a larger closing error is allowable.

A standard autolevel practice is to perform a "two-peg" test of instrument calibration prior to a survey or sporadically throughout the survey season. The two peg test involves firmly driving two pegs (A and B) in approximately level ground separated by approximately 30 m. The tripod and scope should be set in the middle of the two pegs and a rod reading should be recorded on pegs A and B. In your notes, the rod readings should be recorded as "a" and "b," respectively. The next step is to move the tripod and scope as close as possible to one of the pegs, but not so close that the rod cannot be read. From this location, another set of readings will be recorded with the rods back on pegs A and B, but record the shots as "c" and "d" in your notes. The difference between "a" and "b" is a measure of the difference in elevations of the tops of pegs A and B. An independent estimate of the difference

2 Wetland Bathymetry and Mapping

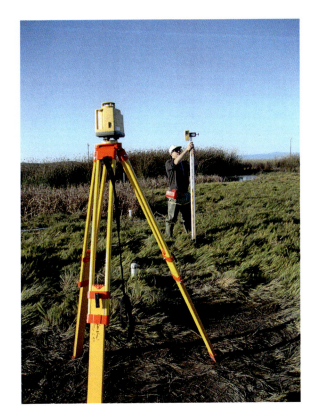

Fig. 2.11 An example of rotating laser equipment, which can be used by a single person

is the difference between c and d. Therefore, absolute value of (a-b) should equal absolute value (c-d). If they vary by more than a few mm, consider instrument calibration prior to surveying.

2.3.2.3 Rotating Level Laser

A direct substitute for an autolevel is a rotating laser level (Fig. 2.11). A rotating laser level performs a similar function, but emits a laser (usually red) that can be used to measure the vertical distance from a level plane created by the spinning laser beam. The same notes and data are taken (e.g., HI, FS, and BS), and horizontal control is still provided by the tape grid, but a rotating laser replaces the autolevel on the tripod. Some rotating level lasers have self-leveling servo motors, while others must be leveled by hand using an integrated bubble level and three leveling screws. Once leveled, the rotating laser emits a laser beam from a lens that is spinning about a vertical axis. The beam describes a horizontal plane that represents HI. There is a laser sensor attached to the survey rod that emits a beeping sound when the beam hits the sensor. The rod end is placed on the BM or wetland surface, and the rod is telescoped up or down until the sensor cuts the laser. Then, the rod reading gives the distance from HI to the ground surface as before.

2.3.2.4 Pool-Point Radial Survey Method

Modern, high-precision survey instruments can collect many three dimensional points with little effort. These types of surveys (e.g., pool-point radial survey, total station, real-time, kinetic GPS (RTK-GPS), and ground-based LiDAR), do not require an external grid for determining locations. Instead, they are able to generate a network of X, Y, and Z locations in a digital format which often leads to data sets being composed of thousands or millions of X, Y, and Z point locations that require specialized software for processing beyond the scope of this chapter.

The pool-point radial survey method relies on mapping the maximum wetted perimeter and radial transects with X, Y, and Z values. In simple wetlands, you can take relatively few measurements to obtain accurate bathymetry estimates with two relatively quick site visits. During the first visit, you insert a stake at the pool-point or deepest part of the wetland and then measure ground surface elevation and location from the pool-point to the perimeter along 3–5 radial lines. More radial lines may be required if the wetland is topographically complex. It is important to obtain an adequate number of elevation readings above the maximum height of the wetland so there are no interpolation errors near the high water mark. Later, when the water is at its highest, you survey the perimeter of the high water edge and the stake at the pool-point, which will provide the elevation of the pool relative to the maximum pool depth. This survey can be conducted with a meter tape and stadia rod, lead line, total station, rotating laser, handheld GPS, or survey-grade GPS. The simplest form of pool-point radial survey design is measuring along the long and short axis of the wetland.

If using a meter tape and stadia rod or lead line, the depth of the water at point locations are recorded along the meter tape. One limitation of using water depth as the measurement for establishing elevation is that you are limited to the area of wetland that is inundated. If the survey is conducted at the highest stage of inundation, you can maximize the bathymetric coverage. A total station or GPS that collects horizontal and vertical position can be used in either wet or dry conditions and collect a complete data set regardless of water stage. Using a handheld laser is quite rapid and efficient, as all the dry and wet measurements can be recorded in less than 20 min for a 50 m diameter wetland (Wilcox and Los Huertos 2005).

2.3.2.5 Total Station

Total station equipment ranges in functionality from basic point and shoot (i.e., aim the station at a prism pole and record the X, Y, and Z values) to fully robotic scanners (Fig. 2.12). The basic principle is that the instrument sends out a laser pulse in a known direction and calculates the position of the ground by analyzing the laser signal that is reflected back to the instrument. Some total stations require a prism on a rod to create the reflection, while others can receive the laser reflected from the

Fig. 2.12 An example of total station equipment, which can be used to rapidly capture precise X, Y, and Z points

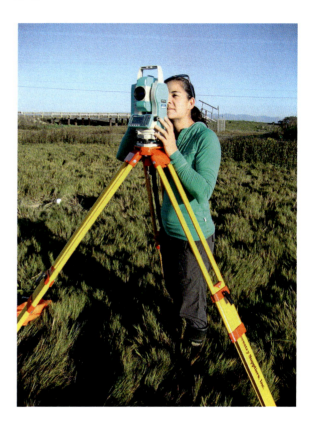

ground or vegetation. In general, the instrument calculates the horizontal and vertical angles of the aimed laser beam and then calculates the distance to the target by analyzing the reflected light. It converts the resulting spherical coordinates into X, Y, Z coordinates for export to a spreadsheet. Although high precision positions can be used to analyze wetland geometry, the geospatial data are not referenced to any external vertical or geographic reference frame unless the survey is intentionally linked into those frameworks by incorporating published benchmarks or local benchmarks that have been referenced by GPS work. We also note that the use of reflectorless total stations is not practical in wetlands where the laser would be reflected from dense vegetation or water rather than the ground.

2.3.2.6 Real Time Kinematic GPS

The RTK GPS is a positioning system that uses two GPS receivers: a base station and rover (Fig. 2.13). The base station can be positioned over a local BM to determine its georeferenced position, or over a published BM with a known georeferenced position. If the base station is set over a local BM, it should be left

Fig. 2.13 An example of real-time, kinetic GPS (RTK GPS) equipment. The base (*right*) and rover (*left*) can communicate with each other up to several kilometers with radio signal booster. RTK GPS is efficient and requires only one operator

there for up to several hours to record GPS signals. This long record will average out most of the error associated with instantaneous GPS positions. The base station position can be further refined by differentially correcting the data to long-term GPS stations in the region. Additionally, NOAA maintains a free web service (OPUS) for correcting base station data (http://www.ngs.noaa.gov/OPUS/). The differentially corrected positions commonly have less than 1 cm of error in horizontal position and less than 2 cm error in vertical position.

The rover is a GPS antenna that is mounted on a hand-held staff of known length. The rover staff can instantly record a position throughout the wetland. The point position is automatically "corrected" in real time by radio communications with the base station. The same satellite errors affect the base station and rover, but the base station knows its position. Thus, it also calculates the time-specific errors and can correct the rover positions. At a rate of one point per 15 s (including moving from one position to another), it is possible to collect several hundred precise survey points in a survey session. Wilcox and Los Huertos (2005) describe a simple and rapid method for bathymetric mapping using a total station and GPS.

The great value of RTK GPS is that each point is independently georeferenced, and will plot precisely on existing regional map data. The limitations of this technology include potentially poor satellite positions, and the inability to operate when a tree canopy or valley walls block satellite reception.

2 Wetland Bathymetry and Mapping

Fig. 2.14 An example of mobile terrestrial LiDAR positioned by RTK GPS and an inertial motion sensor that was used to map large tidal wetlands in central California

2.3.2.7 Light Ranging and Detection Technology

LiDAR technology offers the ability to survey extensive wetlands with great efficiency and precision. LiDAR is an optical system that sends out several thousand laser pulses each second and records various aspects of the reflected light. In topographic surveys, the primary variables are the direction and distance the laser beam traveled before it was reflected back to the sensor. Those variables are used to calculate an X, Y, Z position to the reflecting surface (e.g., plant, ground, building, etc.). LiDAR has the advantage of shooting thousands of X, Y, and Z points per second from a plane or terrestrial platform, but it has a disadvantage of not being able to easily penetrate water or very dense wetland vegetation. There is currently some experimentation to achieve better water penetration using various light wavelengths. Aerial LiDAR data sets are available from public internet sources including the National Center for Airborne Laser Mapping (http://www.ncalm.cive.uh.edu/). Currently, terrestrial LiDAR scanners are not widely available. California State University Monterey Bay has engineered a mobile LiDAR system that can be attached to an all-terrain vehicle (Fig. 2.14) to precisely digitize soil erosion rates. The early results are very promising, but the instrumentation is beyond the budgets of most practitioners.

2.4 Modeling and Visualization of the Bathymetric Surface

Once the field survey work has been completed, the next step is to create a bathymetric surface. The data are entered into a spreadsheet (e.g., Excel) and processed to create a digital representation of the surface. It is important to note that the depths between the sampling points are interpolated using one of a variety of methods. The points are used to create surface models either as a raster or a triangulated irregular network (TIN) surface model. A raster is a grid of evenly spaced elevation values (points) created by interpolation from the survey data. A TIN is created by making a network of irregular, nonoverlapping triangles between the survey data points. By using relatively affordable software such as ArcGIS with extensions (Spatial Analyst and 3D Analyst), the tabulated data (X, Y and Z) can be calculated to develop a bathymetric surface for display and further analysis. Typical analyses can determine water elevation (stage), and the wetted perimeter, area, and volume as a function stage.

GIS software is now used in all walks of academic and professional environmental science. In the following discussion, we assume that the reader has used GIS software such as ESRI ArcMap. Each new version of ArcMap provides slightly different ways of achieving the desired results we want, so some steps we describe below may become outdated with newer software versions. However, we are confident that the general principles will apply far into the future.

The general steps toward wetland visualization and geometric analysis are:

1. Enter or import the X, Y, and Z data into a spreadsheet;
2. Save the file in a format readable by ArcMap;
3. Import that file to ArcMap;
4. Produce an ArcMap point file;
5. Create a digital elevation model (DEM) by interpolating the data into an elevation raster, or by making a TIN; and
6. Use the DEM to visualize and analyze the wetland structure.

We provide a step-by-step process below using a wetland example.

2.4.1 *Molera Wetland GIS Analysis Example*

2.4.1.1 Data Preparation and Import

Molera Wetland, which is located along the central California Coast, was selected to provide an example of surveying and data analysis. The example also serves as an exercise at the end of the chapter. To follow our example, use the data available via this weblink: https://sites.google.com/a/csumb.edu/marc-los-huertos/home/molera-wetland-bathymetry. Download the rtk_gps_wetlands.xlsx file in a folder called Molera.

Fig. 2.15 ArcGIS map of Molera Wetland with the survey points

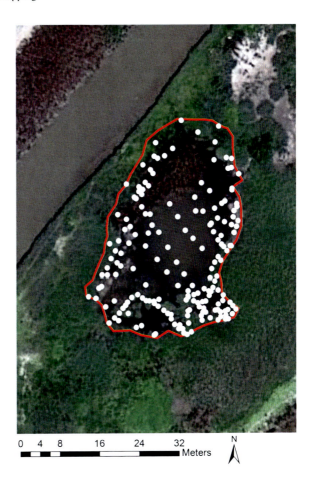

Approximately 270 positions were shot using RTK GPS during one afternoon at Molera Wetland (Fig. 2.15). The base station was placed on a known benchmark. The survey focused on the pond edges because they were more complex than the central part of the wetland pond. In ArcMap, we drew a polygon around the pond perimeter and used the 188 points inside the polygon in the following geometric analysis. Next we entered the data into a spreadsheet with columns labeled "Easting", "Northing", and "Elev". In general, those columns can also represent any X, Y positioning system that was used in the survey, including an arbitrary local survey framework.

In some cases, one must make vertical or horizontal adjustments (e.g., adjust the Z value to relate to stage or NAVD88 for example). We used the GPS system to provide output in WGS 84 UTM, Zone 10 North meters as the horizontal reference and NAVD88 meters as the vertical reference, so no further adjustments were required.

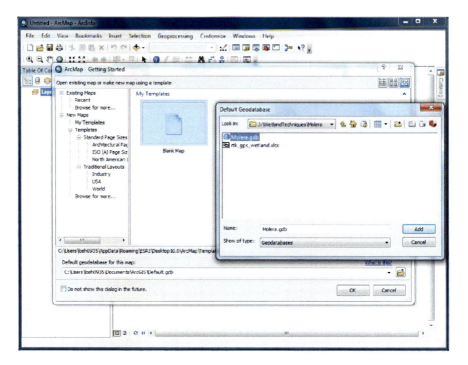

Fig. 2.16 Creating a new geodatabase is the first step when you start ArcMap v10. This screen shows a new directory where we created a new file by selecting the new geodatabase icon (a small grey round cylinder, often used as a database symbol). We then add this file and select OK in the "Getting Started" window

To follow this example, you will need to use ESRI's ArcMap. In the new version of ESRI's ArcGIS, ArcMap v.10 uses a default geodatabase that we will redefine. To begin, you should click on the small folder icon near the bottom of the "Getting Started" window and navigate to the Molera folder using the "Connect to Folder" icon. Once the Molera folder is selected, you should click on the "New File Database" icon and rename it as "Molera" (Fig. 2.16) and click okay. Once this file (geodatabase) is created, you can open the excel file and save the datasheet as a comma delimited ("comma separated values" that is abbreviated as csv) file, which can be done in any spreadsheet software and then add the csv file to the map, using File > Add Data > Add XY Data. You then will select the csv file and assign the X direction as the "Easting" column, Y direction as "Northing" column and Z as the "Elev". You must be sure to assign the appropriate projected coordinate system (Projected Coordinate System > UTM > WGS 84 > Northern Hemisphere > WGS 1984 UTM Zone 10N.prj). ArcMap will give an error (warning) because there is no object-ID field, but you can ignore the warning and proceed by clicking okay. After clicking okay, you should be able to see points displayed on the map. The map should then be saved in the newly created directory with an appropriate name (e.g., Molera). The next step will be to create a shape file, which is a specific file structure

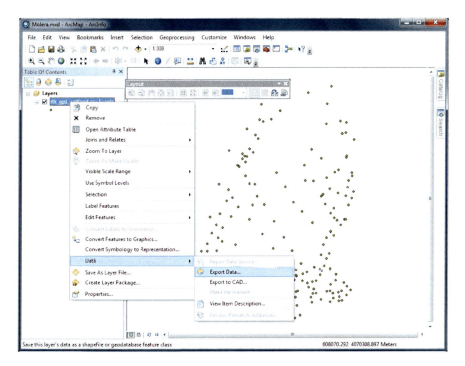

Fig. 2.17 A screenshot of the process of preparing to export the csv file into a shapefile

for maps in ESRI software. You can begin this process by right clicking on the name of the imported file "Events" in the "Table of Contents" in the left panel and choose the Data > Export Data (Fig. 2.17) as a shape file. After you have exported the data as a shape file, you will click on the folder icon and select "shapefile" in the bottom dialog box as the file format. You must be sure that you are saving the file in the Molera directory and as the correct named file; we used "GPSPoints" as the shapefile name. After completing this step, you can remove the original .csv file (by right clicking on it and selecting remove) to clean up the ArcGIS "Table of Contents" and save your Map Document.

Now, you are ready to plot the survey points, which will also allow you to make sure there are no data entry errors (e.g., outlier points located far away from the cluster of survey points). If you observe outlier points, you should check the data entry to make sure the coordinates did not include a typo. By right clicking the database file > plot xy, the data will be prepared for display on the screen. The resulting "point file" will be used in the following analyses. You can also bring up an aerial image, which is available on the website. You should see the points match the extent of the aerial photo. If you do not, then the projection may be incorrectly defined. The exported map should appear as in Fig. 2.15.

2.4.1.2 Creating a Digital Elevation Model

For creating a DEM of the wetland, we recommend using either the "natural neighbor" method or the "kriging" method. Both methods are commonly used for creating a DEM, but there are many choices for grid interpolation methods. Given that we are working with less than thousands of data points, you can create a TIN or you can create a surface by interpolating points to make a raster. We describe kriging in this example. We created a synthetic wetland from our data and sub-sampled it using a variety of sampling grid spacing to synthesize the accuracy achieved by different levels of effort in surveying (Fig. 2.18). We then created both TINs and krig DEMs to illustrate how accurately each one represented the original synthetic wetland in terms of volume, which was analyzed at a variety of depths.

Figure 2.18 shows the rate at which accuracy improves as more survey shots are taken, which results in tighter survey grids being used. However, you should also be aware that there is much less accuracy at lower water stage compared to higher stage. From our experience, we found that kriging improved accuracy between 2 and 10 % in the 5 m grid survey and by 11–30 % in the 10 m grid survey. The results indicate that the advantage of kriging increases markedly when the survey has fewer shots to control elevations in the Molera Wetland.

We now describe the procedure for kriging. Using the "ArcToolbox" (icon with a red tool box), select Spatial Analyst > Interpolation > Kriging with a hammer symbol. Next, you should select the point features created and select the Z value as "Elev". Finally, you must define the results into a DEM folder within your project directory. We used the default options for this kriging. ArcGIS creates a default output cell size, but we rounded the value to 0.1 m. Changes to the "Maximum Distance" for kriging "search radius" might improve the output depending on the bathymetric variability, but we left the value blank (Fig. 2.19). The resulting DEM will appear in the project.

Finally, we created a polygon shape file as a mask that can be used to trim the kriged surface so that we do not extrapolate elevations beyond the GPS collected points. To do this, we opened the ESRI ArcCatalog program and navigated to our project via the Folder Connections where you can right click the folder > new > shapefile and select polygon as file type. We call it "KrigMask," and assign the project's coordinate system. You now add the newly created shapefile to the map in ArcMap and open the editor menu where you select the correct shape file to edit and then go to construction tool and select polygon. After selecting polygon, you can create a polygon on the boundary of the wetland points, and use the aerial photo to help define where to click each point. To end polygon construction, you must double click. You can now save your edits, and then click "stop editing" to complete the process. To further refine the DEM, you should select Spatial Analyst Tools > Extraction > Extract by Mask and create a new trimmed DEM (e.g., Krig_Trim) using the DEM as a raster and the mask shape file (Fig. 2.20).

2 Wetland Bathymetry and Mapping

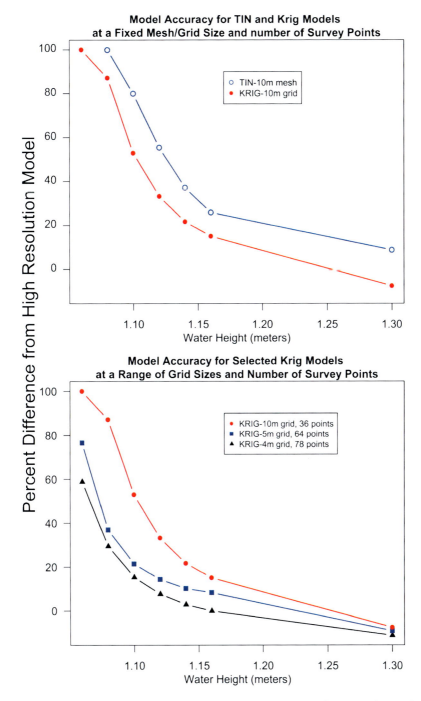

Fig. 2.18 A comparison of the accuracy of krig versus TIN models for calculating wetland volumes at various grid sizes and number of survey points shot

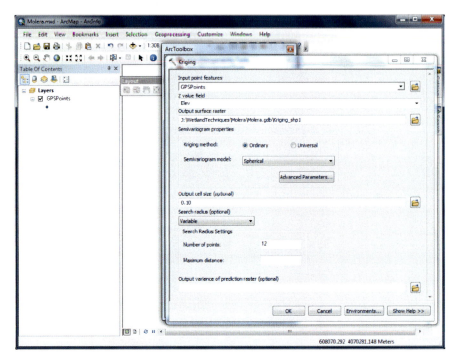

Fig. 2.19 An example of krig dialog box

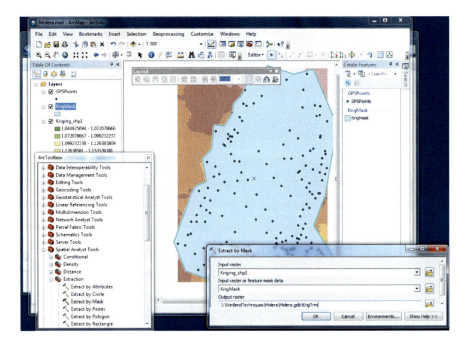

Fig. 2.20 A screenshot of the process of preparing to trim the map

2 Wetland Bathymetry and Mapping

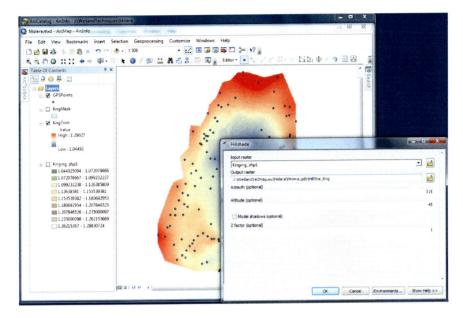

Fig. 2.21 An example of the hillshade dialog box

2.4.1.3 Visualizing Wetland Geometry

The DEM can be displayed so each range of depth can be associated with a color. This operation will color the DEM by elevation. You have many choices, and the selection will be dictated by the information you are seeking. Two standard coloring methods are 1) "stretched" color ramp that gives a continuous gradation of color from high to low elevations or 2) "classified" which gives more information and more control. For stretched color ramp, you will use the following sequence. First, you should right-click the DEM Filename > Properties > Symbology > stretched, and select the color ramp that you think is appropriate. In addition, you can create a shaded image called a hillshade to show topographic variation. From ArcToolBox, you should select 3D Analyst > Raster Surface > Hillshade and make sure the unmasked DEM is the selected file in the dialog box. You will then extract the file using the mask polygon boundary shapefile as before. You will need to control the sun angle and azimuth for illuminating the digital surface (Fig. 2.21). In this example, we used the default values. We suggest you experiment with this option to determine how it influences the results. This process may take some time depending on the speed of the computer, so you should be patient. In this example, the display defaults to a categorical color scheme because the surface is fairly flat, thus, not very useful. To enhance the hillshade, you can manually change the hillshade symbology by setting the high value to 255 and low value to zero. This will create a reasonable hillshade grayscale. To do this, you will right click the hillshade on the left side of the screen and select properties. In the symbology tab,

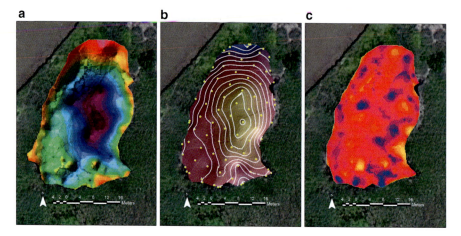

Fig. 2.22 Examples of differences in wetland analyses using kriging. (**a**) Colored hillshade of the synthetic wetland pond created from kriging our 189 GPS shots. (**b**) Colored and contoured hillshade created from kriging the superimposed 4 m grid shots (*yellow*) that were used to subsample the wetland surface in **a**. (**c**) Map of the surface differences between the original model (**a**) and the subsampled model (**b**), with differences mapped as color intensities. *Yellow* indicates areas where the 4 m grid model underestimated the depth, and *blue areas* indicate an overestimate. The average difference was not significant at 95 % confidence level. The maximum local differences were ±0.04 m, indicating a very good model comparison

you will select stretched, set the stretch type to minimum-maximum, check the edit high/low levels box and type in 255 for the high value and zero for the low value.

The DEM can be displayed so each range of depth can be represented. Much more technical information about your wetland can also be displayed by draping (layering) data which may include contour lines, vegetation layers, or elevation coloration on top of a hillshade. The basic technique is to display both the data layer (e.g., colored DEM) and the hillshade in the same map view. For example, Fig. 2.22a illustrates that effect in a Fledermaus project. We can also improve the visual effect by making the layer semi-transparent (Fig. 2.22b), which can be accomplished by right-clicking on the Hillshade > Properties > Display > Transparency and then using trial and error on the transparency level to obtain the desired effect (e.g., in our example, we used 35 %). You may add contours using 3D Analyst > Raster Surface > Contour. We used a 0.2 interval and a base contour of 1.05 (approximately the deepest point). To make the lines more visible, we changed the color to white and simplified the map by turning off other layers (Fig. 2.22b). Figure 2.23b illustrates that effect in an ArcMap project. Further detail can be added as artwork by exporting a map, and using an art program such as Adobe Illustrator. Figure 2.22c demonstrates an analysis of the surfaces mapped in the original and subsampled models.

2 Wetland Bathymetry and Mapping 81

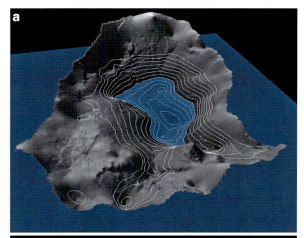
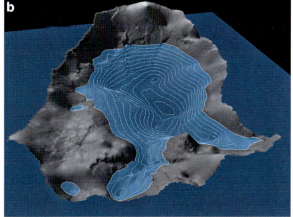

Fig. 2.23 A digital model of Molera Wetland projected using Fledermaus software. The contour interval is 0.02 m and begins at 1.05 m stage. (**a**) Contour lines help to visually define the location of the point of zero volume (pzv). A transparent *blue plane* inserted at an elevation of 1.09 m helps visualize the complexity and general shape of the wetland at that stage. (**b**) A transparent *blue plane* inserted at a stage of 1.16 m indicates that this stage is very near the point of incipient flooding (pif) for this particular depression. If the water surface were higher, it would flood to adjacent landscape elements

2.4.1.4 Quantifying Wetland Geometry

There are many elements that can be measured in a digital model of wetland topography and several ways to calculate their values. Some of the elements that researchers may need to know include what is the deepest point in the wetland, what is the point where water may overflow to the next basin, and what is the water volume and surface area of the wetland under a variety of different water levels. The deepest point in a wetland is referred to as the point of zero volume (pzv). The pzv will be the last refuge for fully aquatic organisms as the wetland dries. It is the lowest elevation value in the DEM (Fig. 2.22b). The point of incipient flooding (pif) is the elevation where water spills from one wetland depression to another, or to the adjacent terrace. It is important to remember that the pif is not considered the highest elevation in the digital model because water will "spill over" at a "saddle", but is a low point between basins. This is clearly illustrated in the Fledermaus project of Molera Wetland (Fig. 2.23a,b).

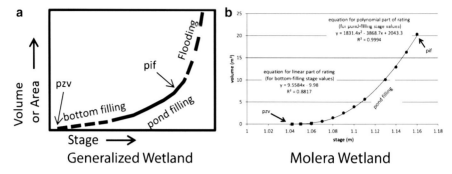

Fig. 2.24 A representation of the relationship between a wetland's water stage and a wetland's volume. (**a**) Wetland ponds can be geometrically complex, so more than one equation is needed to adequately model the volume to stage relation. The simplest wetlands will have a minimum of three volume zones (bottom filling or point of zero volume [pzv], pond filling, and flooding or point of incipient flooding [pif]) as they are analogous to a bathtub with a wide, relatively flat bottom, steep sides, and overflow. (**b**) The bottom of Molera Wetland is relatively flat, and the contour lines are far apart (refer to Fig. 2.23a,b for a representation of its bathymetry), so an increase in stage which inundates the lowest contours results in a very small increase in wetland volume. Nearer to the perimeter, the wetland has relatively steep sides, and the contours are closer together, so that an increase in stage results in a much larger increase in wetland volume

Table 2.3 The relationship of water stage to wetland area and volume in our Molera Wetland example

Water stage (m)	Area (m^2)	Volume (m^3)
1.05	3.54	0.007
1.07	61.5	0.58
1.09	120	2.3
1.11	183	5.3
1.13	256	9.7
1.15	360	15.8
1.17	536	24.7
1.19	748	37.7

Wetland water volume and surface area can be determined at a range of water levels. To calculate water volume and surface area, you will select 3D Analyst Tools > Functional Surface > Surface Volume > to input your trimmed DEM filename, indicate the water elevation for which you want the calculations, and then indicate that the analysis is "below" the plane. You should be sure to create an output file. This output text file stores the analysis results. We placed our file in a folder called "stage_vol directory." If wetland volume or surface area is calculated for a range of stages, a visual graph and mathematical "rating" equation can be created that relates stage to volume or area (Fig. 2.24). Rating equations can be linear, power, or polynomials, and different parts of the data set might require different equations (Fig. 2.24). It is important that you use good judgment in limiting extrapolation beyond the data, given that wetland geomorphology can be very complex as demonstrated by our example of how changes in stage can dramatically change wetland area and volume (Table 2.3). For our calculation of wetland area and volume at different stages, the calculation output was appended to the text file with each new stage we used when we used ArcGIS version 9.3, but in version 10, we needed to create a new output file for each stage, which is a bit tedious.

2.5 Conclusion

Bathymetric data and analysis can refine our understanding of wetland status and the impacts of human activities on wetlands. The mapping and analysis of wetlands requires several distinct steps that include planning, data acquisition (via field surveys or obtaining digital data), importing data into mapping software, visualization, and analysis. Because the precision and accuracy of wetland bathymetry can play an important role in understanding wetland structure and function, the technology used have become increasingly sophisticated. Survey equipment accuracy and efficiency have increased dramatically in the last 20 years, whereas, software and mapping programs have become more powerful, readily available, and user friendly. The ability to understand and effectively use bathymetric mapping and visualization techniques will convey an advantage to anyone who is interested in wetland science and management.

References

Bliss SA, Zedler PH (1997) The germination process in vernal pools: sensitivity to environmental conditions and effects on community structure. Oecologia 113:67–73

Corti D, Kohler SL, Sparks RE (1997) Effects of hydroperiod and predation on a Mississippi River floodplain invertebrate community. Oecologia 109:154–165

Faulkner SP, Patrick WH (1992) Redox processes and diagnostic wetland soil indicators in bottomland hardwood forests. Soil Sci Soc Am J 56:856–865

Gallardo A (2003) Spatial variability of soil properties in a floodplain forest in northwest Spain. Ecosystems 6:564–576

Haag KH, Lee TM, Herndon DC, County P, Water TB (2005) Bathymetry and vegetation in isolated marsh and cypress wetlands in the northern Tampa Bay Area, 2000–2004. U.S. Geological Survey Scientific Investigations Report 2005–5109, 49 p

Haag KH, Lee TM, Water TB (2010) Hydrology and ecology of freshwater wetlands in central Florida: a primer. U.S. Geological Survey Circular 1342:138 p

Harrelson CC, Rawlins CL, Potyondy JP (1994) Stream channel reference sites: an illustrated guide to field technique. General Technical Report, RM-245, U.S. Forest Service, Rocky Mountain Research Station, Fort Collins, CO, 61 p

Karraker NE, Gibbs JP (2009) Amphibian production in forested landscapes in relation to wetland hydroperiod: a case study of vernal pools and beaver ponds. Biol Conserv 142:2293–2302

Keeley JE, Sandquist DR (1992) Carbon: freshwater plants. Plant Cell Environ 15:1021–1035

Lee T, Haag K, Metz P, Sacks L (2009) Comparative hydrology, water quality, and ecology of selected natural and augmented freshwater wetlands in west-central Florida. U.S. Geological Survey Professional Paper 1758:152 p

Pechmann JHK, Scott DE, Whitfield Gibbons J, Semlitsch RD (1989) Influence of wetland hydroperiod on diversity and abundance of metamorphosing juvenile amphibians. Wetl Ecol Manag 1:3–11
Poole GC, Stanford JA, Running SW, Frissell CA (2006) Multiscale geomorphic drivers of groundwater flow paths: subsurface hydrologic dynamics and hyporheic habitat diversity. J N Am Benthol Soc 25:288–303
Ripley BJ, Holtz J, Simovich MA (2004) Cyst bank life-history model for a fairy shrimp from ephemeral ponds. Freshw Biol 49:221–231
Takekawa JY, Woo I, Athearn ND, Demers S, Gardiner RJ, Perry WM, Ganju NK, Shellenbarger GG, Schoellhamer DH (2010) Measuring sediment accretion in early tidal marsh restoration. Wetl Ecol Manag 18:297–305
van der Valk AG (1981) Succession in wetlands: a Gleasonian approach. Ecology 62:688–696
Wilcox C, Los Huertos M (2005) A simple, rapid method for mapping bathymetry of small wetland basins. J Hydrol (Amsterdam) 301:29–36

Student Exercises

Classroom Exercises

Classroom Exercise #1: Evaluating Wetlands in Google Earth

Developing a familiarity with Google Earth is a good starting point for many mapping projects. This exercise guides you through the use of Google Earth to evaluate the Molera Wetland for this and the following exercises.

1. If you do not have Google Earth installed on your computer, follow this link http://www.google.com/earth/index.html to download the program.
2. To find Molera Wetland, paste the following longitude and latitude coordinates into the search window: 36°46'20.93"N, 121°47'18.60"W. The wetland parcel is bounded on the north by the Tembadero Slough and on the west by the historic Salinas River channel. It is bounded on the south and east by roads.
3. Click on the year at the bottom of the screen and you will get a time slider at the top that can be used to choose the dates of various images. In the case of the Molera Wetland and at the time of writing this exercise, there were several images available from 1993 to 2012. The images vary in terms of resolution and number of color bands.
4. Molera Wetland was constructed as a water treatment wetland. To accomplish that goal, it consists of an elongate, sinuous channel in the southern part and an open water wetland in the northern part. Use the time slider to determine when the land use changed from agriculture to wetland. Note how the "wetness" of the open wetland changes through time. Can you determine if the wetness changes are more related to season differences or annual differences?
5. Select an aerial image year that shows the wetland very wet (e.g., May 2011). We will make a rough estimate of open wetland size in that image. Click on the ruler icon at the top of the Google Earth page.

(a) Use the ruler to measure the perimeter of the wetland.
(b) We can estimate the area of the wetland by approximating it as an ellipse. First, measure and record the long and short dimensions of the wetland and then take ½ of those dimensions and multiply them by pi (π) as demonstrated in the following equation:

$$Area \cong \frac{1}{2}(long\ dimension) * \frac{1}{2}(short\ dimension) * \pi$$

Unfortunately, calculating the area this way has limited value because it is based on what you can see and the water depth at that time. You should also select a few other images and calculate how they have changed in different seasons and different years. Can you say anything about the bathymetry of the wetland based on these dimensions?

Classroom Exercise #2: Creating a Bathymetric Surface for Visualization and Analysis

The following exercise uses locally referenced grid of survey points from Molera Wetland. The data for the exercise are available at the following website https://sites.google.com/a/csumb.edu/marc-los-huertos/home/molera-wetland-bathymetry. Download the "exercise_data.csv" file from a folder called Molera. The X, Y coordinates are linked to a piece of rebar we assigned as (50,50) meters, and the elevations are referenced to a local county benchmark elevation in meters. The data were collected using a grid similar to the one shown in Fig. 2.8a, and a subset of the field notes are shown in Fig. 2.5. This data set is coarser than the one used in the chapter, so you can compare the impact of lower resolution on the analysis values. These data have no real-world horizontal coordinates, and ArcGIS will give you a warning message to that effect, which can be ignored for the purpose of this exercise. Use the same steps outlined in the Molera Wetland example in the chapter. The minor differences are noted below in keeping with a survey that has no horizontal georeferencing.

1. Open a new map project in ArcMap. Make the map units meters by clicking the View menu, then Data Frame Properties > General and select "meters" for the units of the map and display.
2. Import the csv file into ArcMap, and create a point shapefile so that the survey points are displayed on the map, and the points have an attribute table.
3. Create a polygon shapefile to be the mask representing the wetland boundary. Draw the boundary using the outer-most points of the survey as a guide.
4. Interpolate the points into a DEM by kriging. For kriging, use the mask to limit the extent of the analysis.
5. Color the DEM to create a map. You have many choices, and the selection will be dictated by the information you are seeking. Two standard coloring methods

are "stretched" color ramp that gives a continuous gradation of color from high to low elevations, or "classified" which gives more information and more control. For stretched color ramp, use the following sequence. Right-click the DEM filename > properties > symbology > stretched, and select the color ramp. For classified coloration use the following sequence. Right-click the DEM filename > properties > symbology > classified and then select a number in the "classes" box to indicate the number of discrete elevation color bands you want and the color ramp. More statistical information, and coloration controls are present if you click "classify."

6. Add contour lines using a base contour of 1.05 m and a contour interval of 0.02 m, or another value of your choice. Color the contours to your liking.
7. You can calculate the perimeter and area of the analysis region by analyzing the perimeter mask shapefile you created. Right click the mask filename > open attribute table > add a field. Select "short integer" and name the field "perimeter." When you click "OK," you will see a new column in the attribute table. Right click the top of the column and select calculate geometry > perimeter. The perimeter value will appear in the column. Try the same steps for determining the area. These are the values that do not correspond to a specific water level, but are values for describing the wetland in general. How did these values compare with the Google Earth measurements you made earlier in this exercise?
8. Quantify Wetland Volume and Surface Area using the following values for wetland water stage: 1.05, 1.07, 1.09, 1.11, 1.13, 1.15, and 1.17 m. Compare the results to those that we created in Table 2.3. Create a graph of the stage and volume relationship and stage and surface area relationship in a spreadsheet. The values you obtained will differ somewhat, because you are using a lower resolution survey than the one presented in the chapter example. In comparing your results with our results in the chapter example (Table 2.3), you can qualitatively evaluate whether a having a large number of points improves accuracy and provides more information.
9. Practice making a hillshade and making it semi-transparent. We find that using a z-factor of four and lowering the sun angle to about 25° improves the visual impact of the hillshade in low relief settings such as this.

Chapter 3
Assessing and Measuring Wetland Hydrology

Donald O. Rosenberry and Masaki Hayashi

Abstract Virtually all ecological processes that occur in wetlands are influenced by the water that flows to, from, and within these wetlands. This chapter provides the "how-to" information for quantifying the various source and loss terms associated with wetland hydrology. The chapter is organized from a water-budget perspective, with sections associated with each of the water-budget components that are common in most wetland settings. Methods for quantifying the water contained within the wetland are presented first, followed by discussion of each separate component. Measurement accuracy and sources of error are discussed for each of the methods presented, and a separate section discusses the cumulative error associated with determining a water budget for a wetland. Exercises and field activities will provide hands-on experience that will facilitate greater understanding of these processes.

3.1 Introduction

The physical, biological, and chemical properties of a wetland all are greatly influenced by water and chemical fluxes, both to and from the wetland, as well as the temporal variability of these fluxes. Therefore, hydrologic processes are central to the character and features of a wetland and to virtually everything that occurs surrounding and within a wetland basin. A question occasionally posed by wetland scientists is whether a wetland "has hydrology." This terminology likely stems from

---

D.O. Rosenberry (✉)
U.S. Geological Survey, Denver Federal Center, MS 413, Bldg. 53, Box 25046
Lakewood, CO 80225, USA
e-mail: rosenber@usgs.gov

M. Hayashi
Department of Geoscience, University of Calgary, 2500 University Dr. NW,
Calgary, AB T2N 1N4, Canada
e-mail: hayashi@ucalgary.ca

a need to determine whether a landscape has characteristics of a wetland setting for regulatory or protection purposes. Hydrology is basically the study of water as it is distributed over, on, and within the earth. All landscapes, and particularly wetlands, have hydrologic properties that are an integration of all the water-related characteristics and processes that occur there. Wetland hydrology encompasses study of the distribution and flow of all water that is added to, lost from, or stored in a wetland.

A wetland is a portion of a landscape that is wet for a period sufficiently long that physical, chemical and biological conditions are indicative of a wet setting. Wetlands occur in a wide range of settings where geological and hydrological processes enhance the accumulation and retention of water (Winter 1988). Water, therefore, is present at or just beneath land surface at a substantial percentage of the time in wetland settings. Given that water is integral to wetland settings, an overarching challenge in determining the type or persistence or quality of a particular wetland setting is to determine the relative contributions of the various components of wetland hydrology (i.e., precipitation or evapotranspiration or surface-water inputs or groundwater inputs or overland flow). A water-budget approach for making this determination is perhaps the best way to categorize and describe the wide range of wetland types that exist in the world (Winter and Woo 1990; Winter 1992) and is the perspective from which this chapter is presented.

3.2 Wetland Hydrology from the Perspective of a Water Budget

Knowledge and understanding of the storage and mass balance of water and chemicals is critical to understanding a wetland ecosystem. This includes quantifying all of the sources, losses, and changes in storage in the wetland. Simply determining the relative magnitude of various hydrologic components can largely determine a wetland type. For example, surface water may be the dominant source and sink of water and solutes for a riparian wetland whereas overland flow and evapotranspiration may dominate in a prairie wetland. One will have greatly different water chemistry and biogeochemical processes than the other, all because of the relative mix of sources and sinks of water and chemicals.

Wetland stage is an integrated response to all source- and sink-terms in a hydrologic budget. It also incorporates temporal variability in the balance of all hydrologic fluxes and is, therefore, strongly linked to wetland hydroperiod and wetland hydrodynamics, both of which are important to most disciplines that encompass wetland science (Euliss et al. 2004). Wetland stage and volume can also provide a direct and often sensitive response when climate change may be affecting the relative magnitude and importance of specific hydrologic components.

For these reasons and more, an accounting of hydrologic components of a wetland water budget should be one of the first items on a wetland-scientist's agenda (LaBaugh 1986). Preliminary estimates of the relative volume associated with each hydrologic component is often a valuable first step. These estimates will allow attention to be focused on the most important hydrologic components of a particular wetland setting or type. The importance of quantifying individual hydrological components also depends on the issues and questions being asked. For example, at an extensively studied wetland in the prairie-pothole region of North Dakota, groundwater discharge was a small component (3.5 %) of the water budget, small enough that it might be ignored. However, groundwater discharge delivered a large percentage of chemicals to the wetland and was an important contributor to wetland chemistry (LaBaugh et al. 2000).

A wetland water budget can be written as

$$\frac{\Delta V}{\Delta t} + R = P + O_f + S_i + G_i - ET - S_o - G_o \qquad (3.1)$$

where $\Delta V/\Delta t$ is the change in volume of surface water in the wetland per time, P is precipitation, O_f is overland flow, S is surface water, G is groundwater, ET is evapotranspiration, and R is the residual, or unaccounted water, in the water budget. Subscripts i and o refer to water flowing into or out of the wetland. This basic equation should be modified to suit specific wetland settings. For example, some wetlands will have dewfall or stem flow that is substantial and quantifiable whereas other wetlands will not have any surface-water inputs or losses. Many wetlands in northern latitudes also have an input term associated with drifting snow (e.g., Hayashi and van der Kamp 2007). Some wetlands will rarely contain surface water, in which case $\Delta V/\Delta t$ can be based on changes in volume of surface water, groundwater, and soil-moisture storage over time. If surface water is not present, hydrologic fluxes are distributed over an area based on criteria other than areal extent of surface water, perhaps the areal extent of wetland vegetation. In this chapter we will restrict discussion primarily to settings where surface water is present.

Equation 3.1 can be rearranged to solve for any of the components provided the others are known. An example is presented later for determining G_i and G_o as the unknown entities of the water-budget equation. ET also can be a difficult value to obtain and is occasionally solved as the unknown of a water-budget equation. However, the uncertainty associated with ET commonly is much smaller than the uncertainty associated with quantifying G_i or G_o. In many wetland settings, errors associated with quantifying groundwater exchange are so large that solving for ET as the residual would be meaningless.

3.2.1 Determining the Accounting Unit

As mentioned earlier, the change in wetland storage, ΔV, integrates all of the input and loss terms of a hydrologic budget. This term can also be approximated as

$$\Delta V \cong \Delta h \left(A + \frac{\Delta A}{2} \right) \qquad (3.2)$$

where A is wetland surface area and h is wetland stage. Details for determination of V in the typical case where A changes with depth are provided in Sect. 3.3.2.

Table 3.1 Errors indicated in % for water-budget components of selected studies conducted on lakes, reservoirs, and wetlands (– indicates parameter was not determined; calc indicates value was calculated as the residual)

	P	E or ET	S	G	Of	ΔV
Winter (1981)	5–10	10–15	5–10	13–36	–	–
LaBaugh (1985)	33	10	5–15	–	–	10
Belanger and Kirkner (1994)	10	10	50	50	–	10
LaBaugh et al. (1995, 1997)	5	10	–	50	–	5
Lee and Swancar (1997)	10	16	–	102–106	–	5
Sacks et al. (1998)	5–9	10	30–100	calc	–	5
Choi and Harvey (2000)	8.5	20	10	10	–	15
Harvey et al. (2000)	15	10	10–15	10	–	15
Motz et al. (2001)	5	20	11–15	50	100	5
Rosenberry and Winter (2009)	5	15	5	25	–	10
Median	9	10	10	36		10
Maximum	33	20	100	106		15
Minimum	5	10	5	10		5

When Δh is small and A is much greater than ΔA, this relation often is simplified by assuming that A is constant (i.e., $\Delta A = 0$). A minimum measurable change in wetland stage is, therefore, a logical accounting unit in a wetland water budget. Precipitation and evapotranspiration already are usually expressed in terms of depth applied over the wetland surface per time (commonly mm/day). Other water-budget components more commonly measured in terms of volume per time, such as surface-water or groundwater inputs and losses, can be expressed as Δh by dividing by A. This seemingly simple task can be a substantial problem at many wetlands, as evidenced by the relatively large errors associated with the ΔV term listed in Table 3.1; errors of 10–15 % are common. Since measuring stage is quite simple and can be done very accurately, often with accuracies of ± 3 mm or better, the estimation of surface area is the source of most of this error.

The shoreline must be identified before wetland area can be determined. Unfortunately, an indistinct shoreline as shown in Fig. 3.1 is common. In some cases, an area of dense emergent vegetation forms an abrupt boundary, not at the shoreline but at the edge of the open-water portion of the wetland, that confounds the determination of the actual shoreline. If this border occurs at a water depth of 0.3 or 0.5 m, an example of which is shown in Fig. 3.1, the actual shoreline, where water depth decreases to zero, can be many meters away and obscured by additional dense emergent vegetation.

For wetlands situated in low-gradient settings, the shoreline can move laterally a large distance in response to a small stage change (e.g., Lee et al. 2009). An accurate bathymetry map, and associated stage-area and stage-volume plots, are particularly important for minimizing error when determining ΔV. Generating an accurate stage-area plot is not nearly as onerous as it once was (see Chap. 2 on wetland bathymetry).

3 Assessing and Measuring Wetland Hydrology

Fig. 3.1 Example of a wetland where the shoreline is not easily distinguishable (Photo by Donald Rosenberry)

Methods that provide high-resolution topographic information, such as a map generated by the light detection and ranging (LIDAR) technique, are particularly useful for determining appropriate areas to assign to specific stages. These methods are best employed when stage is lowest. Also, some wetlands that normally have no surface-water outlet can develop one during extremely wet periods. This process is commonly referred to as "fill-and-spill" (van der Kamp and Hayashi 2009; Shook and Pomeroy 2011; Shaw et al. 2012). Modern approaches based on differential geographic information system (GIS) are capable of determining stage- and scale-dependent contributing areas with regard to net overland flow that contributes to a particular wetland basin. In situations where these relatively new tools and procedures are prohibitively expensive or labor intensive, simplifying assumptions based on general knowledge of wetland shape can provide reasonably accurate stage-area and stage-volume relations (Hayashi and van der Kamp 2000).

Some wetland basins become separate entities during dry periods and then coalesce during wet periods (e.g., Winter and Rosenberry 1998). Water budgets need to be determined for each distinct wetland sub-basin, based on separate stage-area and stage-volume relations, until the wetlands coalesce, at which point a new stage-area relation should be used for the now combined wetland.

Thus far, ΔV has been determined based on the surface area of the open-water or standing-water portion of a wetland. This concept is not appropriate for wetlands that do not contain standing water to any measurable depth; for example, wetlands on hillslopes or wetlands that drain rapidly following rain or flooding events. In those settings, the accounting unit may need to be set based on topography or areal extent of specific types of wetland vegetation. If the wetland surface is considered to be saturated virtually all the time, then one could reasonably assume that ΔV is zero. In this case, any additions of water to the wetland must instantly be balanced by an equal volume of loss terms. The accounting unit also could be the quantification of water stored in the vadose zone or the sum of water contained in the vadose zone and in groundwater beneath the vadose zone. Assumptions may need to be made regarding the level of saturation of the wetland soils to quantify a change in stored volume. If the water table decreases to below the wetland bed, water-volume change could be estimated based on water-level measurements in monitoring wells and assumptions about volumetric storage capacity of the wetland soil. A further complication is associated with the typically small distance of the water table below the land surface or wetland bed. The capillary fringe is a zone of tension saturation that exists above the water table in all settings; the thickness is inversely proportional to the grain size of the soil. In the generally fine-grained sediments found in most wetland settings, the soils may be essentially saturated beneath much to all of the wetland bed. If this is the case, even small rainfall or recharge events can bring the water table directly to land surface and result in surprisingly large amounts of overland flow to the wetland (Gerla 1992).

3.2.2 Determining the Accounting Period

The proper time interval (Δt in 3.1) over which a water budget is determined depends on the questions asked, the duration of the study, or the reasons for quantifying a water budget (Healy et al. 2007). If the question is related to wetland response to climate change, an annual water budget may be all that is necessary. If determining the relative significance of a particular hydrologic component is important, the study may need to extend over several years and quarterly or monthly time steps would be appropriate. If the concern is related to the response of a wetland to individual recharge events or to specific physical, chemical, or biological processes, daily time steps may be the most appropriate. In general, because of technological advances in data collection, scientists are tending to use shorter time steps. Whereas monthly measurements may have been the norm during previous decades, it is more likely that data are collected every minute to every hour and hourly or daily values are then calculated based on those data.

3.3 Water-Budget Hydrology

The volume of water contained in a wetland, V, is an integrated response to all of the hydrological processes that add or remove water. Therefore, if all of the components of a wetland water budget were measured perfectly, the sum of those processes should equal the volume of water stored in a wetland for any given accounting period. By simply measuring the change in the elevation of the wetland surface, and multiplying that change by the surface area of the wetland, we can obtain a change in wetland volume over an accounting period and relate that change to the hydrologic inputs and losses that occurred over that same accounting period. Relative to the complexities associated with measurement of all of the input and loss terms, measurement of wetland stage should be relatively simple and error free. However, even small measurement error, and poor characterization of wetland bathymetry and geometry, can still result in substantial errors (e.g., Winter 1981).

3.3.1 Stage Measurement

The relative height of the wetland water surface commonly is referred to as wetland stage, herein symbolized as h. This is sometimes confused with wetland elevation, which is the height of the wetland water surface relative to a citable datum (reference elevation); for example, North American Vertical Datum of 1988 (NAVD88). This also is not to be confused with wetland water depth, which is the vertical distance from the sediment-water interface (herein, referred to as the wetland bed) to the water surface. Stage typically is determined relative to a local datum, such as a painted mark on a rock outcrop or stable concrete fixture, a pipe or rod driven into the ground, a lag screw placed near the base of a nearby tree, or a benchmark if one is located nearby. Wetland hydrologists commonly make the assumption that the wetland surface is flat and that wetland stage can be measured at any location in a wetland (see Sect. 3.3.3.3 below on how to address seiches for large wetlands). Therefore, measurements typically are made either at a location convenient to the observer or at the deepest point in the wetland if it is expected that the wetland might go dry. In some cases, the wetland bed is artificially deepened at the point of measurement so that the water level can be measured for a short distance below the deepest portion of the wetland during drawdown. A water-table monitoring well installed in the wetland is required to track further water-level drawdowns during prolonged dry periods. Several of the more commonly used methods for measuring stage are described below. Greater detail is provided in a U.S. Geological Survey (USGS) report on methods for making stage measurements (Sauer and Turnipseed 2010).

3.3.1.1 Staff Gage

The simplest and most common method for measuring wetland stage is to visually observe the value where the water surface cuts across a graduated plate placed vertically in the water (Fig. 3.2). Commonly made of fiberglass or enamelled metal, the staff gage is bolted to a stable surface or placed on a pipe or solid rod driven into the wetland bed. Data-collection interval commonly is variable and depends on the

Fig. 3.2 Surveying a wetland staff gage to a local datum. Rod is held on a screw projecting from the plank on which the staff plate is mounted. Water-level is at 26.38

timing and frequency of observer visits to the site. Staff gages are subject to movement because wetland sediments tend to have a relatively large content of organic material and are, therefore, often poorly competent, meaning they are loosely compacted and may readily deform. Pipes or rods to which staff gages are attached should be driven deeper to provide a stable anchor if sediments are soft. If the wetland surface freezes during winter, any change in the elevation of the ice surface over the course of the winter, such as a rising ice surface during snowmelt, can also move the staff gage, either horizontally or vertically. Therefore, the height of the staff gage relative to the local datum needs to be determined at least annually to provide inter-annual continuity of stage data. Staff plates can be stacked vertically if wetland stage varies over a distance greater than the length of a single staff plate. Because this method is so simple and relatively robust, staff-gage values commonly are used as the reference value when automated sensors are used to collect more frequent stage data. Staff plates need to be cleaned regularly to remove chemical or biological accumulations at or near the water surface.

3.3.1.2 Float-Based Gage

A float and counterweight connected to opposite ends of a tape or wire draped over a rotating pulley is another wetland-stage measurement method that has been in use for many decades. The float moves up and down with the wetland water level, which turns a shaft on which the pulley is mounted. The counterweight maintains tension on the system and keeps the tape or wire taut against the measurement wheel. Earlier versions usually were linked to a mechanical chart recorder, but the rotating shaft now more commonly used is attached to an electrical potentiometer or a device that generates an electrical pulse for a specific degree of shaft rotation (shaft encoder), either of which can easily be interfaced with a digital datalogger. Drag or frictional resistance associated with movement of the float, the float wire, and the rotational resistance of the potentiometer or shaft encoder cause the float to ride higher in the water during a falling water table than during a rising water table (instrument hysteresis). The accuracy of this system is related to a large extent to the diameter of the float. The float displaces a greater volume of water during rising than during falling stage. Because displacement volume is equal to float-immersion depth times the cross-sectional diameter of the float, variation in immersion depth becomes smaller as the float diameter is increased.

3.3.1.3 Bubbler System

A bubbler system, also commonly referred to as a bubble gage, measures water level above an orifice submerged beneath the water surface. A very accurate non-submersible pressure transducer is often used to measure the pressure required to push gas through the orifice; the pressure is proportional to the height of the water column above the orifice. Gas (typically nitrogen or air) supplied by a pressurized cylinder or a small pump is pushed through a flexible hose or pipe to the orifice that is affixed at a stable location beneath the water surface. The tubing or pipe often is buried beneath the wetland bed to prevent disturbance, damage or vandalism. Systems can be designed to either pump gas continuously or to intermittently purge the orifice line and then collect a pressure reading once the gas flow has stabilized. The latter design either uses less gas if a compressed cylinder is the supply or requires less power consumption if a pump supplies the pressurized gas. A bubbler system also allows measurements beneath an ice-covered surface. Data of poor quality may result from siltation of the orifice or if the orifice is placed where surface-water currents are substantial.

3.3.1.4 Capacitance Rod

Capacitance is a measure of the charge that builds up between two plates relative to an applied voltage. Capacitance is directly proportional to the area of the plates and to the dielectric property of the material between the plates, and inversely proportional to the distance between the plates. Since water has a greatly different dielectric property than air, output from a capacitance rod that is partially submerged in water is proportional to the submergence distance. Therefore, capacitance rods should be suspended from a fixed point, such as a stilling well placed in a wetland, so that the water level in the wetland does not go below the bottom or above the top of the rod.

Capacitance rods generally are available in lengths ranging from 0.5 to 2 m and some models contain integrated dataloggers as well as a temperature sensor that provides water temperature output and also corrects for the influence of changing temperature on sensor output. Accuracy generally depends on the length of the rod and is commonly about 1 % of the full scale (e.g., ±20 mm for a 2-m rod). Care must be exercised in determining the vertical placement of the rod so a rising water level does not overtop the rod and possibly damage the data-processing hardware.

3.3.1.5 Submersible Pressure Transducer

A submersible pressure transducer measures the pressure of a column of water above the sensor while it is submerged in the fluid. The most common type is a silicon strain gage, in which electrical resistance across a silicon wafer changes in proportion to the slight deflection (strain) that occurs in response to differential pressure applied across the plane of the wafer (Freeman et al. 2004). Sensors that are vented contain a small-diameter tube that extends from the transducer to the point at which the power and signal wires are connected to a computer or datalogger. The vent allows changes in atmospheric pressure to be transmitted to the sensor so that the output reflects only changes in the height of the water column above the sensor. Non-vented sensors measure the combined pressure of the water column and overlying atmosphere and require use of a separate pressure sensor (barometer) to allow atmospheric pressure changes to be subtracted from output from the non-vented sensor. The advantage of vented sensors is that only one measurement is required, minimizing cost, complexity, and eliminating any measurement error associated with a separate sensor. The problem with a vented sensor is the vent itself. The vent needs to remain completely unobstructed; the vent can become blocked if the cable is inadvertently kinked, for example. If moisture condenses inside the vent so that a water drop extends across the cross section of the vent, changes in atmospheric pressure no longer are completely transmitted to the sensor. Furthermore, corrosion and corruption of sensor output is likely if the water or moisture is transmitted to the sensor housing. Non-vented sensors not only eliminate the vent problem, they also commonly contain an on-board datalogger and do not have any electrical wires extending beyond the sensor housing. This minimizes problems with cable-related leaks. The disadvantage, in addition to the requirement of a separate sensor to measure atmospheric pressure, is that the sensor generally needs to be retrieved to download the data. Measurement error can occur if the sensor is not re-suspended at exactly the same elevation below a control datum.

Output in pressure (P) is converted to stage with the relation

$$\psi = \frac{P}{\rho g} \qquad (3.3)$$

where ψ is pressure head (m), P is pressure (Pa), ρ is density of water (kg/m^3), and g is acceleration of gravity (m/s^2). Since pressure is force per area and can be expressed as kg m/s^2/m^2, units for ψ are kg/s^2/m divided by ρg (kg/s^2/m^2), which yields m. As long as the pressure transducer is positioned somewhere within the open column of water in the wetland (i.e., not buried in the sediment beneath the wetland), the pressure head is directly proportional to wetland stage because, within a water column, pressure head is exactly offset by the elevation head (i.e., the height at which the pressure head is being measured). The deeper the transducer is positioned in the wetland, the smaller the elevation head but the larger the pressure head. For example, if the transducer is mounted at a depth 0.1 m above the wetland bed, and the converted output from the transducer is 0.3 m, then wetland depth is equal to elevation head (0.1 m) plus pressure head (0.3 m), which is equal to 0.4 m. If instead the transducer is mounted at 0.25 m above the bed, then the height of the water column above the transducer would be smaller because the transducer would be immersed at a shallower depth in the wetland. In this case, elevation head would be 0.25 m, output from the transducer converted to pressure head would be 0.15 m, and wetland stage would still be 0.4 m.

3.3.1.6 Manual Measurement with a Graduated Rod

In seasonally frozen wetlands, a staff gauge needs to be surveyed annually to account for movement due to frost action or moving ice. Conly et al. (2004) presented a simple and practical method to eliminate this requirement using a water-depth measurement rod. In this method, permanent markers, typically metal stakes, are driven into the wetland sediments at locations where standing water commonly is present. The elevation of the top of the marker stake is surveyed once, which is used to establish the elevation of the wetland bed, which is in turn used to calculate the wetland stage from depth measurements. An observer places a wooden measuring rod, approximately 2 cm in diameter and 1.5 m in length and graduated to 1 mm resolution, on the bed next to each marker stake. A 6-cm diameter metal or plastic disk attached to the base of the rod prevents the rod from being pushed below the sediment-water interface, ensuring greater repeatability of measurements. A notch is cut into the base so the rod can be slid along the side of the metal stake, ensuring consistency of the measurement location among site visits. The observer slides the rod downward until the base touches the bed. Distances from the marker base to the water surface, from the marker top to the water surface, and from the marker top to the wetland bed are recorded, providing redundancy in the water-surface measurement. Conly et al. (2004) reported the accuracy of this method to be on the order of 1–2 cm. While this degree of uncertainty may be too large for daily water budgets, the method provides a useful approach to maintaining inter-annual consistency during long-term monitoring of wetland stage and in determining monthly to seasonal water budgets.

3.3.1.7 Remote Sensors

Several other sensors are capable of measuring wetland stage without coming in contact with the water, some with considerable accuracy. Acoustic sensors transmit an acoustic wave to the water surface and record the time of transmission upon reflection of the acoustic wave back to the sensor. This provides a useful method to monitor the stage of a seasonally frozen wetland when a pressure transducer could suffer damage caused by freezing (Hayashi et al. 2003). Corrections need to be made for air temperature and density to maintain a high level of accuracy. Sensors that transmit and receive a radar pulse operate under the same assumptions. For the radar sensors in particular, the diameter of the water surface over which stage is being determined depends on the transmission beam angle as well as the distance the sensor is mounted above the water surface. Therefore, any object(s) projecting above the water surface that are within the cone of influence can corrupt the measurement. Several acoustic and radar sensors can provide water-level measurements that are within 3 mm of the true value. Laser-based devices also are available, but for water that is particularly clear, the laser beam may penetrate the water rather than reflect off it. Use of a floating reflector positioned inside a stilling well may minimize this problem.

High-resolution satellite images or aerial photographs provide reasonably accurate estimates of inundated wetland areas under ideal conditions. If the relation between wetland area and stage is known (e.g., Eq. 3.4 below), then the wetland stage can be estimated with reasonable accuracy. However, the accuracy of this method depends on the delineation of inundated area, which may be difficult with the presence of emergent vegetation (e.g., Fig. 3.1), and on the accuracy of the stage-area relation.

3.3.2 Converting Stage Change to Volume

Measurement of wetland stage commonly is determined on a short time interval, perhaps every 15 min or once an hour, unless the measurement is made manually. This allows quantification of stage in response to individual precipitation events if precipitation also is determined on a short time interval. However, for the purpose of determining a water budget, change in stage should be determined on the same time interval as the hydrologic component with the longest measurement interval. In most situations, the time-limiting parameter will be evaporation, which rarely is determined on less than a daily interval. Therefore, assuming that all other hydrologic components are determined at least on a daily basis, stage change should be determined based on subtracting wetland stage at midnight from wetland stage during midnight of the subsequent day. In this way, daily change in wetland stage will be integrated over the day, just as is the case for measurement of the rest of the hydrologic components.

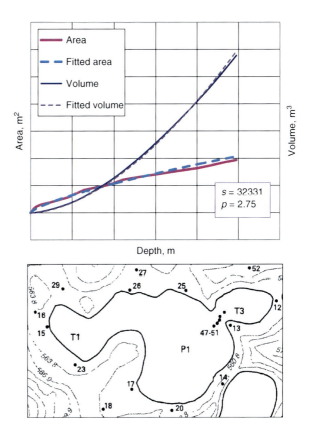

Fig. 3.3 Wetland area and volume related to stage based on a detailed map of wetland bathymetry for an irregularly shaped wetland. D.O. Rosenberry unpublished data for wetland P1, Cottonwood Lake Area, North Dakota

A relation between wetland stage and surface area or volume is needed to determine a volume associated with change in stage. If detailed bathymetry data are available, curves relating wetland area and volume with stage can be generated, from which wetland volume can be determined for any given stage value (e.g., Fig. 3.3). In this case, it is a simple matter of taking the difference between volumes associated with two sequential values of wetland stage to determine change in wetland volume.

Unfortunately, it often is not a simple matter to determine wetland bathymetry. Wetlands commonly are situated in a low-gradient landscape where small changes in stage can result in large changes in surface area. Dense or tall emergent vegetation also can hinder bathymetry determinations based on remote-sensing technology or even on direct observation, as previously noted in Fig. 3.1. It often is necessary to use the brute-force approach and collect high-density measurements of the elevation of the wetland bed at well-determined locations, either with detailed on-site surveying or a combination of surveying and differential global positioning system (GPS). A study of cypress wetlands in Florida, for example, determined location and elevation at 86–145 measurement points/ha in order to generate wetland areas and volumes for every 3 mm increase in wetland stage (Haag et al. 2005).

Lacking such detailed data, it is possible to determine change in volume with reasonable accuracy by making some basic assumptions related to wetland geometry. Assuming a symmetrical wetland basin with the deepest part located at the center of the basin, wetland area can be determined by making an assumption about the change in slope of the wetland basin with distance from the center. Using this approach, Hayashi and van der Kamp (2000) developed the following relation:

$$A = s \left(\frac{H}{H_0} \right)^{\frac{2}{p}} \tag{3.4}$$

where A is wetland surface area, H is wetland depth, H_0 is unit depth (e.g., 1 m), s (m^2) is a scaling factor that is equal to the wetland surface area at H_0, and p is a dimensionless scaling factor that is related to the shape of the wetland basin. For example, if the profile of the wetland bed extending from the center to the perimeter is a straight line, then p is equal to 1. If the wetland is bowl shaped, then p is close to 2. If the wetland has a broad, flat basin that steepens near the wetland edge, then p is somewhere between about 5 and 100, with p increasing to infinity for a rectangular cross section. Wetland volume also can be approximated using the same fitting factors and the equation

$$V = \frac{s}{1 + 2/p} \frac{H^{1+2/p}}{H_0^{2/p}} \tag{3.5}$$

(Hayashi and van der Kamp 2000).

An example of comparing fitted and measured values for area and wetland area and volume is shown in Fig. 3.4. Even with an irregularly shaped wetland basin, Eqs. 3.4 and 3.5 approximate values for A and V reasonably well based on measured bathymetry.

3.3.3 Sources of Error

Measurement of wetland stage is conceptually very simple and any given observation has a high likelihood of being very accurate. However, several sources of error can increase as study duration extends to multiple months or years and greatly diminish the accuracy of wetland stage that is very important to a water-budget analysis. The significance of these errors depends on the accuracy requirements. The U.S. Geological Survey, for example, requires an accuracy of ±0.01 ft (3 mm) for water-level measurements over the range typically encountered in most wetland settings (Sauer and Turnipseed 2010).

If daily water budgets are a goal, then measuring stage to a level of precision and accuracy similar to hydrologic fluxes summed over a day would be appropriate.

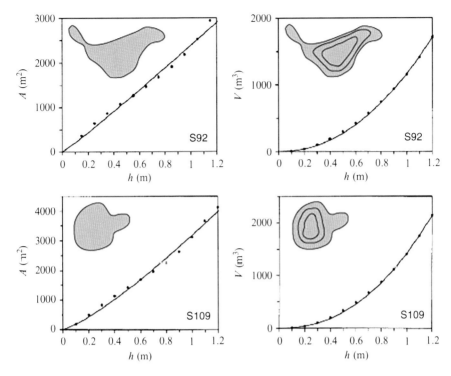

Fig. 3.4 Measured versus modeled areas and volumes of two wetlands in the St. Denis Wildlife Area, Saskatchewan, Canada. Wetland shape and bathymetry shown in *upper left* portion of plots (Modified from Hayashi and van der Kamp (2000)). Published with kind permission of © Elsevier 2000. All Rights Reserved)

Precipitation commonly is measured to the nearest 0.3 mm with an accuracy generally considered to be ±5 to 15 % (Winter 1981). Daily evaporation commonly ranges from 0 to 4 mm and rarely exceeds 6 mm. Accuracy of surface-water measurements depends on the surface-water discharge relative to the wetland surface area. Even if inputs are relatively large and wetland surface area is relatively small, measurement error expressed in terms of wetland stage usually is less than 3 mm. Given these magnitudes of daily hydrologic fluxes common to wetland settings, measuring wetland stage to within 1 mm is not an unreasonable goal, even though it is rarely achieved with current technology.

3.3.3.1 Staff-Gage Errors

Although accuracy of wetland stage to within 1 mm is desirable, it is quite difficult to read a staff gage more accurately than about ±3 mm. Most staff gages are incremented no finer than 3 mm and many display 10-mm increments. Observation errors result from waves that cause the water surface to fluctuate during a gage reading. Clear water makes it difficult to see specifically where the water cuts across the staff plate. Corrosion or algal growth can obscure the values on the staff plate. A tilted staff plate causes the indicated stage change to be larger than the actual stage change.

As mentioned earlier, a staff gage mounted on a pole or stake driven into the bed can move over time, resulting in a bias in time-series trends. This problem is greatly enhanced in locations where the surface water freezes during winter. Staff gages commonly are pulled upward in the spring when recharge to the wetland causes the floating ice to rise before the ice melts enough to release contact with the staff gage. If wind causes the ice to move horizontally during the melting process, the staff gage can be tilted or even sheared from the pole or stake. In these cases, staff gages need to be resurveyed to a stable datum annually.

3.3.3.2 Local Datum Errors

Elevation of the local datum to which the staff gage is related also can change over time. Lag screws at the base of large trees are surprisingly stable, but a strong wind can cause a tree to lean or fall to the ground. Therefore, many installations make use of multiple datums and annual surveys are made from the staff gage to each one. Some studies have used nearby monitoring wells as stable reference marks for maintaining inter-annual elevation control. Wells associated with wetland studies commonly are shallow because the water table is close to land surface. Shallow wells are not well anchored, leaving them susceptible to frost expansion; some have been observed to move 10 cm or more in a single winter. Therefore, use of a well as a stable reference point is not recommended in environments where soil frost occurs (Rosenberry et al. 2008).

3.3.3.3 Automated Sensor Errors

Automated sensors also are subject to error. Each sensor has specifications that indicate sensor resolution and accuracy. In addition to stated sensor limitations, problems can develop that are specific to particular types of sensors.

Systems that include a float are subject to hysteresis, as described earlier. Errors are proportional to the square of the float and pulley diameters and generally will be less than 3 mm if the float diameter is greater than about 60 mm and the pulley diameter is 0.1 m. Floats also can ride deeper in the water or even sink due to a leak. Debris accumulation on the float can cause the float to ride deeper in the water, creating a bias. A potentiometer connected to a float can fail (basically, wear out) over specific depth increments if the float remains at essentially the same height but waves cause the float and potentiometer to move back and forth ("paint") for extended periods. During exceptionally high or low water levels, the counterweight may exceed its length of travel, resulting in faulty measurements.

Bubble gages can drift or fail if sediment covers the bubble orifice. The system will fail if the pressurized-gas source is exhausted or if the battery supply is insufficient to power the pump. Output from submerged pressure transducers can drift due to cable or wire stretch, slippage of the point from which the cable or wire is suspended, or simply due to sensor electronic drift (formerly, a common problem). Non-contact sensors that make use of radar or sonar are sensitive to air temperature and sensor height above the water surface. Floating debris, wind-generated waves, rain or snow, and other falling debris (e.g., pollen) can further degrade data quality.

Errors can result from sensor location as well. If the sensor is not located in the deepest part of the wetland basin, the sensor might indicate that the wetland has gone dry while there is still standing water in a deeper portion of the wetland. For larger wetlands, strong wind can generate a seiche, or oscillation of water stage associated with piling of water on the downwind side of the wetland. The effects of seiche can be minimized by averaging multiple measurements over a period longer than the characteristic period of oscillation, which is roughly proportional to the length of the wetland, and inversely proportional to the square root of the depth of water (Wilson 1972).

3.4 Precipitation

Precipitation is the main driver of most wetland water budgets through direct application to a wetland surface and indirect inputs via surface runoff and groundwater discharge (Winter and Woo 1990). Accurate measurements or estimates of precipitation are essential. Compared to other water-budget components, precipitation measurement at a given location (i.e., a point measurement) using a properly designed precipitation gage is relatively straightforward and accurate. However, there are a host of issues that can introduce error in these simple point measurements, such as poor installation or maintenance of instruments. Scaling up from point data, sparse data, or off-site data to precipitation distributed over a watershed also increases the uncertainty of a representative value. In the following sections, we will present methods for making point measurements, indicate potential sources of errors, and then present methods for scaling from point measurements to determining precipitation on a watershed scale.

3.4.1 General Consideration for Point Measurements

To obtain representative values of precipitation over an area of interest, the choice of measurement site, the type and exposure of the instrument, the prevention of evaporation loss, and the reduction of wind effects and splashing all are important considerations (WMO 1994:91). Ideally, a precipitation gage should be located within an open space in a fairly uniform enclosure of trees, shrubs, or fences, so that wind effects (see below) are minimized. None of the surrounding objects should extend into the volume of an imaginary inverted cone with the apex positioned directly above the sensor and the sides extending from the sensor orifice at a 45° angle (Dingman 2002:112). The gage orifice should be horizontal even on a sloping ground surface. The gage height should be as low as possible to minimize wind effects, which increase with height, but high enough to prevent splash of rain drops from the ground. If the gage is used to measure snowfall, the orifice should be located above the maximum snow height. An orifice height of 0.3 m is used in many countries in areas that receive little snow. A standard of 1 m is suggested for most areas that accumulate larger amounts of snow during winter (WMO 1994:92).

Orifice size needs to be sufficiently large to minimize edge effects and should be known to the nearest 0.5 % for an accurate conversion of volume of water collected (m^3) to equivalent depth of precipitation (mm). An orifice area of 200–500 cm^2 is common (WMO 1994:94). The collection cylinder should be deep enough and the slope of the funnel steep enough to prevent rain from splashing out of the gage. For storage-type gages (see below), a smaller-diameter restrictor should be positioned between the orifice and the collection cylinder, and the cylinder should be covered with a highly reflecting material to minimize loss of water by evaporation. Adding a layer of non-volatile immiscible oil floating on the collected water also minimizes evaporation. Low viscosity, non-detergent motor oils are recommended for this purpose; transformer and silicone oils have been found to be unsuitable (WMO 1994:95).

3.4.2 Type of Precipitation Gages

Precipitation gages can be classified into non-recording and recording types. Non-recording gages generally consist of an open receptacle with vertical sides or a funnel, and a reservoir that stores the collected water. Precipitation is determined by weighing or measuring the volume of water collected in the reservoir, or by measuring the depth of water using a calibrated measuring stick or scale. Care must be taken to minimize observation errors for both graduated-cylinder and weighing-device measurements (see WMO 1994:96–100 for detailed discussion). If measurements are made infrequently, evaporation loss can cause substantial negative bias in the data, or overflow of the collector may occur as a result of unusually heavy storm events. Despite these potential sources of errors, carefully operated non-recording gages present a useful alternative to recording gages because they are simple, accurate, and relatively inexpensive. They are particularly useful for applications that require a large number of points to capture spatial variability of rainfall at a low temporal resolution (e.g., weekly or monthly).

The most commonly used recording devices are the weighing gage and the tipping-bucket gage. Weighing gages measure the weight of the storage reservoir and its contents using electronic sensors, such as a load cell or strain gage, or a spring. Since these gages store water in a reservoir, similar to non-recording gauges, they also are susceptible to evaporation loss or overflow resulting from infrequent site visits. Weighing gages record cumulative precipitation data at a specified frequency (e.g., hourly). Estimating the amount of rainfall or snowfall during specific time intervals may not be straightforward due to instrument noise. In particular, electronic sensors are sensitive to temperature; even after temperature compensation routines are applied, the data may contain substantial temperature-related error. Therefore, it is best to use weighing gages for recording precipitation over a relatively long interval (e.g., weekly, in which case temperature-related effects can be integrated over a longer time), and tipping-bucket gages for studies requiring greater temporal resolution.

Tipping-bucket rain gages introduce the water received in a funnel to one of a pair of identical vessels (buckets) balanced on a fulcrum. When one bucket is filled, it tips and sends an electronic pulse to a recording device, and the other bucket is brought into position for filling (Dingman 2002:105). These instruments are useful for collecting rainfall data at a high temporal resolution, but they also have some disadvantages. For example, during high-intensity rainfall the sensor measures less rainfall than actually occurs. The bucket does not tip instantly and during the first half of its motion, rain is being fed into the compartment already filled with the designed amount of rainfall (WMO 1994:103). This delay results in systematic negative bias in measured rainfall for high intensity events. Events with less than a minimum amount of precipitation required to tip the bucket also are not recorded (these are called "trace" events). Similarly, a small amount of water collected at the end of an event commonly is left in the bucket and subsequently evaporates, resulting in underestimation of precipitation. The sensitive balance of buckets requires periodic calibration of the amount of precipitation per each tip. Without such calibration, the data may contain a substantial degree of positive or negative bias.

Optical devices are less commonly used, but represent promising new technology (Nitu and Wong 2010). When water passes through an optical scintillation gage it alters the frequency of an infrared beam, which can be analyzed to deduce the time, amount, and intensity of precipitation. In the second type, the sensor measures the extinction caused by precipitation droplets falling through a thin sheet of light, from which precipitation type (rain or snow), amount, and intensity are deduced. The third type measures the forward optical scattering by the particles, from which the precipitation type, amount, and intensity are estimated. In comparison to conventional gages, these optical devices tend to have a larger degree of measurement uncertainty, but they provide useful alternatives when tipping-bucket or weighing gages cannot be used, for example, on a ship or a floating platform affected by wave motion (Nystuen et al. 1996).

Recording precipitation gages are commonly used with internal or external datalogging devices that record the data at a fixed interval (e.g., every 30 min) or record the time stamp of individual tips of a tipping bucket. Many data-logging devices can also accommodate other environmental sensors, such as water-level or water-quality sensors, and transmit the data via telephone, satellite, or wireless communication network.

Fig. 3.5 Examples of precipitation gages equipped with a windshield. (**a**) a weighing gage equipped with an Alter shield. (**b**) a non-recording gage equipped with a Nypher shield

3.4.3 Effects of Wind

Precipitation gages that project above the ground surface generate wind eddies that tend to reduce the catch of the smaller raindrops and snowflakes (Dingman 2002:109). This can be a major source of error in precipitation measurements. The amount of measured precipitation relative to "true" precipitation is referred to as gage-catch deficiency. The degree of deficiency generally increases with wind speed and can be on the order of 20 % (Yang et al. 1998). Several types of wind shield are commonly installed around a precipitation gage to reduce wind eddies above the gage orifice (Fig. 3.5). Even with a wind shield, catch efficiency is significantly less than one, especially for snowfall. Empirical correction formulas can compensate for the negative bias caused by wind. Coefficients are usually determined by fitting a regression curve to data that relate precipitation to wind speed (Fig. 3.6). Unfortunately, these curves are fitted to what often are noisy data sets (e.g., Goodison et al. 1998:36–37). Therefore, protection from wind should be a high priority in selecting a site for measuring precipitation, particularly for measurements of snowfall.

3.4.4 Snowfall Measurement

Measurements of snowfall using a precipitation gage have additional challenges even if wind effects are minimized. If a weighing gage is used, the reservoir needs

3 Assessing and Measuring Wetland Hydrology

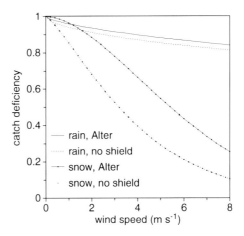

Fig. 3.6 Gage-catch deficiency for rain and snow as a function of wind speed at the orifice height. Data are for the U.S. National Weather Service standard 8-in. gage with and without Alter shields (equations compiled by Dingman 2002:111–112)

to have a sufficient amount of antifreeze solution to melt incoming precipitation, even at very low temperatures. If a heated tipping bucket gage is used, the effects of evaporation due to heating need to be considered. During high-intensity snowfall, snow may pile up at the gage orifice and subsequently blow off, causing negative bias in measured precipitation. If a gage is located in an exposed area, drifting snow may fall into the gage, causing positive bias.

Blowing snow causes redistribution of accumulated snow and substantial sublimation loss, resulting in large local-scale variability of snow density and accumulation. Depending on the surface condition of a wetland and the surrounding upland, the wetland may accumulate higher or lower amounts of snow than recorded by precipitation gages. For example, if a wetland has a smooth-ice surface and the surrounding upland is covered by tall grasses, much of the snow that falls on the wetland may drift to the wetland edge or the upland, where it is trapped by grasses. On the other hand, if a wetland is situated in a relatively deep basin and has extensive emergent vegetation, snow may drift from the upland and accumulate in the wetland. For these reasons, it often is better to quantify and distinguish between snow accumulation on the wetland, and snow accumulation on the surrounding terrain, rather than trying to relate snowfall measured with a precipitation gage to distribution across a landscape of interest.

A snow survey conducted at peak snowpack accumulation, just before the snowpack starts to melt, can integrate, both spatially and temporally, all of the processes that apply and redistribute snow over a landscape, including drifting, loss due to sublimation, and mid-winter melt events. A snow survey should be conducted along an established line, called a snow course, that encompasses the local-scale heterogeneity of landforms and vegetation across the area of interest. At a fixed interval (typically 10–50 m) along the snow course, a snow tube is used to measure the depth of snow and collect a sample for determining snow density by weight. Depending on the scale of survey, the process can be abbreviated by measuring depth at all of the sites and measuring density at a subset of those sites. In this case, the depth-density relation needs to be established to estimate density for those points with only depth measurements. The amount of snow water equivalent, commonly abbreviated as SWE, is calculated from the product of depth and density. Details on snow course measurements are described in WMO (1994:117–122).

In many studies, the amount of precipitation is recorded without differentiating whether it is rain or snow. Local air-temperature data can be used to separate rainfall from snowfall. The transition from snow to rain normally occurs at temperatures between 1 and 3 °C (e.g., Dingman 2002:108).

3.4.5 Calibration and Maintenance

Precipitation gages need to be recalibrated at specific intervals to operate at their stated accuracy. Detailed calibration procedures are specific to the instrument, and are usually found in the manufacturer's operation manual. Calibration is simple for manual gages; it involves accurate measurement of the gage orifice to confirm that orifice dimensions are as stated in the manufacturer's specifications. If not, a custom multiplier can be applied to the data. For weighing gages, known weights of water are added to the gage to validate the gage reading. For tipping-bucket gages, a known volume of water is slowly poured through the gage (e.g., 5–10 tips per minute) and the total number of tips is used to determine the amount of precipitation (mm) per tip, which can be different from the manufacturer's specification by as much as 10 % (Marsalek 1981).

Precipitation gages need to be periodically cleaned (at least once every field season) to prevent clogging of the funnel opening, accumulation of dust in the tipping buckets, and to minimize friction of all moving parts. Gages should be checked to ensure they remain level. Any potential obstruction, such as fallen woody debris or vegetation growing close to the orifice, should be removed. Recording instrument calibrations and servicing in a maintenance record is useful. Such information should ideally be kept in a metadata file accessible in the same area as the data. This advice applies to collection of all automated data related to wetland hydrology and eliminates confusion during subsequent data processing and analysis, or when someone other than the operator uses the data.

3.4.6 Spatial Variability and Interpolation

Precipitation can have large spatial variability due to variation in elevation, topography, prevailing wind direction, and distance from moisture sources or from urban areas. For example, the amount of precipitation may systematically decrease on the lee side of mountain ranges. Urban "heat islands" may have complex effects on the local distribution of precipitation. Convective storms have a relatively small area of influence, often resulting in a large difference in rainfall between sites located

Fig. 3.7 Construction of Thiessen polygons by: Step-1, drawing the dashed lines connecting adjacent gages, Step-2, drawing solid lines perpendicularly bisecting the dashed lines, and Step-3, dividing the area into four polygons defined by the solid lines

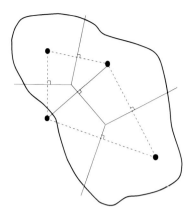

within and beyond the narrow storm track. For these reasons, precipitation gages need to be located as close as possible to the wetlands being studied. Determining precipitation over a large wetland or drainage basin of wetlands may require multiple gages to capture spatial variability. If precipitation data from nearby weather stations are used, instead of an on-site gage, errors resulting from spatial variability should be considered carefully.

To estimate areal precipitation using data from multiple stations, or to estimate local precipitation using data from distant weather stations, data need to be interpolated spatially using one of several surface-fitting method (Dingman 2002:118–130). The weighted-average method estimates precipitation (P_{av}) by the sum of the product of individual station data (p_i) multiplied by a station-specific weighting factor (w_i):

$$P_{av} = \sum_{i=1}^{N} w_i p_i \quad \text{and} \quad \sum_{i=1}^{N} w_i = 1 \qquad (3.6)$$

where N is the number of stations. The simplest averaging scheme is the arithmetic mean that uses $w_i = 1/N$ for all stations. Another commonly used scheme is the Thiessen polygon method, in which the area of interest is divided into N polygons as shown in Fig. 3.7, and w_i is given by the area of each polygon divided by the total area. Other surface-fitting methods are more convenient for constructing precipitation maps from a large or relatively dense network of gages. In these methods, the area is divided into a large number of grid cells and precipitation for each grid cell is computed from station data using a weighted average scheme similar to Eq. 3.6. The most commonly used weighting schemes include the inverse distance method, where w_i is inversely proportional to the distance between the grid cell and the station, and the kriging method, which assigns w_i based on geostatistical correlation among station data (e.g., Kitanidis 1997). These methods are available in popular software packages such as ArcGIS (Environmental Systems Research Institute Inc.) and Surfer (Golden Software Inc.). The software packages can be used to estimate total precipitation over the area, or point values of precipitation for a location that does not have a local precipitation gauge.

3.4.7 Missing Data and Other Issues for Using Archived Weather Data

It is not uncommon to have gaps in precipitation records due to malfunctioning equipment, temporary closure of a station, and various other factors. Missing data are often estimated using data from other precipitation gages, preferably located close to the site. This estimation can be made using the weighted average method (Eq. 3.6), or using the normal-ratio method (Dingman 2002:115) that estimates missing values (p_m) from long-term average precipitation at the station with missing record (P_0), the long-term average precipitation at each station (P_i), and the data for the missing time interval (p_i) at other stations

$$p_m = \frac{1}{N} \sum_{i=1}^{N} \frac{P_0}{P_i} p_i \qquad (3.7)$$

where N is the number of nearby stations with valid data. The long-term average may be annual or monthly depending on the site climatic condition and the nature of analysis (e.g., weekly or monthly water budget).

Archived precipitation data from other sources should to be examined for consistency, particularly if the user does not have first-hand knowledge of the station history and conditions. Changes in measurement method, gage location, or the surrounding environment can induce artificial offsets or trends in the data (Dingman 2002:117). It is important to review the station history, if available, and also use a double-mass curve technique to identify any suspicious data. A double-mass curve is a graph showing cumulative monthly or annual precipitation from a reference station on the horizontal axis and cumulative precipitation from the station of interest on the vertical axis (Searcy and Hardison 1960). The slope of a double-mass curve should be constant if there has been no change in the station of interest. If there is a statistically significant change in slope, the data from the station of interest can be multiplied by a correction factor to compensate for the change. The reference station should have consistent data and be located reasonably close to the station of interest.

In addition to checking the consistency, attention should be paid to gage calibration and correction procedures. For example, some weather stations operated by governmental and municipal agencies may not apply corrections for gage-catch deficiency, resulting in a negative bias in measured precipitation, snowfall in particular. In recent years, gridded precipitation data have become available from government agencies such as the U.S. National Oceanic and Atmospheric Administration (NOAA). These data are generated over a large region (e.g., North America) typically by interpolating observation data and/or refining numerical weather model outputs. While these data sets offer convenient means to estimate precipitation for a given region, the data are not intended as a surrogate for local precipitation measurements. Therefore, gridded precipitation data should be validated using observational data from local stations.

3.4.8 Summary

Precipitation measurement provides critical input data to water-budget analysis. With proper installation and operation of precipitation gages, and sufficient attention to calibration and site maintenance, it is feasible to achieve an accuracy of 5–15 % in point precipitation measurement. Spatial variability adds another degree of uncertainty for larger areas of interest, but the uncertainty can be reduced by using multiple gages and appropriate spatial interpolation techniques.

3.5 Evapotranspiration

Evapotranspiration (ET) is the combined flux from surface water to the atmosphere resulting from evaporation and transpiration from plants. Evaporation is the process of converting liquid water to water vapor, along with the transport of that vapor from the water surface to the atmosphere. Transpiration is a similar process, but one that occurs in plants. Liquid water is pulled through roots from the soil and transported to the plant leaves. The vaporization process occurs within plant leaves and the release of water vapor to the atmosphere occurs via small openings on the leaf surface called stomates. We usually cannot distinguish between evaporation and transpiration so hydrologists generally combine the measurement of these processes (e.g., Shoemaker et al. 2011). Most of the time, this is sufficient from a water-budget perspective. Unless loss of wetland water to groundwater is unusually large, ET usually is the largest water-budget loss term for wetlands that do not have a surface-water outlet and is, therefore, an important term to quantify as accurately as possible.

3.5.1 Commonly Used Methods

Vaporization of water is an energy-intensive process; 4.2 J are required to raise 1 g of water 1 °C whereas approximately 2,500 J are needed to vaporize 1 g of water. Both evaporation and transpiration are dependent on the amount of energy available to drive the process. They also depend on the relative availability of liquid water as well as on the ability of the atmosphere to remove the water vapor once it is formed, thereby allowing for the formation of additional water vapor at the wetland surface. Since wetlands by definition are settings that generally (although not always) have an ample water supply, it is usually assumed that ET occurs at a rate that is unlimited from a water-supply perspective. Therefore, ET is assumed to occur at the maximum potential ET rate based on available energy. This same assumption cannot be made regarding the ability of the atmosphere to remove the recently vaporized water, however. The vapor-removal component of ET depends on the degree of instability of the lower atmosphere. If the atmosphere is stable, in which case, temperature and density both decrease with increasing elevation, there is very little vertical movement that could otherwise aid the vapor-removal process. Therefore, most methods for quantifying ET also measure the vapor-pressure gradient, wind speed, or both, just above the surface supplying water for ET.

3.5.2 Direct Measurements

Of the options available for quantifying ET, only two can be considered capable of directly measuring the process. The evaporation pan is very simple, but requires a coefficient that can be difficult to determine. The eddy-covariance method often is considered a direct method because it has a sound theoretical foundation and requires no assumptions regarding atmospheric stability or the wind-speed velocity profile. It also requires extensive instrumentation and data processing.

3.5.2.1 Evaporation Pan

The evaporation pan is perhaps the most direct method for quantifying ET. It consists of a cylinder nearly filled with water with the top open to the atmosphere. Evaporation is determined by totalling the water added to the system, minus water removals following rainfall, required to maintain the water at a constant level. Pans of a variety of shapes, sizes, and depths have been used over many decades. Since the 1950s and 1960s, many national networks (e.g., Jovanovic et al. 2008) have adopted the class A pan configuration of 1.21-m diameter and 0.25-m height with a stilling well positioned inside the tank to facilitate accurate measurement of the water level (Fig. 3.8). To ensure uniformity, the standard installation is located in an open area at least 20 m by 20 m with close-cropped grass on a wooden platform 15 cm above ground (Allen et al. 1998). Some sites also measure wind speed just above the rim of the pan as well as air and water temperature to make additional empirical corrections for those environmental variables (e.g., Harbeck et al. 1958).

The amount of water lost from an evaporation pan almost always is larger than water lost from a nearby lake or wetland because of enhanced wind flow across the pan surface and radiational heating of the sides of the pan. A pan coefficient of 0.7 is generally applied to convert measured ET to actual ET on an annual basis (Dingman 2002), although values have been reported ranging from 0.64 for the Salton Sea, California, to 0.81 for Lake Okeechobee, Florida (Linsley et al. 1982). Numerous studies have determined site-specific monthly coefficients that vary over a much larger range to achieve greater accuracy (Kohler 1954; Abtew 2001; Masoner et al. 2008). Several studies have employed a floating pan situated in the water body of interest. For small wetlands, this method works well because waves do not become large enough to overtop the floating pan and corrupt the data. Data from a floating pan situated in a small wetland near Norman, Oklahoma, were within 3 % of

3 Assessing and Measuring Wetland Hydrology

Fig. 3.8 Two standard class-A evaporation pans in use near a lake in northern Minnesota. One is serviced daily and the other weekly (Photo by Donald Rosenberry)

evaporation-chamber measurements and were considered to represent actual ET with no correction factor (Masoner and Stannard 2010). Comparisons with the empirical Priestley-Taylor method (discussed in the Combination methods section) indicated that the Priestley-Taylor method over-estimated ET during the day because the air over the hot, dry landscape surrounding the wetland did not represent atmospheric conditions directly over the evaporating water in the wetland. The study also indicated that the floating-pan measurements over-estimated ET during mid to late afternoon and under-estimated ET during nighttime to early morning. This was attributed to the shallower depth of the water inside of the pan being more sensitive to diurnal air-temperature changes than the deeper water column adjacent to the floating pan. However, the errors were largely offset so the floating pan provided daily ET values with little bias.

3.5.2.2 Eddy-Covariance Method

The process of evaporation can be viewed as vapor-rich rotating eddies of various sizes that rise because they are less dense than other volumes of drier air that descend to occupy the volume that the moist, rising air just vacated. The process is 3-dimensional and also occurs on horizontal axes, but for the purpose of determining evaporation from a wetland surface we are most concerned with the vertical axis. Sensors measure the vertical velocity of these air packets, as well as their "concentrations" (either temperature or absolute humidity) to obtain the vertical velocity of the upward or downward flux of these properties. Vertical flux is then represented as a covariance between measurements of vertical velocity and concentration of either humidity (vapor flux) or temperature (sensible heat flux).

Mathematically, the process can be presented for latent heat flux ($\lambda \rho_w E$) and sensible heat flux (H) as:

$$\lambda \rho_w E = \lambda \overline{w' \rho'_v} \tag{3.8}$$

$$H = \rho_a c_p \overline{w' T'} \tag{3.9}$$

where overbars indicate an average (typically a 30-min average), primes indicate departures from the mean, λ is the latent heat of vaporization (J kg^{-1}), ρ_w is the density of water (kg m^{-3}), E is evapotranspiration flux (m s^{-1}), w is the vertical wind speed (m s^{-1}), ρ_v is the absolute humidity (also called vapor density) (kg m^{-3}), ρ_a is the density of air (kg m^{-3}), c_p is the specific heat capacity of air (J kg^{-1} °C^{-1}), and T is the air temperature (°C). Measurements are typically made 10–20 times a second. For both latent and sensible heat fluxes, units are in J m^{-2} s^{-1} or W m^{-2}. E rather than ET is used in Eq. 3.8 to be consistent with other literature that describes the evaporation process. The process is identical in wetland settings, although the source for some of the water is via the stomates of leaves. In this chapter, ET refers to the process of evapotranspiration and E refers to evapotranspiration flux in units of distance per time.

The above description is suitable for flat, open areas with uniform vegetative cover over long distances upwind of the sensors. The process is a bit more complex on sloping land surfaces or where air streams converge or diverge upwind of the sensors, in which case covariances are determined on three axes and coordinate rotations to the data may need to be performed before E is determined (Wilczak et al. 2001). A krypton hygrometer or infra-red gas analyzer, and sonic anemometer, are typically used to measure humidity and wind speed, respectively. Temperature is provided either as a by-product of the sonic anemometer or from a separate sensor. Instrumentation has improved rapidly in this field; newer sensors are much more robust and are now capable of being deployed during rain events. This method is sensitive to misalignment; sensors need to be deployed and maintained on a stable platform and at a constant orientation (Wilczak et al. 2001). Height of deployment is strongly related to the upwind area that the measurement represents. The roughness of the upwind area also greatly affects the sensor signal.

3.5.3 Estimation Methods

Numerous empirical methods have been developed to estimate evaporation, ranging from methods that require only measurement of air temperature to methods that have a sound physical basis and require numerous parameters. Methods can be grouped into those that quantify available energy to determine evaporation as the residual,

3 Assessing and Measuring Wetland Hydrology

those that quantify evaporation based on the aerodynamics of the near-surface atmosphere, and those that combine energy and aerodynamic approaches.

3.5.3.1 Energy-Balance Method

Energy removed in the evaporation process is usually offset by resupply of energy from the wetland and the atmosphere. Solving for evaporation by accounting for all of the other energy terms can be expressed as:

$$Q_n - \lambda \rho_w E - H = Q_x - Q_v \tag{3.10}$$

where Q_n is net radiation, Q_x is increase in energy stored in the wetland water column, and Q_v is the net amount of energy advected to the wetland from the sum of streamflow to and from the wetland, groundwater flow to and from the wetland, and rainfall. Atmospheric terms are on the left side and water and sediment terms are on the right side of the equation. Because of the errors associated with determining Q_x and Q_v, the accounting period for this method historically has been 5 days or longer but newer instrumentation has led some to determine evaporation using this method on a daily basis.

Unfortunately, neither $\lambda \rho_w E$ or H can be directly measured, requiring the use of the Bowen ratio (B):

$$B = H/\lambda \rho_w E \tag{3.11}$$

The Bowen ratio can be determined by measuring differences in temperature and vapor pressure in the atmosphere directly above the evaporating surface:

$$B = \gamma \left(\frac{T_s - T_a}{e_s - e_a} \right) \tag{3.12}$$

where γ is the psychrometric constant, T_s is the temperature at the water surface, T_a is the air temperature, e_s is the saturation vapor pressure at the temperature of the water surface, and e_a is the atmospheric vapor pressure. T_a and e_a are measured at the same height above the water surface, commonly 2 m. The psychrometric constant is not really a constant but is a function of specific heat capacity and atmospheric pressure. It is equal to

$$\gamma = \frac{c_p P}{0.622 \lambda} \tag{3.13}$$

where c_p and λ are as described above, P is atmospheric pressure, and 0.622 is the ratio of the molecular weights of water vapor and air (Perez et al. 1999).

Substituting B into Eq. 3.10, and assuming that Q_v is negligibly small (a reasonable assumption for most wetland settings), we obtain

$$E = \frac{Q_n - Q_x}{\lambda \rho_w (1 + B)} \tag{3.14}$$

This expression is commonly termed the short form of the energy-budget evaporation equation. For a version that includes individual terms for net radiation, as well as Q_v and heat transfer to and from wetland sediments (Q_b), see Parkhurst et al. (1998) or Winter et al. (2003).

Determination of Q_x can be challenging for some wetlands where water depth is sufficiently large that the wetland column is thermally stratified. In that case, several temperature sensors are required at different water depths to represent the change in heat stored for each depth increment, or horizontal slice, of wetland water. The change in heat can be expressed as

$$Q_x = \sum_{i=1}^{n} \rho_{wi} c_{wi} h_i \frac{\Delta T_{wi}}{\Delta t} \tag{3.15}$$

where h_i and T_{wi} are the thickness and average temperature, respectively, of each horizontal slice of wetland water, ρ_w and c_w are defined as before but now apply to the water in each specific depth increment of the wetland, and Δt is the time between two successive temperature measurements. Q_x is the sum of the change in heat for all depth increments. The same procedure can be used in the soil where standing water is not present if temperature sensors are installed at several depths beneath the wetland bed, with the deepest sensor at a depth where temperature does not change considerably over periods less than weeks to perhaps a month. In this case, the term for changes in heat stored in the sediment commonly is referred to as G rather than Q_x.

In drier environments, where soil at the surface is not at or near saturation, the assumption cannot be made that vapor pressure at the surface is at saturation. In these cases, temperature and vapor pressure need to be measured at two heights. Because differences often are very small, and instrument bias could lead to substantial error, it is common to use a device that alternates the positions of the upper and lower sensor so that bias can be subtracted.

Several other empirical formulations are used to determine ET based on estimation of available energy. Most require measurement of either solar or net radiation, or air temperature, or both. Three that use solar radiation and air temperature are Jensen-Haise, Makkink, and Stephens-Stewart (McGuinness and Bordne 1972). Most radiation-temperature models can perform fairly well in the environment for which they were developed but do not transfer well when applied in other climates (e.g., Rosenberry et al. 2004, 2007). Several others require measurement only of air temperature and day length (e.g., Blaney-Criddle and Hamon methods) or air temperature only (e.g., Papadakis and Thornthwaite methods). The Thornthwaite method, in particular, has been widely used because of its simplicity.

However, it has been shown based on comparisons in several settings that these rather simplistic formulations provide ET values with much greater uncertainty than methods that are more physically based (Winter et al. 1995; Dalton et al. 2004; Drexler et al. 2004; Rosenberry et al. 2004, 2007; Elsawwaf et al. 2010).

3.5.3.2 Energy Balance Method with Large Aperture Scintillometer (LAS)

Sensible heat flux from the ground causes variations in the refractive index of the atmosphere, referred to as scintillation. These variations can be detected by a scintillometer using a beam of light emitted by the transmitter and detected by a receiver, from which sensible heat flux is estimated (Hill et al. 1992). A large aperture scintillometer (LAS) measures scintillation over a horizontal line at a scale of several hundred meters. Using sensible heat flux estimated by the LAS method with measurements of net radiation and other components of the energy balance, it is possible to estimate latent heat flux and evapotranspiration from Eq. 3.10. Brunsell et al. (2008) applied the LAS method at a tall grass prairie site in Kansas and obtained ET values that were comparable to measurements by the eddy-covariance method. This method is not as widely used as the eddy-covariance method, but offers a promising alternative for larger-scale measurements of ET.

3.5.3.3 Aerodynamic Methods

Numerous equations have been developed to determine ET based on the removal of water vapor (mass transfer) from the evaporating surface. Many are referred to as Dalton-type equations, named after the English chemist John Dalton who first formulated such an equation in 1802 (Dingman 2002), and are of the form

$$E = N \cdot f(u)(e_s - e_a) \tag{3.16}$$

where N is a locally-determined coefficient (not required for some formulations), $f(u)$ is a function of wind speed, and $e_s - e_a$ is the vapor-pressure difference presented in Eq. 3.12. N, if present, generally is determined by regression of the product of the wind function and the vapor-pressure difference (the mass-transfer product) with another means of determining evaporation.

Rasmussen et al. (1995) presented a comparison of seven versions of this basic equation applied to a number of lakes in Minnesota. Singh and Xu (1997) presented results of 13 mass-transfer equations applied to data from 4 climate stations in Ontario, Canada. Results indicated a general insensitivity to wind speed. In addition, methods with parameters determined for one location did not transfer well when applied to other locations.

A range of methods for quantifying ET, including the mass-transfer method with locally determined values for N, were compared with the energy-budget approach at a small wetland in North Dakota and at a small lake in New Hampshire. Mass-transfer results were within 20 % of energy-budget results 45 and 57 % of the time at the North Dakota and New Hampshire sites, respectively, and were among the poorest methods in comparison to energy-budget results (Rosenberry et al. 2004, 2007).

3.5.3.4 Combination Methods

Combination methods determine ET based on measuring both available energy and aerodynamic efficiency. Probably the best known is the Penman method (1948), which often is written as

$$E = \frac{\Delta}{\Delta + \gamma} \frac{Q_n - Q_x}{\lambda \rho_w} + \frac{\gamma}{\Delta + \gamma} E_a \qquad (3.17)$$

where Δ is the slope of the saturation vapor pressure versus temperature curve at the mean air temperature, γ is the psychrometric constant described in the *Energy-balance methods* section, and E_a is described as the drying power of the air and is basically a mass-transfer product, as described in the *Aerodynamic methods* section. The first and second terms on the right side of the equation are the radiation and aerodynamic terms, respectively; hence, a combination method. Many mass-transfer products have been associated with the Penman method. A form of the equation that includes an often-used mass-transfer product in place of E_a, along with a multiplier to convert to units of mm/day, is (Rosenberry et al. 2007):

$$E = \frac{\Delta}{\Delta + \gamma} \left(\frac{Q_n - Q_x}{\lambda \rho_w} \right) \times 86.4 + \frac{\gamma}{\Delta + \gamma} (0.26(0.5 + 0.54 u_2)(e_{sa} - e_a)) \qquad (3.18)$$

where u_2 (m s^{-1}) is wind speed measured at 2 m above the water surface, e_{sa} (hPa) is the saturation vapor pressure at the air temperature, and e_a (hPa) is the measured vapor pressure at 2 m height. One of the benefits of using the Penman equation is temperature and vapor pressure only need to be measured at one height. Penman originally formulated this method to use T_a as the temperature at which to obtain Δ and e_{sa} because T_s was considered difficult to measure.

The Penman method requires a lot of data: net radiation, temperature of the water body at multiple depths, air temperature, vapor pressure of the air, and wind speed. Numerous simplifying assumptions have been made to reduce the data requirements. The Priestley-Taylor (1972) method is likely the best of these alternate approaches. It assumes that the aerodynamic portion of the equation is 26 % of the energy term and replaces the aerodynamic term with the coefficient 1.26 applied to the energy term. Therefore, only T_a, Q_n, and Q_x need to be determined:

$$E = 1.26 \frac{\Delta}{\Delta + \gamma} \frac{Q_n - Q_x}{\lambda \rho_w} \qquad (3.19)$$

Despite its simplicity, the Priestley-Taylor method applied to data from a wetland in the prairie-pothole region of North Dakota produced better data that the Penman method when compared to results from the energy-budget method (Rosenberry et al. 2004).

Another type of combination method that may work well for small wetlands, or wetlands that are moisture limited, is the complementary relationship method. First proposed in 1963, the complementary-relationship method makes the assumption that actual evapotranspiration (AET) is reduced relative to evapotranspiration in a wet environment (ET_{wet}) to the same extent that potential evapotranspiration (PET) is larger than ET_{wet}; the drier the environment, the greater the difference between AET and PET. Following this logic,

$$AET + PET = 2ET_{wet} \qquad (3.20)$$

By calculating ET_{wet} based on available energy, and measuring PET, one can then determine AET with

$$AET = 2ET_{wet} - PET \qquad (3.21)$$

Any number of methods can be used to determine ET_{wet} and PET (e.g., Morton 1983a). One approach is to determine ET_{wet} using the Priestley-Taylor method and determine PET using the Penman method (Brutsaert and Stricker 1979). As one might expect, the Brutsaert-Stricker method applied over a wetland in North Dakota (Rosenberry et al. 2004) and over a small lake in New Hampshire (Rosenberry et al. 2007) gave results very close to either the Penman or the Priestley-Taylor methods alone because the environment was wet; ET was occurring at the potential rate. However, wetlands do not always have an ample water supply. During times when wetlands are relatively dry, the complementary relationship method likely would produce better results than other methods designed to indicate ET at the potential rate. The method also can be used to determine the extent to which warm, dry air may increase ET along the upwind edge of wetlands, a particular concern for wetlands that are small or situated in arid environments (Morton 1983b). This also is a concern for other ET methods that rely on the assumption that vapor and temperature gradients are adjusted to the wetland surface at the point of measurement. However, at a wetland in North Dakota, the open-water portion of which varied in size from 1.5 to 3 ha, insufficient fetch was found to cause errors in ET estimates of less than 2 % (Stannard et al. 2004).

3.5.4 Measurement Parameters for Estimating ET

Quantification of net radiation is required for most of the ET methods. This involves measurement of downward shortwave and longwave radiation from the atmosphere, upward reflected shortwave and longwave radiation, and upward longwave

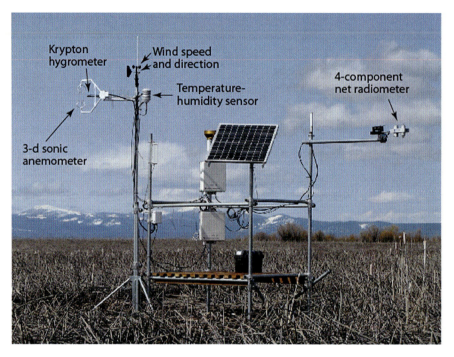

Fig. 3.9 Upward and downward facing 4-component net radiometer deployed over a large wetland in southern Oregon. Other sensors also are labeled (Photo printed with kind permission of © David Stannard, U.S. Geological Survey 2014. All Rights Reserved)

radiation emitted from the land or water surface proportional to the temperature of the surface. Several broad-spectrum radiometers with sensors facing both upward and downward can provide a single net-radiation value. Another option is to deploy four sensors, two facing upward to separately measure shortwave and longwave radiation, and two facing downward to separately measure shortwave and longwave radiation (Fig. 3.9). Although deploying four sensors provides greater accuracy, it comes at a substantially larger cost. If only two upward-facing sensors are deployed, the upward radiation vectors can be calculated with reasonable accuracy (Sturrock et al. 1992; Parkhurst et al. 1998; Winter et al. 2003). Sensors need to be deployed on a stable platform and maintained in a level, horizontal orientation to minimize bias.

All evaporation methods require measurement of air temperature and most also require measurement of humidity at the same location. For gradient-based methods, sensors usually are deployed that provide both temperature and relative humidity. With both parameters, humidity output can be converted to vapor pressure, vapor density, or whatever form of humidity is needed for a specific ET method. Several ET methods require measurement of the water-surface or land-surface temperature and make the assumption that vapor pressure is at saturation based on the surface temperature. Methods that quantify change in heat stored in the water body (Q_x) require a temperature sensor for each horizontal slice of water contained in the

3 Assessing and Measuring Wetland Hydrology

Fig. 3.10 ET raft deployed on Cottonwood Lake Area Wetland P1. *Inset on left* shows raft being put into place just after ice out. Surface-water-temperature sensor is located near the shallow end of the float so it is resting just below the water surface. The white float minimizes heating from solar loading (Photos by Donald Rosenberry)

water body. For settings where substantial horizontal variation in temperature exists, such as in a wetland that has several embayments of differing total depth, profiles of temperature sensors may need to be deployed at several locations to accurately represent change in heat stored in the entire wetland. For other wetlands, a single deployment of sensors at different depths may be sufficient. Rosenberry et al. (1993) provide additional information regarding the errors associated with making measurements at a single location relative to multiple locations. Temperature usually is measured with a thermistor, which has a non-linear, inverse electrical resistance relative to temperature. Polynomial functions are applied to the resistor output to provide temperature that commonly is within ± 0.1 to 0.5 °C of true temperature. Thermocouples are inexpensive and stable temperature-measurement devices based on the principle that an electrical current is generated when a junction is made between two dissimilar metals. The current is proportional to the temperature difference between two electrical junctions at different locations in the same circuit. Output is linear and stable, but small. Temperature difference usually is related to a separate temperature sensor located very near one of the bi-metal junctions. If wires are well shielded with insulation so that temperature gradients are small (or at least remain constant) in the vicinity of the reference thermometer, modern dataloggers are well capable of resolving the small current changes that occur in response to temperature changes.

Wind speed often increases approximately logarithmically with height above a surface. Measurement of wind speed needs to be specified with regard to height above the water surface. Measurement height is somewhat arbitrary, but 2 m is common. Wind speed, air temperature, and humidity should all be measured at the same height (Fig. 3.10). All anemometers have a threshold below which the sensor indicates zero wind speed. A sensor should be selected with the lowest threshold that is affordable. Most sensors have cups or propellers that spin in response to the wind and generate either a pulse per revolution or direct current (DC) volts proportional to the rotational velocity. Ultrasonic anemometers operate based on resolving the difference between times of transmission of ultrasonic pulses sent in both directions between two transducers. Combinations of sensors are commonly oriented along two or three axes to determine horizontal or 3-dimensional wind speed, respectively.

Either dataloggers or field computers can be used to measure and store data from analog and digital sensors that provide data to feed evaporation methods. Sensor scan rate usually is determined based on the suite of sensors connected to a particular datalogger as well as requirements specific to evaporation methods or other products generated from the installation.

The most representative data are collected over the wetland in a location designed to maximize fetch in all directions, or at least in the directions of the prevailing winds. Sensors commonly are deployed from a floating raft or a platform fixed to the bed, depending primarily on the wetland depth. Sensors that need to be deployed at a constant height above the water surface, such as temperature-humidity sensors and anemometers, need to be adjusted as the water level of the wetland rises and falls. The temperature probe at the surface should be installed so it rests as close to the water surface as practically possible. A float is commonly used to keep sensors at the proper height above the water surface (Fig. 3.10). Data from land-based sensors often are used if a raft or platform is not available. In some cases, use of land-based sensors results in little additional error (Rosenberry et al. 1993).

3.5.5 Measurement Errors

ET is difficult to estimate or measure in part because of the numerous sources of error associated with the large number of sensors required. Error varies substantially depending on the chosen measurement method.

The evaporation pan has sources of error associated with the actual physical measurement as well as conceptually. A sensitive depth gage is very important when adding or removing water in response to evaporation or rainfall. For a class-A evaporation pan, evaporation of only 1 mm of water equates to 1.15 L of water that needs to be added back to the pan to maintain a constant water level. Pans also eventually develop leaks. If the pan is buried to minimize the effect of sidewall heating, it would be difficult to diagnose a leak. Similarly, a floating pan is subject to waves splashing over the side of the pan during windy periods. Numerous modifications also have been made to minimize the effect of animals drinking from the pan water. If a wire mesh is placed over the pan, a correction may need to be made to account for the shading effect from the wire mesh. From a conceptual perspective, a pan deployed in an arid environment, and elevated relative to the land surface around it, is subject to what commonly is termed the oasis effect; the pan represents a surface that is evaporating at the maximum potential rate. However, if the surface around the pan has a limited moisture source, and the rate of ET is substantially smaller, then the cooling effect associated with evaporation is reduced. The air mass that passes over the pan is both warmer and drier than what it would be if the surface surrounding the pan was losing water at the same rate as the pan. The warmer, drier air enhances evaporation from the pan, just as described for the complementary-relationship method.

Sources of error, in addition to simple measurement error associated with each sensor, often are the result of prolonged sensor deployment. Some sensors, such as humidity sensors and radiometers, drift and require regular recalibration. Bearings in anemometers wear out and require replacement. Thermistors deployed in water experience algal growth that delay the response time of the sensor. If the wetland water level changes and the sensor heights are not adjusted, then bias can occur in temperature and vapor-pressure differences. Loss of data due to power interruptions, sensor failure, or breaks in the sensor wires also occur. An error that often is overlooked is extrapolation of measured evaporation rates over the evaporating surface. As wetland stage changes, along with the corresponding change in shoreline location and wetland surface area, substantial error can occur by applying measured evaporation rates to an incorrect wetland surface area (see Sect. 3.2.1).

3.5.6 Cost Effectiveness

All scientists would like to quantify fluxes and processes as accurately as possible, but the benefit associated with greater accuracy has to be balanced with the cost of instrumentation and methodology. The eddy-covariance method may yield ET rates with the smallest uncertainty, but sensors are expensive, data processing is lengthy, and installation costs can be large. At the opposite extreme, temperature is one of the least expensive parameters to measure, making methods such as Thornthwaite particularly attractive if the scientist is willing to sacrifice accuracy in the interest of economy. To a large extent, the choice of method depends on the importance of accurate quantification of ET. If ET is a large component of a wetland water budget, then a substantial investment in time and money is warranted. If the wetland is dominated by surface-water flow and ET is a relatively small component, then perhaps a method based on temperature, or temperature and radiation, is sufficient. At a small lake in New Hampshire where all components of the water budget were characterized as accurately as possible, the Priestley-Taylor method was deemed the best compromise between accuracy and cost. It produced data nearly as good as the energy-budget method but did not require measurement of surface-water temperature or quantification of advected energy sources and sinks (Rosenberry et al. 2007). Another study of a reservoir in northern Florida came to a similar conclusion regarding use of the Priestley-Taylor method despite the water budget being dominated by surface-water inputs and losses (Dalton et al. 2004).

3.6 Surface Water Inflow and Outflow

Surface water flow is commonly the largest component of a water budget for wetlands in humid regions, making it an important component to measure accurately. Accurate measurements are possible if flow is confined to well-defined channels. Measurement uncertainty can be very large, however, in wetland settings with exceptionally low topographic gradients where flow commonly occurs through a network of poorly constrained channels. Furthermore, the distinction between surface-water flow and diffuse overland flow can be difficult to determine. In this section, we limit discussion to settings where a well-defined channel exists and briefly discuss low-gradient settings in the section on flow estimation using indirect methods.

Surface-water flow in terms of volume per time (m^3/s) is often referred to as discharge in the stream hydrology literature. A more precise definition is the volume rate of flow through a stream cross-section at a right angle to the flow direction. Since it is impractical to manually measure stream discharge with 15-min to daily temporal resolution over long periods of time, it usually is calculated from stage measured at a stream-gaging station installed in a channel, or stage measured in an artificial control structure (i.e., flume or weir) using a pre-established stage-discharge relation. Often referred to as a stage-discharge rating curve, this empirical function is determined with direct measurements of stream discharge over a range of stages or by calibration of theoretical formula. In the following section, we will briefly describe establishment of a stream-gaging station, measurement of stream discharge and stage, development of a rating curve, and the associated potential problems and sources of error.

3.6.1 Stream Gaging Station

The accuracy of stream discharge measurement is strongly dependent on the accuracy of the rating curve, which is dependent on the degree of flow control. Therefore, whenever possible, it is best to establish gaging stations in a stream section with good natural flow control (see below) or in a specifically designed control structure such as a weir or flume. Flow control can be defined as a feature some distance downstream of a gaging station that controls the stage-discharge relation upstream at least as far as the location of the gaging station.

Ideally, a gaging station should be located in a place where: (1) the channel is straight about 100 m upstream and downstream, (2) the total flow is confined to one channel at all stages, (3) the streambed is not subject to scour and fill and is free of aquatic vegetation growth, (4) banks are permanent, free of brush, and high enough to contain floods, (5) the downstream flow control is unchanging, (6) a pool is present upstream from the control, (7) the site is far enough upstream from the confluence with another stream that flow in the tributary does not affect flow at the gaging station, and (8) a suitable stream section for making manual discharge measurements is available near the gaging station (Rantz 1982:5). It is usually impossible to meet all the above conditions, and judgement must be exercised to select an adequate location despite some shortcomings. For measuring outflow from a wetland, conditions 6 and 7 can easily be satisfied by using the wetland itself as a pool and measuring stage in the wetland near the outlet. However, controlling the areal extent and density of vegetation in the outlet channel is often a challenge in wetland settings.

Some of the conditions above can be improved by modifying a channel; for example, reinforcing the bank or removing flow obstacles, without significantly altering the habitat characteristics for aquatic life. The most accurate data are obtained by installing an artificial control structure, but installation may require an environmental impact assessment if the required channel modifications are extensive.

3.6.2 Characteristics of Flow Control

Some stream reaches are relatively straight for a long distance with constant slope, channel geometry, and bed roughness. This situation, where the control on the upstream stage-discharge relation is the channel itself, is called channel control. Other flow-control settings include a place where the stream flows across bedrock, a reach where the channel narrows, or the point beyond which the downstream stream reach steepens (i.e., the upstream end of a riffle). Some artificial structures that are not designed specifically to be a flow control, such as a bridge or a culvert under a road, may serve as a good flow control. Two attributes of a satisfactory flow control are stability and sensitivity (Rantz 1982:11). If a control is stable, then the stage-discharge rating curve does not require frequent adjustment. For example, a constriction provided by rock-ledge outcrop is not affected by flood events (stable), but upstream boulders and gravel may move during floods (unstable). Regarding sensitivity, a control section ideally should have a relatively narrow width at low discharge condition so that a small change in discharge is reflected by a significant change in stage.

If natural conditions do not provide adequate stability and sensitivity, artificial control should be considered. In natural streams having a wide range of discharge conditions, it is common to use broad-crested weirs (Fig. 3.11a) that conform to the general shape and height of the streambed (Rantz 1982:12). It is generally impractical to build the control high enough to avoid submergence at high discharges. Therefore, broad-crested weirs are effective for low to medium discharge only. In canals and drains where the range of discharge is limited, thin-plate weirs (Fig. 3.11b) or flumes (Fig. 3.11c) may be used to cover the complete range of discharge. Thin-plate weirs are suitable for channels in which the flow has relatively low sediment load and the banks are high enough to accommodate the increase in stage (backwater) upstream of the weir. Flumes are largely self-cleaning and can be used in channels with high sediment load, and do not cause significant backwater. However, flumes are generally more costly to build than weirs (Rantz 1982:13). Other types of control structures include weirs with moving gates installed in canals, commonly referred to as "head gates".

Fig. 3.11 Examples of artificial flow control structures: (**a**) a broad-crested weir in a stream in northern Manitoba, Canada, (**b**) a thin-plate weir in a stream in Banff, Canada, (**c**) a Parshall flume. The streamflow direction is from the left to the right in all photographs (Photos a and b by Masaki Hayashi; photo c by Donald Rosenberry)

Ideally, artificial controls should have structural stability, their crest should be as high as practical to eliminate the effects of variable downstream conditions, and the stage should be sensitive to discharge. As a general rule, a weir is more advantageous than a flume because it is less expensive, can be designed to have greater sensitivity, and its rating curve can be extrapolated beyond the normal operation range without serious errors (Rantz 1982:18). Flumes are more advantageous in streams carrying heavy sediment load or when backwater created by a weir is undesirable. Specific design and characteristics of different types of weirs and flumes can be found in Rantz (1982:294–326).

3.6.3 Stage Measurement

The first step in measuring stream stage is to establish a permanent datum (reference elevation) and a staff gage (Fig. 3.2). The zero for a stream-stage datum should be below the elevation of zero flow on a natural control, and usually at the elevation of zero flow in an artificial control such as a weir. Changing the datum during a

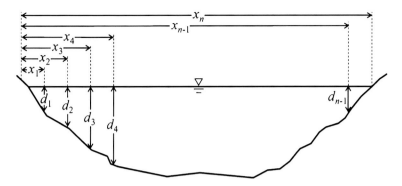

Fig. 3.12 Schematic diagram showing a channel cross section divided into rectangular subsections for the velocity-area method of discharge measurement (Adapted from Rantz 1982:81. Published with kind permission of the U.S. Geological Survey. Figure is public domain in the USA. All Rights Reserved)

monitoring period should be avoided, but if change is unavoidable, the relation between the new and old datum needs to be clearly established and recorded.

Details regarding installation, operation, and maintenance of stage-measuring devices are described in the Sect. 3.3.1 on wetland stage. For stage measurements in streams, a water-level sensor can be protected from flowing debris by installing it in a stilling well, which also dampens waves generated by wind or turbulence and provides a more stable reading. A stilling well is a vertical pipe or culvert that is hydraulically connected to the stream water level. It can be installed in a stream bank and connected to the stream by a subsurface horizontal pipe, or installed directly in a channel. It must be secured to a stable anchor so that its elevation does not change over time. The bottom of a stilling well should be at least 0.3 m below the minimum stage and its top should be above the peak flood level. Details regarding construction of stilling wells and their shelters is described by Rantz (1982:41–52) and Sauer and Turnipseed (2010:6–11).

3.6.4 Discharge Measurement

3.6.4.1 Velocity-Area Method

Among several methods for discharge measurement, the most commonly used is the velocity-area method, which involves direct measurement of flow velocity using a current meter at successive locations along a channel cross section (Fig. 3.12) and summation of measured values to calculate the total discharge. The ideal stream cross section is located within a straight reach having streamlines parallel to each other; the streambed is relatively uniform and free of numerous boulders and extensive aquatic vegetation; flow is relatively uniform and free of

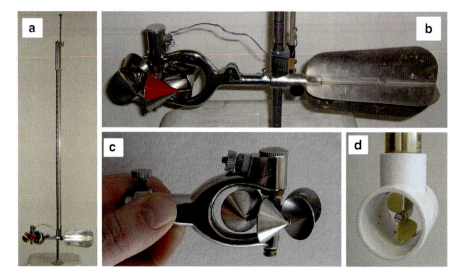

Fig. 3.13 Example of flow-velocity measuring devices that are commonly used in North America: (**a**) top-setting wading rod with a meter attached, (**b**) vertical-axis current meter (Price AA), (**c**) vertical-axis current meter (Price Pigmy), (**d**) horizontal-axis current meter (Global Water FP101)

eddies, slack water, and excessive turbulence; velocity and depth are within the range for which measuring devices give accurate results; and the observer can safely carry out measurements (Rantz 1982:139; Dingman 2002:609). If necessary, the condition can be improved by removing obstructions in, above, and below the channel section without affecting discharge.

Measurements can be made by wading, from a boat, or from bridges. After the selection of a suitable section, the first step is to extend an incremented line across the channel above the water surface and measure the total channel width. A tape measure, a rope marked at constant intervals, a marked tag line for boat measurements, or marking constant intervals on a bridge can be used for this purpose. Except for bridge-based measurements, the line is placed at right angles to the direction of flow. The next step is to determine the measurement interval, or the width of individual rectangular subsections in Fig. 3.12. Rantz (1982:140) recommend 25–30 subsections with no single subsection contributing more than 5 % of total discharge, meaning that smaller widths need to be assigned for subsections in deeper and faster portion of the channel.

Depth of water and average velocity is determined at the middle of each subsection. When flow velocity and depth allow measurements to be made while wading the stream, a flowmeter commonly is mounted to a top-setting wading rod specifically designed to indicate water depth and easily placed at the proper depth for each velocity measurement (Fig. 3.13a). The observer should stand in a position that least affects the velocity being measured; for example, by standing downstream of the tag line and facing either bank, or in the case of a sufficiently

3 Assessing and Measuring Wetland Hydrology

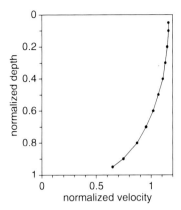

Fig. 3.14 Typical velocity profile plotting dimensionless velocity normalized to the profile-mean value against dimensionless depth normalized to the total depth of profile. At a normalized depth of 0.6, normalized velocity is approximately equal to 1 (Adapted from Rantz 1982:133. Published with kind permission of the U.S. Geological Survey. Figure is public domain in the USA All Rights Reserved)

narrow channel, standing on an elevated plank crossing the channel (Rantz 1982:146). A cable with a heavy depth-sounding weight at the bottom is used to measure depth and suspend the current meter at the proper depth when measurements are made from a boat or bridge or cableway (Rantz 1982:101–102).

Flow velocity in a channel also varies vertically. Within a rectangular subsection, the velocity is very small at the streambed and increases upwards. Figure 3.14 shows the standard velocity profile used by the U.S. Geological Survey (Rantz 1982:133) based on intensive investigation of open channel hydraulics. In this profile, the velocity measured at a normalized depth of 0.6 (six tenths of the distance from the water surface to the bed) is almost identical to the profile-mean velocity, and also the arithmetic mean of two velocity values measured at normalized depths of 0.2 and 0.8 is almost identical to the profile-mean velocity. Based on these observations, it is common practice to represent the mean velocity of a subsection by the arithmetic mean of two measurements at 0.2 and 0.8 depths (2-point method), or a single velocity measurement using a current meter placed at 0.6 depth (six-tenth method). The 2-point method generally gives more reliable results than the six-tenth method, but in shallow streams making a measurement at 0.2 depth may not be possible, depending on the size and specification of the current meter, in which case the six-tenth method is preferred (Rantz 1982:134–135). An alternative approach is to use a current meter that integrates velocity measurements and calculates a profile-mean velocity as the meter is continuously moved up and down the entire profile at a uniform rate for a sufficiently long period (integration method).

Several types of current meters are commonly used to make streamflow measurements. Mechanical current meters have long been the sensor of choice, but electromagnetic current meters and acoustic Doppler velocimeters (ADV; Winter 1981) are rapidly gaining popularity; ADV's may already be the most widely used type of current meter in North America. These sensors are preferable because they output direct velocity values, and in the case of the ADV, they have the ability to measure flow velocities below the limit of mechanical meters (see Turnipseed and Sauer (2010:44–58) for an overview of mechanical and non-mechanical current meters). Mechanical current meters are classified into vertical-axis (Fig. 3.13b, c) and horizontal-axis meters (Fig. 3.13d) depending on the alignment of moving cups or a propeller with respect to the flow direction. In North America, the standard equipment used by U.S. Geological Survey and the Water Survey of Canada are Price[1] AA and Price pigmy vertical-axis current meters. These meters require an operator to place it at a fixed depth, count the number of rotations of the cups, and convert the count to velocity using a calibration table or rating function. This operation, formerly conducted manually, is automated in most meters that are currently available on the market. Horizontal-axis meters also require a count of the rotations of a propeller and conversion to velocity. They are suitable for continuous profiling of flow velocity, and are often equipped with electronics designed for the integration method of velocity measurement (see above).

Regardless of the meter type, it is important to ensure that the meter is correctly functioning and well calibrated. The meter should be visually checked before and after its use, and cleaned if necessary. It should be periodically checked during measurements, when it is out of the water, to ensure that it spins freely. It is not uncommon for debris or aquatic weeds to become trapped between moving parts causing underestimation of velocity. Rantz (1982:93–94) describes the maintenance and care of mechanical meters in detail. A meter should be periodically calibrated to ensure that its rating function (i.e., relation between the speed of rotation and flow velocity) has not changed beyond a specified tolerance. The rating function is normally established by towing the meter at a constant velocity through a long water-filled trough (Turnipseed and Sauer 2010:53). While it is desirable to have current meters calibrated in a dedicated facility operated by experienced staff, it is possible to carry out similar calibrations using a more accessible facility. For example, if access to a swimming pool can be obtained for a short period of time then a flow meter can be attached to a cart and towed at a reasonably constant velocity under completely calm pool conditions (Fig. 3.15).

Once the depth (d_i, m) and mean velocity (u_i, m s^{-1}) of each individual subsection are measured accurately, total discharge (Q, m^3 s^{-1}) is calculated by the summation of discharge in all subsections:

$$Q = \sum_{i=1}^{n} u_i A_i = \sum_{i=1}^{n} u_i d_i \frac{x_{i+1} - x_{i-1}}{2} \qquad (3.22)$$

---

[1] Any use of trade, firm, or product names is for descriptive purposes only and does not imply endorsement by the U.S. Government.

3 Assessing and Measuring Wetland Hydrology

Fig. 3.15 Calibration of a horizontal-axis current meter in a swimming pool. The meter is attached to a cart that is pushed at a constant velocity (Photo by Masaki Hayashi)

where A_i (m^2) is the cross sectional area of each rectangular subsection, and x_i (m) is the distance along the line of measurements. This method of computation is called the midsection method, and is known to provide more accurate values of Q than many other methods (Dingman 2002:611).

In Eq. 3.22, total discharge is calculated using a number of data points measured over the length of time required to quantify flow across the entire cross section. In stream sections undergoing rapid flow transience (for example, during a storm event or water extraction/release upstream), discharge may change substantially during the measurement period. If these conditions are suspected, stream stage should be monitored during the measurement period to assess the potential magnitude of error caused by the transience.

3.6.4.2 Tracer-Dilution Methods

Tracer-dilution methods are useful alternatives to the velocity-area method where it is difficult or impossible to use a current meter due to high velocities, turbulence, debris, rough channels, shallow water, or other physical reasons; or where the cross-sectional area cannot be accurately measured (Kilpatrick and Cobb 1985). However,

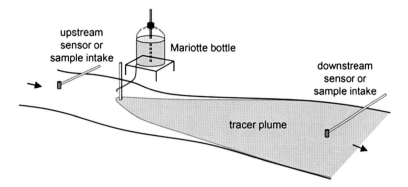

Fig. 3.16 Typical set up for the constant-rate injection method of tracer-dilution gaging. The tracer solution is released in the center of a stream section using a Marriote bottle system. The background value of tracer concentration is monitored at an upstream point, and the fully mixed value of tracer concentration is monitored at a downstream point. The set up is the same for the sudden injection method with the exception that the Mariotte bottle is not used because the tracer is instantaneously poured into the stream (From Dingman (2002) modified with kind permission of © S. Lawrence Dingman 2002. All Rights Reserved. The figure was created based on an illustration originally appearing in Gregory and Walling (1973:134))

tracer-dilution methods are generally more difficult to use than current-meter methods, and under most conditions the results are less reliable. Therefore, these methods should not be used when conditions are favorable for a current-meter measurement (Rantz 1982:212).

In a typical tracer-dilution measurement, a tracer solution is injected into the stream and the stream discharge is estimated from measurements of the tracer-solution concentration, tracer-solution injection rate, and tracer concentrations at a sampling cross section downstream from the injection site (Fig. 3.16). Two methods are commonly used; the constant-rate injection and the sudden injection. For both methods, it is assumed that the tracer is completely mixed at the downstream measurement point. In the constant-rate injection (CRI) method, a tracer of known concentration C_1 (kg/m) is injected at a constant rate q (m^3/s) in the center of a stream channel using a constant-flow device such as Mariotte bottle (e.g., Moore 2004). If the injection is continued for a sufficiently long period, monitoring of concentration at a downstream sampling cross section will show a plateau of constant concentration C_2 (kg/m). Discharge (Q) is estimated from the dilution ratio by:

$$Q = q(C_1 - C_2)/(C_2 - C_b) \qquad (3.23)$$

where C_b (kg/m) is the background tracer concentration in the stream.

In the sudden-injection or slug-injection (SI) method, a slug of tracer solution is instantaneously applied to a stream, and concentration is monitored at a downstream sampling cross section to generate a concentration-time curve (Fig. 3.17). The total mass of the injected tracer must equal the total mass of tracer going

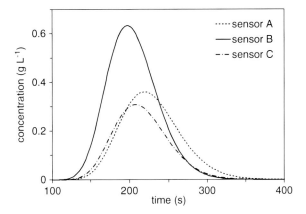

Fig. 3.17 Hypothetical concentration-time curves at a downstream sampling section during a tracer-dilution gaging by the sudden-injection method. *Three curves* represent concentrations recorded by three sensors placed at different locations in the same cross section. The areas under the three curves should be identical if the tracer is completely mixed. In this example, sensor B is recording a much larger mass of tracer passing the section, indicating incomplete mixing

through the sampling section, which is given by the integration of concentration under the curve, assuming that discharge does not vary during the monitoring period. Therefore, discharge is estimated from the mass balance by:

$$Q = C_1 V_1 \bigg/ \int_0^\infty [C(t) - C_b] dt \qquad (3.24)$$

where V_1 (m^3) is the volume of tracer solution introduced to the stream and $C(t)$ (kg m^{-3}) is the time-varying concentration at the sampling cross section. In practice, the integral in Eq. 3.24 is approximated by:

$$\sum_{i=1}^{n} (C_i - C_b) \Delta t \qquad (3.25)$$

where C_i are the concentrations measured at a discrete time interval Δt (s) until C_i becomes indistinguishable from C_b at the nth sample.

Both tracer-injection methods assume that complete vertical and lateral mixing of the tracer with stream water has occurred at the sampling cross section. Vertical mixing usually occurs very rapidly, but a substantial distance is required for lateral mixing. Therefore, it is important to establish the sampling location a sufficient distance downstream of the injection point to ensure complete mixing. It also is important to sample at the downstream location long enough to establish a concentration plateau for the CRI method or to capture the entire concentration-time curve for the SI method. Depending on the site condition and access, it may be difficult to achieve complete mixing, as illustrated in Fig. 3.17, in which case a large degree of

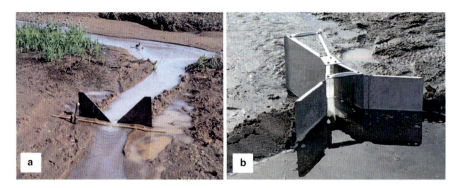

Fig. 3.18 (**a**) Example of a portable V-notch weir (Photo by Masaki Hayashi), (**b**) portable 2.54-cm Baski flume (Photo printed with kind permission of © Kirk Miller, U.S. Geological Survey 2012. All Rights Reserved)

error can be introduced in estimated Q. It is strongly recommended that the methods be tested in the field according to procedures described by Rantz (1982:216–220). Other sources of error include the loss of tracer solution between the injection and sampling points to groundwater or hyporheic exchange, photochemical and other reactions, sorption to streambed materials, and interference of concentration monitoring devices by turbidity and other sensor-calibration issues. Errors due to reaction and sorption, and to some extent sensor issues, can be minimized by the choice of tracer. The ideal tracer has very low natural concentration in the stream, is chemically and biologically conservative (no reaction, absorption, release or uptake), is readily detectable at low concentration, and is harmless to the observer and aquatic life. Sodium chloride and fluorescent dye are commonly used as tracers.

3.6.4.3 Other Methods of Discharge Measurement

For narrow streams with small discharge, it is often possible to divert the entire flow to a container having a known volume, and measure the time it takes to fill the container to calculate discharge (the bucket-stopwatch method). Examples of sites presenting the opportunity for this method are a V-notch weir or a cross-section of natural channel where a temporary earthen dam can be built over a small-diameter pipe (Rantz 1982:263). If a temporary structure is built for volumetric measurement, the stage behind the structure (e.g., a dam) should be allowed to stabilize before the measurement. The measurements should be repeated several times to obtain consistent results. Where a portable V-notch weir or Parshall flume can be installed in the stream (Fig. 3.18), discharge is estimated from the stage measurements in these devices using a laboratory-calibrated formula (see next section).

In situations where no current meter or tracer-dilution equipment are available, or the condition does not permit the use of other methods, estimates of discharge can be obtained using surface floats. Any distinguishable article can be used as a float, such as wooden disks, bottles partially filled with water, or even oranges. Two cross sections are selected along a reach of straight channel, so that the time the float takes to pass from one cross section to the other can be measured accurately. Distance from the upstream to the downstream cross sections needs to be measured and floats should be applied far enough upstream of the upper cross section that they are travelling at the same speed as the surface current when they pass the upper cross section. For best estimates of discharge, a number of floats are distributed uniformly across the stream width, and the position of each with respect to distance from the bank is noted. The stream-channel width can be segmented just as with the velocity-area method described above, depth for each channel section can be measured, and discharge calculated as the sum of each velocity-area product. This method will over-estimate discharge because the surface velocity is faster than the depth-integrated velocity. Therefore, a coefficient of 0.85 is commonly used to convert the surface velocity to mean velocity (Rantz 1982:262).

3.6.5 Stage-Discharge Rating Curve

Rating curves for discharge gaging stations are normally determined empirically with periodic measurements of stage and discharge made over the full range of stage at a particular station. Thereafter, only periodic measurements (commonly a minimum of 10 per year) are needed if it has been demonstrated that the rating curve does not vary with time (Rantz 1982:285). However, shifts in rating curves are common at gaging stations with natural control due to changing channel conditions, including scour and fill, vegetation growth, boulder movements, and ice formation. Rating curves are created by plotting the stage-discharge data and fitting a suitable mathematical function to the entire data set, or fitting a set of functions to separate segments of the data. Power functions are commonly used to fit the data:

$$Q = a(h - h_0)^m \quad (3.26)$$

where Q (m^3 s^{-1}) is discharge, a (m^{3-m} s^{-1}) is a scaling constant, h (m) is stream stage (i.e., water level), h_0 (m) is the stage at zero flow, and m is a dimensionless constant. This function generates a straight line when Q and $h - h_0$ are plotted on logarithmic axes. For gaging stations with no control or natural control, a and m are determined by minimizing the difference between the measured discharge and the predicted discharge using Eq. 3.26.

For gaging stations with an artificial control structure, a formula developed in the laboratory is provided with the device. For example, a rectangular thin-plate weir has a rating curve in the form:

$$Q = Cb(h - h_0)^{3/2} \quad (3.27)$$

where b (m) is the width of the weir through which water is flowing, and C (m$^{1/2}$/s) is a coefficient dependent on the geometry of the weir (Rantz 1982:296–297). A V-notch weir has a rating curve in the form of:

$$Q = C \tan(\theta/2)(h - h_0)^{5/2} \qquad (3.28)$$

where θ is the angle of notch ($\theta = 90°$ for the standard V-notch weir), and C (m$^{1/2}$/s) is a weir coefficient, which has a value of 1.38 in the ideal condition where the weir plate is perfectly made and vertical and water in the pool upstream of the weir has negligible velocity approaching the weir (Rantz 1982:304). The formulas for several types of broad-crested weirs and flumes, including the Parshall flume, are described by Rantz (1982:306–326).

Stage-discharge rating formulas for artificial controls were developed based on laboratory and modeling studies conducted under ideal conditions. To achieve maximum accuracy in the field, data from these instruments should be compared with separate measurements of discharge conducted at the weir or flume installation (Rantz 1982:295) and correction coefficients applied as necessary. The velocity-area method (Fig. 3.11a) or volumetric method (Fig. 3.11b) is most often used for this purpose.

The stage-discharge relation is sensitive to changing conditions of the flow control and the channel reach in the vicinity of the gaging station. Therefore, changes in the channel (e.g., boulder and gravel movement, streambed scouring or filling, bank erosion, vegetation growth) may cause a shift in the rating curve, which can only be detected through periodic stage-discharge measurements. When an observer measures discharge at an existing gaging station, it is a good practice to calculate discharge before leaving the site and plot the new measurement on the rating chart. If the new point deviates noticeably from the established rating curve, a second discharge measurement is carried out to confirm a shift (Rantz 1982:346). If the second measurement confirms a shift, a note should be made to pay particular attention to the site conditions during the next visit. If several consecutive measurements show a consistent shift, the rating curve will need to be adjusted to account for the new channel conditions.

3.6.6 Flow Estimation by Indirect Methods

In low-gradient wetland settings, surface-water flow can be very slow and occur over a broad area without a well-defined channel, often covered with extensive emergent vegetation, making it impossible to measure surface-water flow using any of the methods above. In this case, surface-water flow may have to be treated as the residual of a water-budget equation. Surface-water flow can still be estimated, although often with a large degree of uncertainty, using one of several hydraulic equations with estimated parameters. If the depth of water is sufficiently large compared to the height of submergent vegetation, and if the flow is turbulent, a standard equation for open-channel flow, such as Manning's equation, can be used to estimate average flow velocity (v, m/s):

$$v = y^{2/3} S^{1/2}/n \qquad (3.29)$$

where y (m) is depth of water, S is slope of the water surface (unitless), and n (s m$^{-1/3}$) is a roughness coefficient whose values can be found in the literature (e.g., Kadlec and Wallace 2009:40).

The use of open channel equations such as Eq. 3.29 is inappropriate for those commonly encountered situations in wetlands where flow is laminar, water depth is shallow, stream slope is very small and nearly impossible to measure, and flow resistance exerted by vegetation is strongly related to water depth. In these settings, it is more appropriate to use an empirical equation having the form

$$v = \alpha y^{\beta-1} S^{\gamma} \qquad (3.30)$$

where α (m$^{2-\beta}$ s) is an empirical coefficient representing hydraulic resistance, and β and γ are additional empirical coefficients (Kadlec and Wallace 2009:35). A disadvantage of Eq. 3.30 is that all empirical coefficients are site-specific (and possibly season-specific) and have to be determined locally. Kadlec and Wallace (2009:39) listed ranges of α from 70 to 2,300 m$^{2-\beta}$/s, β from 1.4 to 3.0, and γ from 0.7 to 1 from a survey of the existing literature and recommended the following values be used when no data are available: $\alpha = 120$ and 580 m/s for densely and sparsely vegetated wetlands, respectively, $\beta = 3$, and $\gamma = 1$.

Indirect estimates of discharge using equations such as 3.29 and 3.30 have a very large degree of uncertainty unless the equation is calibrated using site-specific field data. Therefore, they should be used as a last resort to obtain an order-of-magnitude estimate of flow.

3.6.7 Errors and Challenges

Using well-calibrated instruments for velocity and depth measurement in a carefully chosen location with appropriate flow control, it is possible to measure stream discharge with accuracy of 3–6 % using the velocity-area method (Turnipseed and Sauer 2010:80). Similar accuracy can be achieved for the volumetric measurement using a well-calibrated vessel. However, flow often has to be measured in non-ideal locations caused by unsatisfactory flow control, local flow lines not crossing the measurement section at a right angle, obstruction of flow by boulders and vegetation, vertical and lateral leakage of surface water to fluvial sediments, loss of water to side channels, or several other causes. It is advisable to conduct duplicate or triplicate velocity-area discharge measurements on different sections located within the same stream reach to evaluate the uncertainty in flow measurement. Compared to the velocity-area method and volumetric measurements, tracer-dilution methods tend to have a larger degree of uncertainty due to lack of complete mixing, loss of tracer to groundwater, sensor calibration issues, and insufficient time for complete capture of tracer. Other methods likely have greater degrees of uncertainty.

Errors associated with the stage-discharge rating curve are expected to be reasonably small for well-maintained and calibrated artificial control structures such as weirs and flumes. However, rating curves for naturally-controlled gaging stations can be greatly affected by short- and long-term changes in channel conditions. It is not uncommon to have root-mean-squared (RMS) errors of rating curves exceeding 20 %, particularly for those locations that are susceptible to changes in density and extent of vegetation or shifts in boulders or gravel bars. Therefore, as stated earlier, it is important to visit stations frequently to conduct maintenance, make manual discharge measurements, detect any shifts in the rating curve, and make appropriate adjustments. Errors in the stage measurement (see Wetland stage section) also affect discharge data, and should be kept at a minimum.

The formation of channel ice in cold regions subject to freezing temperatures can have a major effect on the stage-discharge relation by causing backwater due to increased flow resistance. The backwater effect is dependent on the quantity and nature of the ice, as well as the amount of discharge, which necessitates frequent discharge measurements, particularly during freeze-up and thaw when the flow is highly variable. Rantz (1982) describes procedures for making discharge measurements in ice-covered streams and adjustments to stage-discharge rating curves to account for the backwater effect (Rantz 1982:360–376). If it is not feasible to maintain ice-free condition or to measure discharge frequently, the record may be regarded as "seasonal", with no data available during the ice-covered period.

3.6.8 Summary

Surface-water flow can be a major component of a wetland water budget, and the uncertainty of surface-flow measurements can dominate the cumulative uncertainty of a water-budget calculation. Measurement accuracy of 10 % or better can be achieved at well-maintained gaging stations with appropriate flow control provided the stage-discharge rating curve is frequently checked and adjusted. However, stream channels in many wetland settings are ill-defined and conditions can be highly variable over time. If it is necessary to measure discharge in an undesirable location, the observer should strive to obtain estimates of measurement error and uncertainty, which can be reflected in the water-budget calculation.

3.7 Diffuse Overland Flow

During snowmelt or storm events, surface runoff generated within the drainage area may directly reach the wetland by flowing over the land surface without entering stream channels. This process is called diffuse overland flow. Since it is practically impossible to measure diffuse overland flow over a large area, this component is frequently neglected or treated as a residual in the water-budget equation, especially when there is evidence indicating that the magnitude of diffuse overland flow is much smaller than stream inflow or groundwater inflow. However, diffuse overland flow is a major water-budget component, at least temporarily, in some wetlands without channelized stream inputs. Examples include many prairie wetlands in the Northern Prairies region of North America (Winter 1989) and ephemeral forest pools in the New England region of the United States (Brooks 2009).

Diffuse overland flow generally moves toward a wetland and is nearly always considered as an input term in wetland water budgets, but this is not always the case. In low-gradient settings where surface-water outflow is very slow and occurs over a broad area with an ill-defined channel, loss of wetland water could be considered either as slow surface-water flow or diffuse overland flow that is moving away from the wetland. Although some have separated diffuse overland flow into separate input and loss terms (LaBaugh 1986), here we will consider any slow-moving flow occurring over a broad area that is leaving a wetland basin to be surface-water outflow, as quantified with Eqs. 3.29 and 3.30.

Since the amount of diffuse overland flow input entering a wetland is proportional to the length of wetland perimeter, the effect of diffuse overland flow is particularly pronounced in relatively small (e.g., $<10^4$ m^2) wetlands that have large perimeter-to-area ratios. A shallow water table in areas adjacent to a wetland can rise to the surface with a relatively small amount of infiltration during storm events (Gerla 1992), which precludes further infiltration and generates runoff (Dunne and Leopold 1978:268). In cold regions that have seasonally or permanently frozen soil, reduced infiltrability of frozen soil causes a large amount of snowmelt runoff in the surrounding uplands, which can be the dominant mode of water input to wetlands (e.g., Winter and Rosenberry 1995; Hayashi et al. 1998). Using the same logic, low-permeability soils also will retard infiltration, resulting in a greater percentage of precipitation reaching the wetland as diffuse overland flow. Although difficult to quantify over a large area, there are a number of methods for measuring flow volume on a local scale or for individual storm events, or for obtaining order-of-magnitude estimates of flow volume using simple models.

3.7.1 Measurement of Diffuse Overland Flow

Flow traps are commonly used to measure diffuse overland flow over a small area. Figure 3.19 shows a very simple flow trap consisting of an isolated area and a pit to collect water. In this example, the pit is emptied after each storm event to determine the overland flow volume for individual events. A pit also can be equipped with a V-notch weir and water-level recorder for continuous monitoring of overland flow. If the area contributing diffuse overland flow to a wetland is delineated with reasonable accuracy, then the data obtained using small flow traps may be extrapolated to a larger area to estimate the total water input to the wetland.

Fig. 3.19 A simple overland flow trap consisting of the contributing area delineated by concrete walls and a pit to collect water (Photo by Masaki Hayashi)

If diffuse overland flow occurs in an area of relatively uniform vegetation and slope, and depth of surface water is measured with reasonable accuracy, approximate flow rates can be estimated from the empirical Eq. 3.30 described in the section on streamflow measurements. This method has a large degree of uncertainty due to uncertainties and errors associated with empirical coefficients, uncertainty in delineation of the area contributing to overland flow, and uncertainty in estimating the water depth of the overland flow.

The methods described above can be applied to areas with uniform characteristics; results from each of those areas are then summed to generate O_f for the entire wetland. Methods described below all provide a single, integrated value for the entire wetland.

If a wetland does not have inflow and outflow streams, and the groundwater flow rate is much slower than the overland flow rate, the amount of diffuse overland flow during a short-lasting storm event can be estimated from the water-balance equation. Omitting stream and groundwater flow terms and assuming negligible evaporation during the event, Eq. 3.1 can be written as

$$O_f = \Delta V / \Delta t - P \tag{3.31}$$

Integrating (3.31) for the entire event, the total overland flow volume (O_{ftot}) is given by

$$O_{ftot} = V_{fin} - V_{ini} - P_{tot} \tag{3.32}$$

where V_{ini} and V_{fin} are the initial and final volumes of water contained in the wetland, respectively, and P_{tot} is the total volume of precipitation applied at the wetland surface. Although this method requires specific conditions and is not applicable to all water-budget calculations or all wetland settings, data obtained using Eq. 3.32 may be used to calibrate or validate models used for estimation of diffuse overland flow (see below).

3.7.2 Estimation Using Simple Rainfall-Runoff Models

If the measurement of diffuse overland flow is impossible, it can be estimated from other variables using a hydrological model. One of the simplest models is the "rational method", which estimates the volume of overland flow (O_{fc}, m^3) generated from a contributing area (A_c, m^2) as a fixed ratio of precipitation (Mitsch and Gosselink 2007:128):

$$O_{fc} = R_c p A_c \qquad (3.33)$$

where R_c is a dimensionless "rational coefficient" taking values between 0 and 1, and p (m) is the amount of precipitation. Equations similar to Eq. 3.33 are commonly used by engineering hydrologists for estimating storm runoff generation in urban areas. However, their applicability to wetlands is limited because R_c is dependent on many factors including soil, vegetation, and the depth to the water table. It is usually difficult to represent the variable conditions in a wetland catchment, both in time and space, with a single parameter.

More sophisticated models consider soil type, vegetation, land use, and numerous other factors, and treat soil moisture (and water-table) conditions as time-dependent variables. A relatively simple example of such models is the "curve number" (CN) method developed by the U.S. Soil Conservation Service in the 1950s and 1960s to estimate storm runoff from agricultural lands. Since then, the CN method has become a standard tool for hydrologists and has been used in numerous computer-based hydrological models such as the Soil and Water Assessment Tool (SWAT) (http://swatmodel.tamu.edu/) and Hydrologic Engineering Center (HEC) (http://www.hec.usace.army.mil/) models. This method computes O_{fc}/A_c in Eq. 3.33 as a non-linear function of precipitation. The non-linear dependence of runoff on precipitation is represented by a CN coefficient, which is determined by soil texture and drainage condition, land use, and the amount of rain during a period prior to the storm event (i.e., antecedent moisture). Step-by-step instructions of the CN method and examples of applications are found in introductory hydrology textbooks such as Dunne and Leopold (1978:291–298), and can be easily implemented in a computer algorithm.

Regardless of model sophistication, estimates made by these methods commonly have a large degree of uncertainty resulting from violations of model assumptions and uncertainty in model parameters and input variables. Therefore, estimated overland flow should be verified with measured data whenever possible using the methods described above.

3.8 Groundwater Inflow and Outflow

Determining exchanges between wetland water and groundwater can be a surprisingly complex task for wetland hydrologists. Low-permeability, organic-rich soils are often situated beneath and adjacent to wetlands; they reduce rates of exchange and can greatly increase residence time of water in pore spaces, enhancing geochemical processes. Flow across the sediment-water interface is variable both in space and time on multiple scales. Directions of flow between groundwater and surface water can reverse seasonally or in response to individual precipitation events or evapotranspiration (e.g., Doss 1993; Rosenberry and Winter 1997).

A variety of tools and methods are available to quantify exchange between groundwater and surface water, the selection of which should include strong consideration of the appropriateness of the scale of the measurement method relative to the scale of the goals of the study. On the larger end of the scale spectrum, suitable methods include watershed-scale rainfall-runoff modeling, groundwater-flow modeling, quantifying changes in streamflow along stream-reach segments (commonly called a seepage run), and making use of aerial imagery to locate areas of focused groundwater discharge. On a local scale, appropriate for smaller wetlands or specific shoreline segments or riparian reaches, methods include measurement of hydraulic properties using piezometers and water-level monitoring wells, use of seepage meters to quantify flow across an isolated portion of submerged bed sediment, and measurements of temperature to determine quantitatively or qualitatively distribution and rates of groundwater discharge to specific portions of wetland beds. Several of the most-commonly utilized methods are described below.

3.8.1 Darcy Flux Method

Use of the Darcy equation to determine flow through porous media is one of the core concepts of hydrogeology and is commonly employed to estimate exchanges between surface water and groundwater in wetland settings. The Darcy equation (e.g., Freeze and Cherry 1979) can be expressed as

$$Q = -KA\frac{h_1 - h_2}{l} \tag{3.34}$$

where Q is the volume of water that flows across the bed of the wetland to enter or leave the wetland, K (hydraulic conductivity, m/s) is a proportionality constant that represents the ease with which water can flow through porous media, A is the area of the sediment-water interface through which water flows to enter or leave a wetland, $h_1 - h_2$, or Δh, is the difference between hydraulic head measured at a nearby monitoring well and the wetland surface, and l is the distance from the monitoring

3 Assessing and Measuring Wetland Hydrology 143

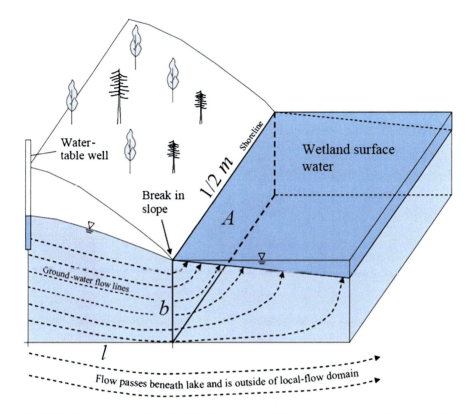

Fig. 3.20 Components required to determine exchange between groundwater and surface water using the Darcy method (Modified from Rosenberry et al. (2008). Published with kind permission of the U.S. Geological Survey. Figure is public domain in the USA. All Rights Reserved)

well to the shoreline of the wetland (Fig. 3.20). The minus sign is included to indicate that the flow (Q) occurs in response to a decrease in hydraulic head along a groundwater flowpath. The ratio $\Delta h/l$ is also called the hydraulic gradient and is commonly indicated with lower-case i, leading to the shorthand version of the Darcy equation:

$$Q = KiA \tag{3.35}$$

Note the absence of the negative sign that appears in Eq. 3.34. In Eq. 3.35, i is formulated so that the difference between heads is positive. The hydraulic gradient is usually the easiest term to quantify, requiring only measurements of the water level in a nearby monitoring well, wetland stage, and the horizontal distance between the well and the shoreline. The elevation of the top of the monitoring well relative to the water surface of the wetland also is needed to relate the water level in the monitoring well to wetland stage.

A as described above is the wetted portion of the wetland bed across which water flows to enter or leave the wetland. By definition, flow across A is perpendicular to the plane of A, the orientation of which ranges from nearly vertical to horizontal at the wetland bed. Therefore, A in Fig. 3.20 is instead oriented vertically and located at the shoreline of the wetland, which simplifies determination of i by allowing it to be made on a horizontal axis. Flow across this vertical plane is assumed to be equal to flow across the sediment-water interface of the wetland.

A is the product of m, the shoreline reach associated with the monitoring well, and b, the thickness of the portion of the aquifer that exchanges with the wetland (Fig. 3.20). Both m and b are conceptually simple but often difficult to determine. The distance, m, of a shoreline segment associated with the gradient between a monitoring well and the wetland is based on how far one can reasonably extrapolate the hydraulic gradient along the wetland shoreline. Because the monitoring well for each shoreline segment is usually located at the center of the segment, m, and therefore A, as shown in Fig. 3.20 are both half of what they should be because the figure does not show the half of the shoreline segment that would extend out of the page. If the shoreline reach is straight and hydrogeological conditions are expected to be uniform, the shoreline segment corresponding to a monitoring well could be quite long and the extent relatively easy to determine. If the shoreline is curvilinear (commonly the case in wetland settings) and hydraulic gradients are expected to vary substantially along the shoreline reach, then additional monitoring wells should be installed and shoreline segments should be correspondingly shorter. The other component needed to determine A, and one that often is the most difficult to estimate, is b, the thickness of the vertical plane through which water has to flow to enter (or leave) the wetland. Water passing through any portion of the aquifer deeper than b will not exchange with the wetland but will instead pass beneath the wetland. This is shown by the two flowlines that extend beyond the wetland in Fig. 3.20. Unless the subsurface geology is known to constrain exchange with the wetland, or unless the wetland depth extends to the base of the aquifer, b has to be estimated. One common approach is to arrive at a reasonable estimate for b through the use of a simplified groundwater flow model.

The terms of the Darcy-flux method are extrapolated along an entire shoreline segment, the extent of which is based on what is determined to be reasonable. The longer the segment, the weaker the assumption that K, A, and $\Delta h/l$ indeed are uniform along the segment. Therefore, the wetland perimeter is divided into several segments, each of which is associated with a specific monitoring well located near the wetland. This commonly is referred to as the segmented-Darcy approach, an example of which is depicted in Fig. 3.21. Once the hydraulic gradient, shoreline-segment length, and estimated K and b are determined for each shoreline segment, all of the information is available to calculate Q for each segment and the entire wetland. The example shown in Fig. 3.22 is of a flow-through wetland, one that both receives groundwater discharge and contributes wetland water to the adjacent groundwater system. This approach allows for estimation of total groundwater discharge, total recharge of wetland water to groundwater, as well as the net (G_i minus G_o) term.

3 Assessing and Measuring Wetland Hydrology   145

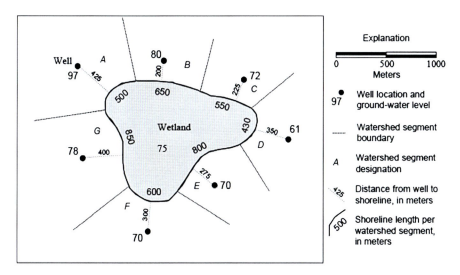

Fig. 3.21 Segmenting of a hypothetical wetland for determination of Darcy fluxes based on locations of seven monitoring wells surrounding the wetland (Modified from Rosenberry et al. (2008). Published with kind permission of the U.S. Geological Survey. Figure is public domain in the USA. All Rights Reserved)

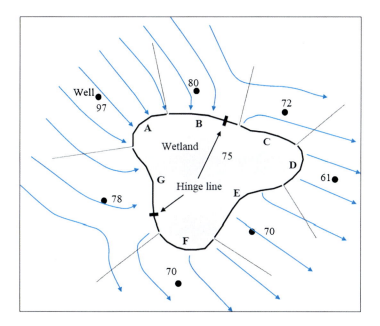

Fig. 3.22 Wetland perimeter divided into segments for determination of Darcy fluxes with groundwater flowlines shown in *blue* (Modified from Rosenberry et al. (2008). Published with kind permission of the U.S. Geological Survey. Figure is public domain in the USA. All Rights Reserved)

3.8.1.1 Location of Monitoring Wells, Assumptions, and Errors

The value of a single monitoring well for determining exchange between groundwater and wetland water is slight. Analysis of sediment and distribution of soil types based on materials removed during the well installation has local value but the lateral extent of these properties is unknown. Difference in hydraulic head between the single well and the wetland can be determined, but little can be known about the actual direction of flow of the ground water with only a single well and wetland stage. Two wells provide additional information about the local-scale geology and hydraulic gradient. If holes augured on opposite sides of a wetland both indicate similar geology, then confidence is increased that geology surrounding the wetland is somewhat uniform. However, information still will be insufficient to characterize hydraulic gradients around the entire wetland perimeter with any certainty.

Unless aquifer-gradient information is known a-priori, the minimum number of wells required to estimate groundwater exchange with a wetland, at least qualitatively, is three. If wells are distributed approximately evenly around a wetland, contour lines of equivalent hydraulic head (equipotential lines) can be drawn based on the head values from the wells and the stage of the wetland. Once equipotential lines are drawn, groundwater flowpaths can be drawn perpendicular to the equipotential lines. Flowpath lines will provide an indication of the direction of groundwater flow. For flow-through wetlands that both receive groundwater discharge and recharge water to groundwater, the locations of hinge lines, defined as those points along a shoreline that separate reaches where groundwater discharges to a wetland from reaches where wetland water flows to groundwater, can be drawn (Fig. 3.22).

This rudimentary analysis forms the beginning of a groundwater flow-net analysis, which is another method for estimating the direction of groundwater flow, described more completely in Rosenberry et al. (2008). Examples are shown in Fig. 3.23 based on a variety of combinations of monitoring wells. For example, heads from any two wells selected from the array of wells shown in Figs. 3.22 and 3.23, with the exception of wells A and D or A and E, will lead to an incorrect interpretation of directions of groundwater flowpaths in the vicinity of the wetland. Data from only wells C and G will lead to the assumption that flow is to the northeast (Fig. 3.23b). Data from wells C and F will lead to the assumption that the wetland is losing water to groundwater at least along the majority of the wetland margin (Fig. 3.23c).

Heads only from wells B, E, and G would result in a correct interpretation of the direction of groundwater flow (Fig. 3.23d). However, without data from wells A or D, the interpretation would be that far less groundwater exchanges with the wetland. Groundwater flow to the wetland would occur only along the shoreline represented by well B. Assuming that segments A, B, and C were assigned to the gradient at well B, that b is 20 m, and that K is 30 m/day, inflow would total 25,500 m^3/day. Outflow, assuming well E is assigned to segments D, E, and F, well G is assigned only to segment G, and b and K remain the same at 20 and 30, respectively, would be $-19,964$ m^3/day. This would result in an imbalance of over 5,000 m^3/day or 20 % of inflow. If wells were located near the three protruding bays of the wetland (wells A,

3 Assessing and Measuring Wetland Hydrology   147

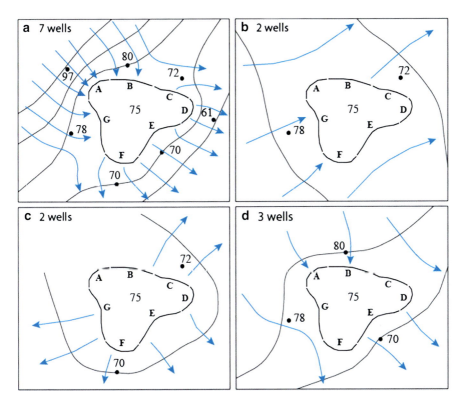

Fig. 3.23 Groundwater equipotential lines and flowlines based on (**a**) seven wells, (**b**) two wells, (**c**) a different combination of two wells, and (**d**) three wells

D, and F), inflow would total just over 62,000 m^3/day and outflow would total just over −43,000 m^3/day. Groundwater discharge to the wetland would be more than double the estimate based on seven monitoring wells and flow of wetland water to groundwater would be only two-thirds of groundwater discharge. It is clear that the location and number of monitoring wells are crucial for determining a reasonably accurate indication of groundwater exchange with a wetland.

The segmented-Darcy approach assumes that groundwater flow vectors are perpendicular to the wetland shoreline. This clearly is not the case for some wetland settings, including the example shown in Fig. 3.22. Groundwater flow is primarily tangential to shoreline segments B and G, where hinge lines are located. Those hinge lines also indicate that, not only is the hydraulic gradient not uniform along the entire segment reach, the gradient is in opposite directions from one end of the segment to the other. Contouring the hydraulic-head data and drawing flowlines provides additional information about groundwater exchange with the wetland. For example, flowlines can provide a much better indication of the locations of hinge lines. Knowing that, shoreline segments can be modified to end in the vicinity of hinge lines and more realistic values of flow can be determined for each shoreline segment.

Errors of interpretation and incorrect assumptions regarding directions of flowlines can be greatly reduced through the use of one of a variety of commonly available flow-modeling techniques. Perhaps the simplest and oldest is the previously mentioned hand-drawn flow-net approach. However, many analytical and numerical computer-based models can provide a quick analysis of likely groundwater flowpaths and estimations of volumes of exchange between groundwater and water in the wetland. The influence of upper and lower bounds of K and b on volumetric exchange also can be determined with numerical simulation.

3.8.1.2 Location and Installation of Monitoring Wells

The interpretation of groundwater-surface-water exchange, whether by segmented-Darcy, flow-net, or analytical or numerical models, depends on data from monitoring wells. Fortunately, installation of wells for the purpose of measuring the elevation of the upper extent of saturated sediments (the water table) is often relatively simple and inexpensive. Shallow water tables and small depths to water, not to mention soft and often nearly saturated sediments in near-shore wetland margins, allow monitoring wells to be installed manually rather than with a drilling rig. Although wells can sometimes be driven to depth with a post driver, sledge hammer, or hydraulic-push rig, it usually is better to auger a test hole, collect, describe, and analyze the sediments removed from the hole, and then install the well in the test hole. This may be difficult in some sediments, either because the sediments are poorly consolidated and slump back into the hole, or because sediments contain a large fraction of cobbles or larger particles, making hand auguring difficult or impossible.

Two types of monitoring wells are used for wetland-hydrology investigations. A water-table monitoring well is designed to indicate the level of the top of the aquifer, where total pressure is equal to atmospheric pressure and below which all the pores in the soil are filled with water; in other words, the water table. Because the water table can fluctuate over a range of several meters in some wetland settings, a water-table well needs to have a well screen that is long enough that it intersects the water table whether the water table happens to be high or low at the time. Note the long well screen for the water-table well shown in Fig. 3.24 that extends above the water table. The other type of monitoring well is often termed a piezometer. A piezometer is designed to represent hydraulic head at a single point in an aquifer. Ideally, such a well would just have an opening at the bottom of the well casing to represent pressure. Because many piezometers also are designed as water-quality sampling points and need to produce some water for sampling, they often have short screened intervals. In such cases, the mid-point of the screened interval is the depth to which pressure head indicated by the piezometer is generally associated. Two piezometers are indicated in Fig. 3.24. One represents a piezometer installed near a wetland and will provide a pressure head to compare with the adjacent water-table well. The other is installed in the sediments beneath the wetland bed and is designed to provide a hydraulic gradient on a vertical axis as

3 Assessing and Measuring Wetland Hydrology

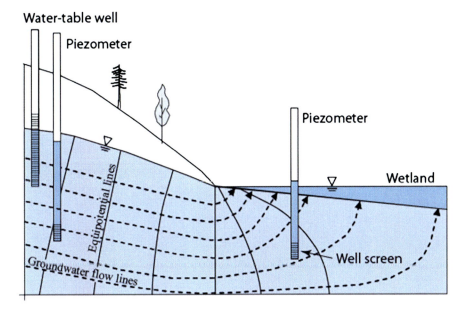

Fig. 3.24 Depiction of the two types of monitoring wells commonly installed in wetland settings. The water level in the piezometer on the left is lower than the water level in the water-table well, indicating that a downward gradient exists in the aquifer at that point. The piezometer to the right positioned in the wetland has a water level higher than the wetland surface, indicating an upward gradient exists in the aquifer beneath the wetland

determined by the pressure difference between the mid-point of the piezometer screen and standing water in the wetland. Actual flow vectors beneath a wetland bed cannot be known with a single piezometer installed in standing water in a wetland. However, because the hydraulic gradient is determined on a vertical axis, the direction of flow in these settings is almost always assumed to be vertical.

Regardless of the preferred installation method or monitoring-well type, well screens should be selected so the width of openings in the screen, commonly called the slot size, is representative of the grain-size distribution of the sediment adjacent to the screen. This ensures that the monitoring well is in good hydraulic connection with the surrounding sediments. The screen length and installation depth should adequately represent the elevation of either the water table or pressure head at a specific depth below land surface. If a water-table well is installed at too great a depth, and the well has a screened interval that is substantially below the water-table elevation, the well will likely function as a piezometer, providing a water level indicative of hydraulic head at some depth beneath the water table rather than the actual water-table elevation. This effect generally is a concern only when vertical hydraulic gradients are large.

Wetland settings, although generally conducive to studies involving monitoring wells, can present unexpected challenges in data interpretation. Evapotranspiration from emergent vegetation and dense riparian vegetation can extract groundwater until the water-table elevation is below the wetland stage. Recharge to groundwater is more rapid and extensive in these near-shore margins where the unsaturated zone is thinnest, and may result in the water table being higher than elsewhere, effectively forming a hydraulic dam between the wetland and groundwater farther from the wetland. Numerous examples of either a water-table trough or ridge between a monitoring well and the wetland shoreline are reported in the literature (e.g., Rosenberry and Winter 1997). If either a trough or a ridge is present between a wetland and nearby monitoring well, water cannot flow from the well to the wetland or vice versa. It often is prudent to install two or more monitoring wells at different distances from shore to determine if transient water-table ridges or troughs occur, are frequent, or persistent.

Once installed, determining the water level in a monitoring well can be accomplished with several methods ranging from something as simple as lowering a chalked steel tape into the well to immersing a pressure transducer that includes a self-contained datalogger for collecting time-series data. Details for measuring water levels in wells are presented in Cunningham and Schalk (2011).

3.8.1.3 Methods for Determining Hydraulic Conductivity (K)

Of all the factors that control the degree of exchange between groundwater and wetland water, K is the most spatially variable and often the most difficult to determine. A complex history of erosion and deposition of organic and inorganic sediments is commonly encountered in many wetland settings where stage and shoreline location can vary by a large amount over time. Organic-rich sediments, typically with small values of K, can be situated next to wave-washed sand and gravel in these dynamic environments, complicating the determination of K on a scale that is relevant to a wetland water budget. Furthermore, determination of K is itself scale dependent (e.g., Rovey and Cherkauer 1995). Point measurements may represent conditions within a few meters of a monitoring well, but will not be representative of a more transmissive portion of the sediments that may route most of the groundwater to or from a wetland. Most sediment is more permeable to horizontal flow than to vertical flow. In addition, K commonly decreases with sediment depth (Hayashi et al. 1998). Reduction in K with depth also is particularly common in peat. An additional complexity of peat is that it is compressible, which also affects K (Surridge et al. 2005; Hogan et al. 2006). For these many reasons, determinations of K require careful consideration and several avenues of investigation.

A single-well slug test provides a reasonable indication of K at a scale comparable to the size of the well screen. This method involves recording the water level within a well, typically with a submerged pressure transducer, while the water level is suddenly increased or decreased (e.g., Fetter Jr 2001). The rate of recovery of the water level in the well is proportional to the hydraulic conductivity of the sediments that surround the well screen. Analysis of the recovery curve assumes that flow to or from the well is primarily horizontal and requires use of one of several analytical methods such as Bouwer (1989), Bouwer and Rice (1976) or Hvorslev (1951) to calculate K.

Obtaining a single value for K that is representative of an entire wetland is possible where wetland stage can be changed rapidly, either by pumping wetland water elsewhere or by altering wetland stage with a control structure. As an example, water was pumped from a 60-m-diameter wetland in Florida until wetland stage was lowered 0.3 m and the recovery of wetland stage was recorded following the end of the pumping period (Wise et al. 2000). Because the recovery occurred during a time of minimal rainfall, and the rate of recovery was much larger than potential effects of rainfall or evapotranspiration, recovery of wetland stage following pumping was attributed to seepage from groundwater. By measuring the vertical hydraulic gradient, i_v, between wetland stage and several piezometers installed within the wetland basin, carefully measuring wetland bathymetry to obtain a good estimate of A for each increment of wetland stage, and knowing the amount of water pumped from the wetland (Q), Darcy's law can be manipulated to calculate the vertical component of hydraulic conductivity, K_v, of the wetland sediments:

$$K_v = Q/(i_v A) \quad (3.36)$$

Once K_v is known, i_v can be monitored with measurements of piezometers installed in the wetland and G_i or G_o can be determined depending on whether i_v is indicating upward or downward flow potential.

3.8.2 Direct Seepage Measurements

Most devices or methods for quantifying exchange between groundwater and surface water are based on indirect measurements. For example, hydraulic gradient and hydraulic conductivity are determined using the segmented-Darcy approach, but the actual quantity of interest is the flux across the sediment-water interface. A seepage meter is an instrument that directly measures flow across the sediment-water interface between groundwater and surface water. Although several early versions developed in the 1950s and 1960s were unwieldy and quite complex (listed and described in Carr and Winter 1980), the meter generally in use since the mid 1970s, the "half-barrel" seepage meter, is very simple and inexpensive. The device consists of an open-ended seepage cylinder placed on the bed to which an attached plastic bag is used to record the time-averaged rate of flow (Lee 1977). The open-ended cylinder isolates a portion of the bed, commonly 0.25 m^2, and all flow across the bed area covered by the cylinder is routed to (or from, depending on the direction of flow) the attached plastic bag (Fig. 3.25). By recording the volume contained in the bag at the times of emplacement and removal, the volumetric seepage rate is determined:

$$Q = \frac{V_{t1} - V_{t2}}{t_1 - t_2} \quad (3.37)$$

Fig. 3.25 Half-barrel seepage meter with seepage bag located inside a bag shelter for protection from currents and waves (Photo by Donald Rosenberry)

where V_{t1} is the volume contained in the bag at the start of the measurement period, V_{t2} is the volume in the bag at the end of the measurement period, and t_1 and t_2 are the times at the start and end of the measurement period. Dividing that result by the area covered by the seepage cylinder gives seepage flux in length per time:

$$q = \frac{Q}{A} \tag{3.38}$$

Although conceptually very simple, the device is not necessarily simple to use. Inferior data have been collected and published, likely because the simplicity and low cost of the meter have resulted in insufficient understanding of sources of error and attention to measurement commensurate with the cost and complexity of the instrument. However, given sufficient measurement care, the half-barrel seepage meter can provide reliable and repeatable data (Rosenberry et al. 2008).

Several modifications to the basic design are commonly employed to reduce measurement error and improve measurement efficiency. Perhaps the most important is to place the seepage bag inside of a shelter to minimize the influence of currents and waves, as shown in Fig. 3.25. Seepage bags exposed to currents can fill with water due to velocity-head effects not normally considered by groundwater scientists (Sebestyen and Schneider 2001; Rosenberry 2008). Other modifications include deploying the bag 1 m or more from the seepage cylinder to minimize local disturbance during attachment and removal of the bag, increasing the diameter of bag-connection hardware to improve meter efficiency, and connecting multiple cylinders to a single bag to reduce measurement time and increase the bed area represented by each measurement (Rosenberry 2005). Additional discussion of modifications and sources of error (and how to minimize them) are presented in Rosenberry et al. (2008).

Seepage meters also have been modified for use in streams (Rosenberry 2008). Such a meter would be useful for riparian wetland settings where flow, although usually relatively slow, could still corrupt seepage measurements made with meters not modified for use in flowing water. For the typically slow flow velocities associated with wetland settings, the most important consideration is to place the bag inside of a bag shelter.

As indicated earlier, many of the errors associated with seepage measurements can be attributed to problems associated with the seepage bag. Furthermore, any variability in seepage rate is integrated over the duration of each bag attachment. To address these concerns, the bag can be replaced with alternate means of quantifying flow ranging from chemical-dilution methods to heat-pulse flow technology to mechanical or electromagnetic flowmeters (Rosenberry et al. 2008). Much finer temporal resolution is possible with these designs that allow quantification of processes that would otherwise be impossible with standard designs (Rosenberry and Morin 2004; Rosenberry 2011).

3.8.3 Determining Groundwater Fluxes as the Residual of a Water Budget

The wetland water budget presented earlier (Eq. 3.1) can be reordered to solve for net groundwater exchange:

$$G_i - G_o \pm R = \Delta V/\Delta t - P + ET - S_i + S_o - O_f \quad (3.39)$$

where the terms are as described earlier. This is a common approach for determining net groundwater contribution to lakes, wetlands, or stream reaches where groundwater fluxes are difficult to determine with more direct measurements. Hood et al. (2006), for example, used Eq. 3.39 to estimate the contribution of groundwater to an alpine lake. Note that this provides only the net groundwater exchange ($G_i - G_o$). For wetlands where either G_i or G_o dominates, determining the net groundwater contribution may be all that is needed, but for many other wetland settings determining the net term may not be sufficient. For example, by only knowing net groundwater exchange, the water residence time cannot be determined. Fortunately, if we also have a chemical constituent of some sort that is associated with each of the water terms, then both G_i and G_o can be determined.

The water-budget equation, including chemical concentrations associated with each term, can be written as a mass-balance equation:

$$\Delta(C_W V)/\Delta t = C_P P + C_{Si} S_i + C_{Gi} G_i + C_{Of} O_f - C_{ET} ET - C_{So} S_o \\ - C_{Go} G_o \pm \varepsilon \tag{3.40}$$

where C is the concentration of the chemical constituent, ε is the total hydrologic and chemical-measurement error, subscripts are related to the various components of the water budget, and $_W$ refers to surface water in the wetland. In many settings, surface water in the wetland is well mixed; therefore, it is reasonable to assume that the concentration associated with G_o and S_o are the same as that of the wetland surface water, C_W. Under this assumption, Eq. 3.39 is rearranged to isolate G_o and this expression is substituted into Eq. 3.40 to obtain

$$G_i \pm \varepsilon = \frac{V \frac{\Delta C_w}{\Delta t} + (C_W - C_P)P + (C_W - C_{Si})S_i + (C_W - C_{Of})O_f + (C_{ET} - C_W)ET}{C_{Gi} - C_W} \tag{3.41}$$

Note that the residual term R in Eq. 3.39 was omitted in this substitution because all errors are now lumped into ε. Lastly, G_i can be inserted into Eq. 3.39, which can be rearranged to solve for G_o:

$$G_o \pm \varepsilon = P + S_i + G_i + O_f - ET - S_o - \frac{\Delta V}{\Delta t} \tag{3.42}$$

Another important assumption is that the chosen chemical constituent is conservative, meaning that it is not altered by any chemical or biological process. Water solutes are commonly used in this analysis and chloride is often considered conservative in many settings. Stable isotopes of water, usually deuterium (^2H) or oxygen-18 (^{18}O), are an excellent choice because they are not a dissolved solute but part of the water molecule. If chloride or another solute is used, the equation is simplified somewhat because the evaporation process distils the water and no solute is lost with the evaporating water; therefore $C_{ET}ET$ is zero. If a stable isotope of water is used, the isotopic value of the evaporating water needs to be determined. This value is rarely available, is relatively difficult to obtain, and often is estimated based on other studies conducted within the area or region (e.g., LaBaugh et al. 1997).

This method is not well suited for wetland water budgets dominated by groundwater discharge. As G_i becomes large, the difference between the two terms in the denominator of Eq. 3.41, $C_{Gi} - C_W$, becomes small, at which point measurement errors can greatly affect the solution. If water isotopes are used, the method is not very robust when the water residence time of the wetland is short or seasonal variation in isotopic composition is large (Krabbenhoft et al. 1994). In such instances, it is better to use a conservative major ion. Errors can be substantial for some of the terms and in some cases the residual term, ε, can approach or exceed

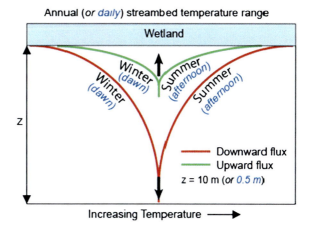

Fig. 3.26 Streambed temperature profiles for downward water flow and upward water flow (Modified from Constantz (2008). Published with kind permission of the U.S. Geological Survey. Figure is public domain in the USA. All Rights Reserved)

either G_i or G_o. A good discussion of errors associated with use of the water budget to determine groundwater exchange terms can be found in LaBaugh and Winter (1984), Krabbenhoft et al. (1990), or Choi and Harvey (2000).

3.8.4 Measurement of Temperature to Quantify Groundwater-Surface-Water Exchange

Diurnal and seasonal temperature changes in wetland water are attenuated with depth beneath the sediment-water interface. Attenuation is controlled by the capacity of the sediments to conduct heat. Direction and rate of groundwater flow modifies the conduction-driven attenuation. Net upward flow reduces, and net downward flow increases, the amplitude of diurnal or seasonal temperature responses with depth beneath the sediment-water interface (Figs. 3.26 and 3.27).

Robert Stallman (Stallman 1965) developed a method that could determine vertical flow of water through sediment based on measurement of temperature. The method required measurement of diurnal (or seasonal) fluctuation of temperature at two depths, the volumetric heat capacity of the water, and estimates of thermal conductivity and volumetric heat capacity of the bulk sediment. If the time series of the diurnal or seasonal temperature data are sinusoidal, this method provides a reasonable indication of the groundwater flow to or from a surface-water body. However, when applied to the typical wetland setting where wetland stage (and, therefore, i) is often highly variable over time, a method is needed that can solve for flow that may be substantially non-uniform.

Fortunately, several numerical models exist that simultaneously solve for heat and fluid flow in porous media. In addition to the parameters listed above, reasonable assumptions regarding boundary conditions, and values for i_v and K_v, are all that is needed to determine the flow of water across the sediment-water interface.

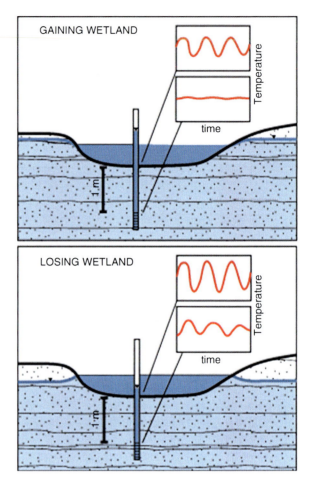

Fig. 3.27 Diurnal temperature response at the wetland bed and at a well screen installed 1 m beneath the wetland for upward and downward groundwater flow beneath the wetland (Modified from Stonestrom and Constantz (2003). Published with kind permission of the U.S. Geological Survey. Figure is public domain in the USA. All Rights Reserved)

The vertical hydraulic gradient can be obtained with the installation of a shallow piezometer in the wetland bed. This installation also allows convenient deployment of a temperature sensor to provide data at depth to compare with temperature at the bed of the wetland. Although K_v can be determined in-situ, K_v often is heterogeneous and is scale dependent. Therefore, K_v usually serves as the model calibration factor that is adjusted until the simulated time-series temperature data generated by the model match the measured time-series temperature data. Once K_v is calibrated so the modeled and measured temperatures are in good agreement, the model produces q, the specific groundwater flux across the sediment-water interface.

Temperature sensors are among the most accurate, robust, and inexpensive devices commonly used in the earth sciences, making this method particularly attractive. Thermal conductivity (K_T), the heat-flow equivalent of hydraulic conductivity (K), is a property that varies over a much narrower range than hydraulic conductivity. It can be reasonably estimated based on the type of sediment present

Table 3.2 Typical values of parameters required to determine q using numerical models that solve for heat and fluid flow through porous media (Modified from Stonestrom and Constantz 2003)

Material	Porosity	Density (10^6 g/m^3)	Volumetric heat capacity (10^6 J/m^3/°C)	Thermal conductivity (W/m/°C)	Thermal diffusivity (10^{-6} m^2/s)
Liquid water	1.0	1.0	4.2	0.6	0.1
Ice		0.9	1.9	2.2	1.2
Quartz		2.7	1.9	8.4	4.3
Soil minerals	0.2–0.4	2.7	1.9	2.9	1.5
Clay minerals	0.4–0.7	2.7	2.0	2.9	1.5
Soil organic matter	0.4–0.9	1.3	2.5	0.25	0.1
Organic estuary[a]	0.8		2.3	0.9	0.2

[a]Values from Land and Paull (2001)

beneath the wetland. Other properties, such as porosity, diffusivity, and heat capacity of the sediments, affect the solution to a lesser extent. These values commonly are estimated based on the type of sediment present at the site of interest (Table 3.2).

Any numerical model that simultaneously solves for fluid flow and heat flow can be used. Two commonly used models are SUTRA and VS2DH (Stonestrom and Constantz 2003). If a substantial horizontal component of flow is suspected, such as near the perimeter of many wetlands, additional wells can be installed near the wetland, temperature measured at several depths, and a 2-d version of the model can be created and calibrated to solve for flow along the wetland bed (Fig. 3.28). Although both Figs. 3.27 and 3.28 indicate that temperature is measured at several depths beneath the wetland bed, a single temperature measurement at depth will suffice. With only one temperature measurement in the sediments, the assumption is that K of the sediments is uniform. Additional temperature measurements provide additional data regarding the variability and distribution of K within the sediments. An example of adjusting K to obtain a good fit of modeled output to measured time-series data is presented by Stonestrom and Constantz (2003:88).

Vertical flow velocity also can be calculated by comparing the ratio of amplitudes or the phase shift of time-series data from temperature measured at different depths (Hatch et al. 2006). The selection of amplitude ratio or phase shift depends on the vertical velocity; very fast seepage rates are better resolved with the phase-shift solution whereas the slower seepage rates common in wetland settings are better determined with the amplitude-ratio method. This procedure has the added benefit of calculating changes in seepage over time and does not require measurement of hydraulic-head gradients. Although the degree and type of filtering of the data can be rather complex, advancement and use of this technique has been rapid and several variations of the original concept have been presented (Keery et al. 2007; Swanson and Cardenas 2010; Vogt et al. 2010). An automated data-processing routine has recently been developed to make the method faster and more user-friendly (Gordon et al. 2012).

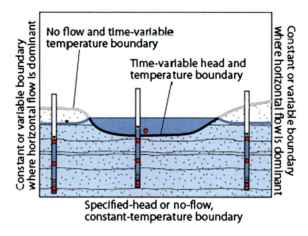

Fig. 3.28 Boundaries associated with a coupled water- and heat-flow model to simulate groundwater exchange with a wetland in two dimensions. *Red dots* are locations where temperature is measured. Head is measured at the wetland and at the screened interval of two monitoring wells (Modified from Stonestrom and Constantz (2003). Published with kind permission of the U.S. Geological Survey. Figure is public domain in the USA. All Rights Reserved)

Measuring and mapping the temperature of the submerged wetland bed also can be used to determine rates and distribution of groundwater discharge, but only for wetlands where groundwater discharge is prominent and pervasive (Schmidt et al. 2007). This method uses the Turcotte and Schubert (1982) solution for steady-state 1-dimensional advection-diffusion heat flow and relates temperature measured at about 20-cm depth in the bed sediment to an assumed constant temperature at greater depth in the sediment. The method requires that the surface-water temperature has small diurnal variability prior to and during the mapping of the temperature of the wetland bed, a condition best met during winter or during prolonged cloudy periods. Although the method was developed for use in streams, it should provide acceptable results for many wetlands that receive groundwater discharge; bed-sediment temperatures should be measured during periods when diurnal fluctuations are minimal. The method should work particularly well for wetlands that are ice covered during winter.

Mapping the bed temperature has become much easier with the growing use of what is now commonly called the distributed temperature system (DTS) (Selker et al. 2006; Fleckenstein et al. 2010). This system uses a device that sends a laser pulse down a length of fiber-optic cable that can be up to several km long. The light signal is reflected back to the sensor from every point along the cable. By timing the return, and resolving the frequency distribution of the light scattering, temperature can be determined to about 0.1 °C resolution and averaged over cable increments of 0.5–1 m. Temperature mapping of a sediment bed can be done as frequently as every minute to several minutes, allowing a qualitative determination of temporal as well as spatial variability of groundwater discharge (e.g., Henderson et al. 2009).

3.9 Subsurface Storage Above the Water Table

The storage term (ΔV) in the water balance equation commonly represents the amount of surface water in a wetland. For those wetlands with seasonal or ephemeral surface water, subsurface water storage also is an important hydrological consideration. Moisture content of the exposed soil influences the transport of oxygen and other gases, thereby affecting redox condition and biogeochemical processes. Moisture conditions affect the viability of soil fauna and the growth of plants adapted to high moisture environments.

After surface water in the wetland dries up, water loss from the wetland soil continues, mainly due to transpiration by wetland vegetation, which causes the water table to drop and the soil to become unsaturated. The volume of sediment between land surface and the water table is called the vadose zone. The soil remains nearly saturated immediately above the water table due to surface tension that holds water in the soil pores. As soil dries and the water table continues to decline, it becomes increasingly difficult for plant roots to extract water from the soil. As a result, the rate of transpiration decreases, the rate of water-table decline decreases, and the water table eventually reaches a relatively stable position. This condition persists until something changes; most often the change is a subsequent recharge event that adds water to the unsaturated sediments. The amount of water required to saturate the soil completely and bring the water table to land surface is called the moisture deficit. A wetland with a small moisture deficit can recover from a dry condition relatively quickly when wet meteorological conditions return. Therefore, subsurface moisture storage is an indicator of the resilience of a wetland to fluctuations in water inputs.

Subsurface moisture storage is determined by the depth to the water table and soil water content in the vadose zone. Methods for determining the position of the water table are described in the section on groundwater flow. Here we describe methods for measuring soil-water content and then introduce the concept of specific yield, S_y, that relates subsurface storage to water-table depth.

3.9.1 Thermo-Gravimetric Method for Measuring Soil Water Content

This method starts with collecting a sample of undisturbed soil in a metal cylinder of precisely known volume (e.g., 100 cm^3) using a soil corer, or inserting the cylinder into the side wall of a soil pit. Care must be taken to fill the cylinder completely while at the same time avoiding soil compaction. The top and bottom of the sample are leveled using a metal scraper so that the soil volume is equal to the volume of the cylinder, and sealed with plastic caps and electrical tape to prevent evaporation. Samples are transferred to the laboratory and weighed using a balance to determine the pre-drying weight. Samples are placed in an oven with a temperature set at 105 °C for 24–48 h to evaporate all liquid water without volatilizing organic components of the soil. The sample weight may also be measured periodically during drying until it does not change any longer. Samples are cooled in a sealed container with desiccant to prevent absorption of atmospheric vapor during cooling, and weighed again. Volumetric soil water content, θ_v (cm^3/cm^3), is given by

$$\theta_v = [(\text{original weight} - \text{dry weight})/\text{density of water}]/\text{sample volume} \quad (3.43)$$

The same sample may be used to determine other soil parameters such as dry bulk density or porosity (see Chaps. 4 and 8 on soil sampling).

The water content determined with this method is commonly used as the reference to test or calibrate other methods. However, it should be noted that the thermo-gravimetric method does not necessarily yield exact results (Topp and Ferré 2002) because the measured value may be affected by the drying temperature and time, vapor absorption during cooling, and most importantly, errors in measurement of sample volume and weight. A major disadvantage of this method is that it requires the removal of the sample and is not suitable for continuous, in situ monitoring of soil water. Therefore, instrumental methods are commonly used for continuous monitoring.

3.9.2 Time Domain Reflectometry

Of the various types of instruments available for continuous monitoring of soil moisture, time domain reflectometry (TDR) and the capacitance method are the most widely used (Ferré and Topp 2002). These methods both make use of the fact that the velocity of electromagnetic (EM) waves is equal to the speed of light ($c = 3.0 \times 10^8$ m s^{-1}) in a vacuum but is lower in other media. EM-wave velocity is determined by a property called dielectric permittivity. The dielectric permittivity of water relative to a vacuum is much greater (≈ 80) than that of air (≈ 1) or soil solids (≈ 3–8). Therefore, volumetric water content can be estimated from measurements of soil dielectric permittivity.

In the TDR method, a very sharp voltage pulse from the signal source travels through the soil along a wave guide, typically consisting of parallel stainless steel rods (Fig. 3.29), and is reflected back to the source. The reflected signal is recorded as a time series of voltage values, commonly called the wave form (Fig. 3.30), in which the time to the negative peak (t_1) indicates the two-way travel time of the EM wave between the source and the top of the wave guide, and the time to a rapid rise (t_2) indicates the two-way travel time between the source and the bottom of the wave guide. If the length of wave guide is L (m), then the apparent velocity of the EM wave (v_{EM}) in the soil is

$$v_{EM} = 2L/(t_2 - t_1) \quad (3.44)$$

3 Assessing and Measuring Wetland Hydrology

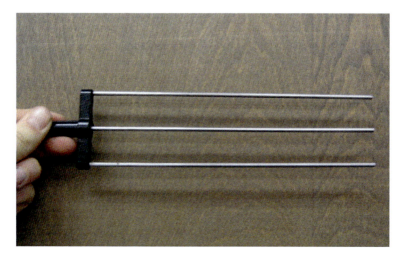

Fig. 3.29 Example of a parallel-rod wave guide for time domain reflectometry (TDR)

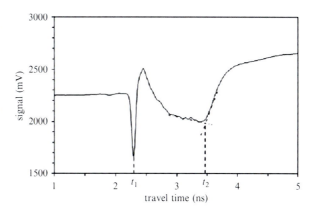

Fig. 3.30 Example of waveform data obtained by a TDR device with a 0.2-m parallel-rod wave guide (see Fig. 3.29) installed in a mineral soil. Times t_1 and t_2 indicate the travel time of signals reflected from the top and bottom of the wave guide

For low-salinity soils, relative dielectric permittivity (ε_r) is given by

$$\varepsilon_r = (c/v_{EM})^2 \qquad (3.45)$$

Equation 3.45 only gives approximate values for soils with high electrical conductivity (Ferré and Topp 2002). Volumetric water content is estimated from a calibration curve relating θ_v and ε_r. For a large variety of agricultural mineral soils, a "universal" formula of Topp et al. (1980) has been found to yield reasonably accurate values of θ_v:

$$\theta_v = -0.053 + 0.0292\varepsilon_r - 0.00055\varepsilon_r^2 + 4.3 \times 10^{-6}\varepsilon_r^3 \qquad (3.46)$$

However, significant deviation has been noted in organic soils. Therefore, it may be necessary to develop soil-specific calibration curves for application of the TDR method in organic soils typically found in wetlands (see next section for calibration methods).

TDR wave guides (or probes) can be installed vertically from the surface or horizontally in a soil pit, which should be refilled very carefully to prevent preferential infiltration affecting the measurements. The measurement is sensitive to soil disturbance and gaps between the probe and soil matrix. Therefore, probes must be inserted straight into the soil, with minimum wobble, to minimize disturbance. A TDR probe measures an average θ_v over the entire probe length in a cylindrical region within a diameter of approximately 1.5 times the rod separation (Ferré and Topp 2002). If the depth to the water table is shallow (e.g., <0.3 m), vertical probes may be used to cover the entire vadose zone. Where the water table is deeper, horizontal probes need to be installed at multiple depths to measure a profile of water content from the surface to the water table.

Probes may be connected to a portable field device for manual recording of wave forms and determination of ε_r, or connected to a digital datalogger. Most of commercially available TDR devices determine θ_v using internal algorithms and output the value of θ_v, which is a convenient feature for long-term monitoring. However, it is prudent to store the raw waveform data and periodically check the accuracy of automatically determined θ_v. Detailed discussion on the TDR method and useful guidance for its application are found in Ferré and Topp (2002).

3.9.3 Capacitance Method

This method also utilizes the large contrast in dielectric permittivity between water and other soil components, but it is based on the principle that frequency of oscillation of a circuit consisting of an electrode-soil capacitor is a function of dielectric permittivity (Starr and Paltineanu 2002). Since the functional relationship is dependent on electrode configuration and soil type, soil-specific calibration is required to calculate θ_v from the frequency measured with a capacitance probe. Compared to the TDR method, which yields reasonably accurate results using the universal formula, the disadvantage of the capacitance method is the necessity of soil-specific calibration. On the other hand, once well calibrated, the capacitance method offers a much more robust and convenient tool for continuous monitoring of soil water content than the TDR method. Most commercially available capacitance probes are designed to work with standard dataloggers and can be used as part of the standard collection of sensors that make up hydrological monitoring stations. This is another advantage over TDR, which typically requires an expensive control unit in addition to a datalogger.

Depending on the monitoring objectives and probe length, capacitance probes may be installed vertically from the surface, or horizontally at multiple depths in a soil pit. Similar to TDR wave guides, measurement with capacitance probes is sensitive to soil disturbance and air gaps between electrodes and the soil matrix. Therefore, probes need to be inserted carefully to minimize sensor error. Detailed discussion on the capacitance method and practical procedures are found in Starr and Paltineanu (2002).

If a large block of undisturbed soil representative of field conditions can be removed intact, capacitance probes may be calibrated in the laboratory. Briefly, the sample is placed in a sealed container, a probe is placed in the sample, and the sample is brought to saturation. The total weight is measured and the output of the probe is recorded. The sample is then drained and dried in several stages with each stage given enough time to establish uniform water content in the container, and total weight and probe output are recorded. At the end of drying, subsamples are collected from the container to determine the dry bulk density of the soil, from which the weight of the soil is converted to volumetric water content. Starr and Paltineanu (2002) describe detailed procedures for laboratory calibration using disturbed and repacked soil, which is suitable for agricultural mineral soils but may not be applicable to organic soils. If it is not feasible to conduct laboratory calibrations, soil samples can be collected from the probe depth at the time of installation, and θ_v determined with the thermo-gravimetric method is then compared to the initial probe data for a single-point calibration.

3.9.4 Specific Yield

Fluctuations of the water table represent changes in subsurface storage. Since it is easy to measure the water-table elevation in monitoring wells, attempts have been made to estimate changes in subsurface storage (ΔS_{sub}) from changes in water-table elevation (Δh_{WT}) using a concept called specific yield (S_y), also known as drainable porosity. When ΔS_{sub} is expressed as depth of water, S_y is the ratio of ΔS_{sub} to Δh_{WT}, or more precisely, it is the volume of water released from or taken into storage per unit cross sectional area following a unit change in water-table elevation (Freeze and Cherry 1979:61).

Despite being conceptually simple, S_y is somewhat complex because soils can retain a sizable and variable amount of water above the water table that is related to the size of void spaces in the soil matrix. The relation between water content and the magnitude of tension force holding water in pores is called the soil water characteristic (SWC) curve. Under static conditions in the absence of vertical flow, the magnitude of tension force is proportional to distance above the water table. Therefore, SWC is commonly shown as a vertical profile of θ_v, which represents the theoretical distribution of water content after complete gravitational drainage of the vadose zone following complete saturation (Fig. 3.31). Suppose that a certain amount of groundwater is extracted, causing the water table to drop. This extraction induces drainage of water from the vadose zone until the new static condition is reached (Fig. 3.31). The amount of extracted water should be equal to the difference between the pre- and post-extraction profiles (Fig. 3.31). In other words,

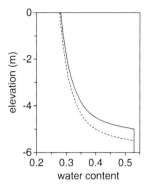

Fig. 3.31 Example of soil water characteristic (SWC) curve of agricultural mineral soil having a clay-loam texture. *Solid line* shows the soil water-content profile for the water table located 5 m below the ground surface, and *dashed line* shows the profile for the water table lowered to 5.5 m below the surface

$$\Delta S_{sub} = \int_{z1}^{0} \Delta \theta_v(z) dz \quad (3.47)$$

where $\Delta \theta_v(z)$ is change in volumetric water content as a function of elevation (z), and z_1 is a reference point below the water table. Figure 3.31 shows an example for a relatively deep water table. For a shallow water table, for example 2 m below the surface, the value of ΔS_{sub} for the same amount of Δh_{WT} is much smaller, indicating the strong dependence of ΔS_{sub} and S_y on the water-table position. In addition, depending on soil hydraulic conductivity, complete drainage of soil water following the water-table drop may take a long time, meaning that S_y is dependent on the time scale of measurement (Healy and Cook 2002).

Despite the variable nature of S_y, a constant value is used in most practical applications. Healy and Cook (2002) reviewed several field and laboratory methods for estimating S_y. If the SWC is available from laboratory analysis of soil samples, S_y can be calculated directly from the comparison of theoretical water-content profiles resulting from a given drop in the water table (e.g., Fig. 3.31). If the water table is sufficiently deep, for example deeper than 3 m for the soil shown in Fig. 3.31, the calculated S_y is little affected by the assumed position of the water table.

While this method is theoretically simple, determination of SWC is time consuming and labor intensive. The column-drainage approach offers an alternative method for laboratory measurement of S_y. In this method, a column is filled with undisturbed or repacked soil taken from a field site and then saturated with water from the bottom. Water is then allowed to drain from the bottom of the column. The top of the column is open to the atmosphere, while avoiding evaporation, and the bottom is placed in a very shallow (e.g., <2–3 mm) water reservoir to maintain a water-table condition. Dividing the amount of drainage (ΔS_{sub}) by the length of the column (Δh_{WT}) gives S_y. Care should be taken to drain the soil column completely, which may take several hours or days, depending on the soil type. The column should be sufficiently long to reduce the influence of column length on S_y. For example, the soil used in constructing Fig. 3.31 would require a column longer than 3 m, although such a long column is not practical.

Complete drainage of soil water, which is assumed in laboratory methods, may take up to several weeks. Since such assumptions do not reflect the dynamic water-table conditions in the field, it is preferable to use field methods to estimate S_y. Several methods have been proposed based on conducting an aquifer pumping test and interpreting the data. However, pumping tests require installation of a network of observation wells and the interpretation is strongly influenced by model assumptions, such as the boundary conditions or time scale of soil-water drainage (Healy and Cook 2002).

An alternative to aquifer pumping tests is the water-balance approach, where ΔS_{sub} over a given time period is estimated from careful measurements of other water balance components:

$$\Delta S_{sub} = P + O_f - E + S_i - S_o + G_i - G_o \qquad (3.48)$$

Since it is usually impossible to measure all components in Eq. 3.48, periods are chosen so that some of the hydrologic components are negligible and can be omitted. For example, if a wetland does not have surface water input and output, overland flow is negligible, and the magnitude of net groundwater input $(G_i - G_o)$ is expected to be much smaller than that of net atmospheric input $(P - E)$, then plotting Δh_{WT} observed in response to rainfall events against the amount of net precipitation $(P - E)$ may yield a linear relation between $P - E$ ($\approx \Delta S_{sub}$) and Δh_{WT}. The slope of this linear relation is equal to S_y. This type of approach is commonly used in wetland studies (e.g., Gerla 1992; Rosenberry and Winter 1997).

Considerable uncertainty and discrepancy is noted in values of S_y estimated using different methods. Field-based methods generally give smaller values than laboratory methods, presumably because laboratory methods usually allow a long time for complete drainage of soil samples compared to the time scale of field processes (Healy and Cook 2002). Therefore, investigators must be aware of the time scale of processes under investigation, as well as the assumptions associated with the definition of S_y.

3.10 Use of Conservative Tracers

In many low-gradient wetlands with extensive areas of vegetation it can be difficult to quantify several of the terms in the wetland water budget. In these situations, water chemistry may provide a separate or perhaps better estimate of some of the water-budget terms. In Sect. 3.8.3 we discussed the procedure of writing two equations, one a water-budget equation (Eq. 3.39) and the other a mass-balance equation for a conservative tracer (Eq. 3.40), for the purpose of determining either G_i or G_o. These equations can be rearranged and this procedure can be used to solve for any of the water-budget components, not just G_i or G_o. Here we discuss further the characteristics of a conservative tracer, assumptions associated with this method, and procedures for proper sample collection.

3.10.1 Evaluation of Water and Mass Balance Equations

An ideal tracer for the mass-balance method is chemically and biologically conservative (no reaction, absorption, release or uptake), is readily measureable, and is harmless to the observer and aquatic life. Chloride and bromide are commonly used as tracers (e.g., Choi and Harvey 2000; Parsons et al. 2004). For relatively shallow and small wetland ponds, water is often assumed well mixed so that tracer concentration is uniform within the pond. When this assumption is justified based on field data, for example by the analysis of samples from multiple points and depths in a pond, Eq. 3.40 is simplified to

$$\Delta(C_W V)/\Delta t = C_P P + C_{Si} S_i - C_W (S_o + G_o) + C_{Gi} G_i + C_{of} O_f \qquad (3.49)$$

(note that $C_W V$ is equal to mass and that $C_E E$ has been neglected as it is assumed that no tracer mass is lost through evaporation or transpiration). If the assumption is not justified, C_{So} and C_{Go} need to be measured for each stratified layer, or for each isolated embayment of the pond, and mass needs to be determined based on the concentration and volume of each separate entity of the pond (e.g., Choi and Harvey 2000). Further simplification to Eq. 3.49 can be made in some settings. For example, if chloride is used in a region that receives very small amounts of chloride in precipitation relative to other components, then C_P is assumed negligible. Tracers also can be applied to an entire wetland in some cases. If bromide is used as an artificial tracer in a system that has very low natural bromide concentrations, then C_P, C_{Si}, C_{Gi}, and C_{Of} are all assumed negligible.

Mass balance calculation requires measurements or estimates of all "known" terms in the water balance equation and all concentrations in the mass-balance equation. Specialized monitoring devices, such as ion-specific electrodes, are available for continuous monitoring of tracer concentration, but these devices are prone to instrument drift and calibration issues. Therefore, best results are obtained by collecting water samples and analyzing them in the laboratory. This process can be conducted as part of a routine sampling program for water-quality monitoring (see Chap. 6 on water quality). Using automated water samplers (Fig. 3.32), concentrations can be determined daily or even hourly and the mass balance calculated on a fine temporal resolution. When a longer sampling interval, such as weekly or monthly, is used the water and mass balance equations need to be evaluated using average values. The equations can be written as:

$$(V_2 - V_1)/(t_2 - t_1) = <P> - <E> + <S_i> - <S_o> + <G_i> - <G_o> + <O_f> \qquad (3.50)$$

$$(C_{w2}V_2 - C_{w1}V_1)/(t_2 - t_1) = <C_P P> + <C_{Si} S_i> - <C_w(S_o + G_o)> + <C_{Gi} G_i> + <C_{Of} O_f> \qquad (3.51)$$

where subscripts 2 and 1 indicate sampling dates 2 and 1, respectively, and $<>$ indicates an average value for the time interval. Definition of average is

3 Assessing and Measuring Wetland Hydrology

Fig. 3.32 Automated sampler for collecting stream water samples at a prescribed time interval. (**a**) A field sampler with the sample intake in a pool above a V-notch weir. (**b**) The same sampler showing the control unit and 24 sample bottles inside the sampler body (Photos by Masaki Hayashi)

straightforward for some terms; for example, $<P>$ is given by the total amount of precipitation divided by $t_2 - t_1$. However, assumptions may have to be made for averages of other terms. For example, if continuous surface inflow data are available but only two values of concentration are available, a reasonable approximation is to use an arithmetic mean for concentration

$$<C_{Si}S_i> \approx (C_{Si1} + C_{Si2})/2 \times <S_i> \qquad (3.52)$$

Fig. 3.33 Application of bromide tracer in a prairie wetland (Parsons et al. 2004). To ensure even distribution of tracer, the tracer solution was slowly released to the wetland from a boat moving around the wetland (Photo printed with kind permission of © David Parsons 2012. All rights reserved)

Once the balance equations are written for measured values, they can be implemented in a spreadsheet or a simple computer program to compute desired components. For example, Choi and Harvey (2000) used chloride to quantify groundwater inflow and outflow in constructed wetlands in Florida, U.S.A. Parsons et al. (2004) used bromide to quantify evaporation and groundwater outflow in a prairie wetland in Saskatchewan, Canada.

3.10.2 Remarks on Water Sample Collection

Successful application of the mass-balance approach depends on how well the simplifying assumptions are satisfied and how well all the concentration terms are represented by measured values. If a single value of C_W is used to represent the whole wetland pond, it is important to verify that the pond is well mixed by periodically sampling and analyzing water from different locations and depths. If an artificial tracer is applied to a pond, it needs to be applied evenly in the entire pond area to ensure a uniform initial concentration (Fig. 3.33). Samples near the water surface can be collected by simply submerging a clean bottle in water. Deeper samples need to be collected using a tube with an intake at the sampling depth connected to a pump or plastic syringe, or using a van Dorn, Kemmerer, or other type of sampler (see Ward and Harr 1990). In water up to approximately 1 m in depth, a sample can be collected from the entire water column using a cylinder inserted vertically into the wetland water (Swanson 1978). If $C_P P$ is a significant component of tracer input, then a specialized sampling device must be used to collect atmospheric deposition (see Allan (2004) for specification). If a specialized device is not available, it may be better to use published data as a proxy rather than using erroneous data collected via inappropriate methods. Precipitation chemistry data are available from national and international atmospheric deposition monitoring networks, such as Global Atmospheric Watch (http://gaw.empa.ch/gawsis/).

Temporal and spatial variability in C_{Si} and C_{Gi} need to be properly represented by collecting multiple samples over time. If there are multiple inflow streams, samples should be collected at all streams and volume-weighted averages of concentrations should be used in the mass-balance equation. Due to geologic heterogeneity, solute concentrations in groundwater may have large spatial variability even within a relatively small area. Therefore, it is necessary to collect groundwater samples from several locations (and depths) in the areas of anticipated groundwater discharge. Water samples usually are collected directly from monitoring wells and piezometers, or springs; they also can be extracted from sediment core samples (see Adams (1994) for methods). Wells should be bailed or pumped prior to sample collection to ensure that the sample represents the composition of groundwater in the aquifer surrounding the well screen. A common purging protocol is to pump 3–5 times the volume of water in the well to ensure that the stagnant water in the casing has been completely removed prior to sample collection. Another option is to slowly pump water through an intake tube that is placed in the screened interval of the well (low-flow sampling method). The pumping rate needs to be slow enough that virtually no drawdown occurs in the well, in which case nearly all of the water supplied during pumping originates from the aquifer. Detailed procedures for groundwater sampling are found in manuals and handbooks on this subject (e.g., Yeskis and Zavala 2002; Wilde 2006). The number of groundwater samples commonly used is insufficient to determine the precise value of an average tracer concentration, $<C_{Gi}>$. Therefore, a recommended practice is to use one standard deviation from the arithmetic mean of all groundwater samples to represent the uncertainty in the mass-balance calculation.

3.10.3 Use of Multiple Tracers

If more than one conservative tracer is available for water-budget determinations, and their concentrations are not correlated, then Eq. 3.51 can be written for each individual tracer. This increases the number of equations and thus, the number of unknowns that can be determined. For example, using naturally occurring chloride and artificially introduced bromide, a set of three equations can be solved for three unknowns. Alternatively, the water-budget equation is solved separately with each mass-balance equation to provide separate estimates of the same unknowns. If resulting values using separate tracers are greatly different, then possible errors in estimation or measurement of known terms or missing terms in the equations are indicated.

The mass balance of a conservative tracer can be used with a potentially reactive tracer to identify a possible reaction and estimate its rate. The mass-balance equation for a reactive tracer can be written as

$$\Delta(C_w V)/\Delta t = C_P P + C_{Si} S_i - C_w(S_o + G_o) + C_{Gi} G_i + C_{of} O_f + R_{xn} \quad (3.53)$$

where R_{xn} (kg/m^3/s) is the rate of reaction per volume. If all other terms in Eq. 3.53 are known from solving the water-budget equation and the mass balance equation for a conservative tracer, then R_{xn} can be determined as the residual of Eq. 3.53. For example, Heagle et al. (2007) solved the water-budget equation with the mass balance equations for naturally occurring chloride and sulphate to estimate the rate of sulphate reduction in a prairie wetland in Saskatchewan, Canada.

3.10.4 Final Remarks

The tracer mass-balance approach provides a useful tool for estimating the water-budget components that are difficult to measure directly. Unlike other methods for estimating groundwater flow, the mass-balance method evaluates the flow averaged over the entire wetland, while giving no information about the spatial distribution of groundwater recharge or discharge within the wetland. Therefore, it is beneficial to use this method in combination with other methods that give local values of flow, such as a seepage meter or mini-piezometer (see Sect. 3.8). The mass-balance method provides a constraint on the possible range of total groundwater flow, whereas a local-scale method is useful for delineating areas of focused recharge or discharge, which may have significant influence on the distribution of wetland flora and fauna (e.g., Rosenberry et al. 2000).

3.11 Estimation of Errors

As introduced in Sect. 3.2, a wetland water budget can be written as the change in wetland volume per time (plus residual) equal to the sum of all inputs and losses (Eq. 3.1). If all of the hydrological components are measured as accurately as possible, it is almost certain that the sum of those components will not equal the change in volume in the wetland over an accounting period. R in Eq. 3.1 can be disturbingly large relative to ΔV for some water budgets. Error stems from (1) incorrect measurement of a parameter (instrument error), (2) misapplying point measurements to specific areas or volumes of a wetland (a common but often neglected error), and (3) misinterpreting the hydrologic setting, usually by not

3 Assessing and Measuring Wetland Hydrology

measuring one or more parameters that are important to the water or chemical budget. Positive errors often are offset by negative errors, so R commonly is smaller than the error associated with one or more individual terms. A first-order error analysis takes this into account when determining error as a function of multiple parameters. When errors are additive, first-order analysis (e.g., Taylor 1982) involves calculating the square root of the sum of the squared values of each of the parameters:

$$\delta = \sqrt{\delta_P^2 + \delta_E^2 + \delta_{Si}^2 + \delta_{So}^2 + \delta_{Gi}^2 + \delta_{Go}^2 + \delta_{Of}^2 + \delta_{\Delta V}^2} \qquad (3.54)$$

where

δ = error,
P = precipitation,
E = evapotranspiration,
Si = surface-water flow to the wetland,
So = surface-water flow from the wetland,
Gi = ground-water discharge to the wetland,
Go = loss of wetland water to ground water,
Of = overland flow,
ΔV = change in volume of water contained in the wetland (positive for increase in volume).

First-order error analysis assumes that errors are independent and randomly distributed. This clearly is a poor assumption. For example, most of the water-budget parameters are dependent on precipitation. If substantial interdependence is suspected, a more rigorous analysis can be conducted where covariances between terms are considered (e.g., LaBaugh 1985). However, a large percentage of water-budget studies, if they present error estimations at all, simply assume parameters are independent and apply an equation similar to 3.54.

By estimating the error associated with each of the water-budget components of a wetland, cumulative error, δ, can be compared with the residual, R, of Eq. 3.53. If differences are large, it is likely that at least one of the components has been determined incorrectly or that errors associated with one or more of the water-budget components have been poorly estimated.

A common question among wetland scientists is just how large are these errors? Estimates vary substantially depending on the setting, goals of the study, and methods of measurements. Errors reported in a selection of publications that provide error estimates for water-budget components of studies of lakes, wetlands, and reservoirs generally are smallest for precipitation and largest for groundwater. Based on values presented in ten such studies in Table 3.1, median estimates for error associated with P, E, S, G, and ΔV are 9, 10, 10, 36, and 10 %, respectively.

3.12 Chapter Summary

After reading this chapter, the reader may come away with the thought that wetland hydrology is complex but there are many different approaches and tools that can be used, or that quantifying hydrologic fluxes in wetland settings is difficult and fraught with error. Either impression would be correct. Perhaps the most important conclusion is that the pursuit of parallel lines of evidence, using multiple methods for achieving the same goals, will lead to a better understanding of these complex processes and a more accurate assessment of the various hydrologic components that constitute the hydrologic setting of a wetland. Armed with the numerous methods at our disposal, and knowledge of the various sources and magnitudes of error associated with each approach, the wetland hydrologist can feel comfortable in pursuing quantification of the various hydrological components with a judicious selection of methods appropriate to the goals and budget associated with the investigation.

References

Abtew W (2001) Evaporation estimation for Lake Okeechobee in south Florida. J Irrig Drain Eng 127:140–147
Adams DD (1994) Sediment pore water sampling. In: Mudrock A, MacKnight (eds) Handbook of techniques for aquatic sediments sampling. Lewis Publishers, Boca Raton
Allan MA (2004) Manual for the GAW precipitation chemistry programme: guidelines, data quality objectives and standard operation procedures. World Meteorological Organization Global Atmosphere Watch No. 160:170
Allen RG, Pereira LS, Raes D, Smith M (1998) Crop evapotranspiration – guidelines for computing crop water requirements. Food and Agricultural Organization of the United Nations FAO Irrigation and drainage paper 56:328
Belanger TV, Kirkner RA (1994) Groundwater/surface water interaction in a Florida augmentation lake. Lake Reserv Manag 8:165–174
Bouwer H (1989) The Bouwer and Rice slug test – an update. Ground Water 27:304–309
Bouwer H, Rice RC (1976) A slug test for determining hydraulic conductivity of unconfined aquifers with completely or partially penetrating wells. Water Resour Res 12:423–428
Brooks R (2009) Potential impacts of global climate change on the hydrology and ecology of ephemeral freshwater systems of the forests of the northeastern United States. Clim Chang 95:469–483
Brunsell NA, Ham JM, Owensby CE (2008) Assessing the multi-resolution information content of remotely sensed variables and elevation for evapotranspiration in a tall-grass prairie environment. Remote Sens Environ 112:2977–2987
Brutsaert WH, Stricker H (1979) An advection-aridity approach to estimate actual regional evapotranspiration. Water Resour Res 15:443–450
Carr MR, Winter TC (1980) An annotated bibliography of devices developed for direct measurement of seepage. U.S. Geological Survey Open-File Report 80-344:38
Choi J, Harvey JW (2000) Quantifying time-varying ground-water discharge and recharge in wetlands of the northern Florida Everglades. Wetlands 20:500–511
Conly FM, Su M, van der Kamp G, Millar JB (2004) A practical approach to monitoring water levels in prairie wetlands. Wetlands 24:219–226

Constantz J, Niswonger R, Stewart AE (2008) Analysis of temperature gradients to determine stream exchanges with ground water. In: Rosenberry DO, LaBaugh JW (eds) Field techniques for estimating water fluxes between surface water and ground water. U.S. Geological Survey Techniques and Methods 4-D2, Denver, pp 117–127

Cunningham WL, Schalk CW (2011) Groundwater technical procedures of the U.S. Geological Survey. U.S. Geological Survey Techniques and Methods 1-A1:151

Dalton MS, Aulenbach BT, Torak LJ (2004) Ground-water and surface-water flow and estimated water budget for Lake Seminole, northwestern Georgia and northwestern Florida. U.S. Geological Survey Scientific Investigations Report 2004-5073:49

Dingman SL (2002) Physical hydrology. Prentice-Hall, Upper Saddle River

Doss PK (1993) Nature of a dynamic water table in a system of non-tidal, freshwater coastal wetlands. J Hydrol 141:107–126

Drexler JZ, Snyder RL, Spano D, Paw UKT (2004) A review of models and micrometeorological methods used to estimate wetland evapotranspiration. Hydrol Process 18:2071–2101

Dunne T, Leopold LB (1978) Water in environmental planning. W.H. Freeman and Company, New York

Elsawwaf M, Willems P, Feyen J (2010) Assessment of the sensitivity and prediction uncertainty of evaporation models applied to Nasser Lake, Egypt. J Hydrol 395:10–22

Euliss NH Jr, LaBaugh JW, Fredrickson LH, Mushet DM, Laubhan MK, Swanson GA, Winter TC, Rosenberry DO, Nelson RD (2004) The wetland continuum: a conceptual framework for interpreting biological studies. Wetlands 24:448–458

Ferré PA, Topp GC (2002) Time domain reflectometry. In: Dane JH, Topp CC (eds) Methods of soil analysis part 4 – physical methods. Soil Science Society of America, Madison

Fetter CW Jr (2001) Applied hydrogeology, 4th edn. Prentice Hall, Upper Saddle River

Fleckenstein JH, Krause S, Hannah DM, Boano F (2010) Groundwater-surface water interactions: new methods and models to improve understanding of processes and dynamics. Adv Water Resour 33:1291–1295

Freeman LA, Carpenter MC, Rosenberry DO, Rousseau JP, Unger R, McLean JS (2004) Use of submersible pressure transducers in water-resources investigations. U.S. Geological Survey Techniques of Water-Resources Investigations 8-A3:50

Freeze RA, Cherry JA (1979) Groundwater. Prentice Hall, Englewood Cliffs

Gerla PJ (1992) The relationship of water-table changes to the capillary fringe, evapotranspiration, and precipitation in intermittent wetlands. Wetlands 12:91–98

Goodison BE, Louie PYY, Yang D (1998) WMO solid precipitation measurement intercomparison final report. World Meteorological Organization Report No. 872, Geneva

Gordon RP, Lautz LK, Briggs MA, McKenzie JM (2012) Automated calculation of vertical pore-water flux from field temperature time series using the VFLUX method and computer program. J Hydrol 420–421:142–158

Gregory KJ, Walling DE (1973) Drainage basin form and processes. John Wiley and Sons

Haag KH, Lee TM, Herndon DC (2005) Bathymetry and vegetation in isolated marsh and cypress wetlands in the northern Tampa Bay area, 2000–2004. U.S. Geological Survey Scientific Investigations Report 2005-5109:49

Harbeck GEJ, Kohler MA, Koberg GE (1958) Water-loss investigations: lake mead studies. U.S. Geological Survey Professional Paper 298:100

Harvey JW, Krupa SL, Gefvert CJ, Choi J, Mooney RH, Giddings JB (2000) Interaction between ground water and surface water in the northern Everglades and relation to water budget and mercury cycling: study methods and appendixes. U.S. Geological Survey Open-File Report 00-168:395

Hatch CE, Fisher AT, Revenaugh JS, Constantz J, Ruehl C (2006) Quantifying surface water – groundwater interactions using time series analysis of streambed thermal records: method development. Water Resour Res 42:W10410. doi:10410.11029/12005WR004787

Hayashi M, van der Kamp G (2000) Simple equations to represent the volume-area-depth relations of shallow wetlands in small topographic depressions. J Hydrol 237:74–85

Hayashi M, van der Kamp G (2007) Water level changes in ponds and lakes: the hydrological processes. In: Johnson E, Miyanishi K (eds) Plant disturbance ecology. Academic, Burlington

Hayashi M, van der Kamp G, Rudolph DL (1998) Water and solute transfer between a prairie wetland and adjacent uplands, 1. Water balance. J Hydrol 207:42–55

Hayashi M, van der Kamp G, Schmidt R (2003) Focused infiltration of snowmelt water in partially frozen soil under small depressions. J Hydrol 270:214–229

Heagle DJ, Hayashi M, van der Kamp G (2007) Use of solute mass balance to quantify geochemical processes in a prairie recharge wetland. Wetlands 27:806–818

Healy RW, Cook PG (2002) Using groundwater levels to estimate recharge. Hydrogeol J 10:91–109

Healy RW, Winter TC, LaBaugh JW, Franke OL (2007) Water budgets: foundations for effective water-resources and environmental management. U.S. Geological Survey Circular 1308:98

Henderson RD, Day-Lewis FD, Harvey CF (2009) Investigation of aquifer-estuary interaction using wavelet analysis of fiber-optic temperature data. Geophys Res Lett 36:L06403. doi:06410.01029/02008GL036926

Hill RJ, Ochs GR, Wilson JJ (1992) Measuring surface-layer fluxes of heat and momentum using optical scintillation. Boundary-Layer Meteorol 58:391–408

Hogan JM, van der Kamp G, Barbour SL, Schmidt R (2006) Field methods for measuring hydraulic properties of peat deposits. Hydrol Process 20:3635–3649

Hood JL, Roy JW, Hayashi M (2006) Importance of groundwater in the water balance of an alpine headwater lake. Geophys Res Lett 33:13405

Hvorslev MJ (1951) Time lag and soil permeability in ground water observations. U.S. Army Corps of Engineers Waterways Experimental Station Bulletin No. 36:50

Jovanovic B, Jones D, Collins D (2008) A high-quality monthly pan evaporation dataset for Australia. Clim Chang 87:517–535

Kadlec RH, Wallace SD (2009) Treatment wetlands. CRC Press, Boca Raton

Keery J, Binley A, Crook N, Smith JWN (2007) Temporal and spatial variability of groundwater-surface water fluxes: development and application of an analytical method using temperature time series. J Hydrol 336:1–16

Kilpatrick FA, Cobb ED (1985) Measurement of discharge using tracers. U.S. Geological Survey Techniques of Water-Resources Investigations Chapter A16:52

Kitanidis PK (1997) Introduction to geostatistics: application to hydrogeology. Cambridge University Press, New York

Kohler MA (1954) Lake and pan evaporation In: Water-loss investigations: Lake Hefner studies, technical report U.S. Geological Survey Professional Paper 269, Washington, DC

Krabbenhoft DP, Bowser CJ, Anderson MP, Valley JW (1990) Estimating groundwater exchange with lakes 1. The stable isotope mass balance method. Water Resour Res 26:2445–2453

Krabbenhoft DP, Bowser CJ, Kendall C, Gat JR (1994) Use of oxygen-18 and deuterium to assess the hydrology of groundwater-lake systems. In: Baker LA (ed) Environmental chemistry of lakes and reservoirs. American Chemical Society, Washington, DC

LaBaugh JW (1985) Uncertainty in phosphorus retention, Williams Fork Reservoir, Colorado. Water Resour Res 21:1684–1692

LaBaugh JW (1986) Wetland ecosystem studies from a hydrologic perspective. Water Resour Bull 22:1–10

LaBaugh JW, Winter TC (1984) The impact of uncertainties in hydrologic measurement on phosphorus budgets and empirical models for two Colorado reservoirs. Limnol Oceanogr 29:322–339

LaBaugh JW, Rosenberry DO, Winter TC (1995) Groundwater contribution to the water and chemical budgets of Williams Lake, Minnesota, 1980–1991. Can J Fish Aquat Sci 52:754–767

LaBaugh JW, Winter TC, Rosenberry DO, Schuster PF, Reddy MM, Aiken GR (1997) Hydrological and chemical estimates of the water balance of a closed-basin lake in north central Minnesota. Water Resour Res 33:2799–2812

LaBaugh JW, Winter TC, Rosenberry DO (2000) Comparison of the variability in fluxes of ground water and solutes in lakes and wetlands in central North America. Verh Int Ver Theor Angew Limnol 27:420–426

Land LA, Paull CK (2001) Thermal gradients as a tool for estimating groundwater advective rates in a coastal estuary: White Oak River, North Carolina, USA. J Hydrol 248:198–215

Lee DR (1977) A device for measuring seepage flux in lakes and estuaries. Limn Oceanogr 22:140–147

Lee TM, Swancar A (1997) Influence of evaporation, ground water, and uncertainty in the hydrologic budget of Lake Lucerne, a seepage lake in Polk County, Florida. U.S. Geological Survey Water-Supply Paper 2439:61

Lee TM, Haag KH, Metz PA, Sacks LA (2009) Comparative hydrology, water quality, and ecology of selected natural and augmented freshwater wetlands in west-central Florida. U.S. Geological Survey Professional Paper 1758:152

Linsley RK Jr, Kohler MA, Paulhus JLH (1982) Hydrology for engineers. McGraw-Hill, New York

Marsalek J (1981) Calibration of the tipping-bucket rain gauge. J Hydrol 53:343–354

Masoner JR, Stannard DI (2010) A comparison of methods for estimating open-water evaporation in small wetlands. Wetlands 30:513–524

Masoner JR, Stannard DI, Christenson SC (2008) Differences in evaporation between a floating pan and class a pan on land. J Am Water Resour Assoc 44:552–561

McGuinness JL, Bordne EF (1972) A comparison of lysimeter-derived potential evapotranspiration with computed values. U.S. Department of Agriculture Agricultural Research Service Technical Bulletin 1452:69

Mitsch WJ, Gosselink JG (2007) Wetlands. Wiley, New York

Moore DA (2004) Construction of a Mariotte bottle for constant-rate tracer injection into small streams. Streamline 8:15–16

Morton FI (1983a) Operational estimates of areal evapotranspiration and their significance to the science and practice of hydrology. J Hydrol 66:1–76

Morton FI (1983b) Operational estimates of lake evaporation. J Hydrol 66:77–100

Motz LH, Sousa GD, Annable MD (2001) Water budget and vertical conductance for Lowry (Sand Hill) Lake in north, central Florida, USA. J Hydrol 250:134–148

Nitu R, Wong K (2010) CIMO survey on national summaries of methods and instruments for solid precipitation measurements at automatic weather stations. World Meteorological Organization Instruments and Observing Methods Report 102:57

Nystuen JA, Proni JR, Black PG, Wilkerson JC (1996) A comparison of automatic rain gauges. J Atmos Ocean Technol 13:62–73

Parkhurst RS, Winter TC, Rosenberry DO, Sturrock AM (1998) Evaporation from a small prairie wetland in the Cottonwood Lake area, North Dakota – an energy-budget study. Wetlands 18:272–287

Parsons DF, Hayashi M, van der Kamp G (2004) Infiltration and solute transport under a seasonal wetland: bromide tracer experiments in Saskatoon, Canada. Hydrol Process 18:2011–2027

Penman HL (1948) Natural evaporation from open water, bare soil, and grass. Proc R Soc Lond A193:120–145

Perez PJ, Castellvi F, Ibañez M, Rosell JI (1999) Assessment of reliability of Bowen ratio method for partitioning fluxes. Agric For Meteorol 97:141–150

Priestley CHM, Taylor RJ (1972) On the assessment of surface-heat flux and evaporation using large-scale parameters. Mon Weather Rev 100:81–92

Rantz SE (1982) Measurement and computation of streamflow, vol 2. Computation of discharge. U.S. Geological Survey Water-Supply Paper 2175:631

Rasmussen AH, Hondzo M, Stefan HG (1995) A test of several evaporation equations for water temperature simulations in lakes. J Am Water Resour Assoc 31:1023–1028

Rosenberry DO (2005) Integrating seepage heterogeneity with the use of ganged seepage meters. Limnol Oceanogr Methods 3:131–142

Rosenberry DO (2008) A seepage meter designed for use in flowing water. J Hydrol 359:118–130
Rosenberry DO (2011) The need to consider temporal variability when modeling exchange at the sediment-water interface. In: Nutzmann G (ed) Conceptual and modelling studies of integrated groundwater, surface water, and ecological systems. IAHS Press, Oxfordshire
Rosenberry DO, Morin RH (2004) Use of an electromagnetic seepage meter to investigate temporal variability in lake seepage. Ground Water 42:68–77
Rosenberry DO, Winter TC (1997) Dynamics of water-table fluctuations in an upland between two prairie-pothole wetlands in North Dakota. J Hydrol 191:266–269
Rosenberry DO, Winter TC (2009) Hydrologic processes and the water budget. In: Winter TC, Likends GE (eds) Mirror Lake: interactions among air, land, and water. University of California Press, Berkeley
Rosenberry DO, Sturrock AM, Winter TC (1993) Evaluation of the energy-budget method of determining evaporation at Williams Lake, Minnesota, using alternative instrumentation and study approaches. Water Resour Res 29:2473–2483
Rosenberry DO, Striegl RG, Hudson DC (2000) Plants as indicators of focused ground water discharge to a northern Minnesota lake. Ground Water 38:296–303
Rosenberry DO, Stannard DI, Winter TC, Martinez ML (2004) Comparison of 13 equations for determining evapotranspiration from a prairie wetland, Cottonwood Lake area, North Dakota, USA. Wetlands 24:483–497
Rosenberry DO, Winter TC, Buso DC, Likens GE (2007) Comparison of 15 evaporation methods applied to a small mountain lake in the northeastern USA. J Hydrol 340:149–166
Rosenberry DO, LaBaugh JW, Hunt RJ (2008) Use of monitoring wells, portable piezometers, and seepage meters to quantify flow between surface water and ground water. In: Rosenberry DO, LaBaugh JW (eds) Field techniques for estimating water fluxes between surface water and ground water, U.S. Geological Survey Techniques and Methods 4-D2, Denver
Rovey CW II, Cherkauer DS (1995) Scale dependency of hydraulic conductivity measurements. Ground Water 33:769–780
Sacks LA, Swancar A, Lee TM (1998) Estimating ground-water exchange with lakes using water-budget and chemical mass-balance approaches for ten lakes in ridge areas of Polk and Highlands Counties, Florida. U.S. Geological Survey Water-Resources Investigations Report 98-4133:52
Sauer VB, Turnipseed DP (2010) Stage measurement at gaging stations. U.S. Geological Survey Techniques and Methods 3-A7:45
Schmidt C, Conant B Jr, Bayer-Raich M, Schirmer M (2007) Evaluation and field-scale application of an analytical method to quantify groundwater discharge using mapped streambed temperatures. J Hydrol 347:292–307
Searcy JK, Hardison CH (1960) Double-mass curves. U.S. Geological Survey Water-Supply Paper 1541-B:66
Sebestyen SD, Schneider RL (2001) Dynamic temporal patterns of nearshore seepage flux in a headwater Adirondack lake. J Hydrol 247:137–150
Selker JS, Thévenaz L, Huwald H, Mallet A, Luxemburg W, van de Giesen N, Stejskal M, Zeman J, Westhoff M, Parlange MB (2006) Distributed fiber-optic temperature sensing for hydrologic systems. Water Resour Res 42:W12202
Shaw DA, Vanderkamp G, Conly FM, Pietroniro A, Martz L (2012) The fill-spill hydrology of prairie wetland complexes during drought and deluge. Hydrol Process, n/a-n/a
Shoemaker WB, Lopez CD, Duever M (2011) Evapotranspiration over spatially extensive plant communities in the Big Cypress National Preserve, southern Florida, 2007–2010. U.S. Geological Survey Scientific Investigations Report 2011–5212:46
Shook KR, Pomeroy JW (2011) Memory effects of depressional storage in Northern Prairie hydrology. Hydrol Process 25:3890–3898
Singh VP, Xu CY (1997) Evaluation and generalization of 13 mass-transfer equations for determining free water evaporation. Hydrol Process 11:311–323

Stallman RW (1965) Steady one-dimensional fluid flow in a semi-infinite porous medium with sinusoidal surface temperature. J Geophys Res 70:2821–2827

Stannard DI, Rosenberry DO, Winter TC, Parkhurst RS (2004) Estimates of fetch-induced errors in Bowen-ratio energy-budget measurements of evapotranspiration from a prairie wetland, Cottonwood Lake area, North Dakota, USA. Wetlands 24:498–513

Starr JL, Paltineanu IC (2002) Capacitance devices. In: Dane JH, Topp CC (eds) Methods of soil analysis part 4 – physical methods. Soil Science Society of America, Madison

Stonestrom DA, Constantz J (2003) Heat as a tool for studying the movement of ground water near streams: U.S. Geological Survey Circular 1260:96

Sturrock AM, Winter TC, Rosenberry DO (1992) Energy budget evaporation from Williams Lake: a closed lake in north central Minnesota. Water Resour Res 28:1605–1617

Surridge BWJ, Baird AJ, Heathwaite AL (2005) Evaluating the quality of hydraulic conductivity estimates from piezometer slug tests in peat. Hydrol Process 19:1227–1244

Swanson GA (1978) A water column sampler for invertebrates in shallow wetlands. J Wildl Manag 42:670–672

Swanson TE, Cardenas MB (2010) Diel heat transport within the hyporheic zone of a pool-riffle-pool sequence of a losing stream and evaluation of models for fluid flux estimation using heat. Limnol Oceanogr 55:1741–1754

Taylor JR (1982) An introduction to error analysis: the study of uncertainties in physical measurements. University Science, Mill Valley

Topp GC, Ferré PA (2002) Thermogravimetric using convective oven-drying. In: Dane JH, Topp CC (eds) Methods of soil analysis part 4 – physical method. Soil Science Society of America, Madison

Topp GC, Davis JL, Annan AP (1980) Electromagnetic determination of soil water content: measurements in coaxial transmission lines. Water Resour Res 16:574–582

Turcotte DL, Schubert G (1982) Geodynamics: applications of continuum physics to geological problems. Wiley, New York

Turnipseed DP, Sauer VB (2010) Discharge measurements at gaging stations. U.S. Geological Survey Techniques and Methods 3-A8:87

van der Kamp G, Hayashi M (2009) Groundwater-wetland ecosystem interaction in the semiarid glaciated plains of North America. Hydrogeol J 17:203–214

Vogt T, Schneider P, Hahn-Woernle L, Cirpka OA (2010) Estimation of seepage rates in a losing stream by means of fiber-optic high-resolution vertical temperature profiling. J Hydrol 380:154–164

Ward JR, Harr CA (1990) Methods for collection and processing of surface-water and bed-material samples for physical and chemical analysis. U.S. Geological Survey Open-File Report 90-140:71

Wilczak JM, Oncley SP, Stage SA (2001) Sonic anemometer tilt correction algorithms. Bound-Layer Meteorol 99:127–150

Wilde FD (2006) Collection of water samples. U.S. Geological Survey Techniques and Methods Book 9, Chapter A4:166

Wilson BW (1972) Seiches. In: Chow VT (ed) Advances in hydrosciences. Academic Press, New York

Winter TC (1981) Uncertainties in estimating the water balance of lakes. Water Resour Bull 17:82–115

Winter TC (1988) A conceptual framework for assessing cumulative impacts on the hydrology of nontidal wetlands. Environ Manag 12:605–620

Winter TC (1989) Hydrologic studies of wetlands in the northern prairie. In: van der Valk A (ed) Northern prairie wetlands. Iowa State University Press, Ames

Winter TC (1992) A physiographic and climatic framework for hydrologic studies of wetlands. In: Robards RD, Bothwell ML (eds) Aquatic ecosystems in semi-arid regions, implications for resource management. Environment Canada, Saskatoon

Winter TC, Rosenberry DO (1995) The interaction of ground water with prairie pothole wetlands in the Cottonwood Lake area, east-central North Dakota, 1979–1990. Wetlands 15:193–211

Winter TC, Rosenberry DO (1998) Hydrology of prairie pothole wetlands during drought and deluge: a 17-year study of the Cottonwood Lake wetland complex in North Dakota in the perspective of longer term measured and proxy hydrological records. Clim Chang 40:189–209

Winter TC, Woo MK (1990) Hydrology of lakes and wetlands. In: Wolman MG, Riggs HC (eds) Surface water hydrology, the geology of north America. Geological Society of America, Boulder

Winter TC, Rosenberry DO, Sturrock AM (1995) Evaluation of 11 equations for determining evaporation for a small lake in the north central United States. Water Resour Res 31:983–993

Winter TC, Buso DC, Rosenberry DO, Likens GE, Sturrock AMJ, Mau DP (2003) Evaporation determined by the energy budget method for Mirror Lake, New Hampshire. Limnol Oceanogr 48:995–1009

Wise WR, Annable MD, Walser JAE, Switt RS, Shaw DT (2000) A wetland-aquifer interaction test. J Hydrol 227:257–272

WMO (1994) Guide to hydrological practices, 5th edn. World Meteorological Organization WMO Publication 168:735

Yang D, Goodison BE, Metcalfe JR, Golubev VS, Bates R, Pangburn T, Hanson CL (1998) Accuracy of NWS 80 standard nonrecording precipitation gauge: results and application of WMO intercomparison. J Atmos Ocean Technol 15:54–68

Yeskis D, Zavala B (2002) Ground-water sampling guidelines for superfund and RCRA project managers. U.S. Environmental Protection Agency Ground Water Forum Issue Paper EPA 542-S-02-001

Student Exercises

Classroom Exercises

Short Exercise 1: Converting Pressure to Water Depth and Stage

Measuring wetland stage and hydraulic head, and determining direction and potential for flow between groundwater and surface water, are among the most basic requirements in wetland hydrology. A sketch of a common monitoring installation appears below (Fig. 3.34). A piezometer designed to indicate hydraulic head beneath the wetland bed is instrumented with a submersible pressure transducer. The sensor is suspended from the surface of the well casing by a metal wire. The distance from the attachment point to the sensor port commonly is described as the hung depth. This particular type of sensor stores the data on a circuit card; the sensor must be retrieved and the data downloaded periodically. Some installations instead have a data cable extending from the sensor to a datalogger that can query and store data from multiple sensors. In some models the cable contains a vent tube that allows changes in atmospheric pressure to be transmitted to the pressure sensor. Venting allows the pressure measurement to be relative to atmospheric pressure. The transducer in this example is not vented to the atmosphere; some would argue this is preferable because there is no associated opportunity for water vapor to reach and damage the sensor electronics. However, without

3 Assessing and Measuring Wetland Hydrology 179

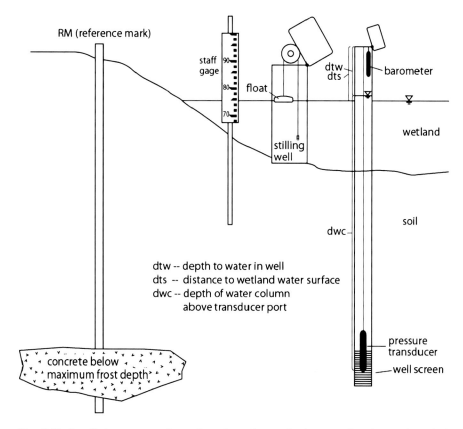

Fig. 3.34 Installations commonly used to determine wetland stage, elevation, and vertical hydraulic-head gradient

venting, the sensor output is the sum of hydrostatic pressure of the water column above the sensor port (the dwc or depth of the water column that we want to know) and atmospheric pressure. Therefore, atmospheric pressure needs to be measured and subtracted from the output of the submerged pressure transducer to obtain the height of the water column above the submerged sensor. A barometer is suspended in the piezometer casing, well above the water level, to provide atmospheric-pressure measurements. If the well is susceptible to occasional flooding, the barometer could instead be located anywhere nearby as atmospheric pressure does not change appreciably over distances of several km.

Output from pressure transducers, as well as many other sensors, commonly is converted to units in which field check measurements are made. In wetland settings, that unit usually is feet or meters of water head. Meters will be used here. To convert output in pressure to head, recall that Pressure $= \rho g h$ where ρ is density of water (kg m^{-3}), g is acceleration due to gravity (m s^{-2}), and h, hydraulic head, is the height to of a column of liquid that would exert a given pressure, in m. Output from pressure transducers commonly is in units of Pascals. Recall that a Pascal is a

Newton per square meter and that a Newton, a unit of force, is determined in terms of mass times acceleration (kg m s^{-2}). Therefore,

$$h = \frac{P_{trans} - P_{bar}}{\rho g} + \mathit{Offset} \tag{3.55}$$

where P_{trans} is the output from the submerged pressure transducer, P_{bar} is the output from the barometer, and *Offset* is a value that equates the sensor output to a local datum or reference elevation.

A stilling well also is displayed in the drawing. Although another submerged pressure transducer could have been used to indicate wetland stage, this stilling well contains a float and counterweight that together rotate a pulley connected to a potentiometer or pulse-counting device. As water level changes, the float moves and the pulley rotates, changing either the electrical resistance if the sensor is a potentiometer, or causing electrical pulses to be sent to a data recording unit if the sensor is a pulse-counting device (often called a shaft encoder). The output of the sensor in the stilling well commonly is set to be equal to the water level indicated by a nearby staff gage.

The staff gage is connected to a metal pipe driven into the wetland bed. This simple device is designed to provide a direct indication of the relative stage of the wetland. The units on the "staff plate" in this example are in meters, but units of feet are perhaps more common in the US. Some wetland sediments are relatively soft, and some wetlands freeze during winter, providing the potential for the staff gage to move over time. To determine whether this occurs or not, we need a stable reference point to which the staff gage can be compared; hence, the reference mark, commonly called an RM. The term RM is used so as to not confuse it with BM (bench mark), which is an official surveying location that is part of a national geodetic survey. This particular RM consists of a pipe that extends into the ground. However, in areas where soil frost is common and can extend a meter or more beneath ground surface, pipes also can move. Therefore, this particular RM was set in a mass of concrete that was installed beneath the deepest expected extent of soil frost.

Our tasks here are to:

1. compare the potentiometer output from the stilling well to the output from the submerged pressure transducer in common units,
2. make separate measurements of water levels inside of the well and of the wetland surface,
3. determine the difference in hydraulic head (Δh) between the wetland and the piezometer, and
4. verify that our sensors are providing the correct output.

3   Assessing and Measuring Wetland Hydrology   181

Field site data

Staff gage	0.750 m (manually read)
Potentiometer	0.755 m
dts	0.198 m (manually measured)
dtw	0.178 m (manually measured)
Barometer	100.510 kPa
Pressure transducer	110.610 kPa
Pressure-transducer offset	−0.250 m

1. What is the dwc in m of water? Assume fresh water at 20 °C. (therefore, density = 998 kg/m^3)_____
2. What does the pressure transducer indicate for head in the piezometer in m relative to the local datum?_____
3. What do the sensors indicate for Δh?_____
4. What is the manually measured Δh?_____
5. Is the potential for flow upward or downward based on the measured values? _____
6. How does the Δh indicated by the sensors differ from the Δh calculated from the manual measurements?_____
7. What is the gradient assuming the midpoint of the well screen is 0.75 m below the wetland bottom?_____
8. If the top of the staff gage plate is at an elevation of 102.550 m, what is the elevation of the water level inside of the piezometer?_____

Short Exercise 2: Wind Correction of Precipitation Data

Table 3.3 shows daily mean air temperature and wind speed, and daily total precipitation recorded by a weighing precipitation gauge with an Alter wind shield (similar to Fig. 3.5a), at a hydrological research station in Calgary, Alberta, Canada, in 2008. There were two precipitation events, on December 7 and 12.

1. Based on the air temperature, determine the form of precipitation (rain or snow).
2. If the precipitation occurs as snow, then a correction must be made to account for the gage-catch deficiency (see Fig. 3.6). Use the following equation (Dingman 2002:111–112) to compute the catch deficiency factor (CD) from wind speed (u, m s^{-1}) for each day.

$$CD = 100 \exp\left(-4.61 - 0.036 u^{1.75}\right) \quad (3.56)$$

3. Divide the uncorrected precipitation by CD to estimated true (i.e., corrected) precipitation.
4. Calculate the total of two precipitation events for both uncorrected and corrected data. What is the degree (percentage) of underestimate by not correcting the data?
5. Many winter precipitation data sets available on the internet have not been corrected. Discuss the potential problem of using such data for a water-budget analysis.

Table 3.3 Daily mean air temperature and wind speed, and daily total precipitation

Date	Air temp. (°C)	Wind spd. (m s^{-1})	Recorded pcp. (mm)	CD	Corrected pcp. (mm)
Dec. 7	−1.4	1.7	13	—	—
Dec. 12	−5.4	3.7	17	—	—
Total			—		—

3 Assessing and Measuring Wetland Hydrology

Short Exercise 3: Spatial Interpolation of Precipitation Data

Table 3.4 shows monthly total precipitation (mm) at three meteorological stations in Alberta, Canada. Olds Station is located between two other stations, approximately 50 km south of Red Deer and 70 km north of Calgary. The first three columns list the long term average for 1971–2000; the last three columns list the data recorded in 2010. The 2010 data for Olds are missing.

1. Using the normal ratio method (Eq. 3.7), estimate monthly total precipitation in Olds for the three missing months.
2. Actual precipitation data recorded at the Olds station were 77 mm for June, 85 mm for July, and 79 mm for August. Discuss the magnitude of uncertainty associated with this method.

Table 3.4 Long-term average monthly precipitation and 2010 monthly precipitation (mm) at three meteorological stations in Alberta, Canada

	1971–2000 average			2010		
	June	July	Aug.	June	July	Aug.
Red Deer	84	92	70	138	144	62
Calgary	80	68	59	64	66	87
Olds	90	87	65	—	—	—

Data source: Environment Canada National Climate Data and Archive (http://climate.weatheroffice.gc.ca/climateData/canada_e.html)

Short Exercise 4: Calculation of Discharge from Tracer Data

Tracer dilution methods were used to estimate the discharge of two small streams flowing into a wetland. The constant injection method was used in the first stream, where chloride solution having a concentration of 60 g L^{-1} was injected at a rate of 12 L min^{-1}. The tracer concentration in the stream reached a steady value of 100 mg L^{-1} by 150 s after the start of injection (Fig. 3.35). The background chloride concentration in the stream was 1 mg L^{-1}.

1. Using Eq. 3.23, estimate the stream discharge from concentration data.

The slug injection method was used in the second stream, where 10 L of tracer solution containing 3 kg of chloride mass was instantaneously injected in the stream. The tracer concentration reached a peak about 40 s after the release and declined quickly afterwards (Fig. 3.35). The background chloride concentration in the stream was 2 mg L^{-1}.
Concentration data are listed in Table 3.5.

2. Using Eq. 3.25 with $\Delta t = 10$ s, estimate the integral in the denominator of Eq. 3.24.
3. Using Eq. 3.24 with $C_1 V_1 = 3$ kg, estimate the stream discharge.

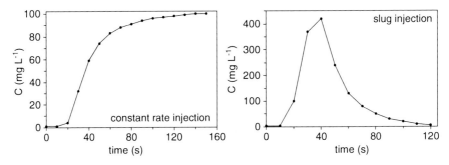

Fig. 3.35 Concentration of chloride tracer in streams. *Left*: constant-rate injection test. *Right*: slug injection test

Table 3.5 Data for slug injection test

t (s)	C (mg L^{-1})	$(C - C_b)\Delta t$ (kg m^{-3} s)	t (s)	C (mg L^{-1})	$(C - C_b)\Delta t$ (kg m^{-3} s)
0	2	____	70	80	____
10	2	____	80	50	____
20	100	____	90	30	____
30	370	____	100	20	____
40	420	____	110	10	____
50	240	____	120	4	____
60	130	____		Total =	____

Short Exercise 5: Calibration of Weir Coefficient

V-notch weirs provide stable and reliable flow measurements, particularly when the coefficient C in the weir formula (Eq. 3.28) is determined to reflect site-specific conditions. Table 3.6 lists measurements of water level (h) and discharge (Q) for the V-notch weir shown in Fig. 3.11b. The water level is measured with respect to the base of the weir. Therefore, $h_0 = 0$ in Eq. 3.28.

1. Compute $h^{5/2}$ and convert Q to m^3 s^{-1}.
2. Plot $h^{5/2}$ and Q in the graph and determine the slope of the plot.
3. Determine C in Eq. 3.28. Note that $\theta = 90°$; thus, $\tan(\theta/2) = 1$. Compare this value to the theoretical value for an ideal weir, $C - 1.38$.

Table 3.6 Water level (h) in a 90° V-notch weir and independently measured discharge (Q) in the weir shown in Fig. 3.11

Date	h (m)	Q (L s^{-1})	$h^{5/2}$ (m$^{5/2}$)	Q (m^3 s^{-1})
Jun. 18, 2009	0.055	0.8		
Aug. 7, 2009	0.096	3.6		
Oct. 30, 2009	0.114	5.5		
May 28, 2010	0.144	9.7		
Jul. 8, 2010	0.156	12.1		
Oct. 5, 2010	0.132	7.5		

Short Exercise 6: Determination of Stage-Discharge Rating Curve

Coefficients for the stage-discharge rating curve (Eq. 3.26) of a stream gauging station can be determined from a series of measurements of stage (h) and discharge (Q) encompassing different flow conditions. Table 3.7 lists the measured h and Q in a small stream in Calgary, Alberta, Canada. The stage at zero flow (h_0) is 0.35 m at this gauging station. Equation 3.28 can be written in a logarithmic form

$$\log Q = \log a + m \log(h - h_0) \qquad (3.57)$$

When the logarithms of data are used to fit a straight line, the intercept and slope of the line give $\log a$ and m, respectively.

1. Compute $\log(h - h_0)$ and $\log Q$ for each measurement.
2. Plot $\log(h - h_0)$ and $\log Q$ in the graph and fit a straight line.
3. Determine the intercept and the slope of the plot, and compute a and m.

Table 3.7 Water stage (h) and discharge (Q) measured in a small stream near Calgary, Alberta, Canada in 2011

Date	h (m)	Q (m³ s⁻¹)	$\log(h - h_0)$	$\log Q$
June 9	0.65	0.56	___	___
June 14	0.59	0.46	___	___
June 21	0.88	1.10	___	___
June 28	0.59	0.45	___	___
July 6	0.50	0.27	___	___
July 13	0.53	0.29	___	___
July 26	0.52	0.24	___	___
Aug. 8	0.47	0.15	___	___
Aug. 24	0.44	0.11	___	___

Short Exercise 7: Estimation of Diffuse Overland Flow

The amount of diffuse overland flow can be estimated using a wetland as a natural overland flow trap. If the wetland does not have inflow or outflow streams, and the contribution of groundwater flow is negligible during a short-duration storm, then the water balance equation for the wetland pond is given by Eq. 3.32. Total overland flow during the storm (O_{ftot}) is estimated from measuring the volume of pond water before (V_{ini}) and after (V_{fin}) the storm. The figure embedded in Table 3.8 shows the pond stage and cumulative precipitation in Wetland 109 in the St. Denis National Wildlife Area in Saskatchewan, Canada, on July 4–5, 1996 (see Hayashi et al. 1998 for a site description). The cumulative precipitation (p_{cum}) during the entire storm was 51 mm. The pond stages recorded at 21:00 and 02:00 are listed in Table 3.8. Water depth (H) at the deepest point in the pond is given by subtracting 551.68 m from the pond stage. The area of pond surface (A) and the volume of pond water (V) can be estimated using Eqs. 3.4 and 3.5 with $s = 3{,}180$ m^2 and $p = 1.61$ (Hayashi and van der Kamp 2000). The effective drainage area (A_{eff}) of Wetland 109 is 20,100 m^2.

1. Calculate the initial (21:00) and final (02:00) pond area and volume from the stage data.
2. Calculate the total amount of precipitation (P_{tot}) falling within the pond by multiplying p_{cum} by the pond area (A_{fin}) at 02:00.
3. Using Eq. 3.32, determine O_{ftot}.
4. Runoff-contributing area to the pond is given by $A_{eff} - A_{fin}$. From O_{ftot}, estimate the areal average runoff (mm) in the contributing area.
5. Estimate the runoff coefficient ($R_c =$ runoff/precipitation) for this storm.

Table 3.8 Pond stage in Wetland 109 in the St. Denis National Wildlife Area, Saskatchewan, Canada on July 4, 1996

Time	Stage (m)	H (m)	A (m^2)	V (m^3)
21:00	552.41			
02:00	552.62			

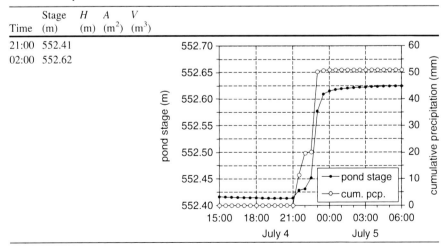

Short Exercise 8: Calculation of Groundwater Flow Using the Segmented-Darcy Method

The segmented-Darcy approach shown in Fig. 3.21 provides values for Q_{In} and Q_{Out} that are based on data from monitoring wells and wetland stage. The figure below (Fig. 3.36) is identical to Fig. 3.21 but heads for three of the wells are changed slightly. Use the data shown in Fig. 3.36, along with the assumptions that K is 30 m/day and b is 20 m, to fill out the data in Table 3.9. Sum the positive values to determine Q_{In} and sum the negative values to determine Q_{Out}. Then answer the following questions.

1. Where is the greatest rate of exchange (Q/A) between groundwater and the wetland? Why?
2. A hinge line is a point along a shoreline that separates a shoreline reach where groundwater discharges to the wetland from a shoreline reach where wetland water flows to the groundwater system. What are the approximate locations of the hingelines?
3. If there is no surface-water exchange with the wetland, and overland flow is negligible, what does this analysis tell you about the other terms of the water budget?

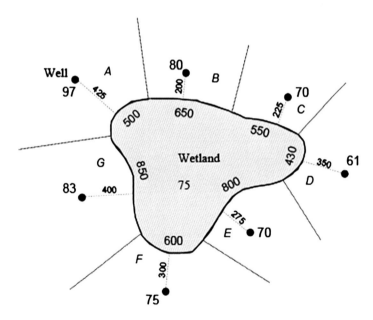

Fig. 3.36 The same wetland setting shown in Fig. 3.21 but with several different head values. Figure legend is shown in Fig. 3.21

Table 3.9 Parameters needed to determine Q_{In} and Q_{Out} using the segmented-Darcy approach

Watershed segment	Horizontal hydraulic conductivity (K), in m/d	Effective thickness of the aquifer (b), in m	Hydraulic head in well minus surface-water stage (h_1-h_2), in m	Distance from the well to the shoreline (L), in m	Length of shoreline segment (m), in m	Water flow (Q), in m^3/d
A						
B						
C						
D						
E						
F						
G						
					ΣQ_{In} =	
					ΣQ_{Out} =	
					In - Out =	
					% imbalance =	

Short Exercise 9: Simple Flow-Net Analysis

We do not need a sophisticated numerical model to give us a good first estimate of groundwater flows to and from wetlands. Reasonable values for exchange between groundwater and a wetland can be calculated with: (1) a map showing the locations of a few monitoring wells and their hydraulic-head values, (2) a value for stage of the wetland, and (3) estimates of hydraulic conductivity. In this brief exercise you will make a flow-net analysis to determine flow between groundwater and a wetland and also compare those values with values that were obtained with the segmented-Darcy approach in short exercise SE 8.

The flow-net analysis is a graphical approach for determining 2-dimensional groundwater flow. The Darcy equation is used to solve for flow through individual "stream tubes" that are drawn based on contour lines drawn from head data. The method assumes steady-state flow is two-dimensional. The flow net can be drawn in plain view, as we did with SE 8, or in cross-sectional view. We will assume that the aquifer is homogeneous and isotropic, although modifications can be made when drawing the flow net if the aquifer is known to be anisotropic. A brief description of how to draw a flow net follows. More detail can be found in Fetter Jr. (2001) and Cedergren (1997).

A flow net consists of equipotential lines (contour lines of equal hydraulic head) that are drawn perpendicular to flow lines that indicate the direction of groundwater flow. The net is bounded by no-flow boundaries or constant-head boundaries. The equipotential lines intersect no-flow boundaries at right angles and the flow lines intersect constant-head boundaries, if present, also at approximately right angles. A simple example is shown in Fig. 3.37. Equipotential head drops consist of the area

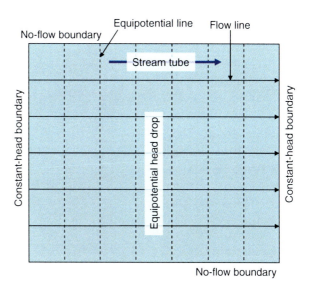

Fig. 3.37 Diagram of a simple rectangular flow net showing boundary conditions, equipotential lines, and stream tubes

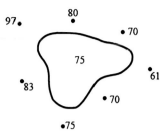

Fig. 3.38 Draw contour lines based on the heads displayed at the monitoring wells and the wetland stage

of the flow net bounded by adjacent equipotential lines and stream tubes consist of the area of the flow net bounded by adjacent flow lines.

The example in Fig. 3.37 contains seven equipotential head drops and six stream tubes. The flow-net equation can be written as

$$Q = \frac{MKbH}{n} \qquad (3.58)$$

where M is the number of stream tubes, n is the number of equipotential head drops, K is the assumed hydraulic conductivity, b is the sediment thickness in the third dimension, and H is the total head drop across the flow net. M is commonly presented as m in most texts, but we use upper-case M here to distinguish it from m, the shoreline length presented earlier in Fig. 3.21. Q is in units of volume per time.

Some basic steps to follow are:

1. Determine boundaries and boundary conditions,
2. Draw equipotential lines by contouring head data from wells and wetland stage,
3. Draw flow lines to create approximate squares (you should be able to draw a circle bounded by the equipotential lines and flow lines),
4. Flow lines cross equipotential lines at right angles (assuming we have isotropic conditions) and flow lines also intersect constant-head boundaries at right angles,
5. You can draw half-equipotential lines for areas with smaller gradients.
6. Five to ten flow lines usually are sufficient,
7. Count up stream tubes and equipotential drops to determine M and n,
8. Determine H, and estimate b and K.
9. Calculate Q for flow to and/or from the wetland.

Let's see how well this can work. The same wetland setting in Short Exercise 8 is displayed in Fig. 3.38. This is the same wetland shown in Fig. 3.21 but with head values changed for three of the seven wells. Your task will be to determine the extent to which changes in head will affect the interpretation of flow of groundwater to and from the wetland. Draw contour lines based on the head data and then draw flow lines based on the instructions provided above. After that, you will count up flow tubes and head drops and calculate flow to the wetland and flow from the wetland. Use K and b values from Short Exercise 8. You will then be able to answer the following questions:

1. How does flow to the wetland compare to flow from the wetland? If the values are different, why are they different?
2. How do the values for flow to the wetland and flow from the wetland compare to those you obtained with the segmented-Darcy approach? Which method do you prefer? Which method provides more realistic results? What might be sources of error for both methods?
3. How do the flowlines you have drawn compare with the flowlines shown in Fig. 3.22? What effect do the different head values have on the positioning of the hinge lines?

References

Cedergren HR (1997) Seepage drainage and flow nets, 3rd edn. Wiley, New York

Fetter CW Jr (2001) Applied hydrogeology, 4th edn. Prentice Hall, Upper Saddle River

Short Exercise 10: Measurement of Groundwater Flow Using a Half-Barrel Seepage Meter

Seepage meters were used to quantify rates and distribution of exceptionally fast flow through a lake bed (Rosenberry 2005). In this exercise you will use data from that report to determine groundwater-surface-water exchange and also compare standard flow measurements with those based on connecting multiple seepage cylinders to a single seepage bag.

Mirror Lake is a small, 10-ha lake in the White Mountains of New Hampshire. A dam built in 1900 raised the lake level by about 1.5 m, increasing the lake surface area and inundating what had previously been dry land. Water leaks out of the lake through a portion of the southern shoreline that, because of the stage rise following dam construction, has been covered by water for only about 110 years. More water is lost as seepage to groundwater than from the lake surface-water outlet (Rosenberry et al. 1999). Seepage meters were used to determine where rapid rates of seepage were occurring and to determine the rates of seepage from the lake to groundwater.

Data shown in Table 3.10 were collected from 18 seepage meters that were installed in the area shown in Fig. 3.39. The photo inset shows the locations of some of the seepage cylinders that were installed prior to the installation of seepage bags and associated bag-connection hardware. Most of the measurements were made from standard seepage meters similar to Fig. 3.25. However, two sets of measurements were made from four seepage cylinders that were all connected (ganged) to one seepage bag. Your task is to fill in the missing data in Table 3.10 for meters 3 and 13 and then answer the following questions. To convert from ml/min to cm/day you will assume that 1 ml = 1 cm^3 of water. You will divide your result in cm^3/min by the area covered by the seepage cylinder (2,550 cm^2) and then multiply by the number of minutes in a day to obtain units in cm/day.

1. What are the averages of seepage measurements made at each of meters 3, 4, 5, and 6? Values for 4, 5, and 6 are already provided. What is the range in seepage rates at these 4 m? How does the variability in seepage among these 4 m compare with the ranges of values at each meter based on repeat measurements?
2. Repeat this analysis for meters 13, 17, 18, and 20. How do these seepage rates compare with meters 3 through 6? How does the range in seepage among meters compare with the ranges of measurements at individual meters?
3. Calculate average values for the two sets of ganged measurements (13, 17, 18, 20 and 3, 4, 5, 6). How do these values compare with the sums of seepage rates based on measurements made at individual meters? What can you say about summed versus ganged measurements for areas of slow versus fast seepage?

Table 3.10 Values collected from Mirror Lake, NH, during July 16–18, 2002

Seepage measurements at Mirror Lake, Campton, New Hampshire

Dates: 7/16/02–7/19/02   Meter area = 2,550 cm^2

Date	Meter	V1 (ml)	V2 (ml)	T1	T2	ΔV (ml)	Δt (min)	ΔV/Δt (ml/min)	ΔV/Δt (cm/d)	Ave.
7/16/02	3	1,000	675	9:26	9:42					
7/16/02	3	1,000	200	9:46	10:28					
7/16/02	3	1,000	860	13:14	13:21					
7/16/02	3	1,000	720	16:07	16:21					
7/16/02	3	1,000	650	16:29	16:47					−4.2
7/16/02	4	1,000	910	9:17	9:30	−90	13	−6.9	−3.9	
7/16/02	4	1,000	780	9:35	10:04	−220	29	−7.6	−4.3	
7/16/02	4	1,000	710	10:24	11:07	−290	43	−6.7	−3.8	
7/16/02	4	1,000	540	11:08	12:12	−460	64	−7.2	−4.1	
7/16/02	4	1,000	625	12:13	12:57	−375	44	−8.5	−4.8	
7/16/02	4	1,000	910	14:55	15:06	−90	1	−8.2	−4.6	
7/16/02	4	1,000	910	16:07	16:24	−90	14	−6.4	−3.6	
7/16/02	4	1,000	870	16:29	16:47	−130	18	−7.2	−4.1	
7/16/02	5	1,000	970	9:42	10:14	−30	32	−0.9	−0.5	
7/16/02	5	1,000	995	10:19	10:49	−5	30	−0.2	−0.1	
7/16/02	5	1,000	940	10:50	11:28	−60	38	−1.6	−0.9	
7/16/02	5	1,000	905	11:29	12:49	−95	80	−1.2	−0.7	
7/16/02	5	1,000	890	13:19	14:57	−110	98	−1.1	−0.6	
7/16/02	5	1,000	990	16:04	16:14	−10	10	−1.0	−0.6	
7/16/02	5	1,000	980	16:27	16:41	−20	14	−1.4	−0.8	−0.6
7/16/02	6	1,000	560	13:19	14:47	−440	88	−5.0	−2.8	
7/16/02	6	1,000	970	16:04	16:14	−30	10	−3.0	−1.7	
7/16/02	6	1,000	940	16:27	16:41	−60	14	−4.3	−2.4	−2.3
7/17/02	13	1,000	210	17:34	17:37					
7/17/02	13	1,000	465	17:43	17:45					
7/17/02	13	1,000	440	18:30	18:52					
7/18/02	17	1,000	800	9:45	9:51	−200	6	−33.3	−18.8	
7/18/02	17	1,000	830	10:09	10:13	−170	4	−42.5	−24.0	

(continued)

3 Assessing and Measuring Wetland Hydrology

Table 3.10 (continued)

Seepage measurements at Mirror Lake, Campton, New Hampshire

Dates: 7/16/02–7/19/02   Meter area = 2,550 cm^2

Date	Meter	V1 (ml)	V2 (ml)	T1	T2	ΔV (ml)	Δt (min)	ΔV/Δt (ml/min)	ΔV/Δt (cm/d)	Ave.
7/18/02	17	1,000	840	10:24	10:29	−160	5	−32.0	−18.1	−20.4
7/18/02	17	1,000	930	10:49	10:51	−70	2	−35.0	−19.8	
7/18/02	17	1,000	940	15:14	15:16	−60	2	−30.0	−16.9	
7/18/02	17	1,000	960	16:31	16:32	−40	1	−40.0	−22.6	
7/18/02	17	1,000	920	16:36	16:38	−80	2	−40.0	−20.6	
7/18/02	18	1,000	280	9:48	9:56	−720	8	−90.0	−50.8	−53.5
7/18/02	18	1,000	550	10:09	10:13	−450	4	−112.5	−63.5	
7/18/02	18	1,000	490	10:24	10:29	−510	5	−102.0	−57.6	
7/18/02	18	1,000	770	10:49	10:51	−230	2	−115.0	−64.9	
7/18/02	18	1,000	510	10:58	11:03	−490	5	−98.0	−55.3	
7/18/02	18	1,000	840	15:19	15:21	−160	2	−80.0	−45.2	
7/18/02	18	1,000	920	16:31	16:32	−80	1	−80.0	−45.2	
7/18/02	18	1,000	840	16:36	16:38	−160	2	−80.0	−45.2	
7/18/02	20	1,000	900	11:05	11:07	−100	2	−50.0	−28.2	−28.9
7/18/02	20	1,000	800	11:26	11:30	−200	4	−50.0	−28.2	
7/18/02	20	1,000	900	15:08	15:10	−100	2	−50.0	−28.2	
7/18/02	20	1,000	945	16:08	16:09	−55	1	−55.0	−31.1	
7/18/02	13,17,18,20	1,000	670	15:27	15:28	−330	1	−330.0	−46.6	−47.6
7/18/02	13,17,18,20	1,000	650	15:32	15:33	−350	1	−350.0	−49.4	
7/18/02	13,17,18,20	1,000	670	15:37	15:38	−330	1	−330.0	−46.6	
7/16/02	13,17,18,20	1,000	660	15:41	15:42	−340	1	−340.0	−48.0	
7/16/02	3,4,5,6	1,000	670	15:17	15:27	−330	10	−33.0	−4.7	−4.5
7/16/02	3,4,5,6	1,000	670	15:46	15:56	−330	10	−33.0	−4.7	
7/17/02	3,4,5,6	1,000	680	16:56	17:06	−320	10	−32.0	−4.5	
7/17/02	3,4,5,6	1,000	680	17:07	17:17	−320	10	−32.0	−4.5	
7/17/02	3,4,5,6	1,000	690	18:14	18:24	−310	10	−31.0	−4.4	

3 Assessing and Measuring Wetland Hydrology

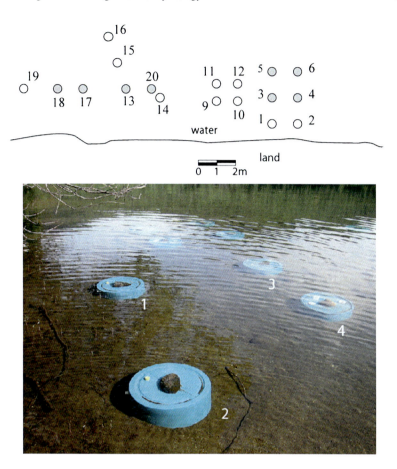

Fig. 3.39 Distribution of seepage meters installed in Mirror Lake, New Hampshire, USA. Seepage cylinders that were ganged for a single, integrated measurement are shown by *shaded circles*. *Numbers* in the photo inset correspond to the numbered seepage meters in the drawing. Note the rocks positioned on top of the seepage cylinders to counteract the buoyancy of the plastic cylinders, and that bag shelters have not yet been attached to the seepage cylinders

References

Rosenberry DO (2005) Integrating seepage heterogeneity with the use of ganged seepage meters. Limnol Oceanogr Methods 3:131–142

Rosenberry DO, Bukaveckas PA, Buso DC, Likens GE, Shapiro AM, Winter TC (1999) Migration of road salt to a small New Hampshire lake. Water Air Soil Pollut 109:179–206

Short Exercise 11: Estimation of Seepage Flux Using Temperature Data

Diurnal oscillation of temperature in wetland-bed sediments can be used to estimate groundwater seepage flux based on mathematical analysis of vertical heat transfer. When the temperature at the sediment-water interface oscillates in a sinusoidal manner with a fixed period (τ), (here we will assume 1 day), and amplitude A_0 (°C), then the temperature T (°C) of the sediment at depth z (m) is given by:

$$T(z,t) = T_m(z) + A_0 \exp(-az) \sin(2\pi t/\tau - bz) \tag{3.59}$$

where $T_m(z)$ is the time-averaged temperature profile representing the effects of a long-term temperature gradient, t is time, and a (m^{-1}) and b (m^{-1}) are constants defined by the thermal properties of the sediment and the magnitude and direction of seepage flux (Stallman 1965, equation 4; Keery et al. 2007, equation 2).

Equation 3.59 indicates that the amplitude of oscillation decreases with depth, and the phase delay of the sinusoidal signal increases with depth. Both amplitude and phase delay are dependent on the thermal properties of the saturated sediment and seepage flux. Suppose that the data recorded at two temperature sensors located at depth z_1 and z_2 ($z_1 < z_2$) have amplitudes of A_1 and A_2, and a phase shift (i.e., time difference of peak temperatures between two depths) of Δt (s). Seepage flux q (m s^{-1}) is positive for downward seepage in this example, which is the opposite of its definition elsewhere in this chapter. Seepage is defined this way in this exercise to be consistent with the construct used by Keery et al. (2007). Seepage flux is related to temperature amplitude by (Keery et al. 2007):

$$\frac{H^3 D}{4(z_2-z_1)} q^3 - \frac{5H^2 D^2}{4(z_2-z_1)^2} q^2 + \frac{2HD^3}{(z_2-z_1)^3} q + \left(\frac{\pi^2 c^2 \rho^2}{\lambda_e^2 \tau^2} - \frac{D^4}{(z_2-z_1)^4} \right) = 0 \tag{3.60}$$

where c (J kg^{-1} °K^{-1}) and ρ (kg m^{-3}) are the specific heat capacity and density, respectively, of bulk sediment, λ_e is the effective thermal conductivity of bulk sediment, and c_w (J kg^{-1} °K^{-1}) and ρ_w (kg m^{-3}) are the specific heat capacity and density, respectively, of water. In addition,

$$H = c_w \rho_w / \lambda_e \quad \text{and} \quad D = \ln(A_1/A_2) \tag{3.61}$$

It also follows that the magnitude of q is related to Δt by (Keery et al. 2007):

$$|q| = \sqrt{\frac{c^2 \rho^2 (z_2-z_1)^2}{\Delta t^2 c_w^2 \rho_w^2} - \frac{16 \pi^2 \Delta t^2 \lambda_e^2}{\tau^2 (z_2-z_1)^2 c_w^2 \rho_w^2}} \tag{3.62}$$

Therefore, q can be estimated from the analysis of temperature signals using Eqs. 3.60, 3.61 and 3.62.

3 Assessing and Measuring Wetland Hydrology

Table 3.11 Temperature measured in sandy sediments underlying a wetland at depths of 0.2 m and 0.4 m over a period of 2 days

	0.2 m	0.4 m
T_{max} Day 1		
T_{min} Day 1		
T_{max} Day 2		
T_{min} Day 2		
Peak time Day 1		
Peak time Day 2		
Amplitude, 0.2 m =	(°C)	
Amplitude, 0.4 m =		(°C)
$\Delta t =$	(h)	(s)

Accurate estimates of q using this method requires pre-processing the signals using Fourier transform or a dynamic harmonic regression algorithm (Keery et al. 2007; Gordon et al. 2012). In this exercise, a simple graphical technique is used for demonstration purposes.

The figure embedded in Table 3.11 shows the temperature data collected in sandy sediments underlying a wetland.

1. Record the maximum and minimum temperature recorded on Day 1 for the 0.2 and 0.4 m sensor depths and enter the values in Table 3.11. Repeat the procedure for Day 2.
2. Record the time of peak temperature on Day 1 at 0.2 and 0.4 m depths and enter the values in the table. Repeat the procedure for Day 2.
3. Estimate the average amplitude of temperature oscillation by calculating $(T_{max} - T_{min})/2$ and taking the average of the 2 days.
4. Estimate the average phase shift Δt by calculating the difference in peak time for each day and taking the average of the 2 days.
5. Calculate D and H in Eq. 3.61 assuming: $c_w = 4{,}160$ J kg^{-1} °K^{-1}, $\rho_w = 1{,}000$ kg m^{-3}, and $\lambda_e = 2.0$ W m^{-1} °K^{-1}.
6. Calculate all constants in Eq. 3.60 assuming $c = 1{,}400$ J kg^{-1} °K^{-1}, $\rho = 2{,}000$ kg m^{-3}. Note that the period of oscillation τ is 86,400 s (24 h).
7. Solve Eq. 3.60 for q. The third-order polynomial equation has three roots, but only one is a real number. Various numerical tools are available; for example, MATLAB[2] software or its freeware equivalents have a line command for solving polynomial equations. The solution also can be obtained graphically by treating the left hand side of Eq. 3.60 as a polynomial function $f(q)$ and

---

[2] Any use of trade, firm, or product names is for descriptive purposes only and does not imply endorsement by the U.S. Government.

plotting $f(q)$ against q on the graph below. Starting with $q = 1 \times 10^{-6} \text{ms}^{-1}$, keep plotting $f(q)$ for increasing values of q until $f(q) = 0$ is reached, which is the solution. A positive value of q indicates downward flow, and a negative value upward flow.
8. Calculate the magnitude of q using Eq. 3.62 and check the consistency of the values calculated from Eqs. 3.60 and 3.62.

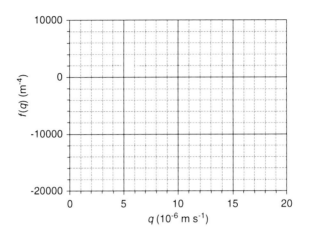

References

Gordon RP, Lautz LK, Briggs MA, McKenzie JM (2012) Automated calculation of vertical pore-water flux from field temperature time series using the VFLUX method and computer program. J Hydrol 420–421:142–158

Keery J, Binley A, Crook N, Smith JWN (2007) Temporal and spatial variability of groundwater–surface water fluxes: Development and application of an analytical method using temperature time series. J Hydrol 336:1–16

Stallman RW (1965) Steady one-dimensional fluid flow in a semi-infinite porous medium with sinusoidal surface temperature. J Geophys Res 70:2821–2827

Short Exercise 12: Estimation of Specific Yield

When inflow to and outflow from a wetland containing no surface water are negligible over a short-duration storm, the change in subsurface storage (ΔS_{sub}) is approximately equal to the net vertical input or loss of water from the wetland ($P - E$) (see Eq. 3.48 and the associated paragraph). Assuming that E is much smaller than P during the storm, specific yield can be estimated as the proportionality constant between ΔS_{sub} ($\cong P$) and increases in the water table (Δh) caused by storms:

$$\Delta S_{sub} = S_y \Delta h \tag{3.63}$$

The figure embedded in Table 3.12 below shows the water-table elevation recorded beneath Wetland 109 in the St. Denis National Wildlife Area in Saskatchewan, Canada (see Hayashi et al. 1998 for the site condition), in July-August 1995 when the water table was mostly below the sediment surface (551.68 m). During this period, there were five storms that caused measurable increases in the water table without bringing it to the surface (see Table 3.12 below).

Table 3.12 Total precipitation and water-table increases during storms recorded in July-August 1995 at Wetland 109. The graph shows the water-table elevation and cumulative precipitation

Date	P (mm)	Δh (mm)
July 30	7.8	193
Aug. 7	10.1	188
Aug. 8	7.6	149
Aug. 26	3.8	94
Aug. 29	10.8	240

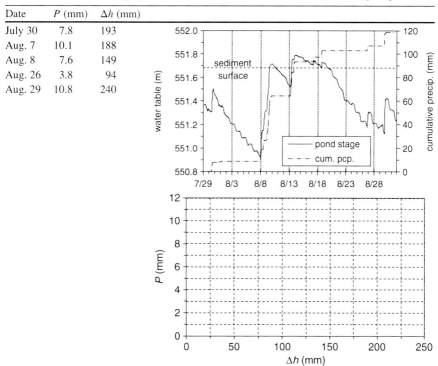

1. Plot P and Δh in the graph.
2. Draw a straight line that goes through the origin and provides the best fit with all five points.
3. Determine the slope of the straight line and estimate S_y.
4. The sediments in this wetland are rich in clay (20–30 % by weight). Discuss the relation between S_y and the texture (i.e., grain size distribution) of the sediments. Would sandy sediments have higher or lower S_y than the value computed in this exercise?

Reference

Hayashi M, van der Kamp G, Rudolph DL (1998) Water and solute transfer between a prairie wetland and adjacent uplands, 1. Water balance. J Hydrol 207:42–55

Short Exercise 13: Influence of Error on the Water Budget

Whatta Wetland is a hypothetical 1.5-ha wetland situated in a humid environment where annual precipitation is nearly three times larger than evaporation (Table 3.13). The stage of Whatta Wetland is controlled by a small dam that increases the water level about 0.3 m. As such, it has a well-defined outlet channel, which allows accurate measurement of surface-water flow from the wetland using a weir. A weir also is used to measure surface-water flow to the wetland. In fact, great care was taken to measure all input and loss terms of the Whatta water budget. Based on a report from the wetland observer indicating that she has never seen overland flow at this sandy location, we assume that overland flow, if any, is insignificant. Maximum errors associated with individual components of the water budget are estimated to be:

Precipitation	P	$\pm 5\%$
Evapotranspiration	ET	$\pm 15\%$
Streamflow into the wetland	S_i	$\pm 5\%$
Streamflow from the wetland	S_o	$\pm 5\%$
Groundwater flow to the wetland	G_i	$\pm 25\%$
Wetland flow to groundwater	G_o	$\pm 25\%$
Change in lake volume	ΔV	$\pm 10\%$

We can write our water-budget equation as

$$R \pm \varepsilon = P + O_f + S_i + G_i - ET - S_o - G_o \qquad (3.64)$$

where R is the sum of all of the water-budget components (except change in wetland volume) and ε is the cumulative error associated with all of the water-budget terms on the right hand side.

We are interested in determining how R compares with our measured value for ΔV, which will tell us if we have any bias in our water budget or whether there are some unknown or missing terms. Ideally, R will be very close to ΔV. If this is not the case, we want to know if the difference between R and ΔV can be attributed to measurement error or if there really is a missing component or some substantial bias in our estimates of one or more of the water-budget terms.

The uncertainty associated with determination of each term also is presented in Table 3.13. After quick calculation, you can confirm that the sum of all the input and loss terms, R, is more than eight times larger than our measured annual change in wetland volume, ΔV. If we make the worst-case assumption that all errors are at the positive extreme and then sum all of the error terms, the value based on a summation of the positive error terms is so large that it encompasses the measured value for ΔV. Alternately, manipulating the sum to obtain a minimal cumulative error cannot be supported either. Thus, simple sums of the error values do not provide a means of discriminating whether R is a valid measure of the residual.

Table 3.13 Water-budget terms of Whatta Wetland, including percent of input our output terms, maximum percent error, and maximum error in m^3 per year

Water-budget term	Volume (m^3/year)	Percent of input or loss	Percent error	Error (m^3/year)
P	18,200	26 %	5 %	±910
S$_i$	46,900	68 %	5 %	±2,345
G$_i$	4,250	6 %	25 %	±1,063
ET	6,540	10 %	15 %	±981
S$_o$	49,730	79 %	5 %	±2,487
G$_o$	6,940	11 %	25 %	±1,735
R	6,140			
ΔV	700		10 %	70

If we can justify making two simple assumptions, we can estimate our cumulative error with far less uncertainty. First, we assume our errors are distributed normally. Given that measurements were made approximately biweekly, making our number of measurements around 26, this assumption appears reasonable. Second, we assume that errors in our measurements are independent. Given that precipitation is measured with a rain gage, streamflow with a flow-velocity meter, evaporation with a suite of sensors, and groundwater with a tape measure of some sort, there is small possibility that any of our sources of measurement error are dependent on another. Assuming errors are normally distributed and independent, cumulative error is reduced based on an equation similar to Eq. 3.54, but without the ΔV term:

$$\varepsilon = \sqrt{\varepsilon_P^2 + \varepsilon_{ET}^2 + \varepsilon_{Si}^2 + \varepsilon_{So}^2 + \varepsilon_{Gi}^2 + \varepsilon_{Go}^2} \qquad (3.65)$$

Using ε as a measure for the cumulative error, Eq. 3.64 indicates that $\Delta V = R \pm \varepsilon$. Based on the above information, answer the following questions:

1. How does R compare with ΔV? Are these values reasonably close? If not, suggest a reason for why they are different.
2. What is the additive error associated with determination of R (what is $R \pm \varepsilon$?) What is the error associated with R based on Eq. 3.65? Based on ε determined with Eq. 3.65, are you comfortable with stating that R is different from ΔV?
3. What if our weir failed and we had to use floating oranges all year to make estimates for the S_i term. Recalculate the maximum error for S_i assuming an error of 20 %. How does this affect R, ε, and your assessment of the water budget relative to ΔV?
4. What if the weir was fine but, instead, we had only air temperature data and were forced to estimate evaporation using the Thornthwaite method, which we decided had a maximum error of 50 %. How would increasing the error associated with evaporation from 15 to 50 % affect the determination of R relative to ΔV?

Field Exercises

Field Activity 1: Installation of a Wetland Staff Gage, Water-Table Well, and Piezometer

With a staff gage to indicate wetland stage and measurement of the depth to water in a nearby water-table well, a wetland scientist can determine whether groundwater has the potential to flow to the wetland or whether the wetland is likely to lose water to the adjacent groundwater system. If we know hydraulic conductivity (K) at the well, and make the assumption that K is uniform in the vicinity of the well and the wetland, we can calculate flow (Q) between the wetland and groundwater in an area for which we think data from the well is representative. Lastly, two additional measurements of Q can be made; one utilizes a seepage meter installed in the wetland bed and the other makes use of changes in temperature gradients in the wetland sediments. The temperature method requires installation of sensors at various depths beneath the wetland bed. Since we have to auger a hole or pound a pipe a meter or two into the sediment to install these sensors, it also makes sense to put a well screen at the bottom, in which case we can determine the hydraulic gradient on a vertical plane as well as K based on a single-well test. With that information, and our measurement of Q from the seepage meter, we can use Darcy's law to calculate K of the wetland sediment on a vertical axis. This will give us an idea of anisotropy, the ratio of horizontal to vertical hydraulic conductivity. With this small investment of time and money, we will have learned a great deal about wetland hydrology and hydrogeology at this site.

This first of three exercises near the wetland shoreline will demonstrate the installation of a monitoring well and a staff gage. Detailed instructions and parts lists presented here, and also those presented in the other field exercises, represent the authors' preferences and describe only one of many different ways to achieve these objectives. Students are encouraged to seek other descriptions and opinions for accomplishing these tasks and then develop their own impressions and methods for collecting data in the field.

Wetland Staff Gage

Figure 3.40 shows a wetland staff-gage installation and illustrates some of the problems that can be associated with their use. First, note that there are two staff gages in the photograph. In settings where wetland stage changes substantially, it may be necessary to have multiple staff gages so that when one gage is completely submerged during periods of high water another situated at a higher elevation can be read to indicate wetland stage. Secondly, note the substantial angle from vertical of the staff gage in the distance. This is the result of ice on the wetland surface having moved at some point during the winter, tilting the staff gage. If the ice moves enough, the staff gage can be completely removed from the wetland bed and sometimes transported a considerable distance. The surveyor holding the rod on the

Fig. 3.40 Staff gages installed in a wetland in the Nebraska sandhills with a surveyor standing on the frozen wetland surface and holding a survey rod at the distant gage. Note that ice movement has tilted the staff gage in the distance. Staff-gage movement is an annual occurrence in locations where ice forms on the wetland surface during winter, requiring re-surveys to maintain year-to-year continuity of wetland stage data

staff gage in Fig. 3.40 will also record the angle from vertical of the staff gage so that corrections can be made to any stage measurements obtained while the gage is tilted. Once straightened, the gage will need to be re-surveyed.

Construction of the staff gage in the foreground is typical of many installations. A steel fence post is attached to a piece of lumber that is treated to resist rot (the example in Fig. 3.40 uses U-clamps to attach a wooden board to the post). An incremented staff section, usually made of enameled metal or fiberglass, is screwed to the wood. The fence post can be attached to the wood and then driven into the wetland bed, or if the wetland sediments are very resistant, the fence post can be driven first and then the board complete with face plate is subsequently attached. A length of steel pipe is often substituted for the fence post. Many installations also have a bolt or screw projecting out of the wood next to the face plate so that a survey rod can be placed on the bolt and held in a constant position relative to the values on the face plate while surveying the relative elevation of the staff gage.

Monitoring Well Installation

Two types of monitoring wells, or piezometers, will be installed as part of this field activity, one constructed to indicate the elevation of the water table adjacent to a

3 Assessing and Measuring Wetland Hydrology

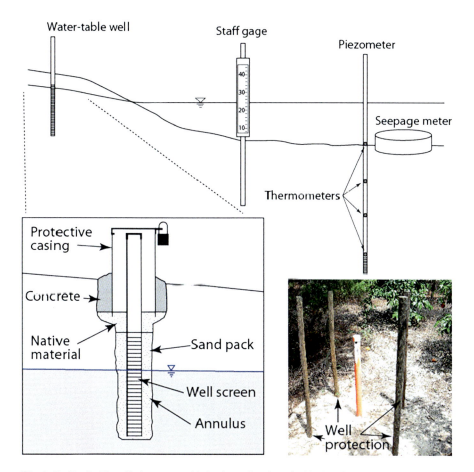

Fig. 3.41 Typical installation to quantify horizontal and vertical hydraulic gradient, seepage rate, and hydraulic conductivity

wetland and the other constructed to indicate hydraulic head at some point beneath the water table (Fig. 3.41). Although both can be considered as piezometers, we will refer to the first as a water-table well.

Water-Table Well Installation

A water-table well is designed to indicate the elevation of the top of the saturated portion of the sediments where pressure head is equal to atmospheric pressure (the water table). Installation of a water-table monitoring well can be simple and inexpensive if the land surface slopes gently away from the wetland edge, in which case the vertical distance from land surface to the water table is usually small. In these shallow, near-shore margins a monitoring well can usually be installed by hand, precluding the need for a large, mechanical drill rig. Such is the assumption for the following field activity describing the installation of a shallow monitoring well. Items you will need include:

- Polyvinyl chloride (PVC) pipe (a wide range of diameters are available but 5.1-cm diameter is very common)
- PVC well screen (see Fig. 3.42c for examples of commercially made screens. See the section on piezometer installation for making screens from regular pipe)
- Associated couplings and caps and PVC cement
- Bucket auger and associated hardware (8.9-cm (3.5-in.) diameter is common)
- Supply of medium sand (approximately 5-L but amount will vary depending on the diameter of the augered hole relative to the diameter of the monitoring well)
- Shovel
- Tamping rod (handle of the shovel or unused sections of auger rod can suffice)
- Hand saw
- Sledge hammer
- Tape measure or folding rule
- Water-level measurement device (e.g., chalked-steel tape, electric tape)
- Notebook, hand lens, sediment-sample bags

First, select a location for installation of the water-table monitoring well. The well should be located so that it is representative of conditions along a specific reach or area of the wetland. Criteria that are commonly considered when locating a water-table well include topographic gradient, vegetative cover, aspect, geology and soil type. Once the location is selected, use a shovel to remove the vegetation from an approximately 0.25-m^2 area surrounding the intended well site. Note the vegetative cover and organic soil type and thickness.

Install an appropriate auger head on a section of rod (Fig. 3.42a) (closed-head for sand and loosely consolidated sediment, open-head for cohesive sediment) and begin turning the auger in a clockwise direction until the auger bucket is full. Remove the bucket from the hole and shake or push the sediment out of the auger head (Fig. 3.42b), allowing the sediment to fall onto a clean surface, such as a board or tarp. Record the depth of the hole with a tape measure. Describe the sediment in the field notebook. Place a sample from the auger in a sample bag for later lab analysis of percent organic matter and grain-size distribution. Repeat this process until you reach the water table or the intended depth. As you auger deeper, you may need to add one or more rod extensions to the soil-auger assembly. You also may encounter large rocks that inhibit continued augering. Persistence will sometimes get you past a rock or rocky layer, but you also may have to abandon the hole and try again a short distance away.

The water table may not necessarily be obvious if the permeability of the sediment is small enough that water does not readily flow into the auger hole. In some cases, squeezing the sediment with your hand can indicate whether the sediment is saturated or not. If the sample was removed from below the water table, water will be released from the sediment as you squeeze the sample. In settings where the sediment is sandy and poorly cohesive, it is likely that saturated sediment will slump back into the hole as sediment below the water table is removed. The

3 Assessing and Measuring Wetland Hydrology

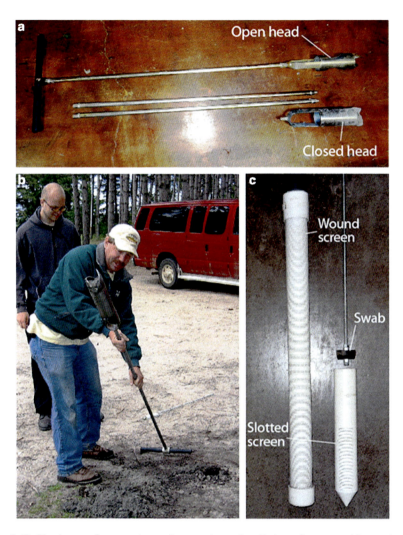

Fig. 3.42 Hand auger for removing sediment prior to installation of a water-table monitoring well. (**a**) Auger head, rod, and handle with two rod extensions and an additional auger head; (**b**) Augering a hole with the bucket inverted for removal of sediment; (**c**) PVC wound well screen, PVC slotted well screen, and well-screen swab. Note the two different types of fittings at the end of the well screen (standard PVC cap and cone-shaped PVC point). If the slotted screen is inverted and the cap is attached to the opposite end, the non-slotted interval becomes the sump

common solution to this problem is persistence. Keep augering through this sediment with strong downward force on the auger handle. You may need to change to an auger head that has solid sides and a narrower opening between the cutting fins so that loose, wet sand is better retained when the auger is pulled from the hole. The hole below the water table will gradually deepen as you continue to remove sediment and the loose slurry occupying the hole will become less and less dense as you continue to remove sediment from the hole. Once the desired depth has been reached, commonly about 1–1.5 m below the water table, it is time to assemble and install the well.

Record the total depth of the hole by marking the auger rod at the point where it is even with land surface when the auger is at the bottom of the hole. Remove the auger from the hole and measure the distance from the mark to the bottom of the auger. Add a distance, commonly 0.6–1 m, for the extent of the well casing that will be above the ground. This is often called the "stickup." The sum of these distances will be the total length of the monitoring well. Assemble the well screen by gluing a cap to the bottom of the well screen and a coupling to the top of the screen (Fig. 3.42c). If available, it is desirable to use a well cap that either is cone shaped or that has the same outer diameter as the well screen to reduce resistance when pushing the assembly into the loose sediments below the water table. The well screen should be sized to be long enough that the water table is usually within the screened interval of the well. The slot size (the width of the openings in the screen) should be selected so that most of the sediment cannot pass through the well screen.

Well screens often have an interval at the bottom of the screen that does not have any slots. This is called the sump, or the volume below the screen where fine sediments that pass through the screen can accumulate without blocking the well-screen openings. Be sure to record the presence of a sump and indicate the length of the sump. This information will be important in determining the precise screened interval of the well. The existence of a sump becomes particularly important if the water table is below the bottom of the screened interval. Measurements of depth to water will indicate an erroneous water level equivalent to the elevation of the bottom of the well screen because water will be trapped in the sump. Drilling small holes in the bottom of the sump prior to well installation may allow trapped water to drain from the sump if the well goes dry.

Cut the PVC casing so that the total well length is the distance of the hole depth plus the desired stickup length. If the hole is relatively deep, you may need to attach another PVC coupling and another length of well casing to reach the desired total assembly length. By now, the sediment in the auger hole may have settled and solidified and it may be necessary to remove several additional buckets full of recently slumped sediment from the hole. Keep removing sediment from the hole until the auger has reached the bottom of the hole and the sediment is once again poorly consolidated. At this point it is important to move rather quickly, especially in sediments that readily slump and solidify, such as medium to fine sand. As soon as the last bucket of sediment is pulled out of the hole, immediately shove the completed well casing and screen into the hole and push it down until it stops. You may need to pound lightly on the top of the well casing with the sledge hammer to drive the well to the intended depth. It is prudent to place a board or drive cap on the well casing to prevent damage to the top of the well casing. While pounding lightly, grab the well casing and push downward, essentially vibrating the well downward through the loose sediment. In most cases, you will be able to reach or get very near the desired well depth. Once the well is in place, it is a simple matter of filling the annular space between the edge of the augered hole and the well casing with sediment that was removed from the hole. Tamp the sediment repeatedly as you fill the hole so the sediment is tightly consolidated. This will prevent any preferential flow of water along the outside of the well casing during recharge events. If unused segments of auger rod are used for this purpose, place duct tape over the end of the rod to prevent damage of the threads.

If the sediment is sufficiently cohesive that the augered hole remains open below the water table, inserting the completed well screen and casing is as simple as placing the assembly into the auger hole. In this case, you will then need to pour sand coarser than the well-screen slot size down the hole so that it surrounds the entire screened interval. This backfill, often called a sand pack, will ensure that the well screen does not become clogged with fine-grained sediment that otherwise would be situated next to the well screen. Once sufficient sand is added to fill the annular space to just above the screened interval, material removed from the auger hole can be added to fill the remainder of the augered hole. As described before, this sediment should be tamped to ensure that the density of the sediment filling the annular space is not less than the undisturbed material. It is common to add soil to create a small mound of soil at the base of the well that will direct rainfall away from the well casing.

Now all that is left is to install a well cap, install well protection, and make several measurements. A well cap can be as simple as a plastic slip cap that stays on the casing via friction and gravity. You might instead wish to glue on an assembly that has a threaded cap or that allows access to the well to be protected with a keyed lock. In either case, make sure that the well cap can easily be removed from the casing for measurements of depth to water. Shallow monitoring wells are not well anchored to the soil because of the smaller contact area with the soil that surrounds the well casing. Some wells can easily be moved, even in an attempt to remove a firmly attached well cap, which may change the vertical positioning of the top of the well and introduce error in determinations of hydraulic gradient. A small hole also may be drilled through the well casing to facilitate equilibration of the pressure inside of the well casing with changes in atmospheric pressure. If air cannot readily enter the well casing, the position of the water table inside of the well may not represent the water table.

In many areas, regulations require some form of protection that will minimize the chance of the well casing being inadvertently broken by a falling tree or branch or a wayward automobile or lawnmower. This may entail placing a steel casing of larger diameter over the top of the well casing and into the ground (Fig. 3.41), or installation of three or four wooden or metal posts positioned so that wayward objects will strike the posts rather than the well casing (Fig. 3.41 photo inset). Lastly, make measurements of the stickup length and the distance to the bottom of the well. Survey to the top of the well casing and determine the spatial coordinates of the well with a global positioning system (GPS) or similar device.

Piezometer Installation

The piezometer will be installed in a location where the wetland bed is beneath the water surface. In this situation, the piezometer will indicate the vertical hydraulic gradient. In order to ensure that the difference in head between the piezometer screen and the wetland stage will be measurable, the screen needs to be placed a considerable distance below the sediment-water interface, often 2–3 m or more below the sediment-water interface. If the sediments are well consolidated and do not readily slump, it may be possible to use a bucket auger to create a hole in which the well screen and casing are placed, as described previously for installation of a water-table well. If augering is possible, the augered hole should not be larger than the outside diameter of the well to prevent vertical preferential flow of water along the outside of the well casing, which could alter hydraulic head at the well screen. However, in most inundated settings the sediments simply collapse into the augered hole and it is extremely difficult to auger a hole deep enough for a piezometer installation. It is much more common to drive a piezometer to depth with a well pounder or post driver. That is what we will do here. The items you will need include:

- Well screen, cap, couplings, and casing (typically steel to withstand the rigors of pounding)
- Device for driving the well and casing to the desired depth
- Cap to protect the top of the well casing
- well swab (a device to shove water through the well screen)
- bailer or pump for removing or adding water to the well
- Measuring tape

You will want to select a well diameter that is small enough to permit the driving of the well to depth but large enough to allow installation of monitoring equipment inside of the well casing, such as a pressure transducer or temperature sensors. A common diameter for these purposes is 1.9–3.2 cm (0.75–1.25 in.). Commercial well screens are preferred because of the large surface area open to the sediments, although holes or slots can be drilled or cut with hand tools to create simple screens in coarser-grained settings. If the latter option is pursued, the much smaller aggregate surface area of the holes and slots relative to a commercial well screen may result in an unacceptable response time of the well to changes in hydraulic head.

Considerable care is needed to ensure that the well screen is not clogged during installation, especially if a well screen is made by cutting or drilling holes in the well casing. To minimize this possibility, a well swab can be constructed to force water through the screen and to clean out the screened interval of the well during and following the well installation. A well swab can be as simple as a rubber washer or washers attached to the end of a metal rod (Fig. 3.42c) so that the rubber washer rubs against the side of the well casing and screen as it is pushed up and down inside of the well casing. By pushing the rod downward, water inside the well casing is forced through the screen. An upward motion pulls water through the well screen into the well casing. Repeated up and down motion generally is sufficient to remove particles that may be stuck in the screened openings, improving the connection with the aquifer sediments and reducing the time required for the head inside of the well to become representative of the adjacent saturated sediments.

Whether a post driver or well-head driver or sledge hammer is used to advance the well assembly, it should not directly strike the top of the well casing if threads are present. Doing so could deform the threads and make it impossible to attach a coupling or additional sections of casing that would otherwise allow the well screen to be driven deeper into the sediment. A drive cap or coupling should be screwed onto the threads at the top of the well casing before striking the top of the casing to drive it farther into the sediment. The drive cap or coupling should be tightened occasionally as the casing is driven into the sediment; not doing so also may result in damaged threads. It is prudent to periodically stop driving the well and swab the well to remove sediment that may have clogged the well screen. It may be necessary to pour water into the top of the well casing so the swab pushes and pulls water, and not air, through the well screen. If additional sections of pipe are required, Teflon tape or pipe dope should be used liberally, and the fittings tightened using pipe wrenches, to ensure that no leaks occur at the junctions between pipe segments. Once the well is driven to depth, it should be thoroughly developed by repeatedly swabbing the well and screen, including periodic removal of water and suspended sediment from the well with a pump or bailer, until the water level inside the well casing recovers readily to the static water level. Once this occurs, the well is considered developed and is functioning as a piezometer.

After well installation and development you will want to measure and record:

1. Distance from the top of casing to the well bottom,
2. Distance from top of casing to the wetland bed,
3. Screened interval, sump interval (if present), and
4. Distance from the water surface to the wetland bed.

With these values determined, the distance from the sediment-water interface to the mid-point of the screened interval can be calculated. Commonly referred to as l in the Darcy equation (or sometimes l_v to indicate that the gradient is distributed on a vertical axis), this is the distance that the head difference is divided by to determine the vertical hydraulic gradient. The head difference can easily be determined by measuring the distance from the top of casing to the wetland water surface and subtracting the distance from the top of casing to the water surface inside of the well. For a small-diameter well completed in low-permeability sediments, measurements of depth to water can be corrupted if a portion of the measuring device needs to be immersed in the water to make a measurement. The volume of the sensor device immersed in the water will cause the water level to rise inside of the well. Low-permeability sediments will not permit the water level inside the well to return to static equilibrium in a sufficiently short time, resulting in a false depth-to-water measurement. Care should be taken to prevent this possibility by using a measurement method that does not require immersion of a large sensor relative to the well-casing diameter during a water-level measurement. The cut-off end of a chalked-steel tape is a particularly good device for this purpose because the volume of the steel tape immersed to make a measurement is very small.

Once the piezometer is installed, GPS coordinates and well-top elevations are determined, and measurements are made to determine the hydraulic gradient. Sensors also can be installed to continuously monitor hydraulic head, and temperature at one or more depths, inside of the piezometer (Fig. 3.41).

Field Activity 2: Single-Well Response Test

In Field activity 1, a piezometer was installed either on the margin of or beneath a wetland bed. Figure 3.43a demonstrates a piezometer in a wetland with the screen (slotted portion in the bottom) in direct contact with the sediments, and panel b demonstrates a piezometer completed in a dry margin of a wetland (the water table is below the ground surface). The latter has been installed in an augered hole with a sand pack around the screen and a clay seal above to prevent "short-circuiting" of water through the annular space. A horizontal line beneath an inverted triangle is a commonly used symbol to indicate surface-water level. This symbol is displayed here to indicate the pond water level in (a) and the water table in (b), as well as the undisturbed water levels (also called static head) in the piezometers.

A single-well response test, often referred to as a slug test, is initiated by changing the water level in a water-table well or piezometer very quickly (within a few seconds) and monitoring the recovery of the water level from the initial disturbed value to the static level. A number of methods are available for creating this near-instantaneous water-level change (Butler 1998). The easiest method is to quickly lower a solid cylinder (typically made of metal or high-density plastic) attached to a length of rope into the piezometer. This solid "slug" displaces a known volume of water as it is rapidly lowered into place and the slug remains stationary for the duration of the test. The water level in the well returns to the static level at a rate that is controlled by the hydraulic conductivity of the porous medium around the well screen. After the static level is reached, a second test can be initiated by rapidly removing the cylinder, thereby causing an instantaneous drop of the water level. It is always good practice to conduct two response tests (positive and negative displacement) and check the consistency of results.

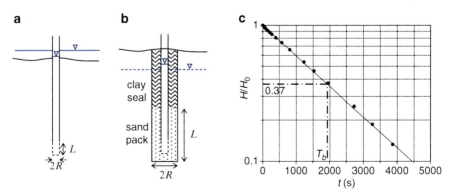

Fig. 3.43 Schematic diagrams of piezometers with screen length L and radius R without (**a**) and with (**b**) a sand pack; (**c**) example of the plotted recovery of a single-well response test conducted in a piezometer located in Wetland 109 in the St. Denis National Wildlife Area

Water level is monitored during the slug test using either a manual water-level sounder or a pressure transducer, depending on the rate of water-level recovery. For low-permeability settings, manual measurements can often be made quickly enough to capture the initial rapid phase of water-level recovery and can easily be made frequently enough during the slower phase of recovery. A pressure transducer is a far better choice for wells installed in sand or coarser sediments where the entire recovery can be completed in a matter of seconds. The transducer is suspended prior to the test at a depth greater than the reach of the slug to avoid damage to the transducer, and early enough that the well has recovered to the static water level following displacement of water during immersion of the transducer. The combined length of the slug and rope needs to be carefully measured to ensure that the slug does not slam into the pressure transducer as it is rapidly lowered into the well. If the slug is completely submerged during deployment, the known slug-displacement volume can be used to estimate the initial rise (or drop) of the water level during the test. Calculation of the maximum water-level change can then be compared with the measured value. A substantial difference between calculated and measured water-level change may indicate a procedural problem or a problem with the piezometer construction. It also is important to ensure that the piezometer water level does not go below the top of the screen or the top of sand pack during the entire test. For this reason, single-well response tests are not recommended for water-table wells.

The average (or bulk) hydraulic conductivity (K_b, m s^{-1}) of the material surrounding the piezometer screen (or sand pack, if present) can be estimated from the recorded water-level data:

$$K_b = \pi r^2/(FT_b) \tag{3.66}$$

where r (m) is the radius of the inside of the well casing, F (m) is a shape factor representing the dimension and geometry of the groundwater flow field around the screen, and T_b (s) is the basic lag time of the piezometer (see below for definition). The "sample volume" of this method is approximately equal to a sphere with a radius similar to the length of the well screen, L (m). F is a function of L and R, the radius of the outer surface of the well screen or the sand pack, if present. Numerous equations have been suggested to estimate F for different types of piezometers under different conditions (see Butler 1998). In most cases, if L/R is not substantially smaller than 4, the formula of Hvorslev (1951) as cited by Freeze and Cherry (1979:341) gives a convenient means to approximate F:

$$F = 2\pi L/\ln(L/R) \tag{3.67}$$

T_b is determined by plotting head versus time on a semi-logarithmic plot (Fig. 3.43c). For convenience, head is normalized as:

$$H/H_0 = (h - h_s)/(h_0 - h_s) \tag{3.68}$$

where h (m) is measured head, h_0 (m) is the water level immediately after the introduction of the slug, and h_s (m) is the static water level prior to introduction of the slug. Once a straight line is fitted to the data, T_b is determined as the time in seconds since the beginning of the introduction of the slug when H/H_0 equals 0.37 ($\cong e^{-1}$) (Fig. 3.43c).

Once the slug test data have been collected and entered in a spreadsheet, you should follow the procedure listed below:

1. Prepare a data table containing time in one column ($t = 0$ at the maximum h value following introduction of the slug) and h in the second column corresponding to each value of t.
2. Compute H/H_0 for each reading.
3. Plot H/H_0 versus t, using a logarithmic axis for H/H_0.
4. Fit a straight line to the data points, and determine the value of t where the straight fitted line crosses $H/H_0 = 0.37$. Em shows an example, in which $T_b \cong 1{,}930$ s.
5. From T_b and the dimensions of the piezometer, compute K_b.
6. In the example shown in Fig. 3.43c, the piezometer is constructed similarly to panel b and has dimensions of $L = 0.73$ m, $R = 0.075$ m, and $r = 0.016$ m. Substituting these values and T_b into Eqs. 3.66 and 3.67 gives $K_b = 2.1 \times 10^{-7}$ m s^{-1}. This test was conducted in a piezometer located in Wetland 109 in the St. Denis National Wildlife Area in Saskatchewan, Canada (see Hayashi et al. 1998 for details).

References

Butler JJ (1998) The design, performance, and analysis of slug tests. Lewis, Boca Raton

Freeze RA, Cheery JA (1979) Groundwater. Prentice-Hall, Englewood Cliffs

Hayashi M, van der Kamp G, Rudolph DL (1998) Water and solute transfer between a prairie wetland and adjacent uplands, 1. Water balance. J Hydrol 207:42–55

Field Activity 3: Installation of a Seepage Meter and Temperature Sensors

The use of multiple methods to determine flow between groundwater and surface water is always a good idea because it improves understanding of the physical setting and it provides independent values representative of multiple spatial scales. Field activities 1 and 2 demonstrated measurement of hydraulic gradients and hydraulic conductivity to determine Q. Field activity three provides two additional methods for determining Q. A seepage meter makes a direct measurement of Q, but over a very small portion of the wetland bed. The piezometer that we installed in the wetland can serve double duty if we suspend temperature sensors inside of the piezometer casing, allowing calculation of Q based on temperature gradients and attenuation of diurnal cycles in temperature with depth.

Seepage Meter Construction and Installation

What you will need:

- 208-L (55-gal) plastic storage drum
- Hand saw for cutting plastic drum
- Permanent marker
- Measuring device
- Power drill (battery-powered or electric)
- Drill bits appropriately sized for the hose-connection hardware
- Hose-connection fittings
- Rubber or cork stopper
- Plastic tub and lid to serve as a seepage bag shelter
- Plastic seepage bag (approximately 3–5 L)
- Tube and fittings to connect plastic bag to hose
- Hose to connect bag shelter to seepage cylinder
- Brick or suitable weight to place on top of seepage cylinder

A seepage meter can be made from many different readily available products. The standard "half-barrel" seepage meter is described as such because it was made by cutting the ends off of a standard 208-L (55-gal) storage drum (Lee 1977). Although many other cylinders have been used as seepage meters, such as coffee cans, cut-off trash cans, trash-can lids, even wading pools, the half-barrel meter is often used because it is rigid, durable, does not readily deform, covers a larger surface area than many of the other devices, is still quite inexpensive, and can be easily obtained from many industrial supply companies. A storage drum will be used in this exercise. First, obtain a storage drum from one of a large number of suppliers. Either metal or plastic drums can be used, but to simplify construction for this exercise, you should obtain a plastic drum. Be sure to order a closed-top drum to eliminate possibilities of leaks associated with an open-top drum where the top can be removed, and order the larger 208-L (55-gal) drum because it covers a larger surface area than the 114-L (30-gal) drum. You will make seepage cylinders from the top and the bottom thirds of the drum.

Mark the side of the drum a consistent distance from one end of the barrel; commonly, a length of 30–35 cm is used. Connect the dots (marks) by drawing a line along the circumference of the drum. Use the hand saw to cut along this line to remove one end from the drum. Repeat this process for the other end of the drum. If vegetation on the wetland bed is tall and dense, you may instead simply cut the barrel in half, essentially making two seepage cylinders, each approximately 45 cm tall. A cross-cut hand saw can be used to cut the plastic drums whereas a cutting torch or reciprocating saw (or a hack saw used with great persistence) are generally required to cut a metal barrel. Carefully measure the diameter or the circumference of the open end of the cut-off cylinder and calculate the area based on either measurement. This open end of the cylinder will equal the area of the wetland bed covered by the seepage cylinder. Most 208-L drums will cover an area of about 0.25 m^2.

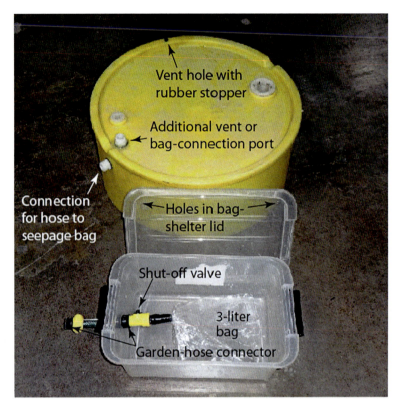

Fig. 3.44 Half-barrel seepage cylinder showing ports installed both in the top and the side of the cylinder. A section of garden hose with female garden-hose connectors on both ends (not shown) is used to connect the bag shelter to the seepage cylinder

Next, you will need to drill a hole in the side of the drum, near the drum end, to which a short hose will be attached (Fig. 3.44). The short hose will extend from the seepage cylinder to a seepage-bag shelter that will protect the bag from wind and waves, curious animals, and diving ducks (Fig. 3.45). The diameter of the hole will depend on the hardware that you use to attach the hose to the seepage cylinder. There are many different options available. Water flows through a seepage meter under very low pressure. The fitting should not leak under small pressures but you do not need to go to the expense of installing a water-tight bulkhead fitting either. Lastly, drill a small hole approximately 0.5–1 cm in diameter at the highest point of the seepage cylinder (vent hole identified in Fig. 3.1). This will be the vent for releasing any gas that is trapped during seepage-meter deployment. This hole will be open during installation of the seepage cylinder and then plugged with a rubber or cork plug during operation. If substantial amounts of gas are generated, a common situation in many wetland settings, you may need to install a vent tube that will extend above the water surface so that gas can be released to the atmosphere during seepage-meter operation (Lee and Cherry 1978).

3 Assessing and Measuring Wetland Hydrology 219

Fig. 3.45 Half-barrel seepage meter installed in sandy sediment. Note side port, to which the hose is connected, and top port with cap and vent hole with rubber stopper

The seepage bag, used to measure the volume of water that flows across the sediment-water interface covered by the seepage cylinder, also can be made from a variety of materials. A convenient bag volume is 3–4 L and thin-walled, flexible bags are preferable. Lightweight freezer-storage bags have often been used. Avoid using bags with thicker walls, such as medical intravenous (IV) bags or solar-shower bags; these bags have a substantial resistance to expansion and contraction in response to being filled or emptied. Use of these bags will substantially reduce the volume of water that otherwise would flow across the bed covered by the seepage cylinder. The opening of the bag can be gathered together around a hose and taped to the hose so the fitting does not leak. Another option is to weld or otherwise seal the bag opening and cut a small slit in one of the corners of the bag, through which you will insert a hose or tube and tape the bag to the hose or tube. As with the seepage cylinder, the bag and fittings should not leak under small pressures but the assembly does not withstand large pressures. It is convenient to install hardware that includes a valve that can be closed while the bag is being transported, attached or removed from the seepage cylinder, and during subsequent handling prior to being weighed or measured.

The bag should be placed in a shelter for several reasons: (1) to prevent the bag from being exposed to currents, (2) to maintain the bag in a proper orientation, and (3) to protect the bag from fish or mammals or waterfowl, a particularly important consideration in many wetland settings. Many different types of bag shelters have been used; examples are provided in Figs. 3.44 and 3.45. Design and build a bag shelter of your choosing, including a section of tubing or hose that will connect to the side opening on the seepage cylinder. The hose or tubing should be approximately 1–2 m long, which ensures that you will not disturb the seepage cylinder while attaching or removing the seepage-collection bag.

Seepage-Meter Installation

Select a location near the piezometer that you installed as part of field activity 1. Wade to the location, making sure to not step on the area that will be covered by the seepage cylinder. The bed should not be covered by any large rocks or debris (i.e., waterlogged sticks) that would alter seepage or prevent insertion of the seepage cylinder. Make sure the rubber plug is removed and the port on the side of the seepage cylinder is open; this allows water to escape as you are pressing the seepage cylinder into the wetland sediments. Press the cylinder into the sediment very slowly, allowing gas and water to escape through the top vent tube. You may need to twist the cylinder to aid in cutting through a vegetative mat, if one is present. If aquatic vegetation is very dense you may need to first cut a slit in the vegetative mat with a long knife to facilitate insertion of the cylinder. The bottom rim of the cylinder typically needs to penetrate the sediment approximately 5–10 cm to ensure a good seal with the sediment. However, if the bed surface is uneven, the insertion depth may need to be increased so no gaps are present beneath the edge of the seepage cylinder. You should probe with your fingers along the interface between the wetland bed and the seepage cylinder. If you can feel the bottom edge of the cylinder, then the insertion depth is not sufficient. In this case, press the cylinder deeper into the sediment until you can no longer feel the bottom edge of the cylinder. The meter also should be inserted with a slight tilt so that the vent hole is at the highest point, allowing any gas released from the sediment to escape. Once the meter is set, place a weight on the meter to counter the buoyant force of the plastic material. A concrete or masonry brick usually is sufficient. Plug the vent tube with the rubber stopper. The stopper will be removed later, prior to seepage measurement, to provide a relative guide for the volume of gas released from the sediment. If the volume is substantial, you will want to install a vent tube to release gas to the atmosphere. If unvented, gas released from the sediment will collect inside of the seepage cylinder, displacing water that will be routed to the seepage-collection bag.

Install the bag shelter and connect the shelter to the seepage cylinder. You may also need to place a small weight inside of the bag shelter to hold it in place and prevent movement in response to waves. The wetland bed has been substantially disturbed during meter installation and it is common for seepage rates to be larger than normal following meter installation. It is common practice to wait for hydraulic conditions at and near the bed to stabilize before measuring seepage. If your field schedule permits, wait until the next day before making the first measurement, or measure seepage directly after installation and compare those values with measurements made the following day.

Seepage-Meter Measurement

Since you do not know whether water is flowing into or from the wetland across the portion of the wetland bed isolated by the seepage cylinder, start your first measurement with the seepage bag approximately half filled with water. Place a known volume of water inside of the bag. Volume can be determined either with a graduated cylinder or by weighing the water and the bag with an electronic scale. If using an electronic scale, knowing that the density of water is 1 g/cm^3 and that 1 ml equals 1 cm^3 allows you to measure change in volume by recording change in weight of the seepage bag. Before making any measurements using an electronic scale, you should weigh the bag empty, then completely full, so you will know the range of volume that can be measured with the bag.

Once an initial volume of water in the bag has been measured (or weighed), you will need to remove all remaining air from inside of the bag prior to connecting the bag to the seepage cylinder. This is commonly called de-airing the bag. Close the valve on the bag that contains a measured volume of water, walk out to the bag shelter, suspend the bag vertically while holding onto the bag fitting, open the valve, and slowly lower the bag into the water, immersing the bag with the valve constantly pointing up and always above the water surface. This process will force air inside of the bag to leave via the open valve located above the water surface. Once the bag is pulled beneath the surface to the point where water inside of the bag is at the same level as the valve, close the valve. The bag is now de-aired and ready for deployment.

Carefully remove the bag-shelter lid and attach the bag to the threaded fitting inside of the bag shelter. Straighten the bag so the bag material is not twisted and the bag is oriented in a relaxed position inside of the bag shelter. Open the valve and record the time of opening. Your measurement has begun. Place the lid on the bag shelter very slowly to avoid forcing water out of the bag during the measurement. Now you wait. Since you do not know the seepage rate a priori, the wait time is somewhat of a guessing game. A half hour to an hour should be sufficient to allow a change in water volume that is large enough to allow you to know whether water is flowing to or from the bag. To remove the bag, repeat the process described above but in reverse. Remove the lid on the bag shelter very slowly, and close the valve on the bag being careful to not touch the bag. Record the time as you close the valve. Remove the bag and measure the final volume of water (or determine the final weight of the bag plus water if an electronic scale was used prior to bag attachment). By the gain or loss in volume or weight, you will know the direction of flow and have an initial assessment of the relative seepage rate. If the bag is full or empty upon removal, you waited too long and your next measurement should be conducted over a shorter period. If there is no measurable change in volume, your next measurement period should be increased. After one or two iterations, you should have a good estimate for the amount of time it will take to make a seepage measurement. Simply divide the change in volume by the time of bag attachment to get seepage results in ml/min. Divide that value by the area covered by the seepage cylinder to report your results in flux units (distance per time).

Installation of Temperature Sensors

Accurate measurements of temperature can be made easily with inexpensive instruments, making its use in quantifying exchanges between groundwater and surface water particularly attractive. Here we will make use of newer technology

Fig. 3.46 Nest of piezometers installed at different depths beneath the wetland bed with pressure transducers and temperature sensors installed in five of the seven wells. All sensors are connected to a digital datalogger positioned on shore to the left of the photo. Note also the four seepage meters, with bags attached directly to the tops of the seepage cylinders, installed near the wells. Attaching the bag directly to the seepage cylinder is sometimes acceptable where wind and currents are minimal

for measuring temperature, along with the concepts presented in Sect. 3.6, to determine a value for Q at the piezometer we installed earlier. This value can be compared to Q determined with the Darcy method described in Field activity 1.

Two basic types of electronic sensors are commonly deployed for this purpose. The thermocouple is a device that consists of two wires made of different metals that are connected together at both ends. A current is generated when two junctions of these wires are exposed to different temperatures. Copper and constantan wires are commonly paired for use in environmental applications. The method requires that one of the junctions be related to a known temperature. Therefore, a separate reference temperature sensor also is required to use this measurement method. The second commonly used sensor, and one that often is used as the reference thermometer for thermocouple installations, is the thermistor. A thermistor is basically a resistor that changes resistance in response to changing temperature. The choice of thermocouple or thermistor often depends on the number of temperature sensors required. If more than 5–10 sensors are required, it may be more cost effective to deploy thermocouples.

Two methods of deploying temperature sensors also commonly are used. One consists of a sensor connected to wires that transmit the signal to a nearby data-collection device (Fig. 3.46), and the other consists of the sensor and datalogger in a single, self-contained unit. Recent versions of the latter device have become very small (e.g., 17 mm diameter) and can be inserted inside small-diameter piezometers.

Either type of sensor can be used for this installation. First, familiarize yourself with the electronic thermometer of choice, making sure that the sensor output is reasonable, that output changes in response to placing the sensor in a warmer or colder environment, and the sensor is logging data. For this application, collecting data at 15-min intervals generally is sufficient to monitor diurnal changes in temperature, although more frequent data collection is certainly acceptable.

Attach one sensor to the outside of the casing of the piezometer that is installed in standing water in the wetland. The sensor should be positioned just above the sediment-water interface. You may also wish to deploy an additional sensor to record changes in air temperature that drive changes in the wetland water temperature. Next, position one sensor at the bottom of the well and another one or two sensors at equal distances between the well bottom and the sediment-water interface. Only one sensor is actually required to be deployed inside of the well; additional sensors allow a determination of the degree of heterogeneity in hydraulic conductivity between the sediment-water interface and the bottom of the well. It is common to suspend sensors on appropriate lengths of string or fine wire from the top of the well (be sure to first check whether the sensors sink or float), or if a signal cable is involved, to affix the signal cable to the top of the well so the sensor hangs at the appropriate depth.

Collect data from the sensors for a period of one to several weeks. Retrieve the sensors, download the data, and plot the time series from all sensors on the same plot.

1. After viewing the data you have collected, is it likely that groundwater is discharging to the wetland or that wetland water is flowing vertically downward to become groundwater? Or is it not possible to make this determination based on your data?
2. Calculate the difference between the daily maximum and minimum temperatures for each sensor. Plot the differences versus time. If you have collected air-temperature data, include daily differences for air temperature as well. Can you make any determination regarding any potential change in the rate of flow across the sediment-water interface?

You can determine the rate of vertical flow across the wetland bed in either direction using the methods described in Short exercise 11. You will also need estimates of thermal conductivity, porosity, dispersivity, and heat capacity of the sediment. Since you also know the vertical hydraulic-head gradient based on measurements you made at this piezometer in field activity 1, you could use one of several methods described in Appendix B of Stonestrom and Constantz (2003) to determine Q. As an additional exercise, you are encouraged to use the free software described in Stonestrom and Constantz to calculate Q based on the temperature data you have collected.

References

Lee DR (1977) A device for measuring seepage flux in lakes and estuaries. Limnol Oceanogr 22:140–147

Lee DR, Cherry JA (1978) A field exercise on groundwater flow using seepage meters and mini-piezometers. J Geol Educ 27:6–20

Stonestrom DA, Constantz J (2003) Heat as a tool for studying the movement of ground water near streams: U.S. Geological Survey Circular 1260:96

Field Activity 4: Stream Gaging Techniques

Stream inflow or outflow may be the dominant component of a wetland water balance, in which case it is important to measure stream discharges as accurately as possible. The following field activities will provide values of stream discharge using three different methods. These measurements are ideally conducted in a relatively small stream with a well-defined channel that is safely accessible by observers.

First, identify a suitable stream reach that satisfies the conditions listed in the first paragraph of the "Discharge measurement" segment of Sect. 3.4. Following the procedures described in "Velocity-area-method" of Sect. 3.4, a measurement section perpendicular to the flow direction should be set up. One observer wades into the stream with a current meter and a device to measure the depth of water (e.g., a wading rod), while the second observer takes notes on the bank and also takes necessary precautions for the safety of the observer in the stream. Depending on the type of current meter used, the velocity is measured at a prescribed depth (e.g., six-tenth point for the Price-type meter), or averaged over the entire depth profile in a subsection. From the depth and velocity data for individual subsections, the total discharge is calculated using Eq. 3.22. Repeat the same measurement two or three times, preferably moving the cross section upstream or downstream by several meters, and compare the results to assess the repeatability and errors of the method.

Next, measure discharge in the same stream reach using the float method described in the section "Other methods of discharge measurement". This method usually is not as accurate as the velocity-area method, but it provides a useful alternative when a current meter is not available. Any floating objects that are clearly visible and are relatively unaffected by wind can be used. Subsections should be determined in a manner similar to the velocity-area method (but usually with coarser spacing of measurement points). Once points are determined, float-velocity measurements simply replace measurements made with a current meter. The profile-averaged velocity can be estimated by multiplying the surface velocity determined with the floats by 0.85.

The tracer-dilution method provides a third value of stream discharge at this stream reach. First, select a suitable location upstream of the measured cross section for release of the stream tracer. This location should be sufficiently far upstream to ensure complete mixing of the tracer solution. This may require preliminary release of tracer at several upstream locations, along with accompanying downstream measurements of tracer concentration at several locations, to confirm complete mixing. You will want to select a tracer that can be released in small quantities but that will not be masked by the background concentration in the stream. The tracer also needs to be one that is not regulated by any stream-management authorities, or one for which you have a permit to release.

It may be convenient to use electrical conductivity (EC) as a surrogate for tracer concentration if a sufficient amount of tracer can be released to create an easily measured increase of the EC of the stream water. In streams that have very low background EC, a strong correlation between tracer concentration (e.g., chloride) and EC can be pre-established, and concentration can be estimated from the measurements of EC. If this is not feasible, water samples will need to be collected and analyzed with a field analyzer or in the laboratory. This will require a large number of samples for slug injection tests.

After the location for tracer release is selected, a choice must be made between the constant-rate injection (CRI) and the slug injection (SI) method. The CRI method requires a device for injecting tracer solution at a constant rate, but only three values of concentration are required (see Eq. 3.23). The SI method does not require a special device, but many concentration values are required to establish the time-concentration curve shown in Fig. 3.17. Here we describe the use of the CRI method. It is assumed that the background concentration is small enough that the tracer concentration can be estimated from the measurement of EC. To establish the relation between EC and tracer concentration, prepare a set of standard solutions from the tracer chemical and the stream water; for example, solutions of 0, 5, 10, 20, ... 1,000 mg of sodium chloride in 1 L of stream water. The EC values of these solutions are plotted against concentration values to establish a calibration curve.

For successful application of the CRI method, the tracer solution should be released at an appropriate rate and concentration to ensure that concentration at the measurement section can be accurately measured relative to the stream background concentration, and that a sufficient volume of tracer solution exists in the tracer-injection reservoir to achieve steady state at the sampling location. The constant release rate of tracer solution can be maintained using a Mariotte bottle or a field-portable pump with controlled flow rate (see Moore 2004 for construction of a simple Mariotte bottle from readily available materials). Once a steady value of EC is established at the sampling location and tracer concentrations are determined, the observer can calculate discharge using Eq. 3.23.

In summary, the suggested field activities for stream gauging are the following:

1. Determine stream discharge using the area-velocity method. If time permits, determine the discharge at multiple locations and assess the errors and uncertainty of this method.
2. Estimate stream discharge using the float method at the same location, and compare the accuracy of this method with the area-velocity method.
3. Determine stream discharge using the tracer dilution method.
4. Compare the values of discharge obtained by all three methods and discuss their advantages and disadvantages for application at this particular location, as well as other possible locations and situations.

Chapter 4
Hydric Soil Identification Techniques

Lenore M. Vasilas and Bruce L. Vasilas

Abstract Conceptually, hydric soils are soils that formed under hydrologic conditions associated with wetlands. Identification of soils as "hydric" is critical to the identification and protection of wetlands. Conditions of saturation and anaerobiosis associated with wetland hydrology create morphological characteristics in soils that can be used to distinguish them from non-hydric (upland) soils. These distinctive morphological characteristics have been used to develop "indicators" to facilitate the rapid identification of hydric soils in the field without relying on chemical assays or long term monitoring. An understanding of how soils form and the soil properties related to hydric soil morphologies such as soil color and texture are needed to field identify indicators of hydric soils. This chapter emphasizes the proper application of field indicators of hydric soils, the process of describing soil morphology inherent to the use of hydric soil indicators, and approaches to address soils suspected to be hydric but do not meet a field indicator.

4.1 Introduction

To fully understand the material in this chapter it should be accompanied by *Field Indicators of Hydric Soils in the United States* (Version 7.0) (USDA, NRCS 2010a) and subsequent errata, the *Army Corps of Engineers Wetlands Delineation Manual* (Environmental Laboratory 1987) and approved Regional Supplements (U.S. Army, COE 2012), and the *Munsell Book of Color* (available from Munsell Color Company, Inc. Baltimore MD).

L.M. Vasilas (✉)
Soil Survey Division, U.S. Department of Agriculture-Natural Resources Conservation Service, Beltsville, MD 20705, USA
e-mail: Lenore.Vasilas@wdc.usda.gov

B.L. Vasilas
Department of Plant and Soil Sciences, University of Delaware,
Newark, DE 19716-2170, USA

Soil morphology refers to field observable soil characteristics that can be assessed visually or by touch. Morphological characteristics addressed in this chapter include horizonation or layers, color, texture, and structure. Soil morphology typically reflects long term hydrologic conditions. Therefore, the ability to identify, document, and interpret soil morphology is critical to many wetland investigations. Expertise in soil morphology and the interpretation of soil morphology assists in (1) determinations and delineation of wetlands subject to federal jurisdiction, (2) assessment of current or past wetland hydrology, and (3) assessment of changes to wetland condition.

Of particular importance for wetland determinations is the ability to apply Field Indicators of Hydric Soils in the United States (hereafter referred to as *Field Indicators*) properly. Hydric soils are routinely identified in the field through hydric soil indicators, which are sets of morphological patterns that are correlated with soils that formed under hydrologic conditions associated with wetlands. Hydric soils are one of three factors needed to identify an area as wetlands subject to federal jurisdiction under the Clean Water Act and Food Security Act. In this chapter, we present soil morphological concepts that are used in the application of the Field Indicators. These same concepts can be used to further characterize site-specific hydrology with respect to hydroperiod (the seasonal pattern of water table depth) and hydrodynamics (the direction and energy of hydrologic inputs).

The goal of this chapter is not to turn the reader into a soil scientist, but to give the individual enough expertise in soil science to allow for routine wetland determinations and delineations, as well as hydrologic assessment. Knowledge of soil morphology also allows the wetlands practitioner to identify difficult situations where a soil scientist should be called in for assistance.

4.2 Overview of Hydric Soils

4.2.1 What Is a Hydric Soil?

Soil is a natural body comprised of solids (minerals and organic matter), liquid, and gases that occurs on the land surface, occupies space, and is characterized by one or both of the following: horizons, or layers, that are distinguishable from the initial material as a result of additions, losses, transfers, and transformations of energy and matter, or the ability to support rooted plants in a natural environment. The upper limit of soil is the boundary between soil and air, shallow water, live plants, or plant materials that have not begun to decompose. Areas are not considered to have soil if the surface is permanently covered by water too deep (typically more than 2.5 m [~8 ft.]) for the growth of rooted plants. The lower boundary that separates soil from the nonsoil underneath is most difficult to define. Soil consists of horizons near the Earth's surface that, in contrast to the underlying parent material, have been altered

4 Hydric Soil Identification Techniques

by the interactions of climate, relief, and living organisms over time. Commonly, soil grades at its lower boundary to hard rock or to earthy materials virtually devoid of animals, roots, or other marks of biological activity. For purposes of classification, the lower boundary of soil is arbitrarily set at 200 cm (~6.5 ft.) (Soil Survey Staff 1999).

The term *hydric soil* was first published in *Classification of Wetlands and Deepwater Habitats* (Cowardin et al. 1979). The initial purpose of the definition was to define a class of soils that were closely correlated with hydrophytic vegetation and to produce a list of soils that could be used with soil surveys to facilitate the development of National Wetland Inventory (NWI) maps. Conceptually, hydric soils are soils that developed under hydrologic conditions associated with wetlands. Because of the role of hydric soil identification in jurisdictional determinations of wetlands, very specific criteria/definitions are applied to distinguish hydric soils from non-hydric soils.

4.2.2 Hydric Soils and Wetland Regulation

Identification of soils as *hydric* is critical to the protection of wetlands under the Clean Water Act (CWA) (Federal Water Pollution Control 2008) and for conservation compliance under the Farm Bill. According to the *US Army Corps of Engineers Wetlands Delineation Manual* (hereafter referred to as the *Delineation Manual*) (Environmental Laboratory 1987), the presence of a hydric soil is one of three factors that must be met in order for an area to meet the definition of a jurisdictional wetland. The other two are the presence of hydrophytic vegetation and wetland hydrology. The use of the Delineation Manual and Regional Supplements (U.S. Army COE 2012) is required for all federal agencies involved in identification of wetlands that may be jurisdictional, as well as for most states that have environmental programs to protect wetlands.

A hydric soil as defined by the National Technical Committee for Hydric Soils (NTCHS) is *a soil that formed under conditions of saturation, ponding, or flooding long enough during the growing season to develop anaerobic conditions in the upper part* (Federal Register, July 13, 1994). For a soil to qualify as a hydric soil for regulatory purposes, it must meet the definition of a hydric soil. It is important to note that a soil meets the definition if it developed under the stated hydrologic conditions. If those hydrologic conditions are altered through drainage or protection (levees), the soil is still considered to be hydric *if the soil in its undisturbed state developed as a hydric soil.*

A hydric soil is defined in the National Food Security Act (USDA, FSA 1985) as *a soil that, in its undrained condition, is saturated, flooded, or ponded long enough during the growing season to develop an anaerobic condition that supports the growth and regeneration of hydrophytic vegetation.* While the definition is slightly different than the definition developed by the NTCHS, the methods (hydric soils list, Field Indicators, and Hydric Soil Technical Standard) that can be used to identify a hydric soil are the same.

Important concepts in the definition to note are:

1. *in its undrained condition* means that the soil formed under wet conditions and the absence of a water table would not preclude the soil from still being considered hydric. In other words, it may be currently in the dry part of the season when it is being observed or it may have been artificially or naturally drained but if the soil formed when the water table saturated the upper part of the soil it is still hydric;
2. *saturated, flooded, or ponded* means that the soil must have water in an unlined bore hole in the upper part of the soil or the water must rise above the surface of the soil;
3. *during the growing season* means that the water must be present during the growing season as determined by the use of the Hydric Soil Technical Standard (NTCHS 2007);
4. *anaerobic condition* means the soil lacks oxygen and is a reducing environment.

If the soil meets all the above mentioned concepts, then it will support the growth and regeneration of hydrophytic vegetation. Hydrophytic vegetation, as defined in the FSA Manual means a plant growing in (A) water; or (B) a *substrate that is at least periodically deficient in oxygen during a growing season as a result of excessive water content* [16 U.S.C. 3801(a)(13)].

4.2.3 Hydric Soil Indicators

Nearly all hydric soils exhibit characteristic morphologies that result from repeated periods of saturation or inundation for more than a few days. Saturation or inundation, when combined with microbial activity in the soil, causes the depletion of free oxygen (O_2). This anaerobiosis (without O_2) promotes certain biogeochemical processes, such as the accumulation of organic matter and the reduction, translocation, or accumulation of iron (Fe) and other reducible elements. These processes result in distinctive characteristics that persist in the soil during both wet and dry periods, making them particularly useful for identifying hydric soils in the field.

Hydric soils are routinely identified in the field through use of the Field Indicators. Most hydric soils are readily identified by observing either a predominance of gray color with redoximorphic concentrations (formerly called "high chroma mottles") near the surface or an accumulation of organically enriched material on the surface. These features indicate that the soil has been chemically reduced and fits the standard saturated soil/wet soil morphology paradigm. These readily observable soil morphologies resulting from oxidation-reduction of principally Fe near the surface and accumulation of organic matter comprise the primary Field Indicators used for jurisdictional determinations of wetlands. The presence of one indicator is evidence that the soil meets the definition of a hydric soil.

4 Hydric Soil Identification Techniques

The hydric soil indicators are "proof positive," i.e., the presence of an indicator is proof that the soil is hydric. The absence of an indicator does not prove that the soil is not hydric ("proof negative"). It is important to remember that a soil that does not contain a hydric soil indicator may in fact be hydric if it meets the definition of a hydric soil. In general, soil morphology reflects long-term hydrologic conditions, which is the basis for the Field Indicators. For a myriad of reasons, some of which are still poorly understood, there are some relatively small but significant areas that are, or appear to be, anomalies to the standard saturated soil/wet soil morphology paradigm. That is, not all hydric soils develop diagnostic redoximorphic features, and some soils have colors that suggest that the soils formed under saturated conditions when, in fact, they did not. It is these anomalous soil morphologies that are so difficult to interpret and are easily misinterpreted by the layperson that have become known collectively as *problem soils*.

Hydric soil lists, Field Indicators, and the Hydric Soil Technical Standard were all created to help identify those soils that meet the definition. If a soil meets the definition of a hydric soil, then it is hydric regardless of whether or not it is a soil series on a hydric soils list or meets an approved Field Indicator.

Currently there are not Field Indicators or soil series mapped that fit every hydric soil condition. These soils are considered problem soils for the purpose of hydric soil identification. Chapter 5 of the Regional Supplements has some suggested methods to assist in making hydric soils determinations in problem soils where Field Indicators may not adequately identify hydric soils. Ultimately, the Hydric Soil Technical Standard may need to be applied to collect data to make a hydric soils determination and/or to develop a field indicator that will work in a problem soil situation.

4.3 Soil Formation

4.3.1 Factors of Soil Formation

Soils develop as a result of the interactions of climate, living organisms, and landscape position as they influence parent material decomposition over time (the five soil-forming factors). Each of these five soil-forming factors also influence the development of morphological patterns on which the Field Indicators are based.

4.3.1.1 Parent Material

Parent material refers to the great variety of unconsolidated organic matter and mineral material in which soil formation begins. Certain parent materials such as red parent material or parent material that weathers to soils with high pH can be problematic because the hydric soils that develop in these parent materials often lack characteristic hydric soil morphologies.

4.3.1.2 Climate

Climate is a major factor in determining the kind of plant and animal life on and in the soil. It determines the amount of water available for weathering minerals. Warm, moist climates encourage rapid plant growth and thus high biomass production (primary productivity). The opposite is true for cold, dry climates. High primary productivity does not necessarily result in high soil organic matter levels as much of the fixed carbon (C) is sequestered in standing biomass. In addition, organic matter decomposition (and the demand for soil O_2) is accelerated in warm, moist climates. In saturated soils, partially decomposed organically enriched material may accumulate, such as in bogs, fens, and swamps.

4.3.1.3 Landscape Position or Topography

Topography in terms of landscape position causes localized changes in moisture and temperature. Even though the landscape has the same soil-forming factors of climate, organisms, parent material, and time, drier soils at higher elevations may be quite different from the wetter soils where water accumulates. Wetter areas may have reducing conditions that will inhibit proper root growth for plants that require a balance of soil O_2, water, and nutrients. Landscape position is an important soil forming factor for hydric soil development. A hydric soil is only going to occur in landscapes that allow for an excessive accumulation of water to cause soil saturation and reduction in the upper part.

Figure 4.1 illustrates a few common landscapes. Older terraces, or soils on second bottom positions, usually have developed B horizons (soil layers characterized by illuviated clay or organic matter). Recent soils deposited in floodplains or first bottom positions usually do not have a developed B horizon. Instead, they may have stratified layers varying in thickness, texture, and composition. Differences in climate, parent material, landscape position, and living organisms from one location to another as well as the amount of time the material has been in place all influence the soil forming process.

4.3.1.4 Organisms

Plants affect soil development by supplying upper layers with organic matter, recycling nutrients from lower to upper layers, and helping to prevent erosion. Microbial activity is the driving force behind the development of soil morphological features that are used as Field Indicators. Soil microbes have adapted to a wide range of soil conditions and are rarely a limiting factor in the development of hydric soil indicators.

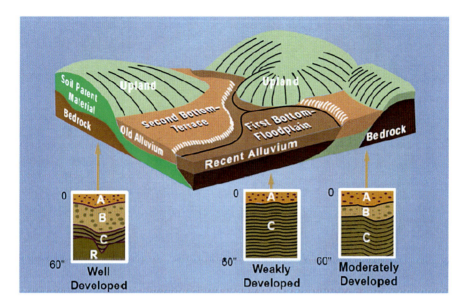

Fig. 4.1 Landscape position influences soil development (Published with kind permission of US Department of Agriculture, Natural Resources Conservation Service (2010b). Figure is public domain in the USA. All Rights Reserved)

4.3.1.5 Time

Time is required for horizon formation. The longer a soil surface has been exposed to soil forming agents like rain and growing plants, the greater the development of the soil profile. Soils in recent alluvial or windblown materials or soils on steep slopes where erosion has been active may show very little horizon development. Soils on older, stable surfaces generally have well defined horizons because the rate of soil formation has exceeded the rate of geologic erosion or deposition. Relatively young soils may lack typical hydric soil morphologies due to lack of time to allow for organic matter accumulation or redoximorphic feature formation.

4.3.2 Soil Forming Processes

The four major processes that change parent material into soil are additions, losses, translocations, and transformations.

4.3.2.1 Additions

The most obvious addition is organic material. As soon as plant life begins to grow in fresh parent material, organic material begins to accumulate. Organic matter gives a black or dark brown color to surface layers. Even young soils may have a dark surface layer. Partially decomposed organic material may accumulate in saturated soils resulting in thick organic surfaces. These thick organic surfaces are one of the features that can be used to identify a hydric soil.

4.3.2.2 Losses

Most losses occur by leaching. Water moving through the soil dissolves certain minerals and transports them into deeper layers. Some materials, especially sodium salts, gypsum, and calcium carbonate, are relatively soluble. They are removed early in the soil's formation. As a result, soil in humid regions generally does not have carbonates in the upper horizons. Quartz, aluminum, Fe oxide, and kaolinitic clay weather slowly. They remain in the soil and become the main components of highly weathered soil.

4.3.2.3 Translocations

Translocation means movement from one place to another. In low rainfall areas, leaching often is incomplete. Water starts moving down through the soil, dissolving soluble minerals such as calcium carbonate as it goes. Saturation promotes the reduction of Fe which helps bridge clay particles together. Reduced Fe and the associated clays will move with the water. When the water stops moving these materials are deposited. Soil layers enriched with clays, calcium carbonate or other salts form this way. Translocation upward and lateral movement is also possible. Translocation is most apparent in seasonally saturated soils as minerals and clay move up and down with the water table.

4.3.2.4 Transformations

Transformations are biogeochemical changes that take place in the soil. Microorganisms that live in the soil feed on fresh organic matter and change it into humus. For example, ferric Fe (Fe^{+3}) commonly present in Fe oxides under aerobic conditions and is readily reduced to soluble ferrous Fe (Fe^{+2}) which is quite easily removed from the soil by leaching. The patterns in the soil as a result of Fe transformations is the most common feature used to identify hydric soils.

4.4 Soil Horizons

The factors of soil formation do not have a consistent impact with depth. For example, plant roots may not extend throughout the entire depth of the soil. Some soils contain more than one type of parent material. Anthropogenic disturbance (such as plowing) is usually restricted to the upper part of the soil. Soil moisture

4   Hydric Soil Identification Techniques

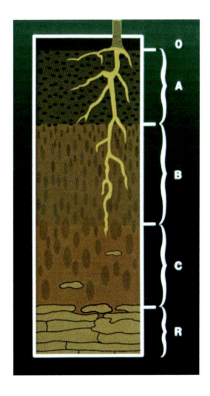

Fig. 4.2 A diagram of soil horizons (Published with kind permission of US Department of Agriculture, Natural Resources Conservation Service (2010b). Figure is public domain in the USA. All Rights Reserved)

typically increases with depth and diurnal fluctuations in soil temperature are minimized with depth. Because of this, soil forming processes are not uniform with depth. As a result, soil morphology typically changes with depth and displays distinct horizontal layers of soil called *horizons* (Fig. 4.2). Horizons can be composed predominately of organic matter (O horizons) or composed predominately of unconsolidated mineral materials (designated as A, E, B, and C horizons). Consolidated bedrock is designated as R.

O horizons form at the soil surface because they are composed primarily of plant roots and leaves in various stages of decomposition. O horizons are dark brown or black. A horizons, commonly called topsoil, are predominately mineral, but distinguished from other mineral horizons by organic matter enrichment. As a result, they tend to be dark brown or black in color. A horizons form at the soil surface or below an O horizon. E horizons represent zones of elluviation, the loss of soil components such as clay, Fe, or organic matter. B horizons represent zones of illuviation, the gain of soil components such as clay, Fe, or organic matter. C horizons display little of the soil forming processes and are similar in composition to parent material. Generally, the horizon you are in does not matter when identifying a hydric soil. However, it is important to understand when you are in an A or E horizon as the requirements for those horizons for some indicators are different than for other horizons.

A soil may lack one or more of these horizons or may have similar horizons at multiple depths. O or A horizons, which form near the soil surface, may be found deeper in the soil due to subsequent formation of horizons above them following

Fig. 4.3 Leaf litter and other recent debris should be removed and not included in the soil description

deposition. An individual horizon may display characteristics of two different types of horizons; its designation reflects this duality. For example, a horizon that is enriched with organic matter but depleted of clay would be designated AE. The presence or absence of specific horizons, and the vertical arrangement of these horizons are used by soil scientists to classify soils.

Soil descriptions document these morphological characteristics. Observations are typically made on a soil profile, a two dimensional vertical slice of soil. For assessing the presence of Field Indicators a slice to a depth of 45 cm (18 in.) is usually sufficient. Shallow soil slices are routinely extracted with a tiling spade hence the phrase *spade slice*. However, spade slices can be extracted with any shovel with a relatively flat blade. It is important to maintain the integrity of the spade slice during the extraction process. This may be facilitated by first digging a pilot hole and extracting and discarding the soil, then cutting a slice from the resulting hole. For deeper observations such as those to accompany monitoring well installment, soil samples are collected with a bucket auger. Samples collected by augering are laid out on the ground in sequence corresponding to the depth of extraction for each sample. Depth of each sample must be documented. A folding carpenter's tape works well for this purpose. Soil descriptions should start directly below the previous year's leaf fall or litter and organic material beneath the layer is considered to be part of the soil (Fig. 4.3).

4.5 Soil Color

4.5.1 Overview

Soil color is an important characteristic of soil morphology as it can be interpreted to provide information on soil mineral composition, distinguish between organic and mineral soil materials, and reflects long term hydrologic conditions. For example,

4 Hydric Soil Identification Techniques

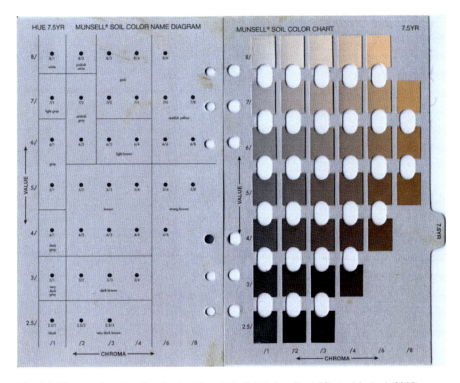

Fig. 4.4 Picture of a color chart in the Munsell™ Soil Color Chart (GretagMacbeth 2009)

in well drained soils Fe oxides usually give soils a yellow, orange, or red color. In soils that are saturated for extended periods, Fe oxides are reduced. The reduced (ferrous) form of Fe is easily removed from the soil by leaching. After the Fe is gone, generally the leached area has a grayish or whitish color. Repeated cycles of saturation and drying create a mottled soil (splotches of color(s) in a matrix of a different color). Part of the soil is gray because of the loss of Fe, and part is red or yellow where the Fe oxides remain.

Therefore, the ability to correctly identify and document soil colors and patterns of soil colors is critical to wetland investigations. Soil scientists rely on the Munsell System of Color Notation in part because it is standardized. The Munsell System includes the entire visible color spectrum using three components: hue, value, and chroma. Colors most commonly found in soils are arranged in books of color chips (Munsell™ Soil Color Charts). One of the Munsell™ soil color charts is presented in Fig. 4.4. Soil is held next to the chips (or better yet, underneath) to find a visual match and assigned the corresponding Munsell™ notation. The notation is recorded in the form: hue, value/chroma – for example, 5Y 6/3. All color chips correspond to an English name in the Munsell™ Soil Color Charts. An example color would be 10YR 4/6, which is called *dark yellowish brown*. 10YR, or 10 yellow-red, is the hue. Four is the value and 6 is the chroma. 10YR means that there are ten parts yellow to one part red.

4.5.2 Components of Soil Color

The four components that have the most affect on soil color are organic compounds (usually black or brown), manganese (Mn) oxides (usually black), iron (Fe) oxides (usually red, orange, or yellow), and the color of the mineral grains (usually clear or neutral gray). Soil color is determined by matching a moist soil to a chip in the Munsell™ Book of Color. Each chip has a specific hue, value, and chroma, identified on the printed page facing each page of chips.

4.5.2.1 Hue

Hue is the chromatic composition (color) of light that reaches the eye. Each Munsell™ page is a different hue that is printed on the upper right corner. Most soils in the Mid-Atlantic Region are on the 10YR (yellow red) page, with redder colors on pages to the left of 10YR and more yellow and grayer colors on pages to the right of 10YR. Additional hues are also used to describe soils on the gleyed pages. These hues include greens, blues, and neutral colors (white, gray, and black).

4.5.2.2 Value

Value is the degree of lightness or darkness of soil color. The value notations are found on the left margin of each page beside each row. The lower values have darker color, while higher values have lighter colors. Value is a continuous scale from 0 to 10. Whenever soil colors do not match a value chip exactly you can round the value to the nearest chip.

4.5.2.3 Chroma

Chroma is the strength or purity of color. The chroma notations are found on the bottom margin of each page under each column. The lower chromas have more neutral (often grayer) color, while highest chromas have the strongest expression of that particular hue. Technically, chroma has no upper limit to the scale, but typically the range found in soils is 0–8. Soil colors that do not match a chroma chip exactly, should be noted as falling between the two color chips. This can be done by estimating a decimal value (10YR 4/2.2) or by using a + (10YR 4/2+). Some Field Indicators require chromas of $\leq x$ while others may require a chroma $<x$. So knowing whether the soil color meets a chroma or is in between a chroma is important.

4.5.3 Conditions for Measuring Soil Color

Ideally, soil color should always be read on a ped (clump of soil) interior, immediately after excavation, in a moist state and under direct natural light. Soil is not smeared prior to reading soil color. Hydric soils, especially when they are saturated, may change color quickly upon exposure to oxygen. Therefore, it is important to describe the colors soon after excavation. If the soil does change color with time, you should also record the color of the soil once it has changed and the amount of time that has passed since excavation.

Although it is best to describe soil color moist, often a hydric soil is saturated and thus it is impossible to acquire a moist sample while in the field. In this case documentation that the soil color was read under saturated conditions is made and a sample may be collected and let dry to a moist state before soil color is read again. A saturated soil may change color as it dries indicating a reduced matrix (Fe is reduced in situ). Changing moisture content may affect soil value, while a change due to oxidation or reduction of Fe will most likely produce a change in chroma and confirms a reduced matrix (reduced Fe was present). If the change in color is only due to moisture state and not Fe reduction, then the moist color only needs to be recorded. However, if it is in fact a reduced matrix both colors and the fact the matrix is reduced should be noted.

Soil color should be read under full natural light with the color book facing the sun at a 90° angle. It is best to do this during mid-day when the sun is high. If soil color is read in a forest, the color should be read in a spot where the sun is shining through the canopy. Morning and evening sunlight makes it much more difficult to distinguish between different colors, especially in the winter.

4.5.4 Describing Soil Colors

Multiple colors are often present in a single horizon or layer of a hydric soil. The color pattern is critical to many of the Field Indicators. Therefore, when describing soil colors it is important to document the pattern of colors according to the following parameters.

4.5.4.1 Matrix Color

The matrix color (dominant color) is the color that occupies the greatest volume of the layer (Fig. 4.5). If there are multiple colors that appear to be equally dominant, the soil is described as having a mixed matrix.

Fig. 4.5 The dominant color of this soil is *gray*. Therefore, the matrix color of this soil would be considered *gray* while the other splotches of color would be considered mottles. In this soil, the mottles are due to wetness in the soil and are a type of mottle called redoximorphic features

4.5.4.2 Mottling Versus Redoximorphic Features

Secondary zones of color less dominant in surface area to the matrix are referred to as mottles. Redoximorphic (redox) features are a type of mottling that is associated with wetness and form as a result of saturation and reduction of Fe and manganese (Mn).

4.5.4.3 Percentages

When assessing a soil layer with multiple colors care should be taken to accurately document percentages as they are critical to many of the Field Indicators. Some of these require a minimum percentage of the matrix and/or redoximorphic features. One example would be A11 Depleted Below Dark Surface, which requires a matrix color that has ≥ 60 % of the layer with a chroma of ≤ 2 starting within 30 cm (12 in.) of the soil surface. Another would be F6 Redox Dark Surface, which requires ≥ 2 % distinct or prominent redox concentrations (F6a) or ≥ 5 % distinct or prominent redox concentrations (F6b).

4.5.4.4 Contrast

Contrast refers to the degree of visual distinction that is evident between associated colors. Three categories of contrast are recognized as faint, distinct, and prominent. Contrast is an important consideration when using the Field Indicators as most indicators require redox concentrations to be either distinct or prominent. Note that currently the only Field Indicator to allow faint contrast is S6 Stripped Matrix. The upper threshold for faint contrast is presented in Table 4.1.

Table 4.1 Upper thresholds for faint contrast. Any feature above the upper threshold for faint features would be considered either distinct or prominent

Upper thresholds for faint contrast		
Δ Hue	Δ Value	Δ Chroma
0	2	1
1	1	1
2	0	0
Hue	**Value**	**Chroma**
Any	3	2

4.5.4.5 Type of Redoximorphic Features

Four classes of redoximorphic features are recognized as defined below. On the data sheet under the category *Type*, they should be noted by their abbreviation. Examples of redox concentrations and depletions are shown in Fig. 4.6.

1. Concentration (C): Bodies of apparent accumulation of Fe-Mn oxides.
2. Depletion (D): Bodies of low chroma (less than or equal to) having values of 4 or more where Fe-Mn oxides alone have been stripped out or where both Fe-Mn oxides and clay have been stripped out.
3. Reduced Matrix (RM): Soil matrices that have a low chroma color *in situ* because of the presence of Fe^{2+}, but whose color changes in hue or chroma when exposed to air as the Fe^{2+} is oxidized to Fe^{3+}. The change in color occurs within 30 min or less after the sample is exposed to air.
4. Masked Sand Grains (CS): This applies to particles masked with coats of organic material.

4.5.4.6 Location of Redoximorphic Features

When noting "Location" there are two categories, which are defined below:

1. Pore Lining (PL): Zones of accumulation are either coatings on a ped or pore surface or impregnations of the matrix adjacent to the pore or ped.
2. Matrix (M): Zones of accumulation that are impregnations within the matrix.

4.6 Soil Texture

Soil texture, or particle size distribution, is the numerical proportion of the mineral particles <2 mm (in.) in size (sand, silt, and clay) and is expressed as percent by weight. These mineral size classes are distinguished by size: sand, 0.05–2 mm; silt, 0.002–0.05 mm; and clay, <0.002 mm. Figure 4.7 shows the relative sizes of sand, silt, and clay. Almost everyone knows what sand and clay feel like, either from playing in a sandbox or sculpting with clay in an art class. Silt feels similar to talcum powder or flour. Typically, a sample of soil will contain all three components in various ratios. Therefore, soil textural classes were created to designate the ratios (Fig. 4.8). For example, a sandy clay contains 35–55 % clay,

Fig. 4.6 Types of redoximorphic features. (a) is a redox concentration as a soft mass on the interior of a ped within the matrix. (b) is a redox concentration along a pore lining. (c) is a redox depletion adjacent to a plant root

4 Hydric Soil Identification Techniques

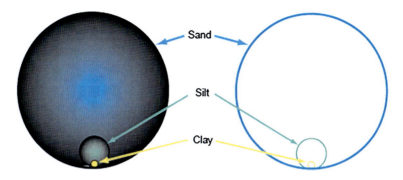

Fig. 4.7 Relative sizes of sand, silt, and clay particles (Published with kind permission of US Department of Agriculture, Natural Resources Conservation Service (2010b). Figure is public domain in the USA. All Rights Reserved)

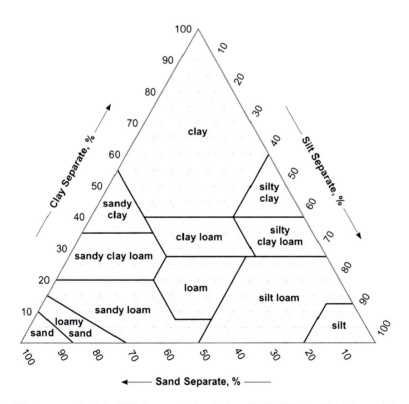

Fig. 4.8 An example of the USDA textural triangle for soils (Published with kind permission of US Department of Agriculture, Natural Resources Conservation Service from Schoeneberger et al. (2002). Figure is public domain in the USA. All Rights Reserved)

45–65 % sand, and 0–20 % silt. Textural classes can be identified in the field using a flow chart (Fig. 4.9). The process is difficult with very dry samples so it is helpful to moisten a dry sample with a spray bottle.

This flow chart is intended only for soil materials that are predominately mineral. For organic soil materials, see *Soil Organic Matter* later in this chapter. A difficult call, even for an experienced soil scientist, is for mineral soil materials enriched with organic material to the extent that it displays characteristics of organic material. In that case, the mineral texture is assigned a mucky modifier, for example, mucky-modified silt loam. This issue is also addressed in *Soil Organic Matter*.

The most important mineral soil texture separation is between loamy fine sand and loamy very fine sand as this break determines which field indicators can be used for identifying the soil as hydric. Based on the diagram, a soil that does not ribbon is generally sandy (loamy fine sand or coarser) and those that do form a ribbon are loamy or clayey (finer than loamy fine sand).

4.7 Soil Structure and Bulk Density

In general, mineral soil particles do not occur as independent units. Instead, multiple particles are grouped into secondary units called peds or aggregates. This aggregation is promoted by oxidized Fe, organic matter, and physical forces associated with wetting and drying cycles, freezing and thawing cycles, or vehicular traffic. Soil structure refers to the shape and distribution of peds and the resistance of peds to physical change. Examples of structural shape classes are presented in Fig. 4.10. Structural units are also rated for strength and are reported as weak, moderate, or strong.

Bulk density is defined as soil dry weight per unit volume. Sand has a higher bulk density than clay. Organic soil materials have a lower bulk density than mineral soil materials. O horizons have lower bulk densities than mineral horizons, and an A horizon generally has a lower bulk density than the underlying B horizon. Bulk density decreases as porosity (% pore space by volume) increases. Compaction is an increase in bulk density; it can be caused by vehicular traffic or by long-term inundation. Structural classes are associated with general ranges in bulk density. For example, granular structure is prevalent in A horizons and is promoted by organic matter; it is associated with low bulk densities. The single grain class is associated with sandy materials, C horizons, and high bulk densities. Blocky structure is associated with B horizons enriched with clay and intermediate bulk densities.

Use of the Field Indicators does not require familiarity with structure or bulk density. However, as addressed later, knowledge of these soil characteristics are critical to the installation of monitoring wells and the assessment of wetland hydrology as they can significantly impact the flow path of water in soil. Figure 4.11 shows the impact of structure on percolation of water through soil.

4 Hydric Soil Identification Techniques

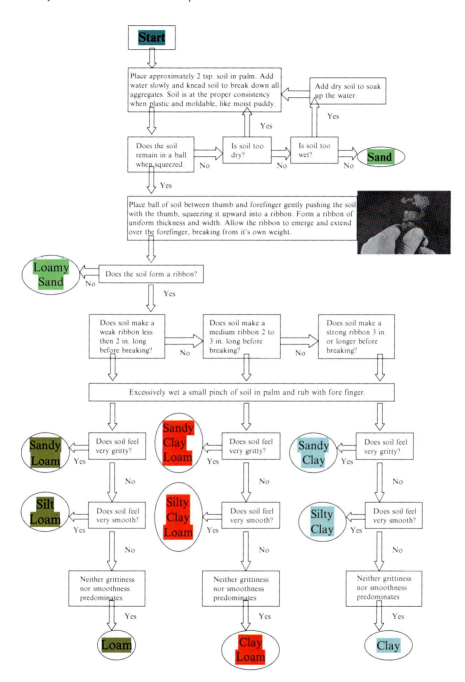

Fig. 4.9 Flow chart for determining soil texture (Modified from Thien (1979), p. 55. Published with kind permission of © American Society of Agronomy, 5585 Guilford Rd., Madison, WI 53711, 1979. All Rights Reserved)

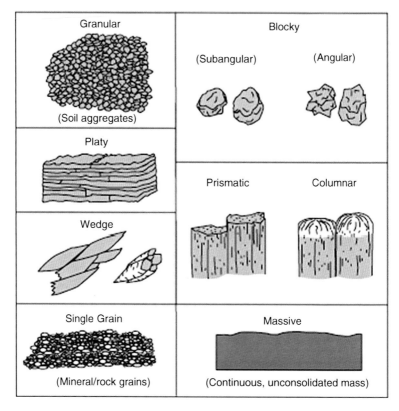

Fig. 4.10 A diagram of the types of soil structure (Published with kind permission of US Department of Agriculture, Natural Resources Conservation Service from Schoeneberger et al. (2002). Figure is public domain in the USA. All Rights Reserved)

4.8 Soil Organic Material

4.8.1 Overview

Soil microbes use carbon (C) compounds found in organic material as an energy source. However, the rate at which organic C is utilized by soil microbes is considerably lower in a saturated and anaerobic environment than it is under aerobic conditions. Therefore, soils that are saturated the entire growing season may accumulate partially decomposed organic material. The result in wetlands is often the development of O horizons of various thicknesses or dark organic-rich mineral surface layers. Three types of O horizons are recognized and distinguished by the level of organic material decomposition. Oa indicates highly decomposed organic material, Oi is slightly decomposed, and Oe is intermediate in decomposition.

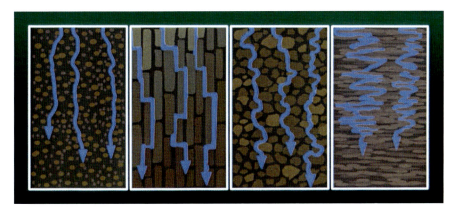

Fig. 4.11 Impact of structure on percolation of water through soil (Published with kind permission of US Department of Agriculture, Natural Resources Conservation Service (2010b). Figure is public domain in the USA. All Rights Reserved)

By definition, organic soil material is saturated with water for long periods or is artificially drained and, excluding live roots, has an organic C content ≥ 18 % with ≥ 60 % clay, or ≥ 12 % organic C with 0 % clay (Soil Survey Staff 1999). Soils with an intermediate amount of clay have an intermediate amount of organic C. Three types of organic soil materials are recognized and distinguished by the degree of decomposition: muck (highly decomposed), peat (very little decomposition), and mucky peat (intermediate decomposition).

To distinguish mineral soil materials that are highly enriched with organic matter, a "mucky" modifier is added to its mineral texture designation; for example, mucky sand. Mucky modified mineral soil with 0 % clay has 5–12 % organic C. Mucky modified mineral soil with 60 % clay has 11–18 % organic C (Soil Survey Staff 1999). Soils with an intermediate amount of clay have intermediate amounts of organic C. The Field Indicators in this category are: A7 5 cm Mucky Mineral, S1 Sandy Mucky Mineral, and F1 Loamy Mucky Mineral.

4.8.2 Field Characterization of Soil Materials High in Organic Carbon

Material high in organic C could fall into three categories: (1) organic, (2) mucky mineral, or (3) mineral. In lieu of laboratory data, the following estimation method can be used for soil material that is wet or nearly saturated with water. The first step is to determine whether the material is mineral or organic. If organic, the second step is to determine the type of soil organic material.

These soil material categories can be determined by gently rubbing wet soil material between forefinger and thumb. For sandy textured soils, if upon the first or second rub the material feels gritty, it is mineral soil material. If after the second rub the material feels greasy, it may be either mucky mineral or organic soil material. If after additional rubbing (2–3×) the sample feels gritty or plastic, it is considered mucky mineral soil material; if it still feels greasy, it is organic soil material. Accumulation of silt residue on fingers after rubbing indicates that the sample is likely mineral or mucky mineral.

Another method is to take equal amounts of known mineral soil and the horizon in question. An organic soil material will be much lighter than equal amounts of mineral material. Mucky mineral would be slightly lighter than equal amounts of mineral material. The reason for the difference in weight is due to the greater bulk density of mineral material compared to organic matter. If the material is organic soil material, a further division should be made to identify the type of organic soil material. Organic soil materials are classified as sapric, hemic, or fibric which correspond to the organic texture classes muck, mucky peat, and peat, respectively. Organic texture class can be determined by rubbing a soil sample about ten times and then visually estimating the proportion of the sample comprised of fibers (excluding live roots). After rubbing, sapric material or muck will have less than 1/6 visible fibers; fibric material or peat will have more than 3/4 fibers; and hemic material or mucky peat will have between 1/6 and 3/4 fibers (Soil Survey Staff 1999).

4.9 Formation of Hydric Soils

Hydric soils are soils that developed under conditions of saturation close to the soil surface. Under saturated conditions, plant roots and microorganisms use O_2 faster than it can be replenished by diffusion from the atmosphere resulting in first anaerobic conditions and then reducing conditions. The change from aerobic conditions to anaerobic conditions causes a shift in the direction or rate of a number of biogeochemical processes, especially those that impact the accumulation or loss of Fe, Mn, sulfur (S), or C compounds. This results in distinct soil morphological characteristics that serve as the basis for the Field Indicators. For more information on this subject refer to Chap. 7 on *Wetland Biogeochemistry Techniques*.

4.9.1 Processes

4.9.1.1 Soil Saturation

A horizon is considered saturated when the soil water pressure is zero or positive (at or above atmospheric pressure). At these pressures, water will flow from the soil matrix into unlined auger holes. Three types or patterns of saturation are defined:

1. Endosaturation-ground water table. Soil is saturated in all horizons below the water table to a depth of 2 m.
2. Episaturation-perched water table. Soil is saturated in a horizon that overlies an unsaturated horizon, and the unsaturated horizon lies within a depth of 2 m from the surface.

4 Hydric Soil Identification Techniques

Table 4.2 Element reduction sequence in inundated soils

Eh threshold, mv	Element	Oxidized form	Reduced form(s)
+350	Oxygen	O_2	H_2O
+220	Nitrogen	NO_3^-	N_2O, NO_2^-
+200	Manganese	Mn^{+4}	Mn^{+2}
+120	Iron	Fe^{+3}	Fe^{+2}
−150	Sulfur	SO_4^{-2}	H_2S
−250	Carbon	CO_2	CH_4

3. Anthric saturation-paddy soil with a created perched water table. Like episaturation but must occur under controlled flooding, for example, wetland rice or cranberries.

It should be noted that the term *water table* is not used in the definition of saturation. Also, horizons within the capillary fringe are technically not considered saturated since this contains soil water that has pressures less than atmospheric pressure. Under ideal circumstances, horizons that are saturated by the above criteria will have all their soil pores filled with water. However, for a horizon to be considered saturated it is not necessary that all pores be filled with water. Horizons that have soil water pressures of zero or positive are considered saturated even if they contain entrapped air in some pores. Saturation can occur at any time during the year.

4.9.1.2 Anaerobiosis

When aerobic conditions exist, bacteria decompose organic matter and consume O_2 in soil pores containing air. Under anaerobic conditions, bacteria decompose organic matter by consuming dissolved O_2 until it is gone. At this point, the soil water is reduced. The bacteria continue to consume organic matter, but at a slower rate. They produce organic chemicals that reduce nitrates (NO_3^-) and minerals, including Fe and Mn oxides. The sequence is shown in Table 4.2. While nitrate reduction is the first indication of anaerobic conditions, it does not leave a visible indicator that can be used for the easy identification of a hydric soil.

4.9.1.3 Iron and Manganese Reduction, Translocation, and Accumulation

Both oxidized Fe and Mn can be chemically reduced under certain soil conditions. Reduction occurs when oxidized forms of Fe (ferric, Fe^{3+}) or Mn (manganic, Mn^{3+} or Mn^{4+}) accepts electrons from another source such as organic matter to produce ferrous Fe (Fe^{2+}) and manganous Mn (Mn^{+2}). When these elements are reduced in a soil, several processes occur: (1) Fe and Mn oxide minerals begin to dissolve in water; (2) the soil colors change to gray; and (3) Fe^{2+} and Mn^{+2} ions diffuse through or move with the soil water to other parts of the soil horizon or may be leached from the soil. When Fe and Mn are in their reduced form, they have much less coloring effect on soil than when they occur in their oxidized forms. Of the two, evidence of Fe reduction is more commonly observed in soils.

4.9.1.4 Sulfate Reduction

Sulfur is one of the last elements to be reduced by microbes in an anaerobic environment. The microbes convert sulfate (SO_4^{-2}) to hydrogen sulfide (H_2S) gas. This results in a very pronounced "rotten egg" odor in some soils that are inundated or saturated for very long periods. In unsaturated or non-inundated soils, SO_4^{-2} is not reduced and there is no rotten egg odor. The presence of H_2S is a strong indicator of a hydric soil, but this indicator is found only in the wettest sites in soils that contain S-bearing compounds. It can sometimes be sensed by simply walking across these areas. This is indicator A4 Hydrogen Sulfide. Caution should be used when using this as an indicator so that other smells such as those associated with the decomposition of organic matter are not mistaken for a sulfidic odor.

4.9.1.5 Organic Accumulation

Soil microbes use C compounds found in organic material as an energy source. However, the rate at which organic C is utilized by soil microbes is considerably lower in a saturated and anaerobic environment than under aerobic conditions. Therefore, in saturated soils, partially decomposed organic material may accumulate. The result in wetlands is often the development of organic surfaces of varying thicknesses, such as peat or muck, or dark organic-rich mineral surface layers. These soils are typically saturated for very long periods of time.

4.9.2 Development of Redoximorphic Features

Redoximorphic features are those formed by the reduction and oxidation of Fe and Mn compounds in seasonally saturated soils. Fe oxide minerals give the soil red, brown, yellow, or orange colors depending on which iron minerals are present. Manganese oxides produce black colors. These oxides tend to coat the surfaces of the soil particles. Without the oxide coatings, the particles are gray. Areas in the soil where Fe is reduced often develop characteristic bluish-gray or greenish-gray colors known as *gley*. Ferric Fe is insoluble but Fe^{2+} easily enters the soil solution and may be moved or translocated to other areas of the soil. Areas that have lost Fe typically develop characteristic gray or reddish-gray colors and are known as *redox depletions*. If a soil reverts to an aerobic state, Fe that is in solution will oxidize and become concentrated in patches and along root channels and other pores where oxygen enters or remains in the soil. These areas of oxidized Fe are called *redox concentrations*. Since water movement in these saturated or inundated soils can be multi-directional, redox depletions and concentrations can occur anywhere in the soil and have irregular shapes and sizes. Soils that are saturated and contain Fe^{2+} at the time of sampling may change color upon exposure to the air, as Fe^{2+} is rapidly converted to Fe^{3+} in the presence of O_2. Such soils are said to have a *reduced matrix* (Vepraskas 1994).

While indicators related to Fe or Mn depletion or concentration are the most common in hydric soils, they cannot form in soils whose parent materials are low in Fe or Mn. Soils formed in such materials may have low-chroma colors that are not related to saturation and reduction. For such soils, features formed through accumulation of organic C may be present.

4.9.3 Types of Redoximorphic Features

4.9.3.1 Iron and Manganese Depletions

Formation is similar for Fe and Mn depletions, and both may occur within the same or adjacent horizons. It is easiest to visualize these features forming around roots that grow along stable macropores. These are required so that features continue to enlarge as succeeding roots grow and die along the same macropore. Roots growing along a structural crack or channel provide an energy source, organic material, that is needed by the microbes for Fe reduction. When the root dies and the macropore is filled with water, the bacteria will consume the root tissue and utilize (reduce) O_2 in the water if soil temperatures are high enough for the bacteria to be active. The newly formed bleached layer where Fe and Mn have been removed along the channels is a redox depletion, specifically an Fe depletion due to its lower content of Fe and Mn.

4.9.3.2 Masses, Nodules and Concretions

When a horizon has been repeatedly saturated, reduced, and drained, Fe masses will form where air penetrates into the horizon slowly to oxidize reduced Mn and Fe ions. Nodules and concretions are believed to form when air penetrates quickly, perhaps at a point into the wet matrix containing Fe^{2+} and Mn^{+2}.

4.9.3.3 Reduced Matrices

A reduced matrix forms simply by the reduction of Fe in the soil. This requires that the soil horizon be saturated to exclude air for a long enough period of time such that Fe reduction occurs. Reduced matrices can only occur where soluble organic matter is present and microorganisms are active.

4.9.4 Location of Redoximorphic Features

Pore linings occur along ped surfaces as well as root channels. They are also found on the roots of living plants that can transport O_2 to their roots in saturated soils (oxidized rhizospheres). These form by diffusion of Fe^{2+} and Mn^{+2} ions toward aerated macropores, where the ions are oxidized adjacent to the macropores and even on root surfaces. If both Fe and Mn are in solution, the Fe tends to precipitate first because it will oxidize at a lower Eh value than will Mn. Therefore, pore linings may appear to consist of clearly separated Mn oxides (in the macropore) and Fe oxides (in the matrix).

In terms of location, reducing conditions will occur near the root channels if the soluble carbon source required by the bacteria comes from dead roots. If the organic compounds are dissolved and dispersed in the soil water, then reduction can occur at any place in a soil horizon where the pores are filled with water.

4.10 Using Field Indicators of Hydric Soils

4.10.1 Overview

To fully understand this section, we recommend that the reader downloads and prints out a copy of the most recent version of *Field Indicators of Hydric Soils in the United States* along with any subsequent errata from the NTCHS website at http://soils.usda.gov/use/hydric/ and a copy of Chapter 3 of the Corps Delineation Manual Regional Supplement for the area of interest at http://www.usace.army.mil/Missions/CivilWorks/RegulatoryProgramandPermits/reg_supp.aspx for reference. The publication *Field Indicators of Hydric Soils in the United States* is a comprehensive list of all the Field Indicators approved for use by the NTCHS. All Field Indicators listed in the Corps Regional Supplements are a subset of the NTCHS national list of indicators. The Regional Supplements also contain the appropriate data sheet for determination and delineation of wetlands and hydric soils in that region.

Not all of the Field Indicators are appropriate for each situation. The Field Indicators are regionalized, and each indicator is only valid in specific Land Resource Regions (Fig. 4.12). In addition, some indicators are restricted by soil texture. There are three categories of Field Indicators which are distinguished by soil texture: All Soils, Sandy Soils, and Loamy and Clayey Soils. *All soils* refers to soils with any USDA soil texture. Examples include A1 Histosol, A4 Hydrogen Sulfide, and A12 Thick Dark Surface, among others. Sandy soils have a USDA texture of loamy fine sand and coarser (sandier). Examples include S1 Sandy Mucky Mineral, S6 Stripped Matrix, and S10 Alaska Gleyed. The loamy and clayey soils category has USDA textures of loamy very fine sand and finer (more clay). Examples include F1 Loamy mucky mineral, F6 Redox Dark Surface, and F9 Vernal Pools.

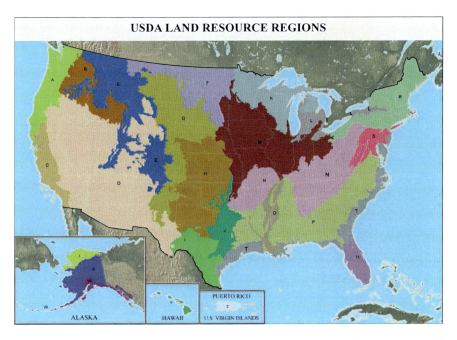

Fig. 4.12 Major land resource regions (Published with kind permission of US Department of Agriculture, Natural Resources Conservation Service (2006). Figure is public domain in the USA. All Rights Reserved)

The descriptions of the Field Indicators are structured as follows:

1. Alpha-numeric listing (A, S, or F Indicators)
2. Short name
3. Applicable land resource regions (LRRs)
4. Description of the field indicator
5. User notes

For example, *A2* is the second indicator in the "all soils" category; the short name is *Histic Epipedon*; the indicator is for use in *all LRRs*; the description is *a histic epipedon underlain by mineral soil material with chroma of 2 or less*. Helpful user notes are added.

4.10.2 Terminology

There are many important definitions that must be understood to properly use the Field Indicators. Many can be found in the glossary of the Field Indicators of Hydric Soils publication (USDA 2010a). Some of these definitions may be slightly different than the use of the same term for other purposes such as for use in soil taxonomy and soil survey. When a term that describes an indicator differs from other soil science references, an asterisk (*) is placed next to the term in the Field Indicators glossary. Familiarity with the following terms (presented in the Field Indicators glossary) is necessary for the identification of Field Indicators.

1. Depleted matrix – This is an important concept used in many of the Field Indicators. Note that all depleted matrices must have values ≥ 4 and chromas ≤ 2, and depending on the value and chroma, may or may not need the presence of redox concentrations.
2. Gleyed matrix – The definition of a gleyed matrix for the purposes of the Field Indicators are different than the definition used in Soil Taxonomy. For the purposes of the Field Indicators, a gleyed matrix has colors found on the gleyed pages of the Munsell Book of Color and also must have a value ≥ 4.
3. Layer(s) – A soil horizon and a layer for the purposes of the Field Indicators are not synonymous. There can be multiple layers in the same horizon if that horizon meets all the requirements of two different layers and is thick enough to meet the combined thicknesses requirements of the both layers. There can also be multiple horizons that meet all the requirements of the same layer except thickness that can be combined to meet the thickness requirements. For a good explanation of combining horizons to meet the thickness requirement, see Chapter 3 of any Corps of Engineers Regional Supplement (http://www.usace.army.mil/Missions/CivilWorks/RegulatoryProgramandPermits/reg_supp.aspx).
4. LRR and MLRA – Field Indicators are regionalized. Identification of the LRR (and in some cases what MLRA the sites occurs in) for the site in question is critical as it limits the Field Indicators valid for that site. Figure 4.10 is a map of the LRRs. MLRA identification can be obtained from: http://soils.usda.gov/survey/geography/mlra/.
5. Organic soil material – Use of the Field Indicators require a distinction between organic soil material and mineral soil material. Some Field Indicators also require the distinction between the grades of decomposition (muck, mucky peat, or peat).
6. Within – When a Field Indicator states that a layer must start within a certain depth, if the layer starts at that depth it is considered to be within that depth.

4.10.3 Concepts and Rules

In addition, a clear understanding of the following concepts is inherent to the proper application of the Field Indicators:

1. The Field Indicators are proof positive. If a soil meets a Field Indicator, it is a hydric soil. If it does not meet an indicator, it is still a hydric soil if it meets the definition.
2. A soil must meet the requirements in the indicator description to meet that Field Indicator. User notes are provided to assist in the interpretation of those requirements.

3. The Field Indicators were developed to locate the hydric soil boundary. Wetter soils may not meet a Field Indicator.
4. Depths and thicknesses are critical in the upper 30–45 cm (12–18 in.) of the soil when using the Field Indicators. It is recommended that a spade, not an auger, be used to excavate the soil.
5. All soil (A) indicators can be used in any layer regardless of texture. Sandy soil (S) indicators can be used in layers that are loamy fine sand or coarser. Loamy and clayey (F) indicators can be used in layers that are loamy very fine sand and finer.
6. Layers are not synonymous with horizons. One horizon may consist of multiple layers or one layer may include multiple horizons.
7. If a soil meets all the requirements of multiple indicators except thickness, you can combine indicators by adding up the thicknesses of each layer that meets the requirements. You can then designate it as hydric if the thickness is as thick as the most stringent thickness requirement of the indicators it meets (see an explanation of this in the introductory information in Chapter 3 of your regional supplement for a more thorough explanation).
8. Chromas should not be rounded. If the chroma appears to be between color chips, indicate that by using a + or a decimal point. Some indicators require a chroma of x or less. Others require a chroma less than x. In the former, if the color is between the required chroma and a higher chroma, it does not meet the requirement. In the latter, if the color is between the listed chroma and the next lower chroma, it does meet the requirement.
9. In LRRs R, W, X, and Y, observations begin at the top of the mineral surface (underneath any and all fibric, hemic, and/or sapric material) except for application of indicators A1, A2, and A3, where observations begin at the actual soil surface. In LRRs F, G, H, and M, observations begin at the actual soil surface if the soil is sandy and for the application of indicators A1, A2, and A3; and at the muck or mineral surface for the remaining Field Indicators. In the remaining LRRs, observations begin at the top of the muck or mineral surface (underneath any fibric and/or hemic material) except for application of indicators A1, A2, and A3 where observations begin at the actual soil surface.
10. Except for indicators A16, S6, S11, F8, F12, F19, F20, and F21 (those indicators that do not require a chroma ≤ 2 to meet the indicator), any soil material above the indicator must be a chroma ≤ 2 or if the chroma is >2 it must be less than 15 cm (6 in.) thick.
11. Both the definition of a depleted matrix and a gleyed matrix require values ≥ 4. This is to separate redox colors from organic matter accumulation colors. A, E and calcic horizons require ≥ 2 % concentrations.
12. Remember to describe organic features such as type (peat, mucky peat, or peat), color, mucky modified mineral, and percent masking of sand grains.

It is critical that the practitioner be familiar with the general rules required for using the Field Indicators. There are situations where a soil may meet all the requirements of the Field Indicator, however, it is not a hydric soil based on that Field Indicator because it has failed one of the general rules. The most common

Fig. 4.13 Two hydric soils. The soil on the *left meets* Field Indicator S7 Dark Surface. The soil on the *right meets* Field Indicator F3 Depleted Matrix (Published with kind permission of US Department of Agriculture, Natural Resources Conservation Service (2010a). Figure is public domain in the USA. All Rights Reserved)

example is a soil that meets all the requirements of F3 Depleted Matrix, but the layer that meets the requirements starts below 15 cm (6 in.) and the matrix chroma above the layer is a chroma >2. While the soil does meet all the specific requirements of the Field Indicator, it fails the rule that any soil material above the indicator must have a matrix chroma ≤2 or, if the matrix chroma is >2 it must be <15 cm thick.

It is helpful for the practitioner to review the Corps Regional Supplement for the geographic area in question and identify those Field Indicators applicable in their LRR or MLRA. It may also be helpful to create a one page cheat sheet of the Field Indicators that only lists the indicator descriptions for those identified for the region in question. However, when learning to use the Field Indicators it is helpful to have the user notes and glossary handy for referral when attempting to use a Field Indicator. The sheer number of Field Indicators that that can be used in a region can be intimidating to the novice. However, with experience, the practitioner will find that a small number of them are used for the majority of hydric soil identifications. The remainder of Field Indicators are used in areas that are obviously wet and not near the hydric soil boundary or for areas or specific situations that did not have commonly used Field Indicators. Nationwide, the commonly used Field Indicators are A11 Depleted Below Dark Surface, F3 Depleted Matrix (Fig. 4.13), F6 Redox Dark Surface, S5 Sandy Redox, and S7 Dark Surface (Fig. 4.12).

4.11 Soil Surveys and Hydric Soil Lists

Soil surveys are available for most areas and can provide useful information regarding soil properties and soil moisture conditions. A list of available soil surveys is located at http://soils.usda.gov/survey/online_surveys/. Soil maps and data are available online at http://websoilsurvey.nrcs.usda.gov/. Soil survey maps divide the landscape into areas called map units. Map units usually contain more than one soil type or component. They often contain several minor components or inclusions of soils with properties that may be similar to or quite different from the major component. Those soils that are hydric are noted in the Hydric Soils List.

Hydric Soils Lists are developed for each detailed soil survey based on criteria to identify soil map unit components that are at least in part hydric (Federal Register [FR Doc. 2012–4733], 2012). These lists rate each soil component as either hydric or non-hydric based on soil property data. If the soil is rated as hydric, information is provided regarding whether the soil meets the definition due to saturation, flooding, or ponding; and on what landform the soil typically occurs. Hydric Soils Lists are useful to identify areas likely to contain hydric soils. However, not all areas within a mapping unit or polygon identified as having hydric soils may be hydric. Conversely, inclusions of hydric soils may be found within soil mapping units where no hydric soils have been identified.

Soil survey information can be valuable during preliminary data gathering and synthesis. Landscape relationships and other information that can help identify the location of the component of the map unit that is hydric vs. non-hydric is also helpful. Local Hydric Soils Lists are available from state or county NRCS offices and over the internet from the Field Office Technical Guide, Section 2 (http://www.nrcs.usda.gov/technical/efotg/index.html) or Soil Data Mart (http://soildatamart.nrcs.usda.gov/). Local Hydric Soils Lists have been compiled into a National Hydric Soils List and are available at: http://soils.usda.gov/use/hydric/.

4.12 Soils That Lack Hydric Soil Indicators

4.12.1 Overview

As stated earlier, the requirements of a hydric soil are those presented in the definition. Field Indicators were created to assist in identifying those soils that meet the definition. However, the Field Indicators do not replace or relieve any of those requirements. If it meets a Field Indicator, it has morphology that indicates that the soil meets the hydric soils definition and therefore is a hydric soil. However, if it does not meet a Field Indicator, it may still be a hydric soil if it meets the requirements in the defintion. Hydric Soils Lists and the Hydric Soil Technical Standard are two approaches that may lead to an assessment of a soil as hydric by definition even though it does not meet a Field Indicator.

4.12.2 Quick Identification of Soils That Lack Field Indicators of Hydric Soils

The following key was created to identify soils that will definitely not meet a Field Indicator. This key does not identify soils that are not hydric soils. However, the identification of soils that cannot meet a Field Indicator saves field time by eliminating the more tedious process of identifying a Field Indicator.

A quick way to identify these soils is to dig a hole to 6 in. (15 cm) and address the following questions:

1. Do organic soil materials or mucky modified layers exist?
2. Do chromas ≤ 2 exist?
3. Are there any distinct or prominent redox concentrations as soft masses or pore linings?
4. Is it a sandy soil with stripped zones?
5. Is there a hydrogen sulfide odor?
6. Are you in red parent material, a depression, on a floodplain, or within 200 m (656 ft.) of an estuarine marsh and 1 m (3.3 ft.) of mean high water?

If answer is no to all five questions, the soil will not meet an indicator. This does not mean the soil is not hydric. If the soil meets the definition of a hydric soil but fails this test it only means it will not meet a Field Indicator.

4.12.3 Problematic Soil Situations

There are many problematic soil situations that currently lack an appropriate Field Indicator. Chapter 5 of the Corps Regional Supplements suggests methods to assist in the identification of a hydric soil in these problematic situations. Also, at the end of Chapter 3 of each Regional Supplement is a list of test indicators that may help in problematic soil situations. Test indicators are Field Indicators that show potential but have not been approved by the NTCHS.

If no hydric soil indicator is present, the additional site information below may be useful in documenting whether the soil is indeed non-hydric or if it might represent a "problem" hydric soil that meets the hydric soil definition despite the absence of indicators. Addressing the following questions can aid in the identification of problematic soil situations.

1. *Hydrology* – Is standing water observed on the site or is water observed in the soil pit? What is the depth of the water table in the area? Is there indirect evidence of ponding or flooding? Is the site adjacent to a downcut or channelized stream? Is the hydrology impacted by ditches or subsurface drainage lines?
2. *Slope* – Is the site level or nearly level so that surface water does not run off readily, or is it steeper where surface water would run off from the soil?

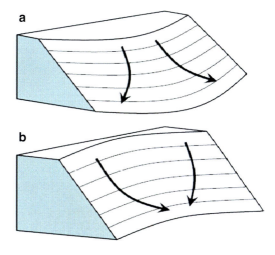

Fig. 4.14 Divergent slopes (**a**) disperse surface water, whereas convergent slopes (**b**) concentrate water. Surface flow paths are indicated by *arrows*

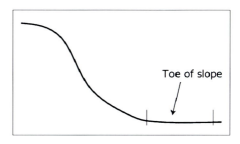

Fig. 4.15 At the toe of a hill slope the gradient is slightly inclined or nearly level

3. *Slope shape* – Is the surface concave (e.g., depressions), where water would tend to collect and possibly pond on the soil surface? On hillsides, are there convergent slopes (Fig. 4.14), where surface or groundwater may be directed toward a central stream or swale? Or is the surface or slope shape convex (e.g., dome shaped), causing water to run off or disperse?
4. *Landform* – Is the soil on a low terrace or floodplain that may be subject to seasonal high water tables or flooding? Is it at the toe of a slope (Fig. 4.15) where runoff may tend to collect or groundwater emerges at or near the surface? Has the microtopography been altered by cultivation?
5. *Soil materials* – Is there a restrictive layer in the soil that would slow or prevent the infiltration of water? This could include consolidated bedrock, compacted layers, cemented layers such as duripans and petrocalcic horizons, layers of silt or substantial clay content, seasonal ice, or strongly contrasting soil textures (e.g., silt over sand). Platy or prismatic soil structure may also result in restrictive layers. Is there relatively loose soil material (sand, gravel, or rocks) or fractured bedrock that would allow the water to flow laterally down slope?
6. *Vegetation* – Does the vegetation at the site indicate wetter conditions than at other nearby sites, or is it similar to what is found at nearby upland sites?

4.12.4 Hydric Soils Technical Standard

For a problematic site that requires monitoring to determine the presence of a hydric soil, the Hydric Soils Technical Standard (NTCHS 2007) is used as guidance for what data must be collected to satisfy the requirements of a hydric soil. In addition, the Hydric Soils Technical Standard can be used to:

1. evaluate the function of wetland restoration, mitigation, creation, and construction,
2. evaluate onsite the current functional hydric status of a soil, and
3. with appropriate regional data, modify, validate, eliminate, or adopt Field Indicators for the region.

The Hydric Soils Technical Standard includes requirements to determine that the soils are saturated, ponded, or flooded through water table monitoring and proof that the soils are anaerobic and reducing. Saturation (or inundation) and anaerobic conditions must be present for at least 14 consecutive days. It should be noted that the growing season is assumed to have started when the soil goes anaerobic since the conditions occur when soil microbes are active. Saturation is confirmed by the presence of free water in a piezometer installed to a soil depth of 25 cm (10 in.). Anaerobic conditions are confirmed by direct measurement of Eh, alpha, alpha-dipyridyl dye, or IRIS tubes. Refer to Chap. 7, *Wetland Biogeochemistry Techniques*, for more details on confirmation of anaerobic conditions. For more information on the use of the Hydric Soil Technical Standard see the NTCHS Technical Note 11 at ftp://ftp-fc.sc.egov.usda.gov/NSSC/Hydric_Soils/note11.pdf.

4.12.5 Normal Rainfall

Any data collected to evaluate hydric soils should be correlated to rainfall. Normal rainfall data, for wetland purposes, are available in NRCS National Weather and Climate Center WETS (wetlands determination) tables. WETS tables are produced for local weather stations throughout the United States. They can be accessed at http://efotg.sc.egov.usda.gov//efotg_locator.aspx. Pick your state and then county of interest. The Field Office Tech Guide menu tree will appear. Pick Section II from the drop down menu, then open the climate tab, and select AgCIS (Climate Information System). Select WETS as the product and then it will give you a list of weather stations that are available for that area. Select the weather station most appropriate for your location and then go and the WETS table will be generated.

To evaluate if a given year has had normal precipitation, local rainfall data (either from a local weather station or from an onsite rain gauge) are compared to data in the geographically appropriate WETS table. Rainfall is normal for any given month if the amount of rain falls between the values for that month in the columns "30 percent chance will have less than" and "30 percent chance will have more than." Water table depths for a given time period are impacted not only by precipitation during that timeframe but also by precipitation in the preceding months; therefore, any evaluation of rainfall data for a given time period should also include consideration of the precipitation patterns prior to the time period of interest. For example, the NTCHS recommends the evaluation of precipitation data for the 3 months prior to the period when the soil in question is most saturated and reduced (NTCHS 2007).

4.13 Other Uses for Soil Morphology Information

4.13.1 Monitoring and Interpreting Wetland Hydrology

4.13.1.1 Field Indicators of Hydric Soils

All wetlands by definition experience saturation in the upper part of the soil for at least part of the year in a majority of years. However, wetlands display a wide range in hydroperiods, from peraquic moisture regimes (continuously saturated) such as tidal marshes to seasonally saturated wetlands such as many mineral soil flats. Some wetlands are inundated for extended periods in most years, others are rarely inundated. In addition, some wetlands such as groundwater driven slope wetlands display a static water table with a consistent depth. Others, such as precipitation driven mineral soil flats display a wide range in water table depths and multiple fluctuations in depth each year.

Soil morphology typically reflects long term hydrologic conditions. Field Indicators were developed to identify soils that developed under hydrologic conditions associated with wetlands. Hydric soils are as diverse as wetlands. Therefore, the Field Indicators represent a range in hydrologic conditions and individual indicators represent a more limited range in hydrologic conditions. Soil scientists recognize this relationship and associate specific indicators with certain hydroperiods. For example, F3 Depleted Matrix is based on the reduction and translocation of Fe, not the accumulation of organic matter. Conversely, A3 Black Histic is based on the accumulation of organic matter. Development of a histic epipedon or Histisols requires longer periods of saturation than the reduction and translocation of iron. For example, A3 Black Histic is found in wetlands that are inundated for extensive periods; whereas, F3 Depleted Matrix is found in wetlands that are rarely inundated and have a very dynamic water table. Therefore, the Field Indicators can be used not only to identify a hydric soil, but also to characterize wetland hydroperiods.

Soil colors can be used to distinguish between episaturation and endosaturation. By definition, episaturation is characterized by two layers of saturated soil separated by an unsaturated zone. Horizons that are saturated for extended periods are typically characterized by low chroma colors. In a peraquic moisture regime, soils will have consistently low chromas with little change with depth. In an endosaturated soil that is rarely inundated, matrix chromas gradually decrease to ≤2 with depth. A horizon with chromas ≤2 directly above a horizon with high chromas indicates episaturation. In most cases, a physical distinction (change in texture or structure) between the two adjacent horizons will be apparent.

4.13.1.2 Monitoring Well Installation

A soil description should accompany the installation of monitoring wells or piezometers. Soil characteristics can impact the proper depth of well installation and may be needed to interpret the well data. For these purposes, the most important characteristics are color, texture, and structure. The identification of soil horizons that may restrict water movement is critical, and episaturation should be distinguished from endosaturation. A common scenario is a precipitation driven depressional wetland which maintains wetland hydrology through episaturation. At times, there will be two water tables-a perched water table and the deeper apparent water table. The water table may drop significantly during the growing season but the soil close to the surface may remain saturated. Installation of a well to a depth below the perched zone will result in misleading data as the wetland will appear to have a dry hydroperiod. If episaturation is suspected, it is best to install two wells, one above and one below the horizon that is restricting water flow. Water may perch directly above an aquitard; a soil layer that transmits water very slowly or not at all. Aquitards can often be identified by high bulk densities or by platy or prismatic structure. Perching can also be due to relatively small differences in texture in adjacent horizons. Free water below the aquitard may have positive pressure. If a well is installed through the aquitard, water will rise in the well to an elevation above the water table, again resulting in misleading data.

4.13.2 Assessing Changes in Wetland Hydrology

4.13.2.1 Field Indicators of Hydric Soils

As stated previously, soil morphology typically reflects long term hydrologic conditions. Draining a wetland will not cause rapid changes in morphology. Some organic matter decomposition will occur and subtle changes in redoximorphic features can occur. However, it is difficult to distinguish between drained and undrained versions of the same soil. This is a limitation to the Field Indicators. However, morphological stability of hydric soils can be used to determine if the hydrology of a wetland has been altered when used in conjunction with direct assessment of hydrology such as monitoring well data. For example, consider a wetland with a static water table that is inundated continuously for several months yet has a soil meeting the Field Indicator F3 Depleted Matrix. Wetland hydrology during the monitoring period does not match the long term hydrology as the site is now wetter. Now consider a wetland with a dynamic water table displaying seasonal saturation and multiple fluctuations in water table depth but the soil meets the indicator A3 Black Histic. That site is now drier. In this process local precipitation data should be considered to distinguish between permanent changes in hydroperiod and unusual short term precipitation patterns.

4.13.2.2 Soil Structure and Horizonation

A number of soil morphological features are associated with dynamic water tables. Redoximorphic features were discussed previously. Subangular and angular blocky structure is believed to be caused by forces created by alternating periods of wetting and drying. We do not expect to see these structural types in soil that consistently stays wet. Similarly, argillic horizons, a B horizon enriched with illuviated clay, do not form in soils with a static water table near the surface as a fluctuating water table is required to transport clay vertically. Therefore, if hydrologic monitoring indicates a static water table near the surface, but the soil has an argillic horizon or strong blocky structure, the site is now wetter.

4.14 Additional Resources

4.14.1 National Technical Committee for Hydric Soils (NTCHS)

The NTCHS is chaired by NRCS and has representation from all federal agencies involved in wetlands work as well as university experts in hydric soils related issues. The NTCHS makes all decisions on issues related to hydric soils.

4.14.2 Documents

National Technical Committee for Hydric Soils 2007. The Hydric Soil Technical Standard. Hydric Soils Tech Note 11. (http://soils.usda.gov/use/hydric/ntchs/tech_Notes/index.html) – The standard required to collect long term data to establish new hydric soil Field Indicators of Hydric Soils or to identify a functional hydric soil in the absence of indicators.

Environmental Laboratory. 1987. Corps of Engineers Wetland Delineation Manual. Tech Report Y-87-1. U.S. Army Engineer Waterways Experiment Station. Vicksburg MS. (http://el.erdc.usace.army.mil/wetlands/pdfs/wlman87.pdf)

Florida Soil Survey Staff. 1992. Soil and Water Relationships of Florida's Ecological Communities. G.W. Hurt, Ed. USDA, SCS, Gainesville FL.

Mausbach, M.J., and J.L. Richardson. 1994. Biogeochemical Processes in Hydric Soils. Current Topics in Wetland Biogeochemistry 1:68–127.

Mid-Atlantic Hydric Soils Committee. 2011. A Guide to Hydric Soils in the Mid-Atlantic Region, Ver. 2.0. Vasilas LM, Vasilas BL, Eds. USDA, NRCS, Morgantown, WV. (ftp://ftp-fc.sc.egov.usda.gov/NSSC/Hydric_Soils/HydricSoilsMidAtlantic.pdf)

Richardson, J.L., and M.J. Vepraskas (editors). 2000. Wetland Soils: Genesis, Hydrology, Landscapes, and Classification. Lewis Sci Publ, Boca Raton, FL.

Schoeneberger, P.J., D.A. Wysocki, E.C. Benham, and W.D. Broderson (editors). 2002. Field Book for Describing and Sampling Soils, Ver. 2.0. NRCS, National Soil Survey Center, Lincoln, NE. (http://soils.usda.gov/technical/fieldbook/)

Soil Science Society of America. 2001. Glossary of Soil Science Terms. Soil Science Society of America, Madison WI. (https://www.soils.org/publications/soils-glossary)

Soil Survey Division Staff. 1993. Soil Survey Manual. Soil Conservation Service. USDA Handbook 18. (http://soils.usda.gov/technical/manual/)

Soil Survey Staff. 2010. Keys to Soil Taxonomy, 11th ed. USDA, NRCS, Washington DC. (http://soils.usda.gov/technical/classification/tax_keys/)

Soil Survey Staff. 1999. Soil Taxonomy: A Basic System of Soil Classification for Making and Interpreting Soil Surveys. 2nd ed. USDA, NRCS, U.S. Gov. Print. Office, Washington DC. (http://soils.usda.gov/technical/classification/taxonomy)

U.S. Army Corps of Engineers. September 2008. Interim Regional Supplement to the Corps of Engineers Wetland Delineation Manual: Midwest Region. (http://www.usace.army.mil/Missions/CivilWorks/RegulatoryProgramandPermits/reg_supp.aspx)

USDA, NRCS. 2006. Land Resource Regions and Major Land Resource Areas of the United States, the Caribbean, and the Pacific Basin. USDA Handbook 296. (http://soils.usda.gov/survey/geography/mlra/index.html)

USDA, NRCS. 2008. National Food Security Act Manual, Fourth Edition.

USDA, NRCS. 2010. National Soil Survey Handbook, title 430-VI. (http://soils.usda.gov/technical/handbook/)

USDA, NRCS. 2010. Field Indicators of Hydric Soils in the United States. Ver. 7.0. Vasilas LM, Hurt GW, Noble CV, Eds. USDA, NRCS in cooperation with the National Technical Committee for Hydric Soils. (ftp://ftp-fc.sc.egov.usda.gov/NSSC/Hydric_Soils/FieldIndicators_v7.pdf)

USDA, NRCS. 2010. From the Surface Down. An Introduction to Soil Surveys for Agronomic Use. (ftp://ftp-fc.sc.egov.usda.gov/NSSC/Educational_Resources/surdown.pdf)

USDA, NRCS. 2010. Hydric Soils Technical Note, Proper Use of Hydric Soil Terminology. (http://soils.usda.gov/use/hydric/ntchs/tech_notes/note1.html)

Vepraskas, M.J. 1994. Redoximorphic Features for Identifying Aquic Conditions. Tech Bull 301. North Carolina Agricultural Research Service, North Carolina State University, Raleigh, NC.

4.14.3 Websites

http://soils.usda.gov/use/hydric/ – This is the NTCHS website and contains all official technical information regarding hydric soils and hydric soil issues.

http://soils.usda.gov/technical/ – This site is the NRCS Soil Survey Division technical resources website and contains all NRCS technical references pertaining to soil survey issues.

http://www.usace.army.mil/Missions/CivilWorks/RegulatoryProgramandPermits/reg_supp.aspx – This is the website to obtain copies of the Corps of Engineers Wetland Delineation Regional Supplements. For information on hydric soils, go to Chapter 3 in your local suppplement and Chapter 5 for problematic soil situations.

http://soils.usda.gov/education/resources/lessons/texture/ – This site is from the USDA, NRCS and includes the Textural Triangle & the Guide to Texture by Feel.

References

Cowardin LM, Carter V, Golet FC, LaRoe ET (1979) Classification of wetlands and deepwater habitats of the United States. FWS/OBS-79/31. U.S. Fish and Wildlife Service, Washington, DC

Environmental Research Laboratory (1987) Corps of engineers wetland delineation manual. Tech Report Y-87-1. U.S. Army Engineer Waterways Experiment Station, Vicksburg. http://el.erdc.usace.army.mil/wetlands/pdfs/wlman87.pdf

Federal Register (1994) Changes in hydric soils of the United States. US Department of Agriculture, Soil Conservation Service, Washington, DC

Federal Water Pollution Control Act (2008) Title 33 – navigation and navigable waters. Chapter 26 – Water pollution prevention and control. [as amended through P.L. 110–288, 29 July 2008] (33 U.S.C. Section 1251 et seq.)

GretagMacbeth (2009) Munsell soil color charts. GretagMacbeth, New Windsor

National Technical Committee for Hydric Soils (2007) The hydric soil technical standard. Hydric Soils Tech Note 11. http://soils.usda.gov/use/hydric/ntchs/tech_notes/index.html

Schoeneberger PJ, Wysocki DA, Benham EC, Broderson WD (eds) (2002) Field book for describing and sampling soils, Ver. 2.0. NRCS, National Soil Survey Center, Lincoln. http://soils.usda.gov/technical/fieldbook/

Soil Survey Staff (1999) Soil taxonomy: a basic system of soil classification for making and interpreting soil surveys, 2nd edn. USDA, NRCS, U.S. Gov. Print. Office, Washington, DC, http://soils.usda.gov/technical/classification/taxonomy

Thien SJ (1979) A flow diagram for teaching texture-by-feel analysis. J Agron Educ 8:54–55
U.S. Army COE (2012) Interim regional supplement to the corps of engineers wetland delineation manual. http://www.usace.army.mil/Missions/CivilWorks/RegulatoryProgramandPermits/reg_supp.aspx
USDA, FSA. National Food Security Act (1985) P.L. 99–198, 99 Stat. 1354–1660, 23 Dec 1985. [16 U.S.C. Section 3801(a) (12)]
USDA, NRCS (2006). Land resource regions and major land resource areas of the United States, the Caribbean, and the Pacific Basin. USDA Handbook 296. http://soils.usda.gov/survey/geography/mlra/index.html
USDA, NRCS (2010a) Field indicators of hydric soils in the United States. Ver. 7.0. In: Vasilas LM, Hurt GW, Noble CV (eds) USDA, NRCS in cooperation with the National Technical Committee for Hydric Soils. ftp://ftp-fc.sc.egov.usda.gov/NSSC/Hydric_Soils/FieldIndicators_v7.pdf
USDA, NRCS (2010b). From the surface down. An introduction to soil surveys for agronomic use. ftp://ftp-fc.sc.egov.usda.gov/NSSC/Educational_Resources/surdown.pdf
Vepraskas MJ (1994) Redoximorphic features for identifying aquic conditions. Tech Bull 301. North Carolina Agricultural Research Service, North Carolina State University, Raleigh

Student Exercises

Classroom Exercises

Classroom Exercise #1: Sources of Hydric Soil Information

Objective: To familiarize students with sources of hydric soils information.

Procedures:

Take a moment to explore the following sources of hydric soils information derived from the soil survey for your selected site location.

1. Go to the NTCHS website (http://soils.usda.gov/use/hydric/) and look up the national Hydric Soil List and find the soil survey area your site is located in.

    (a) Note the soil survey area which is located in column B.
    (b) Column D contains the map unit symbol as it is mapped in the official soil survey.
    (c) Column E contains the map unit name as it is named in the official soil survey.
    (d) Column F contains the component name(s) that are hydric soil components of that map unit. Note that some components are major components that are named in the map unit name and others are minor components (inclusions) within the map unit.
    (e) Column G contains the percentage of that map unit that contains a hydric soil component.

4 Hydric Soil Identification Techniques

(f) Column H contains the landform in which you will find the hydric component.
(g) Column I provides the criteria that was used to establish this component as hydric.

2. Now go to the Soil Data Mart (http://soildatamart.nrcs.usda.gov/) and look up the local Hydric Soil List for your survey area.

   (a) Click on "select state" and select the state you are in.
   (b) Click on "select survey area" and select the one you are interested in.
   (c) Click on "generate reports" and click on "select all" to select all the map units in the survey area. This should highlight all the map units. You can also select individual map units if you do not want a list of the whole county.
   (d) From the pull down menu of reports, select "hydric soils" and click on "generate report".
   (e) It will take a moment for the hydric soil report to come up.
   (f) This report gives you the same information that is presented in the national Hydric Soil List, however, this report is updated any time information is updated in Soil Data Mart, while the national Hydric Soil List is only updated about once a year. There may be differences if an update to Soil Data Mart was made after the national Hydric Soil List was generated. Soil Data Mart should be your source for official (and current) soil survey information.

3. Now go to Web Soil Survey (http://websoilsurvey.nrcs.usda.gov/)

   (a) Click on "start WSS".
   (b) Locate your area of interest. You can do this by: selecting state and county and then zooming in by clicking the icon at the top of the map with a magnifying glass with a + sign in the upper left of the map screen, soil survey area and zooming into your area of interest, typing in an address, using latitude and longitude, or any of the other methods listed (Use the help menu in Web Soil Survey for more detailed instructions.).
   (c) Once zoomed in, click on either the area of interest rectangle icon or the area of interest icon that allows an irregular shape. These icons are located in the upper left of the map screen.
   (d) Outline the area of interest on the map and then click on the "Soil Data Explorer" tab. You should now see a soils map overlaying your area of interest.
   (e) Under the tab "Suitability and Limitations for Use" click on "Land Classification", then "Hydric Rating by Map Unit", and then view rating.
   (f) This will produce a map that shows map units that are hydric, partially hydric (contain components that are both hydric and non-hydric), or not hydric.
   (g) Click on "Add to Shopping Cart" in the upper right and then ok.
   (h) Click on the "Shopping Cart" tab at the top, and then "Check Out" in the upper right and then ok.
   (i) Now you have a customized .pdf soil survey report for your area of interest that contains the Hydric Soils List information from Soil Data Mart along with a map categorizing those map units that are hydric, partially hydric, or non-hydric. You can save this to your computer and/or print the file for use when you do your on-site investigation exercise a little later.

Classroom Exercise #2: Identification of Field Indicators of Hydric Soils

Objective: To familiarize students with the use of Field Indicators of Hydric Soils.

Proceedures:

Below are examples of data sheets that are completely filled out. Go through the latest version of Field Indicators of Hydric Soils and list all indicators that are met for each of the descriptions.

SOIL Sampling Point: _1_

Profile Description: (Describe to the depth needed to document the indicator or confirm the absence of indicators.)

Depth (inches)	Matrix Color (moist)	%	Redox Features Color (moist)	%	Type[1]	Loc[2]	Texture	Remarks
0-8	10YR 3/2	90	7.5YR 5/4	10	C	PL	SL	
8-24	2.5Y 6/1	88	7.5YR 5/4	12	C	M	SL	

[1]Type: C=Concentration, D=Depletion, RM=Reduced Matrix, CS=Covered or Coated Sand Grains.   [2]Location: PL=Pore Lining, M=Matrix.

Hydric Soil Indicators: (Applicable to all LRRs, unless otherwise noted.)

- ___ Histosol (A1)
- ___ Histic Epipedon (A2)
- ___ Black Histic (A3)
- ___ Hydrogen Sulfide (A4)
- ___ Stratified Layers (A5) (LRR F)
- ___ 1 cm Muck (A9) (LRR F, G, H)
- ___ Depleted Below Dark Surface (A11)
- ___ Thick Dark Surface (A12)
- ___ Sandy Mucky Mineral (S1)
- ___ 2.5 cm Mucky Peat or Peat (S2) (LRR G, H)
- ___ 5 cm Mucky Peat or Peat (S3) (LRR F)

- ___ Sandy Gleyed Matrix (S4)
- ___ Sandy Redox (S5)
- ___ Stripped Matrix (S6)
- ___ Loamy Mucky Mineral (F1)
- ___ Loamy Gleyed Matrix (F2)
- ___ Depleted Matrix (F3)
- ___ Redox Dark Surface (F6)
- ___ Depleted Dark Surface (F7)
- ___ Redox Depressions (F8)
- ___ High Plains Depressions (F16) (MLRA 72 & 73 of LRR H)

Indicators for Problematic Hydric Soils[3]:
- ___ 1 cm Muck (A9) (LRR I, J)
- ___ Coast Prairie Redox (A16) (LRR F, G, H)
- ___ Dark Surface (S7) (LRR G)
- ___ High Plains Depressions (F16) (LRR H outside of MLRA 72 & 73)
- ___ Reduced Vertic (F18)
- ___ Red Parent Material (TF2)
- ___ Other (Explain in Remarks)

[3]Indicators of hydrophytic vegetation and wetland hydrology must be present, unless disturbed or problematic.

Restrictive Layer (if present):
Type: _None_
Depth (inches): _____

Hydric Soil Present?   Yes _____   No _____

Remarks:
Landscape: Mineral Flat

4 Hydric Soil Identification Techniques

SOIL — Sampling Point: 2

Profile Description: (Describe to the depth needed to document the indicator or confirm the absence of indicators.)

Depth (inches)	Matrix Color (moist)	%	Redox Features Color (moist)	%	Type[1]	Loc[2]	Texture	Remarks
0-8	10YR4/3	90	7.5YR4/6	10	C	PL	L	
8-24+	2.5Y5/1	82	7.5YR5/4	18	C	M	SL	

[1]Type: C=Concentration, D=Depletion, RM=Reduced Matrix, CS=Covered or Coated Sand Grains. [2]Location: PL=Pore Lining, M=Matrix.

Hydric Soil Indicators: (Applicable to all LRRs, unless otherwise noted.)

- ___ Histosol (A1)
- ___ Histic Epipedon (A2)
- ___ Black Histic (A3)
- ___ Hydrogen Sulfide (A4)
- ___ Stratified Layers (A5) (LRR F)
- ___ 1 cm Muck (A9) (LRR F, G, H)
- ___ Depleted Below Dark Surface (A11)
- ___ Thick Dark Surface (A12)
- ___ Sandy Mucky Mineral (S1)
- ___ 2.5 cm Mucky Peat or Peat (S2) (LRR G, H)
- ___ 5 cm Mucky Peat or Peat (S3) (LRR F)

- ___ Sandy Gleyed Matrix (S4)
- ___ Sandy Redox (S5)
- ___ Stripped Matrix (S6)
- ___ Loamy Mucky Mineral (F1)
- ___ Loamy Gleyed Matrix (F2)
- ___ Depleted Matrix (F3)
- ___ Redox Dark Surface (F6)
- ___ Depleted Dark Surface (F7)
- ___ Redox Depressions (F8)
- ___ High Plains Depressions (F16) (MLRA 72 & 73 of LRR H)

Indicators for Problematic Hydric Soils[3]:
- ___ 1 cm Muck (A9) (LRR I, J)
- ___ Coast Prairie Redox (A16) (LRR F, G, H)
- ___ Dark Surface (S7) (LRR G)
- ___ High Plains Depressions (F16) (LRR H outside of MLRA 72 & 73)
- ___ Reduced Vertic (F18)
- ___ Red Parent Material (TF2)
- ___ Other (Explain in Remarks)

[3]Indicators of hydrophytic vegetation and wetland hydrology must be present, unless disturbed or problematic.

Restrictive Layer (if present):
Type: None
Depth (inches): _____

Hydric Soil Present? Yes _____ No _____

Remarks: Landscape: Mineral Flat

SOIL — Sampling Point: 3

Profile Description: (Describe to the depth needed to document the indicator or confirm the absence of indicators.)

Depth (inches)	Matrix Color (moist)	%	Redox Features Color (moist)	%	Type[1]	Loc[2]	Texture	Remarks
0-8	10YR4/4	90	7.5YR4/6	10	C	PL	SiL	
8-18+	10YR5/6	100					SiL	

[1]Type: C=Concentration, D=Depletion, RM=Reduced Matrix, CS=Covered or Coated Sand Grains. [2]Location: PL=Pore Lining, M=Matrix.

Hydric Soil Indicators: (Applicable to all LRRs, unless otherwise noted.)

- ___ Histosol (A1)
- ___ Histic Epipedon (A2)
- ___ Black Histic (A3)
- ___ Hydrogen Sulfide (A4)
- ___ Stratified Layers (A5) (LRR F)
- ___ 1 cm Muck (A9) (LRR F, G, H)
- ___ Depleted Below Dark Surface (A11)
- ___ Thick Dark Surface (A12)
- ___ Sandy Mucky Mineral (S1)
- ___ 2.5 cm Mucky Peat or Peat (S2) (LRR G, H)
- ___ 5 cm Mucky Peat or Peat (S3) (LRR F)

- ___ Sandy Gleyed Matrix (S4)
- ___ Sandy Redox (S5)
- ___ Stripped Matrix (S6)
- ___ Loamy Mucky Mineral (F1)
- ___ Loamy Gleyed Matrix (F2)
- ___ Depleted Matrix (F3)
- ___ Redox Dark Surface (F6)
- ___ Depleted Dark Surface (F7)
- ___ Redox Depressions (F8)
- ___ High Plains Depressions (F16) (MLRA 72 & 73 of LRR H)

Indicators for Problematic Hydric Soils[3]:
- ___ 1 cm Muck (A9) (LRR I, J)
- ___ Coast Prairie Redox (A16) (LRR F, G, H)
- ___ Dark Surface (S7) (LRR G)
- ___ High Plains Depressions (F16) (LRR H outside of MLRA 72 & 73)
- ___ Reduced Vertic (F18)
- ___ Red Parent Material (TF2)
- ___ Other (Explain in Remarks)

[3]Indicators of hydrophytic vegetation and wetland hydrology must be present, unless disturbed or problematic.

Restrictive Layer (if present):
Type: None
Depth (inches): _____

Hydric Soil Present? Yes _____ No _____

Remarks: Landscape: Closed depression subject to ponding.

SOIL

Sampling Point: 4

Profile Description: (Describe to the depth needed to document the indicator or confirm the absence of indicators.)

Depth (inches)	Matrix Color (moist)	%	Redox Features Color (moist)	%	Type[1]	Loc[2]	Texture	Remarks
0-14	2.5Y 2.5/1	100					S	
14-18	10YR 4/6	100					S	
18-24+	2.5Y 6/1	100					S	

[1]Type: C=Concentration, D=Depletion, RM=Reduced Matrix, CS=Covered or Coated Sand Grains. [2]Location: PL=Pore Lining, M=Matrix.

Hydric Soil Indicators: (Applicable to all LRRs, unless otherwise noted.)

- __ Histosol (A1)
- __ Histic Epipedon (A2)
- __ Black Histic (A3)
- __ Hydrogen Sulfide (A4)
- __ Stratified Layers (A5) (LRR F)
- __ 1 cm Muck (A9) (LRR F, G, H)
- __ Depleted Below Dark Surface (A11)
- __ Thick Dark Surface (A12)
- __ Sandy Mucky Mineral (S1)
- __ 2.5 cm Mucky Peat or Peat (S2) (LRR G, H)
- __ 5 cm Mucky Peat or Peat (S3) (LRR F)
- __ Sandy Gleyed Matrix (S4)
- __ Sandy Redox (S5)
- __ Stripped Matrix (S6)
- __ Loamy Mucky Mineral (F1)
- __ Loamy Gleyed Matrix (F2)
- __ Depleted Matrix (F3)
- __ Redox Dark Surface (F6)
- __ Depleted Dark Surface (F7)
- __ Redox Depressions (F8)
- __ High Plains Depressions (F16) (MLRA 72 & 73 of LRR H)

Indicators for Problematic Hydric Soils[3]:

- __ 1 cm Muck (A9) (LRR I, J)
- __ Coast Prairie Redox (A16) (LRR F, G, H)
- __ Dark Surface (S7) (LRR G)
- __ High Plains Depressions (F16) (LRR H outside of MLRA 72 & 73)
- __ Reduced Vertic (F18)
- __ Red Parent Material (TF2)
- __ Other (Explain in Remarks)

[3]Indicators of hydrophytic vegetation and wetland hydrology must be present, unless disturbed or problematic.

Restrictive Layer (if present):
Type: None
Depth (inches): _____

Hydric Soil Present? Yes ____ No ____

Remarks: Landscape: Mineral Flat

SOIL

Sampling Point: 5

Profile Description: (Describe to the depth needed to document the indicator or confirm the absence of indicators.)

Depth (inches)	Matrix Color (moist)	%	Redox Features Color (moist)	%	Type[1]	Loc[2]	Texture	Remarks
0-2	10YR 3/1	100					MK-PT	
2-14	2.5Y 3/1	100					SiL	
14-28+	2.5Y 6/1	90	7.5YR 4/6	10	C	M	SiL	

[1]Type: C=Concentration, D=Depletion, RM=Reduced Matrix, CS=Covered or Coated Sand Grains. [2]Location: PL=Pore Lining, M=Matrix.

Hydric Soil Indicators: (Applicable to all LRRs, unless otherwise noted.)

- __ Histosol (A1)
- __ Histic Epipedon (A2)
- __ Black Histic (A3)
- __ Hydrogen Sulfide (A4)
- __ Stratified Layers (A5) (LRR F)
- __ 1 cm Muck (A9) (LRR F, G, H)
- __ Depleted Below Dark Surface (A11)
- __ Thick Dark Surface (A12)
- __ Sandy Mucky Mineral (S1)
- __ 2.5 cm Mucky Peat or Peat (S2) (LRR G, H)
- __ 5 cm Mucky Peat or Peat (S3) (LRR F)
- __ Sandy Gleyed Matrix (S4)
- __ Sandy Redox (S5)
- __ Stripped Matrix (S6)
- __ Loamy Mucky Mineral (F1)
- __ Loamy Gleyed Matrix (F2)
- __ Depleted Matrix (F3)
- __ Redox Dark Surface (F6)
- __ Depleted Dark Surface (F7)
- __ Redox Depressions (F8)
- __ High Plains Depressions (F16) (MLRA 72 & 73 of LRR H)

Indicators for Problematic Hydric Soils[3]:

- __ 1 cm Muck (A9) (LRR I, J)
- __ Coast Prairie Redox (A16) (LRR F, G, H)
- __ Dark Surface (S7) (LRR G)
- __ High Plains Depressions (F16) (LRR H outside of MLRA 72 & 73)
- __ Reduced Vertic (F18)
- __ Red Parent Material (TF2)
- __ Other (Explain in Remarks)

[3]Indicators of hydrophytic vegetation and wetland hydrology must be present, unless disturbed or problematic.

Restrictive Layer (if present):
Type: None
Depth (inches): _____

Hydric Soil Present? Yes ____ No ____

Remarks: Landscape: Mineral Flat

Answers

Sampling point 1: This description meets A11 Depleted Below Dark Surface, F3 Depleted Matrix, and F6 Redox Dark Surface.

Sampling point 2: No indicator is met. Note that this description would meet F3 Depleted Matrix. However, it fails the general rule that you cannot have 15 cm (6 in.) or more of a matrix chroma higher than 2 above the depleted matrix (indicator).

Sampling point 3: F8 Redox Depressions. Note that this is one of the landscape specific indicators that allows matrix chromas higher than 2. In this case, it can only be used in soils that occur in closed depressions subject to ponding. Examples are vernal pools, playa lakes, rainwater basins, "Grady" ponds and potholes.

Sampling point 4: S7 Dark Surface. This is an example of where horizons and layers are not synonymous. To meet this indicator, you must have 10 cm (4 in.) with a value of 3 or less and a chroma of 1 or less. Immediately below the 10 cm (still in the same horizon), you meet the next layer requirement with a chroma of 2 or less. However, if you went to the next horizon instead of looking immediately below the 10 cm layer, you would not meet this indicator since it has a chroma higher than 2. S7 is not an approved indicator for all LRRs. When identifying indicators in a real world scenario you would need to identify that you are in an approved LRR before using this indicator.

Sampling point 5: Answer: A11 Depleted Below Dark Surface. Note that if you incorrectly start your measurements at the actual soil surface instead of the mineral surface (at 5 cm), the depleted matrix would be too deep to meet this indicator.

Laboratory Exercises

Laboratory Exercise: Description and Identification of Hydric Soils in the Field

Overview: The following field exercise is intended to allow you to use the skills you have learned to make a hydric soils determination in the field. If you are a novice to writing soil descriptions, you may want to seek out assistance from a soil or wetland scientist with more experience for assistance. Before you go to the field, you will need to gather the soil report you created earlier in Web Soil Survey for the area you will be using for the excercise, a copy of the *Field Indicators of Hydric Soils in the United States* (Version 7.0), a copy of Chapter 3 of your local Corps Regional Supplement, and the Key to Soils that Lack Field Indicators of Hydric Soils.

Objectives: To describe and identify hydric soils in the field

Materials and Equipment Needed:

1. Tiling spade or similar flat bladed shovel
2. Bucket auger if you think you may need to describe your soil to a depth greater than 45 cm (18 in.).

3. Measuring tape
4. Water for estimating soil texture
5. Knife or other tool for picking the soil surface
6. Munsell soil color chart
7. Clipboard
8. Pencil or pen
9. Data sheets
10. You may also want to contact your local Resource Soil Scientist to determine if there are any other tools you may need. For example, in areas where Mn may be used as a Field Indicator, it is useful to carry hydrogen peroxide to determine if dark mottles in the soil are in fact redox concentrations that contain Mn.

Procedures:

Once in the field, locate an area that you feel is on the wet side of the hydric/non-hydric soil boundary and fill out the soils portion of the wetland delineation data sheet completely. Go to the drier side and complete another data sheet filling out the soils portion completely. Make sure you describe all the information that you will need to identify the Field Indicators of Hydric Soils the soil might meet. For example, if you are in a sandy soil with dark matrices, it is important to record an estimate of masked vs. unmasked sand grains because this is a characteristic that can separate hydric soils from non-hydric soils. Once you have completed your descriptions, go through the Field Indicators in the field to identify all those indicators your soil meets. You should record all the Field Indicators met, although only one indicator is required. Once you become familiar with the Field Indicators, you can begin completing your descriptions in the field and going through the Field Indicators following the field visit, but for this exercise, you should attempt to determine the indicators that are met in the field in case you need to go back and identify features you may have forgotten to record. Note the importance of recording colors, soil textures, accurate depths, percentage and location of redoximorphic features, masked vs. unmasked sand grains, and, if it applies, the type of organic soil material (muck, muck peat, peat).

If you have gone through your exercise and were not able to identify a Field Indicator in a site that you feel should contain hydric soils, you may be in a problematic soil situation. The first step in identifying a problematic soil situation is to address the following questions:

1. Look at the big picture.
2. What landscape are you in?
3. Does the vegetative community make sense?
4. Are the soil characteristics what you expect?

Read the information on problematic hydric soils and then go back to the information you have gathered to determine if your site fits any of the problematic hydric soil situations described in your Regional Supplement and whether the information provided can assist you to identify the soil as hydric. If you have not already asked for assistance from the local Resource Soil Scientist, you may want to contact them to discuss whether it is likely that the site is problematic.

Chapter 5
Sampling and Analyzing Wetland Vegetation

Amanda Little

Abstract Effectively sampling and analyzing wetland vegetation is an important part of wetland science, as an indicator of wetland health and quality, and jurisdictional and mitigation success determinations. This chapter explains spatiotemporal vegetation sampling considerations by addressing key questions, such as which wetlands should be sampled and when and at what scale sampling should occur. It also plainly discusses the advantages and disadvantages of basic sampling techniques, such as different types of plot-based, plotless, and relevé systems. Methods of assessing different vegetation and environmental attributes, such as cover and functional groups are discussed in detail. The chapter then describes methods of analyzing wetland vegetation, including simple summary analyses and more complex multivariate methods, such as classification, ordination, and floristic quality indices. Explanations of different types of these analyses and their advantages and disadvantages are provided. Finally, both field and laboratory-based exercises in sampling and analysis are provided for faculty and students studying wetland vegetation.

5.1 The Importance of Wetland Vegetation

There are many reasons to investigate wetland vegetation. Aside from purely scientific interest, wetland vegetation has long been used as an indicator of wetland health and quality (U.S. EPA 2002), a basis of comparison between reference and restored or mitigated states (Matthews et al. 2009a), and as one of the three indicators of jurisdictional wetlands (Environmental Laboratory 1987). It also provides valuable ecosystem services as habitat for fish (Gabriel and Bodensteiner 2011), birds (Valente et al. 2011), amphibians (Hamer and Parris 2011), and insects (Molnar et al. 2009), and is a component of biodiversity in its own right. In addition,

A. Little (✉)
Department of Biology, University of Wisconsin-Stout,
331G Jarvis Hall Science Wing, 712 S Broadway, Menomonie, WI 54751, USA
e-mail: littlea@uwstout.edu

specific properties of wetland vegetation (e.g., carbon storage and uptake) are important in studies of ecosystem function. In this chapter, we will explore common techniques for sampling and analyzing wetland vegetation.

5.2 Considerations of Location and Timing – Which Wetlands and When Should They Be Sampled?

What a scientist finds depends upon where and when they look. This section will explore considerations of sample location at multiple scales: watersheds, wetlands, and zones or communities within wetlands. At each scale, randomization options and pseudoreplication considerations (discussed in Chap. 1) need to be carefully considered and applied.

5.2.1 Which Wetlands?

In many cases, the choice of study wetland is pre-determined by the goals and objectives of the study. However, if the goal is to compare wetlands or generalize about particular wetland types or conditions, the choice of study wetlands becomes the most important decision (Curtis 1959). It is best to begin by identifying the range of possible wetlands within the study's scope. Numerous free resources are available, including the National Wetlands Inventory (U.S. Fish and Wildlife Service), the U.S. Department of Agriculture Web Soil Survey (for locating areas of hydric soil), state or county-level wetlands inventories, and more detailed wetlands inventories created for specific management areas, such as cities, preserves, or forests. These inventories will provide information as to the type of wetland, its size, shape, and geographic location – typically associated with a geographic information system (GIS) map layer. The level of detail provided about wetland type ranges from basic information about the dominant strata and hydrology in the wetland (e.g., emergent, shrub, forested) in large-scale inventories (Fig. 5.1) to species-level and hydrology data provided in more small-scale inventories. In many cases, the inventory will be based upon remotely-sensed data and therefore subject to error, especially for forested wetlands, which are more difficult to detect remotely (Kudray and Gale 2000).

These inventories provide a range of possibilities for study. Practitioners who want to get an unbiased representation of different wetlands across an area could apply a stratified random sampling scheme (see Chap. 1), stratified upon type, to select study wetlands. If wetland-level attributes will be used as samples in statistical analyses, pseudoreplication should be avoided by ensuring that different wetlands are not hydrologically-connected closely within the same watershed. Alternatively, those seeking to identify representative or reference wetlands could

5 Sampling and Analyzing Wetland Vegetation   275

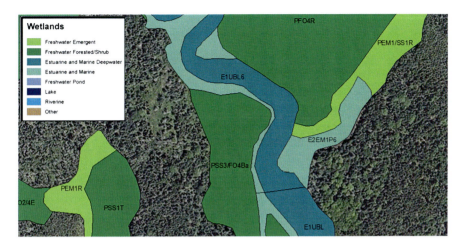

Fig. 5.1 The National Wetlands Inventory map of an estuarine system in Maine. Information includes dominant system (e.g., palustrine, estuarine), vegetation life form (e.g., shrub-scrub, emergent, forested), and limited detailed subclass modifiers (e.g., evergreen type, inundation permanence, and some water chemistry information)

simply use the inventory as a starting point for site visits. In any case, it is critical to keep the purpose of your study in mind while selecting wetlands and to avoid "reinventing the wheel" when possible.

Questions to ask yourself when choosing wetlands:

1. Is it important to have an unbiased sample of wetlands?
2. Is it important to identify representative or reference-type sites using professional judgment?

5.2.2 Where in the Wetland?

With some exceptions, (e.g., monocultures or some large peatlands), wetlands tend to have highly heterogeneous vegetation. This heterogeneity is often due to hydrologic differences at multiple different scales within the wetland. For example, some wetlands have distinctive bands, or zones of vegetation corresponding to large-scale hydrologic gradients (Fig. 5.2a). Each of these zones could be considered a different community within the wetland. Within each zone, vegetation can be further influenced by microtopographic features such as hummocks, tussocks, deep holes, or trees (Fig. 5.2b, (Ehrenfeld 1995)). Disturbance factors, such as herbivory, pollution, fire, or animal trails can lend further heterogeneity to vegetation, often in less predictable patterns.

Wetland vegetation heterogeneity can initially be explored using high-quality aerial photography, which can provide a good idea of large-scale patterns. This

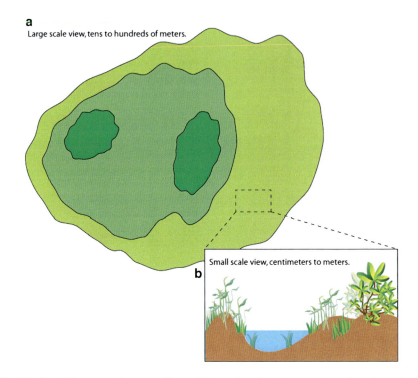

Fig. 5.2 Spatial heterogeneity created by multiple scales of environmental gradients in wetlands. (**a**) Large-scale zonation corresponding to large-scale elevation gradients. (**b**) Small-scale microtopographic heterogeneity corresponding to features like hummocks, pools, and coarse woody debris (Published with kind permission of © M. Kuchta 2014. All Rights Reserved)

photography can be obtained from multiple sources, including the National Wetlands Inventory (aerial imagery viewer), United States Department of Agriculture (USDA) Natural Resources Conservation Service (NRCS) Gateway (http://datagateway.nrcs.usda.gov), and the United States Geological Survey (USGS) EarthExplorer (http://earthexplorer.usgs.gov). Important information to gather from photos includes: wetland size and dimensions, type of strata (aquatic, understory, shrub, and/or tree), access points, and any obstacles to travel within the wetland (e.g. ponds, rivers). After locating and investigating the photography, a preliminary site visit can help identify microtopographic patterns to take into consideration, further obstacles to travel within the wetland, and unknown plant species to identify prior to intensive sampling.

Questions to ask yourself before choosing where to sample within the wetland:

1. Is it important to quantitatively describe the complete vegetation of the wetland or is a description of the vegetation in each community type sufficient?
2. Are communities easily recognizable and discrete, or do they grade into each other so that it is difficult to detect changes without quantitative data?
3. Is pseudoreplication an important consideration? That is, is it important to collect samples only from within one zone or microsite, or across many?

5.2.3 When?

Timing is extremely important in vegetation sampling. Just as there are multiple scales of spatial investigation and variability, temporal variability has multiple scales. The most common consideration in vegetation sampling is season of the year. Within the northern hemisphere, most wetland vegetation is sampled in mid to late summer in order to facilitate sedge, grass, and aster identification, and to capture vegetation at the peak of its growth. However, many early-season sedge, mint, and violet species may be difficult to identify in late summer, and several orchid species seem to disappear entirely after blooming. Within forested wetlands, spring ephemerals may be missed altogether. A good strategy is to visit the wetland site periodically early in the year in order to identify early-blooming and fruiting species and then apply this knowledge to a more comprehensive quantitative sampling later in the growing season when early-bloomers are in a non-flowering state. If a comprehensive species list for the wetland is desired, then returning to sample at multiple times during the year is necessary.

On a larger scale, interannual variability can also be important. When reporting results of vegetation study, include information about whether the climate was typical or unusual that year. Some plant species flourish or become more apparent during times of high or low water levels (Warwick and Brock 2003). Late spring freezes can temporarily eliminate a host of spring herbaceous species. If the sampling year is atypical, an additional year will probably be necessary in order to fully describe the wetland vegetation. Wetlands like prairie potholes or beaver (*Castor canadensis*) meadows experience larger-scale cycles of disturbance that can extend for decades or longer. In order to capture the full variability of these systems, it may be necessary to sample wetlands at different stages in the cycle. It is important to recognize and document the site history and temporal context of your sampling when reporting and applying results.

Questions to ask yourself before choosing when to sample:

1. Is it important to get a general description of the vegetation, or a comprehensive species list?
2. Does the system have a characteristic frequent disturbance regime that will make the results of your sampling only narrowly applicable?

5.3 Basic Vegetation Sampling Techniques

Entire books have been written about how to properly sample vegetation. What follows is an introduction to some basic techniques. For further information, the reader is encouraged to consult Greig-Smith (1983), Bonham (1989), Kent and Coker (1995), Elzinga et al. (1998), Krebs (1998), and Mueller-Dombois and Ellenberg (2003).

Before any sampling protocol can be created, the goals of the project must be clarified. Questions to ask yourself include:

- Is this a one-time assessment, or would you like to track changes in the vegetation over time?
- Are you more interested in one to a few different populations, the plant community as a whole, or in plants as a production component of an ecosystem model?
- What attributes of the vegetation are important to you? Presence or absence of species? Abundance of individual species? Vegetation cover, height or biomass? Dominance or importance of different species?
- Is it important to have quantitative data that can be used in a statistically-valid manner, or is a basic qualitative (e.g., a species list) description of the system adequate?
- Is characterizing the spatial pattern of the vegetation or having precise location information important to answering your question?

Once you have considered and answered these questions, you will be able to determine which techniques are best suited for your purpose.

5.3.1 Attributes of Vegetation

The basic building blocks of any sampling protocol are the attributes to be measured. Once a system for selecting and delimiting sample locations has been established, vegetation characteristics are assessed. Commonly-measured attributes include:

- **Presence:** Does the species occur within the plot or site? This measure can later be used to calculate the **frequency** of a species within a site (the number of plots in which the species occurred).
- **Abundance:** How much of the species occurs? This can be measured in different ways, such as by count of individuals or by visual percent-cover.
- **Production:** How much biomass is produced by different species in the plot? Root, shoot, and or total plant biomass can be measured.
- **Structure:** How tall is the vegetation or how many stems or branching points are produced? How much three-dimensional space is occupied by the plant? These types of measures can be particularly helpful when assessing habitat for animals.
- **Composition:** Which and how many different species occur within the plot?
- **Functional groups:** What is the abundance of species from different functional groups (e.g., perennial graminoid, annual forb, floating-leaf submergent)? Many attributes, such as invasive or wetland indicator status, can be assigned after sampling based upon available information (e.g., USDA Plants Database: http://plants.usda.gov).
- **Morphological characteristics:** What types of traits does each species have? Traits measured in the field include leaf number, leaf shape, specific leaf area, and flower number.

- **Dominance or importance:** Which species is most important in its influence on the community or ecosystem? This attribute is often assessed during the analysis stage by creating composite scores using different attributes, such as relative cover or density. However, attributes like height or basal area (calculated from tree diameters) may be important to collect in the field.
- **Spatial pattern:** Is the vegetation dispersed in a clumped, random, or regular pattern?

5.3.2 The Sample Unit: Plot-Based and Plotless Techniques

The choice of sample unit will depend upon the size of the plants, the resources available (time and money), and the ease of using the subsequent data to meet the specific goals for your analysis or report.

Plot-based techniques assess vegetation within an area of pre-defined size and shape. These techniques have the advantage of leading to relatively straightforward calculations of density and other summary attributes in the analysis stage, because they have a known area. The size of the plot typically relates to the size of the organisms studied. For example, a larger plot will be used to assess trees than to assess understory vegetation. In order to capture a wide variety of species in understory vegetation, plot sizes from 0.01 to 1.0 m^2 are typical and sometimes nested within each other (Elzinga et al. 1998). A common plot size for trees is 100 m^2. Plots can frequently be nested around a common point when investigating multiple strata (e.g., understory, shrubs, trees). Small plots are often called quadrats. As a general rule, the plot should be roughly twice the size of the largest organism in your sample (Greig-Smith 1983). Other researchers suggest that plots be as large as possible given time and effort constraints (Kenkel and Podani 1991). Organism spatial distribution should be considered. If plant populations are clustered, researchers should be sure that plots are not of a size that will result in numerous empty plots when they land in between plant clumps (Elzinga et al. 1998). Another good strategy for choosing plot size is to use previously-published studies. For a detailed discussion of plot size, see Elzinga et al. (1998).

The effects of plot shape have also been debated by ecologists (discussed in (Krebs 1998)). Some argue that a rectangular plot is most effective because it captures the most vegetation heterogeneity (with quadrats oriented to capture variability within plots as opposed to between). Others argue that circular or square plots are best because they minimize edge effects and require the least amount of subjective "in or out of plot" decisions on the part of the worker due to their low perimeter:area ratio. Choose the plot shape based upon the purpose of your study. If the purpose is to capture the maximum heterogeneity, a rectangular plot will be most suitable. If not, choose a square or circular plot in order to maximize accuracy and minimize edge decisions (U.S. EPA 2002).

Nested plots (where smaller plots are located within larger plots) are used for many reasons. The most common use of nested plots is to survey vegetation within

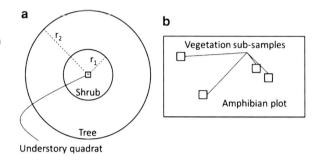

Fig. 5.3 Nested plots. (a) Vegetation of different strata can be sampled within differently-sized nested plots. (b) Different types of data can be sampled within nested plots, corresponding to the organism studied

Table 5.1 Example line intercept data sheet for a line 5.0 m long

Species	Intercepts (m)	Total length (m)	% intercept
Calamagrostis canadensis	0.5–0.8, 1.2–1.7, 2.9–3.5, 4.8–5.0	1.6	32.0
Chamaedaphne calyculata	0.7–1.3, 3.9–4.1	0.6	12.0
Myrica gale	0.0–0.6, 1.6–3.0, 3.4–3.8	2.4	48.0
Typha latifolia	0.0–0.2	0.2	4.0
Spiraea alba	4.0–4.9	0.9	18.0

different strata. In this case, understory vegetation is typically sampled in a small rigid frame at the center of the plot, and then successively larger circular plots are used to survey shrubs and then trees (Fig. 5.3a). Nested plots may also be effective in collecting different types of data at different scales. For example, individual plant count data may be collected in a subsample of small quadrats within a larger plot that is surveyed for trees or non-vegetation-related ecological attributes (Fig. 5.3b). One of the most complex uses of nested plots is to determine the rate of species accumulation over larger and larger areas (Barbour et al. 1998). In this use, the investigator first surveys a series of small quadrats, recording species. Then, within subsequently larger plots, repeats the procedure, adding any new species. The results can be used to determine the most effective plot size for characterizing community or population attributes (the plot size where there is little subsequent change in your estimates). Finally, nested plots can be used to assess the spatial pattern of a plant population or community (Dale 1998; Greig-Smith 1983).

The line-intercept technique is essentially a variation upon the plot technique in which the plot is one-dimensional (very long but with no width). A line is created with a meter tape, and vegetation is measured wherever it intersects the vertical plane created by the tape (both above and below). A data sheet for line-intercept data includes the species that intersect the line, and the distances at which they intersected it (Table 5.1). Line intercept data can be quicker to collect than quadrat data when there are few species and large areas to be covered. A challenge is comparing the intercept of a graminoid (grass-like) species to those of species that have more two-dimensional coverage (like shrubs). Additional challenges include deciding when to start and end the continuous intervals of intercept with plants that are not themselves continuous over the entire interval. The belt-transect method is a combination of the line-intercept and plot-based methods in which field personnel traverse a long, linear plot, counting all individuals within the belt-like plot. This method is most effective for low density plant populations, shrubs, or trees.

Point-based sampling is another variation on the plot method, conducted with a plot of one dimension. A long rod is vertically placed in the ground, and each plant that intersects or touches the rod is recorded. This method is primarily good for assessing understory species presence, but could be combined with a canopy tube to assess tree presence at the point. The point-based method is relatively more objective in that few observer decisions must be made, but it also supplies less information per sample. Since most time in the field is spent moving between locations, it is relatively inefficient. In addition, the technique offers limited statistical flexibility because the wetland site as a whole is quantitatively described (by all of the points), but each point sample is not (no abundance information is collected at each sample). For these reasons, point-based sampling is rarely used to provide a general description of the vegetation in a wetland, although it is used by the Washington State Department of Transportation for wetland monitoring due to its more objective nature (WSDOT Environmental Services 2008).

Plotless methods are location-restricted, but do not have any distinct boundary within which to assess vegetation. They are most commonly used to assess tree populations or forest communities, or relatively rare plant populations. There are a wide variety of plotless methods (Bonham 1989). The point-quarter (PQM) and the Bitterlich methods are commonly used.

The PQM is a distance-based method that can be easily combined with nested plot sampling because it has a central point that smaller quadrats or plots can be centered upon. It also has the advantage of being usable by a single investigator; however, it is most appropriate in forested settings and not in wetlands with bands of herbaceous vegetation. These center points can be positioned randomly or regularly throughout the wetland area, just like other plots. Once the central point is determined, the surrounding region is divided into four 90-degree sectors (typically aligned with compass directions, but not always). The nearest tree or plant of interest within each sector is selected, measured, and the distance back to the center point is recorded (Fig. 5.4a). If trees are of interest, diameter-at-breast-height (DBH, 1.4 m from the ground) is typically measured. The PQM data can be used to calculate the density of different species, the mean basal area, and the frequency of different species in a forest (Barbour et al. 1998). In order to calculate density, a mean point-to-plant distance is calculated for all trees, with:

$$\text{Total density/ha}(10,000\,\text{m}^2) = 10,000/(\text{mean distance in m})^2$$

And the density (no./ha) for each species can be calculated by:

$$\text{Relative density of species A} = (\text{total no. of species A})/(\text{total no. of all trees}) \times \text{total density}$$

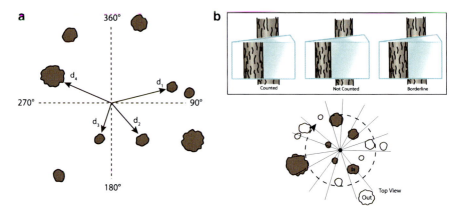

Fig. 5.4 Plotless sampling methods for trees. (**a**) The point-quarter method of dividing the plot space into four quadrants and measuring the distance (d_x) and diameter at breast height of the nearest tree. (**b**) The Bitterlich method in which a central figure counts trees that are "in" and "out" using a prism. If the prism trunk image overlaps actual trunk, the tree is counted. If there is no overlap, the tree is not counted (Published with kind permission of © M. Kuchta 2014. All Rights Reserved)

Basal area (BA) per hectare for each species can then be calculated by:

BA/ha of species A = (Mean BA of species A) × (Mean density of species A)

The Bitterlich method is used to very rapidly calculate the total basal area of trees in a forest, and so has the most application in forestry. Today, hand held glass wedges, called "prisms", are used to carry out Bitterlich sampling. By standing in a central location, the worker uses the prism, which is calibrated to a specific basal area factor (BAF, usually 0.929–1.858 m^2/acre) to determine whether surrounding trees are included in the sample or not. In general, a tree that is closer or larger is more likely to be included than a distant, small tree (Fig. 5.4b). The number of each species included in the sample is then multiplied by the BAF to obtain the number of m^2/acre for each species. Although the inability to obtain density information from this method is a definite drawback, it may be useful in wetlands applications where the primary interest is a general description of the overstory trees with a more specific focus on understory or shrub layers.

Another plotless method frequently used to assess the plant species present within a wetland or community is the timed meander search. The practitioner simply walks around the wetland or community, recording all species that they observe. The time limit means that one can be consistent in order to compare different wetlands to each other. However, the timing should be scaled according to wetland size. An advantage to this technique is that it can capture a larger number of species than plot-based sampling alone because the investigator is free to explore a larger diversity of potentially species-rich microsites wherever they occur. For this reason, it typically leads to larger species counts than plot-based sampling.

5 Sampling and Analyzing Wetland Vegetation 283

This type of sampling is also used to detect rare species (Goff et al. 1982). A disadvantage to this technique is that it can only be used effectively by skilled field botanists who know the likely habitats of different species and can identify them quickly. In addition, the practitioner must be careful to specifically check for the small plants that might be more easily detected using plot sampling.

5.3.3 Locating Samples Within a Wetland

As discussed above, locating sample points within a wetland system is not a simple matter due to the patterned heterogeneity of much wetland vegetation. There are two basic approaches widely used in wetland vegetation assessment today: (1) representative, more subjective sample placement or (2) systematic sample placement based upon a pre-defined objective scheme. These pre-defined schemes have been described in Chap. 1 as random, restricted-random, regular, or haphazard.

Representative sample placement is quick and efficient, but less defensible in scientific or legal settings than systematic sampling. Nonetheless, it is common practice for monitoring wetland mitigation sites and for rapid wetland delineations. Using a combination of on-the-ground and aerial reconnaissance, different plant communities are roughly delineated (frequently on a map or aerial photograph), and a sample is described from one to several representative locations within each community (Fig. 5.5a). Representative sampling is easier to practice when there are relatively distinct and homogenous communities. It is most commonly used when a rapid, general assessment is needed and or there is a high level of trust in the judgment of the practitioner.

Pre-defined sampling schemes frequently use a baseline plus transects, which define a grid system for sample placement within the wetland (Fig. 5.5b). The baseline is established parallel to the dominant hydrologic gradient of the wetland, and transects extend perpendicular to it and the gradient (Fig. 5.5b). Sample sites are then located at specific locations on the transects. Transects can be regularly or randomly arranged on the baseline, and sample sites can be located regularly or randomly on the transects. This method is generally perceived as more objective and accepted by the scientific and legal community in North America. However, it may be overkill in situations where only preliminary descriptions of vegetation are needed. In addition, the method can be difficult to implement in very large wetland complexes with complex or non-obvious hydrologic gradients or in wetlands with large areas of deep water in the middle. Consider the extreme example of placing one 1 m^2 quadrat every 200 m on a 2,000 m long transect. Clearly alternative strategies must be devised. For large complexes, subdividing the region into smaller representative subsections may be more practical. For wetlands ringed with vegetation around large central deep water, a better strategy might be establishing a baseline around the perimeter of the wetland and running transects in toward the center (Fig. 5.5c). However, this method risks over-sampling the wetland's center with more plots than the periphery.

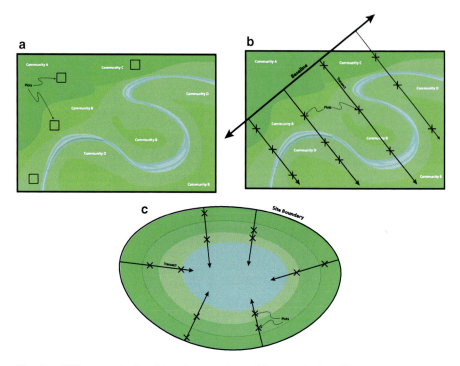

Fig. 5.5 Different methods of locating samples within a wetland. Different shades indicate different plant communities. (**a**) Representative sampling method; squares are samples. (**b**) Systematic sampling method with transects; crosses are samples. (**c**) Sampling with a perimeter baseline tends to oversample the wetland center (Published with kind permission of © M. Kuchta 2014. All Rights Reserved)

5.3.4 Relevé Systems

In general, the more objective system of systematically-placed plots using pre-defined sampling schemes has been adopted by United States ecologists, while the more subjective system of the relevé has been used by European ecologists. Relevés are becoming more common in the United States wetland monitoring community (see Minnesota Department of Natural Resources 2007; U.S. EPA 2002), however, due to the ability to obtain representative and detailed information in a relatively short amount of time. In the hands of a skilled practitioner with expertise in wetland plant identification, relevés can be very effective. The basic procedure involves establishing a 100 m^2 plot (400 m^2 for forested wetlands (Minnesota Department of Natural Resources 2007)) at a representative location within a wetland community. By walking through the plot, practitioners compile a species list, with cover estimates of both life form groups (e.g., evergreen, graminoid) and individual species. Multiple strata are assessed (tree, shrub, understory herbaceous), and heights and sociability (clustered, mat-forming, single) can be estimated. Size information can be taken for trees within the relevé plot to provide a more comprehensive data set. Important species located just outside of the study plot can also be included.

In many ways, the relevé gives a richer picture of the wetland plant community than systematically-placed plots. However, skilled field botanists are needed to execute it properly. Relevé data are more difficult to statistically summarize at smaller spatial scales due to the smaller number of plots. Over large spatial scales, species presence and abundance estimates can be used to assemble solid community descriptions. Practitioners also use permanently-established relevés to detect community change over time. Relevés may not be the best solution in systems with numerous discrete plant communities concentrated within small spatial scales, because collecting data from a high number of relevés can be time-consuming (U.S. EPA 2002).

5.3.5 Number of Samples

The number of samples is always a compromise between the resources available and the desire to collect as much data as possible to adequately characterize the system of interest. Ideally, a pilot study should be conducted prior to implementing a sampling scheme. The pilot study reveals the amount of variability in the wetland vegetation and can therefore give an idea of how many samples are needed to adequately describe that variability. The practitioner will typically vary the number and/or size of samples in the pilot study. The data are then used to create species accumulation curves (for species richness: number of species) or performance curves (for other measures). From these curves, the investigator can estimate the point at which additional samples yield minimal additional information. This point optimizes the efficiency and accuracy of sampling. The pilot study data should also be used in statistical power calculations (see Chap. 1).

A species accumulation curve is obtained by comparing the mean cumulative number of species to the number of samples (or size of plot, Fig. 5.6a). The asymptote of this curve is the point at which an adequate number of samples has been collected to characterize the richness of the system. In practice, this value is tedious to compute (consider: calculating the mean number of species for sample size of one is quite easy, but what about all pairs of samples for sample size of two?). However, there are computer programs (such as EstimateS; Colwell 2009) that will calculate for you based on your data matrix of samples and species. Likewise, the performance curve is obtained by plotting the mean and variability of some attribute against the number of samples (or size of plot, Fig. 5.6b). When there is no further change in the mean (within acceptable limits), the number of samples is adequate.

If there are no resources available for a pilot study, the adequacy of sampling can be assessed post-hoc using these methods. Statistical methods have been developed to estimate actual species richness from inadequate samples (discussed later in this chapter). In the absence of a pilot study, an important rule of thumb is to collect at least 20 samples at each site to meet the demands of some statistical methods (such as linear regression).

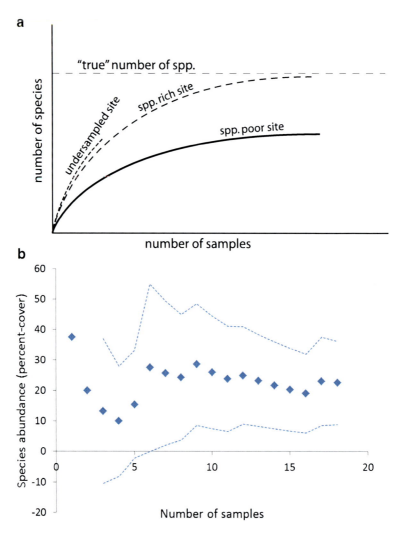

Fig. 5.6 Results from a pilot study. (**a**) Species accumulation curve, spp. = species. (**b**) Performance curve for species-abundance. *Dashed lines* are a 95 % confidence interval

5.3.6 Tracking Change Over Time

Permanent plots and photopoints are techniques for effectively monitoring vegetation change over time. If changes are substantial, it is possible to track changes using random sampling of the same sites at two different times, but maximum statistical power and confidence that change is real comes from sampling *exactly* the same place at different times. In order to install permanent plots, decisions must be made about location, plot type, and monumentation. The locations of permanent plots could be random or representative, depending upon the number of plots and the study purpose. They should be located so as to minimize potential for human or animal disturbance.

Three different types of permanent plots are common: circular (marked by central point), square (marked by four corners), and transect (marked by two endpoints). Researchers should not expect enough accuracy from a compass to be able to mark only one endpoint of a transect. In addition, although global positioning system (GPS) technology has advanced to a stage where most practitioners can get sub-meter accuracy and occasionally sub-foot, this is not high enough accuracy to avoid leaving monuments in the field. GPS can help to narrow down plot location, but then permanent marking structures should be installed to identify exact plot location. Of course, leaving structures in the field brings risks that they will be disturbed by humans, animals, or acts of nature (flood, landslide, fire). Non-visible, ground-level markers minimize the chance that they will be disturbed, but also minimize chances of relocation in the future. The ideal marker is rugged and hard to remove (such as rebar), visible or detectable by metal-detector, and designed to minimize harm to or detection by passers-by. Elzinga et al. (1998) provides a detailed discussion of monumentation considerations. Within wetlands, it can be extremely difficult to use a metal detector in areas of thick litter and vegetation, because the vegetation dampens the signal.

A picture is worth a thousand words. Photopoints not only communicate change over time in a visible way, but they can also help locate permanent plots in the field. A picture is typically located at some type of permanent marker and associated with a compass direction. In order to associate a particular photopoint with a location, record photographs containing location information can be taken in between photographs of the plant community or population. It is also good practice to locate these points using GPS.

5.3.7 Assessment Techniques for Specific Attributes

Substantial research and invention has been invested into how to best assess different vegetation attributes. Provided here is a non-exhaustive list of some commonly-used techniques and considerations for their use.

- **Definitions:** Meaningful definitions are important to establish and consistently use throughout the study. In vegetation studies, understory is typically defined by its height (i.e., <breast height (1.4 m)), the shrub layer defined by its number of stems, height and diameter (i.e., >1.4 m tall, with DBH <4 cm), while a tree is defined by its diameter (i.e., >4 cm DBH). In some cases, definitions of functional groups will need to be created. When assessing plant traits, clear definitions of leaves, stems, and flowers are important.
- **Density:** The number of individuals per area can be estimated using the plot-based counts or plotless techniques described above. Information on density is

Table 5.2 Common cover class systems

Cover class	Braun-Blanquet (1965) (midpoint)	Daubenmire (1959) (midpoint)
+	<1 % (0.5 %)	
1	1–5 % (3 %)	0–5 % (2.5 %)
2	5–25 % (15 %)	5–25 % (15 %)
3	25–50 % (37.5 %)	25–50 % (37.5 %)
4	50–75 % (62.5 %)	50–75 % (62.5 %)
5	>75 % (87.5 %)	75–95 % (85 %)
6		95–100 % (97.5 %)

especially important in population studies, and can be more easily collected for shrubs and trees than other measures of abundance. Difficult decisions must be made when defining an individual, however. Many plants are connected to each other by vegetative reproductive structures (e.g., blanket-like clones of grasses, grass tussocks, branching tree trunks, or multiple shrub stems connected underground). What is an individual in these circumstances? Investigators typically choose a definition that will capture the influence of the individual on the system of investigation. Trees are often considered separate individuals if they are separate at breast height. Shrubs are frequently measured using stem counts with disregard for underground structures. Grasses can be counted using number of shoots, although it is highly tedious. Boundary decisions (is the plant inside or outside the plot?) can also be difficult. Elzinga et al. (1998) provides a good overview of boundary decisions. Comparisons of density between different species are most informative when species are of similar size.

- **Cover:** Percent-cover is a measure of abundance that describes the horizontal area that a plant species or individual occupies. All plot-based techniques can be used to measure cover. It is frequently faster to use than density for very dense populations, and it is more effective than density for mat-like, low-growing vegetation. Some also argue that it is a better measure of the species' influence on the community than density, because it takes the size of the individuals into consideration. A disadvantage of cover measurements is that they change dramatically over the course of a season, so consistency in time-of-year is important.

There are numerous methods of assessing percent-cover, and the boundary decision problem still applies here. One of the most common methods is visual percent-cover. The observer stands over the plot and assigns cover to each species present. Naturally, this process is highly subjective and can differ dramatically between different plots and different observers (Kercher et al. 2003). In order to minimize observer bias, cover classes are typically used (Table 5.2). Mid-points of cover classes are typically used when analyzing the data, although this introduces substantial uncertainty into the data (Podani 2006). Care must be taken when analyzing ordinal data derived from cover classes (Podani 2006).

In order to avoid the observer bias inherent in the visual-percent cover system, some practitioners use a pin-frame system. This is a metal or plastic grid frame with attached vertical pins (10–100 pins) that fits over the top of a plot. Just as in the point-based method, any plant that hits a pin is recorded, and species cover for the plot is the number of pins that a given species intersects. This method is quite tedious, but less subjective. Other investigators use digital cameras suspended above the plot on a frame, which is also more objective and minimizes time in the field (except for camera-leveling). Digital image processing software is then used to differentiate different classes of ground cover (Luscier et al. 2006). The digital image technique is not as effective for plots with high species richness or multiple layers of vegetation. In addition, photographic images may make species identification difficult.

A consistent concern with cover measurements is that they mean different things for plants with different physiognomy. For example, mosses or mat-like plants have substantial horizontal cover, with little vertical structure. Graminoids, on the other hand, have substantial linear, vertical structure and can be under-represented by cover estimates.

- **Biomass:** Biomass assesses the amount of production of different species, and can indicate the above- and below-ground influence of the species on the population, community, or ecosystem. It is a destructive sampling technique, and so is difficult to justify when using permanent plots. Most biomass measurement techniques rely on harvesting the vegetation at the peak of its growth. Above-ground biomass is frequently assumed to represent plant allocation to growth, while below-ground biomass represents allocation to maintenance and mineral nutrient and water acquisition (Gurevitch et al. 2006). Above-ground harvest involves clipping at ground level, air- or oven-drying the harvest in paper bags, and weighing the sample. If species-specific biomass is desired, the species should be sorted in the field and bagged separately. Methods have been developed to visually-estimate biomass, based upon a calibrated clipped subsampling scheme (double-weight sampling (Interagency Technical Team 1996)). Below-ground biomass is substantially more difficult to assess, and typically involves excavating roots from a known and consistent volume of soil. In order to identify roots to the species level, above-ground parts must remain attached. Soil can be removed by washing prior to drying or sieving after drying. Core samplers have been devised for sampling below-ground biomass of submersed aquatic vegetation (Madsen et al. 2007).
- **Dominant Species:** Dominant species assessment can be important in classifying or differentiating wetland community types. Dominance is typically assigned to each vegetation layer (or stratum) separately. The concept of dominant species is complicated, because individual studies or methodologies have their own definitions of dominance (Barbour et al. 1998). A dominant species is one that has a large influence on the community or ecosystem due to its size or abundance. Frequently, this is determined after analyzing data back in the lab. Within forests, dominant tree species are those with the highest basal area. However, it is possible to determine and assign dominant species in the field if there is a clear definition based upon easily-measured attributes. Within a plot, for example, a dominant species could be that with the highest height $\times$ cover value. The height of dominant plant species can be used as an indicator of

wetland productivity or nutrient pollution (Little 2005). Analyzing the environmental tolerances of dominant plant species can also be helpful in modeling the dynamics of wetland plant communities (Squire and van der Valk 1992). The concept of dominance is also important in wetland delineation (Environmental Laboratory 1987). The wetland indicator status of dominant species, as determined by the "50/20" rule, determines whether a plot area is designated wetland.

- **Plant Functional Groups:** In order to effectively model plant communities, it is helpful to reduce the hundreds of species present into a smaller more manageable set. Species are assigned to groups based upon traits that reflect similar function in the ecosystem or community. Groups and traits are defined according to the application at hand. For example, Raulings et al. (2010) used plant response to flooding to create functional groups that they then modeled under varying flooding regimes. Other types of functional groups are based upon growth form (e.g., tussock, rhizomatous) or life history (e.g. annual, perennial) or combinations of these (Bouchard et al. 2007). The wetland indicator status used in wetland delineation (Lichvar and Kartesz 2011), is another example of a plant functional group scheme. Exploring the relations between functional groups and other organisms or environmental variables can yield interesting patterns that help us better understand and make predictions about wetland systems. Using established functional group definitions (such as the wetland indicator status or status from the U.S.D.A. Plants database) makes it easier to connect work to previously published studies, and is more acceptable to the scientific community.

- **Plant traits:** Plant traits are genetically-determined characteristics, like leaf shape, flowering time, seed number, or photosynthetic method that are inherent to the taxa, irrespective of the environment (Violle et al. 2007). They may also include genetically-determined responses to the environment, such as variation in specific leaf area based upon light availability and nutrient status. Relations can be drawn between plant traits and environmental attributes (e.g., carnivory and nutrient-poor wetlands). Practitioners also use plant traits to predict the behavior of individual species (e.g., invasiveness) or their response in wetland restoration settings (e.g., assembly rules, (Matthews et al. 2009a)). Plant traits can frequently be determined from the published literature after field work has been completed. However, if researchers are working with a novel trait-species combination, the trait parameters will need to be assessed in the field using adequate and representative sampling from the population. Use a performance curve to determine sampling adequacy.

5.3.8 Sampling Aquatic Vegetation

Many deep-water aquatic systems are not considered wetland, although they may be surrounded by or grade into wetland systems and so are of interest here. Many of the basic techniques and attributes described above for emergent and terrestrial vegetation can be applied, with modification, to aquatic vegetation. If the submerged vegetation is very shallow, the techniques can be applied directly, but for deeper water, access to the plants can be a problem. In order to sample deep water vegetation, there are two solutions: go to the plants or bring the plants to you. Going to the plants involves SCUBA or snorkeling. Sampling can be accomplished using open-ended polyvinyl chloride (PVC) frames for plots to surround tall vegetation (Parsons 2001). A different method of "going to the plants" involves creating a "viewing tube" out of PVC and clear plexiglass that can be used from a boat. Unless this is permanently-attached to the boat, it must be limited in size in order to penetrate the water. One of the most common methods of sampling aquatic plants is using a simple garden rake to harvest plants from a point, and then estimating percent cover on the rake of different aquatic plant species that are brought up to the surface. If water is very deep, the rake can be attached to a rope instead of the rake handle (Parsons 2001). Wide landscaping rakes used to prepare lawns are often preferred, because they are relatively light and bring up a large quantity of plants. GPS units are essential for locating plots when using a boat to sample.

5.3.9 Practical Considerations

There are several common practices used in field studies that are worth discussing.

- **Trampling the vegetation:** Although this may seem a petty concern, the results of trampling are not petty. When establishing plots and transects, it is important to not trample the vegetation in the area that you will be sampling. Trampled vegetation is more difficult to identify, and visual percent-cover is far more difficult to estimate. Trampling vegetation within a permanent plot can also affect future growth. When walking transect lines, always walk on the side of the tape opposite the side you will be sampling. Always walk outside the plot that you are establishing.
- **Voucher specimens:** It is important to collect a sample specimen of each species in your study. These are pressed in a plant press, identified, and deposited in a local herbarium, where their identities can be verified. The purpose of a voucher collection is to increase the quality of the study so that future researchers can determine the plant species found in the study, even if the names have changed, decades into the future. In situations where there are multiple observers over multiple years, vouchers can ensure consistency in identification. In order to not affect composition and structure of sample plots, when at all possible, voucher specimens should be obtained outside the plot.
- **Site map:** Site maps allow future researchers to return to your site to replicate your study or locate important features, such as monitoring wells or access points. Of course, GIS maps of a site with an aerial photograph for background are the gold-standard in site maps, but even hand-drawn maps with important

features, like access points, labeled permanent plot locations, streams, or different plant community locations can be extremely helpful in the future, or when sharing data collection duties with other workers. They also help ensure that interpretations made in the field align with those assumed back in the lab or stored in the computer file.
- **Multiple observers:** If large amounts of data are collected, it is inevitable that multiple personnel will be involved in vegetation assessment. Working with multiple observers adds additional variation in (1) plot boundary decision interpretation, (2) visual percent-cover estimates, and (3) definitions of individuals, among other aspects. One way to minimize variability is to be clear and consistent about rules and definitions, and document them in standard operating procedures (SOPs). In order to minimize variability in cover estimates, calibrating teams until results are consistent between observers is important (Kercher et al. 2003). This calibration may have to be repeated on a daily basis. Different observers may also have differing levels of expertise in plant identification. If differences in species-richness estimates between individuals are observed, correction factors can be applied post-hoc.

5.3.10 Other Important Data

Some data describing the wetland environment on a small scale can be easily recorded during vegetation sampling.
- **Litter and peat:** Wetlands can produce copious amounts of litter, which eventually may become peat. This litter can potentially suppress plant growth, and so may be an important variable influencing the vegetation. Attributes such as litter depth, percent cover, and type can be easily measured by sampling at one to many locations within a plot. Peat depth and type may also be important in structuring wetland plant populations and communities. Depth is easily measured using >2 m marked plastic rod inserted into the ground. If peat is deeper than the rod, then chances are the extra depth is not biologically significant, and a dummy depth can be used for analysis purposes.
- **Bare ground:** Bare ground within a wetland could signify disturbance, available seed bed, or stressful conditions for plant growth. In any of these cases, it is biologically interesting, and can be easily assessed using the percent-cover method.
- **Elevation or water depth:** Water depth is critically important to wetland plant growth and community structuring. It can be easily measured from the middle of a plot (or subsampled) using a meter stick or tape measure. This type of local measurement is a good supplement to staff gauge or piezometer information, because it is at a smaller scale and may be more relevant to the plants. For more intensive studies, survey equipment (laser level, tripod, and stadia rod) can be used to assess the elevation of each plot. For smaller plots, a single measurement in the plot center is adequate. For larger plots, multiple readings may need to be taken.

- **Microtopography:** Microtopographic features in wetlands (i.e. hummocks, pools, stumps, or tussocks) can exert strong control of the local plant community (Peach and Zedler 2006). Depending upon the study purpose and scope, microtopography can be measured quantitatively (more intense, smaller scope) or qualitatively (less intense, larger scope). Quantitative measurements of high and low points within plots (associated with topographic breaks) can be measured with high-accuracy GPS units associated with local base stations (Werner and Zedler 2002) or using meter sticks to determine tussock height (Peach and Zedler 2006) or maximum height difference within the plot. For studies that are broader in scope, a plot can be assigned qualitative microtopographic scores corresponding to all types within a plot (e.g. stump, high hummock, hummock, low hummock, hollow, flat, or pool). For data analysis purposes, these can be transformed into ordinal scores, and plot microtopographic richness, mean score, or maximum difference can be calculated (Little et al. 2010).
- **Canopy cover:** Canopy cover is an important environmental variable to measure in forested wetlands, because many wetland understory species respond to shade. There are three common ways to measure canopy cover, increasing in accuracy: (1) canopy tube, (2) spherical densiometer, and (3) digital camera with fish eye lens and image-processing software. Canopy tubes are simply vertical tubes with a cross hairs and some type of leveling mechanism. These can be sophisticated tubes with mirrors, or home-made toilet-paper tubes with a dangling level inside. Visual percent-cover of canopy within the tube is recorded from the middle of the plot. Alternatively, the crosshairs can be used to determine presence/absence of canopy at a set of points per plot (Ganey and Block 1994). A spherical densiometer is a small, handheld gridded mirror with a leveling bubble. It is held above the plot, and the observer views how many grid cells are occupied with canopy cover by visualizing a series of dots within the cells (Lemmon 1956). The most sophisticated and accurate measurements of canopy cover use a fish eye (hemispherical) lens with digital image-processing software to calculate canopy cover and light transmission (Englund et al. 2000). However, these cameras are very expensive, and data can only be collected a certain times of day under specific weather conditions – limiting their utility.
- **Spatial data:** A detailed discussion of spatial data collection and autocorrelation is beyond the scope of this chapter, but practitioners should consider whether important spatial relations may exist within or between wetland systems of study. Landscape ecology approaches may be needed to assess relations between wetland sites (consult Turner et al. (2001) for ideas). Within sites, numerous workers have found interesting relations between hydrological features and plant communities using measures as simple as distance of plot from a feature (Grace and Guntenspergen 1999). Since hydrological and dispersal gradients often vary with distance in wetlands, it can be a helpful, and easily-measured surrogate for other variables.

5.3.11 Field Forms

Customized data sheets or files are frequently used in ecology to help streamline data collection and ensure that nothing is accidentally omitted. Data sheets can be as simple as a table on a single page (see Field Labs at the end of this chapter) to a complex multi-page and attribute form like is used in wetland delineation. By listing commonly-encountered plant species in the form in advance, then the data recorder does not have to write them in each time data is collected. In addition, if repeated sampling is planned, consistent data forms can ensure that the same data is collected each time. These forms can also be designed to simplify data entry once fieldwork is completed.

5.4 Basic Analysis Techniques Commonly Used for Vegetation Data

As with sampling techniques, entire books have been written about analyzing ecological and vegetation data. The reader is encouraged to explore McCune and Grace (2002) and Kenkel (2006) for more detailed discussion of multivariate techniques, their assumptions, and the data transformations needed to meet those assumptions.

5.4.1 Basic Calculations

Summarizing the basic attributes of a plant population or community is an important step in the initial stages of data analysis. Exploratory data analysis is critical to understanding the data structure in preparation for more advanced analyses (Kenkel 2006).

- **Frequency:** The number of plots or samples in which a species appears, based upon presence or absence. Frequency is a good measure of how common the species is across the site.
- **Density:** The number of individuals per area. Density measures can be quite variable, spatially. Measures of mean and variability are calculated.
- **Cover or Basal Area:** The areal cover of a plant. Basal area pertains to tree trunks, and is $\pi\left(\frac{d}{2}\right)^2$ where d is the diameter of the tree at breast height (typically measured with a special diameter tape).

Relative values of each of these measures can be calculated, and these are how much each species contributes (as a fraction or percent) of the total frequency, density, or cover of all species. Calculating relative values enables comparisons between sites with dramatically different total cover, for example.

5 Sampling and Analyzing Wetland Vegetation

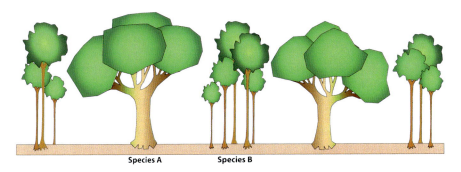

Fig. 5.7 Two species may have different roles in a community, but similar importance values (Published with kind permission of © M. Kuchta 2014. All Rights Reserved)

Relative value of species A = Value of species A/Total value for all species

Frequency, density, and cover each represent a different aspect of the role of a species in a plant community. If a measure of the overall importance of a species is desired, then composite measures, such as an importance value (Curtis 1959) can be calculated.

Importance Value = (relative frequency + relative density + relative cover)/3

Practitioners must use caution when interpreting composite measures, however, because two species with very different biological roles in a community can have the same importance value. If species A has low density but high cover and species B has high density, but low cover, they could have the same importance value (Fig. 5.7).

5.4.2 Assessing Species Richness and Diversity

There are many reasons why it is important to assess wetland plant species richness and diversity. Levels of diversity can indicate the health or status of wetland systems (U.S. EPA 2002). Diversity assessments can be important in establishing conservation priorities. Sometimes practitioners are interested in the factors that contribute to high or low wetland plant diversity or richness (e.g., Bedford et al. 1999, Michalcova et al. 2011). In wetland plant science, richness refers to the number of species found in a defined area. Diversity is defined by a combination of species richness (number) and the evenness of the distribution of abundance among species. For example, a wetland site with 142 different species, but 97 - percent-cover of *Typha* spp. has high richness, but low evenness. Numerous mathematical formulas have been invented and used to describe how diverse such a system is. Regardless of the method used, both richness and diversity give no information about species identity. A very rich community could contain a high number of invasive species, which would be negative from an ecological value perspective.

- **Species Richness:** The number of species is frequently easier to communicate to decision-makers than composite index numbers representing species diversity, and so is frequently used to describe wetland systems. Sampling for species richness can be difficult; pilot studies and species-accumulation curves should be evaluated prior to final sampling (see above). Raw, or observed, species richness straight from the field tends to underestimate the true species richness of a site or system. In order to correct for sampling deficiencies (bias or loss of precision), species richness estimators have been created. Most estimators use a process of generating numerous estimates from randomized resampling of data with different numbers of samples and calculating the mean estimate from the resampling (Michalcova et al. 2011). According to Magurran (2004), the richness estimators with the least bias and highest accuracy are the Chao2, Jack1 and Jack2 methods. Several statistical packages can be used to calculate estimators, including the R package vegan (Oksanen et al. 2011) and EstimateS (Colwell 2009). In practice, the data set with actual observations is entered, yielding an output with several estimates of species richness based upon different estimators. The user then must choose which estimator performs the best for the given data set by examining the output. Some estimators are more conservative than others or will better mirror the observed species accumulation curve.
- **Species Diversity:** Diversity indices are numbers generated from information about species richness and how evenly-distributed the abundance of different species is within the community. These indices provide more information, but they can be open to interpretation and difficult to communicate to decision-makers. There are three general types of diversity: alpha, beta, and gamma. Alpha diversity is the diversity of a single point or site, and is the type most commonly used. Beta diversity is the difference in community composition (change in species and their abundance) over a series of samples. Gamma diversity is the species pool, or the set of species present in the larger regional landscape, and can be important in determining the potential set of propagules available for a restored or disturbed wetland site. There are numerous published diversity indices for describing alpha diversity (see Magurran (2004) for a thorough review), although only a few are widely used. Each index is based upon the proportion of total abundance (p_i) that each species comprises within the community.
    - **Simpson's Index:** This widely-used metric is simply the sum of squares of all species proportions, where S = number of species, and p_i is the proportion of species i:

$$D = \sum_i^S p_i^2$$

Simpson's index of diversity is 1 – D, the probability that any two randomly drawn species will be different. Values range from one (high diversity) to zero (low diversity). It emphasizes common species and de-emphasizes rare species, which means that the measure is not dramatically affected by missing rare species during sampling. The effective number of species using the Simpson's index is 1/D.

- **Shannon-Wiener Index:** This index is also very popular in ecological studies. It is a measure of the "disorder" in a sample. The higher the disorder or uncertainty, the more diverse a system is. The higher the H' value, the more diverse the site. The Shannon-Wiener index is the negative sum of the proportion of each species (p_i) times the log of p_i:

$$H' = -\sum_{i}^{S} p_i \ln p_i$$

Index values typically range from 1.5 (low diversity) to 3.5 (high diversity). This measure is more sensitive to rare species than Simpson's index, but less sensitive than plain species richness. That is, rare species count for more value in the Shannon-Wiener Index than in the Simpson. The effective number of species using the Shannon-Wiener index is $e^{H'}$, the exponent of H'.

- **Effective number of species:** This metric can be calculated from any diversity measure, and describes the equivalent number of equally common species for a data set. That is, if all species were of equal abundance, how many would there be? This number takes into account the evenness of the community, and will always be lower than the actual species richness (unless all species are equally abundant). Sites with higher numbers of effective species are more diverse than sites with lower values. Unlike the index values, using effective number of species makes intuitive sense to a lay audience.

- **Evenness:** An evenness value can also be calculated for each sample. Using the Shannon-Wiener index (H') as a starting point,

$$J = \frac{H'}{\ln S}$$

where S is the species richness. This metric is "Pielou's J," and ranges from one (perfect equitability among species) to zero (no equitability). In terms of evaluating wetland plant communities, higher evenness may mean that a community is more diverse, and less dominated by a few highly competitive species.

5.4.3 Preparation of Multivariate Data

Users must be careful to understand the structure of their multivariate data before beginning. Many parametric data analysis techniques rely on normally-distributed

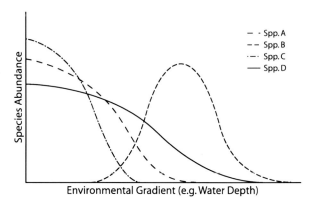

Fig. 5.8 Direct gradient analysis in which species distributions are plotted along an environmental gradient (Published with kind permission of © M. Kuchta 2014. All Rights Reserved)

data sets that have linear relationships with other normally-distributed data sets. Ecological data sets rarely have these characteristics. Species frequently respond to environmental gradients in a non-linear manner (Fig. 5.8), with low abundances at the extremes of their tolerances and high abundances in the center at their ideal conditions (called a Gaussian distribution). In addition, response curves can be solid: even in the most favorable environments, species may not be present due to dispersal restrictions or other factors and species abundances may range from zero to very abundant in the most favorable conditions. In addition, we have no information on species response to conditions beyond the range of their tolerances. Our data sets are truncated at zero, because it is impossible for a species to have a negative abundance (Fig. 5.8). Additional complications arise when species distributions are more skewed or peaked than normal. Finally, many species will exhibit shared absence in numerous sites, creating a species by site matrix that contains numerous zeroes. Just because two species are not present in the same site does not mean that they respond similarly to the same environmental factors. However, this mutual absence may produce a correlation artifact in the data. These characteristics of ecological data can be dealt with by with data preparation and transformation strategies that will minimize variation and maximize expressed data structure. These strategies are beyond the scope of this chapter, but are described in McCune and Grace (2002).

5.4.4 Classification of Wetland Plant Communities

Classifying wetlands, or putting them into categories, is important to effectively manage and restore them. It is typically a first step in any study of a novel system, essential to description. Classification facilitates conservation, and predicting future behavior in response to environmental change. There are numerous methods of classifying wetlands, such as the hydrogeomorphic classification system (Brinson 1993; see Chap. 2 in Vol. 3). Wetlands, or communities within wetlands, can also be classified on the basis of their vegetation. Classifying based upon vegetation can

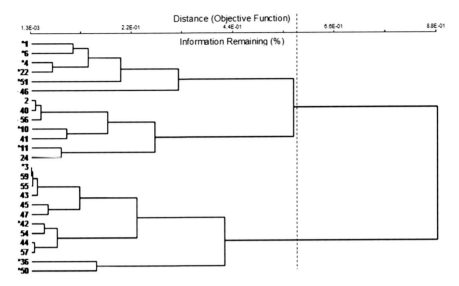

Fig. 5.9 Cluster dendrogram of 25 wetland sites based upon species dissimilarity. *Dashed line* indicates subjective cut-off to create two wetland groups

be useful, because vegetation integrates hydrological, edaphic, and biogeochemical signals (U.S. EPA 2002). It also responds more rapidly to anthropogenic and natural disturbances than hydrogeomorphic setting. On the other hand, this relatively short-term response to disturbance can be a disadvantage if one is attempting to discern response to underlying long-term signals, such as global warming or acid rain.

All classification methods attempt to form groups of like communities out of species data from multiple sites or samples. Early methods relied primarily on relevé descriptions and placement into associations based upon species tables (Barbour et al. 1998), (see Graf et al. (2010) for a recent application). For a while, TWINSPAN (two-way indicator species analysis) was a popular method of forming groups identified by indicator species. However, TWINSPAN should never be used except in simple cases of a single dominant environmental gradient (change in some environmental variable across sites or time (McCune and Grace 2002)). There are multiple methods of distinguishing vegetation groups out of multivariate data. Multivariate data are collected when workers collect multiple measurements (species or environmental variables) at a single sampling location (Kenkel 2006). Complete coverage of methods and their mathematical rationale is given in (McCune and Grace 2002). The most commonly used method in studies of wetland vegetation is hierarchical cluster analysis. Sites or samples can be placed into groups based upon their multivariate vegetation using dissimilarity indices (such as Sorenson or Euclidean distance). It is a hierarchical process, because smaller, more similar groups are combined into larger, less similar groups, with the smaller groups becoming sub-groups of the larger groups. The end product is a dendrogram showing the multivariate similarity between sites or samples (Fig. 5.9). Groups can be defined post-hoc using subjective methods or more objective measures which assess the homogeneity or heterogeneity of groups (Sharma 1996). In general, the practitioner's knowledge of the study system is most important when defining groups that are helpful to modeling the system (not too many or too few groups for understanding). The process of non-hierarchical K-means cluster analysis is becoming more popular (Carr et al. 2010), in which the practitioner first determines the number of groups and then a computer program optimizes a statistical parameter within those groups (McCune and Grace 2002).

The non-parametric multi-response permutation procedure (MRPP) can be used to assess within group homogeneity and to test for significant differences between groups based upon multivariate data (McCune and Grace 2002). However, statistically significant differences are not always ecologically-meaningful.

Currently, classification for mapping purposes is more frequently accomplished remotely using vegetation reflectance from the visual and near infrared spectra. These remotely-detected pixel signals are frequently combined into groups using supervised or unsupervised classification with K-means clustering to identify the spectral signatures of different wetland plant communities (Zhang et al. 2011). Remotely-sensed and classified communities can be mapped very easily (Midwood and Chow-Fraser 2010), however, the level of detail in these classifications is necessarily limited. Numerous statistical packages can perform cluster analyses, including the freeware R package vegan (Oksanen et al. 2011) and PC-ORD (McCune and Mefford 2011).

5.4.5 Ordination

Typically, a practitioner will have multiple sites or samples, with each sample described in numerous ways (the abundances of multiple species, environmental characteristics, etc...). Ordination is a method of discovering patterns and underlying structure in this multivariate data (Kenkel 2006). Because ordination diagrams and the process of ordination itself can be confusing, ordination information is typically not directly presented to lay people or political decision-makers. However, that does not mean that it has no role to play in wetland conservation and management. Ordination has been used to assess the effects of management practices on wetland plant communities (Hall et al. 2008); compare damaged, restored, and reference plant communities (Rooney and Bayley 2011); assess the community-level effects of exotic species invasion (Mills et al. 2009); and generally understand how environmental degradation affects wetland systems (Carr et al. 2010). A complete discussion of ordination techniques, their assumptions, and mathematical background can be found in McCune and Grace (2002), Kenkel (2006), and Legendre and Legendre (1998).

There are several different types of ordination, but all involve reducing the variability in a large data set down to a few axes that express the primary patterns

5 Sampling and Analyzing Wetland Vegetation

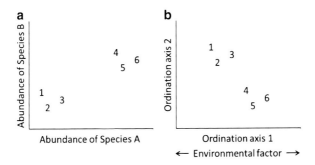

Fig. 5.10 Ordering sites according to their multivariate plant species composition using indirect gradient analysis (ordination). *Numbers* are wetland sites. (**a**) A data set consisting of only two species and six sites. (**b**) A data set consisting of numerous species in six sites. The underlying structure in the species information has been extracted by ordination methods into two axes of variation. Site ordination scores are related to environmental measurements using correlation

in the data using the correlation between the multiple variables (McCune and Grace 2002). Sometimes there is a particular environmental factor of interest. If this is the case, then direct gradient analysis, or positioning samples along axes determined by environmental measurements, is most appropriate (Kenkel 2006). The practitioner can then examine the relations between the species within those sites and the environmental factor of interest (Fig. 5.8).

Indirect gradient analysis does not assume the importance of any particular environmental gradient *a priori*, but rather lets the plant species data order itself. In multivariate speak: "arranging plots in species space." There are several ways of using ordination, but the most common is to plot sites along axes of species composition, and then correlate these axes with environmental variables to determine which environmental variables most strongly influence the plant communities. One can conceptualize this more easily by beginning with a system of multiple sites with only two species. It is easy to plot samples or sites within species space, with each axis representing the abundance of species A or B (Fig. 5.10a). In Fig. 5.10 sites 1, 2, and 3 are similar in their composition of species A and B, and sites 4, 5, and 6 are similar to each other. In a cluster analysis, these two groups would most likely cluster together. Ordination typically involves many more than two species, but it is very difficult to create a graph that contains 100 axes for each different species. In order to make meaning, ordination mathematically sorts through the variation in these different species to draw out the strongest patterns (based on correlation or similarity), and these patterns are reduced to usually two or three axes (Fig. 5.10b). Each site is assigned a score or position along each axis, based upon its species composition. Because each site has associated environmental measurements, the ordination score can be correlated with the environmental factors to determine how these factors affect the community as a whole rather than an individual species.

Various methods exist for indirect gradient analysis: PCA (principle components analysis), Bray-Curtis, NMS (nonmetric multidimensional scaling), CA (correspondence analysis), DCA (detrended correspondence analysis), and CCA (canonical correspondence analysis). McCune and Grace (2002) and Kenkel (2006) thoroughly review the options and mathematical background behind each technique. In general, PCA should be used when there are few primary gradients that relate broadly-linearly to the scope of plant communities studied. CA is best used on categorical contingency table data (optimizing both sites and species variation simultaneously). DCA should be avoided. McCune and Grace (2002) argue that NMS is currently the method of choice because it makes no assumption of linear relations, and performs very well with high diversity data sets. However, Kenkel (2006) suggests using NMS as a last resort only after PCA and CA options have been exhausted. CCA is ordination constrained by environmental variables, and should be used when there are one to few strong environmental gradients of interest. All of these techniques are available in the R package vegan (Oksanen et al. 2011) and in vegetation-specific software, like PC-ORD (McCune and Mefford 2011). Roberts (2011) provides helpful online tutorials for using R to analyze vegetation data.

Ordination is extremely helpful for initial pattern detection and description of novel systems. Ordination using different transformations of the same species dataset (e.g., one using abundance, one using frequency, and one using presence or absence) can be helpful in differentiating between levels of organization in plant communities (Allen and Wyleto 1983). It is critically important to prepare the species dataset for ordination by removing outliers and performing data transformations to meet the assumptions of the technique (McCune and Grace 2002; Kenkel 2006). One criticism of ordination is that it cannot test hypotheses using the philosophy of inferential statistics, although the structure of ordinations themselves can be tested through bootstrapping (testing real data configurations against multiple randomized variations). However, structural equation modeling (SEM) provides a new way of statistically testing relations discovered through ordination (McCune and Grace 2002).

5.4.6 Classification and Regression Trees (CART)

Another method of analysis that addresses the question of how environmental variables affect plant populations or communities is CART (McCune and Grace 2002). Classification trees model which independent variables best differentiate pre-defined groups from each other (e.g., plant community groups from classification or occupied versus unoccupied sites). Classification trees have also been used to assess wetland condition (Cohen et al. 2005). Regression trees have continuous response variables. One advantage of CART is that it is a non-parametric method, meaning that it does not require the same assumptions of data normality that other methods require. The output is a predictive model that resembles a dichotomously-forking tree which separates pre-defined groups based on a threshold value of the best-differentiating environmental variable at each level. Like in hierarchical classification, the initial fork separates two relatively heterogeneous groups from each other, and subsequent divisions can lead all the way down to individual sites or some pre-defined stopping value. The resulting model allows the user to determine, given the values of different environmental variables, the type of plant community likely to occur in a site or whether the site could be suitable habitat for a species. Using this technique, relationships sometimes emerge that are otherwise difficult to detect using other linear or even multivariate models. For more information, see De'ath and Fabricius (2000) and McCune and Grace (2002).

5.4.7 Mantel Test

A Mantel test is simply a method of correlating two similarity or distance matrices with each other. It is also especially helpful when evaluating the effect of spatial proximity on plant community similarity or the strength of plant community – environment relationships. For example, a goal of a project may be to determine whether plant communities respond to wetland disturbance or restoration in a similar fashion to macroinvertebrate communities across a set of wetlands. With the Mantel test, the similarity in response can be compared by correlating the plant and macroinvertebrate distance matrices. A Mantel test can also be used to assess the significance of the correlation between geographic distance and community distance (McCune and Grace 2002).

5.4.8 Indicator Species Development and Analysis

Although wetland plants can be used successfully as indicators of wetland health (U.S. EPA 2002) and wetland status (Environmental Laboratory 1987), this section does not focus solely on those particular applications. Indicator species analysis (Dufrêne and Legendre 1997) is a mathematical technique that can determine indicators for different groups of sites or plant communities. Therefore, it can be used to develop wetland condition indicators, but that is not its sole purpose. Once groups have been established either *a priori* or using the techniques described above, indicator species analysis determines how faithful a given species is to a particular group (whether it is always present), and how exclusive the species is to the group (never occurring in other groups, (McCune and Grace 2002)). A species abundance or presence data set is input, and the output is a table of indicator values (percent of perfect indication, with 100 % being perfect), and an associated P-value based on a Monte Carlo (randomization) test with a null hypothesis of no difference between groups. The R package labdsv (Roberts 2010) and PC-ORD (McCune and Mefford 2011) both calculate indicator species values.

Indicator species analysis has been used to better describe plant community groups (Rooney and Bayley 2011), differentiate wetlands invaded by non-native plant species (Johnson et al. 2010), and associate plant species with different environmental conditions for wetland condition assessment (Johnston et al. 2007). The combination of cluster analysis, NMS ordination, and indicator species analysis is commonly used to describe and differentiate plant communities in the wetland literature.

5.4.9 Floristic Quality Assessment Indices

Floristic quality indices (FQAIs) are frequently used to determine the condition of a wetland based upon the ecological "conservatism" of the plant species (U.S. EPA 2002). Some plant species are more sensitive to human disturbance and therefore more conservative in terms of their growth requirements. These species are indicators of high quality systems. By sampling an area, one can assess its quality using the plant species scores (i.e., coefficients of conservatism). Species are ranked with values from one (not conservative) to ten (conservative and highly ecologically sensitive) by experts. The ranking needs to be done on a regional basis, because species behave differently in different regions. Therefore, one cannot apply a FQAI developed in the Upper Midwest to New England wetlands.

Once a wetland has been sampled, various formulas can be used to summarize the wetland conditions based upon the C of Cs (coefficients of conservatism) of the plants found there. One commonly used formula is

$$I = \sum_{i}^{S} CC_i \Big/ \sqrt{N_{native}}$$

Where I = the FQAI for the site, CC_i is the C of C for species i, and N is the total number of native species found at the site (Andreas et al. 2004). A simple mean C of C ($\bar{C}$) for all species can also be calculated:

$$\bar{C} = \sum_{i}^{S} CC_i / N$$

Both of these measures rely only on species presence data, which is an advantage in that it is faster to inventory species presence than abundance. A weighted average measure can incorporate species abundance, with relativized species abundance as the weights (see Exercise 4). Rooney and Rogers (2002) discuss some problems with FQAI and pose alternative calculations. Matthews et al. (2009b) does a thorough assessment of the performance of different vegetation indicators, including FQAIs, when tracking wetland restoration trajectories in comparison with reference systems. They warn that using any single metric to assess wetland restoration success provides an incomplete picture, that multiple methods should be used, and that there is no simple metric that adequately assesses restoration success (Matthews et al. 2009b).

5.4.10 Using the Wetland Indicator Status of Vegetation

One of the most frequent applications of vegetation sampling and analysis in a wetland setting is for wetland delineation purposes. Very specific sampling and analysis protocols are used, according to the 1987 U.S. Army Corps of Engineers manual (Environmental Laboratory 1987) and the newer Regional Supplements. In this system, plant species are assigned an indicator status (obligate wetland = 1, facultative wetland = 2, facultative = 3, facultative upland = 4, or upland = 5) for different regions according to expert opinion. The indicator status of species also can be used for other purposes aside from wetland delineation protocols. One application is to calculate a weighted average of indicator scores with weights based upon species importance value, cover, or frequency in order to track the relative wetness of a site. This application is especially helpful when conducting repeated studies to assess wetland mitigation success, for example (Atkinson et al. 1993).

5.4.11 More Resources

The subject of vegetation sampling and analysis has generated a vast and rich literature. This chapter is intended to expose the reader to a variety of sampling considerations and basic analysis techniques. The following excellent resources should be consulted for further information:

- Sampling and analysis for plant population studies: Elzinga et al. (1998)
- Plant community data analysis, especially of multivariate data: McCune and Grace (2002) and Kenkel (2006)
- Using vegetation as an indicator of wetland quality: U.S. EPA (2002)
- U.S. Fish and Wildlife Service National Wetlands Inventory: http://www.fws.gov/wetlands/
- U.S. Department of Agriculture Web Soil Survey: http://websoilsurvey.nrcs.usda.gov

References

Allen TFH, Wyleto EP (1983) A hierarchical model for the complexity of plant-communities. J Theor Biol 101:529–540

Andreas BK, Mack JJ, McCormac JS (2004) Floristic Quality Assessment Index (FQAI) for vascular plants and mosses for the State of Ohio. Ohio Environmental Protection Agency, Division of Surface Water, Wetland Ecology Group, Columbus

Atkinson RB, Perry JE, Smith E, Cairns J (1993) Use of created wetland delineation and weighted averages as a component of assessment. Wetlands 13:185–193

Barbour MG, Burk JH, Pitts WD, Gilliam FS, Schwartz MW (1998) Terrestrial plant ecology. Benjamin Cummings, Menlo Park

Bedford BL, Walbridge MR, Aldous A (1999) Patterns in nutrient availability and plant diversity of temperate North American wetlands. Ecology 80:2151–2169

Bonham CD (1989) Measurements for terrestrial vegetation. Wiley, New York

Bouchard V, Frey SD, Gilbert JM, Reed SE (2007) Effects of macrophyte functional group richness on emergent freshwater wetland functions. Ecology 88:2903–2914

Braun-Blanquet MM (1965) Plant sociology: the study of plant communities. Hafner, London

Brinson MM (1993) A hydrogeomorphic classification for wetlands. WRP-DE-4. U.S. Army Corps of Engineers Waterways Experiment Station, Vicksburg

Carr SC, Robertson KM, Peet RK (2010) A vegetation classification of fire-dependent pinelands of Florida. Castanea 75:153–189

Cohen MJ, Lane CR, Reiss KC, Surdick JA, Bardi E, Brown MT (2005) Vegetation based classification trees for rapid assessment of isolated wetland condition. Ecol Indic 5:189–206

Colwell RK (2009) EstimateS: statistical estimation of species richness and shared species from samples 8.2. http://purl.oclc.org/estimates

Curtis JT (1959) The vegetation of Wisconsin: an ordination of plant communities. University of Wisconsin Press, Madison

Dale MRT (1998) Spatial pattern analysis in plant ecology. Cambridge University Press, New York

Daubenmire RF (1959) A canopy coverage method of vegetational analysis. Northwest Sci 33:43–64

De'ath G, Fabricius KE (2000) Classification and regression trees: a powerful yet simple technique for ecological data analysis. Ecology 81:3178–3192

Dufrêne M, Legendre P (1997) Species assemblages and indicator species: the need for a flexible asymmetrical approach. Ecol Monogr 67:345–366

Ehrenfeld JG (1995) Microtopography and vegetation in Atlantic white cedar swamps: the effects of natural disturbances. Can J Bot 73:474–484

Elzinga CL, Salzer DW, Willoughby JW (1998) Measuring and monitoring plant populations. 1730-1. Bureau of Land Management, Denver

Englund S, O'Brien J, Clark D (2000) Evaluation of digital and film hemispherical photography and spherical densiometry for measuring forest light environments. Can J For Res 30:1999–2005

Environmental Laboratory (1987) Corps of engineers wetland delineation manual. Y-87-I. U.S. Army Engineer Waterways Experiment Station, Vicksburg

Gabriel AO, Bodensteiner LR (2011) Ecosystem functions of mid-lake stands of common reed in Lake Poygan, Wisconsin. J Freshw Ecol 26:217–229

Ganey JL, Block WM (1994) A comparison of two techniques for measuring canopy closure. West J Appl For 9:21–23

Goff FG, Dawson GA, Rochow JJ (1982) Site examination for threatened and endangered plant species. Environ Manage 6:307–316

Grace JB, Guntenspergen GR (1999) The effects of landscape position on plant species density: evidence of past environmental effects in a coastal wetland. Ecoscience 6:381–391

Graf U, Wildi O, Feldmeyer-Christe E, Kuechler M (2010) A phytosociological classification of Swiss mire vegetation. Bot Helv 120:1–13

Greig-Smith P (1983) Quantitative plant ecology. University of California Press, Berkeley

Gurevitch J, Scheiner SM, Fox GA (2006) The ecology of plants. Sinauer Associates, Sunderland

Hall SJ, Lindig-Cisneros R, Zedler JB (2008) Does harvesting sustain plant diversity in Central Mexican wetlands? Wetlands 28:776–792

Hamer AJ, Parris KM (2011) Local and landscape determinants of amphibian communities in urban ponds. Ecol Appl 21:378–390

Interagency Technical Team (1996) Sampling vegetation attributes. BLM/RS/ST-96/002+1730. Bureau of Land Management, Denver

Johnson TD, Kolb TE, Medina AL (2010) Do riparian plant community characteristics differ between Tamarix (L.) invaded and non-invaded sites on the upper Verde River, Arizona? Biol Invasions 12:2487–2497

Johnston CA, Bedford BL, Bourdaghs M, Brown T, Frieswyk C, Tulbure M, Vaccaro L, Zedler JB (2007) Plant species indicators of physical environment in Great Lakes coastal wetlands. J Great Lakes Res 33:106–124

Kenkel NC (2006) On selecting an appropriate multivariate analysis. Can J Plant Sci 83:663–676

Kenkel NC, Podani J (1991) Plot size and estimation efficiency in plant community studies. J Veg Sci 2:539–544

Kent M, Coker P (1995) Vegetation description and analysis: a practical approach. Wiley, West Sussex

Kercher SM, Frieswyk CB, Zedler JB (2003) Effects of sampling teams and estimation methods on the assessment of plant cover. J Veg Sci 14:899–906

Krebs CJ (1998) Ecological methodology. Benjamin Cummings, Menlo Park

Kudray GM, Gale MR (2000) Evaluation of National Wetland Inventory maps in a heavily forested region in the upper Great Lakes. Wetlands 20:581–587

Legendre P, Legendre L (1998) Numerical ecology. Elsevier Science, Amsterdam

Lemmon RE (1956) A spherical densiometer for estimating forest overstory density. For Sci 2:314–320

Lichvar RW, Kartesz JT (2011) North American digital flora: national wetland plant list, version 2.4.0. U.S. Army Corps of Engineers, Engineer Research and Development Center, Cold Regions Research and Engineering Laboratory. https://wetland_plants.usace.army.mil. Accessed 2011

Little AM (2005) The effects of beaver inhabitation and anthropogenic activity on freshwater wetland plant community dynamics on Mount Desert Island, Maine, USA. Dissertation or Thesis, University of Wisconsin-Madison

Little AM, Guntenspergen GR, Allen TFH (2010) Conceptual hierarchical modeling to describe wetland plant community organization. Wetlands 30:55–65

Luscier JD, Thompson WL, Wilson JM, Gorham BE, Dragut LD (2006) Using digital photographs and object-based image analysis to estimate percent ground cover in vegetation plots. Front Ecol Environ 4:408–413

Madsen JD, Wersal RM, Woolf TE (2007) A new core sampler for estimating biomass of submersed aquatic macrophytes. J Aquat Plant Manage 45:31–34

Magurran AE (2004) Measuring biological diversity. Princeton University Press, Princeton

Matthews JW, Peralta AL, Flanagan DN, Baldwin PM, Soni A, Kent AD, Endress AG (2009a) Relative influence of landscape vs. local factors on plant community assembly in restored wetlands. Ecol Appl 19:2108–2123

Matthews JW, Spyreas G, Endress AG (2009b) Trajectories of vegetation-based indicators used to assess wetland restoration progress. Ecol Appl 19:2093–2107

McCune B, Grace J (2002) Analysis of ecological communities. MjM Software Design, Gleneden Beach

McCune B, Mefford MJ (2011) PC-ORD. Multivariate analysis of ecological data, Version 6. MjM Software Design, Gleneden Beach

Michalcova D, Gilbert JC, Lawson CS, Gowing DJG, Marrs RH (2011) The combined effect of waterlogging, extractable P and soil pH on alpha-diversity: a case study on mesotrophic grasslands in the UK. Plant Ecol 212:879–888

Midwood JD, Chow-Fraser P (2010) Mapping floating and emergent aquatic vegetation in coastal wetlands of eastern Georgian Bay, Lake Huron, Canada. Wetlands 30:1141–1152

Mills JE, Reinartz JA, Meyer GA, Young EB (2009) Exotic shrub invasion in an undisturbed wetland has little community-level effect over a 15-year period. Biol Invasions 11:1803–1820

Minnesota Department of Natural Resources (2007) A handbook for collecting vegetation plot data in Minnesota: the relevé method. 92. Minnesota County Biological Survey, Minnesota Natural Heritage and Nongame Research Program, and Ecological Land Classification Program. Minnesota Department of Natural Resources, St. Paul

Molnar A, Csabai Z, Tothmeresz B (2009) Influence of flooding and vegetation patterns on aquatic beetle diversity in a constructed wetland complex. Wetlands 29:1214–1223

Mueller-Dombois D, Ellenberg H (2003) Aims and methods of vegetation ecology. The Blackburn Press, Caldwell

Oksanen J, Blanchet FG, Kindt R, Legendre P, O'Hara RB, Simpson GL, Solymos P, Stevens MHH, Wagner H (2011) Vegan: community ecology package R package version 1.17-12. http://vegan.r-forge.r-project.org. Accessed 2011

Parsons J (2001) Aquatic plant sampling protocols. 01-03-017. Washington State Department of Ecology, Olympia

Peach M, Zedler JB (2006) How tussocks structure sedge meadow vegetation. Wetlands 26:322–335

Podani J (2006) Braun-Blanquet's legacy and data analysis in vegetation science. J Veg Sci 17:113–117

Raulings EJ, Morris K, Roache MC, Boon PI (2010) The importance of water regimes operating at small spatial scales for the diversity and structure of wetland vegetation. Freshw Biol 55:701–715

Roberts DW (2010) Labdsv: ordination and multivariate analysis for ecology R package version 1.4-1. http://ecology.msu.montana.edu/labdsv/R. Accessed 2011

Roberts DW (2011) R labs for vegetation ecologists. http://ecology.msu.montana.edu/labdsv/R Accessed 2011

Rooney RC, Bayley SE (2011) Setting reclamation targets and evaluating progress: submersed aquatic vegetation in natural and post-oil sands mining wetlands in Alberta, Canada. Ecol Eng 37:569–579

Rooney TP, Rogers DA (2002) The modified floristic quality index. Nat Areas J 22:340–344

Sharma S (1996) Applied multivariate techniques. Wiley, New York

Squire L, van der Valk AG (1992) Water-depth tolerances of the dominant emergent macrophytes of the Delta Marsh, Manitoba. Can J Bot 70:1860–1867

Turner MG, Gardner RH, O'Neill RV (2001) Landscape ecology in theory and practice: pattern and process. Springer, New York

U.S. EPA (2002) Methods for evaluating wetland condition: using vegetation to assess environmental conditions in wetlands. EPA-822-R-02-020. Office of Water, U.S. Environmental Protection Agency, Washington, DC

Valente JJ, King SL, Wilson RR (2011) Distribution and habitat associations of breeding secretive marsh birds in Louisiana's Mississippi Alluvial Valley. Wetlands 31:1–10

Violle C, Navas M, Vile D, Kazakou E, Fortunel C, Hummel I, Garnier E (2007) Let the concept of trait be functional! Oikos 116:882–892

Warwick NWM, Brock MA (2003) Plant reproduction in temporary wetlands: the effects of seasonal timing, depth, and duration of flooding. Aquat Bot 77:153–167

Werner KJ, Zedler JB (2002) How sedge meadow soils, microtopography, and vegetation respond to sedimentation. Wetlands 22:451–466

WSDOT Environmental Services (2008) WSDOT wetland mitigation site monitoring methods. Washington State Department of Transportation. http://www.wsdot.wa.gov/NR/rdonlyres/C211AB59-D5A2-4AA2-8A76-3D9A77E01203/0/Mon_Methods.pdf. Accessed 2011

Zhang Y, Lu D, Yang B, Sun C, Sun M (2011) Coastal wetland vegetation classification with a Landsat Thematic Mapper image. Int J Remote Sens 32:545–561

5 Sampling and Analyzing Wetland Vegetation

Field Labs

Field Lab 1: The Effect of Quadrat Shape on Plant Density and Spatial Pattern Estimates

Objectives: Be able to...

- Discuss how method of observation (quadrat shape) can influence your results.
- Establish a sampling grid for randomly-placed plots in the field using a tape and compass.
- Use a spreadsheet program to summarize your data.
- Use a statistical program to analyze your data.

Questions

- Which quadrat shape will have more variation between quadrats, leading to a higher variance:mean ratio?
- Do different quadrat shapes yield significantly different plant population density measurements?

Hypotheses

Write down hypotheses pertaining to the questions above. Think about how the quadrat shape relates to plant shape and any environmental variation in the site.

Study system: This exercise is best conducted in a setting that has easily-recognizable plants with somewhat aggregated (clumped) distributions. Alternatively, sampling could include two different plant species, each with a different spatial pattern (clumped, randomly, or regularly-dispersed). In any case, even if it is a clonal plant, you will be counting individual stems (ramets). These stems should be easily-recognizable for all students in the class, so choose the species with care.

The Set-up: Students will be collecting data at randomly-placed points within a grid. Plan enough space for a 10 × 10 m plot for each pair of students in the course, with a buffer in between each plot (Fig. 5.11). Students will establish a grid in the field using meter tapes and a compass. Plant flags or stakes every 1 m to demarcate the grid. Students can either identify pairs of points from a random number table and work within their own plot for the lab, or they can be assigned sets of random numbers (0–10 or 0–20 if using ½ m spacing), and sample all the plots in the class using those same numbers. The lower left corner or center of the frame should be placed at the random grid coordinates. Boundary decisions (how to deal with plants on the edge of the quadrat) should be made and consistently applied within the class. Each of the quadrats in Fig. 5.11 has a total area of 1 m^2. Quadrats of ½ m^2 could also be used and an exploration of quadrat size effect could also be made.

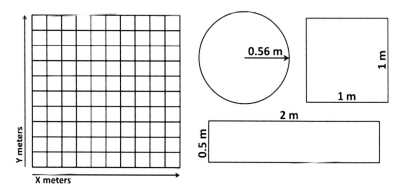

Fig. 5.11 Random sampling grid and quadrats of different shapes (each 1 m² in area)

At each set of random coordinates, assess the number of stems of each species within quadrats of all shapes, and record in the table below. As always, remember to avoid stepping within the plots.

Plot	Quadrat shape	Random Coordinates	Number of Stems Species 1	Number of Stems Species 2	Comments

Data Summary and Analysis

1. There were two response variables in this exercise: plant population density and spatial pattern. What was the independent variable?
2. Using the class data and spreadsheet program, prepare a table that contains the means, standard deviations, and standard errors for number of stems/m² for each of the shapes. Also include the sample size that you are using in each calculation. Discuss whether these results seem to match expectation. If you assessed two species, then create two different tables. Do not forget to include units!

Shape	Mean	Standard Deviation	Standard Error	Variance	n
Square					
Circle					
Rectangle					

3. Using a statistical package and coded data (e.g. 1 = square, 2 = circle, 3 = rectangle), conduct a one-way ANOVA to determine if the different shapes yielded different mean densities.

Shape code	Density

4. In order to detect differences in population spatial pattern, calculate the Variance: Mean Ratio (VMR) of the plant density in quadrats of different shapes. If variance is high compared to the mean, then the population is clumped in pattern. If variance is low compared to the mean, then the population is regularly-distributed. If the VMR $\approx$ 1, then the population is randomly-dispersed. Does spatial pattern change with quadrat shape?

Shape	Mean	Variance	VMR	Spatial pattern
Square				
Circle				
Rectangle				

5. If your statistical package allows, conduct a non-parametric Levene's test to determine whether the three different quadrat sizes gave significantly different variances.
6. Interpret your results. Did quadrat shape significantly affect plant population density or spatial pattern estimates? Why do you think this is?
7. What type of quadrat shape would you use in future studies and why?

Field Lab 2: Tree Populations & Succession

Objectives

- Use plotless methods for assessing population size.
- Use a compass to establish transects.
- Map plant populations using GPS and GIS.
- Interpret population data in order to predict future successional trends.

Background

In this lab, we will assess a forest stand containing interacting populations of trees, which form a community, in order to determine how it will change in the future. This skill is important to many natural resource agencies, which need to predict the future composition of the land.

- A population is a group of individuals of the same species in the same place at the same time. At any moment in time, a population has the attributes of population size and spatial distribution.
- A community contains interacting species in the same place at the same time. The species composition of communities can change over time – a process called succession.

One of the fundamental parameters of interest to ecologists is the density of organisms in a given area. However, in nature it is either impossible or impractical to count all organisms, and so we *estimate* density. For relatively small, immobile organisms, quadrat sampling is used to estimate density. For large, immobile organisms, remote-sensing, plot-based, or plotless techniques can be used. For mobile organisms, ecologists use mark-recapture techniques.

Factors controlled by the investigator that can affect the density estimate:

- the experience of the observer
- method of observation (instrument or chosen sampling technique)
- the number of samples taken

Factors beyond the control of the investigator:

- organism density
- organism spatial arrangement

Plot-based techniques frequently rely upon frames to isolate a sample area. These frames are called quadrats: arbitrarily-sized and -shaped sampling units. There are alternative techniques that are especially useful for large plants (trees). These are commonly called plotless sampling methods. During this laboratory, you will use the plotless Point Quarter Method (PQM) to estimate tree density and basal area

Regular Sampling Scheme

It takes time to establish a random grid and locate plots on it. Although totally random plot placement is the statistical "gold standard," it may be infeasible due to resource constraints. In addition, sometimes you want to ensure an even distribution of plots across a site, in which case totally random sampling may not be appropriate.

Regular sampling consists of using a set spacing between plots. Like random sampling, it typically precludes intentional and unintentional observer bias.

Although not technically statistically sound, ecologists often ignore statistical assumptions in favor of a more representative sample. Sampling schemes including combinations of regular and random sampling are typically favored by ecologists.

> In this exercise, we will implement regular sampling with a random start so as not to bias our samples and save time.

Global Positioning Systems (GPS)

GPS allows ecologists to locate their position on the earth. It relies upon a network of 30+ satellites that encircle the planet, sending signals down to GPS receiver antennas. The receivers differ in quality, some capable of sub-foot accuracy. You will use GPS units to map the center of each plot by establishing waypoints. Be careful to wait until you get roughly 10 m accuracy before plotting a waypoint. Label your waypoint with the plot number. Later, you may download your points into a GIS according to instructor-provided instructions.

Number of Plots

Each group will sample along transects in one of the forests using meter tape and a compass. Take point measurements (as described below) every 20 m until you have sampled at least five points.

Tree Identification

Your instructor will provide you with a tree identification guide and a list of common trees and their abbreviations.

The Point Quarter Method

At each point, divide the surroundings into four quarters along the principal compass directions (N, S, E, W). Use the data sheets provided to record the distance (d, expressed in meters) from the center point to the nearest tree that has a DBH (diameter at breast height) >4 cm in each of the four quarters (Fig. 5.12). Also record the DBH (in cm) and species of each of the four trees. These four measurements constitute data for one point sample. Do not count dead trees. Trees that have multiple trunks, but are separated at breast height are considered multiple trees.

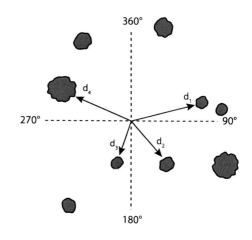

Fig. 5.12 Sampling trees using the Point Quarter Method. The area around a central point is divided into four quadrants, and the closest tree within each quadrant is sampled for distance from point and DBH

Site: _____ Group: _____ Date: _____
Compass bearing: _____ Plot distance apart: _____

Tree Layer

Plot	Tree 1	Tree 2	Tree 3	Tree 4
	Sp, Dist-m, DBH-cm	Sp, Dist-m, DBH-cm	Sp, Dist-m, DBH-cm	Sp, Dist-m, DBH-cm

Notes:

Lab Part 2: Analyzing Point-Quarter Data

Objectives

- Analyze point-quarter data using MSExcel.
- Interpret population data in order to predict future successional trends.

5 Sampling and Analyzing Wetland Vegetation

Analysis of Point-Quarter Data

The final product of your calculations should be a table that looks like this (Table 5.1):

Species	Frequency (no. of plots)	Relative frequency	Density (trees/ha)	Relative density	Mean basal area per tree (m^2)	Mean basal area/ha (m^2/ha)	Relative basal area
Total							

Use the questions and formulas below to fill in the table using the class data.

How common is each species?

1. We can answer this question by simply looking at the number of points that each species occurs in.

    Frequency = no. points that the species occurs at

How frequent is each species relative to the total?

2. If you counted 40 plots total, and 4 of these had white pines, white pines would represent 4/40, or 0.10 of the total points.

    Relative frequency = no. of plots containing species A / total no. of plots

What was the total density of all trees in the site?

3. The first step in analyzing point quarter data is to determine the mean point-to-plant distance for all of the trees on each transect. This value represents the mean distance between trees in the site. Compute this value and write it here:

    Mean point-plant distance for ALL trees = _____ m

4. Next we need to compute tree densities. The mean point-to-plant distance squared (d^2) gives the mean area per tree.

    Mean area per tree over all species = _____ m^2

    By knowing the mean area per tree, we can figure out how many of them are contained in a defined area (usually a hectare (ha), which contains 10,000 m^2).

The average tree density (in trees per ha) on each site = 10,000 m² per ha/(mean m² per tree)

Mean tree density over all species (total density) = _____ trees/ha

What was the mean density of each different tree species?

Mean density for Species A = (no. of trees of Species A)/(total no. of trees) × total density

5. If the total tree density on the site was 800 trees/ha, then the density of white pine trees would be 0.10 × 800/ha = 80/ha. Compute the density for each tree species.

Are some species bigger than others?

6. Foresters are often concerned with how big each tree is and how much wood is on each site as a measure of profitability. Ecologists care about this, because bigger trees can potentially exert more influence on an ecosystem. Tree size is often represented by basal area, which is the cross-sectional area of each tree (usually at breast height).

    Calculate the basal area for each tree by using $BA = \pi r^2$. Use the diameter at breast height (DBH) data to determine the radius (r) of each tree. Once you have computed the basal area of each tree, find the mean basal area per tree of each species on the site.

7. Next, compute the total basal area per hectare of each tree species. This is:

    Mean basal area per tree (in m²) × no. of trees per ha (density)

    For example, if the mean cross-sectional area of a white pine tree was 2,000 cm² you would first divide this by 10,000 to convert it to 0.2 m². Then multiply this by 80 trees/ha (the density of white pines that we calculated above) to find the total basal area. In this case it is 16 m²/ha. A high basal area can be achieved by either having a high basal area per tree or a high density of trees.

8. Finally, compute the relative basal area of each species by dividing that species' basal area per tree by the total basal area per tree for the site.

Questions

Use the data in your tables to answer the following questions in complete sentences:

1. What tree species is present in the highest density and lowest density?
2. What tree species is present in the highest basal area and the lowest basal area?

5 Sampling and Analyzing Wetland Vegetation   317

3. How do species rankings by density compare to rankings by basal area?
4. Draw a forest stand in which species A has high density and low basal area, while species B has low density and high basal area.
5. In order to determine the importance or overall magnitude of a species impact on an ecosystem, we sometimes calculate importance values (IVs). IVs combine all aspects of a species influence into a single number.

   IV = relative density + relative frequency + relative basal area

   Relative values are simply the value of the species divided by the total for all species (taken from Table 5.1). Create a second table of importance values for the different species in your site:

Species	Relative Density	Relative BA	Relative Freq.	IV

6. Use the data in Table 5.2 to answer the following questions:

   A. Which species had the highest importance value?
   B. Which species had the lowest IV?

7. Draw a forest stand in which Species A has a very high IV and Species B has a very low IV.
8. If two species have the same IV, does that mean that they influence the ecosystem in the same ways? Why or why not?

Size-Class Distributions

One way to investigate successional trends in a forested wetland or any forested system is to construct size-class distributions for the different important species. Size-class distributions can be graphically represented by plotting the number of trees in different size classes (e.g., 1, 2, 5, 10 cm classes, Fig. 5.13).

9. Create size-class distribution plots for the three species with the highest IVs.
10. What do these size-class distribution plots tell you about the future of the forest?

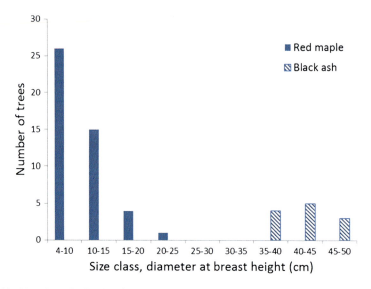

Fig. 5.13 Size-class distribution for red maple and black ash in a forested wetland site

Homework

Exercise 1: Devise a Sampling Strategy

Your goal is to construct a sampling scheme based upon a pilot study (in the case of the provided data set, this is reed canarygrass (*Phalaris arundinacea*)). Using your own data or the data provided below, devise a sampling strategy based upon the (1) species accumulation curve and (2) performance curve of abundance of the species of interest. If you plan to use your own data, download the free program EstimateS (Colwell 2009), to calculate your own species accumulation curve.

Provided data set (calculate a performance curve):

Sample	*P. arundinacea* percent-cover	Cumulative mean percent-cover	95 % Confidence Interval
1	37.5		
2	2.5		
3	0		
4	0		
5	37.5		
6	87.5		
7	15		
8	15		
9	62.5		

(continued)

(continued)

10	2.5		
11	2.5		
12	37.5		
13	2.5		
14	2.5		
15	0		
16	0		
17	87.5		
18	15		

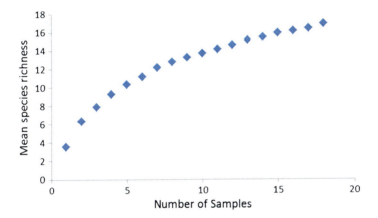

Fig. 5.14 Species accumulation curve

1. Create a performance curve + 95 % confidence interval using the *P. arundinacea* data above and calculating a cumulative mean.
2. Did you collect enough data to accurately estimate the abundance of *P. arundinacea*? Why or why not?
3. If you were trying to maximize efficiency while estimating an accurate abundance of *P. arundinacea*, how many samples would you collect at this site and sites similar to this?
4. According to the species accumulation curve (Fig. 5.14), was the sampling adequate to characterize species richness at this site?
5. How many samples would you need to collect to most accurately and efficiently estimate species richness at this site and sites like it?

Exercise 2. Species Diversity Assessment

Compare the two plant communities below using diversity statistics. Determine which statistics are most helpful, and why.

Community data

Species	Community 1 abundance (percent-cover)	Community 1 abundance (percent-cover)
A	30	12
B	30	12
C	15	12
D	15	12
E	2	12
F	2	12
G	2	12
H	2	12
I	2	4
Total	100	100

1. Simply by inspecting the data, compare the two communities in terms of their species richness and your opinion of their evenness.
2. Calculate Simpson's Index

Species	Comm1 p_i	Comm1 p_i^2	Comm2 p_i	Comm2 p_i^2
A				
B				
C				
D				
E				
F				
G				
H				
I				
Total		D =		D =

Simpson's diversity index = 1 − D: Comm 1: _____ Comm 2: _____
Effective number of species = 1/D: Comm 1: _____ Comm 2: _____

3. Calculate Shannon-Weiner Index

Species	Comm1 p_i	Comm1 ln p_i	Comm1 $p_i \times$ ln p_i	Comm2 p_i	Com21 ln p_i	Comm2 $p_i \times$ ln p_i
A						
B						
C						
D						

(continued)

(continued)

E						
F						
G						
H						
I						
Total			$H' = -$			$H' = -$

Shannon-Wiener index = H': Comm 1: _____ Comm 2: _____
Effective number of species = $e^{H'}$: Comm 1: _____ Comm 2: _____

4. Compare the interpretation of the Simpson's and Shannon-Wiener diversity indices. (A) Which seems to be more effective at distinguishing between the two communities and why? (B) If you were trying to communicate your results to a lay audience, which statistic is easier to interpret and why?
5. Inspect the effective number of species derived from the Simpson's and Shannon-Wiener indices for the two communities. (A) Do the results from the two communities make sense to you? Why or why not? (B) Is there a difference between the Simpson's and Shannon Wiener effective number of species? Why do you think this is?
6. Calculate Pielou's evenness from the Shannon-Wiener index. (Recall that $J = H'/\ln(S)$ where S is the species richness.

   Pielou's J: Comm 1: _____ Comm 2: _____

7. Do the evenness statistics make sense given the initial data? Why or why not?

Exercise 3. Calculating an FQAI

Using either data that you collected yourself, or the data provided below, calculate the floristic quality index and mean C of C for the site. If using the provided data set, refer to the University of Wisconsin – Stevens Point herbarium http://wisplants.uwsp.edu/namesearch.html for the coefficient of conservatism (the wetland site is located in Wisconsin). After entering the species name, select the "more information" link for the species C of C.

Provided data set:

Species	Mean abundance (percent-cover)
Agrostis gigantea	15
Carex atherodes	42
Carex lacustris	13
Carex utriculata	8
Eupatorium perfoliatum	21
Phalaris arundinacea	52
Typha latifolia	10

Calculation table (use your own or provided data set). A typical FQAI does not include abundance data, but only species presence. However, you may have abundance data that you may want to use to weight your findings.

Species	Coefficient of conservatism	Mean abundance	Relative abundance	Weighted C of C (CC'_i)
Sum	A	B	1.00	D

$A = \sum_i^S CC_i$
$B = \sum_i^S x_i$ where x_i is the mean abundance of species i
Relative abundance of species i $= x'_i = x_i/B$
Weighted C of C for species i $= CC'_i = x'_i \times CC_i$
D (Weighted C of C of site) $= \sum_i^S CC'_i$

1. FQAI = _____
2. Mean C of C = _____
3. Weighted C of C of site = _____
4. What does the FQAI tell you about the quality of the wetland site?
5. Do the mean C of C or the weighted C of C provide similar or different interpretations to the FQAI? How are they similar or different?

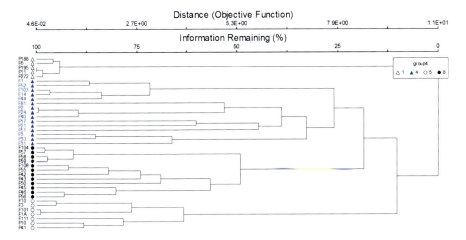

Fig. 5.15 Cluster dendrogram of 39 wetlands based upon *Sphagnum* community dissimilarity. Four groups have been constructed based upon interpretability

Exercise 4. Interpreting Multivariate Data

The following figures are output from a multivariate data analysis of 25 *Sphagnum* species found in 39 different wetlands. Wetlands were clustered into groups based upon their species dissimilarity using hierarchical cluster analysis (Fig. 5.15) and were ordinated within *Sphagnum* species abundance space using non-metric multidimensional scaling (Fig. 5.16).

1. Draw a line on the cluster dendrogram where the group cut-off occurs. What percent of information is remaining at this point?
2. If you were to divide the black circle group into four sub-groups, which wetlands would be included in each group?
3. Which wetland group is the least tightly clustered in this ordination diagram?
4. The red/gray lines are correlations of axes with environmental data collected in each wetland. Which wetland group contains the oldest wetlands? Which wetland group contains wetlands with the highest average groundwater specific conductivity?
5. Which group of wetlands is closest to the centroid for *S. inundatum* on the ordination diagram?
6. What species (three letter abbreviation) is most negatively correlated with Axis 3? Which species is most positively correlated with Axis 3?
7. Which wetland sites (numbers) most likely have the most *S. flavicomans*?

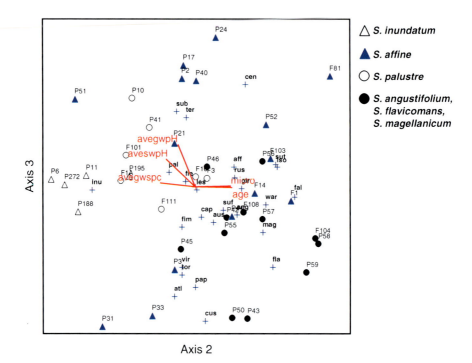

Fig. 5.16 NMS ordination of wetlands (labeled P or F) within *Sphagnum* species space. Wetlands were classified into four groups, named after group indicator species. Centroids of species abundance are labeled by *crosses*, with the three-letter species abbreviation (e.g., cap= *S. capillifolium*). *Lines* are vectors of correlation with environmental variables; *longer lines* indicate stronger correlation. Micro = microtopographic score, age = time since most recent beaver inhabitation, avegwpH and aveswpH are groundwater and surface water pH, respectively, and avegwspc is mean groundwater specific conductivity

Exercise 5. Indicator Species

The table below contains data about the distribution of two species in degraded and non-degraded wetlands. Given these data, which species would be a better indicator of degradation and why?

	Mean abundance in group/ Mean abundance overall		% of sites within group in which species occurs	
	Degraded	Non-degraded	Degraded	Non-degraded
Typha angustifolia	0.92	0.08	100	10
Alnus incana	0.45	0.55	60	80

Chapter 6
Physical and Chemical Monitoring of Wetland Water

Joseph R. Bidwell

Abstract The physical and chemical attributes that comprise the "water quality" of a wetland have a significant influence on the system's biotic structure and function. Assessments of wetland water quality can be used for reference-based monitoring in the development and implementation of wetland water quality standards and to provide ancillary information in support of biotic surveys. While the methods used to evaluate water quality in wetlands are generally the same as those used for other surface waters, wetlands may differ in their dominant source of water, often have greater heterogeneity in habitat types, and can exhibit significant variability in water permanence. These characteristics can lead to spatial and temporal variability within and among wetlands that can make it difficult to use water quality data to detect human impacts. This chapter reviews some of the major sources of variability in wetland water quality and discusses approaches and general sampling considerations for characterizing basic water quality in wetland monitoring studies.

6.1 Introduction

The "water quality" of a wetland encompasses a range of physical and chemical attributes that largely determine its biotic composition and function. This chapter focuses on some of these physical and chemical variables (Table 6.1), with the goal of highlighting issues that should be considered when collecting these data. Even though a significant amount of "chemistry" goes on in wetland soils as they become anoxic after inundation (see Boon 2006; Mitsch and Gosselink 2007), the emphasis here will be on the water column as chemical and physical elements of the water are more commonly measured in routine monitoring of wetlands. This discussion will also primarily focus on inland wetlands, although coastal/tidal systems may exhibit significant variability in water quality as well.

J.R. Bidwell (✉)
Discipline of Environmental Science and Management, School of Environmental and Life Sciences, University of Newcastle, Callaghan, NSW 2308, Australia
e-mail: joseph.bidwell@newcastle.edu.au

Table 6.1 Overview of common water quality parameters measured in wetland monitoring studies

Parameter	General notes/reasons for measuring the parameter	Potential sources of variability to consider when monitoring wetlands
Temperature	Temperature has a major influence on chemical and biological processes including reaction rates, saturation constants of dissolved gases, water density, and the metabolic rates of animals	*Hydrology:* Water input from overland flow, precipitation or groundwater can increase or decrease temperature *Internal and other processes:* Diurnal and seasonal variation in water temperature occurs due to changing incident thermal radiation Differences in thermal regimes can exist between wetlands due to position on the landscape (e.g., shading effects, exposure to prevailing winds) Spatial variability within the wetland can occur due to differences in water depth, presence of macrophytes, and shading from shoreline vegetation Vertical stratification is possible in open water zones
Dissolved oxygen	As the terminal electron acceptor in aerobic respiration, the concentration of dissolved oxygen in water can influence the breakdown of organic matter in wetlands and the presence and distribution of many aquatic organisms that inhabit these systems Oxygen has a major affinity for electrons and determines the status of oxidation/reduction reactions that influence the chemical form of elements in biogeochemical and other chemical processes in wetland soils and water When coupled with other chemical indicators such as levels of nutrients, reduced oxygen levels in wetlands can be indicative of eutrophication	*Hydrology:* Water input from overland flow, precipitation or groundwater can influence dissolved oxygen levels either directly or indirectly by influencing temperature or introducing nutrients (overland flow and groundwater) into the system that stimulate biological productivity and ultimately biological oxygen demand *Internal and other processes:* Diurnal and seasonal variation in water temperature due to changing thermal regimes influences oxygen solubility Diurnal fluctuations in dissolved oxygen levels can result from the combined effects of photosynthesis and respiration Spatial variability in oxygen levels may result from differences in biological activity between stands of macrophytes and open water Vertical stratification of dissolved oxygen levels may also occur in open water zones

pH	pH is a measure of the hydrogen ion (H$^+$) concentration of water with values less than neutral (pH = 7) indicating acidic conditions and those above 7 indicating alkaline conditions. The pH of natural waters is significantly influenced by the carbonic acid equilibrium that develops when carbon dioxide dissolves in water: $$CO_2 + H_2O \rightleftharpoons H_2CO_3 \rightleftharpoons HCO_3^- + H^+ \rightleftharpoons CO_3^{-2} + 2H^+$$ The implications of this are that uptake of CO_2 by photosynthesizing plants can cause pH to increase while release of CO_2 due to respiration causes pH to decrease pH can influence chemical solubility and biogeochemical cycles. For example, acidic pH can increase the solubility and toxicity of sediment-bound metals, while increases in pH above 8 can lead to the loss of ammonia nitrogen (Catallo et al. 1995; Boon 2006) pH has been used as a water quality modifier in wetland classification systems (see Cowardin et al. 1979; Wetzel and Likens 2000)	*Hydrology:* Water input from overland flow, precipitation or groundwater can influence pH- systems with precipitation as the primary water source often have a pH <7 because precipitation tends to be slightly acidic Depending on basin geology and soil type, groundwater input could be acidic or alkaline, with overland flow similarly influenced *Internal and other processes:* Diurnal and spatial variability in pH can result from those same factors that influence dissolved oxygen Seasonal variation is also possible due to temperature effects on respiration
Alkalinity	Alkalinity of water is a measure of the capacity of constituent dissolved chemicals to accept and neutralize protons and is mostly attributed to the presence of carbonates (HCO_3^- and CO_3^{2-}) Alkalinity can indicate the geochemistry of the wetland basin and catchment since it commonly results from carbon dioxide and water attacking limestone or dolomite formations Alkalinity levels in wetlands can influence reactions associated with nutrient cycling (e.g., oxidation of ammonium (Parkes et al. 2007) and so monitoring this parameter can be important in assessing treatment wetland efficiency and understanding wetland nutrient profiles	*Hydrology:* Baseline alkalinity levels in wetlands are influenced by source water and local geochemistry *Internal and other processes:* Diurnal and seasonal variation in wetland water alkalinity has been associated with iron and sulphate reduction, denitrification, and assimilation of ammonium and nitrate (Eser and Rosen 1999; Sisodia and Moundiotiya 2006)

(continued)

Table 6.1 (continued)

Parameter	General notes/reasons for measuring the parameter	Potential sources of variability to consider when monitoring wetlands
Total dissolved solids	Total dissolved solids (TDS) is a measure of those solids in water that pass through a 2.0 μm filter and include ions such as sodium, chloride, and calcium While effects of elevated TDS on water quality are mostly related to impacts on domestic or industrial use, there could be physiological implications for freshwater organisms if levels become sufficiently elevated TDS and salinity are often used interchangeably and conversion factors are available to estimate one value from the other (see salinity)	*Hydrology:* Water source is the major source of between-wetland variability in TDS. For example, groundwater sources often have higher TDS levels due to contact with geologic material while wetlands that fill mostly via precipitation usually have lower TDS levels than those receiving groundwater or other surface input Temporal/seasonal variation in TDS could result from dilution effects due to precipitation events or evapoconcentration associated with water drawdown at the end of the hydroperiod (Charkhabi and Sakizadeh 2006) Spatial differences in TDS could occur within a wetland due to differences in water quality of tributaries flowing into the system.
Salinity	The sum of all dissolved ions in water determines the 'salinity', with the salinity of most inland waters dominated by Ca^{+2}, Mg^{+2}, Na^+, K^+, CO_3^{-2}, HCO_3^-, SO_4^{-2}, and Cl^- Salinity has been identified as a "keystone" chemical parameter because it differentiates freshwater from brackish and marine systems and so significantly influences plant and animal assemblages and wetland structure and function Salinity has been used as a water quality modifier in wetland classification systems (see Cowardin et al. 1979), although moderate variation in salinity common in most inland wetlands may not represent a significant influence on wetland organisms (see Mendelssohn and Batzer 2006)	*Hydrology:* See influences for TDS

Specific conductance (Conductivity)	Conductivity is a measure of the electrical resistance in an aqueous solution and depends on the total concentration of dissolved electrolytes	*Hydrology:*
		See influences for TDS
	Conductivity can serve as a general indicator of productivity of freshwater systems (with highly productive systems usually having higher conductivity than less productive systems) and water source	*Internal and other processes:*
		Seasonal variation in conductivity has been associated with increased biological activity and chemical reactions that result from changing oxidation/reduction characteristics (Eser and Rosen 1999; Stratford et al. 2004)
	Conductivity has been used as a surrogate for total dissolved solids (Trebitz et al. 2007) and factors for converting between the two parameters are available. (e.g., Conductivity × 0.67 ~ total dissolved solids (mg/L) (Dickerson and Vinyard 1999). More complex equations are available to estimate salinity from conductivity (APHA, AWWA and WEF 2005)	Increased biological activity may also lead to spatial differences within the wetland (although this is not extensively reported in the literature)
Total Hardness	Total hardness is a measure of divalent cations in water, primarily calcium and magnesium.	*Hydrology:*
		Water source and associated basin and catchment geochemistry represent the dominant source of between-wetland variation in water hardness
	Total hardness is not often evaluated in wetland monitoring studies although may be a useful modifier for some wetland classification schemes (e.g., Warner and Rubec 1997) or to establish wetland water quality criteria for metal contaminants (Gordon et al. 1997; Nimmo et al. 2006)	Also see influences for TDS
Total suspended solids	Total suspended solids (TSS) is a measure of those solids in water that are retained on a 2.0 µm filter and include both organic and inorganic and living and dead material	*Hydrology:*
		Water source can influence TSS levels due to the import of particulate matter associated with overland flow and tributaries entering the wetland
	Elevated levels of suspended material in the water column can reduce light availability and lead to elevated water temperatures due to increased absorbance of thermal radiation. Light availability and temperature affect bacterial, algal, and zooplankton physiology, as well as the feeding and movements of macroinvertebrates and fish.	Temporal/seasonal variation in TSS could result from dilution effects due to precipitation events or evapoconcentration associated with water drawdown at the end of the hydroperiod (Boeckman and Bidwell 2007)
		Internal and other processes:
	Wetlands may reduce TSS of through-flowing water by facilitating settling of particulate material (Evans et al. 1996)	Blooms of phyto and zooplankton can lead to increased levels of suspended solids and result in spatial differences in TSS within a wetland
	When coupled with other chemical indicators such as levels of nutrients, TSS levels in wetlands can be indicative of eutrophication	

(continued)

Table 6.1 (continued)

Parameter	General notes/reasons for measuring the parameter	Potential sources of variability to consider when monitoring wetlands
		Internal generation of solids occurs through fragmentation of detritus and litter, algal cells, and bioturbation by fish and invertebrates can further contribute to the concentration of wetland TSS (USEPA 2008a)
		Wind-driven mixing of wetland water can cause suspension of particulate material and result in elevated TSS. As such, landscape topography and wetland position on the landscape may also lead to differences between wetlands
Turbidity	Turbidity is a measure of water clarity and results from particulate matter (living and non-living) and dissolved color. See additional discussion for total suspended solids (TSS)	*Hydrology:*
		See influences discussed for TSS, although color and dissolved organic carbon in source water can influence turbidity readings as well (Trebitz et al. 2007)
	Trebitz et al. (2007) found that turbidity could serve as a possible surrogate for total suspended solids, although they state that determining water transparency through the use of a "Secchi transparency tube" would also serve as a surrogate for TSS	*Internal and other processes:*
		See influences discussed for TSS. Biological activity and seasonal variation of dissolved organic carbon in the wetland water column (e.g. Waiser and Robarts 2004) could also influence turbidity readings
Nutrients	Nutrients include the macronutrients (e.g., phosphorus, nitrogen, sulphur, potassium, magnesium, and calcium) and micronutrients or trace elements (e.g., iron, copper, silicon). Nitrogen and phosphorous are often measured in routine water quality monitoring since they may be limiting and so could have a significant impact on productivity of the system. Nitrogen and phosphorus can both occur in the water column as organic and inorganic forms, with each having dissolved and particulate fractions. Different fractions of phosphorous are also designated based on reaction with molybdate (see Wetzel (2001) and APHA, AWWA and WEF (2005) for a more extensive description of these categories). Generally, the dissolved inorganic forms (soluble reactive in the case of phosphorous) are most readily available to plants, but particulate forms can indicate the overall available pool in the system.	*Hydrology:*
		Recent precipitation events may result in increased nutrient loading from overland flow while flooding from rivers may decrease or increase nutrient loads them depending on the nutrient status of the river (Weilhoefer et al. 2008)
		In seasonal wetlands, nutrient levels in the water column may increase shortly after inundation due to release from sediments. Nutrients may also increase during wetland drawdown due to evapoconcentration
		Internal processes:
		Seasonal variation in nutrient levels may occur due to temperature effects on biota and nutrient fluxes influenced by changing oxygen levels (Trebitz et al. 2007; USEPA 2008a)

Important chemical forms:

Nitrogen:

Dissolved inorganic nitrogen: Ammonium: NH_4^+-N, Nitrate: NO_3^--N, Nitrite: NO_2^--N

Total nitrogen (TN): All digestible forms of nitrogen (dissolved and particulate organic and inorganic nitrogen)

Total Kjehdahl Nitrogen (TKN): The sum of organic nitrogen and ammonium nitrogen. Commonly measured as part of permitting requirements for wastewater treatment plant discharges

Phosphorous:

Reactive phosphorous: Phosphates that react with molybdate without preliminary digestion–mostly orthophosphate, PO_4^{3-}. Total reactive phosphorous is derived from water samples that have not been filtered, while dissolved or soluble reactive phosphorous (SRP) is derived from samples passed through a 0.45 μm filter

Dissolved inorganic phosphorous: orthophosphate– basically the same as SRP

Total phosphorous (TP): All digestible forms of phosphorous (dissolved and particulate organic and reactive phosphorous)

The forms of nitrogen and phosphorous most commonly measured in water quality monitoring are total nitrogen and phosphorous, ammonium, nitrate, and orthophosphate/soluble reactive phosphate. However, due to the potential for rapid uptake of dissolved forms of inorganic nutrients, measurement of total fractions has been proposed as a better option for monitoring and characterizing nutrient dynamics (Norris et al. 2007; Trebitz et al. 2007)

Wetland sediments play a critical role in the cycling of nitrogen and phosphorous and the reader is referred to Boon (2006) and Mitsch and Gosselink (2007) for excellent overviews of wetland biogeochemistry

Spatial variation within a wetland may be observed between macrophyte beds and open water zones due to differences in biological activity (Rose and Crumpton 1996)

Vertical profiles may be observed with higher levels of ammonia near-bottom sediments and release of orthophosphate from anoxic sediments (Ryder and Horwitz 1995; Glińska-Lewczuk 2009)

Unless indicated otherwise, information for general notes was derived from Wetzel and Likens (2000), Wetzel (2001) and APHA, AWWA and WEF (2005)

6.2 Uses of Water Quality Data in Wetland Monitoring

One of the more common approaches to assess wetland water quality is to compare measurements of current conditions with that of an ecologically similar but undisturbed or less-disturbed reference site (Norris et al. 2007). Brinson (1988) used this reference-based concept to help define "water quality", stating that good water quality represents the normal unaltered chemical condition with departures representing a deterioration in quality. The biogeochemical assessment module for wetlands developed by the United States Environmental Protection Agency (USEPA 2008a) is an example of a reference-based application of water quality data for wetland monitoring. A number of states in Australia have also incorporated various physical and chemical water quality parameters as a component of the reference-based *Framework for Assessing River and Wetland Health* (Norris et al. 2007; Alluvium Consulting 2011).

A second approach that uses chemical data to evaluate water quality is the comparison of specific chemical concentrations derived from site measurements with those deemed to support designated uses of the water or habitat (Norris et al. 2007; Chapelle et al. 2009). In the United States, this forms the basis for chemical-based wetland water quality standards as mandated by Section 303 of the Clean Water Act (USEPA 1990, 2008b; ELI 2008; Kusler 2011a, b). As stated in USEPA (1990), the original objective for US states was to have either narrative or chemical-based wetland water quality standards in place by 1993. However, as of 2011, only 14 states had adopted standards that were specific to wetlands (Kusler 2011a; also see ELI (2008) for a broader discussion of state-based wetland protection regulations in the US).

Physical and chemical data collected from wetlands may also provide important ancillary information to help understand the distribution of organisms in both basic and applied wetland studies. Water quality has a major influence on what organisms occur in a wetland (e.g., Dunson et al. 1997; Batzer et al. 2004; Euliss et al. 2004; Longcore et al. 2006; Ginocchio et al. 2008; Chen et al. 2011; Bojkova et al. 2011), with parameters such as salinity considered to be "keystone" variables that control plant and animal assemblages and drive wetland structure (Mendelssohn and Batzer 2006). Temperature, pH, dissolved oxygen, and alkalinity also dictate the extent and rate of important wetland functional processes such as nutrient transformations (Kadlec 1999). Physicochemical data have also been used to modify classification systems for wetlands (e.g., Cowardin et al. 1979; Warner and Rubec 1997).

6.3 What Makes Wetlands Different from Other Surface Waters When Monitoring Water Quality?

Mitsch and Gosselink (2007) state that while no particular biogeochemical processes are unique to wetlands, their flooding frequency (permanent or intermittent) can make certain chemical processes more dominant in wetlands than in other types of aquatic systems. In particular, the anaerobic conditions that prevail in wetland sediments can lead to the reduction of chemicals that influences their ultimate fate in the system.

Fluctuations in the presence and depth of water can also lead to greater variability in the physical, chemical and biological attributes of wetlands as compared to other surface waters. For example, Reeder (2011) identified the generally shallow nature of the wetland water column as a key driver of temporal and spatial variability in parameters such as dissolved oxygen. This can pose a challenge for wetland monitoring programs aimed at detecting anthropogenic disturbance since natural variation in biotic and abiotic assessment metrics can make it difficult to detect human effects. Within and between-site variability has also been identified as a key challenge in developing wetland water quality standards (Trebitz et al. 2007; Kusler 2011b).

6.4 Key Factors That Determine Wetland Water Quality

An understanding of the factors that influence the physical and chemical nature of wetland water can help identify sources of variation in these characteristics and assist in the design and implementation of wetland monitoring studies and interpretation of data derived from those studies. For the purposes of discussion, these factors are considered separately, although it is important to realize that these often work together and influence each other. As such, landscape effects on wetland water quality can occur due to effects on the chemistry of water entering the wetland as well as localized effects on biological processes within the system.

6.4.1 Hydrologic Influences

Surface water quality of a wetland integrates the combined influences of geologic setting, hydrology, presence and activity of biota, and human activity within or near the system (Carter 1996; Boon 2006; Azzolina et al. 2007). Of these, hydrology stands as the dominant factor that establishes the "baseline" levels of dissolved and particulate materials in wetland water. As presented by Bedford (1996) and further discussed by Boon (2006), climate and hydrogeologic setting establish key "hydrologic variables" that drive biogeochemical properties of wetlands. These variables include water source, the mineral and nutrient status of that water, and the spatial and temporal dynamics of water in the system.

Wetland water sources include precipitation, groundwater, and overland flow and, as would be expected, the mineral and nutrient status of each is often quite different. For example, vernal pools and other depressional wetlands that fill largely from precipitation commonly have low levels of dissolved solids and are more acidic (pH <7) (Whigham and Jordan 2003; Colburn 2004), while wetlands receiving mostly groundwater input may have variable pHs and dissolved ion levels based on underlying basin and catchment geology (LaBaugh 1989; Bedford 1996; Carter 1996; Winter et al. 2001; Whigham and Jordan 2003; Cabezas et al. 2009; Nelson et al. 2011). Groundwater-fed wetlands may also have lower temperatures than those fed by precipitation and run-off (Korfel et al. 2010). The influence of local geology on groundwater and associated wetland water quality is nicely illustrated by the pH and nutrient content of fen wetlands which vary from quite alkaline (pH 8.4) and nutrient rich (rich or minerotrophic fens) to acidic (pH 3.5) and nutrient poor (acidic or poor fens) depending on the nature of the glacial deposits the source water comes in contact with (Bedford and Godwin 2003; Kolka and Thompson 2006; Nelson et al. 2011). Similarly, wetlands that receive significant input from overland flow often have water quality characteristics that reflect the soil properties of their catchments. This can be observed on a seasonal basis in some coastal wetlands of Lake Huron (Laurentian Great Lakes) that have coloured, acidic water with low conductivity and elevated phosphorous in spring due to input from upland watersheds (deCatanzaro and Chow-Fraser 2011).

In many cases, the chemical profile and associated variability in wetland water is driven by multiple water sources and/or by seasonal changes in water source. In the Lake Huron coastal wetlands mentioned above, upland inflow decreases in summer while input from seiche-derived lake water increases. This leads to increased alkalinity, higher conductivity, and reduced phosphorous levels as compared to springtime when inflow from upland catchments dominates (deCatanzaro and Chow-Fraser 2011). Euliss et al. (2004) state that while precipitation is generally the most significant water source for prairie pothole wetlands, input from groundwater can increase the concentrations of solutes and other dissolved materials. The relative solute concentration in these wetlands ultimately represents the combined effects of groundwater and precipitation. The scenario is further complicated by the length of groundwater flow paths between wetlands that may be influenced by precipitation patterns (Euliss et al. 2004). A comparable interplay between groundwater and surface water sources has been observed in some Australian wetlands (Boon 2006; Jolly et al. 2008). In some cases, groundwater input to depressional wetlands may also serve to buffer increases in dissolved solutes caused by evaporative water loss (Rains et al. 2006; Korfel et al. 2010). Variation in water source and connectivity through groundwater can have important implications for jurisdictional regulation of depressional wetlands that are considered isolated because they lack a clear connection to surface water (see Whigham and Jordan (2003) for further discussion).

Riparian wetlands that are subject to pulse flooding by rivers can have very distinct water quality profiles between the times they are flooded by the river and when they are isolated from it (Gell et al. 2002; Weilhoefer et al. 2008). Flood water can dilute levels of dissolved constituents in floodplain wetland water as observed by Weilhoefer et al. (2008) who report lower levels of conductivity, total phosphorous, and total nitrogen during and just after a flood event. They also concluded that the magnitude and duration of the flood was an important determinant of how much the wetland water quality changed and that flooding could increase nutrient levels in a wetland depending on the nutrient status of the river. Cabezas et al. (2009) also studied the water quality of floodplain wetlands and found that the seasonal chemical profiles of the systems were largely influenced by the relative inputs of river and groundwater. Wetlands receiving major input from the river during flooding had lower conductivity but higher turbidity and nitrate levels, while those receiving mostly groundwater had higher conductivity and lower turbidity.

Wetland hydroperiod (seasonal wetting and drying) has an obvious influence on the abiotic and biotic features of wetlands (Brooks 2000; Jackson 2006) and differences in water between wetlands at different stages of the hydroperiod could be a significant source of between-site variability. Oxidative metabolism of organic matter in the substrate of a dry wetland can result in a pulse of nutrients to the water column when the wetland refloods (Euliss et al. 2004), although Boon (2006) states that nutrient release in newly-flooded wetlands may be derived from other sources as well. Regardless, a succession of chemical reactions occurs in saturated wetland soils as oxygen becomes depleted and new substrates are used as electron acceptors by respiring microorganisms (Boon 2006; Mitsch and Gosselink 2007). Concurrent changes in variables such as dissolved oxygen, pH, dissolved organic carbon, conductivity and water clarity would be expected to occur in the wetland water column as photosynthetic and respiratory processes become established and suspended material begins to settle out, although few studies have evaluated this with sufficient sampling frequency to effectively document the changes that do occur.

Most discussion of wetland hydroperiod and water quality focuses on how water loss influences water quality parameters. Euliss et al. (2004) discuss the "drought" versus "deluge" phases in prairie pothole wetlands in reference to changing solute concentrations that can influence wetland biota. Similarly, evaporative water loss and associated concentrating effects were used to explain stable isotope signatures and increased summertime cation levels in New Zealand peatlands (Chague-Goff et al. 2010), seasonal variation in salinity of arid and semi-arid zone wetlands in Australia (Jolly et al. 2008), and spatial differences in parameters such as conductivity, dissolved organic carbon, and levels of specific dissolved ions between different wetland zones of the Okavango delta (Mackay et al. 2011). Increases in other water quality parameters including total nitrogen, total phosphorous, pH, alkalinity, and hardness have also been associated with evapoconcentration as seasonal wetlands proceed through the hydroperiod (Gell et al. 2002; Boeckman and Bidwell 2007). Water loss and associated evapoconcentration effects on water quality variables can also vary considerably between wetlands based on factors such as basin size, localized landscape position of the wetland, and the presence of plants which can enhance water loss by evapotranspiration (Euliss et al. 2004; Jackson 2006).

Interestingly, increases in the levels of water quality variables due to evapoconcentration may have only limited influence on biotic communities in inland wetlands unless these exhibit extremely elevated conditions as has been described for prairie pothole wetlands and some other systems (Euliss et al. 1999; Mendelssohn and Batzer 2006). Batzer et al. (2004) report variation of up to two orders of magnitude in a suite of water quality parameters they measured in a series of forested wetland ponds that differed in hydroperiod and landform and found these factors had a relatively minor influence on macroinvertebrate assemblages. Babbitt et al. (2003) used plant assemblages and site visits to group a series of forested depressional wetlands according to hydroperiod duration and found permanently inundated wetlands were slightly warmer and had slightly higher pH, higher dissolved oxygen, and lower conductivity than wetlands with the shortest hydroperiod. Based on ordination analysis, dissolved oxygen was the only water quality parameter found to have a strong influence on amphibian assemblages in the wetlands. As such, in some cases, differences in plant and animal assemblages between wetlands with different hydroperiod durations may be driven more by the actual presence of water and available habitat rather than changing water quality (see Jackson (2006) for further discussion).

Finally, water movement through a wetland can establish important within-site spatial differences in water quality, with contrasting levels of parameters such as dissolved oxygen, suspended solids, nutrients, and contaminants observed between the inflow and outflow zones (Ibekwe et al. 2007; Diaz et al. 2012). Trebitz et al. (2005) evaluated the influence of hydrology and geomorphology on water quality (temperature and dissolved oxygen) and other habitat parameters in Lake Superior coastal wetlands and in some cases, found within-wetland spatial differences due to seiche action and tributary inputs that were as large or larger than differences between wetlands.

6.4.2 Landscape Influences

Attributes of the landscape in which a wetland occurs can influence water quality through effects on hydrology, characteristics of water entering the system, and localized effects on wetland microclimate. A number of studies have linked differences in parameters such as levels of suspended solids and nutrients between wetlands with landscape-level differences in agricultural intensity, human population density, and point source pollution (Trebitz et al. 2007; Morrice et al. 2008; Cabezas et al. 2009). In their study, Trebitz et al. (2007) combined individual water quality parameters via principal components analysis into a wetland water quality metric that was more responsive to agricultural intensity than any single parameter alone. Physical characteristics such as size and drainage slope of watersheds that feed wetlands have also been found to significantly influence chemical parameters such as pH and levels of suspended solids and nutrients (deCatanzaro and Chow-Fraser 2011). Differences in parameters such as temperature and pH between forested depressional wetlands have been found to result from subtle differences in landscape topography, forest canopy cover, and tree age and size (Batzer et al. 2000; Skelly and Freidenburg 2000; Hossack and Corn 2008). In an investigation of what constitutes appropriate buffer zones adjacent to wetlands, Houlahan and Findlay (2004) found that water column nutrient levels in the systems studied were influenced by forest cover quality at over 2,000 m from the wetland edge.

6.4.3 Internal Influences

While the hydrologic variables discussed above are significant determinants of wetland water quality, it is important to also consider the effects of internal biological processes on chemical parameters and the physical influences that factors such as temperature and light availability have on these processes. This is clearly illustrated by the differences in wetland water quality that can exist between different habitat types in a single wetland. For example, due to the link between CO_2 and the carbonic acid buffering system in water (Wetzel and Likens 2000, also see Table 6.1), uptake of CO_2 by photosynthesizing aquatic plants increases pH while release of CO_2 by respiring organisms decreases it. The competing effects of photosynthesis and respiration similarly influence dissolved oxygen levels which can have effects on other important parameters such as the oxidation and reduction (redox) potential in water and sediments.

Stands of aquatic macrophytes in particular can have a major influence on commonly measured water quality variables. Dense mats of floating plants and high levels of algal biomass have both been associated with localized increases in water temperature (in some cases by as much as 11 °C, see Reeder (2011) for discussion), while shading from emergent vegetation may locally reduce water temperature (Rose and Crumpton 1996). High microbial respiration associated with decaying plant biomass and reduced diffusion of atmospheric oxygen often leads to near anoxic conditions and reduced pH of the water surrounding beds of emergent and submergent plants as compared to open water zones (Chimney et al. 2006; Rose and Crumpton 2006).

Distinct vertical profiles in water quality variables due to thermal stratification of the water column are well described for deeper ponds and lakes, and may also be observed in the wetland water column. Ryder and Horwitz (1995) report significant differences in temperature, pH, conductivity, dissolved oxygen and redox potential between the surface and bottom of the permanently inundated zone of a depressional wetland in Australia. This stratification was observed in water less than 1.5 m deep, was most pronounced near stands of macrophytes, and exhibited a diel pattern of formation from early afternoon to early evening. Diel cycles of thermal stratification and associated vertical profiles of dissolved oxygen and dissolved methane were also observed in a shallow Australian floodplain wetland (Ford et al. 2002). In this study, surface water oxygen levels were highest in late afternoon and would sometimes be near zero by morning due to high respiratory demand. Boeckman and Bidwell (2007) also observed summertime thermal stratification across a maximum depth of 30 cm in an Oklahoma depressional wetland that resulted in vertical profiles of dissolved oxygen, pH, and suspended solids. The shallow nature of wetlands can make stratification of the water column quite transient as it is easily disrupted by wind (Boeckman and Bidwell 2007). However, stratification may still have an influence on wetland functional processes as indicated by Ryder and Horwitz (1996) who attributed reduced leaf processing in certain areas of the wetland they studied to limitations on microorganisms and invertebrates imposed by the diurnal stratification of the water column.

6.4.4 Temporal Influences

Temporal changes in wetland water quality can be driven by changing hydrologic conditions (see previous discussion of wetland water sources and hydroperiod) and biological activity in the system. Temperature, dissolved oxygen, and pH are among those parameters most prone to diurnal fluctuations as a result of the combined effects of solar heating, radiant cooling, photosynthesis, and respiration, although diurnal variation in parameters such as conductivity and alkalinity have also been observed (Ryder and Horwitz 1995; Stratford et al. 2004; Sisodia and Moundiotiya 2006; Tuttle et al. 2008; Reeder 2011). Cornell and Klarer (2008) report dissolved oxygen levels in a Lake Erie coastal wetland varied between 20 and 150 % saturation over the course of the day and comparable dissolved oxygen fluctuations have been observed in stands of emergent vegetation in restored and natural floodplain wetlands (Boon 2006; Reeder 2011). High levels of photosynthetic activity can also lead to significant increases in pH, although these may be attenuated by other chemical characteristics of the system. Boon (2006) discusses studies of an Australian wetland that exhibited daily pH changes of up to 2 pH units and observed that this fluctuation could influence nitrogen dynamics in the system if the pH were to rise above 8 and convert ammonium to the more volatile ammonia. In contrast, Reeder (2011) did not observe a significant influence of primary productivity or respiration on the pH of restored floodplain wetlands, and attributed these results to buffering by divalent cations in the sediments and/or reduced effects of microbial respiration due to low levels of sediment organic matter.

The intensity of diurnal fluctuations in wetland water quality variables may vary based on habitat type and water source. As compared to open water zones, diurnal fluctuations in temperature may be dampened near stands of emergent vegetation, although daily fluctuations in dissolved oxygen and pH may be greater in beds of both emergent and submergent plants (Rose and Crumpton 1996; Chimney et al. 2006; Reeder 2011). Diurnal fluctuations in temperature and dissolved oxygen were reduced during hydrologic pulses in created riparian wetlands (Tuttle et al. 2008), while groundwater input in vernal pools has also been found to reduce daily temperature fluctuations (Korfel et al. 2010). Daily fluctuations in conductivity and water color in Great Lakes coastal wetlands was attributed to seiche-induced inflow of lake water that increased levels of dissolved ions and diluted color (deCatanzaro and Chow-Fraser 2011).

In temperate zones, seasonal changes in thermal input can lead to significant seasonal variation in wetland water temperature, with differences of 30 °C or more between minimum and maximum temperatures not uncommon (Black 1976; Boeckman and Bidwell 2007). These temperature differences drive seasonal changes in biotic and abiotic processes that influence other water quality parameters. For example, dissolved oxygen levels are often higher during cooler months owing to higher gas solubility and lower respiration (Boeckman and Bidwell 2007). Diel fluctuations in dissolved oxygen, pH, and alkalinity may be greater in summer due to increased rates of photosynthesis and respiration and reduced solubility of oxygen and CO_2. Levels of nitrate and orthophosphate in wetland water may be lower in summer due to greater uptake by plants or, in the case of nitrate, increased rates of denitrification (Mitsch and Reeder 1992; deCatanzaro and Chow-Fraser 2011). However, Eser and Rosen (1999) report seasonal maxima for nitrate and ammonium in late summer which was attributed to nutrient release from the breakdown of organic matter coupled with reduced plant uptake as the peak growing season begins to trail off. Summer increases in wetland orthophosphate levels have also been observed and related to possible release of the nutrients from anoxic sediments (Glińska-Lewczuk 2009). Increased water concentrations of other elements, including potentially toxic metals such as cadmium, have been reported in some temperate wetlands and have been attributed to release of these chemicals from sediments due to increasing temperatures and changing redox conditions in spring (Olivie-Lauquet et al. 2001).

6.5 The Role of Wetlands in Improving Water Quality

Discussions related to wetlands and water quality often focus on "improving water quality" as one of the key services wetlands provide and studies on both natural and created wetlands have demonstrated clear effects on the chemical and physical characteristics of water moving through these systems (see Barnes et al. 2002; O'Geen et al. 2010; Dotro et al. 2011 for basic examples). Some of the proximate mechanisms that underlie these water quality effects have been reviewed by Hemond and Benoit (1988) and Verhoeven et al. (2006). However, there are a few general qualifiers that are worth keeping in mind when considering the "water quality" function of wetlands. First, some wetlands (e.g., some types of vernal pools) have little influence on water quality due to limited hydrological linkage to other waters (Rains et al. 2006). Second, the capacity of wetlands to enhance water quality may ultimately depend on the area of wetland available relative to the total catchment area. For example, Verhoeven et al. (2006) state that wetlands can significantly affect catchment water quality if wetland habitat makes up at least 2–7 % of the catchment area. Finally, in some cases, activities in the surrounding landscape and wetland alterations can actually lead to wetlands becoming sources of sediments, nutrients, and toxic chemicals that could actually degrade downstream water quality (Brinson 1988; Whigham and Jordan 2003; Verhoeven et al. 2006).

6.6 General Study Design and Approaches

6.6.1 Study Design

The objectives and questions to be addressed in collecting water quality data from a wetland should be clearly defined prior to the start of the study since this will inform the study design and associated statistical analyses. USEPA (2002a) discusses some of the key objectives that should be considered in a wetland monitoring study and provides guidance on site selection and sampling designs. Common sampling designs used for wetland water quality monitoring include stratified random sampling, targeted/tiered sampling, and before-after, control-impact (BACI), with the choice of approach dependant on the project objectives (USEPA 2002a). The number of replicate measurements or samples to be taken for analyses of specific wetland water quality parameters is also an important consideration, although often appears to be arbitrarily determined in monitoring studies. An excellent discussion of approaches to determine effective sample number based on study design is available in USEPA (2002b).

6.6.2 Wetland Classification

Wetland classification is aimed at controlling some of the natural variability in the monitoring data derived from wetlands by grouping them according to common physical or biological characteristics such as hydrology, hydrogeomorphology, and/or vegetative assemblages (e.g., Cowardin et al. 1979; Brinson 1993; Reinelt et al. 2001; Jackson 2006). Water quality has also been used to group wetlands or modify classification systems (e.g. Cowardin et al. 1979; Warner and Rubec 1997). Brinson (1988) provides a conceptual discussion of how landscape position and associated flow characteristics, both of which are attributes used for geomorphological classification of wetlands, would influence elemental cycles and wetland water quality.

Studies that have specifically evaluated the extent to which classification helps control variability in water quality parameters among wetlands are limited, although those that are available indicate that broad scale classification may not be particularly effective in this regard. For example, Trebitz et al. (2007) and Morrice et al. (2008) found that grouping Great Lakes coastal wetlands according to relatively coarse hydromorphic types or biogeographic region did not enhance their ability to relate wetland water quality to land use and Trebitz et al. (2007) concluded that finer hydrologic classification would have been desirable in their study. Similarly, Azzolina et al. (2007) were unable to detect significant differences in surface water quality between wetlands grouped according to a modification of Brinson's hydrogeomorphic (HGM) classification (specifically, the LLWW approach as described by Tiner (2003) and Tiner and Stewart (2004)). Euliss et al. (2004) state that while regional landscape position often explains many of the chemical and biological properties of wetlands, finer-spatial scale and temporal influences are also significant determinants of wetland water quality. This is clearly supported by studies that have demonstrated how localized landscape factors and basin characteristics can influence water quality in individual wetlands (e.g., Skelly and Freidenburg 2000; Batzer et al. 2000; Hossack and Corn 2008). The development of regional wetland subclasses as described in the HGM approach (Brinson 1993) or even finer-scale modifiers of wetland classes may therefore be necessary to effectively reduce natural variation in water quality among wetlands.

6.6.3 Temporal and Spatial Considerations

As described above, wetland hydroperiod could influence water quality and lead to divergence between sites due to differences in water loss and associated effects of evapoconcentration. Ideally, water quality would best be compared between wetlands at comparable stages of their hydroperiod. Jackson (2006) provides a discussion of approaches that could be used to characterize wetland hydroperiod, including regular site visits, and use of devices such as staff gauges, automatic water level monitors and piezometers. Unfortunately, this type of intensive surveying may not be possible for studies with limited resources or that have the goal of sampling a large number of wetlands. Species composition of resident plant assemblages has also been used to provide a general indication of the duration of wetland hydroperiod (Babbitt et al. 2003; Sharitz and Pennings 2006) and this may assist in either developing finer-scale groupings of wetlands or as ancillary data to help with interpretation of observed patterns in water quality data.

If the objective of a wetland monitoring study is to obtain water quality data that represent the system as a whole, potential spatial differences between different habitat types should be considered (Fig. 6.1). This could be addressed by first mapping the wetland to identify major habitat types and tributaries running into the system, and then using these areas as strata in a stratified random sampling approach. Vertical stratification of open water zones of the wetland may also need to be considered, particularly if functional parameters such as carbon processing are being assessed (e.g. Ryder and Horwitz 1996). Representative values for the wetland as a whole could then be generated by simply calculating the arithmetic mean of the data derived from each habitat type or by using weighted averages to provide proportional representation for each habitat. Another alternative is to develop composite samples of water derived from different wetland strata and conduct all chemical analyses on those samples. If representative sampling of different habitats in the wetland is not possible, limiting sampling to only the major habitat type may be the next best option. Some effort should also be made to record water quality measurements from comparable habitat types in each wetland visited. Unfortunately, published wetland studies that include collection of water quality data often fail to indicate the type of habitat the data were derived from or if potential habitat variability in the parameters being measured was considered in the study design.

Diurnal variation in water quality driven by changing temperatures, photosynthetic activity and respiration may best be addressed by sampling wetlands within a defined time period (e.g., 10 AM to 2 PM) and some published studies do indicate the time interval during which water quality measurements were made in wetlands (e.g., Trebitz et al. 2007; deCatanzaro and Chow-Fraser 2011). However, this may again pose a challenge for studies aimed at sampling a large number of wetlands or in cases where travel time between sampling sites is significant. This is a good reason to include the time of day the wetland was sampled on field data sheets since this information could be used in later data analysis to determine if any diurnal trends in the data are apparent. Routine water quality monitoring of wetlands often

Fig. 6.1 A freshwater wetland with open water, emergent and floating vegetation. These different habitat zones may differ significantly in water quality parameters such as temperature, dissolved oxygen, pH, and alkalinity

occurs during warmer summer months, although some consideration should be given to the seasonal sampling window since differences in key biological processes that influence water quality could exist depending on when in the growing season individual wetlands are visited.

6.7 General Methodology

The methods used to collect chemical and physical water quality data from wetlands are largely the same as those used to collect these data from other surface waters. Detailed treatments of the chemistry of running and standing waters can be found in any number of basic freshwater ecology or limnology textbooks (e.g., Wetzel 2001), while Mitsch and Gosselink (2007) provide specific overview of wetland biogeochemistry. Similarly, excellent texts that provide in-depth discussions of water sampling and analytical methods are available. These include *Limnological Analyses* by Wetzel and Likens (2000) and *Methods in Stream Ecology* by Hauer and Lamberti (2007). *Standard Methods for the Examination of Water and Wastewater* (APHA, AWWA and WEF 2005) is also a critical reference for those interested in evaluating abiotic and biotic conditions in natural waters. Additional sampling methods and analytical procedures for wetland water quality are available in USEPA (2011).

The range of chemical and physical variables that could be measured in wetland water is extensive and ultimately depends on the objectives of the study. Those listed in Tables 6.1 and 6.2 were selected because they include those variables commonly measured in wetland monitoring programs and because the methodology for most involve basic equipment such as electronic water quality meters or relatively simple wet chemistry procedures. A number of suppliers also offer "environmental laboratory" kits that provide step-by-step methodology using pre-packaged reagents for the analyses of a number of these variables in the field. In most cases, the methods employed in these kits are derived from protocols described in APHA, AWWA and WEF (2005) and listed in Table 6.2. If water quality monitoring data are to be used for regulatory purposes, attention should be paid to whether analytical techniques are acceptable to state or federal regulatory agencies. Some of the methods presented in Table 6.2, such as those for nutrient determinations, have also been adapted for laboratory-based autoanalyzers that allow high through-put of samples.

Other water quality parameters such as forms of organic carbon and individual anions and cations are also often measured in wetland studies but are not discussed in detail here since current methods for their determination involve more expensive types of instrumentation. This is also the case for metals and organic contaminants. A general overview of this instrumentation is provided by Wetzel and Likens (2000). APHA, AWWA and WEF (2005) also provide general overviews and basic methodology for analyzing selected metals and organic contaminants including collection and processing of samples.

In addition to the sampling considerations already discussed, other issues must often be considered when collecting water quality data from surface waters including wetlands. Electronic water quality meters can greatly enhance data collection from the field, with some models allowing the determination of a number of parameters at one time. However, if these instruments are to provide reliable data, attention must be paid to their calibration and maintenance. Without proper calibration, a water quality meter can become an expensive random number generator. As such, all personnel using the device should become acquainted with the manufacture-prescribed frequency for calibration and the calibration procedure. Maintaining records on when the meter was calibrated and by whom is also an important element for data quality assurance. Similarly, regular inspection of the probes of the meter for damage or fouling can help ensure reliable output during use. Measurements taken with a water quality meter often rely on the passage of gasses or ions across an electrode membrane and it may be necessary to provide gentle agitation of the probe to enhance this exchange. Some probes are fitted with small impellers to maintain this water flow. In addition to the handheld water quality meter, a range of other electronic devices are available to collect water quality data from surface waters including temperature loggers and multi-parameter sondes that can be left on site to facilitate data collection over longer time frames. These instruments are particularly useful for characterizing diurnal patterns, although their use may be limited in shallower wetlands.

When using a water quality meter to take measurements from the wetland water column, care must be taken to avoid stirring bottom sediments or allowing the probe

Table 6.2 Summary of methods for the determination of water quality parameters commonly measured in wetland monitoring studies

Parameter (common units)	Example instrumentation/method	Collection container	Minimum sample size (mL)	Preservation	Maximum storage time	Method reference/discussion[a]
Temperature (°C)	Thermometer, Temperature sensor/Water quality meter, data logger	In situ, plastic, glass	N/A	Analyze immediately	NA	1, 2
Dissolved oxygen (mg/L or % saturation)	Oxygen electrode/Water quality meter, Winkler titration	In situ, glass	300	Winkler titration-acidification step	Immediately if by electrode, Winkler titration- 8 h after acidification	1, 3
pH	pH electrode/Water quality meter	In situ, plastic, glass	50	Analyse immediately	NA	1, 4
Alkalinity (as mg CaCO$_3$/L)	Acid titration with indicator solution	Plastic, glass	200	Refrigerate	24 h	1, 8
Total dissolved solids (mg/L)	Sample filtration, drying and weighing	Plastic, glass	200	Refrigerate	7 days	10
Salinity (g/L)	Some water quality meters derive salinity from conductivity measurements, hand-held refractometers can provide a general indication	Glass	240	Analyze immediately	NA	1, 6
Specific Conductance (microSiemens/cm = µS/cm)	Conductivity probe/Water quality meter	In situ, plastic, glass	500	Refrigerate	28 days	1, 5
Total Hardness (as mg CaCO$_3$/L)	EDTA titration	Plastic, glass	100	Add HNO$_3$ or H$_2$SO$_4$ to pH <2	6 months	1, 7
Total suspended solids (mg/L)	Sample filtration, drying and weighing	Plastic, glass	200– depends on level	Refrigerate	7d	11
Turbidity (Nephelometric turbidity units, NTU)	Nephelometer/Turbidity meter	Plastic, glass	100	Refrigerate in dark	24 h	1, 9

Ammonia (mg/L)	Phenate method, Salicylate method, Ammonia selective electrode	Plastic, glass	500	Reduce pH <2 by addition of H_2SO_4 and refrigerate	7 days	1, 12, 13
Nitrate (mg/L)	Cadmium reduction method	Plastic, glass	100	Refrigerate	48 h	14
Nitrite (mg/L)	Diazotization colorimetric method	Plastic, glass	100	Refrigerate	48 h	15
Total Nitrogen (mg/L)	Persulfate digestion followed by nitrate determination	Plastic, glass	500	Refrigerate	48 h	16
Total phosphorous (mg/L)	Persulfate digestion followed by ortho-phosphate determination	Plastic, glass	100	Reduce pH <2 by addition of H_2SO_4 and refrigerate	28 days	17
Orthophosphate (mg/L)	Ascorbic acid method	Acid washed glass	100	For dissolved– filter immediately and refrigerate	48 h	18

All information on sample collection, sample size, preservation, and storage from APHA, AWWA and WEF (2005). Method summaries based on those listed here or alternative procedures can also be accessed on the USEPA website at http://water.epa.gov/scitech/methods/cwa/methods_index.cfm

[a] Manufacturers manual for meters; 1: Wetzel and Likens (2000); 2: Hauer and Hill (2007); 3: APHA, AWWA and WEF (2005) Method 4500-OA-G; 4: APHA, AWWA and WEF (2005) Method 4500-H[+]; 5: APHA, AWWA and WEF (2005) Method 2510 A-B; 6: APHA, AWWA and WEF (2005) Method 2520B; 7: APHA, AWWA and WEF (2005) Method 2340C; 8: APHA, AWWA and WEF (2005) Method 2320; 9: APHA, AWWA and WEF (2005) Method 2130; 10: APHA, AWWA and WEF (2005) Method 2540C; 11: APHA, AWWA and WEF (2005) Method 2540D; 12: APHA, AWWA and WEF (2005) Method 4500-NH_3; 13: Reardon et al. (1966); 14: APHA, AWWA and WEF (2005) Method 4500-NO_3^- E; 15: APHA, AWWA and WEF (2005) Method 4500-NO_2^- ; 16: APHA, AWWA and WEF (2005) Method 4500-N C; 17: APHA, AWWA and WEF (2005) Method 4500-P B,E; 18: APHA, AWWA and WEF (2005) Method 4500-P E

Fig. 6.2 A long-handled sampling pole with a collection bottle on the end can be used to avoid disturbing sediments and/or facilitate collecting water samples from deeper areas of the wetland

to contact sediments since this can significantly influence readings for parameters such as dissolved oxygen. This can sometimes be challenging in shallower systems or if it is necessary to wade into the wetland to access open water. If the wetland is large and deep enough, accessing open water by boat can help avoid these issues although attention must still be paid to where the water quality probe is located in the water column. The use of a long-handled sampling pole with a collection bottle on the end (Fig. 6.2) can be used to pull grab samples of water that can be analysed on shore with a water quality meter or transferred to appropriate containers for determination of wet chemistry parameters. In wetlands with deeper basins, depth-specific samples for wet chemistry analyses can also be collected using a commercial sampling device such as a Van Dorn sampler (Wetzel and Likens 2000). Basic quality assurance for wet chemistry procedures includes following appropriate sample storage and preservation techniques and conducting the analyses within the prescribed time frame to ensure sample viability (Table 6.2). The use of properly cleaned glassware to avoid inaccurate readings due to contamination is also a key issue if conducting sample analyses in house. Appropriate sample labelling and tracking methods are also important considerations. Guidance on analytical quality assurance and data handling is provided by Briggs (1996) and APHA, AWWA and WEF (2005)

References

Alluvium Consulting (2011) Framework for the assessment of river and wetland health: findings from the trials and options for uptake. Waterlines Report No. 58. Australian National Water Commission, Canberra

APHA, AWWA and WEF (2005) Standard methods for the examination of water and wastewater, 21st edn. American Public Health Association, Washington, DC

Azzolina NA, Siegel DI, Brower JC, Samson SD, Otz MH, Otz I (2007) Can the HGM classification of small, non-peat forming wetlands distinguish wetlands from surface water geochemistry? Wetlands 27:884–893

Babbitt KJ, Baber MJ, Tarr TL (2003) Patterns of larval amphibian distribution along a wetland hydroperiod gradient. Can J Zool-Rev Can De Zool 81:1539–1552

Barnes K, Ellery W, Kindness A (2002) A preliminary analysis of water chemistry of the Mkuze Wetland System, KwaZulu-Natal: a mass balance approach. Water SA 28:1–12

Batzer DP, Jackson CR, Mosner M (2000) Influences of riparian logging on plants and invertebrates in small, depressional wetlands of Georgia, USA. Hydrobiologia 441:123–132

Batzer DP, Palik BJ, Buech R (2004) Relationships between environmental characteristics and macroinvertebrate communities in seasonal woodland ponds of Minnesota. J N Am Benthol Soc 23:50–68

Bedford BL (1996) The need to define hydrologic equivalence at the landscape scale for freshwater wetland mitigation. Ecol Appl 6:57–68

Bedford BL, Godwin KS (2003) Fens of the United States: distribution, characteristics, and scientific connection versus legal isolation. Wetlands 23:608–629

Black JH (1976) Environmental fluctuations in central Oklahoma temporary ponds. Proc Okla Acad Sci 56:1–8

Boeckman CJ, Bidwell JR (2007) Spatial and seasonal variability in the water quality characteristics of an ephemeral wetland. Proc Okla Acad Sci 87:45–54

Bojkova J, Schenkova J, Horsak M, Hajek M (2011) Species richness and composition patterns of clitellate (Annelida) assemblages in the treeless spring fens: the effect of water chemistry and substrate. Hydrobiologia 667:159–171

Boon PI (2006) Biogeochemistry and bacterial ecology of hydrologically dynamic wetlands. In: Batzer DP, Sharitz RR (eds) Ecology of freshwater and estuarine wetlands. University of California Press, Berkeley, pp 115–176

Briggs R (1996) Analytical quality assurance, Chapter 9. In: Bartram J, Balance R (eds) Water quality monitoring – a practical guide to the design and implementation of freshwater quality studies and monitoring programmes. Published on behalf of the World Health Organization by E & FN Spon, London, pp 400

Brinson MM (1988) Strategies for assessing the cumulative effects of wetland alteration on water-quality. Environ Manage 12:655–662

Brinson MM (1993) A hydrogeomorphic classification for wetlands technical report WRP-DE-4. U.S. Army Corps of Engineers, Waterways Experiment Station, Vicksburg

Brooks RT (2000) Annual and seasonal variation and the effects of hydroperiod on benthic macroinvertebrates of seasonal forest ("vernal") ponds in central Massachusetts, USA. Wetlands 20:707–715

Cabezas A, Garcia M, Gallardo B, Gonzalez E, Gonzalez-Sanchis M, Comin FA (2009) The effect of anthropogenic disturbance on the hydrochemical characteristics of riparian wetlands at the Middle Ebro River (NE Spain). Hydrobiologia 617:101–116

Carter V (1996) Wetland hydrology, water quality, and associated functions. United States Geological Survey Water-Supply Paper 2425. In: Fretwell JD, Williams JS, Redman PJ (eds) National water summary of wetland resources. United States Government Printing Office, Washington, DC, pp 35–48

Catallo WJ, Schlenker M, Gambrell RP, Shane BS (1995) Toxic-chemicals and trace-metals from urban and rural Louisiana lakes – recent historical profile and toxicological significance. Environ Sci Technol 29:1436–1445

Chague-Goff C, Mark AF, Dickinson KJM (2010) Hydrological processes and chemical characteristics of low-alpine patterned wetlands, south-central New Zealand. J Hydrol 385:105–119

Chapelle FH, Bradley PM, McMahon PB, Lindsey BD (2009) What does "water quality" mean? Ground Water 47:752–754

Charkhabi AH, Sakizadeh M (2006) Assessment of spatial variation of water quality parameters in the most polluted branch of the Anzali wetland, Northern Iran. Polish J Environ Stud 15:395–403

Chen PY, Lee PF, Ko CJ, Ko CH, Chou TC, Teng CJ (2011) Associations between water quality parameters and planktonic communities in three constructed wetlands, Taipei. Wetlands 31:1241–1248

Chimney MJ, Wenkert L, Pietro KC (2006) Patterns of vertical stratification in a subtropical constructed wetland in south Florida (USA). Ecol Eng 27:322–330

Colburn EA (2004) Vernal pools: natural history and conservation. The McDonald and Woodward Publishing Company, Blacksburg

Cornell LP, Klarer DM (2008) Patterns of dissolved oxygen, productivity and respiration in old woman Creek Estuary, Erie County, Ohio during low and high water conditions. Ohio J Sci 108:31–43

Cowardin LM, Carter V, Golet FC, LaRoe ET (1979) Classification of wetlands and deepwater habitats of the United States. FWS/OBS-79/31. U.S. Fish and Wildlife Service, Washington, DC

deCatanzaro R, Chow-Fraser P (2011) Effects of landscape variables and season on reference water chemistry of coastal marshes in Eastern Georgian Bay. Can J Fish Aquat Sci 68:1009–1023

Diaz FJ, O'Geen AT, Dahlgren RA (2012) Agricultural pollutant removal by constructed wetlands: implications for water management and design. Agric Water Manag 104:171–183

Dickerson BR, Vinyard GL (1999) Effects of high levels of total dissolved solids in Walker Lake, Nevada, on survival and growth of Lahontan cutthroat trout. Trans Am Fish Soc 128:507–515

Dotro G, Larsen D, Palazolo P (2011) Treatment of chromium-bearing wastewaters with constructed wetlands. Water Environ J 25:241–249

Dunson WA, Paradise CJ, VanFleet RL (1997) Patterns of water chemistry and fish occurrence in wetlands of hydric pine flatwoods. J Freshw Ecol 12:553–565

ELI (2008) State wetland protection: status, trends & model approaches. Environmental Law Institute, Washington, DC, pp 67. Online accessed at: www.eli.org

Eser P, Rosen MR (1999) The influence of groundwater hydrology and stratigraphy on the hydrochemistry of stump bay, South Taupo Wetland, New Zealand. J Hydrol 220:27–47

Euliss NH, Wrubleski DA, Mushet DM (1999) Wetlands of the Prairie Pothole Region: in-vertebrate species composition, ecology, and management. In: Batzer DP, Rader RD, Wissinger SA (eds) Invertebrates in freshwater wetlands of North America: ecology and management. Wiley, New York, pp 471–514

Euliss NH, Labaugh JW, Fredrickson LH, Mushet DM, Laubhan MRK, Swanson GA, Winter TC, Rosenberry DO, Nelson RD (2004) The wetland continuum: a conceptual framework for interpreting biological studies. Wetlands 24:448–458

Evans R, Gilliam JW, and Lilly JP (1996) Wetlands and water quality. Online accessed at: www.bae.ncsu.edu/programs/extension/evans/ag473-7.html. North Carolina Cooperative extension service, Publication Number AG 473-7

Ford PW, Boon PI, Lee K (2002) Methane and oxygen dynamics in a shallow floodplain lake: the significance of periodic stratification. Hydrobiologia 485:97–110

Gell PA, Sluiter IR, Fluin J (2002) Seasonal and interannual variations in diatom assemblages in Murray River connected wetlands in north-west Victoria, Australia. Mar Freshw Res 53:981–992

Ginocchio R, Hepp J, Bustamante E, Silva Y, De La Fuente LM, Casale JF, De La Harpe JP, Urrestarazu P, Anic V, Montenegro G (2008) Importance of water quality on plant abundance and diversity in high-alpine meadows of the Yerba Loca natural sanctuary at the Andes of north-central Chile. Rev Chil Hist Nat 81:469–488

Glińska-Lewczuk K (2009) Water quality dynamics of oxbow lakes in young glacial landscape of NE Poland in relation to their hydrological connectivity. Ecol Eng 35:25–37

Gordon CC, Flake LD, Higgins KF (1997) Trace metals in water and sediments in wetlands of the rainwater basin area of Nebraska. Proc South Dakota Acad Sci 76:253–261

Hauer FR, Hill WR (2007) Temperature, light and oxygen, Chapter 5. In: Hauer FR, Lamberti GA (eds) Methods in stream ecology, 2nd edn. Elsevier, Burlington, pp 103–118

Hauer FR, Lamberti GA (eds) (2007) Methods in stream ecology, 2nd edn. Elsevier, Burlington

Hemond HF, Benoit J (1988) Cumulative impacts on water-quality functions of wetlands. Environ Manage 12:639–653

Hossack BR, Corn PS (2008) Wildfire effects on water temperature and selection of breeding sites by the boreal toad (*Bufo boreas*) in seasonal wetlands. Herpetol Conserv Biol 3:46–54

Houlahan JE, Findlay CS (2004) Estimating the 'critical' distance at which adjacent land-use degrades wetland water and sediment quality. Landsc Ecol 19:677–690

Ibekwe AM, Lyon SR, Leddy M, Jacobson-Meyers M (2007) Impact of plant density and microbial composition on water quality from a free water surface constructed wetland. J Appl Microbiol 102:921–936

Jackson CR (2006) Wetland hydrology. In: Batzer DP, Sharitz RR (eds) Ecology of freshwater and estuarine wetlands. University of California Press, Berkeley, pp 43–81

Jolly ID, McEwan KL, Holland KL (2008) A review of groundwater-surface water interactions in arid/semi-arid wetlands and the consequences of salinity for wetland ecology. Ecohydrology 1:43–58

Kadlec RH (1999) Chemical, physical and biological cycles in treatment wetlands. Water Sci Technol 40:37–44

Kolka RK, Thompson JA (2006) Wetland geomorphology, soils, and formative processes. In: Batzer DP, Sharitz RR (eds) Ecology of freshwater and estuarine wetlands. University of California Press, Berkeley, pp 7–42

Korfel CA, Mitsch WJ, Hetherington TE, Mack JJ (2010) Hydrology, physiochemistry, and amphibians in natural and created vernal pool wetlands. Restor Ecol 18:843–854

Kusler J (2011a) State water quality standards for wetlands. Discussion paper submitted to Maryland Department of Environment as part of USEPA Wetland Program Development Grant BG 973027-03, 91 pp. Online accessed at: http://aswm.org/pdf_lib/state_water_quality_standards_for_wetlands_061410.pdf

Kusler J (2011b) How wetlands differ from traditional waters; what this means to wetland water quality standards. Discussion paper submitted to Maryland Department of Environment as part of USEPA Wetland Program Development Grant BG 973027-03, 10 pp. Online accessed at: http://aswm.org/pdf_lib/how_wetlands_differ_from_traditional_waters_061410.pdf

LaBaugh JW (1989) Chemical characteristics of water in northern prairie wetlands. In: van der Valk A (ed) Northern prairie wetlands. Iowa State University Press, Ames, pp 56–90

Longcore JR, McAuley DG, Pendelton GW, Bennatti CR, Mingo TM, Stromborg KL (2006) Macroinvertebrate abundance, water chemistry, and wetland characteristics affect use of wetlands by avian species in Maine. Hydrobiologia 567:143–167

Mackay AW, Davidson T, Wolski P, Mazebedi R, Masamba WRL, Huntsman-Mapila P, Todd M (2011) Spatial and seasonal variability in surface water chemistry in the Okavango Delta, Botswana: a multivariate approach. Wetlands 31:815–829

Mendelssohn IA, Batzer DP (2006) Abiotic constraints for wetland plants and animals. In: Batzer DP, Sharitz RR (eds) Ecology of freshwater and estuarine wetlands. University of California Press, Berkeley, pp 82–114

Mitsch WJ, Gosselink JG (2007) Wetlands, 4th edn. Wiley, Hoboken

Mitsch WJ, Reeder BC (1992) Nutrient and hydrologic budgets of a Great Lakes coastal freshwater wetland during a drought year. Wetl Ecol Manag 1:211–222

Morrice JA, Danz NP, Regal RR, Kelly JR, Niemi GJ, Reavie ED, Hollenhorst T, Axler RP, Trebitz AS, Cotter AM, Peterson GS (2008) Human influences on water quality in great lakes coastal wetlands. Environ Manage 41:347–357

Nelson ML, Rhoades CC, Dwire KA (2011) Influence of bedrock geology on water chemistry of slope wetlands and headwater streams in the southern rocky mountains. Wetlands 31:251–261

Nimmo DWR, Johnson RW, Preul MA, Pillsbury RW, Self JR, Bergey EA (2006) Determining site-specific toxicity of copper to daphnids and fishes in a brown-water ecosystem. J Freshw Ecol 21:481–491

Norris RH, Dyer F, Hairsine P, Kennard M, Linke S, Merrin L, Read A, Robinson W, Ryan C, Wilkinson S, Williams D (2007) A baseline assessment of water resources for the national water initiative level 2 assessment, river and wetland health theme, assessment of river and wetland health: potential comparative indices. Australian National Water Commission, Canberra

O'Geen AT, Budd R, Gan J, Maynard JJ, Parikh SJ, Dahlgren RA, Sparks DL (2010) Mitigating nonpoint source pollution in agriculture with constructed and restored wetlands. Adv Agron 108:1–76

Olivie-Lauquet G, Gruau G, Dia A, Riou C, Jaffrezic A, Henin O (2001) Release of trace elements in wetlands: role of seasonal variability. Water Res 35:943–952

Parkes SD, Jolley DF, Wilson SR (2007) Inorganic nitrogen transformations in the treatment of landfill leachate with a high ammonium load: a case study. Environ Monit Assess 124:51–61

Rains MC, Fogg GE, Harter T, Dahlgren RA, Williamson RJ (2006) The role of perched aquifers in hydrological connectivity and biogeochemical processes in vernal pool landscapes, Central Valley, California. Hydrol Process 20:1157–1175

Reardon J, Foreman JA, Searcy RL (1966) New reactants for colorimetric determination of ammonia. Clin Chim Acta 14:403

Reeder BC (2011) Assessing constructed wetland functional success using diel changes in dissolved oxygen, pH, and temperature in submerged, emergent, and open- water habitats in the Beaver Creek Wetlands Complex, Kentucky (USA). Ecol Eng 37:1772–1778

Reinelt LE, Taylor BL, Horner RR (2001) Morphology and hydrology. In: Azous AL, Horne RR (eds) Wetlands and urbanization: implications for the future. Lewis Publishers, Boca Raton, pp 221–235

Rose C, Crumpton WG (1996) Effects of emergent macrophytes on dissolved oxygen dynamics in a prairie pothole wetland. Wetlands 16:495–502

Rose C, Crumpton WG (2006) Spatial patterns in dissolved oxygen and methane concentrations in a prairie pothole wetland in Iowa, USA. Wetlands 26:1020–1025

Ryder DS, Horwitz P (1995) Seasonal water regimes and leaf litter processing in a wetland on the Swan Coastal Plain, Western Australia. Mar Freshw Res 46:1077–1084

Ryder DS, Horwitz P (1996) The diurnal stratification of Lake Jandabup, a coloured wetland on the Swan Coastal Plain, Western Australia. J R Soc West Aust 78:99–101

Sharitz RR, Pennings SC (2006) Development of wetland plant communities. In: Batzer DP, Sharitz RR (eds) Ecology of freshwater and estuarine wetlands. University of California Press, Berkeley, pp 177–241

Sisodia R, Moundiotiya C (2006) Assessment of the water quality index of wetland Kalakho Lake, Rajasthan, India. J Environ Hydrol 14:1–11

Skelly DK, Freidenburg LK (2000) Effects of beaver on the thermal biology of an amphibian. Ecol Lett 3:483–486

Stratford CJ, McCartney MP, Williams RJ (2004) Seasonal and diurnal hydro-chemical variations in a recreated reed bed. Hydrol Earth Syst Sci 8:266–275

Tiner RW (2003) Dichotomous keys and mapping codes for wetland landscape position, landform, water flow path, and waterbody type descriptors. U.S. Fish and Wildlife Service, National Wetlands Inventory Program, Northeast Region, Hadley

Tiner RW, Stewart J (2004) Wetland characterization and preliminary assessment of wetland functions for the Delaware and Catskill watersheds of the New York City water supply system. U.S. Fish and Wildlife Service, National Wetlands Inventory, Ecological Services, Region 5, Hadley

Trebitz AS, Morrice JA, Taylor DL, Anderson RL, West CW, Kelly JR (2005) Hydromorphic determinants of aquatic habitat variability in lake superior coastal wetlands. Wetlands 25:505–519

Trebitz AS, Brazner JC, Cotter AM, Knuth ML, Morrice JA, Peterson GS, Sierszen ME, Thompson JA, Kelly JR (2007) Water quality in great lakes coastal wetlands: basin-wide patterns and responses to an anthropogenic disturbance gradient. J Great Lakes Res 33:67–85

Tuttle CL, Zhang L, Mitsch WJ (2008) Aquatic metabolism as an indicator of the ecological effects of hydrologic pulsing in flow-through wetlands. Ecol Indic 8:795–806

USEPA (1990) Water quality standards for wetlands national guidance. United States Environmental Protection Agency, Office of Water, Washington, DC. EPA 440/S 90 011

USEPA (2002a) Methods for evaluating wetland condition: study design for monitoring wetlands. United States Environmental Protection Agency, Office of Water, Washington, DC. EPA-822-R-02-015

USEPA (2002b) Guidance on choosing a sampling design for environmental data collection for use in developing a quality assurance project plan. United States Environmental Protection Agency, Office of Environmental Information, Washington, DC. EPA/240/R-02/005

USEPA (2008a) Methods for evaluating wetland condition: biogeochemical indicators. United States Environmental Protection Agency, Office of Water, Washington, DC. EPA\440\-822-R-08-022

USEPA (2008b) Nutrient criteria technical guidance manual for wetlands. United States Environmental Protection Agency, Office of Water, Washington, DC. EPA-822-B-08-001

USEPA (2011) National wetland condition assessment laboratory operations manual. United States Environmental Protection Agency, Office of Water, Washington, DC. EPA-843-R10-002

Verhoeven JTA, Arheimer B, Yin CQ, Hefting MM (2006) Regional and global concerns over wetlands and water quality. Trends Ecol Evol 21:96–103

Waiser MJ, Robarts RD (2004) Photodegradation of DOC in a shallow prairie wetland: evidence from seasonal changes in DOC optical properties and chemical characteristics. Biogeochemistry 69:263–284

Warner BG, Rubec CDA (eds) (1997) The Canadian wetlands classification system, 2nd edn. Wetlands Research Centre, University of Waterloo, Waterloo, pp 68

Weilhoefer CL, Pan YD, Eppard S (2008) The effects of river floodwaters on floodplain wetland water quality and diatom assemblages. Wetlands 28:473–486

Wetzel RG (2001) Limnology: lake and river ecosystems, 3rd edn. Academic, San Diego

Wetzel RG, Likens GE (2000) Limnological analyses, 3rd edn. Springer, New York

Whigham DF, Jordan TE (2003) Isolated wetlands and water quality. Wetlands 23:541–549

Winter TC, Rosenberry DO, Buso DC, Merk DA (2001) Water source to four US wetlands: implications for wetland management. Wetlands 21:462–473

Student Exercises

Laboratory Exercises

Laboratory Exercise #1: Spatial Variation in Water Quality

1. Select a study wetland with different habitat types that can be accessed to determine water quality variables and develop a hand-drawn habitat map of the wetland to delineate major habitat types and structural features (e.g., inlets and outlets, stands of emergent, submergent, floating plants, open water). These habitat types will be used to develop hypotheses and as sampling strata in Step 2.
2. Based on the different habitat types in the wetland identified in Step 1 and an understanding of how wetland processes may influence water quality parameters, develop a series of hypotheses or predictions related to how the water quality variables being measured would be expected to differ between the habitat types. For example, how would you expect temperature, dissolved oxygen, and pH to compare between open water and stands of macrophytes?
3. Using a calibrated multi-parameter water quality meter, determine temperature, dissolved oxygen, pH, and conductivity for each wetland habitat. Other water quality parameters (e.g., turbidity, nutrients) may also be determined depending on availability of equipment**. If the wetland has an open water zone that is accessible by boat, use the water quality meter to generate a vertical profile of the water quality parameters from just above the sediment surface to just under the water surface. Space the measurements so that data are collected from at least three depths.
4. Provide a brief summary of whether the data collected supported your hypotheses and a brief discussion of the basis for the results observed.
5. Questions to consider:
   - What are some reasons for any observed differences in the water quality parameters measured between the wetland habitat types?
   - How would you determine if any relationships exist between the parameters measured? (i.e., dissolved oxygen vs. temperature vs. pH)
   - If relationships between parameters are observed, what is the basis for these?

**Water samples may also be collected in clean 1-L plastic bottles to take back to the laboratory for determination of "wet chemistry" parameters such as nitrate, orthophosphate, and alkalinity.

Laboratory Exercise #2: Temporal Variation in Water Quality

1. Select two or three different habitat types within a wetland and develop a sampling schedule that allows collection of water quality data (temperature,

6 Physical and Chemical Monitoring of Wetland Water 353

dissolved oxygen, pH, and conductivity) using a calibrated water quality meter at 3 h intervals from dawn until dusk.
2. Develop a graph for each variable measured by plotting the level of the parameter measured against time.
3. Questions to consider:

- Explain the basis for any observed fluctuations. What biotic and/or abiotic processes underlie the observed changes in water quality parameters over time?
- Do any relationships exist between the water quality parameters measured? If so, what are they? Explain the basis for these relationships.

Laboratory Exercise #3: Land Use Influences on Water Quality

Select a series of depressional or other wetland class that exist across different land use types. For example, crop versus pasture, urban versus park land.

1. Use available resources (topographic maps, digital orthophotographs, etc.) to characterize the major land use types within 1 km of each wetland. Based on this analysis, develop a series of testable hypotheses related to land use and water quality of the wetlands. For example, how would nutrient levels or other water quality parameters of a wetland surrounded by crop land or within a golf course be expected to differ from systems within less disturbed landscapes?
2. What key considerations should be addressed when attempting to compare water quality data between different wetlands?
3. Construct basic habitat maps for each wetland when on site. Also determine the existence of any undisturbed buffer zones around each wetland.
4. Use a calibrated water quality meter and/or collect water samples in clean 1 L bottles for later analysis to obtain water quality data from representative habitats in each wetland. Use these data to address the hypotheses/predictions about the differences between wetlands.
5. Questions to consider:

- Do any observed differences in water quality parameters match what you would expect based on land use types? If so, briefly explain the basis for these differences.
- If no real differences are detected, what are some reasons for this?
- What internal wetland processes may influence wetland water quality and potentially mask differences related to land use effects?

Chapter 7
Wetland Biogeochemistry Techniques

Bruce L. Vasilas, Martin Rabenhorst, Jeffry Fuhrmann,
Anastasia Chirnside, and Shreeam Inamdar

Abstract Biogeochemistry is the scientific discipline that addresses the biological, chemical, physical, and geological processes that govern the composition of the natural environment, with particular emphasis placed on the cycles of chemical elements critical to biological activity. Biogeochemical assays may measure a specific elemental pool, determine the rate of a pathway, or address a surrogate of a biogeochemical process or an elemental pool. In this chapter, we have attempted to emphasize field techniques; however, some of the techniques have relatively standard laboratory components that are beyond the scope of this chapter. This chapter is not meant to be all inclusive. We have chosen to emphasize the cycling of carbon, nitrogen, phosphorous, sulfur, manganese, and iron. Some of these techniques are not appropriate for all types of wetlands, or may be appropriate for a seasonally saturated wetland only during part of the season. Some of the techniques are simple and rely on equipment available to most wetlands practitioners. Others, which utilize isotopic methodologies, require expensive sophisticated equipment. Some techniques, such as soil organic matter determination by loss on ignition, have been accepted as standard methods for decades. Others, such as the determination of dissolved organic matter represent recent advances in a rapidly evolving field of ultra-violet and fluorescence technology. Some techniques rely solely on direct field measurements; others rely on the

---

B.L. Vasilas (✉) • J. Fuhrmann • S. Inamdar
Department of Plant and Soil Sciences, University of Delaware,
Townsend Hall, Newark, DE 19716-2170, USA

M. Rabenhorst
Department of Environmental Science and Technology,
University of Maryland, College Park, MD 20742, USA

A. Chirnside
Department of Entomology and Wildlife Ecology, University of Delaware,
Townsend Hall, Newark, DE 19716-2170, USA

incorporation of published data with field data. Apparent strengths and weaknesses of the various approaches, and wetland scenarios that would preclude the use or compromise the accuracy of a given technique are addressed.

7.1 Overview of Techniques

Biogeochemistry is the scientific discipline that addresses the biological, chemical, physical, and geological processes that govern the composition of the natural environment. Particular emphasis is placed on the study of the cycles of chemical elements such as carbon (C), nitrogen (N), and phosphorous (P) which are critical to biological activity. Biogeochemical assays may measure a specific elemental pool (e.g., soil organic carbon), determine the rate of a pathway (e.g., denitrification), or address a surrogate of a biogeochemical process or an elemental pool. The surrogate approach is popular for rapid assessment to characterize ecosystem health, functional capacity, nutrient loading, or water quality. In each case the practitioner must be aware of the exact nature of the parameter in question as well as limitations to the method. Attempts to quantify individual pools of C or N at best, produce representative estimates. On a wetland scale, it is not realistic to believe that the pool can be quantified with 100 % certainty. There is too much variability in the field and input sources which cannot be completely accounted for. Accuracy is compromised due to precision limits inherent to the technique and due to field variability. Results are often expressed on a per area basis (e.g., m^{-2}). Extrapolation of the values to a larger spatial area or to represent an entire wetland further increases the error. Therefore, the practitioner should consider these methods to be estimates. They are most useful for comparing wetlands, not for deriving absolute values. Also, the wetland concept encompasses a wide variety of ecosystems. So these techniques will be most reliable when comparing wetlands within a given class (e.g., piedmont slope wetlands). This chapter is not meant to be all inclusive. We have chosen to emphasize the cycles C, N, P, sulfur (S), manganese (Mn) and iron (Fe). Since many of these processes are microbially mediated or there is an exchange between the water column and the soil, there is inherent overlap with other chapters.

Some of these techniques will not be appropriate for all types of wetlands, particularly with respect to hydroperiod class. Nitrification levels will be difficult to detect and quantify in a permanently-inundated freshwater marsh as nitrification is an aerobic process. However, nitrification certainly could be measured in a seasonally saturated mineral soil flat as long as the measurements are not taken during a wet phase. Conversely, methane (CH_4) emissions could be detected in a marsh but not in a mineral soil flat. In addition, because some of these processes are strictly aerobic and others are strictly anaerobic, care must be taken to determine the time of season to run a field assay as the target process may not be occurring at detectable levels. This is primarily an issue with seasonally saturated wetlands where the practitioner must take into account seasonal variability in hydrologic conditions.

We have tried to emphasize field techniques. However, some of the techniques have relatively standard laboratory components that are beyond the scope of this chapter. In some cases the reader will be advised to check additional documents for the laboratory techniques. In addition, there are a number of commercial labs or university soil testing labs that will perform some of these assays at cost. In some cases, both a field assay and a lab assay are available to measure the same process. The field assay may be presented in this discussion, but the reader will be referred to documentation that covers the lab assay if the latter is considered to be more accurate.

7.2 Quality Control

7.2.1 Sample Collection

Many of the biogeochemical assays require collection of a field sample which is subsequently analyzed in the laboratory. The quality of the lab data is inherently limited by the quality of the field sample. Whether monitoring wetlands for regulatory purposes or for research studies, it is important to have a sampling program that employs proper field monitoring techniques and accurate laboratory analytical procedures. There are many publications that outline the proper methods for environmental monitoring. The National Wetlands Research Center (http://www.nwrc.usgs.gov) of the United States (U.S.) Geological Survey is a great source of information on wetland assessments. Many states have developed wetland monitoring guidelines; therefore, it is important to check with each state's environmental department for the most current monitoring strategies.

Samples collected for wetland assessment must be representative of the environmental variability that occurs both spatially and temporally within an ecosystem. The Environmental Monitoring Systems Laboratory of the U.S. Environmental Protection Agency's Office of Research and Development has published a guideline on soil sampling of any site under investigation (Mason 1992). Today, different geostatistical evaluations are used to design the monitoring approach and to evaluate the collected data in such a way as to minimize the inherent variability found within soils. Random selection techniques should be used to determine the actual location where the soil samples will be taken. Areas that should be considered before sampling include (1) maximizing the accuracy and precision of collection; (2) selecting sample locations that represent the wetland under study; (3) determining when, how often and how deep to sample; and (4) considering how the size of the wetland will affect the accuracy of sampling. Many field guides describe how these issues are addressed (Barth et al. 1989; Barth and Mason 1984; Brown 1987).

Collection and preservation of samples is dependent on the type of sample required and on the analytical procedures that will be performed on the sample. Each method of analysis requires specific collection methods, sampling containers and storage requirements. These requirements are designed so that no significant changes in the composition of the sample occur before the tests are performed. When sampling for organic compounds and trace metals, special precautions are needed to ensure detection. Often these parameters are present at such low concentrations, that they may be totally or partially lost if the proper procedures are not used. Standard Methods for the Examination of Water and Wastewater (Eaton et al. 2005) is a comprehensive reference book that covers all facets of water and wastewater analytical techniques. Standard Methods is a joint publication of the American Public Health Association, the American Water Works Association, and the Water Environment Federation. The Soil Science Society of America (SSSA) has published the book series, Methods of Soil Analysis, containing the following volumes: Part 1: *Physical and Mineralogical Methods*, Part 2: *Microbiological and Biochemical Properties*, Part 3: *Chemical Methods*, Part 4: *Physical Methods* and Part 5: *Mineralogical Methods*. This series is one of the primary references on methodology in soil science. Part 4 contains information on sampling procedures. Another good reference on soil sampling is a SSSA Special Publication entitled *Soil Testing: Sampling, Correlation, Calibration, and Interpretation* (Brown 1987).

Sampling plans determine the type of sample required for each particular project. The objective of the sampling plan is to ensure that the number and type of samples collected is representative of the "population" under study. The plan designates how many samples are needed, the locations of the samples and the sample depth at each location. The plan may include simple random samples, stratified random samples, systemic samples, sub-sampling and composite samples. The best plan must consider the overall cost and precision (lack of error) of sampling.

Once in the field, it is important to document the sampling operation. Field log books are used to record all information pertinent to field sampling such as: the purpose for sampling; location of the sampling point; date and time the sample was collected; name and address of field contact; procedure for field decontamination of sampling tools between samples to prevent cross contamination, field measurements, and any observations worth noting. A chain-of-custody record should also accompany each sample or group of samples. This record includes: the sample number; signature of the collector; date, time, and address of collection; sample type; signatures of persons involved in the chain of possession; and inclusive dates of possession. Once collected, most samples need to be kept on ice until delivered to the analytical laboratory.

7.2.2 Quality Control and Detection Limits

Analytical laboratories have quality assurance and quality control plans that ensure data accuracy and precision. Quality assurance (QA) is the system that uses procedures and assessments that ensure reliable data. The plan ensures that the best available sample preparation, handling, preservation and storage methods are used as recommended by the appropriate authority. In addition to the above, QA also includes control of the following: calibration and standardization of instruments, preventive and remedial maintenance, proper instrument selection and use, quality laboratory water, clean laboratory environment, replicate analysis, spiking of samples, holding facilities for samples, responsible evaluation of data, and recording and maintaining a quality control (QC) database.

Quality control is the system of practices and procedures that provides the measure of precision, accuracy, detection limits and completeness of the testing facility. Precision measures the degree of agreement among replicate analyses of a sample. It quantifies the repeatability of a given measurement. The precision is calculated as relative percent difference of duplicates. The precision for three or more replicates is estimated by calculating the relative standard deviation (RSD) as

$$RSD = 100 \frac{s}{X}$$

where: s = standard deviation of replicate analysis, and X = mean of replicate analysis.

The best mechanism to evaluate precision is the examination of relative percent difference of duplicate samples in the analytical run. This is expressed in the formula:

$$RPD = 100[(X1 - X2)/\{(X1 - X2)/2\}]$$

where: RPD = relative percent difference, X1 = first observation of unknown X, and X2 = second observation of unknown X.

In analytical chemistry, the detection limit (also called the lower limit of detection or limit of detection) is the minimum concentration of a substance that can be determined with a given level (usually 99 %) of confidence. That is, the true concentration of the substance in question is greater than zero. There are several types of detection limits. For example, some detection limits are set by the manufacturer of a specific piece of analytical equipment. The method detection limit (MDL) is unique in that it is designed for each individual laboratory. A sample containing a known amount of the compound being measured is analyzed by the laboratory seven or more times and the standard deviation of those measurements is determined. The MDL is calculated according to the formula: MDL = Student's t value × the standard deviation.

Sample unknowns are duplicated based on the assay, commonly at the rate of 1 per every 10–20 unknowns, depending on the standard operating procedure. Relative percent differences of 10 % are expected at levels of ten times the method detection limit (MDL) and above. Another mechanism to evaluate precision involves a comparison of a check sample run daily with each batch of samples. If the check sample is run several times during the analytical run, then an estimate of replicability of the run can be obtained. The standard deviation of these results is an estimate of daily precision. The repeatability of the procedure over time can be evaluated by the comparison of the results of this check sample on a day-to-day basis. The pooled standard deviation of the check sample over many days and analyses gives an evaluation of the precision of the method over time.

Accuracy measures the bias in a measurement and can be defined as the degree of agreement of a measurement, X, with an accepted or true value, T. It is usually expressed as the difference between the two values, or as a percentage of the reference value 100 $(X - T)/T$. Accuracy of laboratory measurements are usually defined as percent recoveries of the analyte of interest from matrix spikes, or spike reference material introduced into selected samples of a particular matrix, or by the use of appropriate internationally certified materials. For many projects, percent recoveries of the spiked samples and the laboratory control standards are set at 80–120 %.

The method detection limit is the analyte concentration derived from the method that yields a signal which is large enough to be considered significantly different from the blank with a statistical 99 % probability. The method detection limit is determined by analyzing reagent water fortified at a concentration considered to be two to three times the estimated detection limit. At least seven replicates of this fortified blank are analyzed by the same procedure followed in the determination of unknown samples. The MDL is then calculated using the equation $MSDL = (t) \times (S)$, where $t = 3.14$ (for seven replicates) and $S =$ the standard deviation of the replicate analysis.

Completeness refers to the percentage of valid data received from actual analyses performed in the laboratory. Completeness (C) is calculated as follows: $C = 100 \, (V/T)$; where $V =$ number of measurements judged valid, and $T =$ total number of measurements.

7.3 Background

7.3.1 Characteristics of Wetlands That Promote Biogeochemical Processes

Wetlands are diverse ecosystems and variation is found in topographic position (e.g., slope vs. depression), substrate (organic soils or mineral soils), plant community composition, dominant water source, and hydroperiod. Each of these characteristics affects biogeochemical cycles. One characteristic common to all wetlands is the presence of a water table close to the soil surface for at least part of the growing season. The shallow water table leads first to anaerobic soil conditions and then to reduced soil conditions. A number of biogeochemical pathways proceed only under anaerobiosis or reducing conditions. These pathways play a greater role in biogeochemical cycles in wetlands than in uplands. Many freshwater wetlands display significant temporal variability in water table depth so that anaerobic or reduced soil conditions are present for only a portion of the growing season. In these wetlands, the dominant pathways switch during the year from anaerobic processes to aerobic processes. This temporal variability in soil oxygen (O_2) content promotes some processes such as denitrification as explained below. Water source and landscape position influence inputs and outputs. Sediment loading is a dominant process in riverine wetlands subject to frequent overbank flooding and much of the P inputs will be in particulate form as opposed to ground water driven slope wetlands in which most of the P inputs will be soluble orthophosphates. Hydrodynamics and surface roughness dictate water resonance time. Sedimentation is a more dominant process in a depressional wetland subject to surface runoff than a groundwater driven slope wetland. Mineral soil flats are associated with seasonally saturated hydroperiods and alternating periods of aerobic and anaerobic conditions.

7.3.2 Role of Plants and Microbes

Plants are the dominant source of organic C which supplies the energy for microbially-mediated processes. Microbes are critical to the decomposition of detritus and leaf litter and mediate pathways in the C, N, S, Fe, and Mn cycles. Microbial populations vary with respect to total numbers and species composition according to soil depth and distance from plant roots. These differences are primarily in response to a gradient of available C. The rhizosphere refers to the zone of soil close to and impacted by plant roots. The rhizoplane is the surface of plant roots. Microbial numbers are substantially higher (10- to 100-fold) (Paul and Clark 1996) in the rhizosphere than in bulk soil and are inversely proportional to distance from the roots. The highest microbial numbers, by far, are on the rhizoplane. Plant roots supply most of the C that drives microbial activity in soils. Up to 90 % of fine roots may die and decompose annually in forest soils. In addition, dead root cap cells slough off and supply organic C, and exudates from live roots include readily available C sources (sugars, organic acids), a readily available source of N (amino acids), and growth promoting (and sometimes inhibiting) compounds (Vasilas and Fuhrmann 2011).

7.3.3 Importance of Wetting and Drying Cycles

Soil microbes are critical to the development of anaerobic and reducing conditions in wetland soils. Their activity in turn is impacted by soil moisture conditions. Following the onset of soil saturation, respiration by plant roots and microbes produces anaerobic conditions. Further respiration by microbes produces reducing conditions. For purposes of this discussion we consider reducing conditions to be present when ferric iron (Fe^{3+}) is reduced to ferrous iron (Fe^{2+}). Increased microbial numbers and activity subsequent to rewetting a dry soil are commonly observed and are thought to reflect a temporary increase (pulse) of readily available organic C

(Butterly et al. 2009). The C pulse is thought to result from both the presence of dead microbial cells that accumulated during soil desiccation and the release of previously unavailable organic C sources that resided in the interior of soil aggregates and similarly protected areas. The C released is typically readily available to soil microorganisms and results in increased microbial respiration. Provided O_2 diffusion is restricted as a result of rewetting and sufficient nitrate (NO_3^-) is present, these C pulses can produce sharp spikes in respiratory denitrification (Myrold 2005). Rates of denitrification drop rapidly once C or NO_3^- availability decreases or O_2 availability increases. In fact, N removal from soils due to denitrification is typically greatest when alternating aerobic and anaerobic soil conditions occur frequently. This is because the nitrifying bacteria responsible for converting NH_4^+ to NO_3^- are active only under aerobic conditions, whereas denitrification is dependent on NO_3^- availability (produced during aerobic conditions), presence of easily decomposable C compounds, and lack of O_2 (Vasilas and Fuhrmann 2011). Therefore, the practitioner must be cognizant of these when designing exercises to quantify soil N or C pools, or to quantify rates of processes that contribute to these pools. It is recommended that soil redox potential (Eh) be measured when conducting investigations on biogeochemical processes affected by Eh. The methodology for measuring Eh is presented in *Oxidation-Reduction Processes in Soils*. If nothing else soil moisture conditions during the field assay period should be noted.

7.4 Carbon

7.4.1 Overview

There are six principal C reservoirs in wetlands: plant biomass C, microbial biomass C, soil C (both organic and inorganic), particulate organic C in the water column, dissolved organic C, and gaseous C compounds such as carbon dioxide (CO_2) and methane (CH_4). Often, C in microbial biomass and C in soil organic matter are combined into the category soil organic C (SOC). Carbon is also a major constituent of sedimentary rocks such as coal and limestone. In minerals, it is found predominantly as carbonates, salts of the carbonate ion (CO_3^{2-}) such as calcite ($CaCO_3$). Significant quantities of free carbonates may accumulate in high pH soils in arid climates. In some soils, extensive quantities of C are stored as carbonates (CO_3^{2-}). Public awareness of the C cycle has recently increased due to concerns over global warming which is attributed to the atmospheric increase in greenhouse gases including CO_2 and CH_4. Wetlands can serve as both a source and a sink for C (Kayranli et al. 2010) depending on their age, type, and condition. Some wetlands produce CH_4 (see *Methane Emissions* below). However, most wetlands are characterized by a net retention of organic matter and plant detritus (Mitsch and Gosselink 2000). As such, a critical wetland service is C sequestration-the removal of C (primarily CO_2) from the atmosphere and subsequent long-term storage in a reservoir such as soil organic matter. Disturbance to a wetland, especially in the forms of artificial drainage or deforestation, reverses the net C flow so that disturbed sites initially serve as a source of CO_2.

Because of the impact of O_2 availability on the direction or rate of many biogeochemical reactions, some of the C processes are compartmentalized in specific zones in the soil or water column. For example, CH_4 oxidation occurs in aerobic zones, while methanogenesis is restricted to anaerobic zones (Knight and Wallace 2008). Furthermore, since many of these processes are driven by microbial activity, compartmentalization is further promoted by the availability of organic C as an energy source. For example, the highest decomposition rates are found in close proximity to the wetland surface where there are high inputs of fresh litter and recently synthesized labile organic matter (Sherry et al. 1998) and the highest duration of aerobic conditions.

7.4.2 Primary Productivity

7.4.2.1 Overview

Carbon sequestration refers to the removal of C from the atmosphere and subsequent storage in C sinks such as oceans, forests, and soils. Primary production is the production of organic compounds from CO_2 (atmospheric or aquatic) principally through the process of photosynthesis. Therefore, photosynthesis is integral to C sequestration. The primary producers in wetlands are mainly plants and algae. Net photosynthesis (gross photosynthesis-respiration) can be approximated by assessing biomass. In this section we present methods for determining above-ground biomass for trees and herbs, abscised leaves, and fine roots. Conversion of biomass to C requires a C content value which is obtained from the literature or by chemical analysis of the sampled biomass. Direct chemical analysis will be more accurate as published values will represent averages across species and may not reflect the specific growing conditions of the individual plants in question. Chemical analysis for C content of plant tissue is not presented here. We also do not address biomass and C assessment of shrubs. For this topic we refer the reader to Chojnacky and Milton (2008).

7.4.2.2 Tree Biomass-Allometric Equations

Direct calculations of tree biomass to determine primary productivity or C sequestration is not an option as it requires destructive sampling, determination of dry weight, and chemical analysis for C. However, there are indirect methods that allow for the estimate of tree biomass and C. Above-ground tree biomass can be estimated using a single field measurement, published data, and simple allometric equations frequently in the form of "$M = aD^b$", where M = dry weight of the biomass component, D = diameter at breast height (dbh) (see *Diameter at Breast Height* below), and "a" and "b" are parameters whose specific values are presented in a number of publications. So the practitioner needs to determine dbh and plug its value into the equation with the appropriate parameter values obtained from the literature. Two of the more extensive sources for allometric equations and the parameters are Ter-Mikaelian and Korzukhin (1997), and Jenkins et al. (2003). Ter-Mikaelian and Korzukhin (1997) presented biomass equations for 65 North American tree species based on a literature review. Furthermore, they present equations that address the following biomass components: foliage, branches, stem wood, stem bark, total stem (wood + bark), and total aboveground biomass. The geographic region that generated the data from which the parameter values were derived is presented. Therefore, the practitioner has several equations available for each tree species and should select the equation most closely associated with the site of interest. These equations were developed primarily for timber species, and as such, some wetland tree species may not be represented. It is also likely that the relation between dbh and biomass will be different between an upland situation and a wetland situation. So it should be understood that these indirect methods will give approximate values. However, they are useful for comparing sites.

Biomass can be converted to C either by determining C content on specific samples or by using published values of C content. Carbon analysis should be conducted on tree cores taken at breast height. If bark represents a significant proportion of aboveground biomass, bark should be partitioned from bole wood both in the C analysis and in allometric equations.

7.4.2.3 Diameter at Breast Height

Diameter at breast height is a standard method of expressing the diameter of the trunk or bole of a standing tree. Tree trunks are measured at the height of an adult's breast, which is defined differently in different countries. In the U.S., breast height diameter is measured at a height of 1.4 m. On slopes or in wetlands with pit and mound topography, the soil surface reference point to determine the 1.4 m above ground sampling height may not be obvious. In those situations, the reference point can be set as the highest point on the ground touching the trunk, or set as the average between the highest and lowest points of ground. A consistent approach to setting the reference point is critical.

Diameter at breast height is measured with a diameter (or girthing) tape or calipers. A diameter tape measures the circumference (girth) of the tree; it is calibrated in divisions of π centimeters (3.14 cm) and gives a directly converted reading of the diameter. To determine tree diameter, the tape is wrapped (diameter side facing user) around the tree. Tree diameter is indicated by the alignment of the number "0" aligns with the rest of the tape. Calipers consist of two parallel arms; one is fixed, the other slides along a scale. Calipers are held at right-angles to the trunk with the arms on either side of the trunk. Diameter is directly calculated as the distance between the two arms. Diameter at breast height can also be measured with a Biltmore stick. Although not as accurate as a diameter tape, it is quicker to use. Diameter at breast height is measured by holding the stick at a set distance, usually 64 cm, from the eye, and at breast height. The left side of the stick is lined up with the left side of the tree. The number on the stick that lines up with the right side of the tree is the approximate dbh.

7.4.2.4 Tree Age Determination

In some cases it is of interest to determine the annual rate of woody biomass production. This requires an estimate of tree age. Many tree species increase trunk diameter by producing a single layer of wood each year between the previous year's growth and the bark. In a horizontal cross section cut through the trunk of a tree, these growth bands appear as concentric rings, referred to as growth rings, tree rings, or annual rings. Each ring represents 1 year of growth, so that the tree's age can be determined by counting the rings. The least invasive way to see the tree rings is with an increment borer which takes a small (5 mm diameter) straw-like radial core sample from the tree. An increment borer consists of three parts: handle, steel shaft (core tube or auger), and extractor. Increment borers come in different sizes; the length of the shaft should be at least 75 % the diameter of the tree you are boring. Tree rings should be counted near the base of the tree. For consistency, the boring is commonly taken at breast height (1.4 m). To extract a tree core, the screw tip of the shaft is pressed against the tree and the handle is turned clockwise until the screw bit reaches the center of the trunk. This action forces the core of wood into the tube. The core is extracted by first breaking the core with a counterclockwise one-half turn of the handle. The extractor is then fully slipped through the tube. The core will be removed with the extractor.

There are situations that reduce the accuracy of using ring counts to determine tree age (Avery and Burkhart 2002). One year's growth includes both spring wood (rapid-growing, lighter colored wood) and summer wood (slower-growing, dark colored wood). The method is most reliable when there is a sharp contrast between spring wood and summer wood and for fast-growing coniferous species in northern temperate zones. In tropical or southern temperate zones, tree growth generally does not produce distinctive rings. Some deciduous species produce limited contrast between spring wood and summer wood. Adverse growing conditions, such as drought, result in very narrow rings that are difficult to distinguish. Conversely, a period of favorable growing conditions following a drought can result in *false rings* which represent small growth spurts.

7.4.2.5 Coarse Root Biomass

Destructive sampling of coarse roots (>10 mm diameter) is labor intensive and requires heavy equipment. As such it is not conducive to most wetland investigations. We recommend an alternative approach to estimating coarse root biomass using published allometric equations that depend on dbh. Sources of these equations include Whittaker et al. (1974) and Vadeboncoeur et al. (2007). Carbon content can be determined from coarse roots excavated on site or from core samples taken with increment corers from roots leaving the tree base.

7.4.2.6 Fine Root Biomass

Fine roots (<2 mm diameter) typically contribute <5 % of the total tree biomass (DeAngelis et al. 1981; Vogt et al. 1996). However, it has been estimated that fine root production constitutes about 30–50 % of the C being cycled annually through forest ecosystems (Grier et al. 1981; Vogt et al. 1996). Therefore, fine roots constitute a small but functionally critical fraction of ecosystem biomass. Hertel and Leuschner (2002) consider fine root production and root exudation as the least known processes of the C cycle of forests. However, both processes supply organic C to microbes critical to biogeochemical cycling. In addition, fine root production and turnover may be a sensitive indicator of changing soil environments, and therefore, ecosystem health (Bloomfield et al. 1996).

The sequential root coring method is commonly used to collect fine root biomass data. In this approach, roots are collected from soil cores taken sequentially throughout the year, typically at 1–2 month intervals. It employs a metal tube sharpened on one end which is manually driven into the ground to collect the soil cores. There is no set size for the corer. It should be at least as long as the depth of the soil to be sampled. The wider the corer, the easier it is to extract the soil from the corer while maintaining the integrity of the soil core. This is critical if the soil core is to be divided into sub-samples based on depth. Vogt and Persson (1991) used corers with a diameter of 33 mm and a length of 150 mm. Persson (1978) used a corer with an internal diameter at the hardened steel cutting edge of 6.7 cm while the upper part of the tube had an internal diameter 2 mm larger which facilitates removal of the soil cores by inverting the corer. Typically, the soil core is divided into sub-samples based on horizon. Horizon thickness should be noted prior to sampling so that a sampling volume can be determined. Horizon thickness is determined from an adjacent spade slice as core insertion can compact the soil. This is especially a problem with organic horizons. Thickness of the sub-samples is also determined to allow for the correction of volume lost due to compression when calculating root density. Soil sub-samples are transferred to plastic bags, sealed, and transported to the laboratory for processing. Prior to processing, samples are stored at 4 °C. Live roots can be distinguished from dead roots (necromass) under a dissecting microscope after staining. The necessary techniques are presented by Hertel and Leuschner (2002) and Knievel (1973).

Combining the sequential root coring method with the 'minimum–maximum method' (SC – MM) allows for an estimate of biomass production without distinguishing between biomass and necromass (Edwards and Harris 1977). With the minimum–maximum method, biomass production is calculated as the difference between the minimum value and maximum value of fine root biomass and necromass obtained in the measuring period. Hertel and Leuschner (2002) evaluated four methods that assessed fine root production in a *Fagus* spp. and *Quercus* spp. forest and compared the results with C budget data. Twenty samples each were collected at 4 week intervals over 1 year. The sequential coring or minimum–maximum approach showed the best agreement with the C budget data with an overestimation of 25 %.

7.4.2.7 Quantifying Litterfall

Leaf cages are used to catch abscised leaves from deciduous woody plants so that an estimate of foliage biomass or C can be obtained. The cages can also be utilized to collect abscised flowers and fruit. Techniques for quantifying branchfall are presented by Bernier et al. (2008). It should be understood that senescing leaves typically export simple carbohydrates prior to abscission, so that biomass or C estimates based on abscised leaves will not necessarily equate to foliage biomass or C. Foliage biomass can also be estimated for trees by using allometric equations (see *Tree Biomass* above). Litterfall can also be estimated using the cohort layered screen method (see below). However, the leaf cage approach will provide an estimate of C returned to the forest floor via leaf abscission annually. Commercially available laundry baskets can be used as leaf cages. Holes are drilled into the bottom of the basket and the basket is elevated above the soil surface to promote drainage. In addition, depending on the target species, fine mesh screening may need to be used to cover openings on the sides of the basket. An alternative design consists of an open wooden frame holding window screening. Screening is stapled to the frame to form a box shape open at the top and with four wooden legs. The legs are hammered into the ground to keep the cage in place. The screen bottom is positioned above the soil to allow for air flow and to prevent the leaves from picking up moisture from the soil. The height of the sides should be great enough to prevent leaves from moving out of the frame via wind. Rainfall can leach out soluble carbohydrates from the leaves so that samples should be collected at least weekly or before significant rainfall events. Variability in leaf fall increases with distance from the cage to the target canopy. For shrubs, two cages each halfway between the main stem and the drip line are usually sufficient. For trees, more cages are needed; the appropriate number depends on canopy diameter, and the contents of each cage should be analyzed separately to determine a measure of variability. Sampling designs are addressed by Bernier et al. (2008). The samples are placed in pre-weighed paper or mesh bags and dried to a constant weight at 80 °C. Pre-weighing the bags allows sample dry weight to be obtained without removing the sample from the bag.

Implicit to this technique is the assumption that leaves collected from the cage represent the portion of the canopy equivalent to the surface area of the cage bottom. Therefore, total canopy biomass can be estimated by: (biomass of collected leaves) × (horizontal surface area of canopy) ÷ (surface area of cage bottom). A major limitation to this method is the error associated with extrapolating to a larger area. Wind currents can transport leaves for a considerable distance from the parent tree. The number of cages needed to give a relatively accurate estimate of leaf biomass will depend on the spatial area of the target site and the number of woody species present in the tree and shrub strata. To increase the accuracy, the value for each cage can be weighted by the percentage of the site area shaded by that particular tree or shrub associated with each cage.

7.4.2.8 Herbaceous Biomass

Herbaceous shoot biomass samples can be obtained by clipping at ground level all of the plants within a delineated area. The sample area can be delineated with a square or rectangular frame built from 2.5 cm diameter polyvinyl chloride (PVC) pipe; each piece is connected to the next by a right angle PVC connector. The frame is light in weight and can be easily constructed and broken down in the field. If the site is inundated, the frame should be raised to the surface of the water. This can be facilitated by using three-way connectors and attaching PVC 'legs' to the frame. Another option is to mark the plot with pin flags, but that method does not provide a continuous delineation edge. The samples should be placed in pre-weighed paper or mesh bags and dried to a constant weight at 80 °C. The bags should be pre-weighed so that sample dry weight can be obtained without removing the sample from the bag. Fresh harvested plant tissue will continue to respire and lose weight so samples should be kept on ice until they can be placed in a dryer.

7.4.3 Soil Organic Matter

7.4.3.1 Loss on Ignition

Loss on ignition (LOI) is a relatively simple method of determining soil organic matter (SOM) content as follows (Nelson and Sommers 1996). Pyrex beakers (20 ml) are heated in a muffle furnace at 400 °C for 2 h and then weighed to determine tare weight of the beaker. Air-dried soil samples are ground to pass a 0.4 mm screen. One to 3 mg of dried and ground sample are placed into a tared beaker and heated at 105 °C for 24 h in a drying oven. The beaker is cooled in a dessicator over $CaCl_2$, then weighed to determine the dry mass of the sample. The samples are ignited in a muffle furnace at 400 °C for 16 h, cooled in a dessicator over $CaCl_2$, then weighed to determine weight of the beaker and the ignited sample. All weights should be taken to within 0.1 mg. Organic matter content (%) is calculated as $100 \, (Wt._{105} - Wt._{400})/Wt._{105}$, where $Wt._{105}$ is the sample weight after heating at 105 °C, and $Wt._{400}$ is the sample weight after ignition.

The high temperatures used in this method can cause the loss of structural water from inorganic soil constituents such as hydrated aluminosilicates resulting in weight losses in excess of organic matter content (Nelson and Sommers 1996).

This source of error is most pronounced in subsoils high in clay but low in organic matter (Howard and Howard 1990). This error can be corrected using a series of samples of known C content as described by Nelson and Sommers (1996). For purposes of this discussion, we recommend restricting the LOI method to organic horizons and topsoil where the error is not as large.

The determination of C content of SOM requires complex laboratory assays and or expensive equipment. Several methods are presented by Nelson and Sommers (1996). In lieu of running these assays, the practitioner has two options. Carbon content of SOM has been reported to range from 52 to 58 % (Sparks 1995), but is generally considered to be 58 % (Wolf and Wagner 2005). Therefore, a rough estimate of SOC can be obtained by multiplying SOM values by 0.58. The conversion factor can be fine-tuned by submitting selected samples to a commercial lab for organic C determinations. The selected samples should represent the range in SOM found in the entire suite of samples. It should be clear that the lab analysis is for organic C and not total C as soils formed from calcareous parent materials under arid conditions may contain large quantities of inorganic C (e.g., carbonates).

7.4.4 Organic Matter Decomposition

7.4.4.1 Overview

Biological oxidation of plant tissue and SOM is the principal process that returns terrestrial fixed C to the atmosphere. In the absence of anthropogenic influence, decomposition of plant material before burial is the major pathway for the return of nutrients to the water column. Therefore, decomposition not only returns C to the atmosphere but also supplies nutrients to macrophytes. In this section, we present two approaches (three methods) for assessing organic matter decomposition. One approach is to directly measure biomass losses in leaf litter over time (litter bags and the cohort layered screen method). A second approach (cotton-strip assay) is to measure the loss in tensile strength of cotton fibers. Although an indirect assay, this method has the benefit of using a standardized substrate that allows for the detection of soil effects independent of differences in substrate characteristics. Organic matter decomposition rates are much lower under anaerobic conditions than under aerobic conditions. Therefore, when utilizing any of the following methods, we recommend that soil Eh be documented as temporal differences or spatial differences may solely be a soil moisture effect.

7.4.4.2 Litter Bags

Litter bags are commonly used to determine decomposition rates of plant tissue in terrestrial ecosystems (Aber and Melillo 1980; Wieder and Lang 1982). Plant material of a known mass is placed in mesh bags which are placed in the field and randomly retrieved at predetermined intervals. Bag size is commonly 20 cm × 20 cm. Nylon mesh or fiberglass screening material is used so the bags themselves are not subject to decomposition. Harmon and Lajtha (1999) recommend fiberglass bags for light intensive sites where UV light can degrade nylon. Litter bags typically have a mesh size of 1–2 mm (Robertson and Paul 1999), although mesh sizes from <1 mm to >10 mm have been used. If the mesh is too small, access to some macroinvertebrates may be denied. Large mesh sizes facilitate the loss of small particulate matter. One option is to use a small mesh, but staple the edges of the bags at relatively large intervals (e.g., 5 cm) to provide openings along the periphery for access of macroinvertebrates. Regardless of mesh size, low molecular weight organic compounds can be lost through leaching. The practitioner must be aware of implications to leaching losses and the bag deployment period should reflect that. Decomposition rates will be highest when the soil is moist but not saturated. Leaching losses will be greatest during periods of soil saturation. Therefore, the bags should not be deployed in continuously saturated soils.

In general, at least five replicate litter bags are collected at each sampling interval during year 1 of the study (Karberg et al. 2008). Sites displaying significant microclimate variability may require more replicates. For example, wetlands with pit and mound topography exhibit spatial variability in soil moisture and soil temperature and would require greater replication. Certainly, the researcher has the option of adjusting replicate number in subsequent years. Sample material is chopped into 2–5 cm lengths and a known amount of fresh plant tissue is placed in the bags. Subsamples of the plant material are dried (70 °C, 48–72 h) to obtain their water content. The organic matter of interest (e.g., leaves or fine roots) is placed in nylon mesh bags. The bags are then placed where the organic material would normally be found. For example, abscised leaves would be placed on the soil surface, fine roots would be placed in organic soil horizons or in the topsoil, and detrital tissue would be placed on the soil surface or in the detrital layers. One advantage of this system is that material can be collected from the site in question and returned to its natural environmental conditions. The filled bags should be returned to the site soon after sample collection to ensure representative environmental conditions. Bags are pre-weighed (tared) so that sample weight can be determined in the bag. Loss of biomass due to decomposition is calculated as the difference between initial biomass and remaining biomass. All values are expressed on a dry weight basis. Average rate of decomposition (per day) is determined by dividing biomass loss by the incubation period. However, since biomass decreases over time, a more accurate estimate of decomposition rate is produced with exponential decay equations (see Karberg et al. 2008). If a chemical analysis is conducted on the tissue before and after the incubation period, N and P mineralization rates can also be determined.

The litter bag method also can be used to estimate fine root decomposition rates (Fahey et al. 1988). Fine roots can be collected by the sequential coring method (above) or with a spade. Soil residues are removed by rinsing, and root samples (2 g fresh weight per bag) are placed in litter bags (nylon, 10 × 10 mm, mesh size 1.2 mm). Subsamples of the root material are dried (70 °C, 48 h) to obtain the water content of the fresh roots. Bags are typically left in the field for 2–3 months, although the incubation period will depend on soil and air temperature, soil moisture, and biomass composition. After collection, bags are transported to the laboratory and the samples are gently rinsed to remove soil residues and fungal hyphae, and then dried and weighed.

There are two general approaches to analyzing the data. If the intent is to compare treatments such as litter species composition, or to compare sites, an analysis of variance is performed on the percentage of initial dry mass remaining at time t. If the intent is to determine decomposition rate constants, mathematical models are fitted to the data to describe biomass loss over time. Both single exponential decay models and double exponential decay models have been frequently used to describe organic matter decomposition. The single exponential decay model is based on the assumption that the relative decomposition rate remains constant over time. The double decomposition decay model is based on the assumption that litter has two distinct components, an easily decomposed (labile) fraction and a more recalcitrant fraction. Therefore, each fraction requires a separate decay rate constant. Wieder and Lang (1982) present a critique of these analytical methods.

7.4.4.3 Cohort Layered Screen

The cohort layered screen method is an inexpensive approach to assessing long-term (≥ 3 years) litter decomposition. As presented by Karberg et al. (2008) aluminum or fiberglass window screening (1 m × 1 m, 2–3 mm mesh) is placed over the forest floor following the major annual litterfall. An additional layer of screen is placed over the screen from the previous year following each annual litterfall. The litter is held in place by the screens and decomposes *in situ*. Sections of the screen are cut out to supply samples of the decomposing litter which is then dried and weighed. Dry weights are then compared to stand level estimates (see *Quantifying Litterfall*) for the year in question. One benefit to this approach is that the litter sample is naturally representative of the site, as opposed to the litter bag method where the practitioner chooses the litter sample. The cohort layered screen method does have several limitations in common with the litter bag method. Certain macrofauna may be denied access to the litter which can alter decomposition rates. Also both techniques do not allow for the separation of true decomposition from losses attributed to leaching and comminution. Leaching losses are especially a concern in wetlands. Since the cohort layered screen method is intended for long-term deployment, it is not appropriate for wetlands that exhibit long-term periods of inundation.

7.4.4.4 Cotton-Strip Assay

The decomposition of cellulose strips has been used extensively as a surrogate for plant organic matter decomposition including in a variety of wetlands (Newman et al. 2001). The assay quantifies cellulose decomposition on the basis of the reduction in tensile strength of cellulose fibers, referred to as cotton tensile strength loss (CTSL), of a standardized cotton fabric. Since the assay uses a standardized cotton fabric (97 % holocellulose; Shirley Institute Test Fabric, Didsbury, England) (Latter and Harrison 1988), it provides a method for normalizing substrate quality between sites (Harrison et al. 1988).

Cotton strips can be inserted vertically into the soil with a flat spade or sharpshooter shovel that is at least as wide as the strip. One end of the strip is trapped between the blade edge and the soil surface. The spade is then inserted into the soil, pulling the strip with it. Two sets of strips are used. One set is inserted and removed immediately. These serve as control or reference strips. The remaining strips are left in the soil for one to several weeks, depending on the expected rate of decomposition. Upon removal from the soil, strips are immediately washed in freshwater to remove debris and soil and then washed again in deionized water. The sample strips are dried at room temperature and then stored in plastic bags. The strips are cut into 3 cm wide horizontal segments and reduced to 2 cm segments by fraying. Segments are used to accommodate soil variability with depth that may impact decomposition rates. Tensile strength is measured from each segment with a tensometer (e.g., Monsanto Type-W) equipped with 7.5 cm wide jaws adjusted to 3 cm spacing. Temperature and humidity affect the results so all measurements should be are carried out at 18–22 °C and 100 % relative humidity (facilitated by soaking the strips in deionized water). Individual losses in tensile strength are calculated relative to the reference strips for each site.

Walton and Allsopp (1977) presented the benefits of this assay: (i) cellulose is a major component of plant remains; (ii) the decomposition of dead plant remains is a major biological process; (iii) cellulose provides a major food source for a wide variety of soil organisms; (iv) cotton is a natural substrate; and (v) degradation of any organic material begins with bond breaking, leading to changes in tensile strength. However, different litter constituents do not decompose at the same different rate (Minderman 1968) and Howard (1988) concluded that the rate of breakdown of pure cellulose added to soil cannot reflect litter decomposition rate. Walton and Allsopp (1977) concluded that this technique is best employed for comparative assessments of biological activity in different soils.

7.4.5 Soil Respiration

7.4.5.1 Overview

Soil respiration, or more accurately soil surface CO_2 efflux, is the release of CO_2 from the soil surface to the atmosphere. It results primarily from respiration by plant roots and soil microorganisms and may comprise 50–80 % of ecosystem respiration (Davidson et al. 2002; Giardina and Ryan 2002). Conceptually, respiration reflects substrate decomposition in soils and in many texts soil respiration is included in

SOM decomposition sections. Also, soil surface CO_2 efflux has been used to compare decomposition rates in different soils. But, soil surface CO_2 efflux is an index of respiration by soil organisms and plant roots (Zibilske 1994). As such, it is not directly comparable to SOM degradation.

There are three main approaches to assessing soil respiration-closed chamber systems, open chamber systems, and flux gradient sensors. Chamber-based approaches are simple, economical, and portable. A chamber is placed over the soil to create a headspace of air, which can be sampled repeatedly over a short time period. Closed chamber systems are the most commonly used and commercially available. They are classified as "closed" because there is no exchange of air between the chamber and the outside atmosphere during measurements. Closed chamber systems may be "active" or "static". In dynamic systems, air is continuously circulated between the chamber and an infrared gas analyzer. In static systems, air samples from the chamber are collected with a syringe for laboratory analysis or CO_2 is absorbed by soda lime in the chamber. In open chamber systems, air is exchanged between the chamber and the outside atmosphere. In the flux gradient approach, infrared sensors are inserted into the soil at various depths. The CO_2 concentration gradient over soil depth and additional soil characteristics are used to calculate CO_2 diffusivity. Bradford and Ryan (2008) present an evaluation of the relative benefits and challenges to each system. In this section, we present the soda lime method, a common and relatively simple method of measuring soil respiration that utilizes a static closed chamber approach. In the section *Methane Emissions*, we include methods that can also be used for quantifying CO_2 fluxes.

7.4.5.2 Carbon Dioxide Detection by Soda Lime Absorption

In this method, an open dish containing soda lime is placed on or just above the soil surface and covered with a container to restrict airflow between the soda lime and the atmosphere. Carbon dioxide and water vapor released by the soil microorganisms during decomposition are absorbed by the soda lime. After drying to remove water, the gain in soda lime dry weight during the exposure period reflects the amount of CO_2 evolved. The following specifics were detailed by Zibilske (1994).

The soda lime jar must consist of oven-safe glass with air-tight screw caps. A 5.5 cm diameter jar is suggested. Jar supports (for stability) consist of 12 cm square pieces of galvanized mesh, bent down at each corner to form four 2-cm legs. For jar covers, they used cylindrical cans (28 cm diam., 25 cm height), open at one end. The exact dimensions are not critical, but each container should have an opening at least 600 cm^2 and a total volume of at least 15,000 cm^3. To prevent the jar covers from direct exposure to sunlight in the field, they can be painted white, covered with aluminum foil, or by placing a 50 cm^2 flat board on top of them in the field. The amount of soda lime (6–12) mesh needed for each jar should be slightly greater than 0.06 g for each cm^2 of soil surface area covered by the jar cover container.

Each jar with the cap on is weighed. Soda lime is placed in the jars and dried to a constant weight in a drying oven at 100 °C for 24 h. The jars are re-capped, cooled, and reweighed to determine the amount of dried soda lime in each jar. A minimum of five replicate chambers are used at each location to accommodate spatial variability. Any vegetation or debris that would interfere with the formation of a tight seal between the cover and the soil surface is removed, but without disturbing any leaves and the soil surface under the chamber. The wire mesh legs are pressed into the soil to produce a stable surface. After opening, each soda lime jar is immediately placed on a mesh stand and covered. The lip of the cover is forced into the soil with a twisting motion. A weighted object such as a rock may be placed on the cover to keep it in place and maintain the soil surface seal. Controls are needed to account for any CO_2 absorbed during this part of the procedure. To construct controls, a jar of soda lime is left open for the same amount of time required to deploy the sample jars (from opening to covering) and then covered. For every ten sample jars, two control jars are used. The control jars will be used to produce blanks. Incubation is commonly for 24 hours (h). Retrieved sample jars are tightly and quickly capped; then dried without caps in a 100 °C oven for 24 h, capped, cooled, and weighed. The absorption of CO_2 generates water which is removed during drying of the soda lime. To account for this, the weight gain determined after drying is multiplied by 1.4. All necessary calculations are presented by Zibilske (1994). The data are commonly expressed as mass per unit area per unit time (e.g., g $CO_2/m^2/h$).

Absorption of water vapor is needed to activate the soda lime after drying. However, it should not come in direct contact with surface water. For this reason, this technique is not appropriate for sites with deep inundation. If need be, the jar supports can be constructed with larger legs to keep the jars above surface water. In addition, to compare soils they should have similar water content when sampled as soil moisture content impacts respiration. Therefore, deployment should coincide with a period of stable soil moisture conditions. We recommend that soil moisture content (dry weight basis) be determined for topsoil adjacent to each sampling point. Take four soil samples equally spaced from each other and 0.5 m from each sampling point and combine for soil moisture determinations. Temperature also affects respiration so air temperature should be taken at a height of 0.5 m and soil temperature should be taken at a depth of 5 cm.

7.4.6 Methane Emissions

7.4.6.1 Overview

Methane is of environmental concern as it has been implicated in global warming. Methanogenesis is the utilization of CO_2 as a terminal electron acceptor to produce CH_4, sometimes referred to as "swamp gas". Methane is produced by a distinct group of obligate anaerobic bacteria (methanogens) only under very reduced (redox potential less than -100 mV) conditions. Methane is oxidized to CO_2 by a group of aerobic bacteria (methanotrophs). Although the two competing microbial processes require extremely different redox potentials, both can occur simultaneously in the same soil. For example, CH_4 produced in an anaerobic subsurface horizon can diffuse into an aerobic surface horizon where it is converted to CO_2. Therefore, field attempts to quantify CH_4 emissions in actuality measures net CH_4 fluxes.

Many of the techniques for assessing gaseous emissions from wetlands work equally well for CO_2 as for CH_4. Most of the techniques depend on chamber-based approaches introduced in *Soil Respiration* above. One major difference is in the options for measuring the concentrations of the respective gases in the sample. Portable infrared gas analyzers are commercially available for the instantaneous measurement of CO_2 in the field. Equivalent technology is not readily available for CH_4, so gas samples must be stored and transported to the lab for analysis of the gas through gas chromatography.

7.4.6.2 Static Closed Chambers

The method presented here employs a static-chamber approach to measure CH_4 evolution from soils. It is discussed in detail by Weishampel and Kolka (2008). This approach may be used for other gases (e.g., CO_2) evolving from soil. However, there are subtle differences in the design, construction and deployment of chambers depending on the target gas. The reader is referred to Livingston and Hutchinson (1995) for a discussion of these factors. It is conducive to spatially intensive sampling exercises in wetlands or uplands. This method employs a static enclosure system comprised of collars that are permanently installed in the ground and portable chambers designed for syringe sampling that fit over the collars. The permanent nature of the installed collars maintains a tight seal with the soil surface and minimizes disturbance effects associated with collar installation during periods of gas flux measurement. Both collars and chambers are made of 25 cm diameter, schedule 40 PVC pipe. Methane flux is calculated from the change in concentration during the incubation period (period of chamber deployment in the field). The required calculations are presented by Weishampel and Kolka (2008).

It is critical that sediments are not disturbed during instrumentation or sampling as it can impact gas fluxes so wooden pallets may be needed to accommodate foot traffic. This is especially a concern for saturated or organic soils. In some instances, the collars are hammered partially into the ground. However, hammering can cause significant compaction to organic soils or very wet soils and should be avoided in those instances. Collars should also be installed at least 1 week prior to sampling to minimize any impact from soil disturbance.

Altor and Mitsch (2008) described a static chamber design for measuring CH_4 and CO_2 emissions from freshwater marshes. Chambers were constructed of PVC chamber frames and circular, high density polyethylene (HDPE) bases (0.27 m^2), and transparent 4-mil (0.1 mm thick) polyethylene bags. The frame consisted of a circular top and three legs. Frame heights were 50 cm for sampling points without macrophytic vegetation and 150 cm for sampling points with macrophytic vegetation. Polyethylene bags were in place only during the sampling process. The bags were pulled down over the chamber frames and attached to the base with elastic straps. The bags should be constructed so that they fit snugly over the frame so that the volume of air sampled is consistent. The top of each bag was equipped with a butyl rubber sampling septa and a 3 m Tygon vent tube (1.6 mm inside diameter [i.d.]).

7.4.6.3 Scaling CH$_4$ Fluxes

Measurements of CH$_4$ fluxes from wetland soils typically reveal high spatial and temporal variability. The number of chamber measurements needed and the labor required to carry out this degree of sampling may preclude this approach to characterizing seasonal CH$_4$ fluxes on a landscape scale. Another approach is to rely on modeling or remote sensing methodologies that link CH$_4$ fluxes to more easily measured ecosystem properties or processes. For example, soil temperature, water table depth and range, community structure, and net primary productivity have been used in CH$_4$ flux models for wetlands (Potter 1997; Potter et al. 2006).

7.4.7 Dissolved Organic Matter

7.4.7.1 Overview

Dissolved organic matter (DOM) is operationally defined as the fraction of organic matter that passes through a 0.45 μm filter and is a heterogeneous mixture of compounds including carbohydrates, proteins, lignins, organic acids and other humic substances (Herbert and Bertsch 1995; Kalbitz et al. 2000). The fractions of DOM that contain functional groups with C and N molecules are generally classified as dissolved organic C (DOC) and N (DON). Wetlands have typically been identified as the largest sources of DOM in watersheds (Aitkenhead-Peterson et al. 2003; Mulholland 2003). Concentrations of DOC for wetlands have been observed to range from 3 to 400 mg/L with an average of 30 mg/L (Thurman 1985). Among wetland types, bogs (3–400 mg/L) have been found to yield the highest DOC values, while marshes represent the lower range (3–15 mg/L) (Thurman 1985). The elevated contents of DOM in wetlands can be attributed to a variety of factors including: (a) high primary productivity of wetlands compared to upland and aquatic ecosystems (Thurman 1985); low decomposition rates of organic matter in wetlands due to acidic conditions and anaerobic or low O$_2$ contents of wetland soils and surface waters (Kalbitz et al. 2000); reducing redox conditions that result in reductive dissolution of Fe and aluminum (Al) oxides that could otherwise have served as sorption sites for DOM (Kalbitz et al. 2000); flooding and hydrologic conditions that facilitate the slow continuous leaching of DOM; and surficial hydrologic flowpaths that bypass mineral rich sorption surfaces present in subsurface soil horizons (Inamdar et al. 2011, 2012).

Past studies have generally focused on determining the bulk concentrations of DOC and DON in wetland soils and watershed runoff (Hinton et al. 1997; Inamdar and Mitchell 2006; Raymond and Saiers 2010). While these observations have been important in highlighting the significant role of wetlands for DOM, bulk DOM concentrations provide little information on the reactivity, bioavailability, molecular size, and mobility of DOM. To get an idea of these ecologically relevant characteristics of DOM, we need to know the functional groups or the individual constituents such as carbohydrates, proteins, carboxylic acids, lignins that make up DOM. For example, labile fractions of DOM that are easily consumed by microbes are found to be rich in carbohydrates and proteins (Benner 2003). Aromatic and humic-rich fractions of DOM play a preferential role in the complexation and transport of metals such as cadmium, arsenic, and mercury. Similarly, aromatic compounds of DOM are predisposed to forming carcinogenic disinfection by-products when water is chlorinated for drinking purposes (Nokes et al. 1999). Hydrophobic DOM compounds are preferentially sorbed on Fe and Al oxides in soils while hydrophilic DOM molecules remain in solution and move with runoff waters (Jardine et al. 1989; Kaiser and Zech 1998; Ussiri and Johnson 2004). Thus, to truly understand the fate and transport of DOM in watersheds and its implications for terrestrial and aquatic ecosystems we need to move beyond bulk determinations to characterizing the chemical constituents of DOM.

7.4.7.2 Characterizing Dissolved Organic Matter Using UV and Fluorescence Spectroscopy

In the past, DOM composition or characterization of functional groups have usually been performed using traditional chemical techniques that are labor-intensive, time-consuming, involve high analytical costs, and require large sample volumes. These challenges have precluded the routine characterization of DOM composition for many ecosystem and watershed studies. The recent advances in ultra-violet (UV) (Weishaar et al. 2003) and fluorescence technology (Coble et al. 1990; McKnight et al. 2003), however, overcomes some of these challenges and promises to be a useful tool for characterizing DOM chemistry, especially for studies that generate a large number of DOM samples.

Ultra-violet and fluorescence techniques rely on the property of DOM that different organic molecules absorb and reflect light at differing wavelengths. Thus, investigation of the absorption and fluorescence spectra can provide critical insights into the composition of DOM. It needs to be emphasized here that these spectrofluorometric procedures do not provide information on the concentration or chemical structure of the DOM functional groups but the proportion of fluorescence contributed by specific DOM moieties can be determined through post-processing of the spectra. Furthermore, only a small of fraction of the DOM pool responds to UV and fluorescence measurements (McKnight et al. 2003). Despite these constraints, UV and fluorescence approaches to characterizing DOM have exponentially increased in recent years (Cory and McKnight 2005; Fellman et al. 2009; Helms et al. 2008; Inamdar et al. 2011, 2012; Jaffé et al. 2008; Miller and McKnight 2010; Wilson and Xenopoulos 2009) including some excellent reviews on the subject (Cory et al. 2011; Fellman et al. 2010).

The UV absorption spectra for DOM is generally obtained using a standard spectrophotometer equipped with a 1 cm path-length quartz cuvette (volume of 4 ml) over the 190–1,100 nm wavelength range at 1-nm intervals. Prior to the sample spectra, the instrument is set up and corrected for scattering and baseline fluctuations by using deionized (DI) water. The absorption spectrum for DOM follows an exponential pattern with a decrease in absorption with increasing wavelength. Some of the key UV metrics that are derived from this spectrum and which have been used to characterize DOM are reported in Table 7.1. The UV metric that has been most commonly reported is the specific UV absorbance (SUVA) which is computed by dividing the decadic UV absorbance at 254 nm by the concentration of DOC (mg C/L) (Weishaar et al. 2003). SUVA has been found to be strongly and positively correlated with aromatic content of DOM as determined by ^{13}C-NMR (Weishaar et al. 2003). SUVA values can however be influenced by the pH, nitrate and dissolved iron (Fe) content of the sample and appropriate screening and corrections need to be applied (Weishaar et al. 2003). Since Fe absorbs light at 254 nm, elevated concentrations of Fe (>0.5 mg/L) in the DOM sample can lead to incorrect (high) SUVA values (Weishaar et al. 2003). A metric similar to SUVA, the absorption coefficient at 254 nm (a_{254} in m) is also calculated by using the naperian UV absorption coefficient (Green and Blough 1994). The a_{254} also provides a measure of aromaticity but without normalization to DOC (Helms et al. 2008). Another UV index, the spectral slope ratio, S_R, is calculated as the ratio of the slope of the shorter UV wavelength region (275–295 nm) to that of the longer UV wavelength region (350–400 nm) (Helms et al. 2008) and is obtained using linear regression on the log-transformed spectral ranges (Yamashita et al. 2010). The spectral slope ratio, S_R has been found to be inversely related to the molecular weight of DOM (Helms et al. 2008).

In fluorescence spectroscopy, DOM samples are exposed to light in a fluorometer for a range of excitation wavelengths and the corresponding emitted wavelength and light intensity is recorded (Lakowicz 1999). The matrix of the fluorescence intensities that is generated is referred to as the excitation-emission matrix (EEMs), an example of which is illustrated in Fig. 7.1. Prior to generating fluorescence scans and deriving meaningful indices from the EEMs a number of important steps need to be performed such as correcting for instrument bias, diluting samples with high absorbance values (e.g., A254 $\geq$ 0.2) and applying corrections to account for inner-filter effects (McKnight et al. 2003). Once the EEMs are generated a variety of fluorescence indices can be generated by using the ratios of fluorescence intensities from specific regions (wavelengths) of the EEM matrix. In addition, EEMs can be further analyzed using rigorous multivariate statistical tools such as parallel factor analysis (PARAFAC, Stedmon et al. 2003) that decomposes the EEMs matrix into chemically and mathematically distinct components with the

Table 7.1 Selected UV and fluorescence metrics that have been commonly used to characterize the composition of dissolved organic matter (DOM)

UV and fluorescence indices	Reference	Definition and significance
Specific UV absorbance (SUVA$_{254}$) [L mg C^{-1} m^{-1}]	Weishaar et al. (2003)	UV absorbance at 254 nm divided by DOC concentration in mg C/L; provides a measure of aromaticity of DOM. High values of SUVA indicate more aromatic material
Absorption coefficient a_{254} [m^{-1}]	Green and Blough (1994)	(UV absorbance at 254 nm) × 2.303 × 100 Measure of aromaticity of DOM
Slope ratio S$_R$	Helms et al. (2008)	Ratio of the slope of the shorter UV wavelength region (275–295 nm) to that of the longer UV wavelength region (350–400 nm); Can be used as a proxy for molecular weight (MW) S$_R$ decreases with increasing MW
Fluorescence Index (FI)	McKnight et al. (2003)	Ratio of fluorescence intensities at 470 and 520 nm at excitation of 370 nm; Used to distinguish between terrestrial and microbial sources of DOM; Terrestrial or allochthonous DOM: 1.2–1.5; Microbial or autochthonous DOM: 1.7–2.0
Humification Index (HIX)	Ohno (2002)	HIX = aI435 – 480/(aI300 – 345 + aI435 – 480) Used to characterize humification status of DOM; Ranges from 0 to 1 and increases with increasing degree of humification
Freshness Index (β:α)	Wilson and Xenopoulos (2009)	Ratio of emission fluorescence intensity at 380 nm by the maximum emission fluorescence intensity observed between 420 and 435 nm, calculated at excitation wavelength of 310 nm. A measure of recently produced DOM, where β represents DOM of recent origin and α represents more decomposed DOM
Redox Index (RI)	Miller et al. (2006)	$Q_{red}/(Q_{red} + Q_{ox})$, where Q_{red} is the sum of the reduced components and Q_{ox} is the sum of the oxidized components; provide a measure of the redox state of DOM
Humic-like fluorescence	Coble et al. (1998)	Excitation <260; Emission 448–480 Indicates DOM from vascular plant sources, high molecular weight and humic in nature
Protein-like fluorescence	Coble et al. (1998)	Tyrosine-like: Excitation <270–275; Emission 304–312; Tryptophan-like: Excitation <270–280; Emission 330–368; Indicates protein-like DOM moieties, bioavailable DOM (Fellman et al. (2009)), and DOM of microbial origin

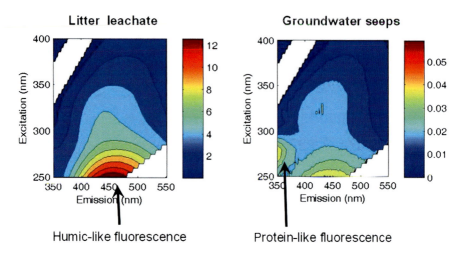

Fig. 7.1 Fluorescence excitation-emission matrices (EEMs) for dissolved organic matter (DOM) from two mid-Atlantic, USA watershed sources (forest floor litter leachate and groundwater seeps) illustrating humic-like and protein-like fluorescence peaks (Modified from Inamdar et al. (2012). Published with kind permission of © Springer Science + Business Media B.V 2012. All Rights Reserved)

relative contribution of each component to the total DOM fluorescence (Cory and McKnight 2005).

Some of the more common fluorescence metrics and PARAFAC components that have been used in recent studies are reported in Table 7.1. The fluorescence index (FI) is calculated using the ratio of fluorescence emission intensities at 470 and 520 nm at an excitation wavelength of 370 nm (Cory and McKnight 2005). McKnight et al. (2003) have used the FI to differentiate between DOM derived from vascular plants (FI: 1.3–1.4) versus microbial or planktonic sources (FI: 1.7–2.0). Another fluorescence metric, the humification index (HIX) was defined (Zsolnay et al. 1999) as a ratio of the peak integrated area under the emission spectra 435–480 nm by the peak integrated area under the emission spectra 300–445 nm; obtained at excitation wavelength 254 nm. This definition was later modified by Ohno (2002) to the integrated areas under the emission spectra 435–480 nm divided by the peak integrated area (300–345 nm + 435–480 nm); again obtained at excitation wavelength 254 nm. The revised equation of Ohno (2002) constrained HIX values to the range of 0–1 with higher values indicating more humified material. Another metric referred to as the freshness index ($\beta:\alpha$; Wilson and Xenopoulos 2009) is computed as the ratio of emission fluorescence intensity at 380 nm by the maximum emission fluorescence intensity observed between 420 and 435 nm, calculated at excitation wavelength of 310 nm. The freshness index is a measure of recently produced DOM where β represents DOM of recent origin and α represents more decomposed DOM. Miller et al. (2006) proposed the redox index (RI) to explain the oxidation states of DOM fluorescence defined as $Q_{red}/(Q_{red} + Q_{ox})$, where Q_{red} is the sum of the reduced components and Q_{ox} is the sum of the oxidized components. Obviously, high values of RI represented DOM components of reduced origin while low values represented oxidized DOM components. In addition to these indices, EEMs have also been used to identify fluorescence peaks in specific regions indicating humic-like and-or protein-like fluorescence (Coble et al. 1990; Cory and McKnight 2005). The humic-like fluorescence is typically assumed to represent DOM from vascular plants, with high molecular weight and aromatic in nature (Coble et al. 1998). In contrast, the protein-like fluorescence is assumed to represent DOM of low molecular weight, representative of amino acids like tryptophan and tyrosine (Yamashita and Tanoue 2003), composed of DOM that is more bioavailable (Fellman et al. 2009), and DOM that may be of microbial origin (Hood et al. 2009).

7.5 Nitrogen

7.5.1 Overview

Inorganic forms of N prevalent in wetland biogeochemical cycles include dinitrogen gas (N_2), ammonium (NH_4^+), ammonia (NH_3), nitrate (NO_3^-), nitrite (NO_2^-), nitric oxide (NO), and nitrous oxide (N_2O). Wetlands receive N inputs via atmospheric deposition, transport in the water column, and biological dinitrogen fixation. Dinitrogen fixation is the conversion of N_2 to NH_3 which is then rapidly converted to organic forms. Dinitrogen fixation occurs in algae, free living bacteria, and bacteria in symbiosis with macrophytes such as legumes. Nitrogen mineralization is the conversion of organic forms of N to inorganic forms (primarily NO_3^- and NH_4^+) during SOM decomposition. The reverse process is referred to as immobilization. Nitrification is the conversion of NH_4^+ to NO_3^-. Inorganic N is removed from wetlands via leaching losses, lateral transport in the water column, and gaseous losses to the atmosphere via denitrification and NH_3 volatilization. Denitrification is the microbial reduction of NO_3^- to gaseous products (primarily N_2O and N_2) which are returned to the atmosphere. Denitrification represents a significant path of N loss from wetlands and it is considered to be one of the more important wetland functions as it contributes to water quality by removing nitrates. Under high pH conditions (pH >8) NH_4^+ is converted to NH_3 which may be volatilized.

Many of the N techniques presented below contain a laboratory component to determine the N content of soil, water, or plant tissue. These chemical assays are beyond the scope of this chapter. The reader is referred to the following references: Bremner (1996), for total N; Mulvaney (1996), for inorganic N; and Stevenson (1996), for organic N.

7.5.2 Stable Nitrogen Isotopes

Many elements of biological interest, including C, H, O, N, and S have at least two or more stable (non-radioactive) isotopes. For a given element, the lightest isotope is present naturally in much greater abundance than the others. Nitrogen isotopes ^{15}N and ^{14}N have similar chemical characteristics and therefore behave almost identically in biological systems. The mass differences, however, result in partial separation of the two isotopes during chemical reactions and during physical processes such as diffusion. This separation of isotopes is referred to as isotope fractionation and results in a higher $^{15}N/^{14}N$ ratio in soils and in water systems than in the atmosphere. Levels of ^{15}N are commonly expressed as atom % excess relative to a standard or baseline level. The background level or baseline for each isotope is usually considered to be equal to atmospheric levels and expressed on a atom percent (At.%) basis: At.%$^{15}N = 0.3663$, At.%$^{14}N = 99.6337$. Atom % excess ^{15}N is any quantity of ^{15}N above this background level. For example, soil organic matter with an At.% ^{15}N value of 0.4773 would have an At.% excess ^{15}N value of 0.1110 (0.4773–0.3663). Individual components of the N cycle can be labeled by enriching it with ^{15}N. This allows one to trace the fate of N from individual pools. Fertilizer enriched with ^{15}N is commercially available. Adding ^{15}N-enriched fertilizer to the soil will label the soil N pool. The addition of organic matter with a high C/N ratio or the addition of sucrose along with the fertilizer will quickly result in a labeled soil organic N pool. Plants grown on the labeled substrate will also become labeled. Labeled plant tissue and labeled soil N can then be used to determine N_2 fixation or N mineralization rates.

7.5.3 Dinitrogen Fixation

7.5.3.1 Stable Isotope Techniques

Dinitrogen fixation in macrophytes such as legumes can be determined by two techniques that rely on the presence of the two stable N isotopes. The natural abundance technique is based on the naturally ^{15}N enriched soil N pool. With the isotope dilution technique, the soil N pool is labeled by additions of ^{15}N enriched fertilizer or ^{15}N enriched organic materials (Vasilas and Ham 1984). With both methods dinitrogen fixation is calculated by comparing the $^{15}N/^{14}N$ ratio in the fixing species with the $^{15}N/^{14}N$ ratio in a control (non-fixing) species. In the fixing species N derived from the soil is diluted with respect to ^{15}N by N derived from the atmosphere via N_2 fixation. The isotope dilution technique is considered to be more sensitive (Weaver and Danso 1994). However, it is not appropriate for non-managed ecosystems where control of the soil N pool is not an option. In wetlands, a further drawback to the isotope dilution method is the presence of a

7 Wetland Biogeochemistry Techniques

periodically high water table which would result in significant losses of the fertilizer N through denitrification.

In both methods, tissue samples from both species (N_2-fixing and non-fixing) are collected, dried, and ground. Nitrogen in the ground tissue subsamples is converted to NH_4^+–N through acid digestion which is then oxidized to N_2. The N isotope ratio of the N_2 sample is determined by mass spectrometry or emission spectrometry. There are commercial labs that will run N isotope analyses on ground tissue samples. The only data needed from the control species is the N isotope ratio. To determine quantities of N fixed, biomass and total N content must be determined for the fixing species. Because the differences between the $^{15}N/^{14}N$ ratios in the atmosphere and that found in soils are very small, there are procedural differences between the isotope dilution method and the natural abundance method (Weaver and Danso 1994). First, in the natural abundance method, isotopic composition is expressed as $\delta^{15}N$ instead of At.% ^{15}N, where $\delta^{15}N = [(R_{sample} - R_{standard})/R_{standard}] \times 1{,}000$, and $R = {}^{15}N/({}^{14}N + {}^{15}N)$, so that 1 $\delta^{15}N$ unit = 0.00037 At.% excess ^{15}N. Second, it is recommended that a member of the fixing species be grown hydroponically on N-free medium to provide an estimate of discrimination that occurs during N_2 fixation. Third, multiple non-fixing plant species should be sampled to provide a mean of the isotopic composition of the soil. With the natural abundance method, the percentage of plant N derived from fixation (%Ndfa) is calculated as follows: %Ndfa = $(x - y)/(x - f) \times 100$, where $x = \delta^{15}N$ of the non-fixing plants; $y = \delta^{15}N$ of the fixing plant grown in soil; and $f = \delta^{15}N$ of the fixing plant grown hydroponically.

7.5.3.2 Acetylene Reduction

Nitrogenase, the enzyme that reduces N_2 to NH_3 (N fixation), will also reduce acetylene (C_2H_2) to ethylene (C_2H_4). The acetylene reduction method has been used to provide a point in time assessment of N_2 fixation in both symbiotic systems (e.g., nodulated legumes) and free living organisms. The following method was presented by Carpenter et al. (1978) to assess N_2 fixation by free living bacteria and algae in tidal marshes. They used surface cores taken from vegetated marsh areas to target algae and sediment slurry samples taken from pannes to target cyanobacteria. A plastic corer is used to take a surface core (0.5 cm^2 dia., 0.25 cm deep); a pipette is used to remove 1 ml of slurry. Each sample is placed in a 6.5 ml wide mouth serum bottle, the bottles are capped, and injected with 1 cc high-purity C_2H_2 gas. The samples are gently rotated to facilitate solution of the C_2H_2, and incubated for 1 h. A gas sample is than removed and analyzed for C_2H_4 by gas chromatography.

Limitations to the acetylene reduction method include the point in time nature of the assay and the indirect nature of the assay which makes extrapolation to the amount of N_2 fixed questionable. Seasonal rates of N fixed cannot be determined from a point in time measurement because the rate may vary over time. Quantifying N_2 fixed from the amount of C_2H_2 reduced requires a conversion factor. Hardy et al. (1968) first suggested a theoretical conversion factor of 3 mol C_2H_2 reduced to

1 mol N_2 reduced (equivalent to 1.5 mol C_2H_4 per mol NH_3). More commonly, a conversion factor of 4 mol C_2H_2 reduced to 1 mol N_2 reduced is used (Boddey 1987). Hardy et al. (1973) suggested that a conversion factor should be experimentally determined for each system.

7.5.4 Nitrogen Mineralization and Nitrification

7.5.4.1 Overview

Nitrification is an aerobic process and N mineralization is more rapid under aerobic conditions than under anaerobic conditions. Field assays for estimating net rates for N transformations were developed for upland conditions. However, they are appropriate for some types of wetlands if care is taken so that the incubation period does not correspond to large fluctuations in water table depth. The results will be most reliable if soil moisture content is relatively constant during the incubation period. They are not appropriate for situations of continuous soil saturation as the rates of these transformations will be too low to be accurately measured.

Precise estimates of N mineralization or nitrification in the field are difficult to obtain and beyond the scope of most field exercises. Nitrate can be removed from the soil by plant assimilation, leaching, and dissimilatory nitrate reduction including denitrification. Plant uptake can be eliminated by using some type of soil containment device that keeps roots away from the sampled soil. Denitrification and other dissimilatory nitrate reduction effects can be minimized by avoiding saturated soil. Leaching effects can be minimized by preventing percolation during the incubation period. However, N mineralization and immobilization occur simultaneously. Therefore, most reported figures for N mineralization or immobilization reflect net gains or losses to a soil N pool. For example, net N mineralization = $(NH_4^+-N + NO_3^--N)_{t+1} - (NH_4^+-N + NO_3^--N)_t$, and net nitrification = $(NO_3^--N)_{t+1} - (NO_3^--N)_t$ (Hart et al. 1994). Gross rates of these processes can only be determined through isotopic techniques.

7.5.4.2 Buried Bag Method

The buried bag method is a simple technique for quantifying net N mineralization and nitrification in the field. It is most appropriate for assessing surface soils; for the assessment of subsurface soils the reader is referred to the closed-top, solid cylinder method (Hart et al. 1994). Intact soil cores are taken with a coring device (PVC or metal tubes), placed in polyethylene bags, sealed, returned to the original hole, and incubated in the field for 1–2 months. The plastic bags used in this procedure are permeable to gases but not to liquids (Gordon et al. 1987). One advantage to this method in comparison to lab incubation assays is that the soil samples are subjected to on-site temperature regimes. However, since the bags are impermeable to water, the technique integrates the on-site soil water dynamics only if the soil water content at the beginning of the incubation period is representative of the entire incubation period. The following specifics to the assay were drawn from Hart et al. (1994).

Soil cores are extracted with thin-walled PVC or metal cylinders, sharpened at 1 end to provide a cutting surface. Recommended cylinder dimensions are 5 cm i.d. and 12 cm length. Longer cylinders allow for deeper soil samples but are more difficult to remove intact cores from the cylinders and require larger incubation bags. Cylinders are traditionally handled in groups of three. Cylinders are hand driven into the soil until 2 cm of the cylinder is above the soil surface. Three cylinders should be used for each 50 m^2 of area; these serve as replicates. Cylinders are removed and the soils are removed from the cylinders resulting in a 0–10 cm sampling depth. Soils from three cylinders are composited for pre-incubation analysis. These will serve to provide baseline (time 0) data for initial concentrations of NH_4^+ and NO_3^-. Additional soil samples are taken (pre- and post-incubation) to determine gravimetric soil water content. As the laboratory analysis is conducted on field moist soil, soil moisture content is needed to express the data on a dry weight basis. Each of the remaining soil samples are enclosed in polyethylene bags (15–30 μm thick), the bags are sealed with plastic ties, and returned to their respective bore holes. The bags are covered with leaf litter, if present. If not, the bags should be covered with a small amount of similar soil to prevent temperature extremes. Bags are removed after a 1 or 2 month incubation period. Samples with perforated bags should be discarded. Consecutive buried bag incubations will provide seasonal patterns of net N mineralization and nitrification. Microbial activity will continue after the bags are retrieved. Therefore, the samples should be kept cool (2–5 °C) during short-term storage (≤2 d) and frozen for long-term storage. Inorganic N is extracted from soil subsamples with 2M KCl in the lab. See Hart et al. (1994) for the sample preparation and extraction procedures. Filtered extracts are analyzed for NH_4^+–N and NO_3^-–N (see Mulvaney 1996). Net N mineralization is calculated as the change in inorganic N (NH_4^+–N and NO_3^-–N) content over the incubation period; net nitrification is calculated as the change in NO_3^-–N content over the incubation period. Net P mineralization can also be determined with this method.

7.5.4.3 Resin Core Method

The *in situ* intact-core resin-bag method was developed to measure N mineralization and nitrification under the existing field soil conditions with respect to temperature and water content (Distefano and Gholz 1986). This system consists of an intact soil core within a PVC or metal cylinder with an ion exchange resin bag at each end. This allows water but not ions to flow through the soil core and eliminates the static moisture regime inherent to the buried bag method. No resin bag is needed at the top of the cylinder if the soil core does not receive leachate from overlying soil horizons or is exempt from inundation.

Wienhold et al. (2009) presented a modified version of the ion resin exchange technique in which they use a resin bag at the bottom only. A metal cylinder

(4.75 cm i.d.) was inserted 17 cm into the soil and immediately extracted, encasing the soil core. Two centimeter of the encased soil were removed from the bottom of the cylinder and replaced with a nylon resin. The bottom of the cylinder was then covered with sturdy nylon cloth to prevent root entry and the cylinder assembly was reinserted into the original hole. They found that incubation periods greater than 60 days resulted in resulted in loss of inorganic N from resins, and recommended 28- to 40-day incubations. However, they conducted their study in uplands; loss of inorganic N from resins would probably be greater under saturated conditions. They also cautioned against compacting the soil below the cylinder during installation as that can impede water movement through the soil core.

7.5.5 Denitrification

7.5.5.1 Overview

Several approaches have been used to estimate denitrification, including *in situ* measurements using natural N isotopic abundances (Søvik and Mørkved 2008), *in situ* measurement of N_2O evolution without an C_2H_2 block (Jordan et al. 2007; Whalen 2000; Wray and Bayley 2007), *ex situ* measurement of N_2 and-or N_2O evolution from intact cores without an C_2H_2 block (Horwath et al. 1998; Wray and Bayley 2007), *ex situ* measurement of N_2O evolution from intact cores using an C_2H_2 block (Bohlen and Gathumbi 2007; Horwath et al. 1998; Hunt et al. 2003), *ex situ* measurement using intact cores coupled with isotopic approaches (Racchetti et al. 2011; Rückauf et al. 2004), and *ex situ* measurement of denitrifying enzyme activity (DEA) in homogenized soil with or without using an C_2H_2 block and optimized incubation conditions (Bruland et al. 2009; Hunt et al. 2003, 2007; Jordan et al. 2007; Sirivedhin and Gray 2006). The methods perhaps best suited to routine estimations, and the two considered here, are (1) *ex situ* measurement of N_2O evolution from intact cores using an C_2H_2 block, and (2) *ex situ* measurement of DEA in homogenized soil with using an C_2H_2 block and optimized incubation conditions.

Using C_2H_2 block eliminates problems with measuring N_2 evolution against high atmospheric background levels of the gas or the necessity of using a non-N_2-containing atmosphere during incubations. Acetylene blocks the conversion of N_2O to N_2, thus simplifying analyses and providing a more sensitive assay. Intact cores can arguably give more realistic estimates of denitrification than does DEA, but are typically more variable both spatially and temporally. In comparison with using intact cores, estimates of denitrification based on DEA using homogenized soil are generally less variable but may give unrealistically high values, especially if the usual practice of optimizing incubation conditions is used (i.e., anaerobic atmosphere, presence of excess C and NO_3^-). It is for this reason that the values obtained are commonly considered potential rates of denitrification and are viewed as representing primarily maximum relative rather than absolute estimates. Thus, the investigator will need to decide which of the two approaches best meets the needs of the situation at hand.

We have included a third method, water column analysis for inorganic N, which represents a very different approach. Nitrogen removal from the water column can be calculated by comparing the inorganic N level in the input water from the N level in the output water. This is an indirect approach; it does not directly assess denitrification and there are other pathways for inorganic N removal (i.e., macrophyte assimilation). Furthermore, total N loss can only be determined if a hydrologic budget is created. However, relative differences in water quality services capacity between wetlands can be characterized by this approach.

7.5.5.2 Measurement of Denitrification in Intact Soil Cores using an Acetylene Block

This method is based on Bohlen and Gathumbi (2007), modified from Horwath et al. (1998) and Mosier and Klemedtsson (1994). A number of 2-cm-diameter soil cores are taken with a hammer corer fitted with cylindrical plastic inserts to the desired depth, the number of cores dictated by the resources available and the expected experimental variability. The plastic insert should be only partially filled, leaving room for sampling headspace gasses. The bottom end of the plastic insert is immediately sealed with a rubber stopper, and the cores kept at 4 °C until analyzed (<24 h). The cores are allowed to equilibrate briefly at ambient or some standard temperature. The top of the plastic cylinder is sealed with a stopper fitted with a rubber septum and C_2H_2 (See comments) is added to a concentration of ~10 kPa using a syringe. The C_2H_2 is distributed throughout the cylinder by repeated pumping of the headspace gasses with a large (~60-ml) syringe. Samples of headspace gasses are taken at intervals (e.g., after 2 and 6 h; see comments) and placed in evacuated vials fitted with rubber septa. Gas samples, including blanks and standards, are analyzed by gas chromatography using a Poropak Q (or equivalent) packed column in combination with an electron capture detector. For examples of specific chromatographic run conditions, refer to Hunt et al. (2007), Jordan et al. (2007), and Sirivedhin and Gray (2006). The amount of N_2O produced between selected sampling times is used to estimate denitrification.

7.5.5.3 Denitrification Enzyme Activity (DEA)

The following method is taken from Jordan et al. (2007) as modified by Tiedje et al. (1989). A weighed amount of soil (~25 g) is placed in a 125-ml flask containing 25 ml of a solution of 1 mM glucose, 1 mM KNO3, and 1 g/L chloramphenicol. The chloramphenicol is used to prevent *de novo* synthesis of denitrification enzymes. The flask is flushed with N_2 or another inert gas such as Ar and closed with a rubber stopper fitted with a gas-sampling septum. The flask is injected with 11 ml of C_2H_2 and incubated for a standard period of time (e.g., 2 h; see comments) at ambient or some standard temperature. Samples of headspace gasses are then taken by syringe and transferred to evacuated vials fitted with rubber septa. Gas samples, including blanks and standards, are analyzed as described for intact cores.

7.5.5.4 Comments

Core dimensions, soil sample size, and sampling times are given as guidelines only and can be modified to meet specific needs and situations. Acetylene used for the C_2H_2 block should not be obtained from commercial (e.g., welding) tanks, as there are reports that these can contain acetone (Jordan et al. 2007). Analytical grade C_2H_2 and C_2H_2 produced by reacting CaC_2 with water have been used successfully. Regarding measurement of DEA specifically, some investigators have chosen to agitate the assay flasks during incubation rather than using static flasks (e.g., Hunt et al. 2007). Additionally, some investigators (Hunt et al. 2003, 2007) used DEA approaches under both optimized (added NO_3^- and glucose) and non-optimized conditions and with and without an C_2H_2 block, in an effort to obtain estimates of denitrification having varying degrees of relatedness to true denitrification rates; the authors conceded that the values they obtained even under non-optimized conditions may still have been overestimates because an O_2-free atmosphere was used during the assay incubations (Hunt et al. 2007).

7.5.5.5 Water Column Analysis

Critical to this approach is the assessment of the hydraulic gradient to identify hydrologic input sources and outputs for sampling. For constructed wetlands, identification of hydrologic inputs and outputs is simple and straightforward. For natural wetlands, the identification of hydrologic input and output sources is more problematic and may require the expertise of a soil scientist. For groundwater driven slope wetlands or depressions, a hydraulic gradient is easily characterized. Slope inputs are easily identifiable as seeps (side slope or toe slope). For depressions, a soil investigation is required to identify the dominant hydrologic source. Often a gravel or sand lens representing a zone of high hydraulic conductivity will extend from the upland into the depression. Slope wetlands usually have apparent outlets; depressions may or may not have outlets. For mineral soil flats, elevations should be shot with the assumption that surface elevations correspond to water table elevations. For surface water driven systems, a flume is needed to collect overland flow.

Inflow points and outflow points are the water sampling points. At each point, a water sampling well is installed. Slotted PVC pipe (i.d. 5 cm) is inserted into a bore hole which is then backfilled with coarse sand or pea gravel and capped with bentonite (see Chap. 3). In some cases, it may be desirable to separate surface water from sub-surface water, or to separate two discontinuous water columns (i.e., episaturation). In those cases, two sampling wells must be installed at each sampling point.

Chemistry of water in the well casing may not be identical to that of the water column. Prior to sampling water, the wells are purged to remove stagnant water. Wells are emptied and the water discarded prior to sampling the wells after they refill. This is known as "purging the wells". As a general guideline, 3–5 casing volumes are purged, where a "casing volume" is defined as the volume of water standing in the well (Barackman and Brusseau 2004). In some cases, recharge of the well is so slow that the water sample must be collected a day after purging. A number of portable battery driven pumps are available for purging and sampling the wells. These pump the water through tubing inserted into the well. To avoid contamination, deionized water should be pumped through the tubing between samples. This method is not appropriate for all wetlands, and it should not be used to compare different types of wetlands. For seasonally saturated wetlands, there may not be sufficient well water to sample during parts of the year.

7.6 Phosphorous

7.6.1 *Overview*

Phosphorous is generally considered to be the major limiting nutrient in most freshwater wetlands (Mitsch and Gosselink 2000) determining the rate of primary production and decomposition. However, when present at very high concentrations, P stimulates macrophyte and algae growth and the resulting increase in the rates of decay of the vegetation depletes O_2 levels in the water. Wetlands represent both a source of P and a sink for P. Furthermore, an individual wetland can switch from a P sink to a P source as its trophic state changes. The trophic state of an aquatic ecosystem is defined by its rate of primary production, the concentration of nutrients and minerals and the rate of supply of organic matter (determined by the balance of primary production and decomposition) (Correll 1998). Oligotrophic systems have low rates and concentrations, while eutrophic conditions exist when the ecosystem is over-enriched with nutrients and minerals. As the flux of P into the wetland increases, primary production and trophic state increase while dissolved O_2 and biodiversity decrease (Correll 1998). These changes in ecosystem processes interfere with the wetland's ability to retain P and wetland soils can become saturated with orthophosphate (OP), after which P may be released to the water column resulting in eutrophication of receiving surface waters (Champagne 2008; Vepraskas and Faulkner 2001).

The major pool of P in natural wetlands is found in the soil sediments (fixed mineral P) and the litter (organic P) comprising approximately 80–90 % of the P within the wetland (Champagne 2008; Richardson and Craft 1993; Vepraskas and Faulkner 2001). The remaining fraction of P is found in the water column and the pore water as soluble OP, dissolved organic P and total dissolved P (Vepraskas and Faulkner 2001). Phosphorus in wetlands can be found as soluble and insoluble complexes in both organic and inorganic forms. The principle inorganic form is OP, which is present in the pH-dependent ionic species; $H_2PO_4^-$ (2 < pH < 7), HPO_4^{2-} (8 < pH < 12) and PO_4^{3-} (pH >12) (Kadlec and Knight 1996). However, OP can react with soil constituents to form both insoluble and soluble compounds. In acidic soils (4–7 pH) under aerobic conditions, OP forms insoluble precipitates with ferric Fe and aluminum oxides and hydroxides. As these soils become first anaerobic and then reduced, dissolution of the precipitates releases P into the soil pore water and water column. In alkaline soils (pH >7), precipitation of insoluble Ca or Mg phosphates or OP adsorption to carbonates occurs (Richardson and Craft 1993; Vepraskas and Faulkner 2001). The soluble inorganic OP is considered the biologically available form of P. Organic forms of P are found in plants, partially decomposed plant tissue, and found as OP bound in organic matter.

Phosphorous has a complex role in wetland ecology. Because of its dual role as both a limiting nutrient and as a major pollutant, especially with its role in eutrophication of surface waters, a number of approaches have been developed to assess different forms of P or different P pools in wetlands. Most of the assays are specific to soil or water. Sequestration is addressed by assays for soil P content. To address P as a limiting nutrient, two approaches are commonly used-a chemical extraction scheme that separates soil P into various fractions representing a range in bioavailability and a mineralization assay. A similar fractionation approach is used to evaluate water quality. As the major eutrophic surface water nutrient, both soil and water assays have been developed that take a bioavailable P approach and target OP. We will not go into detail on the wet chemistry techniques for assessing P. These methods are presented in detail by Kuo (1996) for soils and Eaton et al. (2005) for water. Kovar and Pierzynski (2009) (http://www.sera17.ext.vt.edu/Documents/P_Methods2ndEdition2009.pdf) present methods for soil and water including methods specific to runoff water and flooded soils.

7.6.2 Soil Phosphorous Content

7.6.2.1 Total P

Many methods have been developed to extract and analyze total P in soil; four of the more common include: sodium carbonate (Na_2CO_3) fusion, perchloric acid ($HClO_4$) digestion, sulfuric acid-hydrogen peroxide-hydrofluoric acid (H_2SO_4–H_2O_2–HF) digestion, and sodium hypobromite (NaOBr) oxidation followed by H_2SO_4 dissolution. All four methods convert soil organic P to inorganic P (Kuo 1996), but the Na2CO3 fusion method is considered to be the most reliable (Syers et al. 1967). It is believed that the acid digestion methods fail to extract P from apatite inclusions or imbedded in the matrix of silicate minerals (Syers et al. 1967) and therefore underestimate total P. Three of the methods (Na_2CO_3, H_2SO_4–H_2O_2–HF, and $HClO_4$) are described by Kuo (1996).

7.6.2.2 Total Inorganic P

Inorganic P is extracted with 1 M hydrochloric acid (HCl) or H_2SO_4. Not all of the mineral forms of P may be solubilized with this extraction leading to an underestimation of inorganic P.

7.6.2.3 Total Organic P

Three methods are commonly used to determine total organic P in soils (Reddy and DeLaune 2008) as follows. The methods for quantifying P in extracts are found in Kuo (1996).

1. Difference method: Organic P is calculated as the difference between total P and inorganic P. This approach is best suited for soils high in organic matter; it tends to overestimate organic P in samples high in mineral forms of P.
2. Acid-alkaline extraction method: First P_i is extracted with acid (1 M HCl or H_2SO_4); this is followed by an extraction with alkali (0.5 M NaOH). The alkali extracts are digested to solubilize organic P. This method may underestimate organic P as the extraction procedure may not be sufficient to solubilize all forms of organic P.
3. Ignition method: Organic P is removed by ashing the sample at 550 °C. Inorganic P in the residue is extracted with acid. Organic P is calculated as the difference between total P in the unignited soil and inorganic P in the residue.

7.6.3 Soil Phosphorous Fractionation Schemes to Assess Availability

Orthophosphorus is considered the most bioavailable form of P and it constitutes the bulk of P found in soil pore water. Orthophosphate is readily adsorbed onto surfaces of clay particles and organic matter, and OP can also substitute for silicate within clays (Mitsch and Gosselink 2000). Adsorption of OP onto wetland soils and organic matter is considered long-term P retention. Phosphorus availability is characterized by the concentration of P in the soil solution and by the P buffering capacity that governs the distribution of P between the solution and solid phases. Some of the adsorbed P can be released (desorbed) into soil solution where it is readily bioavailable. Therefore, it is useful to separate exchangeable P, which is potentially bioavailable, from the non-exchangeable P. Because of the complex processes that govern the distribution of P between the solution and the solid phase within wetland soils, P tests have been developed in order to evaluate P availability.

A sequential chemical extraction scheme is commonly used to determine soil P pools in soil to approximate bioavailabilty. This approach is based on differential solubilities in a series of chemical extractants. A number of extraction schemes have been developed for soil P. We offer a simplified scheme below that was presented by Reddy and DeLaune (2008). Most are more complex than the example presented here but they allow for further fractionation. For example, some schemes utilize a filter to separate soluble P from particulate forms of P. Also, extraction schemes have been developed to target soils with specific characteristics (e.g., calcareous soils or organic soils). In each step the resulting extract is assayed for OP. Details on these analytical procedures are presented by Kuo (1996).

1. Extraction with salt solutions (e.g., KCl): This extraction removes soluble and exchangeable P which are considered bioavailable.
2. Extraction with alkali: Residue from Step 1 is treated with 0.1 M NaOH which removes P bound to Fe and Al, and some organic P. This P pool is generally considered to be unavailable but Fe bound P may become available under anaerobic conditions. A more complex extraction sequence can be used to separate Al-P from Fe-P and organic P.
3. Extraction with acid: Residue from Step 2 is treated with 0.5 M HCl which removes Ca and Mg bound P which is considered to be unavailable. The residue contains organic P and stable forms of inorganic P that are considered unavailable.

Soluble P in water and soil extracts can be determined without pretreatment colorimetrically or through instrumentation methods including inductively coupled plasma (ICP) spectrometry and ion chromatography. Ion chromatography measures OP; ICP measure inorganic P and organic P and it is used for soil extraction extracts. Colorimetric methods measure primarily OP. Most commonly used is the ascorbic acid method which measures OP and a small fraction of the condensed phosphates; standard methods calls them "reactive phosphorous". In the ascorbic acid method (Eaton et al. 2005), OP reacts with ammonium molybdate and potassium antimonyl tartrate in an acid medium to form phosphomolybdic acid, which is reduced to an intensely colored molybdenum blue by ascorbic acid. The intensity of the blue color is measure on a spectrophotometer and the concentration is determined from an individually prepared calibration curve.

7.6.4 Phosphorus Fractionation Methods for Reduced Soils

Phosphorous fractionation methods were developed primarily for agricultural soils in an attempt to estimate fertilizer needs. Wetland soils may present an additional challenge because P chemistry in soils and sediments is strongly influenced by the oxidation-reduction status (redox potential). Ferric and manganic oxides and hydroxides provide major adsorption sites for P under oxidized conditions. In addition, ferric and manganic phosphate minerals which form and persist under oxidized conditions become unstable under reduced conditions releasing soluble P into the soil solution (Moore and Reddy 1994). Oxidation of anaerobic sediments results in the rapid conversion of Fe^{2+} to Fe^{3+}. Within minutes, $Fe(OH)_3$ precipitates out of solution and has a tremendous capacity to sorb P, causing soluble P levels in the porewater to be reduced by orders of magnitude. Therefore, it is critical that samples collected from reduced soils and sediments are kept reduced during the sampling procedure, transport to the lab, storage, and the initial phases of P fractionation. Moore and Coale (2009) present methods that specifically address these concerns.

7.6.5 Phosphorous Mineralization

Chemical extraction procedures produce an estimate of available P at one point in time. P mineralization studies produce estimates of available P over time. Phosphorous mineralization rates can be assessed either *in situ* with litter bags (see *Litter Bags* above) or in the lab using an incubation approach. In the incubation approach, soil samples are kept under controlled climate conditions in the lab and leached at set time intervals. The two mineralization approaches would not be expected to produce the same results. The benefit to the litter bag approach is that mineralization is assessed under natural fluctuating conditions of moisture and temperature. It does not totally reflect real life conditions as the substrate is usually fresh plant material and the material is physically separated from the soil. Incubation studies typically use soil samples, a more realistic substrate, but employ optimum temperature and moisture conditions. In addition, soil structure is disturbed as the soil samples are typically mixed with sand to facilitate drainage during leaching. The benefit of the incubation approach is that it allows the comparison of different soils under identical environmental conditions so that inherent differences in mineralization potential can be determined.

Bridgham et al. (1998) utilized the following incubation procedure to estimate P mineralization potential for a series of wetland soils. One advantage to their approach is that they could estimate mineralization potential under both aerobic and anaerobic conditions. This allowed them to determine maximum and minimum mineralization rates for the given temperature. Field moist soil was mixed with acid washed sand to promote drainage during leaching. Samples were placed in 150 ml Falcon filter units and incubated at 30 °C. For aerobic incubations, the samples were exposed to ambient air. For anaerobic incubations, the duplicate samples were placed in filter units which in turn were placed in 500 ml Mason jars filled with water. Samples were leached with 0.001 M $CaCl_2$ at ten dates ranging from 2 to 59 weeks. Both the leachate and the Mason jar incubation water were analyzed for PO_4^{3-}.

7.6.6 Phosphorous in Water

There are many tests for P in water, most use a chemical or physical fractionation scheme as presented below. The P fraction analyzed in each step is based on whether the sample is digested and/or filtered, and the nature of the digestion. In each step, P concentration is determined by the ascorbic acid method. Particulate P is separated from the dissolved (or soluble) fraction by passing the water sample through a filter, typically a 0.45 μm cellulose (Millipore) filter. In each step, the resulting extract is assayed for OP by the ascorbic acid method. Phosphorous measured on undigested samples is considered to be inorganic, predominantly OP. Acid hydrolysis digestion converts inorganic P (primarily condensed phosphates) to OP. Some of the phosphate from organic compounds found in the water may be released but this can be reduced by selecting the right acid strength, the right boiling temperature and the hydrolysis time. Oxidative digestion converts both organic and inorganic forms of P to OP. Organic P is calculated as the difference between total P and inorganic P. Analytical details for this scheme are presented in Eaton et al. (2005).

1. Dissolved acid hydrolyzable P (DAHP): OP determined on filtered samples after acid hydrolysis.
2. Total acid hydrolyzable P (TAHP): OP determined on unfiltered samples after acid hydrolysis.
3. Total dissolved P (TDP): OP determined on filtered samples after oxidative digestion.
4. Total P (TP): OP determined on unfiltered samples after oxidative digestion.
5. Dissolved organic P (DOP): DOP = TDP − DAHP.

7.6.7 Biologically Available Phosphorous

7.6.7.1 Overview

The term "biologically available P" (BAP) is used in a number of disciplines without a standardized meaning. However, in part due to the recognized importance of P in eutrophication of surface waters, BAP has been operationally defined as "the amount of inorganic P a P-deficient algal population can utilize over a period of 24 h or longer" (Sonzogni et al. 1982). The amount of BAP in soil, sediment, and water has been routinely quantified by algal assays or chemical extractions (see Sharpley (2009) for a review of methods). Algal assays require long incubation periods and chemical extraction times. Results from the chemical extraction approach are impacted by the nature of the extractant. Weaker chemical extractants (e.g., NH4F and NaOH) approximate P that is bioavailable under aerobic conditions, whereas stronger extractants (citrate-dithionite-bicarbonate) represent P that may become bioavailable under reducing conditions (Sharpley 2009). Because of these limitations, there has been increased interest in a different approach-one that utilizes P sinks.

The P-sink approach to assessing BAP is based on a chemical sink that attracts only P in forms that would be available to macrophytes and algae. In theory because the P sinks continuously remove dissolved P from the soil solution (Kuo 1996) which creates a concentration gradient, they simulate P removal from soil or water by plant roots and algae. This is not a completely accurate analogy because a chemical sink cannot approximate the rhizosphere influence as root exudates and mycorrhizal fungi can alter P availability. Two of the more common P-sink methods utilize anion exchange resins and Fe oxide impregnated filter paper. These are addressed below.

7.6.7.2 Anion Exchange Resins

Anion exchange resins represent the most common P-sink approach for soils. Anion exchange resins remove dissolved phosphates from the soil solution via surface adsorption (Kuo 1996). The rate of P adsorption is a function of solution P concentration. The resin promotes a low level of solution P concentration to maintain P desorption from the soil. Anion exchange resins can be added directly to soil suspensions, placed in a polyester bag, or the resin may be impregnated onto a plastic membrane. Typically, the procedure utilizes chloride-saturated resin, bicarbonate (HCO_3^-) saturated resin, or a combination of the two. The soil sample and resin are shaken together in deionized water or weak electrolyte 16–24 h. Adsorbed P is extracted with HCl and the P concentration is determined by the ascorbic acid method. For details on this procedure see Kuo (1996).

7.6.7.3 Iron Oxide Method-Runoff

The FeO method is a unique approach to assess the potential for runoff to increase fresh-water eutrophication. The use of iron-oxide (FeO) coated paper to test soil was first reported by Sissingh (1983), who sought to develop a P test to estimate plant-available P in tropical soils without mobilizing other forms of phosphates. Sissingh (1983) created a strip of filter paper impregnated with iron hydroxide which adsorbed mobile P from solution. The main advantage of this approach over standard soil P tests is that the FeO paper functions as an ion sink and doesn't react with soil as will chemical extractants. It has been shown that P extracted by this method (FeO-P) from runoff sediment is correlated to algal growth (Sharpley 1993a, b). A major benefit of the FeO method is its capability to differentiate between soluble inorganic P from FeO-P in sediment of runoff. Sediment FeO-P is considered bioavailable particulate P (BPP) and is calculated as follows: BPP = total BAP – SP; where total BAP is total FeO-P from unfiltered runoff, and SP is soluble inorganic P in filtered runoff (0.45-μm filter).

The FeO method has a stronger theoretical justification for estimating P availability of runoff for plants and algae than do chemical methods (Sharpley 1993a) as chemical methods may mobilize additional forms of P which are not available to plants or algae. Although algae uptake of P is restricted to OP, organic forms of P can undergo mineralization and become available (Correll 1998) and thus be considered a latent source of BAP. There has been discussions focused on methods to restrict hydrolysis of organic P adsorbed onto FeO paper (Robinson and Sharpley 1994). However, adsorption and hydrolysis of organic P is not considered a problem with the FeO method as organic P may be classified as latent BAP which may be readily mineralized, thereby becoming available for algal use.

The FeO paper is made from filter paper immersed first in $FeCl_3 \cdot 6H_2O$ and then in NH_4OH. Paper preparation is described in detail by Myers et al. (1997). The water of interest is sampled and taken to the lab. FeO paper is enclosed within two polyethylene screens to maintain integrity during shaking. The polyethylene screens hold the FeO paper in a fixed orientation during shaking. This helps to prevent contamination from soil particles lodging in the pores of the paper (Myers et al. 1995, 1997) and prevents the paper from sticking to the walls of the shaking vessel, which could reduce adsorption effectiveness of the paper. The FeO paper-screen assembly is inserted into a bottle containing the runoff sample. Both deionized water and a $CaCl_2$ solution (0.01 M) has been used as the shaking matrix for the FeO paper and soil. Deionized water can disperse soil, causing it to lodge in the pores of the filter paper (Sissingh 1983), leading to errors in P analysis (Myers et al. 1995). The need for $CaCl_2$ probably depends upon the clay content of the sediment, the P content of the clay, and the amount of sediment in the runoff. The bottles are shaken on a reciprocating shaker to facilitate adsorption of P to the paper. Phosphorous is extracted from the papers by the addition of H_2SO_4 and further shaking. An aliquot of the H_2SO_4 solution is analyzed for P after neutralization of acidity. For further details on this procedure, see Myers et al. (1995, 1997).

7.7 Sulfur

7.7.1 Overview

Sulfur exists in both organic and inorganic forms. Inorganic forms can be classified as gaseous (e.g., sulfur dioxide-SO_2, and hydrogen sulfide-H_2S), reduced (e.g., elemental S-$S°$, sulfide-S^{2-}), or oxidized (e.g., sulfate-SO_4^{2-}, and sulfite-SO_3^{2-}). Common S containing minerals are gypsum ($CaSO_4$) and pyrite (FeS_2). In wetland soils, S is found mainly in organic forms (Reddy and DeLaune 2008). Under highly reduced conditions (Eh <100 mV), obligate anaerobes use SO_4^{2} as a terminal electron acceptor to produce H_2S. Sulfate reduction dominates anaerobic decomposition in brackish marshes, inhibiting CH_4 production and regulating soil C storage (Megonigal et al. 2003). Soluble S^{2-} at levels found in estuarine marshes can be detrimental or toxic to many organisms, and it has been shown to limit the growth of common marsh grasses including *Spartina alterniflora* Loisel (Koch et al. 1990; Mendelssohn and McKee 1988).

Detrimental environmental impacts occur when reduced S compounds (like sulfides) are oxidized to form sulfuric acid (H_2SO_4). Acid mine drainage refers to the outflow of acidic water from mines. The acidity, due to oxidation of metal sulfides like FeS_2, and subsequent dissolution of heavy metals can cause fish kills and loss of vegetation. Acid sulfate soil is the common name given to soils and sediments exhibiting these problems. Soils containing Fe sulfides (the most common being FeS_2) are referred to as potential acid sulfate soils, but when they oxidize and begin to generate acidity, they are referred to as active acid sulfate soils (Fanning et al. 2010).

Potential acid sulfate soils form under continuously anaerobic conditions and are commonly found in coastal wetlands. When these wetlands are drained, exposure of the soil to O_2 results in the formation of H_2SO_4 creating very acidic conditions (pH <2) and often releasing toxic quantities of Fe, aluminum and heavy metals such as arsenic. Because of the toxic effects of free S^{2-} and the potential for detrimental effects upon oxidation of FeS_2, the rapid assessment of soluble S^{2-} in soil pore water is valuable to wetland scientists and managers dealing with estuarine systems. Below we present two approaches to this assessment.

7.7.2 Measurement of Soluble Sulfide in Marsh Pore Water

7.7.2.1 Sippers and Peepers

Sulfide in pore water is usually measured using two basic methods that collect pore water in the field with sippers or peepers and analysis of water S^{2-} content in the lab using an ion-selective electrode and a set of standards (Eaton et al. 2005). Pore water extractors or "sippers" are inserted into the marsh soil, suction is applied and the pore water is collected in a syringe (e.g., Marsh et al. 2005; Keller et al. 2009). This approach is relatively rapid, but provides poor resolution as the sample is drawn from a soil volume of uncertain dimensions. A second approach uses equilibrium dialysis samplers or "peepers" (Hesslein 1976). In this method, a device containing a vertical series of chambers is filled with deoxygenated, distilled water, covered with a semi-permeable membrane, inserted into the marsh soil, and allowed to equilibrate. When the soluble constituents in the pore water reach equilibrium with those in the chambers, the device is extracted and the water in the chambers is analyzed in the laboratory. Both sippers and peepers are used to obtain vertical profiles of pore water chemistry within the sediment column. By virtue of their placement in the sediment, sippers allow for some vertical segregation of data. Peepers are designed for the collection of discrete water samples at a smaller spatial resolution by preventing vertical mixing of adjacent water masses. Peepers have superior vertical resolution to the sippers (1–2 cm vs. 5–10 cm), but are limited by a relatively long equilibration period (usually 1 week or longer) (Teasdale et al. 1995).

7.7.2.2 IRIS Panels

IRIS tubes were developed to assess the presence of reducing conditions in soils (Castenson and Rabenhorst 2006; Jenkinson and Franzmeier 2006) (see *Techniques to Measure Eh/Assess Reducing Conditions* below). IRIS tubes are PVC tubes coated with an Fe oxide paint. When inserted into soil, the paint dissolves when reduced exposing the bare PVC pipe. It was inadvertently discovered that when IRIS tubes were exposed to S^{2-}, Fe monosulfide (FeS) coatings formed on the tubes

(Stolt 2005) as evidenced by a dark gray to black coating. Rabenhorst et al. (2010) modified the IRIS tube technology to produce IRIS panels for assessing soluble S^{2-} in brackish marshes. The technology can be used for a qualitative assay to detect the presence of soluble S^{2-}, or a quantitative assay to determine S^{2-} concentrations. A simple color change indicates the presence of soluble S^{2-}, and the tube shape is sufficient. A quantitative assay requires a set of standards and image analysis capabilities. The standards correlate the intensity of the color change to a known S^{2-} concentration. Image analysis is used to quantify the surface area of the tube or panel that represents an individual intensity of color change. Rabenhorst et al. (2010) decided to use large flat panels rather than the cylindrical tubes to facilitate quick recording of the images with a flatbed scanner. In contrast to the pore water sampling approach (sippers and peepers), this new technology shows the potential for generating quantitative information on S^{2-} concentrations with millimeter-scale spatial resolution. An additional benefit is the time required to obtain data is restricted to a couple of hours.

Using specially prepared Fe oxide paint (Rabenhorst and Burch 2006), the lower 50 cm of PVC panels or tubes (usually 60 cm in length) are painted. These are then inserted into the marsh soil for periods of 5 or 60 min. The FeS phase is unstable under oxidizing conditions, and the dark color fades over a period of minutes to hours when exposed to the air. Therefore, collection of the images (either by scanning or by photography) must be done quickly. Images are then compared with standard images prepared from painted PVC chips placed into Na_2S solutions (adjusted to pH 7.5) of known concentration (range: 3–300 mg/L S^{2-}) for set periods of time (usually 5 or 60 min.) For more detail, the reader is referred to Rabenhorst et al. (2010).

7.8 Oxidation-Reduction Processes in Soils

7.8.1 Overview

The formal definition of wetlands explicitly refers to the prevalence of "saturated soil conditions" (U.S. Army Corps of Engineers Environmental Laboratory 1987). These "saturated soil conditions" of wetlands are specifically known as hydric soils, which are defined as having "formed under conditions of saturation, flooding, or ponding long enough during the growing season to develop anaerobic conditions in the upper part" (Federal Register 1994). Organic materials in soils (mostly plant remains) are routinely decomposed by heterotrophic microorganisms as an energy source, and during this oxidation reaction where electrons are lost from the oxidized C, some other compound must serve as an electron acceptor and receive or "gain" the electrons (thus, being electrochemically reduced). When available in the soil solution, O_2 is usually the preferred electron acceptor. In saturated soils, however, dissolved O_2 can become depleted during microbial oxidation of organic

7 Wetland Biogeochemistry Techniques

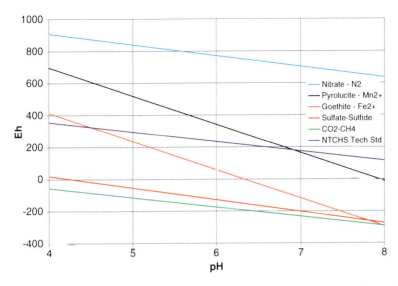

Fig. 7.2 Eh-pH stability phase diagram illustrating the redox conditions under which various common electron acceptors in soils become reduced

matter because the diffusion of O_2 into and through liquid water is quite slow. Such conditions where O_2 has become depleted are referred to as anaerobic.

In order for microorganisms to continue to oxidize organic materials under anaerobic conditions, they must use some alternate electron acceptor to O_2. Chemical thermodynamics determine which compounds that are common in soils can most easily accept electrons and be reduced. Those chemical ions and compounds readily found in soils that function as electron acceptors once O_2 is depleted are: NO_3^-; manganic manganese (Mn^{4+}); Fe^{3+}; sulfate (SO_4^{2-}); and CO_2. These are listed in order of their ease of being reduced which means that the soil environment must become increasingly reduced in order to proceed through this list of electron acceptors. This is sometimes shown in an Eh-pH stability diagram such as Fig. 7.2 where higher Eh values represent more oxidizing conditions and lower Eh represents more reducing conditions. Each line in Fig. 7.2 represents a "redox pair" and above the line, the oxidized form is stable and below the line, the reduced phase is stable. Thus, as the soil becomes progressively more reducing (lower Eh), various reduced phases would be predicted to be stable. First, NO_3^- would be expected to be reduced (function as an electron acceptor), and then Mn oxides (such as pyrolusite), Fe oxides (such as goethite), SO_4^{2-} and eventually (under highly reducing conditions) CO_2 can be reduced to methane.

Although this stepwise change or progression seems very systematic and orderly, soil systems are typically highly complex and also highly variable. Not all soils contain all these compounds. Also, the various proportions among these compounds can be very different among soils or ecological settings. Small scale variability in redox potential (Eh) in soils can be very great over short distances (even millimeters). Nevertheless, if one can measure soil Eh and pH, graphs such as Fig. 7.2 can be useful in helping to predict which electrochemical phases might be stable and which phases might be expected to be reduced or oxidized. For such reasons, soil and wetland scientists are often interested in documenting or measuring the redox status or condition of a soil.

7.8.2 Techniques to Measure Eh/Assess Reducing Conditions

7.8.2.1 Direct Assessment: Pt Electrodes

One approach to documenting soil redox status is to measure the voltage generated between a Pt electrode and a reference electrode (such as a calomel or Ag-AgCl) inserted into the soil and then correcting the measured voltage for the difference between the reference electrode and the standard hydrogen electrode, which is reported as Eh (Bohn 1971; James and Bartlett 2000; Patrick et al. 1996). The Pt electrode is placed into the soil at the depth where one intends to measure the Eh; the reference electrode can be placed nearer the soil surface if that is more convenient. If measurements are to made at soil depths of more than a few cm, it may be useful or necessary to make a pilot hole with some instrument prior to inserting the Pt electrode to the desired depth. Best results are obtained when the soil is moist or wet. Otherwise, a salt bridge may be used to ensure good electrical contact between the reference electrode and the soil (Veneman and Pickering 1983). This may be particularly important if upper part of the soil is not wet at the time of measurement.

It is of particular importance to make the voltage measurement using a device with a high internal resistance (Rabenhorst et al. 2009). This can either be done by using a research grade voltmeter, or by using a low cost multimeter that has been modified to effectively increases the internal resistance to >200 Gohms (Rabenhorst 2009). When an unmodified multimeter (resistance approximately 10 Mohms) is used to measure the voltage, too large a current is permitted to flow during the measurement which alters the electrochemical environment in the vicinity of the Pt electrode causing substantial drift and an unreliable measurement. Also, because there is a great deal of small scale microsite variability in soils in the field, multiple (often as many as five or more) replicate measurements are made. If multiple measurements are to be made over the course of a field season, electrodes are sometimes left in place and then removed at the end of the field season and checked to be sure they are still functioning properly. Automated data loggers may also be programmed to collect repeated measurements. In other situations where it would be impractical to leave the electrodes installed in the field, they may be removed and reinstalled each time measurements are made.

7.8.2.2 Indirect Assessment: IRIS Tubes

During the last decade, another approach has been developed to assess reducing conditions in soils known as IRIS tubes (Indicator of Reduction In Soils) (Castenson and Rabenhorst 2006; Jenkinson 2002). In this approach, an Fe oxyhydroxide paint is applied to (usually 60 cm) sections of 1.25 cm schedule 40 PVC tubing and allowed to dry. After making pilot holes in the soil, the tubes are inserted into the soil for a period of approximately 1 month. If the soils are saturated and microbes are actively oxidizing soil organic matter, some of the Fe oxide paint functions as an electron acceptor, becomes reduced and solubilized, and stripped from the PVC tubing. The degree to which the paint is removed from the tubes is an indication of the degree or magnitude of reduction in the soil (Castenson and Rabenhorst 2006; Rabenhorst 2008). As expected, soil temperature has also been shown to affect the rate at which paint is removed from the tubes with less removal occurring during periods when soil temperatures are low and approaching biological zero (Rabenhorst 2005; Rabenhorst and Castenson 2005). Although the paint is sometimes referred to as being comprised mainly of ferrihydrite (Jenkinson 2002; Jenkinson and Franzmeier 2006), it is important in the synthesis of the Fe oxides to ensure that 40–60 % of the ferrihydrite has been converted to goethite, as otherwise, the paint will not adhere well to the PVC tubing (Rabenhorst and Burch 2006).

One of the perceived benefits of using IRIS tubes for assessing soil reduction is that it integrates the conditions of the soils over the period during which it is installed. So while Eh measurements or indicator dyes can tell you what is happening at the moment, the IRIS tubes provide a better indication of what the redox status of the soil has been over the course of several weeks. They also have the benefit of not requiring specialized lab equipment such as electrodes and volt meters. IRIS tubes are available commercially, and one possible limitation to their use might be the costs involved in purchase of the devices. However, the paint for making the tubes can also be manufactured fairly easily in a laboratory following published methods (Rabenhorst and Burch 2006).

Because the soil redox properties are considered to be quite variable, normal protocols for using IRIS tubes call for using five replicate tubes (Rabenhorst 2008). If a majority (three-fifth) of the tubes exhibits a minimum level of paint removal, the soil is considered to be reducing. To meet the requirement for reducing conditions that is specified in the Technical Standard of the NTCHS, at least 30 % of the paint must be substantially removed from a 15 cm zone, somewhere along the upper 30 cm of the tube (representing the upper 30 cm of the soil) (National Technical Committee for Hydric Soils 2007). However, other work has suggested that a less dramatic removal of the paint might still indicate reducing soil conditions (Castenson and Rabenhorst 2006; Rabenhorst 2008).

7.8.2.3 Indirect Assessment: Alpha-Alpha Dipyridyl Dye

Various dyes can be used as indicators of particular soil chemical conditions and alpha-alpha dipyridyl dye can be utilized for demonstrating the presence of Fe^{2+} in

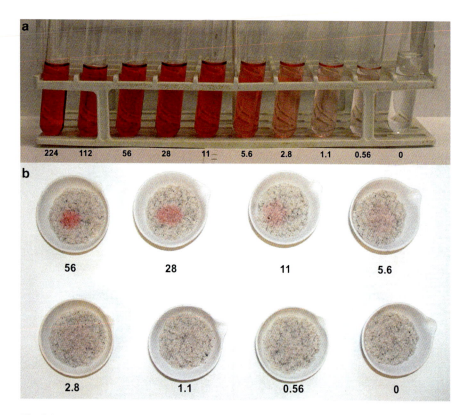

Fig. 7.3 *Pink color* developing in solutions and in saturated sand to which alpha, alpha dipyridyl dye has been added. Numbers represent concentrations of Fe^{2+} (μg/ml). In the *top figure* (**a**), 0.5 ml ααd dye was added to 6 ml of solution. In the *lower figure* (**b**), one drop of ααd dye was added to the saturated sand

the soil solution (Childs 1981). Except for some very unusual acid sulfate soils where oxidation of pyrite can (at least temporarily) form Fe^{2+} under oxidizing conditions, the presence of in soils usually indicates that the soil has been saturated and sufficiently reducing to cause Fe^{3+} oxides to be reduced to Fe^{2+} (Fig. 7.2). When alpha-alpha dipyridyl dye is applied to soils containing ferrous iron, a pink color develops, the intensity of which is function of how much Fe^{2+} is in solution. In clear solutions, as little as 0.5 μg/ml Fe^{2+} can be detected using the dye (Fig. 7.3a). However, when applied to soils, usually a minimum of 3–5 μg/ml is required for the pink color to be detected against the background soil color, and may require even higher concentrations to be seen if the soil is dark colored (Fig. 7.3b).

To test a soil using the dye, a core must usually be extracted and the dye is then applied to an interior broken face of the core. Care should be taken not to apply the dye to portions of the soil that have been in direct contact with a steel auger or spade as reduced iron from the tools has sometimes give a false positive reaction to the dye. Another way that the dye can be used is to extract some fresh soil solution from a saturated soil using a piezometer or suction lysimeter and adding a few drops of the alpha-alpha dipyridyl dye to a few ml of the extracted solution.

7.9 Additional Soil Characteristics

7.9.1 Soil pH

Soil pH refers to the pH of water in equilibrium with soil. pH is defined as the negative logarithm of the hydrogen ion activity in solution: pH = $-$Log (H^+). Activity and concentration (moles/L) are nearly equivalent in dilute solutions, so that concentration is often used instead of activity. Soil pH impacts many of the chemical reactions in soil, determining the rate and or direction of a reaction. pH tends to move toward the neutral value as a soil becomes anaerobic (Mitsch and Gosselink 2000). Soil pH is routinely measured on a soil slurry (1:1 soil to water ratio) in the lab with commercially available pH meters and combination electrodes. The method described below was presented by Thomas (1996).

Ten g of air-dry soil is weighed out in a 50 ml beaker. After adding 10 ml of deionized water, the slurry is mixed well. Let stand for 10 min. After gently swirling the suspension the electrode is inserted. The sample may or may not be stirred during the readings. If stirred, the electrode is inserted into the suspension. If not stirred, the electrode is inserted either into the sedimented soil or into the supernatant above the soil. Readings taken in the supernatant are generally slightly higher than readings taken in the stirred suspension (Thomas 1996). All 3 approaches are acceptable; the important thing is that a consistent approach is used. The value should be recorded as pH_w. The electrodes should be rinsed with distilled water between readings.

Alternative methods for measuring soil pH include pH papers and test kits. Both are colorimetric methods; contact of the soil sample with the pH paper or, in the case of test kits, reagents causes a color change. That color is compared to a reference chart that equates pH values to a specific color. The primary limitations to these methods are a lack of accuracy and a limited pH range. For example, the Hellige-Truog soil pH test kit has long been used for routine soil investigations. It has a pH range of 4.0–8.5 and is accurate to within 0.5 pH units. Relatively inexpensive, rugged pH meters are now available for field work and are more appropriate for biogeochemical assessment of wetland soils.

7.9.2 Soil Oxygen Demand

Biochemical oxygen demand (BOD) is a measure of the quantity of O_2 used by microorganisms in the oxidation of organic matter. Eutrophication is the process where water bodies receive excess nutrients that stimulate excessive plant growth. Traditionally, BOD has been used to assess eutrophication in surface waters but a similar approach can be used to assess nutrient loading in wetlands. For wetlands that do not exhibit continuous inundation, soil O_2 demand (SOD) can provide an indication of nutrient loading. Soil oxygen demand can be easily measured in the lab using a soil incubation procedure and an O_2 electrode. A weighed wet soil sample (5–10 g) is placed into a biological oxygen demand (BOD) bottle (250 ml) containing a stir bar. An additional soil sub-sample is weighed, dried, and re-weighed to determine water content. The bottle is placed on a magnetic stirrer and filled with deionized distilled water while stirring continuously. The contents are stirred for 15 min and dissolved O_2 (DO) is measured with an O_2 electrode. The bottle is sealed with a glass stopper and incubated in the dark for 24 h at 25 °C. Then, DO is again measured under continuous stirring. The analysis should be repeated with smaller sample size if DO levels decrease by more than 50 % (U.S. EPA 2008). Soil O_2 demand is calculated as follows:

$$SOD(mg/kg\text{-hour}) = [(DO_{t0}, mg/L) - (DO_{t24}, mg/L)] \\ \times (\text{water volume}, L)/\text{dry weight of soil}, kg$$

where: DO_{t0} and DO_{t24} = DO at time 0 and DO after the 24 h incubation period, respectively.

7.9.3 Bulk Density

7.9.3.1 Overview

To convert from SOM (or soil N, P, etc.) on a percentage basis to a weight basis, the bulk density (dry weight per unit volume) of the soil must be determined. This entails extracting a known volume of soil and determining the dry weight. Obtaining an accurate estimate of the volume of soil is the most difficult step in this exercise. Bulk density can be determined on intact soil cores obtained at known depth and volume (volume is determined from core diameter and depth), soil clods pulled from a pit (volume is determined by water displacement in the lab), or on unconsolidated materials obtained through excavation (volume obtained by filling the excavation hole with water). The soil is weighed after drying at 105 °C to a constant weight. Bulk density (g/cm^3) is calculated as "dry weight (g)/volume (cm^3)".

Depending on the intended use of the bulk density values, the contribution of rocks in the sample to sample weight and volume may or may not be accounted for. In most cases, the objective is to determine the bulk density of "fine earth" representing particles with diameters less than 2 mm, so coarser material is removed by screening. Mineral particles larger than very coarse sand (>2.0 mm) are considered rock fragments. If the bulk density value is to be used to extrapolate another parameter value to a soil volume or area basis, rocks are left in the sample. Otherwise, rocks in the sample should be accounted for by removing them, weighing them, and determining their total volume through water displacement. The rocks are placed in a graduated cylinder containing a known volume of water. The change in water volume after adding the rocks equals the volume of the rocks. The weight and volume of the rocks is then subtracted from the total sample values before calculating bulk density.

7.9.3.2 Excavation Method

The excavation method is most appropriate for sampling topsoil. Leaf litter is removed from the soil surface. The top of the excavation hole is leveled by removing soil with a trowel or knife and checking with a carpenter's level. Soil is removed with a trowel. The hole is lined with a single piece of Saran wrap or similar waterproof material. The hole is then filled with water from a graduated container. The excavated volume is determined by subtracting the final water volume in the container from the initial water volume. Another approach, which does not require Saran wrap, is to fill the hole with dried sand of a known bulk density. Sand is poured into the hole from a container with a known weight of sand. After filling the hole, the container of sand is weighed again. The volume (cm^3) of the soil cavity is calculated as "difference in sand weight (g)/sand bulk density (g/cm^3)".

7.9.3.3 Can Core Method

With a core method, a cylindrical metal sampler with a known volume is used to extract soil samples for bulk density determinations. Core samplers may be a hydraulically driven probe mounted on a vehicle (e.g., Giddings Probe), a double-cylinder, hammer-driven core sampler, or a simple metal can open at both ends. Use of the first two types of samplers is discussed by Blake and Hartge (1986). In this section, we discuss use of a small can. The can dimensions are not critical (except in calculations) but the diameter of the can should exceed the height of the can. Horizons should be sampled individually as they can vary significantly in bulk density. Tall cans make it more difficult to sample individual soil horizons. A representative size is 6.5-cm diameter and a 4.5-cm height. Three samples are collected from each horizon starting with the surface horizon and working down. At the sampling point, leaf litter is removed from the soil surface which is then smoothed to create a flat horizontal surface. The can is placed end down onto the soil surface and pressed into the soil by hand. This step can be facilitated by laying a board across the top of the can and tapping it lightly with a rubber mallet or hammer. However, care should be taken to avoid compacting the core. For horizons thinner than the height of the can, care must be taken to preclude simultaneously sampling multiple horizons. Dig out the inserted can plus a little of the surrounding soil and cut off the excess soil so the soil sample is flush with the top of the can. Empty the contents of the can into a sample bag. In each case, the thickness of the sampled horizon and the thickness of the core sample must be recorded. To sample subsurface horizons, dig down to the top of the target horizon and repeat the steps presented above.

7.9.3.4 Clod Method

The clod method is the most appropriate method for soil samples deep in the profile where it is difficult to determine volume *in situ*. In this method, clods are coated with a water-repellent substance (e.g., Saran solution) and weighed twice, first in air and then in water. For a complete description of this procedure, the reader is referred to Blake and Hartge (1986). A major limitation to this approach is that typically the clods are extracted from a pit face which is not always feasible in wetland settings.

7.10 Sediment Gains and Losses

7.10.1 Overview

Because of ambiguity in terminology, it is important to preface this section with working definitions for the processes addressed. Accretion is the accumulation of sediment at a particular point; it could be sediment transported from an upslope point or sediment derived at the point itself. The accumulation of detrital material is commonly referred to as organic matter accretion, while sediment accretion refers to the accumulation of both organic and inorganic materials (U.S. EPA 2008). Deposition refers to sediment accumulation at a particular point due to sediment dropping out of flow when the amount of sediment in runoff exceeds the transport capacity of runoff. Erosion is the removal of sediment from a particular point. Subsidence is the decrease in the sediment surface elevation caused by an increase in the bulk density of sediment or a loss of sediment mass through oxidation of organic matter. Loss of sediment through erosion is not considered to be subsidence.

For the most part, the techniques presented in this section were developed for use in tidal marshes where significant sediment gains and losses are inherent to ecological functioning. However, with at most minor modification, they could be used in less hydrodynamic wetlands. The approaches inherent to all of the techniques presented here are based on either sampling and weighing sediment, or measuring the deviation in elevation relative to a fixed elevation point. All of the methods except for Cs137 techniques require multiple visits to the field sites.

7.10.2 Artificial Marker Horizons

Marker horizons provide an easy way to assess marsh accretion over a period of months to years (Cahoon and Turner 1989). White feldspar clay is thinly spread over the marsh surface in replicated plots creating an artificial horizon. Over time sediments accumulate above the artificial horizon through the process of accretion. The depth of sediment accumulation is determined by collecting a core from the sample plot and measuring the distance from the current marsh surface to the artificial horizon. This method is less expensive than isotopic techniques, the cores are simple to collect and process, and sampling success or failure is known at collection time (Cahoon and Turner 1989). White feldspar clay is the marker horizon material of choice because it is conducive to submerged systems, and its bright white appearance is readily distinguishable from the surrounding sediment.

Typical plot size is 50 cm × 50 cm, with a layer thickness of 5 mm. Care should be taken during spreading the clay to ensure uniform horizon thickness. Plots are marked with pipes or rods sufficient in height to rise above the water column or vegetation to facilitate future sampling. Samples can be collected with a thin-walled core tube. Collected cores are refrigerated and taken to the laboratory in a vertical position. If processing is delayed, the cores are stored in the freezer. Melting can compromise core integrity, especially those cores high in organic materials. Therefore, the cores should remain frozen during processing. The frozen core is sectioned to determine the thickness of the material above the feldspar marker. The thickness of the sediment located above the feldspar marker is measured with calibrated calipers. Sectioning the core facilitates sampling for bulk density measurements and organic matter determinations.

A mass estimate of accretion can easily be determined by collecting a known volume of the accreted material and determining bulk density (see *Bulk Density* above). This estimate can be further refined into organic accretion and inorganic accretion by subjecting the sample to LOI assay (See *Soil Organic Matter: Loss on Ignition* above). Shallow compaction can be assessed by utilizing SET's data (See *Sedimentation-SETS* below) with accretion data collected from short-term accretion marker horizons as follows: compaction = sediment accretion − elevation change. For this approach, feldspar plots must be established before taking the SET baseline.

7.10.3 137Cs Dating

Marker integrity can be compromised by bioturbation, tidal action, resuspension, or erosion. Therefore, the reliability of the artificial marker horizon technique suffers in highly dynamic systems (DeLaune et al. 1989). An alternative approach to assessing long-term sediment accumulation is to use cesium-137 tracer methods (DeLaune et al. 1989; Milan et al. 1995). Production of 137Cs (half-life 30 year) resulted from aboveground thermonuclear weapons testing. Atmospheric deposition of 137Cs began in 1954, but peak fallout occurred in 1964 (Ritchie and McHenry 1990). Both dates are used to measure sediment accretion rates, but the 1954 sediment horizon is often difficult to discern (Ritchie and McHenry 1990), so the 1964 peak is most often used as the marker layer in wetland investigations (Reddy et al. 1993). In effect, the zone in the soil profile with the highest 137Cs level represents a "marker horizon", and all accretion above that zone occurred after 1964. The 137Cs signature may be compromised in areas with high erosion rates or where large amounts of sediment washed-in from other areas (Milan et al. 1995). In such situations, Milan et al. (1995) recommend that both 137Cs and 210Pb methods be used (see DeLaune et al. (1989)). The following 137Cs method was presented by DeLaune et al. (1989). Sampling sites are established along a representative transect of the marsh. Sediment cores (15 cm in diameter, ≥ 50 cm in length) are taken along the transect with a thin wall cylinder, with care taken to prevent compaction of the cores. The retrieved core is sectioned, dried, and 137Cs activity is counted. Sediment accretion is measured by counting the 137Cs activity relative to distance down into the core. Bulk density, percent C, and percent organic matter should also be determined by depth in each core to allow for calculations relative to total mass, C, or organic matter.

7.10.4 SETS

Sediment erosion tables (SETS) measure net changes in sediment surface elevation relative to a benchmark (Cahoon et al. 2002a). A SET can be used to determine either the impact of a single meteorological event on sediment surface elevation or a long-term trend in elevation change. SETS consist of a supporting base pipe permanently placed at each site and a portable portion with four components: horizontal arm, vertical arm, flat table or plate, and pins. The base pipe, an aluminum pipe that is driven into the soil to the point of refusal, serves as a benchmark. The table provides a horizontal reference plane. The pins pass through the table at right angles until they touch the marsh surface. Marsh surface elevation relative to the benchmark is determined by measuring the length of pin above the table. Changes in the elevation of the marsh surface are indicated by changes in the distance between the marsh surface and the table. An integrated measure of elevation (accretion minus subsidence) can be calculated by combining data obtained from SETs with that provided by feldspar marker horizons.

SET stations should be located where the sediment surface is plainly visible and not obscured by vegetation or deep water. Data should be collected at times of lowest water level. Efforts should be made to minimize disturbance from foot traffic during SET installation and data collection. One option is to construct platforms with removable planks. The planks should be removed each time you exit the site. Data collection frequency should be based on expected sedimentation rates at the site. In areas with high sediment fluxes, a frequent collection schedule will produce a higher resolution of elevation changes over time. In areas with low sediment fluxes, a similar collection schedule may produce results within the range of methodological error so that actual changes in elevation will not be detected. The conservative approach is for each practitioner to construct a data collection schedule based on preliminary work onsite. For further details on SETS, including design specifics, the reader is referred to Cahoon et al. (2002a, b) and Callaway et al. (2001).

7.10.5 Sediment Collection Tiles

Sediment collection tiles provide a simple and inexpensive approach to assessing deposition in both tidal and non-tidal wetlands. Sediment is deposited on the upper surface of the tile; the sediment is collected, dried, and weighed to produce a mass estimate. Ceramic tiles produced for home use (e.g., glazed kitchen tiles) are appropriate. Ceramic tiles placed on the soil surface will suffice for wetlands subject to low energy flooding. For high energy floodplains or tidal marshes, the sediment collection tile should be anchored to the soil. For sites that are subject to relatively deep flooding such as tidal marshes, tiles may be placed at several elevations relative to the soil surface to determine sediment availability within the water column. This is particularly important if the goal is to separate re-suspended sediments (defined as the amount of sediments lifted 6 cm or more above the sediment surface) from unsuspended, shifting surface material. Tiles may be easily positioned at specific heights relative to the soil surface by attaching PVC pipe to the base of the tile and inserting the pipe into the soil. Tiles are retrieved when the soil surface is exposed to air, usually at 2 week or 1 month intervals. Plant tissue is first removed from each tile. Sediment is scraped and rinsed off the tiles and into sample cups using a rubber scraper and deionized water. Samples are returned to the lab, filtered through pre-weighed paper filters, dried and weighed. Sediment deposition is commonly reported as g dry weight/m^2. Organic matter is an important constituent of sediment in many depositional settings; its content in a sediment sample can be determined by measuring weight loss in subsamples after burning at selected temperatures (see *Soil Organic Matter: Loss on Ignition* above).

The following approach was presented by Pasternack and Brush (1998) and at http://pasternack.ucdavis.edu/sedtiles.htm. The major advantages of this sturdier design is that it lends itself to high energy tidal marshes and can be used to assess erosion as well as accretion. It utilizes an aluminum pipe (1 m long × 2 cm dia.) 'anchor' sunk into the ground and capped with a detachable ceramic tile flush with the marsh surface. A plastic tube is attached to the bottom of each tile; the tube drops over the anchor pipe. This provides a stable tile surface that is not susceptible to motion unless subjected to extreme lift forces. Therefore, it is conducive to high energy tidal marshes. They constructed the tile and tube assembly by attaching (with plastic cement) a 7.5 cm long hard plastic pipe with an inner diameter of 2 cm to the center of a plastic square with plastic cement. Schedule 77 plastic is appropriate for the pipe and the square. The other side of the plastic square was attached to the bottom of a 20 cm × 20 cm glazed-ceramic tile with silicone glue.

Each sampling point should be marked by GPS. Even so, it may be difficult to find the anchor in subsequent visits. One approach is to mark each anchor with a second anchor as follows: Lay a meter stick on the ground with one end on the anchor and the other pointing to a pre-determined compass direction (e.g., due east) and hammer a second anchor ~0.5 m down into the ground at the east end of the meter stick. Flag the top of this second anchor (marker anchor) with a long strip of flagging attached with duct tape. Attach the tile assembly over the sunken auger by placing the plastic pipe over the metal pipe. The metal pipe may need to be filed to accommodate the plastic pipe.

The most appropriate sampling interval for sediment tiles depends on the hydrodynamics and sediment loading rate of the system in question. Extra care is needed when sampling very small or very large amounts of deposition. Pasternack and Brush (1998) recommend the following sampling intervals: tidal freshwater wetland, 2–4 weeks; salt marsh, 4–8 weeks; floodplain, varies with flood regime. These should be taken as initial guidelines and adjusted as needed with experience.

Erosion monitoring requires that the height of each side of the tile above the marsh surface be determined at installation. Also during installation attempts should be made to keep the tile level with the bottom of the tile just touching the marsh surface. If not, any deviations should be recorded. If the tile was flat on the ground upon installation, erosion depth is determined by measuring the height of each tile side above the marsh surface and averaging the measurements. If the tile was not flat, measurements need to be adjusted for the known deviations. A mass estimate of erosion can be determined by multiplying erosion depth by sediment bulk density (see *Bulk Density* above).

7.11 Biogeochemical Indicators for Evaluating Wetland Condition

Many of the biogeochemical assays are expensive, require a high level of technical expertise, and require multiple site visits. Therefore, a number of wetland scientists have evaluated a surrogate approach in which readily measureable indicators of a biogeochemical process or pool are identified and used in lieu of the more complex assay. The surrogate approach may address an individual wetland service, the impact of a specific type of anthropogenic disturbance, or overall wetland condition. A major benefit of this approach is that since the required information can be collected in one visit, a greater number of wetlands can be assessed.

As an example of the surrogate approach to individual wetland services, researchers at the University of Delaware developed an assessment of denitrification capacity for slope wetlands based on soil morphology. In this procedure, wetlands are classified as having high, medium, or low denitrification capacity on the presence of specific hydric soil indicators. Field Indicators of Hydric Soils (USDA, NRCS 2010) are a series of individual soil morphology characteristics that were developed to identify hydric soils in the field. The majority of the indicators are visual and require only a ruler, a shovel, and a Munsell Color Chart. Differences between indicators reflect in part differences in hydroperiods which impact denitrification potential.

There is presently a national wetland monitoring effort based on the use of indicators to assess the impact of a specific type of anthropogenic disturbance or overall wetland condition. Natural and anthropogenic disturbances that negatively impact the functional capacity of a wetland are called "stressors". Widespread anthropogenic disturbances include artificial drainage (e.g., ditches), timber harvesting, and nutrient loading. Commonly, wetland condition is assessed as the degree of deviation from the undisturbed condition. This requires the collection of data from reference sites to identify baseline levels for comparison. Ideally, the reference wetland would be pristine (i.e., un-impacted by anthropogenic activities). However, in many regions such as the mid-Atlantic region of the U.S., it is difficult to find undisturbed wetlands for each class so that minimally impacted sites are used as reference sites. Reference sites should represent the same class of wetlands as the wetlands in question. For example, a mineral soil flat would be compared to a mineral soil flat reference site.

The U.S. EPA recognizes three levels of indicators for wetland monitoring – Levels I, II, and III (U.S. EPA 2008). Level I indicators are used for routine wetland monitoring for condition or impacts. Level II and III indicators are often used to test or support the validity of Level I indicators. The classes are based on the ease of measurement and sensitivity to respond to change as follows:

1. Level I: easily measureable, low cost, less sensitive to stress, long response time, low spatial variability
2. Level II: intermediate in complexity and sensitivity, intermediate response time, moderate spatial variability
3. Level III: highly complex and sensitive, short response time, high spatial variability

Because of the recognized importance of wetlands to water quality, one focus of these monitoring efforts has been to develop methods to evaluate nutrient enrichment of wetlands, which is one of the primary stressors to wetlands in many parts of the country. Eutrophication is commonly considered to be the enrichment of bodies of fresh water by inorganic plant nutrients (e.g., NO_3^-, PO_4^{2-}). This may result in increased primary productivity and a subsequent impact to water quality. Eutrophication of wetlands is usually attributed to nutrient loading, increased external inputs of nutrients from point and non-point sources. However, eutrophication can also result from an increase in nutrient cycling within the wetland itself without an increase in external inputs. For example, artificial drainage or increased development in the surrounding watershed will impact hydrologic characteristics of the wetland. These changes will impact SOM decomposition rates, soil mineralization rates, and denitrification rates. Destruction of vegetation in the wetland will diminish a primary sink for N and P resulting in an increase in those nutrients in the water column.

Not all of the monitoring indicators have a biogeochemical basis. However, because of the interdependence of biogeochemical cycles and the nutrient status of soil and the water column, biogeochemical characteristics are well suited to serve as indicators of wetland condition, especially with respect to nutrient enrichment. Table 7.2 presents some examples of potential biogeochemical indicators to assess nutrient impacts to wetlands. Additional information on using indicators to assess wetland condition can be found in the following references:

Reddy and DeLaune (2008), U.S. EPA (2008),
http://www.epa.gov/waterscience/criteria/wetlands/, and
http://www.epa.gov/owow/wetlands/bawwg/publicat.html.

Table 7.2 Examples of potential biogeochemical indicators to assess nutrient impacts to wetlands (Reddy and DeLaune 2008)

Level I indicators	Level II indicators	Level III indicators
Water column		
Dissolved O_2	Dissolved organic C	Microbial diversity
pH	Enzyme assays	Cellular fatty acids
Temperature	UV absorbance	
Turbidity		
Detritus and soil		
Soil bulk density	Soil O_2 demand	Microbial diversity
Soil Eh	Cation exchange capacity	Substrate-induced respiration
Total N	Soil respiration	Cellular fatty acids
Soil texture	Denitrification potential	Organic matter accretion rates

References

Aber JP, Melillo JM (1980) Litter decomposition: measuring state of decay and percent transfer in forest soils. Can J Bot 58:416–421

Aitkenhead-Peterson JA, McDowell WH, Neff JC (2003) Sources, production, and regulation of allochthonous dissolved organic matter inputs to surface waters. In: Findlay S, Sinsabaugh R (eds) Aquatic ecosystems: interactivity of dissolved organic matter. Academic Press, Amsterdam, pp 26–70

Altor AE, Mitsch WJ (2008) Pulsing hydrology, methane emissions and carbon dioxide fluxes in created marshes: a 2-year ecosystem study. Wetlands 28:423–438

Avery EA, Burkhart HE (2002) Forest measurements, 5th edn. McGraw-Hill Higher Education, New York

Barackman M, Brusseau ML (2004) Groundwater sampling. In: Artiola JF, Pepper IL, Brusseau ML (eds) Environmental monitoring and characterization. Elsevier Academic Press, Burlington, pp 121–141

Barth DS, Mason BJ (1984) The importance of an exploratory study to soil sampling quality assurance. In: Schweitzer GE, Santolucito JA (eds) Environmental sampling for hazardous wastes. American Chemical Society, Washington, DC, pp 97–104

Barth DS, Mason BJ, Starks TH, Brown KW (1989) Soil sampling quality assurance user's guide, 2nd edn. Environmental Monitoring and Support Laboratory. USEPA Report No. EPA/600/8-89/046, Las Vegas

Benner R (2003) Molecular indicators of the bioavailability of dissolved organic matter. In: Findlay SEG, Sinsabaugh RL (eds) Aquatic ecosystems: interactivity of dissolved organic matter. Academic Press/Elsevier Science, San Diego, pp 121–137

Bernier P, Hanson PJ, Curtis PS (2008) In: Hoover CM (ed) Field measurements for forest carbon monitoring. Humana Press, New York, pp 91–101

Blake GR, Hartge KH (1986) Bulk density. In: Klute A (ed) Methods of soil analysis, Part 1. Physical and mineralogical methods, SSSA Book Series No. 5. Soil Science Society of America, Madison, pp 363–376

Bloomfield J, Vogt K, Wargo PM (1996) Tree root turnover and senescence. In: Waisel Y, Eshel A, Kafkafi A (eds) Plant roots: the hidden half, 2nd edn. Marcel Dekker, New York, pp 363–381

Boddey RM (1987) Methods for quantification of nitrogen fixation associated with Gramineae. CRC Crit Rev Plant Sci 6:209–266

Bohlen PJ, Gathumbi SM (2007) Nitrogen cycling in seasonal wetlands in subtropical cattle pastures. Soil Sci Soc Am J 71:1058–1065

Bohn HL (1971) Redox potentials. Soil Sci 112:39–45

Bradford JB, Ryan MG (2008) Quantifying soil respiration at landscape scales. In: Hoover CM (ed) Field measurements for forest carbon monitoring. Humana Press, New York, pp 143–162

Bremner JM (1996) Nitrogen-total. In: Sparks DL (ed) Methods of soil analysis, Part 3. Chemical methods, SSSA Book Series No. 5. Soil Science Society of America, Madison, pp 1085–1122

Bridgham SD, Updegraff K, Pastor J (1998) Carbon, nitrogen, and phosphorous mineralization in northern wetlands. Ecology 79:545–561

Brown JR (ed) (1987) Soil testing: sampling, correlation, calibration, and interpretation, SSSA Spec Publ No 21. Soil Science Society of American, Madison, p 144

Bruland GL, Richardson CJ, Daniels WL (2009) Microbial and geochemical responses to organic matter amendments in a created wetland. Wetlands 29:1153–1165

Butterly CR, Bunemann EK, McNeill AM, Baldock JA, Marschner P (2009) Carbon pulses but not phosphorus pulses are related to decreases in microbial biomass during repeated drying and rewetting of soils. Soil Biol Biochem 41:1406–1416

Cahoon DR, Turner RE (1989) Accretion and canal impacts in a rapidly subsiding wetland: II. Feldspar marker horizon technique. Estuaries 12:260–268

Cahoon DR, Lynch JC, Hensel PF, Boumans RM, Perez BC, Segura B, Day JW Jr (2002a) High precision measurement of wetland sediment elevation: I. Recent improvements to the sedimentation-erosion table. J Sediment Res 72:730–733

Cahoon DR, Lynch JC, Perez BC, Segura B, Holland R, Stelly C, Stephenson G, Hensel PF (2002b) High precision measurement of wetland sediment elevation: II. The rod surface elevation table. J Sediment Res 72:734–739

Callaway JC, Desmond JS, Sullivan G, Williams GD, Zedler JB (2001) Assessing the progress of restored wetlands: hydrology, soil, plants, and animals. In: Zedler JB (ed) Handbook for restoring tidal wetlands. CRC Press, Boca Raton, pp 271–335

Carpenter EJ, van Raalte CD, Valiela I (1978) Nitrogen fixation by algae in a Massachusetts salt marsh. Limnol Ocean 23:318–327

Castenson KL, Rabenhorst MC (2006) Indicator of reduction in soil (IRIS): evaluation of a new approach for assessing reduced conditions in soil. Soil Sci Soc Am J 70:1222–1226

Champagne P (2008) Wetlands. In: Ong SK, Surampalli RY, Bhandari A, Champagne P, Tyagi RD, Lo I (eds) Natural processes and systems for hazardous waste treatment. American Society of Civil Engineers, Reston VA, pp 189–256

Childs CW (1981) Field test for ferrous iron and ferric-organic complexes (on exchange sites on in water-soluble forms) in soils. Aust J Soil Res 19:175–180

Chojnacky DC, Milton M (2008) Measuring carbon in shrubs. In: Hoover CM (ed) Field measurements for forest carbon monitoring. Humana Press, New York, pp 45–72

Coble PG, Green SA, Blough NV, Gagosian RB (1990) Characterization of dissolved organic matter in the Black Sea by fluorescence spectroscopy. Nature 348:432–435

Coble PG, Del Castillo CE, Avril B (1998) Distribution and optical properties of CDOM in the Arabian Sea during the 1995 SW monsoon. Deep-Sea Res II 45:2195–2223

Correll DL (1998) The role of phosphorus in the eutrophication of receiving waters: a review. J Environ Qual 27:261–266

Cory RM, McKnight DM (2005) Fluorescence spectroscopy reveals ubiquitous presence of oxidized and reduced quinones in dissolved organic matter. Environ Sci Technol 39:8142–8149

Cory RM, Boyer EW, McKnight DM (2011) Spectral methods to advance understanding of dissolved organic carbon dynamics in forested catchments. In: Levia DF, Carlyle-Moses DE, Tanaka T (eds) Forest hydrology and biogeochemistry: synthesis of past research and future directions. Ecological Studies Series No. 216. Springer-Verlag, Heidelberg, Germany, pp 117–135

Davidson EA, Savage K, Bolstad P, Clark DA, Curtis PS, Ellsworth DS, Hanson PJ, Law BE, Luo Y, Pregitzer KS, Randolph JC, Zak D (2002) Belowground carbon allocation in forests estimated from litterfall and IRGA-based soil respiration measurements. Agric For Meteorol 113:39–51

DeAngelis DL, Gardner RH, Shugart HH (1981) Productivity of forest ecosystems studied during the IBP: the woodland data set. In: Reichle DE (ed) Dynamic properties of forest ecosystems. Cambridge University Press, Cambridge, pp 567–659

DeLaune RD, Whitcomb JH, Patrick WH Jr, Pardue JH, Pezeshki SR (1989) Accretion and canal impacts in a rapidly subsiding wetland: I. 137Cs and 210Pb techniques. Estuaries 12:247–259

Distefano JF, Gholz HL (1986) A proposed use of ion-exchange resins to measure nitrogen mineralization and nitrification in intact soil cores. Commun Soil Sci Plant 17:989–998

Eaton AD, Clesceri LS, Rice EW, Greenberg AE (eds) (2005) Standard methods for the examination of water & wastewater: centennial ed. American Water Works Association, Hanover, p 1368

Edwards NT, Harris WF (1977) Carbon cycling in a mixed deciduous forest floor. Ecology 58:431–437

Fahey TJ, Hughes JW, Pu M, Arthur MA (1988) Root decomposition and nutrient flux following whole-tree harvest of northern hardwood forest. Forest Sci 34:744–768

Fanning DS, Rabenhorst MC, Balduff DM, Wagner DP, Orr RS, Zurheide PK (2010) An acid sulfate perspective on landscape/seascape soil mineralogy in the U.S. Mid-Atlantic region. Geoderma 154:457–464

Federal Register (1994) Changes in hydric soils of the United States. US Department of Agriculture Soil Conservation Service, Washington, DC

Fellman JB, Hood E, Edwards RT, D'Amore DV (2009) Changes in the concentration, biodegradability, and fluorescent properties of dissolved organic matter during stormflows in coastal temperate watersheds. J Geophys Res – Biogeosci 114. doi:10.1029/2008JG000790

Fellman JB, Hood E, Spencer RGM (2010) Fluorescence spectroscopy opens new windows into dissolved organic matter dynamics in freshwater ecosystems: a review. Limnol Oceanogr 55:2452–2462

Giardina CP, Ryan MG (2002) Total belowground carbon allocation in a fast-growing eucalyptus plantation estimated using a carbon balance approach. Ecosystem 5:487–499

Gordon A, Tallis M, Van Cleve K (1987) Soil incubations in polyethylene bags: effect of bag thickness and temperature on nitrogen transformations and CO_2 permeability. Can J Soil Sci 67:65–75

Green SA, Blough NV (1994) Optical absorption and fluorescence properties of chromophoric dissolved organic matter in natural waters. Limnol Oceanogr 39:1903–1916

Grier CC, Vogt KA, Keyes MR, Edmonds RL (1981) Biomass distribution and above- and belowground production in young and mature *Abies amabilis* zone ecosystems of the Washington Cascades. Can J For Res 11:155–167

Hardy RWF, Holsten RD, Jackson EK, Burns RC (1968) The acetylene-ethylene assay for N_2-fixation: laboratory and field evaluation. Plant Phys 43:1185–1207

Hardy RWF, Burns RC, Holsten RD (1973) Applications of the acetylene-ethylene assay for measurement of nitrogen fixation. Soil Biol Biochem 5:47–81

Harmon ME, Lajtha K (1999) Analysis of detritus and organic horizons for mineral and organic constituents. In: Robertson GP, Bledsoe CS, Coleman DC, Sollins P (eds) Standard soil methods for long-term ecological research. Oxford University Press, New York, pp 143–165

Harrison AF, Latter PM, Walton DWH (eds) (1988) Cotton strip assay: an index of decomposition in soils. In: ITE symposium no. 24. Institute of Terrestrial Ecology, Grange-Over-Sands, Great Britain, 176 pp

Hart SC, Stark JM, Davidson EA, Firestone MK (1994) Nitrogen mineralization, immobilization, and nitrification. In: Weaver RW, Angle JS, Bottomly PS (eds) Methods of soil analysis, Part 2. Microbiological and biochemical properties, SSSA Book Series No. 5. Soil Science Society of America, Madison, pp 985–1018

Helms JR, Stubbins A, Ritchie JD, Minor EC (2008) Absorption spectral slope ratios as indicators of molecular weight, source, and photobleaching of chromophoric dissolved organic matter. Limnol Oceanogr 53:955–969

Herbert BE, Bertsch PM (1995) Characterization of dissolved and colloidal organic matter in soil solution: a review. In: Kelley JM, McFee WW (eds) Carbon forms and functions in forest soils. Soil Science Society of America, Madison, pp 63–88

Hertel D, Leuschner C (2002) A comparison of four different fine root production estimates with ecosystem carbon balance data in a Fagus-Quercus mixed forest. Plant Soil 239:237–251

Hesslein RH (1976) An in situ sampler for close interval pore water studies. Limnol Oceanogr 21:912–914

Hinton MJ, Schiff SL, English MC (1997) The significance of storms for the concentration and export of dissolved organic carbon from two Precambrian Shield catchments. Biogeochemistry 36:67–88

Hood E, Fellman JB, Spencer RGM, Hernes PJ, Edwards RT, D'Amore DV, Scott D (2009) Glaciers as a source of ancient and labile organic matter to the marine environment. Nature 462:1044–1048

Horwath WR, Elliott LF, Steiner JJ, Davis JH, Griffith SM (1998) Denitrification in cultivated and noncultivated riparian areas of grass cropping systems. J Environ Qual 27:225–231

Howard PJA (1988) A critical evaluation of the cotton strip assay. In: Harrison AF, Latter PM, Walton DWH (eds) Cotton strip assay: an index of decomposition in soils. ITE Symposium No. 24. Institute of Terrestrial Ecology, Grange-over-Sands, pp 34–42

Howard PJA, Howard DM (1990) Use of organic carbon and loss-on-ignition to estimate soil organic matter in different soil types and horizons. Biol Fert Soils 9:306–310

Hunt PG, Matheny TA, Szögi AA (2003) Denitrification in constructed wetlands used for treatment of swine wastewater. J Environ Qual 32:727–735

Hunt PG, Matheny TA, Ro KS (2007) Nitrous oxide accumulation in soils from riparian buffers of a coastal plain watershed – carbon/nitrogen ratio control. J Environ Qual 36:1368–1376

Inamdar SP, Mitchell MJ (2006) Hydrologic controls on DOC and nitrate exports across catchment scales. Water Resour Res 42:W03421. doi:10.1029/2005WR004212

Inamdar S, Singh S, Dutta S, Levia D, Mitchell M, Scott D, Bais H, McHale P (2011) Fluorescence characteristics and sources of dissolved organic matter for stream water during storm events in a forested mid-Atlantic watershed. J Geophys Res Biogeosci 116. doi:10.1029/2011JG001735

Inamdar S, Finger N, Singh S, Mitchell M, Levia D, Bais H, Scott D, McHale P (2012) Dissolved organic matter (DOM) concentrations and quality in a forested mid-Atlantic watershed, USA. Biogeochemistry 108:55–76

Jaffé R, McKnight D, Maie N, Cory R, McDowell WH, Campbell JL (2008) Spatial and temporal variations in DOM composition in ecosystems: the importance of long-term monitoring of optical properties. J Geophys Res 113:G04032. doi:10.1029/2008JG000683

James BR, Bartlett RJ (2000) Redox phenomena. In: Sumner ME (ed) Handbook of soil science. CRC Press, Boca Raton, pp 169–184

Jardine PM, Weber NL, McCarthy JF (1989) Mechanisms of dissolved organic carbon adsorption on soil. Soil Sci Soc Am J 53:1378–1385

Jenkins JC, Chojnacky DC, Heath LS, Birdsey RA (2003) National-scale biomass estimators for United States tree species. Forest Sci 49:12–35

Jenkinson B (2002) Indicators of Reduction in Soils (IRIS): a visual method for the identification of hydric soils. PhD dissertation, Purdue University, West Lafayette

Jenkinson BJ, Franzmeier DP (2006) Development and evaluation of Fe coated tubes that indicate reduction in soils. Soil Sci Soc Am J 70:183–191

Jordan TE, Andrews MP, Szuch RP, Whigham DF, Weller DE, Jacobs AD (2007) Comparing functional assessments of wetlands to measurements of soil characteristics and nitrogen processing. Wetlands 27:479–497

Kadlec RH, Knight RL (1996) Treatment wetlands. CRC Press, Boca Raton, p 893

Kaiser K, Zech W (1998) Soil dissolved organic matter sorption as influenced by organic and sesquioxide coatings and sorbed sulfate. Soil Sci Soc Am J 62:129–136

Kalbitz K, Solinger S, Park JH, Michalzik B, Matzner E (2000) Controls on the dynamics of dissolved organic matter in soils: a review. Soil Sci 165:277–304

Karberg NJ, Scott NA, Giardina CP (2008) Methods for estimating litter decomposition. In: Hoover CM (ed) Field measurements for forest carbon monitoring. Humana Press, New York, pp 103–112

Kayranli B, Scholz M, Mustafa A, Hedmark A (2010) Carbon storage and fluxes within freshwater wetlands: a critical review. Wetlands 30:111–124

Keller JK, Wolf AA, Weisenhorn PB, Drake BG, Megonigal JP (2009) Elevated CO2 affects porewater chemistry in a brackish marsh. Biogeochemistry 96:101–117

Knievel DP (1973) Procedure for estimating ratio of living and dead root dry matter in root core samples. Crop Sci 13:124–126

Knight RL, Wallace SD (2008) Treatment wetlands, 2nd edn. CRC, Boca Raton

Koch MS, Mendelssohn IA, McKee KL (1990) Mechanism for the hydrogen sulfide-induced growth limitation in wetland macrophytes. Limnol Oceanogr 35:399–408

Kovar JL, Pierzynski GM (eds) (2009) Methods of phosphorus analysis for soils, sediments, residuals, and waters, 2nd edn. Southern Coop Series Bull No. 408, Virginia Tech University, Blacksburg, VA

Kuo S (1996) Phosphorus. In: Sparks DL (ed) Methods of soil analysis, Part 3. Chemical methods, SSSA Book Series No. 5. Soil Science Society of America, Madison, pp 869–920

Lakowicz JR (1999) Principles of fluorescence spectroscopy, 2nd edn. Springer, Heidelberg, pp 725

Latter PM, Harrison AF (1988) Decomposition of cellulose in relation to soil properties and plant growth. In: Plant G, Harrison AF, Latter PM, Walton DWH (eds) Cotton strip assay: an index of decomposition in soils. ITE Symposium No. 24. Institute of Terrestrial Ecology, Grange-Over-Sands, pp 68–71

Livingston GP, Hutchinson GL (1995) Enclosure based measurement of trace gas exchange: applications and sources of error. In: Matson PA, Harriss RC (eds) Biogenic trace gases: measuring emissions from soil and water, Methods in ecology series. Blackwell Science, Oxford, pp 14–51

Marsh AS, Rasse DP, Drake BG, Megonigal JP (2005) Effect of elevated CO_2 on carbon pools and fluxes in a brackish marsh. Estuaries 28:694–704

Mason BJ (1992) Preparation of soil sampling protocols: sampling techniques and strategies. USEPA Office of Research and Development, Washington, DC. EPA/600/R-92/128. Available at: http://www.sera17.ext.vt.edu/Documents/P_Methods2ndEdition2009.pdf

McKnight DM, Hood E, Klapper L (2003) Trace organic moieties in dissolved organic matter in natural waters. In: Findlay SEG, Sinsabaugh RL (eds) Interactivity of dissolved organic matter. Academic Press, San Diego, pp 71–93

Megonigal JP, Hines ME, Visscher PT (2003) Anaerobic metabolism: linkages to trace gases and anaerobic processes. In: Schlesinger WH (ed) Treatise on geochemistry, vol 8, Biogeochemistry. Elsevier, Amsterdam, pp 317–424

Mendelssohn IA, McKee KL (1988) *Spartina alterniflora* die-back in Louisiana: time-course investigation of soil waterlogging effects. J Ecol 76:509–521

Milan CS, Swenson EM, Turner RE, Lee JM (1995) Assessment of the 137Cs method for estimating sediment accumulation rates: Louisiana salt marshes. J Coastal Res 11:296–307

Miller MP, McKnight DM (2010) Comparison of seasonal changes in fluorescent dissolved organic matter among aquatic lake and stream sites in the Green Lakes Valley. J Geophys Res (in press), 115, G00F12, doi:10.1029/2009JG000985

Miller MP, McKnight DM, Cory RM, Williams MW, Runkel RL (2006) Hyporheic exchange and fulvic acid redox reactions in an alpine stream/wetland ecosystem, Colorado front range. Environ Sci Technol 40:5943–5949

Minderman G (1968) Addition, decomposition and accumulation of organic matter in forests. J Ecol 56:355–362

Mitsch WJ, Gosselink JG (2000) Wetlands, 4th edn. Wiley, New York

Moore P, Coale F (2009) Phosphorus fractionation in flooded soils and sediments. In: Kovar JL, Pierzynski GM (eds) Methods of phosphorus analysis for soils, sediments, residuals, and waters, 2nd edn. Southern Coop Series Bull No. 408 pp 61–70

Moore PA Jr, Reddy KR (1994) Role of Eh and pH on phosphorus geochemistry in sediments of Lake Okeechobee, Florida. J Environ Qual 23:955–964

Mosier AR, Klemedtsson L (1994) Measuring denitrification in the field. In: Weaver RW, Angle JS, Bottomly PS (eds) Methods of soil analysis, Part 2. Microbiological and biochemical properties, SSSA Book Series No. 5. Soil Science Society of America, Madison, pp 1047–1066

Mulholland PJ (2003) Large scale patterns in dissolved organic carbon concentration, flux, and sources. In: Findlay SEG, Sinsabaugh RL (eds) Interactivity of dissolved organic matter. Academic Press, San Diego, pp 139–159

Mulvaney RL (1996) Nitrogen-inorganic forms. In: Sparks DL (ed) Methods of soil analysis, Part 3. Chemical methods, SSSA Book Series No. 5. Soil Science Society of America, Madison, pp 1123–1184

Myers RG, Pierzynski GM, Thien SJ (1995) Improving the iron oxide sink method for extracting soil phosphorus. Soil Sci Soc Am J 59:853–857

Myers RG, Pierzynski GM, Thien SJ (1997) Iron oxide sink method for extracting soil phosphorus: paper preparation and use. Soil Sci Soc Am J 61:1400–1407

Myrold DD (2005) Transformations of nitrogen. In: Sylvia DM, Fuhrmann JJ, Hartel PG, Zuberer DA (eds) Principles and applications of soil microbiology. Pearson Education Inc., Upper Saddle River, pp 333–372

National Technical Committee for Hydric Soils (2007) Technical note 11: technical standards for hydric soils. USDA-NRCS, Washington, DC, Available from: ftp://ftp-fc.sc.egov.usda.gov/NSSC/Hydric_Soils/note11.pdf

Nelson DW, Sommers LE (1996) Total carbon, organic carbon, and organic matter. In: Sparks DL (ed) Methods of soil analysis, Part 3. Chemical methods, SSSA Book Series No. 5. Soil Science Society of America, Madison, pp 961–1010

Newman S, Kumpf H, Lang JA, Kennedy WC (2001) Decomposition responses to phosphorus enrichment in an Everglades (USA) slough. Biogeochemistry 54:229–250

Nokes CJ, Fenton E, Randall CJ (1999) Modeling the formation of brominated trihalomethanes in chlorinated drinking waters. Water Res 33:3557–3568

Ohno T (2002) Fluorescence inner-filtering correction for determining the humification index of dissolved organic matter. Environ Sci Technol 36:742–746

Owens PR, Wilding LP, Lee LM, Herbert BE (2005) Evaluation of platinum electrodes and three electrode potential standards to determine electrode quality. Soil Sci Soc Am J 69:1541–1550

Pasternack GB, Brush GS (1998) Sedimentation cycles in a river-mouth tidal freshwater marsh. Estuaries 21:407–415

Patrick WH, Gambrell RP, Faulkner SP (1996) Redox measurements of soils. In: Sparks DL (ed) Methods of soil analysis, Part 3. Chemical methods, SSSA Book Series No. 5. Soil Science Society of America, Madison, pp 1255–1273

Paul EA, Clark FC (1996) Soil biology and biochemistry. Academic Press, New York

Persson H (1978) Root dynamics in a young Scots pine stand in Central Sweden. Oikos 30:508–519

Potter CS (1997) An ecosystem simulation model for methane production and emission from wetlands. Global Biogeochem Cycles 11:495–506

Potter C, Klooster S, Hiatt S, Fladeland M, Genovese P, Gross P (2006) Methane emissions from natural wetlands in the United States: satellite-derived estimation based on ecosystem carbon cycling. Earth Interact 10:1–12, Article 22

Rabenhorst MC (2005) Biologic zero: a soil temperature concept. Wetlands 25:616–621

Rabenhorst MC (2008) Protocol for using and interpreting IRIS tubes. Soil Surv Horiz 49:74–77

Rabenhorst MC (2009) Making soil oxidation-reduction potential measurements using multimeters. Soil Sci Soc Am J 73:2198–2201

Rabenhorst MC (2010) Visual assessment of IRIS tubes in field testing for soil reduction. Wetlands 30:847–852

Rabenhorst MC (2012) Simple and reliable approach for quantifying IRIS tube data. Soil Sci Soc Am J 76:307–308

Rabenhorst MC, Burch SN (2006) Synthetic iron oxides as an indicator of reduction in soils (IRIS). Soil Sci Soc Am J 70:1227–1236

Rabenhorst MC, Castenson KL (2005) Temperature effects on iron reduction in a hydric soil. Soil Sci 170:734–742

Rabenhorst MC, Hively WD, James BR (2009) Measurements of soil redox potential. Soil Sci Soc Am J 73:668–674

Rabenhorst MC, Megonigal JP, Keller J (2010) Synthetic iron oxides for documenting sulfide in marsh porewater. Soil Sci Soc Am J 74:1383–1388

Racchetti E, Bartoli M, Soana E, Longhi D, Christian RR, Pinardi M, Viaroli P (2011) Influence of hydrological connectivity of riverine wetlands on nitrogen removal via denitrification. Biogeochemistry 103:335–354

Raymond PA, Saiers JE (2010) Event controlled DOC export from forested watersheds. Biogeochemistry 100:197–209

Reddy KR, DeLaune RD (2008) Biogeochemistry of wetlands: science and applications. CRC Press, Boca Raton

Reddy KR, DeLaune RD, DeBusk WF, Koch M (1993) Long-term nutrient accumulation rates in the Everglades wetlands. Soil Sci Soc Am J 57:1145–1155

Richardson CJ, Craft CB (1993) Effective phosphorus retention in wetlands: fact or fiction? In: Moshiri GA (ed) Constructed wetlands for water quality improvement. CRC Press, Inc., Boca Raton, pp 271–282

Ritchie JC, McHenry JR (1990) Application of radioactive fallout cesium-137 for measuring soil erosion and sediment accumulation rates and patterns: a review. J Environ Qual 19:215–233

Robertson GP, Paul EA (1999) Decomposition and organic matter dynamics. In: Sala OE, Jackson RB, Mooney HA, Howarth RW (eds) Methods of ecosystem science. Springer, New York, pp 104–116

Robinson JS, Sharpley AN (1994) Organic phosphorus effects on sink characteristics of iron-oxide-impregnated filter paper. Soil Sci Soc Am J 58:758–761

Rückauf U, Augustin J, Russow R, Merbach W (2004) Nitrate removal from drained and reflooded fen soils affected by soil N transformation processes and plant uptake. Soil Biol Biochem 36:77–90

Sharpley AN (1993a) An innovative approach to estimate bioavailable phosphorus in agricultural runoff using iron oxide-impregnated paper. J Environ Qual 22:597–601

Sharpley AN (1993b) Estimating phosphorus in agricultural runoff available to several algae using iron-oxide paper strips. J Environ Qual 22:678–680

Sharpley AN (2009) Bioavailable phosphorus in soil. In: Kovar JL, Pierzynski GM (eds) Methods of phosphorus analysis for soils, sediments, residuals, and waters, 2nd edn. Southern Coop Series Bull No. 408. Virginia Tech University, Blacksburg, VA, pp 38–43

Sherry S, Ramon A, Eric M, Richard E, Barry W, Peter D, Susan T (1998) Precambrian shield wetlands: hydrologic control of the sources and export of dissolved organic matter. Clim Chang 40:167–188

Sirivedhin T, Gray KA (2006) Factors affecting denitrification rates in experimental wetlands: field and laboratory studies. Ecol Eng 26:167–181

Sissingh HA (1983) Estimation of plant-available phosphates in tropical soils. A new analytical technique. Nota 235. Institute for Soil Fertility Research, Haren

Sonzogni WC, Chapra SC, Armstrong DE, Logan TJ (1982) Bioavailability of phosphorus inputs to lakes. J Environ Qual 11:555–563

Søvik AK, Mørkved PT (2008) Use of stable nitrogen isotope fractionation to estimate denitrification in small constructed wetlands treating agricultural runoff. Sci Total Environ 392:157–165

Sparks DL (1995) Environmental soil chemistry. Academic Press, San Diego

Stedmon CA, Markager S, Bro R (2003) Tracing DOM in aquatic environments using a new approach to fluorescence spectroscopy. Mar Chem 82:239–254

Stevenson FJ (1996) Nitrogen-organic forms. In: Sparks DL (ed) Methods of soil analysis, Part 3. Chemical methods, SSSA Book Series No. 5. Soil Science Society of America, Madison, pp 1185–1200

Stolt MH (2005) Development of field protocols for three-tiered assessments of coastal wetlands in Rhode Island. Final Rep. USEPA Region 1 Wetlands Office. Dep Nat Resour Sci, Univ Rhode Island, Kingston

Syers JK, Williams JDH, Campbell AS, Walker TW (1967) The significance of apatite inclusions in soil phosphorous studies. Soil Sci Soc Am Proc 31:752–756

Teasdale PR, Batley GE, Apte SC, Webster IT (1995) Pore water sampling with sediment peepers. Trends Anal Chem 14:250–256

Ter-Mikaelian MT, Korzukhin MD (1997) Biomass equations for sixty-five North American tree species. For Ecol Manag 97:1–24

Thomas GW (1996) Soil pH and soil acidity. In: Sparks DL (ed) Methods of soil analysis, Part 3. Chemical methods, SSSA Book Series No. 5. Soil Science Society of America, Madison, pp 475–490

Thurman EM (1985) Organic geochemistry of natural waters. M Nijhoff/Dr W. Junk Publishers, Dordrecht, p 516

Tiedje JM, Simkins S, Groffman PM (1989) Perspectives on measurement of denitrification in the field including recommended protocols for acetylene based methods. Plant Soil 115:261–284

U.S. Army Corps of Engineers Environmental Laboratory (1987) Corps of engineers wetland delineation manual. Technical report Y-87-1. U.S. Army Engineer Waterways Experiment Station, Vicksburg

U.S. EPA (2008) Methods for evaluating wetland condition: biogeochemical indicators. Office of Water, U.S. Environmental Protection Agency, Washington, DC. EPA-822-R-08-022

USDA, NRCS (2010) Field indicators of hydric soils in the United States. Ver. 7.0. In: Vasilas LM, Hurt GW, Noble CV (eds) USDA, NRCS in cooperation with the National Technical Committee for Hydric Soils

Ussiri DAN, Johnson CE (2004) Sorption of organic carbon fractions by Spodosol mineral horizons. Soil Sci Soc Am J 68:253–262

Vadeboncoeur MA, Hamburg SP, Yanai RD (2007) Validation and refinement of allometric equations for roots of northern hardwoods. Can J For Res 37:1777–1783

Vasilas BL, Fuhrmann JJ (2011) Microbiology of hydric soils. In: Vasilas LM, Vasilas BL (eds) A guide to hydric soils in the Mid-Atlantic region. Ver. 2.0. U.S. Department of Agriculture Natural Resources Conservation Service, Morgantown, pp 41–47

Vasilas BL, Ham GE (1984) Nitrogen fixation in soybeans: an evaluation of measurement techniques. Agron J 76:759–764

Veneman PLM, Pickering EW (1983) Salt bridge for field redox potential measurements. Commun Soil Sci Plant Anal 14:669–677

Vepraskas MJ, Faulkner SP (2001) Redox chemistry of hydric soils. In: Richardson JL, Vepraskas MJ (eds) Wetland soils: genesis, hydrology, landscapes, and classification. Lewis Publishers, Boca Raton, pp 85–105

Vogt KA, Persson H (1991) Root methods. In: Lassoie JP, Hinckley TM (eds) Techniques and approaches in forest tree ecophysiology. CRC Press, Boca Raton, pp 477–502

Vogt KA, Vogt DJ, Palmiotto PA, Boon P, O'Hara J, Asbjornsen H (1996) Review of root dynamics in forest ecosystems grouped by climate, climatic forest type and species. Plant Soil 187:159–219

Walton DWH, Allsopp D (1977) A new test cloth for soil burial trials and other studies on cellulose decomposition. Int Biodeterior Bull 13:112–115

Weaver RW, Danso SKA (1994) Dinitrogen fixation. In: Weaver RW, Angle JS, Bottomly PS (eds) Methods of soil analysis, Part 2. Microbiological and biochemical properties, SSSA Book Series No. 5. Soil Science Society of America, Madison, pp 1019–1045

Weishaar JL, Aiken GR, Depaz E, Bergamaschi B, Fram M, Fujii R (2003) Evaluation of specific ultra-violet absorbance as an indicator of the chemical composition and reactivity of dissolved organic carbon. Environ Sci Technol 37:4702–4708

Weishampel P, Kolka R (2008) Measurement of methane fluxes from terrestrial landscapes using static, non-steady state enclosures. In: Hoover CM (ed) Field measurements for forest carbon monitoring. Humana Press, New York, pp 163–172

Whalen SC (2000) Nitrous oxide emission from an agricultural soil fertilized with liquid swine waste or constituents. Soil Sci Soc Am J 64:781–789

Whittaker RH, Bormann FH, Likens GE, Siccama TG (1974) The Hubbard Brook ecosystem study: forest biomass and production. Ecol Monogr 44:233–252

Wieder RK, Lang GE (1982) A critique of the analytical methods used in examining decomposition data obtained from litter bags. Ecology 63:1636–1642

Wienhold BJ, Varvel GE, Wilhelm WW (2009) Container and installation time effects on soil moisture, temperature, and inorganic nitrogen retention for an in situ nitrogen mineralization method. Commun Soil Sci Plant 40:2044–2057

Wilson HF, Xenopoulos MA (2009) Effects of agricultural land use on the composition of fluvial dissolved organic matter. Nat Geosci 2:37–41

Wolf DC, Wagner GH (2005) Carbon transformations and soil organic matter formation. In: Sylvia DM, Fuhrmann JJ, Hartel PG, Zuberer DA (eds) Principles and applications of soil microbiology. Pearson Education Inc., Upper Saddle River, pp 285–332

Wray HE, Bayley SE (2007) Denitrification rates in marsh fringes and fens in two boreal peatlands in Alberta, Canada. Wetlands 27:1036–1045

Yamashita Y, Tanoue E (2003) Chemical characterization of protein-like fluorophores in DOM in relation to aromatic amino acids. Mar Chem 82:255–271

Yamashita Y, Maie N, Briceno H, Jaffé R (2010) Optical characterization of dissolved organic matter in tropical rivers of the Guayana Shield, Venezuela. J Geophys Res 115:G00F10. doi:10.1029/2009JG000987

Zibilske LM (1994) Carbon mineralization. In: Weaver RW, Angle JS, Bottomly PS (eds) Methods of soil analysis, Part 2. Microbiological and biochemical properties, SSSA Book Series No. 5. Soil Science Society of America, Madison, pp 835–863

Zsolnay A, Baigar E, Jimenez M, Steinweg B, Saccomandi F (1999) Differentiating with fluorescence spectroscopy the sources of dissolved organic matter in soils subjected to drying. Chemosphere 38:45–50

Student Exercises

Classroom Exercises

Classroom Exercise 1: Understanding pH

The term pH is derived from the French term for hydrogen power, *pouvoir hydrog'ne*. In chemistry, pH is a measure of the acidity or basicity of an aqueous solution. Pure water is considered to be neutral, with a pH close to 7.0 at 25 °C. Solutions with a pH less than 7 are considered to be acidic; solutions with a pH greater than 7 are considered to be basic or alkaline. From a computational standpoint, pH is the negative logarithm (base 10) of the *active* hydrogen ion (H^+) concentration (in moles/L), or $-\log [H^+]$. pH does not precisely reflect H^+ concentration, but incorporates an activity factor to represent the tendency of hydrogen ions to interact with other components of the solution. For purposes of this exercise, consider the active hydrogen ion concentration to be equal to the hydrogen ion concentration. Because a pH value represents the negative log of a concentration, a pH of 4 corresponds to 10^{-4} mol/L, which corresponds to 0.0001 mol/L. What is the

Table 7.3 Ortho-P control chart

OP standards concentrations (mg/L)			
0.005	0.020	0.050	0.100
% transmittance			
99.1	96.8	91.8	84.8
99.2	96.7	91.6	84.6
99.2	96.8	92.6	84.3
99.3	96.2	91.0	84.0
99.8	96.8	92.6	86.9
98.9	96.2	92.8	85.4
98.8	96.6	91.8	84.8
Calculated OP concentrations (mg/L)			
0.005636	0.020227	0.053180	0.102463
0.005009	0.020869	0.054535	0.103930
0.005009	0.020227	0.047789	0.106138
0.004383	0.024090	0.058618	0.108353
0.001262	0.020227	0.047789	0.087264
0.006891	0.024090	0.046448	0.098082
0.007520	0.021512	0.053180	0.102463
Mean			
SD			
CV			
RSD			
MDL			

pH of a 0.1 M hydrochloric acid solution? What is the molar concentration of H^+ in a hydrochloric acid solution at pH 3? Given a 1 L container of pure water, how many hydrogen ions are present?

Classroom Exercise 2: Calculating Detection Limits

An analytical laboratory ran quality control charts to determine the method detection limit for their ortho phosphorus test method. They used the ascorbic acid method and read the color change on a spectrophotometer. The lab assistant made four standard solutions containing 0.005, 0.020, 0.050, and 0.100 mg/L PO_4^{-3}-P. The results are given in Table 7.3. The standard curve was determined before the control samples were run and the regression equation is as follows: $LN(Y) = -1.6094X + 4.6052$, where Y represents the % transmittance values and X represents the concentration. The curve was used to determine the associated concentration of the control sample results.

Calculate the mean (x), standard deviation (SD), coefficient of variation (CV), relative standard deviation (RSD) and the MDL for each of the control standard

Table 7.4 DBH data and allometric parameters for selected trees

Species	Dbh (cm) of individuals						Allometric parameters	
	1	2	3	4	5	6	a	b
Fagus grandifolia	43	28	33	27	26	31	0.0842	2.5715
Carya laciniosa	24	26	18				0.0792	2.6349
Acer rubrum	15	23	28	25	19		0.0910	2.5080
Quercus rubra	36	31	25	22			0.1130	2.4572
Quercus alba	19	36	34	29			0.0579	2.6887

concentrations. The standard deviation from a single standard concentration can be used to determine the MDL. However, the standard deviation is sometimes estimated from measurements made at a minimum of three levels (low, mid-range and high range) of standards. The value for the method standard deviation is calculated by plotting the standard deviation vs. concentration for the different concentrations. The method standard deviation is extrapolated from the curve, the value of the standard deviation as the concentration goes to zero. Plot the calculated standard deviations of the control standards versus concentration and determine the new MDL from the extrapolated SD. See *Quality Control and Detection Limits* for the required equations.

Classroom Exercise 3: Estimating Tree Biomass Using Allometric Equations

Tree biomass can be calculated using the allometric equation, $M = aD^b$, where M is the oven-dry weight of the biomass component of a tree (kg), D is diameter at breast height (DBH) (cm), and a and b are parameters unique to each species. The allometric parameters can be derived experimentally or selected from the literature.

A field investigation of a small wetland revealed 22 trees representing five species. Diameter breast height for each individual and allometric parameters for each species are presented in Table 7.4. The allometric parameters are for total aboveground biomass (Brenneman et al. 1978). Based on the data in Table 7.4, determine the total aboveground biomass for the tree stratum in the plot.

Reference

Brenneman BB, Frederick DJ, Gardner WE, Schoenhofen LH, Marsh PL (1978) Biomass of species and stands of West Virginia hardwoods. In: Pope PE (ed) Proceedings of Central Hardwood Forest Conference II. Purdue University, West LaFayette, pp 159–178

Laboratory Exercises

Laboratory Exercise 1: Measurement of Soil Respiration

Introductory Comments

Laboratory investigations of soil respiration generally monitor carbon dioxide production in either gas-tight, static microcosms or by using a dynamic, flowing-gas system (Zibilske 1994). The former approach is simpler and recommended except when experimental considerations dictate use of a dynamic system. Static systems make use of microcosms containing soil and possessing sufficient headspace to permit gas accumulation and simultaneously avoid the development of excessively low oxygen levels if maintenance of aerobic conditions is desired. If anaerobic conditions are to be investigated, the soil can be flooded with water and the headspace purged with N_2 or another suitable gas prior to incubation. Carbon dioxide produced by aerobic or anaerobic respiration, or by certain anaerobic fermentative processes, either is captured in alkali traps for measurement by acid-base titration or is measured by collecting gas samples for analysis by gas chromatography. The choice between the 2 CO_2 measurement methods is a matter of preference and instrument availability, as both provide good sensitivity. The reader is directed to Zibilske (1994) for details and a complete discussion of the various options.

Containers to be used as microcosms must be air-tight. Canning jars work well, and their lids can be equipped with septa for gas sampling if CO_2 is to be measured by gas chromatography. The nature of the soil placed in the microcosms may range from mixed, sieved soil to intact soil cores. In any case, the ratio of headspace volume-to-soil mass is an important consideration in aerobic studies; a ratio of 10:1 is recommended (e.g., 500 ml container for 50 g of soil), although lower ratios can be used, provided the jars are opened more frequently to allow for aeration. Regardless of the ratio used, it is important in aerobic studies that the oxygen levels of the microcosm be replenished on a regular basis. Presence of volatile fatty acids or reduced S-containing gasses in the headspace (e.g., a rotten egg smell) is an indication that the microcosms need to be opened more frequently. Mixing a bulking agent such as vermiculite or perlite with the soil can help prevent the development of anaerobic microsites (Thien and Graveel 2003). Additionally, for aerobic studies the moisture content of the soil should not exceed field capacity, and it may be desirable especially with fine-textured soils to use a lower moisture content that is still conducive to microbial activity. Regardless of the moisture content chosen, it should be standardized across treatments unless moisture content is being examined experimentally. Another variable affecting aeration status is the rate of any organic matter addition to the soil. A commonly used rate is 1 % by mass, although this will vary with the experimental objectives.

There are additional considerations when alkali traps are being used to collect CO_2. Care must be taken that excess alkali is used in the traps so that the CO_2-trapping ability is not exhausted between samplings. A reasonably conservative starting point when working with soil amended with organic matter at a 1 % rate is the equivalent of 15 ml of 1.5 M NaOH per 50 g of soil (Thien and Graveel 2003). It is absolutely essential that the alkali used in the traps and the acid used for titration be accurately standardized; the use of potassium hydrogen phthalate or a commercially available primary standard is strongly recommended. The concentration of NaOH used in the traps should be determined each time they are replenished. Additionally, it is critical that microcosm blanks (i.e., microcosms containing alkali traps only) be employed to account for any CO_2 in the headspace originally or added when the microcosms are opened for aeration; the CO_2 measured in the blanks should be subtracted from the values obtained from the microcosms containing soil.

Objective: To monitor soil respiration under laboratory conditions using acid-base titration to measure CO_2 evolution.

Materials and Equipment Needed

500-ml soil microcosms (16-oz canning jars with air-tight lids)
Balance
1.5 M NaOH, standardized
Vials for alkali traps (plastic, 50-ml centrifuge tube with air-tight cap)
1.0 M HCl, standardized
1.0 M $BaCl_2$
Phenolphthalein indicator solution
Burets
125-ml Erlenmeyer flasks

Procedures

Overview: The general approach outlined here is suggested for its simplicity and adaptability to a wide range of experimental objectives and is based on the procedure described by Thien and Graveel (2003) for use in instructional soil science laboratories.

1. Weigh 50 g of soil into each microcosm. Adjust moisture content and add a bulking agent and or organic amendment if desired. For anaerobic studies, the soil can be flooded with water at this time.
2. Dispense a precisely measured volume (~15 ml) of standardized NaOH into vials to be used as alkali traps. Place uncovered alkali traps into the microcosms

taking care not to spill the NaOH. Also, prepare microcosm blanks containing alkali traps only.
3. Securely attach the covers of the microcosms, first purging the headspace with N_2 gas for anaerobic treatments. Place the microcosms under the desired incubation conditions.
4. Collect samples for analysis at time intervals dictated by the experimental objectives, typically sampling more frequently early in the study in order to obtain an estimate of CO_2 evolution rate and thereby avoid exhausting the alkali traps by too infrequent sampling. Retrieve the alkali traps from the microcosms and attach air-tight covers to the vials to stop CO_2 collection.
5. After a brief period of time to allow for aeration, place fresh uncovered alkali traps in the microcosms. As noted previously, the NaOH used in the traps should be newly standardized. Replace microcosm lids.
6. For analysis of alkali traps, the solution in a trap is transferred quantitatively, using water rinses, to a titration flask; this procedure and the titration itself should be performed on one sample at a time to minimize spurious CO_2 collection. Add 25 ml of 1 M $BaCl_2$ to the flask, the contents of which should form a white precipitate of $BaCO_3$. A few drops of phenolphthalein indicator are added and should result in a pink coloration, indicating that the NaOH originally added to the trap was not exhausted during the soil incubation (i.e., excess NaOH remains). This excess NaOH is titrated from a pink suspension to a milky white endpoint using standardized 1.0 M HCl.
7. Carbon dioxide evolution is calculated as mg CO_2 = (meq base – meq acid) (22), where meq base is the milliequivalents of NaOH originally present in the traps, meq acid is the milliequivalents of HCl required to titrate the excess NaOH, and 22 is the milliequivalent weight of CO_2. Milliequivalents of acid or base are calculated as the product of concentration expressed as normality and volume expressed as ml [e.g., meq acid = (concentration of HCl as normality used in titration)] (ml of HCl used in titration). For NaOH and HCl, molarity concentration is equal to normality concentration (e.g., 1.5 M NaOH = 1.5 N NaOH).
8. Commonly calculated parameters in respiration studies are cumulative CO_2 evolution with time and mean rate of CO_2 evolution between sampling times. Depending on experimental objectives, it may be desirable to convert values from a CO_2 basis to an elemental carbon basis.

References

Thien SJ, Graveel JG (2003) Laboratory manual for soil science, 8th edn. McGraw Hill, Boston

Zibilske LM (1994) Carbon mineralization. In: Weaver RW et al (eds) Methods of soil analysis. Part 2: Microbiological and biochemical properties. SSSA Book Series No. 5. SSSA, Madison, pp 835–863

Laboratory Exercise 2: Documenting Redox Conditions in Soil Mesocosms Using Pt Electrodes and Alpha-Alpha Dipyridyl Dye

Introduction and Background

When soil pores are filled with water and the soil becomes saturated, the movement of oxygen into the soil is greatly inhibited by the slow diffusion of gasses through liquids. If oxidizable carbon compounds are present and temperatures are sufficiently warm, aerobic heterotrophic microorganisms will begin to oxidize the organic matter using oxygen as the electron acceptor. Once the oxygen has been depleted, various anaerobic microbes will begin to use alternate compounds as electron acceptors. Some of the important and common compounds that serve as electron acceptors in saturated soils and thus can be reduced, include nitrate NO_3^- to nitrite NO_2^- (or eventually to dinitrogen N_2 gas), solid phase manganese oxides MnO_2 to soluble Mn^{2+}, solid phase iron oxyhydroxides FeOOH to soluble Fe^{2+}, and sulfate $SO_4^=$ to sulfide $S^=$. When water is removed from the saturated soil through drainage or evapotranspiration, the reduced species can become oxidized. This reoxidation is often microbially mediated, but some reactions can occur chemically.

The redox potential, or Eh, can be measured using Pt electrodes with a reference electrode and the Eh, together with the pH, can be used to determine how reducing or oxidizing a soil is in order to predict whether particular compounds would be expected to be reduced or oxidized. This is done by comparing the measured Eh and pH values to a line calculated using the Ksp for particular compounds and phases based upon thermodynamic data. There are a number of variables which can affect the precise calculation of the line (such as the concentration of soluble components). Diagrams showing the redox stability lines (fields) for many compounds have been created and published.

Objectives

1. To make Eh measurements in a soil mesocosm
2. To interpret soil redox conditions by using Eh and pH measurements and redox stability diagrams
3. To use, and interpret the use of, alpha-alpha dipyridyl dye
4. To compare the Eh-pH data collected with the alpha-alpha dipyridyl dye reactions observed

Materials and Equipment Needed

Soil Mesocosm 7.5 cm diam by 40 cm high[1]
Six, 40 cm Pt wire electrodes

---

[1] A 7.5 cm schedule 40 PVC pipe 50 cm in length should be sharpened on one end (bevel out) so that the pipe can be driven into the soil to a depth of 40 cm and then excavated to collect the mesocosm.

One reference electrode (with salt bridge)
One, high impedance voltmeter
pH meter and buffers
Light's Solution
Bucket
Duct tape
alpha-alpha dipyridyl dye[2]

Procedures

1. *Selection and testing of electrodes.* Electrodes must be tested to ensure they are working properly. Each group of students will require 6 Pt electrodes. Electrodes can be made at fairly low expense if done according to the procedure described by Owens et al. (2005). To test the Pt electrodes, place a group of electrodes into a beaker containing Light's solution (Light 1972), which contains Fe(II) and Fe(III) in a sulfuric acid solution (be careful). This solution is poised and stable. Electrodes should be tested in groups using the same reference electrode, and all individual electrodes within the group should be within 1–2 mV the mean. Typically, they should read around Eh = 675 (raw voltage reading of about 430 if using a calomel reference). However, it is most important that the electrode readings group together and sometimes values for the Light's solution can drift. Groups of calomel reference electrodes can be similarly tested, although they typically vary a bit more.
2. *Selection of mesocosms.* Each group of (2–4) students will work with one mesocosm. See footnote #1 regarding collection of the mesocosm. A piece of geotextile fabric should be attached across the bottom of the mesocosm using duct tape, in order to help keep the soil in the mesocosm (Fig. 7.4). Care should be taken to give support to the soil within the mesocosm so that it does not accidentally slide out.

    The depth from the top of the mesocosm to the top of the soil should be measured and then marked on the outside of the core. After this, mark the core at the 2 depths at which the Pt electrodes will be later installed. These depths will be at 10 and 25 cm below the top of the soil (NOT below the top of the PVC cylinder) as shown in Fig. 7.4.
3. *Instrumentation of the mesocosms.* An overview of the instrumentation is shown in Fig. 7.5 which illustrates how the mesocosm will appear when all the electrodes have been installed. Probably, the best order in which to install the electrodes is the following: (1) the deep (25 cm) Pt electrodes, (2) the salt bridge and calomel electrode; and (3) the shallow (10 cm) Pt electrodes.

---

[2] This should be made according to the procedure of Childs (1981).

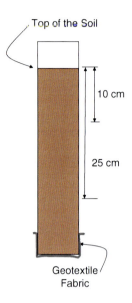

Fig. 7.4 Schematic of collected mesocosm with approximately 40 cm of soil within a 7.5 cm diameter, schedule 40 PVC pipe, 50 cm in length. A piece of geotextile fabric is taped to the bottom to prevent the soil from sliding out. Marks should be placed on the outside of the PVC at depths of 10 and 25 cm below the soil surface

4. *Installing the deep electrodes.* Note that the three deep Pt electrodes will be placed in an equally spaced arrangement around the mesocosm, optimizing the distance between the edge of the mesocosm and the future location of the salt bridge (see Fig. 7.5). Using colored tape, place a mark on each electrode that is 25 cm above the Pt point. Then, using a narrow sharpened stainless steel rod (slightly larger than the electrode), carefully make a VERTICAL pilot hole that is approximately ½ cm shallower than the depth where you will place the Pt electrode. Remove the rod and carefully slide a straight Pt electrode into the hole, and push the electrode to the exact depth where measurements are to be made (25 cm). Carefully install two additional deep electrodes in the same manner being sure to distribute them around the mesocosm (Fig. 7.5). Label these electrodes 25A, 25B, and 25C (for 25 cm below the soil surface).
5. *Installing the salt bridge.* The salt bridge is made from a piece of PVC pipe that has an OD of approximately 22 mm. It is filled with an agar which is saturated with KCl (similar to the calomel electrode itself) (Veneman and Pickering 1983). Prepare to use the salt bridge by inserting your previously selected (and tested) calomel reference electrode into the agar at the top of the salt bridge until the electrode is approximately 2–3 cm into the tube and the agar has encompassed the end of the electrode. Wipe any agar or salt from the electrode and PVC tube and carefully wrap with parafilm to produce a water tight seal that will hold the electrode in place. This will help to keep the electrode from drying out over time.

   A pilot hole approximately 5–10 cm deep should be made using a section of sharpened 2.5 cm dia. pipe, and the soil should be removed to make room for

7 Wetland Biogeochemistry Techniques

Fig. 7.5 Diagrams showing *side view* (**a**) and *top view* (**b**) of mesocosm indicating the spatial placement of the salt bridge and the Pt electrodes

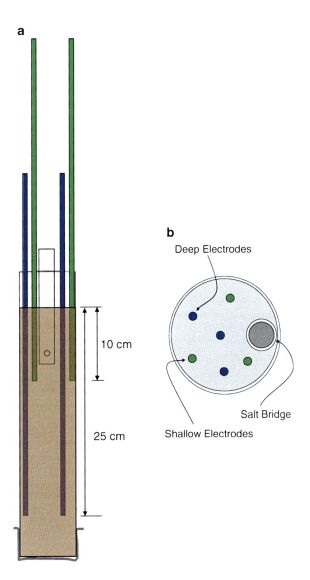

the salt bridge. Then insert the salt bridge firmly into the pilot hole, being careful that the agar does not slide out of the tube. The salt bridge should extend about 5–10 cm above the top of the soil.

6. *Installing the shallow Pt electrodes.* Mark each of the other three electrodes at a depth of 10 cm using colored tape. The three shallow Pt electrodes should be placed in an equally spaced arrangement around the mesocosm, much like the deep electrodes and following a similar installation procedure, but being careful not to disturb the other Pt electrodes or the salt bridge and calomel

electrode (Fig. 7.5). Electrodes should be installed to a depth of 10 cm below the soil surface (which will be 15 cm above the deep electrodes) and should be labeled as electrodes 10A, 10B, and 10C.

7. *Taking initial Eh measurements (prior to saturation).* Using either a lab grade Eh meter or a multimeter in conjunction with a device to create a high resistance circuit (Rabenhorst et al. 2009; Rabenhorst 2009), the voltage should be measured in the circuit created between the calomel reference electrode and each Pt electrode. The positive (red) wire should attach to the Pt electrode and the black wire should connect to the reference electrode. When the electrodes have been recently installed, there may be some slight drift during the measurement, but this drift should become less apparent on subsequent days. Typically, these measurements are recorded to the nearest 0.001 V (note there is too much variability to warrant recording with any greater precision than this so make sure you are NOT reading to tenths of a mV). Commonly, students will occasionally reverse the wires on the volt meter when making measurements. This will result in VERY LARGE ERRORS, because the voltage will have the opposite sign (-300 mV vs. 300 mV). Be very careful to ensure that the Pt electrode leads to the red (+) pole on the volt meter and the reference electrode is connected to the black ($-$) pole. Your initial readings will probably be somewhere in the range of 200–400 mV (before correction for the reference electrode). Over time, the voltages will likely become lower (especially for the deep electrodes installed below the water table.) Note that pH measurements must also be obtained at the same 2 depths where the electrodes will be placed. It is recommended that soil pH at 10 and 25 cm be collected from a replicate mesocosm so that the instrumented mesocosm does not need to damaged.

8. *Saturating the mesocosm.* Mesocosms will be saturated to a depth of 15 cm below the soil surface. In order to saturate each mesocosm, stand the mesocosm vertically in a container (bucket) where the water can be adjusted to the proper height, and secure it using duct tape as shown in Fig. 7.6. Distilled water should then be added slowly to the bucket which will result in filling the soil pore space in the mesocosm from below, which should help minimize the entrapment of air during saturation. The water level should be raised until it is at the appropriate height (Fig. 7.7). An alternate arrangement is also shown in Fig. 7.7 where the mesocosm is saturated to the soil surface. This can be used to illustrate differences in properties of contrasting soil horizons (such as OM content in A vs. B horizons) if different soil materials occur at 10 vs. 25 cm.

9. *Eh measurements after saturation.* Approximately 30–60 min after saturation, collect the second set of Eh measurements from the mesocosm as described previously. After this, Eh measurements should be made on the mesocosms daily for the first week, and then every other day through subsequent weeks. This should be continued for a minimum of 2 weeks and may produce better results if extended for 3 or 4 weeks. The level of the water in the bucket will need to be checked and maintained at the proper height.

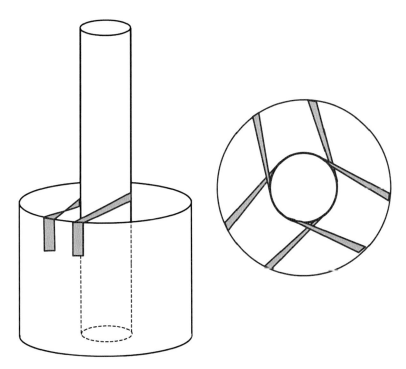

Fig. 7.6 Each mesocosm should be secured in a small bucket using three pieces of duct tape which wrap around the mesocosm and are affixed at approximately 120° from each other

10. *Disassembly of the mesocosms.* At the end of 2 (or more) weeks, remove the mesocosm from the water reservoir and place it on some absorbent material (newspaper) and allow it drain. Empty the water from the bucket into appropriate containers in the lab (do not empty soil down the sinks!)

    Carefully remove each of the Pt electrodes trying not to disturb the soil too much. Rinse off the Pt electrodes and set aside in a group being careful to keep them together and labeled with your mesocosm number. Check each electrode by placing them in the Light's solution and determining the voltage measured using a common calomel reference. This should be done to determine if any appear to be malfunctioning (in which case, data from those electrodes should not be included in the analysis).

    Remove the calomel reference electrode from the salt bridge, rinse off the electrode and make sure it is adequately filled with saturated KCl before storing. Using a rod, pole, or other device, extrude the soil from the PVC sleeve onto newspaper, trying to keep the core as intact as possible.

11. *Testing for Ferrous Fe using alpha-alpha dipyridyl (αad) dye (or strips).*[3]
    Using a knife or spatula, split open the extruded core lengthwise, trying mainly

---

[3] Alpha-alpha dipyridyl strips are available at http://www.ctlscientific.com/cgi/display.cgi?item_num=90725

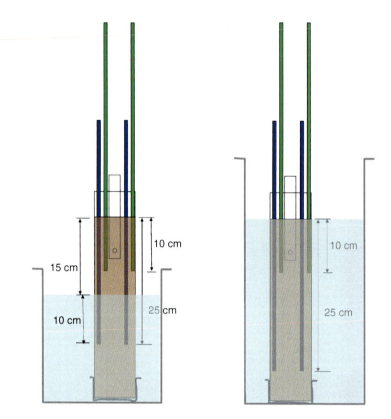

Fig. 7.7 On *left*, distilled water is added to the bucket until it reaches a level 15 cm below the soil surface. The tips of shallow and deep Pt electrodes should be located 5 cm above the water level and 10 cm below the water level, respectively. Water should be added periodically to preserve the desired level. An alternate arrangement (shown on *right*) is to use a taller bucket and to saturate the mesocosm to the soil surface. This can be especially informative if soil horizons with contrasting properties (such as OM content) occur at depths of 10 and 25 cm

to break open and expose the fresh soil surface (rather than a knife-cut soil surface). Note that you will need to test the freshly exposed and broken soil surface with the ααd (sometimes a false positive reaction to ααd is detected from where the low valence Fe from the steel in a knife blade or shovel has contacted the soil.)

Place a few drops of ααd (Childs 1981) on one-half of the core at various depths along the length of the mesocosm and note whether there is a reaction. Look for a pink color to develop. (If there is a lot of Fe^{2+} in the soil solution, this can sometimes occur quite rapidly. Other times, if there is very little Fe^{2+} in solution, it may take a few minutes for a subtle reaction to be observed.) In particular, pay attention to whether there is a reaction in the vicinity of where you had placed the electrodes (10 and 25 cm below the soil surface). Also keep in mind, the depth at which the water table was maintained in your mesocosm.

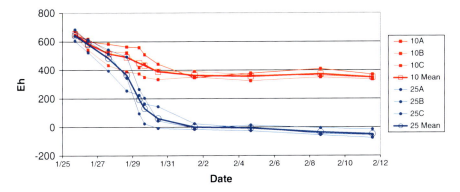

Fig. 7.8 Plotting of Eh measurements as a function of time. Note that replicate Eh measurements sometimes vary by more than 100 mV

Record your observations (especially at the depths of the electrodes), noting whether the reaction to ααd was negative, or positive, and if positive, indicate whether you think the reaction was weak, moderate or strong. Document any reaction to ααd at various other places up and down the core to try to obtain a better sense of where in the soil, Fe^{3+} has been reduced to Fe^{2+}. If you observe a positive reaction to ααd in the soil core, be careful to document where in the soil that reaction was observed, especially in relation to where the water table was located within the soil.

12. *Measuring soil pH.* In order to be able to plot data on an Eh-pH diagram, you will need to measure the pH of the soil at the same depths where you measured Eh. Use the OTHER HALF of the core to which you did NOT apply the ααd for measuring pH. Collect a few (10) ml of soil material from each of the two depths where Eh was measured and make a thick slurry by adding distilled water and stirring (the goal is the equivalent of 1:1 soil:water, but as you are not starting with dry soil, this will be an approximation). Allow the slurry to sit for 10–15 min and then mix again. Measure the pH by using a calibrated pH meter with a combination electrode and record to the nearest 0.1 units.

13. *Data Analysis.* Eh data should be reported in three ways. First, these data should be plotted as a function of time. This will allow you do evaluate whether any of the readings from any of the electrodes was spurious. Normally, a given electrode will show trends over time, and not provide erratic readings. If all the readings for a given electrode follow a trend and then one reading is way off, there is a good chance that the one reading is faulty. An example of Eh data plotted in this way is shown in Fig. 7.8.

The second way in which the mesocosm data should be evaluated is with regard to an Eh-pH stability diagram. This is to determine whether the soil conditions were (theoretically) reducing with respect to Fe at any time during the experiment. You will only have pH measurements from your mesocosm at beginning and end of the experiment, and the pH is not be expected to change dramatically over the course of a couple of weeks (often one unit or less). For the sake of simplification, we will make the assumption that the pH changed

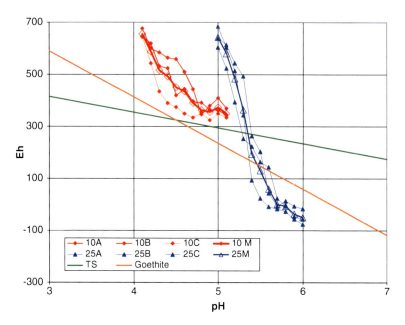

Fig. 7.9 Example data from mesocosms plotted on an Eh-pH diagram. Also shown is the "technical standard" line (TS) from the National Technical Committee for Hydric Soils which they use to define reducing soil conditions and also the stability line for goethite-Fe^{2+}

gradually (linearly) through the period of the experiment and thus use interpolated data for the dates in between. When plotted, these data may look something like those in Fig. 7.9. Also shown in Fig. 7.9 is the line for the equation Eh = 595 − (60 pH). This is the equation from the technical standard of the NTCHS (sometimes referred to as the "Technical Standard" line) (National Technical Committee for Hydric Soils 2007). This is an empirically derived line (shown in green), and data that plot above this line are considered to be oxidizing and those below the line indicate reducing conditions. The Eh-pH line showing the stability field for the crystalline iron oxide mineral goethite is also shown on the diagram (brown). Some people prefer to discuss iron reduction and Eh-pH data with respect to predictions using thermodynamically based equations such as this one. Above the line, goethite would be predicted to be stable and below the line, it would be predicted to unstable with Fe^{3+} being reduced to Fe^{2+}.

A third, and perhaps even more useful way to view the data is to plot the data over time, with the Eh values relative to the TS line (or this could also be done with respect to the a line like the goethite line). This is done by using pH data (for each date and depth) to calculate the corresponding Eh along the TS line, using the equation Eh = 595 − (60 pH). For example, if the pH ranged from 4.1 to 5.1 over a 6 day period, we would calculate the corresponding Eh values as shown in Table 7.5 (these change from 349 to 289 as the pH changes from 4.1 to 5.1).

7 Wetland Biogeochemistry Techniques

Table 7.5 Calculation of Eh data with respect to the Technical Standard Line of the NTCHS

Day	pH	Eh of TS line	Measured Eh	Eh relative to the TS line
1	4.1	349	655	306
2	4.3	337	342	5
3	4.5	325	371	46
4	4.7	313	342	29
5	4.9	301	209	−92
6	5.1	289	−30	−319

These calculated values can then be subtracted from the measured Eh values (made using Pt electrodes). Positive values indicate that they are above the TS line (oxidizing) and if the values are negative, this means the data would plot below the TS line (reducing). If the data shown in Fig. 7.8 are plotted relative to the TS line, we obtain the graph shown in Fig. 7.9. On the graph in Fig. 7.9, the "zero" of the Y axis (Eh relative to the TS) represents the TS line itself. So any values that are greater than zero imply that the data plot above the TS line (in an Eh pH diagram) and are oxidizing with respect to Fe, and any negative values imply that the data plot below the TS line and are reducing with respect to Fe. This plot allows a quick visual evaluation of how these conditions change with time and at what point in time they become reducing.

Submission of Lab Report

Each student will write and submit a laboratory report based upon the data collected by the group (in the table below, data collected by students are shown by the shaded boxes). Each student should then calculate the Eh (considering the correction for the reference electrode). A graph similar to Fig. 7.8 should be constructed. The Eh of the TS line should be calculated for the pH values measured (and interpolated) for the soil. Then you should calculate, for each electrode, the Eh relative to the TS.

Using data measured and calculated in the student's version of Table 7.6, each of the following figures should be prepared:

- A figure showing Eh over time (similar to Fig. 7.8)
- A figure showing Eh and pH in relation to the TS line and to the goethite stability line (similar to Fig. 7.9)
- A figure showing Eh relative to the TS as a function of time (similar to Fig. 7.10)

Each student should provide a discussion of the data and figures that should include some reference to the following:

- Reproducibility and replication of the data measurements;
- Comparison of redox data collected at the two soil depths;
- The meaning of the Eh/pH data collected, implications of the data based on Eh/pH stability diagrams and where data plot relative to the TS line; and
- Comparisons between methods used – in particular the Eh/pH data and the reaction of the soil with alpha-alpha dipyridyl dye.

Table 7.6 Draft data sheet for recording measurements and calculations to be used in later analyses and figures

	Raw mV			Eh					Eh relative to the TS		
	Elect	Elect	Elect	Elect	Elect	Elect		TS	Elect	Elect	
Date	#	#	#	#	#	#	pH	Eh	#	#	Elect #

A similar sheet should be used for collecting data both at 10 cm and at 25 cm

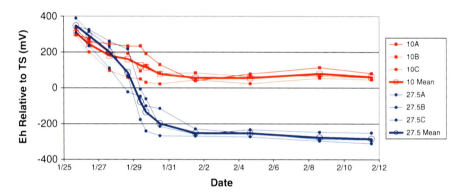

Fig. 7.10 Plot showing how Eh values are related to the Technical Standard line of the NTCHS, as a function of time. Positive values (above zero) indicate that the Eh is above the TS line (and thus is oxidizing). When values drop below zero, it indicates reducing soil conditions

References

Childs CW (1981) Field test for ferrous iron and ferric-organic complexes (on exchange sites on in water-soluble forms) in soils. Aust J Soil Res 19:175–180

Light TS (1972) Standard solution for redox potential measurements. Anal Chem 44:1038–1039

National Technical Committee for Hydric Soils (2007) Technical note 11: technical standards for hydric soils. USDA-NRCS, Washington, DC. ftp://ftp-fc.sc.egov.usda.gov/NSSC/Hydric_Soils/note11.pdf

Owens PR, Wilding LP, Lee LM, Herbert BE (2005) Evaluation of platinum electrodes and three electrode potential standards to determine electrode quality. Soil Sci Soc Am J 69:1541–1550

Rabenhorst MC (2009) Making soil oxidation-reduction potential measurements using multimeters. Soil Sci Soc Am J 73:2198–2201

Rabenhorst MC, Hively WD, James BR (2009) Measurements of soil redox potential. Soil Sci Soc Am J 73:668–674

Veneman PLM, Pickering EW (1983) Salt bridge for field redox potential measurements. Commun Soil Sci Plant 14:669–677

7 Wetland Biogeochemistry Techniques

Field Exercises

Field Exercise 1: Use and Interpretation of IRIS Tubes

Objectives

1. To understand the principles behind the use of IRIS tubes.
2. To understand how to use IRIS tubes in the field.
3. To understand how to interpret the data from IRIS tubes.
4. To compare the functioning of IRIS tubes with other measures of reduction in soils.

Part I: IRIS Tube Installation

Materials and Equipment Needed

Five IRIS tubes
7/8″ push probe for making pilot holes
Spade or shovel
Tape measure
Transparent mylar grids
Equipment for making Eh measurements
Five, 40 cm Pt wire electrodes
One reference electrode (with salt bridge)
One, high impedance voltmeter
pH meter and buffers (This can be done in the field using a portable meter, or else samples can be returned to the lab for pH measurement.)
Alpha, alpha dipyridyl dye solution or test strips

Procedures

Overview: IRIS tubes will be installed following protocols spelled out in the Rabenhorst (2008) article. Tubes will remain in the soil for 4 weeks, after which they will be examined and paint removal will be quantified. On the dates on which the tubes are installed and extracted, water table levels will be documented. Also on these dates, soil reduction will be assessed using Eh (and pH) measurements and also by testing with alpha, alpha dipyridyl dye. Comparisons will be made among all three methods for assessing soil reduction.

1. Students should work in teams of 3–4 persons.
2. Each team will install a set of five IRIS tubes in the field (at a site provided by the instructor), following protocols spelled out in the Rabenhorst (2008) article.

3. At the location where the IRIS tubes were installed, an estimation of the level of the ground water table should be obtained by digging a small hole to a depth slightly greater than the depth to the water table and allowing it to equilibrate in the hole.
4. At the location where the IRIS tubes were installed, Eh should be measured using five replicate Pt electrodes at depths of 12.5, 25, and 40 cm. A stainless rod should be used to make pilot holes for the electrodes to the specified depths. First the five electrodes should be placed at 12.5 cm and measurements made. Then the pilot holes should be deepened and electrodes placed at 25 cm and measurements made a second time. A third set of measurements should be made after electrodes have been placed at 40 cm. Because electrodes will not remain in the field between the two measurement dates, the reference electrode does not need to be placed within a salt bridge. Rather, a small amount of water should be added to the soil surface (if not already saturated) and stirred to make a paste, into which the reference electrode should be placed. The reference electrode should be situated within 25–75 cm of the location of the Pt electrodes.
5. Using either the push probe, a spade or an auger, samples should be collected from depths of 12.5, 25 and 40 cm so that pH can be determined at the same depths where Eh was measured.
6. Using either soil on cores removed when making pilot holes for IRIS tubes or soil collected with a spade or auger, apply a few drops of alpha-alpha dipyridyl dye at various depths in the soil to see whether there is a positive reaction to the dye, indicating the presence of Fe^{2+}.
7. Make a brief soil description at the location where the IRIS tubes were installed making special note regarding whether or not the soil meets any of the approved field indicators of hydric soils (USDA-Natural Resources Conservation 2010).
8. After approximately 4 weeks have elapsed, on the date when the IRIS tubes are extracted and examined, you should again make measurements of Eh, pH and water table height, and evaluate whether the soil gives a positive reaction to the alpha-alpha dipyridyl dye.

Part 2: IRIS Data Collection and Analysis

1. Four weeks after the IRIS tubes were installed, they should be extracted from the soil. Adhering soil can be removed by rinsing with very gentle brushing (as needed) under a stream of water, and then placed aside to dry.
2. Quantification will be accomplished using the mylar grid method (Rabenhorst 2012)because this method is more accurate and more reproducible than visual estimations (Rabenhorst 2010). A 15 cm by 6.7 cm grid containing 390 50 mm by 51 mm sectors should be printed on transparent mylar sheet so that it can be wrapped around an IRIS tube and held in place using rubber bands (Fig. 7.11).
3. The Technical Standard (National Technical Committee for Hydric Soils 2007) requires that paint be removed from 30 % of the IRIS tube within a 15 cm zone

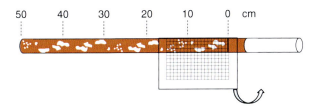

Fig. 7.11 Illustration showing placement of mylar grid around a 15 cm portion of an IRIS tube. By marking and counting sectors where paint has been removed, paint removal can be easily quantified (Modified from Rabenhorst 2012 with kind permission of © The Soil Science Society of America, Inc. 2012. All Rights Reserved)

that occurs somewhere within the upper 30 cm of the soil.[4] Therefore each group must determine which 15 cm zone within the upper 30 cm of each IRIS tube shows the maximum paint removal, and quantification should proceed for this zone.

4. A clear mylar grid should be wrapped around that portion of each IRIS tube showing the greatest degree of paint removal and should be secured using rubber bands. Using a fine point permanent marker, each sector with greater than 50 % paint removal should be marked. Once all the sectors are marked, the sectors should be counted and the percentage of the area can be calculated.

Part 3: Comparison of IRIS Data with Eh, pH and αα d Data

1. Using Eh and pH data collected and plotted on an Eh-pH stability diagram you should be able to evaluate whether the soil is reducing (or at what depths, the soil is reducing).
2. Based on the data you have collected, you now have 3 independent assessments of whether the soil is reducing: (1) Eh-pH; (2) ααd; (3) IRIS tubes.
3. Provide a thorough discussion of your data, being sure to include:

   - A summary of whether or not the soil was reducing according to each of the three assessment tools;
   - How results from each of these assessment methods were similar to or different from the others;
   - What factors might account for any differences that you observed; and
   - Your own conclusions (based upon your data) regarding whether or not the soil is reducing. Note that according to the Technical Standard of the NTCHS, in order for a soil to be reducing, it must meet the reducing specifications of one of the three methods, within the zone of 0–30 cm for a period of 14 days. For Eh and IRIS, a majority of the instruments (three-fifth) must fall within the specified range (not the mean).

---

[4] Other work has suggested that 20 % removal from a 10 cm zone somewhere within the upper 30 cm might be a better assessment of reducing conditions in soils (Castenson and Rabenhorst 2006; Rabenhorst 2008).

4. Your submitted lab report should include the following:

    - Data tables showing all data collected on each of the two dates (Eh, pH, αad, temperature, water tables);
    - Your discussion of water tables and temperatures over the course of the month (using your data and the continuous class record);
    - Completed IRIS data forms with data from yourself and from others in your group;
    - Eh-pH diagrams showing your data plotted for each of the two dates;
    - Your soil description and assessment regarding whether the soil meets any of the Field Indicators; and
    - Your complete discussion of all the data (including morphological data) with your conclusions.

References

National Technical Committee for Hydric Soils (2007) Technical note 11: technical standards for hydric soils. USDA-NRCS, Washington DC. Available from: ftp://ftp-fc.sc.egov.usda.gov/NSSC/Hydric_Soils/note11.pdf

Rabenhorst MC (2008) Protocol for using and interpreting IRIS tubes. Soil Surv Horiz 49:74–77

Rabenhorst MC (2010) Visual assessment of IRIS tubes in field testing for soil reduction. Wetlands 30:847–852

Rabenhorst, MC (2012) Simple and reliable approach for quantifying IRIS tube data. Soil Sci Soc Am J 76:307–308

USDA, NRCS (2010) Field indicators of hydric soils in the United States. Ver. 7.0. In: Vasilas LM, Hurt GW, Noble CV (eds) USDA, NRCS in cooperation with the National Technical Committee for Hydric Soils

Field Exercise 2: Assessing Leaf Litter Decomposition with Litter Bags

Objectives

1. To understand the principles behind the use of litter bags.
2. To understand how to construct and deploy litter bags.
3. To understand the impact of soil moisture content on decomposition rates.

Materials and Equipment Needed

Leaf litter samples
Litter bags – fiberglass screening material, mesh size of 1–2 mm
Drying oven
Balance
Pin flags

Procedures

Overview: There are several options with this exercise with respect to the type of organic materials used and the location of bag deployment in the field. In the procedures below, we suggest collecting fallen leaves from one deciduous tree species. This is the simplest approach and promotes uniformity in the samples. However, the exercise can be modified to compare multiple species or even root samples. With respect to deployment location, we suggest two sites, a wetland and an adjacent upland. The main point to this exercise is that wet conditions slow down decomposition rates. To further illustrate this point, the class may choose to place some litter bags in inundated areas. We suggest avoiding permanently inundated areas as it may be difficult to retrieve the bags.

9. *Litter bag construction*. Ten litter bags will be needed for each team of students. For each litter bag cut a 20 cm × 40 cm section of screen. Double the screen section over and staple the sides together at 5 cm intervals. You should now have a screen pouch open at one end. The open end will be stapled shut after the bag is filled with leaf litter. Each bag should be numbered with a permanent marker for future identification. Weigh each bag with three additional staples. Record the weight. The weight of the additional staples is needed to account for the staples used to close the open end of the bag after the leaf sample is added.
10. *Leaf litter sample collection*. Collect fallen leaves from the chosen field site. To promote uniformity in the samples; collect all samples from the adjacent upland, sample only the top layers of the litter, and sample only one deciduous tree species.
11. *Sample preparation*. Chop the leaves into 2–5 cm lengths. Randomly collect five sub-samples of the chopped leaves. Determine fresh weight of the sub-samples, then dry them (70 °C, 48–72 h) and re-weigh to obtain their water content.
12. *Bag filling*. Fill the bags with the prepared samples. Try to place a similar amount of material in each bag (It does not need to be exactly the same). The material should be relatively loose in the bag to allow for air movement. Staple the open end closed with three staples. Weigh each bag and record. Sample fresh weight is calculated as the difference between the filled bag weight and the empty bag weight.
13. *Litter bag deployment*. Place five bags in the upland and five bags in the wetland. For placement in the wetland, choose deployment points with similar hydrologic conditions as soil moisture will impact decomposition rates. For example, if the site has pit and mound topography, do not place 1 bag on top of a mound and another bag in a pit. Avoid permanently inundated areas. Clear away any duff and place the bags directly on top of the mineral soil surface. Mark the location of each bag with a pin flag.
14. *Litter bag retrieval and analysis*. Retrieve the bags after 2–3 months. Dry the retrieved bags (70 °C, 48–72 h) and weigh. Calculate sample dry weight.

15. *Calculations.* Loss of biomass due to decomposition is calculated as the difference between initial biomass and remaining biomass. All values are expressed on a dry weight basis. Initial dry weight (pre-deployment) of the samples is calculated from the fresh weight of each sample and the average water content of the five sub-samples determined in Step 3. Average rate of decomposition (per day) is determined by dividing biomass loss by the incubation period. Compare the mean biomass loss for the upland samples to that of the wetland samples.

Index

A
Accounting period, 92, 93, 115, 170
Accounting unit, 89–92
Accuracy, 5, 24, 34, 41, 51, 52, 54, 60, 76, 77, 83, 86, 90, 95–98, 100–102, 108, 111, 112, 120, 123, 124, 136–140, 162, 225, 279, 285, 287, 293, 296, 313, 356–360, 365, 368, 403
Acetylene block, 386–388
Acetylene reduction, 383–384
Acid mine drainage, 396
Acid sulfate soil, 396, 397, 402
Acoustic sensors, 98
Actual evapotranspiration (AET), 119
Adaptive resource management, 7
Additions, 1, 6, 14, 16, 19, 23, 28, 29, 32, 35–37, 40, 50, 64, 79, 92, 96, 102, 110, 123, 150, 155, 162, 164, 198, 228, 232–234, 252, 254, 260, 261, 273, 281, 283, 287, 289, 294, 298, 312, 343, 345, 356, 357, 359, 361, 366, 367, 374, 378, 381, 382, 393, 396, 423
Aerial photographs, 55, 75, 76, 98, 275, 283, 291
Aerobic conditions, 46, 234, 246, 248–250, 361–363, 369, 384, 389, 393, 394, 423
Aerobic process, 356, 361, 384
Aerodynamic methods, 117–118
 estimation method, 117–118
 mass-transfer method, 117–118
AET. *See* Actual evapotranspiration (AET)
Algal biomass, 337
Alkalinity, 327, 332, 334, 335, 338, 342, 344, 352
Allsopp, D., 372
Alpha, alpha dipyridyl dye, 260, 401–402, 426–438

Alpha diversity, 296
Alpha level, 36
Alternative hypothesis, 3, 4, 8, 9, 20
Altor, A.E., 375
Ammonia (NH_3), 331, 338, 345, 381, 383, 384
Ammonium (NH_4^+), 327, 331, 338, 362, 381, 383–385, 392
Amphibians, 10, 12, 18, 19, 26, 27, 32, 273, 336
Anaerobic conditions, 46, 229, 230, 248, 249, 260, 332, 360–362, 369, 384, 392, 393, 397–399, 423
Anaerobic microbes, 250, 426
Anaerobiosis, 230, 249, 360
Analysis techniques, 294–305
Analyte concentration, 360
Analytical techniques, 343, 358
Anderson, D.R., 9
Anion exchange resins, 394, 395
Annual cycle, 366
ANOVA. *See* 2-Way analysis of variance (ANOVA)
Anthropogenic activities, 22, 411
Aquatic invertebrates, 19, 23
Aquatic plants, 291, 337
Aquatic vegetation, 124, 127, 220, 289–291
ArcGIS, 52, 72–76, 83, 85, 109
Archived weather data, 110
ArcMAP, 72–74, 76, 80, 85
Army Corp of Engineers (USACOE), 227, 229, 305, 398
Artificial control structure, 124–126, 135, 138
 flumes, 124
 weirs, 124
Artificial marker horizons, 406–407
Ascorbic acid, 345, 392, 393, 395, 421
Ash, 15, 318

Ash-free dry mass (AFDM), 15
Assessment studies, 18, 21, 22
Atmospheric deposition monitoring networks, 169
Atmospheric pressure, 96, 115, 148, 179, 207, 211, 248
Audience, 297, 321
Autolevel, 54, 58, 63–67
Azzolina, N.A., 340

B
Babbitt, K.J., 26, 335
BACI. *See* Before-after/control-impact (BACI)
Bacteria, 249, 251, 252, 362, 374, 375, 381, 383
Balance, 88, 105, 133, 159, 232, 389, 424, 440
Balcombe, C.K., 13
BAP. *See* Biologically available phosphorous (BAP)
Barometer, 96, 179, 180
Basal area, 279, 281, 282, 289, 294, 312, 315, 316
Baseline, 11, 63, 283, 284, 327, 333, 378, 382, 385, 407, 411
Bathymetry, 49–83, 90, 93, 99–101, 151
  data, 53
  map, 90
Batzer, D.P., 335
Beaver, 277, 324
Bedford, B.L., 333
Bed roughness, 125
Before-after/control-impact (BACI), 21–23, 340
Belanger, T.V., 90
Benham, E.C., 264
Bennett, J., 10
Benoit, J., 339
Bentonite, 388
Bernier, P., 367
Bias, 3, 4, 14, 17, 19, 24, 30–32, 40, 102, 104–107, 110, 113, 116, 120, 123, 203, 288, 296, 312, 313, 360, 378
Bidwell, J.R., 325–353
Bioavailability, 377, 390
Bioavailable particulate P (BPP), 395
Biochemical oxygen demand (BOD), 394, 403
Biodiversity, 273, 389
Biogeochemical cycling, 327, 360, 366, 381, 411
Biogeochemical indicators, 410–412
Biogeochemical processes, 88, 159, 230, 248, 332, 356, 360–361, 410

Biogeochemistry, 50, 248, 260, 331, 355–442
Biologically available phosphorous (BAP), 394–396
Biological oxygen demand (BOD) bottles, 404
Biological population, 11
Biological significance, 35
Biomass, 13, 15, 20, 31, 232, 278, 289, 337, 362–371, 383, 422, 442
Bioturbation, 330, 407
Birds, waterfowl, 2
Bitterlich method, 281, 282
Blake, G.R., 405, 406
Blough, N.V., 379
BOD. *See* Biochemical oxygen demand (BOD)
Boeckman, C.J., 337
Bohlen, P.J., 387
Bonham, C.D., 277
Boon, P.I., 333, 335, 338
Bottomland wetlands, 17
Bouwer, H., 150
Bouwer analytical method, 150
Bouwer and Rice analytical method, 150
Bowen ratio, 115
BPP. *See* Bioavailable particulate P (BPP)
Brackish marsh, 29, 396, 398
Bradford, J.B., 373
Braun-Blanquet, M.M., 288
Bray-Curtis Index, 301
Breeding birds, 39
Bremner, J.M., 381
Brennen, E.K., 14
Bridgham, S.D., 393
Briggs, R., 346
Brinson, M.M., 332, 340
Broad-spectrum radiometers, 120
Broderson, W.D., 264
Brooks, R.P., 16
Brown, M.L., 5
Brunsell, N.A., 117
Brush, G.S., 409, 410
Brushing, 438
Bubble gage, 95, 103
Bubbler system, 95
Bucket-stopwatch method, 134
Bulk density, 163, 244–246, 262, 404–408, 410, 412
Buried bag method, 384–385
Burning, 14, 15, 409

C
Cabezas, A., 334
Cahoon, D.R., 408

Index

Calibration, 33, 34, 66, 67, 105, 108, 110, 111, 123, 124, 130, 131, 134, 138, 156, 161–163, 166, 225, 292, 343, 358, 359, 392
Calipers, 364, 407
Callaway, J.C., 408
Can core method, 405
Canonical correspondence analysis (CCA), 302
Canopy cover, 293, 336
Capacitance method, 160, 162–163
Capacitance rod, 95–96
Capillary fringe, 92, 249
Carbon (C), 15, 232, 246–249, 252, 274, 330, 335, 341, 343, 356, 362–381, 425, 426
   analysis, 364
   sequestration, 363
Carbonates, 234, 326, 362, 369, 390
Carbon dioxide (CO_2) detection, 373–374
Carpenter, E.J., 383
CART. *See* Classification and regression trees (CART)
Cartesian coordinate system, 60
Cattail, 8
CCA. *See* Canonical correspondence analysis (CCA)
Cedergren, H.R., 191
Census, 23, 24
Chamberlain, T.C., 9
Channel control, 125
Channel geometry, 125
Channel ice, 138
Chemical methods, 358, 395
Cherry, J.A., 215
Childs, C.W., 427
Chirnside, A., 355–442
Chlorophyll, 15, 23
Choi, J., 90, 155, 168
Chojnacky, D.C., 363
Chroma, 230, 237–241, 251, 253–256, 258, 261, 262, 271
Circular plots, 279, 280
Classification, 38, 229, 298–300, 302–303, 327–329, 332, 340
Classification and regression trees (CART), 302–303
Class I hypotheses, 10
Class II hypotheses, 10
Clean Water Act of 1972, 2, 228, 229, 332
Climate, 10, 16, 21, 22, 88, 92, 116, 117, 183, 229, 231, 232, 260, 277, 333, 362, 393
   change, 21, 22, 88, 92
Clod method, 405–406
Cluster analysis, 299–301, 304, 323

Coale, F., 392
Coarse root biomass, 365–366
Coastal marsh, 17, 20, 21, 29, 38
Coastal wetland(s), 9, 334, 336, 338, 340, 397
Coastal wetland loss, 9
Coble, P.G., 379
Cochran, W.G., 31, 32
Cohen, J., 27
Cohen's *d*, 37
Cohort layered screens, 367, 369, 371
Coker, P., 277
Collins, J.P., 10
Colorimetrically, 392
Color scheme, 79
Combination methods, 118–119
   estimation method, 118–119
   Penman method, 113, 118–119
Communication, 53, 56, 70
Community, 10, 12, 17, 18, 23, 26, 37, 50, 274, 275, 279–285, 287–290, 292–305, 312, 319–321, 323, 335, 376
Community type, 289
Comparative experiments, 17
Completeness, 34, 359, 360
Conceptual hypothesis, 15
Conceptual models, 3, 15–17
Concretions, 251
Conductivity, 142, 150–151, 155–157, 164, 191, 192, 198, 205, 207, 214–216, 223, 324, 329, 334, 335, 337, 338, 344, 352, 353, 388
Conly, F.M., 97
Conservation, 2, 10, 18, 33, 38, 233, 235, 295, 298, 300
Conservative tracers, 165, 169, 170
Constantz, J., 157, 223
Construction equipment, 76
Contrast, 51, 162, 228, 240–241, 259, 336, 338, 365, 381, 398, 430, 432
Controlled experiments, 17
Cook, P.G., 164
Cores, 7, 22, 23, 30, 142, 169, 289, 364–366, 383–388, 402, 404–408, 423, 427, 431–433, 438
   invertebrate samplers, 23
   plexiglass, 291
   steel pipe, 206
Cornell, L.P., 338
Correspondence analysis (CA), 302
Cost effectiveness, 123, 222
Cotton-strip assay, 369
Covariates, 14, 39, 40
Cover class, 288

Cowardin, L.M., 38
Cowardin Classification System, 38, 229
Cox, D.R., 19
Created wetlands, 339
137Cs dating, 407–408
Culverts, 61, 125, 127
Cunningham, W.L., 150
Current meters, 127–130, 132, 134, 222, 224
   acoustic Doppler velocimeter (ADV), 129, 130
   mechanical meter, 130
Cyanobacteria, 383

D

Daily evaporation, 101
Dalton-type equations, 117
   estimation method, 117
3-D analyst, 72, 79, 80, 82
Darcy Flux method, 142–145, 222
Data forms, 40, 41, 294, 440
Data management, 32, 34
Daubenmire, R.F., 288
dbh. *See* Diameter at breast height (dbh)
DCA. *See* Detrended correspondence analysis (DCA)
DEA. *See* Denitrification enzyme activity (DEA)
De'ath, G., 303
Decomposition, 231, 232, 235, 246, 247, 250, 254, 262, 361, 363, 369–373, 376, 381, 389, 411, 441, 442
Decomposition rates, 363, 369–373, 376, 411, 441
Deductive reasoning, 8
Degree of agreement (X), 359, 360
DeLaune, R.D., 391, 407, 411
DEM. *See* Digital elevation models (DEM)
Deming, W.E., 26
Dendrogram, 299, 323
Denitrification, 327, 338, 356, 361, 362, 381, 383, 384, 386–389, 410–412
Denitrification enzyme activity (DEA), 386–388
Density, 13, 14, 54, 62, 97–99, 107, 108, 112, 114, 120, 125, 138, 157, 160, 163, 179, 198, 211, 214, 221, 244–246, 248, 262, 279, 281, 282, 287, 288, 293–295, 309–312, 315, 316, 326, 336, 375, 404–408, 410, 412
Dependent variables, 3, 5, 6, 13–15, 17, 22, 29, 31, 35, 36, 39, 40
Depleted matrix, 254–256, 261, 263

Depletion (D), 241
Depressional wetlands, 262, 333–337, 361
   playa lakes, 271
   prairie potholes, 16, 17, 21, 23, 38, 89, 119, 277, 334, 335
Describing soil color, 239–241
Descriptive inference, 3, 12
Descriptive studies, 18, 34
Detection limits, 358–360, 421–422
Detrended correspondence analysis (DCA), 302
Devito, K.J., 16
Diameter at breast height (dbh), 281, 282, 294, 313, 316, 364–365, 422
Diffuse overland flow, 124, 138–141, 188
Digital datalogger, 95, 162, 222
Digital elevation models (DEM), 50, 52, 55, 56, 72, 76–82, 85, 86
Digital image technique, 289
Dingman, S.L., 132
Dinitrogen fixation, 381–384
Dinitrogen gas (N_2), 381, 426
Direct current (DC), 122
Direct seepage measurements, 151–153
Discharge, 38, 89, 101, 103, 124–138, 142, 144, 146, 147, 154, 158, 169, 170, 184–187, 189, 224, 225, 331
   measurement, 124, 127–137, 185–187, 224
Dissimilarity indices, 299
Dissimilatory nitrate reduction, 384
Dissolved inorganic phosphorus, 331
Dissolved organic carbon (DOC), 257, 330, 335, 376–379
Dissolved organic matter (DOM), 376–381
Dissolved organic nitrogen (DON), 331, 376, 377
Dissolved oxygen (DO), 326, 327, 332, 333, 335–338, 342, 346, 352, 353, 404
Distributed temperature system (DTS), 158
Dithionite, 394
Diurnal temperature, 113, 155, 156, 198, 216, 223, 235, 326, 338, 341
Diversity assessments, 295–298, 318–321
DO. *See* Dissolved oxygen (DO)
Dobbie, M.J., 27
DOC. *See* Dissolved organic carbon (DOC)
DOM. *See* Dissolved organic matter (DOM)
Dominant species, 289–290
Dominant water source, 360
DON. *See* Dissolved organic nitrogen (DON)
Double exponential decay models, 371
Double-mass curve, 110
Drag, 95

Index 447

Drawdowns, 46, 47, 93, 169, 328–330
Dredges, invertebrate samplers, 9
Drying cycles, 244, 361–362
Dry mass, 15, 368, 371
DTS. *See* Distributed temperature system (DTS)
Dunne, T., 141

E
Eaton, A.D., 394
Ecosystem function, 274, 290
Ecosystem health, 356, 366
Ecosystem services, 273
Eddy-covariance method, 112–114, 117, 123
   direct measurement, 112
EEMs. *See* Excitation-emission matrix (EEMs)
Effect size, 280
Electromagnetic waves, 160
Elevation, 47, 50, 52–66, 68, 72, 76–82, 85, 86, 93, 94, 96, 97, 99, 102, 108, 112, 126, 127, 143, 148, 149, 163, 164, 179, 180, 201, 205–207, 210, 213, 232, 262, 276, 292, 388, 406–409
Elevation head, 97
Ellenberg, H., 277
Elser, J.J., 15
Elzinga, C.L., 277, 279, 287, 288, 305
Emergent vegetation, 90, 98, 99, 107, 136, 149, 337, 338, 342
Emphasis, 325, 356
Empirical determination, 135
Empirical equation, 137, 140
Energy-balance method, 115–118
   estimation method, 115–117
Environmental factors, 301
Environmental Monitoring Systems Laboratory, 358
Ephemeral wetlands (ponds), 139
Episaturation, 248, 249, 261, 262, 388
Equilibrium dialysis samplers, 397
Erosion, 71, 136, 150, 232, 233, 406–410
Errors, 3, 4, 19, 20, 24, 28–35, 40, 41, 53, 60, 66, 68, 70, 74, 75, 80, 89, 90, 93, 96, 100–106, 109, 113, 115, 116, 119, 122–124, 126, 131, 134, 137–138, 140, 146–148, 152–155, 160, 163, 169–172, 203–204, 211, 224, 225, 274, 310, 311, 356, 358, 367, 369, 396, 408
Estimates, 285, 296, 318
Ethylene, 383, 384
Euclidean distance, 299
Euliss, N.H., 16, 334, 335, 340

Evaporation loss, 103–105
Evaporation pan, 112–113, 122
   direct measurement, 112–113
Evenness, 295, 297, 320
Excavation method, 405
Excitation-emission matrix (EEMs), 378, 380, 381
Experimental error, 3, 19, 20, 31
Experimental unit, 3, 4, 12–29, 31
Explanatory inference, 12

F
Falcons, 393
Farm bill, 229
Fen wetlands, 334
Ferré, P.A., 162
Ferric iron (Fe^{+3}), 234, 249, 361
Ferrous iron (Fe^{+2}), 234, 237, 249, 361, 402, 436
Fertilizer (fertilizing), 382, 383, 392
Fetter, C.W. Jr., 191
Field indicators, 438
Field log books, 358
Field studies, 9, 25, 28, 36, 37, 47, 291
Fill-and-spill, 91
Filters, 8, 157, 328, 329, 331, 345, 376, 378, 385, 391, 393–396, 409
Filtration, 344
Findlay, C.S., 336
Fine-mesh seines, 367
Fine root biomass, 366–367
Fire, 14, 21, 22, 275, 287
Fish, 23, 26, 219, 273, 293, 329, 330
Fish and Wildlife Service (USFWS), 33, 38, 274, 305
Fisher, R.A., 19
Fixed effect, 3, 13, 16
Fledermaus, 52, 80, 81
Float-based gage, 94–95
Floating pan, 112, 113, 122
Flood duration, 125
Flood frequency, 38, 332
Floodplain wetlands, 334, 337, 338
Floodwater storage, 16, 50
Floristic Quality Assessment Indices (FQAIs), 304, 321–323
Floristic Quality Index (FQI), 304, 321
Flow control, 124–126, 137, 138
Flowpaths, 143, 146, 148, 377
Flow rate, 140, 225
Flow traps, 139, 140, 188
Flow velocity, 127–130, 137, 153, 157, 204

Flumes, 124–126, 134, 136, 138, 388
Fluorescence spectroscopy, 377–381
Forested wetlands, 274, 277, 284, 293, 317, 318, 335
FQAIs. *See* Floristic Quality Assessment Indices (FQAIs)
FQI. *See* Floristic Quality Index (FQI)
Freeware R package vegan, 300
  cluster analyses, 300
Freeze, R.A., 215
Frequency, 38, 46, 94, 105, 158, 162, 278, 281, 294, 295, 302, 305, 315, 317, 332, 335, 343, 408
Freshwater wetlands, 342, 360, 389, 410
Fuhrmann, J., 355–442
Functional capacity, 356, 410
Functional processes, 332, 337

G
Galatowitsch, S.M., 10
Gas chromatography, 375, 383, 387, 423
Gathumbi, S.M., 387
Geographically isolated wetlands, 21
Geographic information system (GIS), 38, 50, 51, 54, 72–83, 91, 274, 291, 311, 313
Geomorphic setting, 38
Geomorphology, 82, 336
Georeferencing, 53, 62, 85
Geospatial Data Gateway, 276
Geostatistical correlation, 109
Geostatistical evaluations, 357
Giardina, 372
Gleyed matrix, 254, 255
Global Atmospheric Watch, 169
Global positioning system (GPS), 51–54, 60, 62, 68–71, 73, 76, 80, 99, 211, 213, 287, 291, 293, 311, 313, 409
Google Earth, 55, 56, 84–86
Gosselink, J.G., 332, 342
GPS. *See* Global positioning system (GPS)
Grace, J.B., 300, 302, 303, 305
Graduated rod, 97
Graminoid species, 280, 284
  grass-like, 280
Graveel, J.G., 424
Gravity acceleration, 97, 179
Gray, K.A., 387
Grazing, 20, 26
Great Lakes wetlands, 334, 338, 340
Greenhouse gases, 362
Green, S.A., 379
Gregory, K.J., 132

Greig-Smith, P., 277
Gross error, 32
Groundwater
  discharge, 38, 89, 103, 142, 144, 146, 147, 154, 158, 169, 171, 189
  fed wetlands, 333
  flow modeling, 142, 144
  flow-net analysis, 146
  fluxes, 153–156
  inflow, 139, 142–158, 168
  outflow, 142–158, 168
  recharge, 144, 146, 170
  wells, 146
Groundwater-surface-water exchange, 142, 143, 148, 155–158, 194
Groups, functional, 278, 287, 290, 376, 377
Growth forms, 290
Grubb's test for outliers, 34
Guthery, F.S., 8
Guy, C.S., 5

H
Haag, K.H., 54
Half-barrel seepage meter, 151, 152, 194–197, 219
Hannaford, M.J., 23
Haphazard sampling, 25, 26
Hardness, 329, 335, 344
Hardwood bottomlands, 17
Hardy, R.W.F., 383, 384
Harmon, M.E., 370
Harrelson, C.C., 57
Hartge, K.H., 405, 406
Hart, S.C., 385
Harvey, J.W., 90, 155, 168
Hauer, F.R., 342, 345
Haukos, A., 1–47
Hayashi, M., 87–225
Head gate weirs, 125
Heagle, D.J., 170
Healy, R.W., 164
Helms, J.R., 379
Hemond, H.F., 339
Herbaceous biomass, 368
Herbaceous vegetation, 281
Herbacous plants, 368
Herbicide, 8, 29
Herbivory, 276
Hertel, D., 366, 367
Heterogeneous vegetation, 275, 276, 279
Heterotroph, 398
HGM subclass, 340

Index 449

High density polyethylene (HDPE), 375
Hill, A.R., 16
Hill, W.R., 345
Hood, J.L., 153
Hopfensperger, K.N., 27
Horizon formation, 233–236
Hornung, J.P., 26
Horwath, W.R., 387
Horwitz, P., 337
Houlahan, J.E., 336
Howard, P.J.A., 372
Hue, 237, 238, 241
Humidity, 113, 114, 120–123, 203, 372
Humification index (HIX), 379, 380
Hunt, P.G., 387
Hunting, 11
Hutchinson, G.L., 375
Hvorslev, M.J., 150, 215
Hvorslev analytical method, 150
Hydraulic gradients, 143, 144, 146–149, 151, 156, 157, 179, 205, 207, 211, 213, 216, 223, 388
Hydraulic resistance, 137
Hydric soil(s)
 formation, 248–252, 266–267
 identification, 227–272
 indicators, 227, 228, 230–231, 233, 250, 252–262, 264, 268–270, 272, 410
 lists, 229, 231, 257
 technical standard, 230, 231, 257, 263
Hydrodynamics, 38, 88, 228, 406, 409
Hydrogeology, 16, 142, 205
Hydrogeomorphic (HGM), 299
Hydrogeomorphic approach, 340
Hydrogeomorphic classification, 38, 298, 340
Hydrologic conditions, 50, 228, 229, 231, 236, 261, 376, 441
Hydrologic Engineering Center (HEC), 141
Hydrologic gradients, 275, 283
Hydrology, 9, 17, 21, 22, 29, 38, 39, 87–225, 228, 229, 244, 258, 261–263, 274, 326–330, 333, 336, 340
Hydroperiod, 26, 50, 51, 88, 228, 261–263, 328, 329, 335–337, 341, 356, 360, 361, 410
Hydrophytes, 38
Hydrophytic vegetation, 229, 230
Hyphae, 371
Hypothesis testing, 30, 35, 41
Hypothetico-deductive approach, 8
Hyrdraulic conductivity, 142, 150–151, 156, 164, 191, 192, 205, 207, 214–216, 223, 388

I
Ice cover, 95, 138, 158
Identification techniques, 227–272
Impact assessment studies, 21
Impact studies, 21–23
Impairment, 27
Importance value, 295, 305, 317
Inamdar, S., 355–442
Independent variables, 3, 13–16, 39, 41, 302, 310
Index
 classification, 296–297
 floristic quality, 321
Index of relative abundance, 322
Indicator species analysis, 299, 303–304
Indirect gradient analysis, 301
Induction, 7, 8
Inductively coupled plasma (ICP) spectrometry, 392
Inductive reasoning, 7
Infrastructure design, 59
Inland wetlands, 325, 328
Inorganic phosphorous, 331
Inorganic soil constituents, 368
Insects, 23, 273
 invertebrate taxa, 38
In situ measurements, 344, 386
Instrument exposure, 103
Instrument hysteresis, 95
Instrument type, 56, 160
Interaction, 10, 14, 50, 229, 231
Interannual variability, 277
Interpretation, 25, 30, 35, 41, 146, 148, 149, 165, 192, 228, 254, 292, 296, 321, 322, 341, 358, 437
Inundation, 68, 230, 244, 260, 275, 325, 330, 371, 374, 385, 403
Invasive species, 10, 21, 296
Invertebrates, 13, 19, 23, 26, 37, 46, 330, 337
 aquatic, 19, 23
Ion chromatography, 392
IRIS panels, 397–398
IRIS tubes, 260, 397, 401, 437–439
Iron (Fe)
 accumulation, 230, 248–250
 depletion, 251
 reduction, 234, 239, 240, 249–251, 261, 327
 translocation, 249–250, 261
Iron oxides, 238
 method-runoff, 395–396
Isopods, invertebrate taxa, 38

J

Jackson, C.R., 341
Jenkins, J.C., 364
Jensen-Haise method, 116
Jordan, T.E., 387
Judgment sampling, 25–27

K

Kadlec, R.H., 137
Kantrud, H.A., 15
Karberg, N.J., 371
Keery, J., 198
Kelly, R.L., 10
Kemmerer bottle, 168
Kenkel, N.C., 300, 302, 305
Kent, M., 277
Kentula, M.E., 12
Kirkner, R.A., 90
Klarer, D.M., 338
Klemedtsson, L., 387
Knievel, D.P., 366
Kolka, R.K., 375
Korzukhin, M.D., 364
Krabbenhoft, D.P., 155
Krebs, C.J., 277
Krig, 76–78
Krigging, 76, 80, 85, 109
Krypton hygrometer, 114
Kuo, S., 390, 391, 395

L

LaBaugh, J.W., 90, 155
Laboratory studies, 15, 25
Lajtha, K., 370
Lake Erie wetlands, 338
Lake Huron wetlands, 334
Lamberti, G.A., 342
Landform, 107, 257, 259, 267, 335
Land, L.A., 157
Landscape position, 231–233, 335, 340, 361
Lang, G.E., 371
Large Aperture Scintillometer (LAS), 117
LAS. *See* Large Aperture Scintillometer (LAS)
Latent heat flux, 114, 117
Leaf litter decomposition, 440
Lee, D.R., 90
Legendre, L., 300
Legendre, P., 300
Leopold, L.B., 141
Leuschner, C., 366, 367

Levees, 9, 229
LiDAR. *See* Light detection and ranging technology (LiDAR)
Light detection and ranging technology (LiDAR), 52, 53, 58, 68, 71, 91
Likens, G.E., 331, 342, 345
Linear regression, 285, 378
Line-intercept technique, 280
Litter, 236, 287, 292, 330, 361, 369–372, 380, 385, 389, 393, 405, 440, 441
Litter bags, 369–371, 393, 441
Little, A., 273–324
Livingston, G.P., 375
Logistical considerations, 37–39
LOI. *See* Loss on ignition (LOI)
Long-term studies, 41
Longwave radiation, 119
Los Huertos, M., 49–86
Losses, 88–90, 93, 123, 170, 228, 233, 234, 368–372, 381, 383, 384, 406–410
Loss on ignition (LOI), 355, 368, 409
Luo, H.R., 20

M

Mackay, D.S., 15
Macroinvertebrate, 26, 303, 329, 335, 370
Macrophyte assimilation, 387
Macrophytes, 326, 337, 352, 369, 381, 382, 394
Magurran, A.E., 296
Main effects, 14
Maintenance, 33, 103, 108, 111, 127, 138, 289, 343, 359, 423
Makkink estimation method, 116
Manganese (Mn), 238, 240, 249–251, 356, 399, 426
  accumulation, 249–250
  depletion, 251
  oxides, 238
  reduction, 249–250
  translocation, 249–250
Manipulative treatments, 18
Manning's equation, 137
Mantel test, 303
Maple, 318
Mariotte bottle, constant-flow device, 132
Marker integrity, 407
Marsh pore water, 397–398
Masked sand grains (CS), 241
Mass-balance equations, 154, 166–170
Masses, 251, 258, 397, 407

Index

Matrix (M), 162, 163, 237, 239–242, 248, 251, 252, 254–256, 261–263, 271, 285, 298, 360, 378, 390, 396
  color, 239–240
  spikes, 360
Matthews, J.W., 304
Maurer, D.A., 13
McCune, B., 300, 302, 303, 305
McKnight, D.M., 379, 380
Measurement errors, 122–123
Measurement site, 103, 332
Measuring soil color, 239
Mechanical error, 31, 32
Mechanical meter
  horizontal-axis meter, 130
  vertical-axis meter, 130
Megonigal, J.P., 9
Mesocosm, 13, 426–434
Meta-analysis, 15
Metapopulation, 3, 11
Methane (CH_4), 337, 356, 362, 373–375, 399
  emissions, 356
Methanogenesis, 363, 374
Methanotrophs, 375
Method detection limit (MDL), 359, 360, 421, 422
Methods of soil analysis, 358
Microbes, 232, 246, 250, 251, 260, 361, 366, 377, 401, 426
Microbiological and biochemical properties, 358
Microhabitats, 26
Microtopographic features, 275, 293
Microtopography, 259, 275, 276, 293, 324
Migration, 14, 47
Milan, C.S., 407
Miller, M.P., 379, 380
Milton, M., 363
Mineral grain color, 238
Mineralogical methods, 358
Mineral soil flats, 261, 361, 388
Missing data, 31, 41, 110, 194
Mississippi River, 9
Mitsch, W.J., 332, 342, 375
Mixed model, 14
Monitoring wells, 92, 102, 142, 144–150, 158, 163, 189, 191, 192, 206, 211, 244, 262, 291
  piezometer, 142, 148, 149, 169, 206, 262
  water-table monitoring well, 148, 207–209
Monocultures, 275
Monumentation, 286, 287
Moore, P., 392

Morphology, 228, 230, 231, 235, 236, 261–263, 410
Morrice, J.A., 340
Morrison, M.L., 28
Mosier, A.R., 387
Mosses, 289
Motz, L.H., 90
Mueller-Dombois, D., 277
Multiple observers, 292
Multivariate data, 297–300, 305, 323–324
Multivariate methods, 35
Multivariate statistics, 13
Mulvaney, R.L., 381
Myers, R.G., 395, 396

N
National Food Security Act, 229
National Geodetic Service (NGS), 58
National Geodetic Vertical Datum (NGVD), 59
National Oceanic and Atmospheric Administration (NOAA), 110
National Technical Committee for Hydric Soils, 229, 263, 401, 434, 438
National Wetland Inventory (NWI), 38, 229
National Wetlands Research Center, 357
Natural Resources Conservation Service (NRCS), 227, 257, 260, 263, 376, 410
Nelson, D.W., 369
Nested plots, 279, 280
Newton, W.E., 15
Nitrate (NO_3^-), 249, 327, 331, 334, 338, 345, 352, 362, 378, 381, 384, 399, 426
Nitric oxide (NO), 381
Nitrification, 356, 381, 384–386
Nitrifying bacteria, 362
Nitrite (NO_2^-), 331, 381, 426
Nitrogen (N), 15, 16, 46, 95, 249, 327, 330, 331, 334, 335, 338, 346, 356, 381–389, 426
Nitrogenase, 383
Nitrogen mineralization, 384–386
Nitrous oxide (N_2O), 381
Nodules, 251
Non-hierarchical K-means cluster analysis, 300
Nonmetric Multidimensional Scaling (NMS), 301, 302, 304, 324
Non-parametric multi-response permutation procedure (MRPP), 300
Normal rainfall, 260–261
Normal-ratio method, 110
North American Datum (NAD), 62
North American Vertical Datum, 93

Nuisance variables, 14, 17, 19, 22, 24, 28, 39, 41
Numerical models, 148, 155, 157
   SUTRA, 157
   VS2DH, 157
Nutrient content, 334
Nutrient cycling, 327, 411
Nutrient limitation, 15
Nutrient loading, 331, 356, 403, 410, 411
Nutrient transformations, 332

O

Observational studies, 18
Ogden, J.C., 16
Ohno, T., 379, 380
Open-ended PVC frames, 291
Ordination, 13, 300–302, 304, 323, 324, 336
Organic accumulation, 250
Organic carbon, 247–248, 340
Organic compounds, 238, 358, 363, 370, 394
Organic content, 247
Organic matter, 208, 228, 230–236, 244, 248–250, 255, 261, 262, 326, 335, 338, 339, 362, 363, 368–372, 376–382, 389–391, 401, 403, 406–409, 412, 423, 424, 426
Organic matter decomposition, 232, 262, 369–372
Organic P, 389–392, 394, 395
Organism(s), 3, 11, 19, 81, 229, 231, 232, 279, 280, 290, 312, 326, 327, 332, 337, 372, 373, 383, 396
Organismic treatments, 18
Orifice, 95, 103, 104, 106–108
Orphophosphate, 331, 338, 339, 345, 352, 361, 389, 391
Orthophosphorous, 391, 421
Overland flow, 88, 89, 91, 92, 124, 138–141, 165, 171, 188, 189, 203, 239, 326, 327, 330, 333, 334, 388
Overland flow volume, 139–141
Owens, P.R., 427
Oxygen, 46, 154, 159, 230, 239, 249, 251, 326–328, 332, 333, 335–338, 344, 346, 352, 353, 361, 403–404, 423, 426

P

Paltineanu, I.C., 163
Paradigms, 230, 231
PARAFAC. *See* Parallel factor analysis (PARAFAC)
Parallel factor analysis (PARAFAC), 378
Parent material, 228, 231–235, 251, 258, 369
Parkhurst, R.S., 116
Parsons, D.F., 168
Particle size distribution, 241
Pasternack, G.B., 409–410
Paull, C.K., 157
PC-ORD, 300, 302, 303
   cluster analyses, 300
Peat, 150, 247, 248, 250, 254, 255, 272, 292
Peatlands, 17, 275, 335
Peepers, 397, 398
Percentages, 88, 89, 139, 171, 182, 240, 272, 360, 368, 371, 383, 404, 439
Performance curves, 285, 286, 290, 318, 319
Permanent plots, 286, 287, 289, 291
Permits, 34, 56, 220, 225, 252, 254
Persson, H., 366
Pfeiffer, D.U., 15
pH, 231, 324, 327, 333–338, 342, 344, 347, 352, 353, 362, 378, 381, 389, 390, 397, 398, 400, 403, 420–421, 426, 427, 430, 433–440
Phosphorous (P), 15, 23, 39, 46, 330, 331, 334, 335, 345, 356, 389–396, 421
   fractionation, 391–392
   mineralization, 385, 393
   in water, 393–394
Photopoints, 286, 287
Photosynthesis, 326, 337, 338, 363
Physical and mineralogical methods, 358
Physical methods, 358
Pierce, C.L., 26
Pilot study, 5, 36, 40, 41, 285, 286, 318
Pin-frame system, 288
Plant communities, 18, 284, 285, 290, 297–300, 302–304, 319
Plant community composition, 295
Plant cover, 17
Plant diversity, 295
Plant functional groups, 290
Plant productivity, 232
Plant response, 46
Plant traits, 287, 290
Platt, J.R., 9
Playa wetlands, 20, 21
Plot-based techniques, 279–283, 288, 312
Plotless methods, 281, 282, 311
Plot location, 287, 292
Plot shape, 279
Plot type, 286
Point-based sampling, 281

Index 453

Point measurements, 64, 103, 150, 154, 313
2-Point method, 129
Point-quarter method (PQM), 281, 282, 312–314
Pollutants, 390
Polyethylene bags, 375, 376, 384, 385
Pool-point radial survey method, 68
Popper, K.P., 9
Population, 2–4, 10–12, 14, 18, 19, 22–31, 36, 37, 40, 41, 46, 47, 278–281, 287–290, 292, 294, 302, 305, 309–312, 314, 358, 361, 394
Pore lining (PL), 241, 242, 252
Pore water extractors, 397, 398
Potential evapotranspiration (PET), 119
Potentiometer, 95, 102, 180, 181
Power, 6, 13, 30, 31, 35, 36, 82, 83, 95, 96, 103, 118, 123, 135, 217, 285, 286, 420
PQM. *See* Point-quarter method (PQM)
Prairie pothole region (PPR), 89, 119
Prairie potholes, 16, 17, 21, 23, 38, 89, 119, 277, 334, 335
Precipitation, 8, 14, 16, 21, 22, 38, 88–90, 98, 101, 103–111, 139–142, 165–167, 169, 171, 182, 183, 188, 201, 203, 204, 260–263, 326–330, 333, 334, 390
  measurement, 106, 110, 111
Precipitation gages, 104–111, 182
  non-recording, 104–106
  recording, 104, 108
Precision, 5, 24, 25, 28, 36, 40, 53, 54, 57, 59, 63, 66, 69, 71, 83, 100, 296, 356–360, 430
Prescribed fire, 14
Preservation, 344–346, 357, 358
Preservation techniques, 346
Pressure head, 98, 148, 149, 207
Pressure output, 96, 120, 179
Priestley-Taylor method, 113, 118, 119, 123
Primary producers, 363
Primary production, 363, 389
Principle components analysis (PCA), 301, 302
Probability sampling, 25, 28
Probes, 122, 162, 163, 220, 343, 344, 346, 405, 437, 438
Procedural error, 32
Propagules, 296
Proxy variables, 15
Pseudoreplication, 20, 22, 26, 274, 276
P-sink approach, 394, 395
Psychrometric constant, 115, 118
Pt electrodes, 400, 426–432, 435, 438
Pulse disturbance, 21

Q
QA. *See* Quality assurance (QA)
QAPP. *See* Quality assurance project plan (QAPP)
Quadrat sampling, 312
Qualitative methods, 158
Quality assurance (QA), 33, 343, 346, 358, 359
Quality assurance and quality control program (QA/QC), 32, 33
Quality assurance project plan (QAPP), 33
Quality control (QC), 33, 34, 357–360, 421, 422
  database, 359
Quantifying litterfall, 367–368, 371
Quantitative methods, 6, 281
Quivira National Wildlife Refuge, 2

R
Rabenhorst, M., 355, 442
Radar sensors, 98
Radiation-temperature models, 116
Rainfall, 92, 104, 105, 108, 112, 115, 122, 151, 165, 211, 234, 260–261, 367
Random effect, 3, 13, 16
Randomization, 3, 17–21, 274, 303
Randomized experiments, 17
Random selection techniques, 357
Rantz, S.E., 127, 128, 130, 134, 136, 138
Rasmussen, A.H., 117
Raulings, E.J., 290
Real-time kinematic (RTK), 68–71, 73
  GPS, 68–71, 73
Reardon, J., 345
Recharge, 92, 102, 144, 146, 149, 159, 170, 211, 389
Recommended techniques, 359, 383, 385, 423
Recording, precipitation gages
  optical devices, 105
  tipping-bucket gage, 104–105
  weighing gage, 104–105
Recovery curve, 150
Reddy, K.R., 391, 411
Redoximorphic features, 231, 240–242, 250–252, 263, 265, 272
Redox pair, 399
Redox potential, 337, 362, 375, 392, 399, 426
Reduced matrix (RM), 180, 239, 241, 251
Reduced soils, 360, 392
Reduction, 1, 103, 150, 170, 230, 232, 234, 239, 240, 249–252, 261, 326, 327, 329, 332, 337, 345, 362, 372, 381, 383–384, 392, 396, 398–402, 434, 437, 440

Reed canary grass, 13, 318
Reeder, B.C., 333, 338
Reference
   sites, 22, 332, 411
   standard, 60, 358, 360, 364, 400
   system, 304
Refuges, 2, 81
Regional supplements, 227, 229, 231, 254, 264, 265, 271, 272, 305
Regulations, 211, 229–230, 332, 334
Relative abundance, 322
Relative standard deviation (RSD), 359, 421
Relative value, 294, 295, 317
Relevance characteristics
   meaningful, 41
   personal, 41
Relocations, 41, 287
Remotely-sensed data, 274
Remote sensors, 98
Removal sampling, 35, 366
Repeated measure, 25
Replication, 1, 3, 17–23, 370, 435
Representative sample placement, 283
Resh, V.H., 23
Residual water, 89, 136, 139, 153–155, 170, 171, 203
Resin core method, 385–386
Respiratory parameters, 335
Response curves, 298
Response variables, 3, 13, 16, 18, 302, 310
Resuspension, 407
Retroduction, 8, 18
Retroductive reasoning, 1
Retrospective power, 31
Rhizoplane, 361
Rhizosphere, 252, 361, 394
Rice, C.L., 26
Rice, R.C., 150
Richardson, J.L., 264
Riparian wetlands, 88, 153, 334, 338
Riverine wetlands, 55, 361
RM. *See* Reduced matrix (RM)
Roberts, D.W., 302
Robson, D.S., 22
Rogers, D.A., 304
Romesburg, H.C., 8, 18
Rooney, T.P., 304
Root-mean-squared (RMS) errors, 1387
Rosenberry, D.O., 87–225
Rotating laser level, 53, 63, 64
RTK. *See* Real-time kinematic (RTK)
Ryan, M.G., 373
Ryder, D.S., 337

S
Sacks, L.A., 90
Salinity, 9, 21, 29, 161, 328, 329, 332, 335, 344
Salinity gradients, 29
Salt marsh, 410
Sample
   distribution, 30, 35
   number, 340
   preservation, 346, 357, 358
   processing, 366
   size, 5, 15, 28, 30, 35–37, 41, 285, 310, 344, 388, 404
   spacing, 309, 372
   storage, 6, 346
   timing, 274–277, 282
   transport, 366
   transportation, 28
Sample collection
   large volume sampling, 366
   small volume sampling, 366
Sampled population, 4, 12, 14
Sampling, 1, 3, 4, 19, 22–32, 35, 36, 39–41, 47, 54–56, 63, 72, 76, 132–134, 148, 160, 166, 168, 169, 225, 251, 271, 273–305, 309, 310, 312–314, 318–319, 335, 339–343, 346, 352, 357, 358, 363–367, 370, 374–376, 385, 387–389, 392, 398, 405–407, 409, 410, 423–425
   design, 1, 23–25, 27, 30, 339
   duration, 341
   error, 4, 20, 31
   frame, 25, 28
   intensity, 26, 398
   methods, 27, 169, 282, 284, 312, 342, 357
   procedures, 358, 392
   qualitative, 293, 398
   quantitative, 25, 276–278, 293, 398
   techniques, 277–294, 312
Sampling devices, water-column, 169, 346
Satellite images, 98
Saturated soil, 230–232, 234, 239, 250, 252, 261, 262, 370, 384, 398, 402, 426
Saturation, 92, 115, 116, 118, 120, 163, 229, 230, 234, 240, 248–249, 251, 257, 260, 261, 263, 326, 338, 344, 361, 370, 384, 398, 430
Sauer, V.B., 127
Scaling
   CH_4 fluxes, 376
   factor, 100
Schalk, C.W., 150
Schoeneberger, P.J., 243, 246, 264
Schubert, G., 158

Index 455

Scientific method, 4–6
Scrub-shrub wetlands, 275
SCUBA, 52, 291
Sea-level rise, 58–60
Search sampling, 25, 27
Season, 9, 17, 29, 39, 66, 84, 108, 137, 229, 230, 246, 260, 262, 277, 288, 339, 342, 356, 360, 398, 400
Seasonal changes, 334
Seasonal temperature, 155
Sediment
  collection tiles, 408–410
  gains, 406–410
  input, 50, 159, 361
  losses, 137
Sedimentation, 51, 55, 361, 407, 408
Sediment erosion tables (SET), 407, 408
Seed(s), 46
  bank, 10
  production, 46
Seepage
  bags, 152, 194, 217–221
  flux, 152, 198–200
  meter, 142, 151–153, 170, 194–197, 205, 216–223
Segmented-Darcy approach, 144, 147, 151, 189–191, 193
SEM. *See* Structural equation modeling (SEM)
Sensible heat flux, 114, 117
Separation methods, 371
SET. *See* Sediment erosion tables (SET)
Shaft encoder, 95, 180
Shallow depths, 97, 113
Shannon-Wiener Index, 297, 320, 321
Shoreline, 44, 64, 66, 90, 91, 123, 142, 143, 146, 147, 150, 189, 192, 194, 205, 326
Shortwave radiation, 119, 120
Shrubs, 104, 274, 276, 279–282, 284, 287, 288, 363, 367, 368
Sieving, 289
  invertebrate samplers, 289
Significance level, 30, 35
Sign surveys, 143
Silicon strain gage, 96
Simple rainfall-runoff models, 141
  curve number (CN) method, 141
  rational method, 141
Simpson's Index, 296–297, 320
Singh, V.P., 117
Single exponential decay models, 371
Single velocity method, 129
Sippers, 397, 398
Sirivedhin, T., 387

Sissingh, H.A., 395
Site map, 291–292
Size-class distribution, 317, 318
Skalski, J.R., 22
Slope
  shape, 259
  wetland, 261, 356, 361, 388, 410
Slugs, 132, 150, 184, 214–216, 225
Small mammals, 219
Snorkeling, 291
Snowfall measurement, 106–108
Snowpack accumulation, 107
Snow survey, 107
SOC. *See* Soil organic carbon (SOC)
SOD. *See* Soil oxygen demand (SOD)
Soda lime absorption, 273–274
Sodium carbonate fusion, 390
Sodium hypobromite oxidation, 390
Soil(s)
  color, 236–241, 249, 261, 272, 402
  cores, 366, 384–387, 404, 423, 433
  formation, 431–434
  horizonation, 228, 263
  horizons, 234–236, 250–252, 254, 262, 370, 377, 385, 405, 430, 432
  mesocosms, 426
  microbes, 232, 246, 250, 260, 361
  morphology, 228, 230, 231, 233, 235, 236, 261–263, 410
  profile, 233, 236, 407
  respiration, 372–375, 412, 423–425
  structure, 228, 244–246, 259, 263, 393
  subsidence, 408
  surveys, 229, 253, 257, 266, 267
  texture, 141, 241–245, 252, 259, 272, 412
  types, 17, 18, 29, 39, 141, 162, 164, 208, 257, 327
  water content, 159–160, 162, 164, 385
Soil and water assessment tool (SWAT), 141
Soil materials
  mineral, 236, 244, 247, 248, 253, 254
  mucky mineral, 248
  organic, 244, 246–248, 254, 258, 272
Soil organic carbon (SOC), 356, 362
Soil organic material, 244, 246–248, 254, 258, 272
Soil organic matter (SOM), 157, 232, 244, 362, 363, 368–369, 373, 381, 401, 404, 407, 409, 411
Soil oxygen demand (SOD), 403–404
Soil phosphorous content, 390–391
Soil redox potential (Eh), 362

Soil saturation
  anthric saturation, 249
  endosaturation, 248
  episaturation, 248, 249
Soil Science Society of America (SSSA), 358, 439
Soil Testing: sampling, correlation, calibration, and interpretation, 358
Soil water characteristic (SWC) curve, 164
Soil-water drainage, 165
Soluble reactive phosphorus (SRP), 231
Soluble sulfide, 397–398
SOM. *See* Soil organic matter (SOM)
Sommers, L.E., 369
Sorenson distance, 299
Spatial analyst, 72, 76
Spatial data, 62, 293
Spatial interpolation, 111, 183
Spatial patterns, 278, 279, 309–311
Spatial variability, 104, 108–109, 111, 158, 159, 326, 327, 333, 370, 411
Spatiotemporal vegetation, 273
Species abundance, 278, 286, 298, 304, 323, 324
Species accumulation curves, 285, 286, 296, 318, 319
Species diversity, 296, 319–321
Species occurrence, 26, 28
Species richness, 13, 285, 289, 292, 295–297, 318–321
Specific conductance, 329, 344
Specific objectives, 11, 16, 39, 47, 284
Specific yield, 163–165, 201–202
Spectrophotometer, 378, 392, 421
Splashing effects, 103, 104, 122
SRP. *See* Soluble reactive phosphorus (SRP)
SSSA. *See* Soil Science Society of America (SSSA)
Stable isotope analyses, 154, 335
Stable isotope techniques, 382–383
Stable nitrogen isotopes, 382
Staff gage, 93–94, 101–102, 126, 180, 181, 205–213
Staff plate, 59–62, 64, 94, 102, 180
Stage-discharge rating curve, 124, 135–136, 138, 187
Stage measurement, 93–98, 126–127, 134, 138, 206
Stallman, R.W., 155
Standard deviation, 169, 310, 311, 359, 360, 421, 422
Standardized methods, 369, 372, 394
Standard methods, 342, 358, 364, 392, 422

Standard Methods for the Examination of Water and Wastewater, 342, 358
Starr, J.L., 163
Static closed chambers, 373, 375–376
Statistical analyses, 6, 19, 22, 23, 34, 35, 37, 41, 339
Statistical error, 1, 19, 20, 30, 31, 34, 35
Statistical parameters, 11, 19, 24
Stevenson, F.J., 381
Stewart, J., 340
Stomates, 111, 114
Stonestrom, D.A., 157, 223
Storage term, 159
Storfer, A., 10
Stratified random sampling, 27–29, 274, 339, 341
Stream gaging station, 124–125
Structural equation modeling (SEM), 302
Structured decision making, 7
Study design, 1–42, 46, 47, 339–342
Study objectives, 7, 8, 39, 46, 47
Subclass modifiers, 275
Submersible pressure transducer, 95–97, 178
Sub-sampling, 289, 358
Substrates
  composition, 360
  types, 360
Subsurface storage, 159–165, 201
Succession, 20, 311–314, 335
Suction devices, 397
Sulfate reduction, 250, 396
Sulfide, 396–398, 426
Sulfur (S), 248–250, 356, 396–398
Sulfuric acid, 390, 396, 427
Sulfuric acid-hydrogen peroxide-hydrofluoric acid digestion, 390
Suren, A.M., 23
Surface float method, 134
Surface water
  inflow, 124–138
  outflow, 124–138
Surfer, 109
Swancar, A., 90
System(s)
  dominant, 274, 275
  plot-based, 279–283
  plotless, 279–283
Systematic sample placement, 283
Systematic sampling, 27–30, 283, 284

T

Targeted/tier sampling, 340
Target element, 12

Index 457

Target population, 3, 4, 11, 12, 14, 18, 19, 22–29, 31
TDR. *See* Time domain reflectometry (TDR)
TDR wave guides, 162
Temperature, 8, 96, 232, 326, 368
Temperature sensors, 96, 116, 120, 121, 156, 198, 212, 216, 221–223, 344
Temporal patterns, 23
Temporal variability, 87, 88, 277, 360, 361, 376
Temporal variance disturbance, 21
Tension saturation, 92
Tensometer, 372
Ter-Mikaelian, M.T., 364
Terrestrial buffers, 8
Theoretical error, 30
Thermal conductivity, 155–157, 198, 223
Thermal stratification, 337
Thermistors, 121, 123, 222
Thermocouples, 121, 222
Thermo-gravimetric method, 16, 159–160
Thien, S.J., 427
Thiessen polygon method, 109
  surface-fitting methods, 109
Thin-plate weirs, 125, 126, 135
Thomas, D.H., 10
Thomas, G.W., 403
Tidal action, 407
Tides, 21
Tiedje, J.M., 387
Time, 3, 50, 88, 229, 277, 334, 356
Timed meander search, 282
Time domain reflectometry (TDR), 160–162
Tiner, R.W., 340
Topographic maps, 38, 54–56, 353
Topographic position, 360
Topography, 49–51, 55, 62, 81, 232, 330, 336, 364, 370, 441
Topp, G.C., 161, 162
Total dissolved solids, 328, 329, 344
Total hardness, 328, 344
Total reactive phosphorus, 331
Total station survey, 52
Total suspended solids, 329, 330, 344
Tracer(s), 132–134, 138, 165–170, 184, 224, 225, 407
Tracer-dilution methods, 131–134, 138, 184, 224, 225
  constant-rate injection (CRI) method, 132, 133, 225
  sudden injection (SI) method, 132, 133, 225
Tracer mass-balance approach, 170
Tracer solution, 132–134, 168, 224, 225

Trade-off analysis, 58
Transect(s), 27, 52, 62–64, 68, 283, 284, 287, 291, 311, 313, 315, 408
Transect plots, 25, 27
Transformations, 35, 41, 228, 233, 234, 294, 298, 302, 332, 384
Translocations, 233, 234, 249–250, 261
Transpiration, 89, 111, 159, 166
Trap(s), 139, 140, 188, 423–425
Treatment, 3, 4, 8, 9, 12–14, 16–22, 25, 26, 28–31, 35, 37, 40, 46, 47, 327, 331, 342, 371, 423, 425
Trebitz, A.S., 336, 340
Trees, 61, 70, 93, 102, 104, 211, 260, 275, 276, 279–282, 284, 287–289, 294, 302, 311–317, 336, 363–368, 422, 441
  age determination, 365
  biomass-allometric equations, 363–364, 422
True value (T), 3, 25, 98, 360
Turbidity, 23, 134, 330, 334, 344, 352, 412
Turcotte, D.L., 158
Turner, R.E., 9
Turnipseed, D.P., 127
TWINSPAN, 299
Type I error, 30
Type II error, 30

U
Ultrasonic anemometers, 122
Unbiased sample, 275
Underwood, A.J., 22
Unit depth, 100
United States Geological Survey (USGS), 38, 58, 59, 93, 100, 120, 127, 129, 130, 134, 143, 145, 155, 156, 158, 357
Univariate statistics, 19
USDA Natural Resources Conservation Service (NRCS), 38, 438
USDA NRCS Geospatial Data Gateway, 276
USDA Plants Database, 278, 290
U.S. Department of Agriculture, 233, 235, 243, 246, 247, 253, 256, 274, 305
U.S. Department of Agriculture Web Soil Survey, 274, 305
U.S. Environmental Protection Agency, 33, 273, 279, 284, 285, 295, 299, 303–305, 404, 406, 411
U.S. Fish and Wildlife Service, 33, 38, 274, 305
USFWS. *See* Fish and Wildlife Service (USFWS)

USGS. *See* United States Geological Survey (USGS)
USGS. EarthExplorer, 276
UV Spectroscopy, 377–381

V

Vadeboncoeur, M.A., 366
Vadose zone, 92, 159, 162, 163
Value, 2, 60, 89, 237, 285, 329, 359
van der Kamp, G., 100
van der Valk, A.G., 10
Vapor flux, 114
Vaporization, 111, 114
Vapor-pressure gradient, 112
Vasilas, B.L., 227–272, 355–442
Vasilas, L.M., 227–272
Vegetation
  abundance, 278
  composition, 278
  dominance, 278, 289
  frequency, 305
  functional groups, 278
  importance, 278
  morphological characteristics, 278
  presence, 278
  production, 278
  reflectance, 300
  sampling techniques, 277–294, 305
  spatial pattern, 279
  structure, 278
Velocity-area method, 127–131, 135, 137, 138, 224
Vepraskas, M.J., 264
Verhoeven, J.T.A., 339
Vernal pools, 17, 252, 271, 333, 338, 339
Vertical-axis meter
  Price AA, 128, 130
  Price pigmy, 128, 130
Vertical flux, 113
Video cameras, 289, 293
Vogt, K.A., 366
Voucher specimens, 291

W

Wallace, S.D., 137
Walling, D.E., 132
Walton, D.W.H., 372
Wastewater treatment plants, 331
Water budget, 39, 50, 51, 88–92, 97, 98, 100, 103, 110, 111, 123, 124, 136, 138, 139, 141, 150, 153–155, 165, 169–171, 182, 203–204
Water budget hydrology, 92–103
Water chemistry, 88, 165, 275, 397
Water column analysis, 387–389
Water density, 326
Water depth (WD), 9, 14, 16, 19, 21, 53, 54, 64, 66, 68, 85, 90, 93, 97, 116, 128, 137, 140, 178–181, 188, 292
Water flow, 123, 124, 136, 142, 146, 155, 218, 262, 343
Waterfowl, 2, 12, 38, 219
Water levels, 46, 50, 54, 56, 61, 64, 81, 86, 92 95, 98, 100, 102, 105, 112, 122, 123, 127, 135, 139, 142, 149, 150, 179–181, 185, 186, 203, 208, 210, 213–216, 277, 341, 408, 430, 432
Water motion, 212
Water movement, 38, 251, 262, 336, 386
Water quality, 16, 19, 29, 37, 148, 166, 325–344, 346, 352, 353, 356, 381, 387, 390, 411
  meter, 343, 344
  monitoring, 330, 331, 339, 341, 343
Water resonance, 361
Water sample collection, 168–169
Watershed-scale rainfall-runoff modeling, 142
Water source, 38, 326–329, 333, 334, 337, 338, 360, 361
Water source data, 328, 329
Water storage capacity, 16, 51
Water table, 55, 92, 93, 95, 102, 139, 141, 148–150, 159–165, 201, 205–212, 214, 215, 228, 230, 234, 248, 249, 258–263, 360, 376, 383, 384, 388, 430, 432, 433, 437, 438, 440
2-Way analysis of variance (ANOVA), 311
Weighted-average method, 109
  surface-fitting methods, 109
Weighting schemes, 109
  inverse distance method, 109
Weilhoefer, C.L., 334
Weirs, 124–126, 134–139, 167, 185–186, 203, 204
Weishaar, J.L., 379
Weishampel, P., 375
Well screens, 148, 150, 156, 169, 181, 205, 208–215
Wetland(s)
  algae, 363
  animal communities, 37, 50
  area, 13, 14, 16, 29, 50, 54, 82, 90, 98–100, 281

assessment, 357
assessment area, 357
basins, 87, 91, 100, 103, 139, 151, 327
bathymetric maps, 50
buffers, 8, 334, 336
characteristics, 360–361
classification, 38, 327, 329, 340
complex, 37, 68, 390
condition, 39, 50, 125, 228, 302–304, 410–412
conservation, 300
declines, 2
degradation, 16
delineation, 272, 283, 290, 294, 305
depth, 97, 100, 122, 144
design, 50
development, 47, 250, 361
drainage, 54
ecosystem services, 273
elevation, 60, 93
enhancement, 2
fish communities, 26
forested, 277, 284, 293, 317, 318, 335
functions, 22, 332, 337, 381
geology, 144, 146, 333
geometry, 69, 72, 79–81
geomorphology, 82
health, 303, 332
hydrological regimes, 50
hydrology, 9, 29, 38, 39, 50, 87–172, 178, 205, 228, 261–263, 335, 341
infrastructure, 18
inventories, 274
location, 37, 38, 93, 122
losses, 88, 89, 170
management, 21, 33, 83
mitigation, 283, 305
monitoring, 281, 284, 326, 327, 332, 333, 339, 341, 343, 344, 410, 411
perimeter, 54, 139, 144–146
physical anthropogenic features, 39
plant communities, 23, 284, 285, 290, 292, 295, 297, 298, 300, 303
principles, 389
processes, 352, 353
protection, 332
quality, 26, 305
reference, 13, 274, 411
regulation, 229–230
representative, 122, 151, 284, 341, 357
reserve program, 362
restoration, 16, 260, 304
sampling, 27, 31, 63, 273–305, 343
sampling considerations, 343
science, 2, 9–11, 18, 20, 26, 83, 88
soils, 92, 159, 325, 326, 335, 361, 376, 377, 389, 391–393, 403
stage, 88–90, 93, 94, 97–101, 103, 123, 127, 138, 143, 149, 151, 178–180, 189, 192, 205, 206, 212
stressors, 16
subclass, 340
surface area, 89, 100, 101, 123
topography, 62
unit design, 47
values, 2, 10
vegetation, 64, 71, 83, 92, 159, 273–305
vegetation assessment, 283
volume, 77, 82, 86, 93, 99, 100, 170, 203
water depth, 18, 54, 93, 116, 292
wildlife, 2, 46, 47, 101, 201, 305
Wetland types, 6, 11, 13, 16–18, 21, 38, 39, 88, 274, 376
    depressional, 262, 333–337, 353, 361
    estuarine fringe, 52
    riverine, 55, 361
    slope, 261, 356, 361, 388, 410
Wetting cycles, 277, 340, 360, 381
Wetzel, R.G., 331, 342, 345
Whittaker, R.H., 366
Whitten, S.M., 10
Wieder, R.K., 371
Wienhold, B.J., 385
Wilcox, C., 70
Wildlife response to management, 214
Wilson, H.F., 379
Wind effects, 103, 104, 106
Wind speed, 107, 112, 114, 117, 118, 121, 122, 182
Winter, T.C., 90, 116, 155
Woody debris, 108, 276
Woody vegetation, 108
Wysocki, D.A., 264

X

Xenopoulos, M.A., 379
Xu, C.Y., 117

Z

Zedler, J.B., 13
Zibilske, L.M., 373, 374, 423
Zimmer, K.D., 23

FOREWORD AND ACKNOWLEDGEMENTS

The journey through Exodus offered by Fabrizio Ficco speaks for itself. It reveals a relevant common thread in the narrative's plot: the Book of Exodus recurrently depicts ritual actions in which blood is shed. These include circumcision in Exod 4:24-26, the blood of the lamb in 12:6-7, 21-23, the blood of the covenant in 24:3-8, and the blood in the high priest's (future) consecration in chapter 29. These episodes all present a liminal quality: that is, they are associated with a threshold, and with the (perilous) journey beyond this threshold – into a life shared with God. When applied or sprinkled, blood is the vehicle of a pulsating, circulating life.

This exegetical and hermeneutical enterprise is to be welcomed. It is carried out with rigourous historical-critical analysis, and finds important analogies in Ancient Near Eastern literature (for example, in vassal treaty documents). At the same time – and this is undoubtedly its profound originality – this work is conducted with the help of narrative exegesis. Following a sequential and cumulative perspective, the texts are examined in the same order in which they appear in Exodus. The act of reading creates a bond not only between the narrative's protagonists (Yhwh, Moses, the people), but also with the reader.

Let us take, for example, the covenant ritual in Exod 24. The people's entry into the $b^e r\hat{\imath}t$ occurs initially as a bond created by a speech act – the illocutionary act reported in vv. 3.7 (see v. 7: "All that Yhwh has uttered, we will do and we will listen"). When the people are subsequently sprinkled with the sacrificial blood, the bond symbolically becomes twofold through the creation of a bond in the very life of God. Although the reader does not experience the sprinkling, she participates via her imagination in the founding event. Nothing better than blood expresses the intimacy of a shared life. We must thank Fabrizio Ficco for unveiling this.

Jean-Pierre Sonnet, SJ

This monograph would have never been possible without the encouragement of a master and true friend: Jean-Pierre Sonnet, SJ. May I take this opportunity to express my deep gratitude to him, because it was thanks to his witness, his expertise, and his support that this essay came to being.

I also want to thank the Faculty of Theology of the Pontifical Gregorian University – where I have been teaching for more than ten years now – which has contributed to the publication of this work. My sincere thanks go especially to the Dean, Fr. Philipp Gabriel Renczes, SJ. I would also like to thank Fr. Michael Kolarcik, SJ, Rector of the Pontifical Biblical Institute, for his help, and for agreeing to publish this study in the prestigious series "Analecta Biblica Studia". Finally, let me mention affectionately my colleagues in the Department of Biblical Theology, and especially its Director, Prof. Nuria Calduch-Benages.

The English text of *Blood and Liminality* is the fruit of my own writing, but it has been proofread and edited by several people whose assistance has been invaluable. First of all, I want to mention Fr. Davide Lees, whose meticulous and insightful suggestions have deeply improved the quality of the text; I would also like to thank some seminarians from the Redemptoris Mater College, where I have been serving as a formator for years: Paul Moore, Cody Merfalen, James Rice, and Joseph Deegan. In conclusion, I remember with affection and gratitude the whole Redemptoris Mater College in Rome, and in particular, the other formators and faculty.

Fabrizio Ficco

ABBREVIATIONS

AAR.SR	Americana Academy of Religion. Studies in Religion
ABD	*The Anchor Bible Dictionary.* I–VI (ed. D.N. FREEDMAN) (New York, NY: Doubleday, 1992)
AKM	Abhandlungen für die Kunde des Morgenlandes
Alonso	L. ALONSO SCHÖKEL, *Dizionario di Ebraico biblico* (Cinisello Balsamo: San Paolo, 2013)
AnBib	Analecta Biblica
ANET	J.B. PRITCHARD, *Ancient Near Eastern Texts Relating to the Old Testament* (Princeton, NJ: University Press, 1950)
Ant. Jud.	*Antiquitates Judaicae*
Aristeas	*Letter of Aristeas*
ARM	Archives Royales de Mari
AThANT	Abhandlungen zur Theologie des Alten und Neuen Testaments
AUSS	*Andrews University Seminary Studies*
BA	*Biblical Archaeologist*
BASOR	*Bulletin of American Schools of Oriental Research*
BBR	*Bulletin for Biblical Research*
BiblInterp	*Biblical Interpretation*
BiblInterpS	Biblical Interpretation Series
BKAT	Biblische Kommentar. Alten Testament
bNed	*Talmud Bavli. Nedarim*
bShab	*Talmud Bavli. Shabbat*
BTB	*Biblical Theology Bulletin*
BWANT	Beiträge zur Wissenschaft vom Alten und Neuen Testament
BZAW	Beihefte zur Zeitschrift für alttestamentliche Wissenschaft
CBOTS	Coniectanea Biblica Old Testament Series
CBQ	*Catholic Biblical Quarterly*
CSSC	Centro Studi Sanguis Christi
CEJL	Commentaries on Early Jewish Literature
Ch./Chs.	Chapter/Chapters
CISR	Conférence Internationale de sociologie des religions
CLV	Centro liturgico vincenziano
COS	W. HALLO – K. LAWSON YOUNGER Jr., *The Context of Scripture.* I. Canonical Compositions from the Biblical World. II. Monumental Inscriptions from the Biblical World (Leiden: Brill, 2003)
DJD	Discoveries in the Judean Desert
DSBP	Dizionario di Spiritualità Biblico-Patristica

EBR	H.-J. KLAUCK – al., ed., *Encyclopedia of the Bible and its Reception*. I–XX (Berlin – Boston, MA: De Gruyter, 2009-2020)
EDB	Edizioni Dehoniane Bologna
EstB	*Estudios Bíblicos*
EstR	*Midrash Rabba. Esther*
ExR	*Midrash Rabba. Exodus*
EstE	Libro di Ester, versione in ebraico
ETR	*Études théologiques et religieuses*
FAT	Forschungen zum Alten Testament
FAT.2	Forschungen zum Alten Testament 2. Reihe
FoSub	Fontes et Subsidia ad Bibliam pertinentes
FRLANT	Forschungen zur Religion und Literatur des Alten und Neuen Testaments
GenR	*Midrash Rabbah. Genesis* (ed. H. FREEDMAN) (London: Soncino Press, 1939)
Gilgamesh	*The Babylonian Gilgamesh Epic*. I (ed. A.R. GEORGE) (Oxford: University Press, 2003)
GK	H.W.F. GESENIUS, *Hebräische Grammatik* (ed. E. KAUTSCH) (Leipzig: Vogel, 1909)
HALOT	L. KOEHLER – al., *The Hebrew and Aramaic Lexicon of the Old Testament* (Leiden – Cologne – New York, NY: Brill, 1994-2000)
HAR	*Hebrew Annual Review*
HBI	The Heritage of Biblical Israel Series
HBM	Hebrew Bible Monographs
HBT	*Horizons in Biblical Theology*
HeBAI	*Hebrew Bible and Ancient Israel*
HebStud	*Hebrew Studies*
HEx	ORIGENES, *Homiliae in Exodum*
HSS	Harvard Semitic Studies
HThKAT	Herder theologischer Kommentar zum Alten Testament
HUCA	*Hebrew Union College Annual*
Ibn Ezra	M. CARASIK, ed., *The Commentators' Bible. The Rubin JPS Miqra'ot Gedolot. Exodus* (Philadelphia, PA: JPS, 2005)
ICC	International Critical Commentary
IECOT	International Exegetical Commentary on the Old Testament
IG XII	E. ZIEBARTH, ed., *Inscriptiones Graecae*. XII, 9. Inscriptiones Euboeae insulae (Berlin: Georgium Reimerum, 1915)
IGT	R. KÖRNER, ed., *Inschriftliche Gesetzestexte der frühen griechischen Polis* (Köln: Vandenhoeck & Ruprecht, 1993)
ISBL	Indiana Studies in Biblical Literature
JBL	*Journal of Biblical Literature*
JBQ	*Jewish Bible Quarterly*
JEA	*Journal of Egyptian Archaeology*

JM	P. JOÜON – T. MURAOKA, *A Grammar of Biblical Hebrew* (SubBib 27; Roma: PIB, 2006)
JNES	*Journal of Near Eastern Studies*
JPS	Jewish Publication Society
JPS.TC	Jewish Publication Society. Torah Commentary
JSOT	*Journal for the Study of the Old Testament*
JSOT.S	Journal for the Study of the Old Testament. Supplement Series
KAI	H. DONNER – W. RÖLLIG, *Kanaanäische und aramäische Inschriften*. I-III (Wiesbaden: Harrassowitz, 1962-1964)
KTU	M. DIETRICH – O. LORETZ – J. SANMARTÍN, *Die keilalphabetischen Texte aus Ugarit. Einschliesslich der keilalphabetischen Texte ausserhalb Ugarits*. I. Transkription (Alter Orient und Altes Testament 24/1; Kevelaer – Neukirchen-Vluyn: Ugarit-Verlag, 1976)
LAR	D.D. LUCKENBILL, *Ancient Records of Assyria and Babylonia* (Chicago, IL: University Press, 1926-1927)
LeDiv	Lectio Divina
LThK	W. KASPER, ed., *Lexicon für Theologie und Kirche*. I-X (Freiburg – Basel – Rom – Wien: Herder, 1993-2001)
MekhY	*Mekhilta de-rabbi Ishmael* (ed. J.Z. LAUTERBACH) (JPS Classic Reissues; Philadelphia, PA: JPS, 2004)
MekhSh	*Mekhilta de-rabbi Shimon bar Yoḥai* (ed. W.D. NELSON) (Philadelphia, PA: JPS, 2006).
mMen	*Mishna. Menaḥot*
MoBi	Le monde de la Bible
mPes	*Mishna. Pesaḥim*
MT	Masoretic Text
mYoma	*Mishna. Yoma*
Nahmanides	M. CARASIK, ed., *The Commentators' Bible*. The Rubin JPS Miqra'ot Gedolot. Exodus (Philadelphia, PA: JPS, 2005)
NIBC	New International Biblical Commentary
NICOT	New International Commentary on the Old Testament
NICNT	New International Commentary on the New Testament
NIGTC	New International Greek Testament Commentary
NJPS	*Hebrew-English Tanakh. The Jewish Bible* (Skokie, IL: JPS, 2009)
NRSV	New Revised Standard Version
NTS	*New Testament Studies*
NVBTA	Nuova Versione della Bibbia dai Testi Antichi
OBO	Orbis biblicus et orientalis
OTL	Old Testament Library
OTS	*Oudtestamentische studiën*
PIB	Pontificio Istituto Biblico
PIHANS	Publications de l'Institut historique-archéologique néerlandais de Stamboul

PRE	Midrash Pirkê de rabbi Eliezer (ed. G. FRIEDLANDER) (New York, NY: Hermon Press, 1965)
PRU IV	Le palais royal d'Ugarit. IV (ed. J. NOUGAYROL) (Mission de Ras Shamra IX; Paris: Imprimerie Nationale & Klicksieck, 1956)
PSam	Pentateuco Samaritano
PUG	Pontificia Università Gregoriana
Rashi Gen	Y.I. ZVI HERCZEG, *The Torah.* With Rashi's Commentary Translated, Annotated, and Elucidated. I. Bereshis / Genesis (The Sapirstein Edition of Rashi; Brooklyn, NY: Artscroll, ²1996)
Rashi Exod	M. CARASIK, ed., *The Commentators' Bible.* The Rubin JPS Miqra'ot Gedolot. Exodus (Philadelphia, PA: JPS, 2005)
SAA	State Archives of Assyria
SAA 2	S. PARPOLA – K. WATANABE, ed., *Neo-Assyrian Treaties and Loyalty Oaths* (SAA 2; Helsinki: Eisenbrauns, 1988)
SANT	Studien zum Alten und Neuen Testament
SBL	Society of Biblical Literature
SBL.AB	SBL. Academia Biblica
SBL.AIL	SBL. Ancient Israel and Its Literature
SBL.MS	SBL. Monograph Series
SBL.SCS	SBL. Septuagint and Cognate Studies
SBLSymS	SBL. Symposium Series
SBL.WAWSup	SBL. Writings from the Ancient World Supplement Series
SCh	Sources Chrétiennes
SJLA	Studies in Judaism in Late Antiquity
SJOT	*Scandinavian Journal of the Old Testament*
SubBib	Subsidia Biblica
TgN	*Targum Neophiti*
TgO	*Targum Onkelos*
TgPsJ	*Targum Pseudo-Jonathan*
THAT	E. JENNI – C. WESTERMANN, ed., *Theologisches Hand-wörterbuch zum Alten Testaments.* I-II (München: Gutersloher, 1971, 1976)
ThWAT	G.J. BOTTERWECK – al., ed., *Theologisches Wörterbuch zum alten Testaments.* I-VIII (Stuttgart: Kohlhammer, 1970-2000)
TTB	R. PENNA – G. PEREGO – G. RAVASI, ed., *Temi teologici della Bibbia* (Dizionari San Paolo; Cinisello Balsamo: San Paolo, 2010)
UBC	Understand the Bible Commentary Series
VT	*Vetus Testamentum*
VTE	D.J. WISEMAN, "The Vassal-Treaties of Esarhaddon," *Iraq* 20 (1958) i-ii.1-99.
WBC	World Biblical Commentary
WO	B.K. WALTKE – M. O'CONNOR, *An Introduction to Biblical Hebrew Syntax* (Winona Lake, IN: Eisenbrauns, 1990)
ZAW	*Zeitschrift für die Alttestamentliche Wissenschaft*

INTRODUCTION

What, then, o Moses? He says: "Immolate a lamb without blemish and anoint the gates with its blood." What are you talking about? Is the blood of an animal without reason able to save men with reason? Yes, he says, not because it is blood, but because it is figure of the blood of the Lord.[1]

The quoted passage is a suitable introduction for this monograph, dedicated to the study of some texts taken from the book of Exodus in which the blood shed within a ritual action constitutes a significant element of the narrative. These words are pronounced by John Chrysostom in one of his *Baptismal Instructions*; they state that the fluid used in Exod 12 is a figure of the salvific blood of Christ. Chrysostom imagines a dialogue with Moses in which the prophet himself recognises that the extraordinary effectiveness of blood was already active when the Israelites' doors were anointed with it.

Although the text of John Chrysostom is so promising, the choice of this theme generates ambiguous first impressions. Indeed, when referring to blood, what immediately comes to mind are the studies of Réne Girard, who promoted an overcoming of sacrificial language.[2] He attempted to unmask the mechanism of the scapegoat which, in his view, lies behind the sacrifice, in order to finally free the field from the "sacrificial ghost."[3] According to his approach, the Revelation of Jesus Christ is in direct opposition to the sacrificial worship.[4] Therefore, even if the theology of sacrifice has been recently rethought and the more dialectical positions reconsidered,[5] many are the questions that the reader may have when confronted just with the theme and

---

[1] JOHANNES CHRYSOSTOMUS, *Ad illuminandos catecheses*, III, 14 (Sources Chrétiennes 50, 159): Τί οὖν ὁ Μωϋσῆς; «Θύσατε, φησίν, ἀμνὸν ἄμωμον καὶ τὸ αἷμα αὐτοῦ ἐπιχρίσατε ταῖς θύραις». Τί λέγεις; Αἷμα ἀλόγου σῷζειν ἀνθρώπους τοὺς λογικοὺς οἶδε; Ναί, φησίν, οὐκ ἐπειδὴ αἷμα ἐστιν ἀλλ᾽ἐπειδὴ τοῦ αἵματος τοῦ δεσποτικοῦ τύπος ἐστί.

[2] Regarding "sacrifice" in the New Testament, see R. GIRARD, *Des choses cachées*, 247, "Je crois que la définition sacrificielle de la passion et de la rédemption ne mérite pas de figurer parmi les principes qu'on peut légitimement tirer du texte néo-testamentaire."

[3] The metaphor, "sacrificial ghost," is used by M. RECALCATI, *Contro il sacrificio*, 13.

[4] A certain modesty regarding blood is also made manifest in the removal of the liturgical memorial of Jesus' circumcision (Lk 2:21) from the Missal of the Second Vatican Council, celebrated in the *Vetus Ordo* the day after the Octave of Christmas. See J.-H. TÜCK, ed., *Die Beschneidung Jesu*, where the restoration of this feast is highly recommended.

[5] See G. CANOBBIO, "La morte di Gesù," 141-164; S. PASSERI, "Rinunciare al sacrificio?", 165-188.

with the title of this work. Does the manipulation of blood have a specific use in the book of Exodus? How do the different passages fit into its plot?

The intention of this work will be precisely to demonstrate that blood appears in significant texts for the structure, plot and theology of Exodus. Thereupon, a clearer light will be shed on the shadow created by the sacrificial language, in order to show its full potential. In order to illustrate the working hypothesis of this study, a prior examination of the current state of enquiry seems necessary.

1. *Status quaestionis*

Much has been elaborated on the theme of blood in recent scholarship, especially in regards to the Old Testament sacrificial praxis.[6] A relevant contribution to research appeared in the proceedings of a conference entitled *Sangue e antropologia biblica* (Blood and Biblical Anthropology. 1980). The articles devoted to blood in the Bible and in the Old Testament texts, are an important starting point for this enquiry.[7] Recent studies on biblical sacrificial practice have also been useful,[8] together with Girard's survey on the symbolic use of the term "blood" in the Old Testament.[9] Analysing the use of דם in the Old Testament, many authors reached similar conclusions: since the term דם is used as a synonym of נפש, "soul, life" (Gen 9:4-5; Lev 17:11-14; Deut 12:23; see Ezek 22:27), blood seems to be a powerful symbol of life.[10] This idea has been taken as the key concept to understand sacrificial worship[11] (Lev 17:11).

---

[6] The contributions of the lexicons has proved to be significant for an initial approach to the subject of our research (see especially KEDAR-KOPFSTEIN, B., "דם," *ThWAT*, II, 253-267); see also S.D. SPERLING, "Blood," 761-763; N. FÜGLISTER, "Blut," 531-534; C.A. EBERHART, "Blood. I. Ancient Near East and Hebrew Bible/Old Testament," 201-212; F. MANZI, "Sangue," 1232-1237. For the greek αἷμα, see A. WEISSENRIEDER – K. DOLLE, ed., *Körper und Verkörperung*, 733-798.

[7] See F. VATTIONI, "Sangue: vita o morte nella Bibbia," 367-378; A. PENNA, "Il sangue nell'Antico Testamento," 379-402 (with bibliography); N.M. LOSS, "Carne, anima e sangue," 403-412. In addition to these articles, there are also specific essays dedicated to some Old Testament texts: J.A. SOGGIN, "Il sangue nel racconto biblico delle origini," 413-423; P.P. ZERAFA, "Il significato del sangue nella Pasqua biblica," 453-465; R. GELIO, "Il rito del sangue," 467-476; ID., "Sangue e vendetta," 515-528; F. VATTIONI, "'Sposo di sangue' (Es 4,24-26)," 477-496; ID., "Il sangue dell'alleanza (Es 24,8)," 497-513.

[8] See, in particular, A. MARX, *Les systèmes sacrificiels* (with bibliography).

[9] See M. GIRARD, *Symboles bibliques*. II, 1454-1488.

[10] A. PENNA, "Il sangue nell'Antico Testamento," 394-402; N.M. LOSS, "Carne, anima e sangue," 403-412; J.A. SOGGIN, "Il sangue nel racconto biblico delle origini," 413-423.

[11] J. MILGROM, *Leviticus 1–16*, 711-712.

The monograph *Blood Ritual in the Hebrew Bible* (2004), written by W.K. Gilders, analyses the meaning of blood manipulation and the power attributed to this element. The author criticises the belief that Lev 17:11 is a crucial passage to understand the ritual use of blood.[12] Instead, he adopts N. Jay's approach, according to which ritual gestures are not actions that need to be interpreted according to the context where they are created, but *indexes*, namely signs that by their very external nature refer to a meaning that can be understood without the need for any cultural mediation.[13]

The conclusions of Gilders' work are as follows: the connection between blood and life, expressed in Gen 9:4-6, Lev 17:11-12 and Deut 12:20-25, is not found elsewhere in the Bible and only Lev 17 (interpreted as an H text) links this relation to cultic practice. Therefore, such an association does not seem relevant for the comprehension of blood manipulation in P and in sacrificial cult, since the author assumes that H is secondary to P.[14] The non-P texts,[15] older than P (he includes Exod 12:21-23 and 24:3-8), are in fact less elaborate and do not present an explicit reference to the cultic use of blood; in these texts, the meaning of manipulation is understandable simply from an examination of ritual gestures, interpreted as *status indexes* which manifest the special position of priests and exalt their prerogatives. Gilders' work, then, remains an important point of reference, even if his basic hypothesis has been criticised[16] and his analysis shows a lack of consideration of the narrative nature of the Bible. This absence is evident considering the texts that will be the object of our interest (Exod 4; 12; 24; 29), which are not analyzed respecting the order of the narration, but considering their hypothetical date of composition.

Owiredu's more recent study, *Blood and Life in the Old Testament* (2004), considers the role of fluids in the life of the body, assessing the place of blood in the physiology of the OT: the link between "blood" and "life" is explained with the association of death and the loss of body fluids. In examining the use of blood, he pays special attention to its ritual capability to reverse the condition of "death" caused by the disease of

---

[12] See W.K. GILDERS, *Blood Ritual*, 8.49.
[13] See N. JAY, *Throughout Your Generations Forever*, 6-7.
[14] Gilders refers to the studies of I. KNOHL, *The Sanctuary of Silence*.
[15] Before dealing with the texts belonging to P literature, Gilders analyses the food prohibition mentioned in the "Covenant Code" (Exod 23:18; see 34:25), the manipulation of blood found in the conclusion of the Sinai covenant (24:3-8), the narration of the first Passover (12:6-7, 21-23), the ritual of Deut 12:20-27 and 2 Kings 16:10-18.
[16] See CH. NIHAN, "The Templization of Israel," 94-130.

leprosy.[17] In the last chapter, the texts of Exodus are addressed, especially Exod 12 and 4:24-26; in them, blood has the power to preserve life by means of a ritual substitution.[18]

In a recent work (2011) Feder examines the significance of the ritual use of blood in the sin offering and its purifying value in Hittite and biblical literature. In his opinion, this cultic practice derives from a background which is common to both cultures.[19] The symbol of blood is explained by the frequent association between the physiological fluid and violent death:[20] the sacrifice, then, has a "substitution" purpose and achieves a certain "restitution" and compensation.[21] Although many of Feder's considerations can be found rather useful in this investigation, the texts analysed here present a broader perspective, because the ritual use of blood is not a part of the rite designated as a חטאת.

The literature referring to blood, then, is quite rich, and the topics covered are manifold. However, a study entirely devoted to the motif of blood in the book of Exodus has not been encountered. The main purpose of this inquiry, therefore, is to fill this gap.

2. Blood in the Book of Exodus: A Working Hypothesis

The outline of previous research regarding the subject of blood creates the opportunity to describe the position of the present work, which is limited to those texts of Exodus which allude to a ritual manipulation of blood. Three of the passages contain ritual actions that are unique in the biblical *panorama* (the circumcision in 4:24-26; the blood of the lamb in 12:6-7, 21-23; the covenant in 24:3-8) and therefore cannot fail to stimulate considerable interest in the reader. The novelty of this work will consist, therefore, in the recognition of a *common thread* that runs through the book of Exodus and determines it in a significant way.[22]

---

[17] C. OWIREDU, *Blood*, 108.
[18] C. OWIREDU, *Blood*, 133-172.
[19] See Y. FEDER, *Blood Expiation*, 1.
[20] See ch. I, § 1.1.
[21] Y. FEDER, *Blood Expiation*, 265.
[22] The hypothesis that prompted me to write this monograph stems from the suggestion and the encouragement of a colleague and friend, Jean-Pierre Sonnet. The sentence, which I quote here, together with his advice, provided the basis for developing this work; see J.-P. SONNET, "Generare, perché?" 174, "Blood thus provides Exodus with its *fil rouge*. It is the blood of the nocturnal episode in Exodus 4, which links God, Moses, his wife and son; it is the blood on the doorposts of the houses in Exodus 12; it is the blood of the conclusion of the Covenant in Exodus 24, which unites the altar and the people, that is, God and the

Secondly, this analysis will attempt to express that the *narrative order* of the episodes is relevant to grasp their meaning. The blood that transforms Moses' condition on a personal and family level (4:24-26), is a vital element for the families of the Israelites engaged in the celebration of the Passover (12:6-7, 21-23), and for the people gathered on mount Sinai for the stipulation of the covenant with Yhwh (24:3-8). Finally, with the blood, Moses will "separate" Aaron and his sons from the people, assigning them to the priestly ministry (29:12, 20-21). Consequently, the sequence of episodes reveals that Moses' personal experience (4:24-26), where the first appearance of ritual blood is found, enables the reader to have a better understanding of the subsequent episodes where blood manipulations appear; moreover, the series of narratives shows that the handling of blood, first reserved to the heads of families (12:7, 13, 22-23) and then to Moses on Sinai (24:3-8), will ultimately be entrusted to Aaron and his sons, multiplying and at the same time limiting the mediators.

The first chapter is devoted to some preliminary questions (c. I): a lexicographical examination of the use of דם in the Hebrew Bible; a more detailed presentation of the working hypothesis; a commentary on two passages of Exodus (4:9; 7:14-24) where the word "blood" is used, but which do not fall directly within the field of this study.

The next four chapters consist of the *lectio cursiva* of the passages that represent the main subject of the work, in the light of their immediate context. Thus, chapter II is formed by the study of Exod 4:24-26, in connection with 4:18-31, which narrates the return of Moses to Egypt after his calling (3:1–4:17). Chapter III is dedicated to the reading of the proclamation of the last plague (Exod 11:1-10), the institution of the Passover (12:1-27) and the liberation from Egypt (12:28-42). Chapter IV analyses Exod 24 in order to describe the ritual of 24:3-8, and finally chapter V reads Exod 29 and 30:1-10, trying to grasp the complex articulation of the rite of Aaron's consecration. The monograph concludes with a summary of the path, followed by some hermeneutical and theological considerations.

3. Methodological Clarifications

Exegetical research on Exodus has devoted a great deal of energy to clarifying the various editorial layers which compose the final text.[23] For

---

people; it is the blood in the ritual of the consecration of the high priests in Exodus 29. In this sense, Exodus is the story of the consanguinity that progressively unites God and the people." (My translation)

[23] See T.B. DOZEMAN – C.A. EVANS – J.N. LOHR, ed., *The Book of Exodus*; T.E. LEVY – T. SCHNEIDER – W.H.C. PROPP, ed., *Israel's Exodus in Transdisciplinary Perspective*.

this reason, the historical-critical method will remain in the background of every reading and will accompany it at all times.[24] Before reading the texts, the main hypotheses for the reconstruction of their editorial history will be examined. This step will be important above all in order to grasp the textual tensions that may be the sign of a layered composition. By highlighting the most plausible reconstructions, the elements obtained from the reading will be used as a starting point to appreciate the editors' work and the final form of the composition.[25]

Thereupon, aware of the textual problems of each passage, the literary coherence of the whole text will be verified, in order to find the meaning that can be derived by reading in sequence even those parts that the critics tend to divide. So, the main sensitivity of the work will be *discourse-oriented* and the inquiry will always ponder the textual coherence in its final form.[26] The exegetical method that, more than any other, will help to explain the literary value of the texts is certainly narrative analysis[27] and for this reason the procedures of this approach will be often used in the *lectio cursiva*.

Finally, this approach will permit a harmonious reading of the various fragments and illuminate all the elements in the light of the ensemble. Moreover, this approach will attempt to embody a typical requirement of *biblical theology*, which is not limited just to the study of the single pericopes, but seeks to obtain a comprehensive understanding of the Scriptures considered as a whole.[28]

---

[24] Regarding the historical-critical method, see: J. BARTON, *The Nature of Biblical Criticism*; CH. NIHAN, "L'analisi redazionale," 121-166.

[25] The inner-biblical exegesis manifests that biblical literature was formed through a process of revision of ancient texts that were preserved and re-interpreted in the light of new conditions; the final text shows the traces of this constant revision. See M. FISHBANE, *Biblical Interpretation*; B. LEVINSON, *Legal Revision*. See also, J.-P. SONNET, "La Bible et l'histoire, la Bible et son histoire," 1-23.

[26] For the concepts of *source oriented* and *discourse oriented*, see ch. II, § 1.2. Paul Beauchamp defined the historical-critical sensibility as "archaeological" and the attention to the text in its final form as "teleological." See P. BEAUCHAMP, *L'un et l'autre Testament*. I, 174, "On comprendra mieux le sens d'une attitude "résolument" téléologique en l'opposant aux présupposés qui ont dominé la recherche exégétique à l'étape précédente.... L'orientation de la lecture exégétique donne de faibles signes de devenir, sous nos yeux, téléologique, après avoir été *archéologique* en ce sens que, jadis et naguère, la fascination des commencements la guidait."

[27] For narrative analysis, see M. STERNBERG, *The Poetics*; R. ALTER, *The Art of Biblical Narrative*; J.-P. SONNET, "L'analisi narrativa," 45-85; J.-L. SKA, *"I nostri padri."*

[28] See P. BOVATI, "Teologia Biblica e ispirazione," 285-289.

CHAPTER I

Preliminary Considerations

Blood, pumped by the heart muscle, is a fluid which circulates in the animal and human body[1] (Deut 12:27). The sight of blood can generate a certain dismay, due to the association of this liquid with the trauma of the open wound[2] and with death.[3] At the same time, blood can have a very positive value, since it is connected with life[4] (see Lev 17:11), and is involved in its transmission.[5] In Mesopotamian culture, blood plays a significant role in the myth of creation: for both the *Atra-ḫasīs* and the *Enūma eliš*, man was created through the blood of the gods, so that the divine fluid may flow through his veins as a sign of his special condition.[6]

In the following pages, the first steps of the investigation are taken by presenting a synthetic overview of the contexts in which the term דם appears most frequently. This outline will allow the reader to gather and synthesize the conclusions of the works that have devoted special attention to the term "blood," in order to introduce the heart of this chapter, where the working hypothesis that guides this study is illustrated in more detail. The chapter ends with an *excursus* dedicated to the first recurrence of דם in

---

[1] The דם is a physiological fluid, it can come out of the nose (Prov 30:33), flow out of the body (1 Kings 22:35; also 18:28), be lapped by dogs (21:19); the women have a continuous experience of bloodloss, due to menstruation (Lev 15:19; 20:18) and childbirth (Lev 12:4, 5, 7; Ezek 16:6, 22).

[2] J. DE LA ROCHETERIE, *La symbologie des rêves*, 184-185. Dreaming something connected with blood is a sign that often reveals the presence of a deep soul wound.

[3] Regarding the link between violence and blood in the Sumerian world, see G. PETTINATO, "Il sangue nella letteratura sumerica," 41-43. See also E. TAGLIAFERRO, "Sangue: area lessicale," 173-221.

[4] A. WEISSENRIEDER – K. DOLLE, ed., *Körper und Verkörperung*, 733-798.

[5] So M. GIRARD, *Symboles bibliques*. II, 1458-1462.

[6] G. PETTINATO, "Il sangue nella letteratura sumerica," 44-45 e L. CAGNI, "Il sangue nella letteratura assiro-babilonese," 78-85. See *Atra-ḫasīs*, I, 208.210-213, "From his flesh and blood / Let Nintu mix clay / That god and man / May be thoroughly mixed in the clay" (see W.G. LAMBERT – A.R. MILLARD, *Atra-ḫasīs*, 59). In the *Enūma eliš*, when Marduk decides to create man, Ea states that to achieve this goal, one of the gods must die. Kingu is chosen, the groom who has stirred up Tiamat, causing the war of the gods; he is brought before Ea who "de son sang, il créa l'humanité" (VI, 33; P. TALON, *Enūma eliš*, 99).

the book of Exodus (4:9) and to the narrative of Exod 7:14-24; although these texts do not fall within the specific interest of this research, they represent a sort of "first impression" that the red colour of blood offers to those who open the Book of Exodus.

1. The Term דם in the Hebrew Bible

The lexicographical study is a preliminary operation that allows the identification of the semantic fields in which דם is used in the Hebrew Bible.[7] The first step is to distinguish the occurences where דם refers to human blood from those where animal blood is involved.

1.1 *Human Blood: Violence and Death*

When the term דם denotes the human blood, the context is frequently violent.[8] This consideration is confirmed by the very first occurrence of דם in the Hebrew Bible, Gen 4:10-11.

> [10] "What have you done? The voice of your brother (קול דמי אחיך) cries[9] to me from the ground. [11] Now, you are cursed from the ground which has opened its mouth to receive your brother's blood (את־דמי אחיך) from your hand."

Yhwh declares the guilt of Cain by means of a unique expression: the blood gives voice to Abel, brutally murdered by his brother Cain and unable to cry for justice.[10] The term "blood" can be used to designate the human person in general, especially indicating his or her fragile and mortal condition[11] (Ps 72:14; Prov 1:18). Whenever דם, as in Gen 4, is *plural*, it designates the blood dispersed (Ezek 16:6, 9) and shed[12] (שפך, 1 Kings 2:31), the bloodguilt[13] (Exod 22:1-2; Num 35:27; Deut 19:10; Hos 4:2) or

---

[7] So C.A. EBERHART, "Blood. I. Ancient Near East and Hebrew Bible/Old Testament," 202-212.

[8] R. GELIO, "Sangue e vendetta," 515-528; B. KEDAR-KOPFSTEIN, "דם," 254-255; Y. FEDER, *Blood Expiation*, 173-176; M. GIRARD, *Symboles bibliques*. II, 1465-1470.

[9] The form of דם is plural, but the meaning singular (see LXX, Vulgata, Peshitta). PSam has דם and צעק instead of the plurals; the reading seems a *lectio facilior* that seeks to achieve full concordance between these terms and the singular קול; G.J. WENHAM, *Genesis 1–15*, 94.

[10] Not only is צעק used for the invocation of those in need of help, but it can also express the appeal for justice made by the innocent to the judge (2 Kings 8:3; Job 19:7).

[11] The expression "flesh and blood" is used in Sirach and in the NT referring to man (see 1 Macc 7:17; Sir 14:18; 17:31; Matt 16:17; John 6:53-56; 1 Cor 15:50; Gal 1:16; Eph 6:12).

[12] See JM § 136*b*; WO § 7.4.1*b*.

[13] The murderer has hands full of blood (דמים + מלא, Isa 1:15; see Ezek 9:9).

even the sentence for the committed crime.[14] The murderer is defined as a "man of blood"[15] (איש הדמים, 2 Sam 16:7; Pss 5:7; 26:9; 55:24; 59:3; 139:19; Prov 29:10). war is the main scenario that provokes bloodshed (1 Kings 2:5; Is 9:4).

The *singular* of דם reflects a similar pattern: when the term is used in violent contexts, it is often linked to the verb שפך, "to pour."[16]

> Gen 9,5-6, "The blood for your lives (ואך את־דמכם לנפשתיכם) will I require (אדרש), at the hand of every beast will I require it; at the hand of man, at the hand of each one [the life of] his brother,[17] will I require (אדרש) the life of man. Whoever sheds man's blood (שפך דם האדם), by man/for this man[18] shall his blood be shed (באדם דמו ישפך). For in the image of God he made man."

Murder, which can also be indicated just with the noun דם (Deut 17:8; 2 Sam 3:27), is committed with the shedding of "innocent blood,"[19] disrespecting the precious value of life.[20] When the misconduct is ascertained, the perpetrator is held to account; the procedure against the murderer is expressed, as in Gen 9:5, through the syntagm דם + דרש as a direct object[21] (Gen 42:22; Ezek 33:6; Ps 9:13), or with the synonym בקש (2 Sam 4:11; Ezek 3:18, 20; 33:8; also 1 Sam 20:16). The judgement is

---

[14] The expression "the blood is on him" (Lev 20:9, 11-13, 16, 27; 1Kings 2:33; Ezek 18:13; see also 2 Sam 1:16; 21:1) indicates that the person is considered responsible for the blood; see P. BOVATI, *Re-establishing Justice*, 359.

[15] If murder is a collective guilt, the city is designated as a "city of blood" (עיר הדמים, Ezek 22:2; 24:6, 9; Na 3:1; see also Mi 3:10, and the expression הארץ הדמים, Ezek 9:9).

[16] Gen 37:22; Lev 17:4; Num 35:33; Deut 21:7; 1 Sam 25:31; 1 Kings 2:31; 2 Kings 21:16; 24:4; Isa 59:7; Jer 7:6; Ezek 18:10; 22:3, 13, 27; 36:18; Joel 4,19; Zeph 1:17; Ps 79:3; 106:38; Prov 1:16; 6:17; Lam 4:13; 1 Chron 22:8.

[17] The Hebrew איש מיד אחיו designates the relationship of brotherhood between each human being and his neighbour; GK § 139c; G.J. WENHAM, *Genesis 1–15*, 154.

[18] The sentence in Gen 9:6 begins with a participle, a *casus pendens* (GK § 143c) which indicates the contingency that provokes the subsequent consequence; GK § 116w; JM § 156g; WO § 37.5a. Here, the preposition ב introduces the agent of the passive verb, the instrument by which justice is carried out; GK 121f; JM § 132e; WO § 11.2.5d; G.J. WENHAM, *Genesis 1–15*, 151; V.P. HAMILTON, *Genesis 1–17*, 151. Fokkelman proposes a different reading and translates the preposition ב "for" (Deut 19:21; 2 Sam 3:27; 1Kings 16:34) as a *beth pretii* (JM § 133c; WO § 11.2.5d); J.P. FOKKELMAN, *Narrative Art*, 35.

[19] The syntagm נקי + דם, "innocent blood," is extremely common in the Hebrew Bible: Deut 19:10, 13; 21:8-9; 27:25; 1 Sam 19:5; 1 Kings 21:16; 24:4; Is 59:7; Jer 7:6; 19:4; 22:3, 17; 26,15; Joel 4:19; Jona 1:14; Ps 94:21; Prov 6:17; also Jer 2:34.

[20] A. PENNA, "Il sangue nell'Antico Testamento," 394-395; the expression "precious in the eyes (יקר בעיני) of God" is used only to designate human blood (Ps 72:14) and the death of the righteous (Ps 116:15).

[21] The verb דרש can be translated "to require" even in other circumstances: דרש + רשע, Ps 10:15; also Deut 18:19; Ps 10:4, 13.

entrusted to the גאל הדם, "the avenger of blood,"[22] a family member appointed for the punishment of the murderer.

The alleged perpetrator could take shelter in a "city of refuge,"[23] if he has killed without having full responsibility; in this case, the גאל had the power to pursue him up to the city gates and demand a trial conducted by the elders (Num 35:12, 24-25). The bloodguilt, therefore, cannot be covered (כסה, Gen 37:26; Is 26:21; Job 16:18) and justice must be restored through the just "revenge" (דם + נקם, Deut 32:43; 2Kings 9:7; Ps 79:10; see also 2Kings 9:26; Joel 4:21); no ransom may enable the murderer to avoid the death penalty[24] (Num 35:31).

1.2 The Prohibitions of Consuming Blood

The connection between blood and life is generally regarded as an original feature of biblical literature,[25] although this belief has recently been reconsidered.[26] Firstly, this connection appears in texts where it is prohibited to consume animal blood; in the book of Genesis, this rule is linked to the prohibition of shedding human blood, while in the book of Leviticus it is connected to the use of blood in sacrificial worship. Thus, the association between blood and נפש allows the transition from the domain of violence (Gen 9:5-6) to the sphere of cultic manipulation of animal blood (Lev 17:11), which will be the subject of the next paragraph.

Gen 9:4, "Only[27] flesh with[28] its life (אך־בשר בנפשו), its blood[29] (דמו), you shall not eat."

---

[22] Num 35:19; Deut 19:6, 12; Jos 20:3, 5, 9; 2 Sam 14:11.

[23] See F. Cocco, *The Torah as a Place of Refuge*. The author analyses the legislation of Num 35:9-34 on the "cities of refuge" where the person who had killed without full responsibility could find asylum. The material executor of the sentence, the "avenger of blood" (גאל הדם; *Ibid.*, 69-72), could be the son or another close relative of the victim. The use of the root גאל, "to redeem," shows that the intervention was intended to restore a broken balance and, ultimately, to protect the victim's family.

[24] So Y. Feder, *Blood Expiation*, 175.

[25] D.J. McCarthy, "The Symbolism," 166-176; Id., "Further Notes," 205-210. See also Y. Feder, *Blood Expiation*, 196-207, with bibliography. See also C.A. Eberhart, "Blood. I. Ancient Near East and Hebrew Bible/Old Testament," 206; the author notes that in Ezek 37:1-14 blood is not listed among the bodily components of the person (the text lists only bones, nerves, flesh, skin), as if blood were connected to life on a deeper level.

[26] See W.K. Gilders, *Blood Ritual*.

[27] The adverb אך often introduces a restriction with respect to a concession stated previously (WO § 39.3.5*d*).

[28] See GK § 119*n*. In this case, the preposition ב suggests the idea of proximity. See also V.P. Hamilton, *Genesis 1–17*, 314.

The noun נפש basically means "throat" (Ps 69:2) and it refers to the oral cavity, in connection with respiration (1Kings 17:21-22) used during the ingestion of food (Is 5:14), often related to thirst (Pr 25:25; Ps 107:9), and linked with appetite and desire[30] (see Ps 42:2). The noun, especially when in parallel with the term "heart" (see Deut 4:29; 6:5; 10:12; etc.), denotes the "soul" as the center of the inner life, the vast terrain of feelings (Ps 31:8; Is 53:11) and thoughts (Pss 13:3; 139:14; Prov 2:10). However, in Gen 9:6 נפש indicates the principle of life[31] in a general sense, because the passage deals with animal blood without focusing on the human condition: it is blood, circulating in the "flesh,"[32] that makes the body "animate" and "vital,"[33] being a kind of vehicle[34] for the נפש (Lev 17:11, נפש הבשר בדם).

The identification of נפש and דם in the prohibition of eating blood is explained in several ways[35] and its meaning is elucidated by the context. Since in Gen 1:29 God imposes a vegetarian diet on mankind, in order to promote a peaceful relationship between living beings,[36] similarly, in Gen 9:2-4 God forbids the consumption of blood because it is not something at one's disposal, since it is the center of vitality;[37] the law therefore "seeks to warn against those thoughts that move violence in the depths of the human heart"[38] and then has a dissuasive character. The near context confirms this interpretation, since in Gen 9:5-6 there is an explicit reference to the absolute condemnation of the shedding of human blood[39] (Exod 20:13).

---

[29] The term דמו is an apposition that modifies the preceding בנפשו (GK § 131*k*).

[30] C. WESTERMANN, "נפש," 71-96; H. SEEBASS, "נפש," 532-555. Westermann (*Ibid.*, 75) lists a series of expressions related to נפש that presuppose the basic meaning, "throat" (פה, "mouth," Eccles 6:7; גרגרת, "neck," Prov 3:22). Other parallel terms justify the translation of נפש with "appetite" (שבע, "satiate," Ps 63:6 and Is 58:11; מלא, "fill," Jer 31:25 and Prov 6:30) and "desire" (when נפש is the subject of אוה, "to desire," Deut 12:20; 14:26; Isa 26:9; Prov 13:4).

[31] A recent study explored the "reflexivity" of נפש: E. JONES, "Direct Reflexivity," 411-426. See also É. GRENET, *Unité du "Je,"* 20-21.

[32] The human person, as a fragile and mortal being (Isa 40:6-7), is defined by the term בשר, normally translated "flesh" (Gen 2:23). The pair "flesh and blood" can be used to indicate human nature: Deut 32:42; see M. GIRARD, *Symboles bibliques*. II, 1470-1471.

[33] See M. VERVENNE, "The Blood is the Life," 453.

[34] So C. OWIREDU, *Blood*, 6; F. MANZI, "Sangue," 1232.

[35] Since blood is life, it is not available as an aliment, because it belongs to God (N.M. SARNA, *Genesis*, 61; J. MILGROM, *Leviticus 1–16*, 417); the prohibition of consuming blood means "to do honour to the principle of life" (U. CASSUTO, *Genesis*. II, 126).

[36] So V.P. HAMILTON, *Genesis 1–17*, 141. Gen 1:29 has been interpreted as an instruction to teach men a dominion over violence; see A. WÉNIN, *Da Adamo ad Abramo*, 32; see also P. BEAUCHAMP, *Pages exégétiques*, 105-144.

[37] C. OWIREDU, *Blood*, 10.

[38] A. WÉNIN, *Da Adamo ad Abramo*, 140.

[39] J.A. SOGGIN, "Il sangue nel racconto biblico delle origini," 418-421.

The same prohibition appears again in legal texts (see Lev 3:17; 7:26-27; 19:26; see also 1 Sam 14:32-34; Ezek 33:25); in two of them the rule is justified through the connection between דם and נפש. In the book of Deuteronomy, desctruction of all the other altars, required by Deut 12:2-5, complicated the practice of slaughtering domestic animals.[40] Deut 12:13-27 therefore states that meat may also be slaughtered locally, but without consuming the blood,[41] because it is the נפש. This rule establishes a clear difference between the sacrificial blood intended for the altar and the blood of the animal slaughtered for domestic use, which must simply be poured on the ground[42] (Deut 12:26-27).

The second text is taken from the book of Leviticus and has been considered a pivotal passage to explain blood manipulations.

Lev 17:10-11, "If anyone[43] of the house of Israel or of the strangers who sojourn among them eats any blood (כל־דם), I will set my face against[44] the person[45] who eats the blood (בנפש האכלת את־הדם) and will cut him off from among his people. For the life of the flesh (נפש הבשר) is in the blood[46] (בדם), and I have given it to you for making atonement (לכפר) for your lives[47] (על־נפשתיכם) upon the altar; for it is blood (הדם הוא) which, through life (בנפש), makes atonement."

In Lev 17:3-4 it is stated that the domestic animal shall not be slaughtered away from the Tabernacle, but must be handed over to the priest so that he may smear its blood on the altar and burn its fatty parts (vv. 5-6). Thereafter,

---

[40] So E. OTTO, *Deuteronomium 12,1 – 23,15*, 1141.

[41] G. PAPOLA, *Deuteronomio*, 174; Y. FEDER, *Blood Expiation*, 201-202.

[42] See W.K. GILDERS, *Blood Ritual*, 52, "The pouring of blood onto the ground indexes away from Yahweh and by an act of ritualization ... establishes noncultic slaughter as substantively different from cultic slaughter."

[43] The expression ואיש איש designates the whole with an accentuation of the distributive aspect: "everybody and everyone" (GK § 123c; Exod 36:4).

[44] The syntagm נתן + פנים in Lev 20:3 and Ezek 14:8 indicates a new orientation of one's gaze, focused on the object that has attracted attention. The use of the preposition ב specifies that the attitude is hostile, "he orients himself against." See J. MILGROM, *Leviticus 17–22*, 1471.

[45] Milgrom translates נפש with the term "person." J. MILGROM, *Leviticus 17–22*, 1471.

[46] The phrase נפש הבשר בדם היא can be rendered either "the life of the flesh is in the blood" or "the life of the flesh is the blood," depending on whether the preposition ב is translated as a locative *beth* or as a *beth essentiae* (GK § 119i; JM § 133c). See W.K. GILDERS, *Blood Ritual*, 169.

[47] The construction על נפשתיכם means "for your lives." So J. MILGROM, *Leviticus 1–16*, 707-708; W.K. GILDERS, *Blood Ritual*, 170. In fact, the repetition of נפש in v. 11 creates an obvious correlation between the blood shed, which is the "life/life force," and its favourable effect on the "life" of men. The expression נפש + על + כפר is often connected to the concept of ransom (*kōper nepeš*, Exod 30:12, 15-16; Num 31:50; 35:31; Prov 13:8); B.J. SCHWARTZ, "Prohibitions," 51; Y. FEDER, *Blood Expiation*, 203.

Yhwh forbids the consummation of blood and introduces, in that which is a singular recurrence in Scripture, a connection between the terms דם, נפש and the practice of blood manipulation. Yhwh, in fact, institutes the application of the blood upon the altar and donates it[48] ("I have given it to you") because blood, through life[49] (בנפש), can make atonement for the Israelites. The text can be interpreted as a late revision of the known practice of shedding blood on the altar[50] or as a connection between the prohibition of consuming blood and blood manipulation interpreted as a ritual "substitution."[51] Reading the texts of Exodus in which we find a manipulation of blood will allow us to reflect further on this issue.

1.3 *The Ritual Use of Animal Blood*

The positive role of blood, in the cultic sphere, has long been considered to be a distinctive characteristic of Israel.[52] This belief has been reconsidered recently, especially thanks to Feder's study, which highlights some significant contacts between Hittite rituals and biblical texts.[53] Nevertheless, it must be recognised that blood manipulation has a unique function in the Israelite cultic practice.[54]

Leaving aside Exodus, it is essentially the book of Leviticus that reports the greatest number of occurrences of דם in connection with worship. Blood is employed primarily for the *burnt offering* (עולה), the most frequent and perfect sacrifice in Israel,[55] where the victim is completely burned with fire and

---

[48] See B.J. SCHWARTZ, "Prohibitions," 51.

[49] The preposition ב is commonly explained as a *beth instrumenti* (GK § 119o; WO § 11.2.5d; JM § 125m) that introduces the means or the instrument by which the action described by the verb is performed (Gen 32:21; Lev 5:16; 1 Sam 3:14; see also Prov 16:6); J. MILGROM, *Leviticus 17–22*, 1478; Y. FEDER, *Blood Expiation*, 204.

[50] After Knohl (I. KNOHL, *The Sanctuary of Silence*, 112-113) there is a tendency to recognise Lev 17:10-11 as redaction of H, namely a late revision of P. See also J. MILGROM, *Leviticus 17–22*, 1472-1479; W.K. GILDERS, *Blood Ritual*, 12-13.

[51] Milgrom thinks that the blood mentioned in Lev 17:11 is considered as a vital force that can cleanse from impurity; J. MILGROM, *Leviticus 1–16*, 711-712. Feder, instead, thinks that the blood serves rather as a "substitution": it can be given in place of the culprit's life, as is said in Lev 17:3-4; Y. FEDER, *Blood Expiation*, 206.

[52] D.J. MCCARTHY, "The Symbolism," 166-176; ID., "Further Notes," 205-210; see also J. MILGROM, *Leviticus 1–16*, 706. The testimony of Ugarit's texts, shows that there is no evidence of the ritual use of blood in the Syro-Palestinian area during the Late Bronze Age; see P. XELLA, "Il sangue," 123-124.

[53] Y. FEDER, *Blood Expiation*.

[54] So A. MARX, *Les systèmes sacrificiels*, 101-103.

[55] F. DI GIOVANBATTISTA, *Il sistema sacrificale*, 133-134.

offered to God with generosity (2 Chron 29:31, "[...] all that were of a generous/noble heart brought burnt offerings"). Before the burning, the offering requires that the animal blood (bovine, Lev 1:5; ovine or caprine animals, 1:11; birds, 1:15) be thrown upon the altar (Lev 7:2; Ezek 43:18, 20).

The *sacrifice of well-being*[56] (זבח שלמים) follows the same procedure: after the slaughter of the animal (bovine, Lev 3:2; ovine, 3:8; caprine, 3:13) at the entrance of the Tabernacle of Yhwh, the blood is thrown upon the altar. Unlike the burnt offering, only the fatty parts of the animal are burnt in the fire (the kidneys and the fat enveloping the liver, 3:3-4, 9-10, 14-15) and offered to God (see Lev 7:14, 33; 17:6).

The *sin offering* (חטאת) is required for unintentional sin[57] (Lev 4:2, 22, 27; 5:2-3; see also Lev 6:23; 10:18; Ezek 45:19) and includes a complex use of blood (see Lev 4:5-7, 16-18). The *Yom Kippur* is a special type of חטאת (Lev 16) that combines the sin offering with the "scapegoat" rites[58] (Lev 16:20-22); the blood of the bull and the goat are sprinkled on the propitiatory (Lev 16:14-15), on the Tent of Metting (16:16) and then put on the horns of the altar seven times (16:18-19).

Even the *ordination offering*, perfomed during the consecration of Aaron and his sons[59] (Exod 29 and Lev 8), and the *purification of the person affected by leprosy* (צרעת) need blood manipulation. In the case of the leper, he is sprinkled seven times with the blood of an immolated bird[60] (Lev 14:6-7) and, after eight days, the priest takes the blood of a guilt offering and daubs it on the lobe of his right ear, the thumb of his right hand and the big toe of his right foot (Lev 14:14, 25).

The use of דם in cultic context, then, reveals a major characteristic of this fluid: although the explanatory value of the connection between blood and life, exposed in Lev 17:11 ("for it is blood that through life makes atonement"), has been reconsidered,[61] the quick overview presented shows that it is impossible not to recognise some "vital power" in blood. This idea is confirmed by the association between דם and Hebrew verbs correlated with the domain of life and death.

---

[56] A. MARX, *Les systèmes sacrificiels*, 37-38.
[57] CH. NIHAN, "The Templization of Israel," 94-130.
[58] F. DI GIOVANBATTISTA, *Il sistema sacrificale*, 280.
[59] See ch. V.
[60] The house that had a stain of "leprosy" could be sprinkled with the blood of an immolated bird in order to be purified (see Lev 14:51-52).
[61] See W.K. GILDERS, *Blood Ritual*, 186-187; Y. FEDER, *Blood Expiation*.

In fact, since the blood of sacrifices is, like water,[62] an instrument of *purification*[63] (טהר), it is clearly related to life, considering that the domain of impurity is linked to death (Lev 15:31) and that the contact with a corpse is one of the most common causes of an impure condition (Num 19:13, 18-19). The connection between דם and the root טהר is attested in the ritual for the cleansing of the leper (Lev 14:14, 17, 25) and in the sin offering of *Yom Kippur*[64] (in Lev 16:19, the theological heart of Leviticus[65]). These rituals presume a certain nexus between blood and life. Leprosy and sin, in fact, are both characterised by a dangerous contact with death, since the first condition was considered as a sort of anticipation of one's decease,[66] while sin was the mortal power *par excellence* (Ezek 18:21, 24; 33:14-15).

Another root linked to blood and life is the verb כפר which expresses *expiation-atonement*.[67] This association between the term דם, the verb כפר and life is explicitly attested to in Lev 17:11, "for it is blood that through life makes atonement." Moreover, the term דם is connected with the verb כפר with regard to the sin offering (חטאת; Lev 6:23; see 2Chron 29:24) in the consecration of the altar during the ordination of Aaron (Lev 8:15) and in the rite of Atonement Day (Lev 16:18, 27). The meaning of the root כפר has been the subject of numerous discussions.[68] The etymological translation, "to wipe off," is usually accepted, even if it has often been criticised.[69] Levine maintains that the verb has two different meanings, "to

---

[62] There are many biblical texts in which the purpose of bathing is to purify objects (Lev 11:32) or persons, in the case of a skin disease (Lev 13:6,34; 14:8-9), gonorrhoea (Lev 15:13), in the case of ingestion of something impure (Lev 17:15) or contact with a corpse (Num 19:12, 19). See J.D. LAWRENCE, *Washing in Water*, 35-38.

[63] A. PENNA, "Il sangue nell'Antico Testamento," 386.

[64] The contact between impurity and sin is also attested by those texts in which the forgiveness of guilt is expressed with the verb טהר (Jer 33:8; Ezek 36:25, 33).

[65] See G. DEIANA, *Levitico*, 162. In a study (*Sainteté et pardon*), Luciani explores the hypothesis that Lev 16 is the structural nucleus of Leviticus; Rendtorff even thinks that Lev 16 is the heart of the entire Torah: R. RENDTORFF, "Leviticus 16 als Mitte der Tora," 252-258. See also B.D. BIBB, "Blood, Death, and the Holy," 137-146.

[66] See C. GRANADOS GARCÍA, *La nueva alianza*, 160. Ezek 36:25 presents a correlation between the act of purification and the vivification. This aspect is clear in the story of 2Kings 5, in which Naaman's leprosy is cured with a bath in the Jordan River (5:10, 12, 14); when the Israel's king responds to the request of Aram's king, he manifests his conviction that the condition of the leper is equivalent to death: v. 7, "Am I God to give death or life (להמית ולהחיות)?" See also C. OWIREDU, *Blood*, 92-108.

[67] A. PENNA, "Il sangue nell'Antico Testamento," 387.

[68] Y. FEDER, *Blood Expiation*, 167-173 and B. JANOWSKI, *Sühne als Heilsgeschehen*.

[69] See Y. FEDER, "On *kupurru, kippēr*," 535-545. The explanation of *kippēr* as a verb derived from the Akkadian *kupurru* ("to dry, to rub") and the translation "to eliminate, to

expiate" and "to ransom,"[70] and Gilders translates it in a very literal way, "to effect removal."[71] The use of the verb in non-cultic contexts can be a valuable aid to clarifying the different nuances:

– Divine subject + כפר *piel* + a term taken from the semantic field of sin, "to forgive (to remove) the sin."[72]

– Divine subject + כפר *piel* + על + a term taken from the semantic field of sin, "to forgive (to remove) the sin" (Jer 18:23; Ps 79:9).

– Divine subject + כפר *piel* + ל + the culprit, "to forgive the sinner/to make atonement for the sinner" (Deut 21:8; Ezek 16:63; see also Deut 32:43[73]).

– Human subject + כפר *piel* + בעד + the sin, "to make atonement for a sin/to obtain forgiveness," as in Exod 32:30.[74]

– Human subject + כפר *piel* + על + the culprit, "make atonement for," as when Aaron intercedes for the people who murmur against God (Num 17:11).

– Human subject + כפר *piel* + the consequence of sin, the righteous wrath of God (Prov 16:14; see Gen 32:21), "to appease."[75]

In cultic context, however, the syntactic framework is much more stable: normally it is the priest who performs the action described with כפר *piel* and the verb is followed by the preposition על (or בעד, Lev 9:7; 16:24) which introduces the person[76] or the element for which atonement is made (the altar, Lev 8:15; 16:18; the house, 14:53; the sanctuary, 16:16; see also 16:33). In this sense, the same syntagma used to indicate the intermediary of reconciliation

---

erase by rubbing" is untenable. The theoretical prerequisites of this article are those developed by J. BARR, *The Semantics*, 117-119.

[70] B. LEVINE, *In the Presence*, 61-62.

[71] W.K. GILDERS, *Blood Ritual*, 28-29.

[72] The syntagmas that express the forgiveness of sin through the verb כפר are various: divine subject + כפר *piel* + עון (Ps 78:38; Dan 9:24; see 1 Sam 3:14; Is 22:14; 27:9); כפר *pual* + חטאת (Is 6:7); divine subject + כפר *piel* + פשע (Ps 65:4).

[73] In Deut 32:43 the subject is divine and the object is not introduced by the preposition ל, "he will forgive/cleanse his people's land (וכפר אדמתו עמו)."

[74] In Exod 32:30, Moses decides to implore Yhwh for the reconciliation between Yhwh and the people. Moses is the subject of the verb כפר *piel* and the term חטאת is preceded by the preposition בעד; therefore, the verbal root presents an action performed by the intercessor and can be translated "to obtain forgiveness / to make atonement for."

[75] Y. FEDER, *Blood Expiation*, 171-172. In Gen 32:21 Jacob sends a gift to Esau with these words, "He thought, 'I will appease him (אכפרה פניו) with the present that goes ahead of me.'" Rashi translates אכפרה פניו, "I will neutralize his wrath" (אבטל רגזו, Rashi Gen, 368); see *TgO*: אניחיניה לרגזיה, "I will make his wrath cease." See Ps 21:10.

[76] The priest can atone (כפר *piel*) for a person or for the people (Lev 1:4; 4:20, 26, 31, 35; 5:6, 16, 13, 16, 18, 26; 8:34; 10:17; 12:7-8; 14:18-21; etc.).

between two parties is employed for the ritual context.[77] Therefore, the priest is not the main agent that removes sin, but acts as a mediator of reconciliation, which is effectively realized by God through the sacrificial victim.

The third crucial association, useful to recognise the vital force of cultic blood, is that between the root קדש and דם. This connection is detectable in the consecration of Aaron (Exod 29:21; Lev 8:30) and in two other passages where the priest pours the blood on the altar (Lev 8:15; 16:19); in all these cases the blood used has the power to *sanctify*.[78] The root קדש refers to the domain of God's perfection[79] and the *qal* expresses the special destination of the verbal subject: the "consecrated" is a person who is set apart for Yhwh. The forms *piel* and *hifil* are used for the act of consecrating, performed by God (Lev 20:8; 21:8.15; 22:9; Num 3:13; 8:17; 1Kings 9:7; Jer 1:5), when Yhwh reserves someone for Himself, or by a man, when he dedicates an object or a person to God.[80] Consecration, then, signals a separation, a destination and, therefore, a new sense of belonging. This idea is explicit in Num 3:11-13 where Yhwh establishes that the Levites are chosen so that they may share in the special consecration of the first-born: Yhwh, who "has consecrated for himself" (Num 3:13, הקדשתי לי) every firstborn of Israel, will replace them with the Levites, who "shall be his" (3:12) and will help Aaron in worship[81] (8:18-19).

Thus, blood manifests the capacity that worship has to purify, preventing "death" and giving life; to make atonement, abolishing the division between God and man; and to consecrate, connecting the person to the sphere of God.

---

[77] See CH. NIHAN, "The Templization of Israel," 100-111. The author presents a complex lexicographical analysis of כפר *piel* in Lev 4 and 16.

[78] A. PENNA, "Il sangue nell'Antico Testamento," 384-385; the rite of Aaron's consecration in Exod 29 and Lev 8 includes the anointing with blood and with oil of the priest (Exod 29:7; Lev 8:12, 30); the consecrations of the king (Judg 9:8-15; 1 Sam 9:16; 16:22) and the prophet (1Kings 19:16), instead, simply consist of an anointing with oil.

[79] See W. KORNFELD – H. RINGGREN, "קדש," *ThWAT*, VI, 1179-1204. The adjective קדוש is used attributively for Yhwh (Josh 24:19; 1 Sam 6:20; Is 5:19); it can refer to a "place" especially related to the sphere of God (the Tabernacle, Lev 6:19; also Lev 7:6; 10:13; 16:24; 24:9). The term קדוש is also a predicate of God (Lev 11:44-45; 19:2; Ps 22:4; etc.) or, especially in Isaiah, a title to indicate Yhwh himself, "the Holy One" (Isa 30:12, 15; 40:25; 41:20; 43:14; 45:11; Hos 11:9; Hab 3:3).

[80] The subject of the verb קדש *piel* is often Moses and the consecrated object may vary: the firstborn (Exod 13:2), Aaron and the priests (30:30; 40:13; Lev 8:12, 30), Aaron's clothes (Exod 28:41), the animals for sacrifice (29:27), the altar and its utensiles (30:28-29), the Tabernacle (40:9; Lev 8:10-11; Num 7:1). For קדש *hifil*, see Lev 27:14; Deut 15:19.

[81] See P. BOVATI, "'Non so parlare' (Ger 1,6)," 68. The author notes a connection between Jer 1:5 and Num 8:18-19, "If being set apart corresponds to a particular relationship with God ('to be for God'), this same act is identified with a destination, a task ('to be for the nations')." (My translation).

2. The Working Hypothesis: Blood in the Book of Exodus

The lexicographical clarification of the different domains in which the term דם is used, permits to draw crucial consequences for the passages that will be objects of this study. This research will manifest that the motif of "blood" is employed in the book of Exodus in a unique way (with the exception of Exod 29-30) and has a peculiar function for its entire plot.

Blood constitutes an actual *Leitmotif* that determines the Book of Exodus in significant moments of its trajectory (the return of Moses to Egypt, ch. 4; the Passover, ch. 12; the stipulation of the covenant, ch. 24; the consecration of Aaron, ch. 29; the rite described in ch. 30). The term דם also appears in 4:9 and 7:17, 19-21, but, as seen above, these passages use it symbolically. In addition, the present work will not consider three legal texts — the rule concerning blood vengeance towards a thief (22:1-2) and the law regarding the offering of leavened bread together with the sacrificial blood[82] (23:18; 34:25) — because they are not directly relevant to our main interest. Among the passages which are going to be analysed, there is a harmonious use of the blood motif and thus, the main goal of this paragraph will be precisely to demonstrate how they are linked thanks to lexical contacts.

The first episode to be examined is the assault on Moses (4:24-26) in his journey to Egypt (4:18-31). During the night Moses is attacked by God, and becomes a "bridegroom of blood" thanks to the intervention of Zipporah, through the shedding of blood that unites the spouses and, by means of circumcision, creates a bond between Moses' family and the people of Israel. The association between 4:24-26 and the second text that will be studied is noted by many,[83] because also in ch. 12 blood (although in this case it is

---

[82] In Exod 23:18 and 34:25 two practically identical norms are enunciated. In 23:18 it is stated, "you shall not sacrifice (לא־תזבח) with leavened bread the blood of my sacrifice (דם־זבחי)," while 34:25 says "you shall not immolate (לא־תשחט) with leavened bread the blood of my sacrifice (דם־זבחי)." The traditional interpretation of 23:18 refers the law to the Passover: leavened bread should not be mixed "with the blood of my Passover" (*TgO* Exod 23:18; Rashi Exod, 201: לא תשחט הפסח, "do not slaughter the Pesach"; Ibn Ezra, 201); see Sarna, 146. Cassuto translates the expression דם־זבחי of 23:18 "my bloody sacrifice," interpreting דם as an adjective (*Exodus*, 212), since the term דם is never the object of the verb זבח. Many, however, believe that דם designates the blood manipulation on the altar and that the law refers to the prohibition of burning leavened items (especially cereal offerings; see Lev 2:11) on the same altar: see Childs, 485; Propp, II, 284; Priotto, 452. Evidently, the law establishes a certain incompatibility between the leaven and the blood used in sacrifices; see W.K. GILDERS, *Blood Ritual*, 34-37.

[83] The connection between Exod 4:24-26 and Exod 12 is already noted by rabbinic exegesis, exposed in the Targums. See G. VERMES, "Baptism and Jewish Exegesis," 312, who quotes *TgPsJ* Exod 4:25 "and brought the circumcision of the foreskin to the feet of the

animal's blood) plays a rather decisive role. A further link between the two passages is established by the repetition of the verb נגע.

Exod 4:25	Exod 12:22
Then Zippora *took* a flint and cut off her son's foreskin and touched his feet (ותגע לרגליו) ...	*You will take* a bunch of hyssop and dip it in the blood that is in the basin and touch (והגעתם) the lintel ...

Moreover, the context that frames the two sections demonstrates even more clearly the correlation between these stories. The Passover narrative is set between the announcement of the tenth plague (11:1-10) and its fulfilment, the killing of the first-born of Egypt (12:29-34). This dramatic epilogue is already announced by Yhwh in 4:21-23, immediately before the meeting with Moses (4:24-26); in this passage, Yhwh justifies the extreme intervention precisely on the basis of the special bond that unites him to the people: Israel is his first-born son (4:22). Finally, the two texts are also thematically similar: both Moses and the people live an experience in which blood marks "the boundaries between life and death."[84] These references attest that blood creates and signals the special relationship of the people and of Moses with the "family" of God: Moses is made "consanguineous" by virtue of the blood of circumcision; the people is recognised in terms of its *status* of first-born son, as a result of the blood applied on the doorposts (12:22).

The sound wave of these two passages creates a third strong reverberation in Exod 24. The connection between Exod 4 and 24 is guaranteed by the choice of the root כרת in Exod 4:25 to indicate circumcision; the verb in fact belongs to the semantic field of the covenant[85] and is linked to circumcision only in this verse (the most common root used for it is מול).

Exod 4:25	Exod 24:8
Then Zippora *took* a flint and cut off (ותכרת) her son's foreskin and touched his feet [...]	Behold the blood of the covenant (דם־הברית) that Yhwh "cut" (כרת) with you on the basis of all these words.

In addition to this lexical contact, the correlation between the covenant domain and circumcision is already evident on the basis of Gen 17, since

---

destroying angel," referring explicitly to *TgPsJ* Exod 12:12. See also B.P. ROBINSON, "Zipporah to the Rescue," 453; Sarna, 25; Nepi, I, 131-132; Priotto, 113; Dohmen, I, 177.

[84] Nepi, I, 132.
[85] Of the 128 recurrences of the verb כרת *qal* in the Hebrew Bible, 77 are related to the term ברית (e.g. Gen 15:18; 21:27.32; Exod 34:10; Deut 4:23; 5:2).

circumcision has been designated as "the sign of the covenant" from the very beginning[86] (see 17:10-14).

The connection between ch. 4, ch. 12 and ch. 24, moreover, has an important consequence: the hypothesis of a comparison between the Exodus account and the raising of a child, formulated by Ilana Pardes, finds new confirmation.[87] The "firstborn son" (4:22-23), with whom Moses is associated "through blood" (4:24-26), is spared by the angel of death thanks to blood (12:22), while the firstborns of Egypt are killed. After the crossing of the Sea the people are "born" in ch. 14–15, they lament like infants in Exod 15:22–17:16 and, "carried" on eagle's wings (Exod 19:4; see Deut 1:31), they acquire the possibility of a more mature filial[88] relationship with God thanks to the covenant (Exod 19–24), sealed in Exod 24:3-8 through the blood.[89]

The fourth echo has a two-beat rhythm, which includes in the same movement Exod 29 and 30. Many lexical relationships can be noted especially between ch. 29, where Yhwh gives to Moses his instructions for the consecration of Aaron and the priests, and ch. 24.

Exod 24:6, 8	Exod 29:16, 20
[6] Moses took (ויקח) half of the blood and put it in basins, while throwing (זרק) half of the blood upon the altar (על־המזבח). [8] Moses took (ויקח) the blood, and threw (ויזרק) [it] on the people and said, "Behold the blood of the covenant that Yhwh has done with you on the basis of all these words"	[16] You shall slaughter the ram and take (ולקחת) its blood and throw (וזרקת) [it] upon the altar (על־המזבח), all around. [20] You shall slaughter the ram and take (ולקחת) from its blood and put it on the tip of the right ear of Aaron […]; then you will throw (וזרקת) the blood upon the altar (על־המזבח), all around.

In the rite of consecration, the blood manipulation creates a connection between the altar and Aaron (with his sons); similarly, the covenant rite of Exod 24 bonds together the altar and the people. Therefore, the two rites signify and realise a relationship between the altar, which represents the

---

[86] See J.J. KRAUSE, "Circumcision and Covenant," 151-165, with bibliography.

[87] I. PARDES, *The Biography of Ancient Israel*, 26.

[88] See F. FICCO, *"Mio figlio sei tu,"* 83-138.

[89] The connections between blood and covenant are common in the Ancient Near East; see D. CHARPIN, *"Tu es de mon sang,"* 74-78. The author examines the Near Eastern covenant treaties, basing his inquiry on the most recent archaeological discoveries. Especially in some 2nd millennium texts, the positive role of blood in a covenant is noted. See e.g. A.2730 in which a counsellor suggests to the king of Mari how to respond to a request for help in the war against King Šarraya, "*Blood relations* and strong ties are established between me and between Šarraya […]. I wrote Šarraya, '*You are [of] my blood*! Give me troops of yours and let your troops camp with my troops!'" (Italics mine; W. HEIMPEL, *Letters*, 509) Regarding the rite of Exod 24:3-8, see ch. IV, § 2.4.

sphere of God, and a group, the people in ch. 24,[90] the priests in ch. 29; through this ritual, Aaron and his sons are made especially close to God. A further confirmation of the contact between ch. 24 and ch. 29 is offered by the studies dedicated to the connection between priesthood and covenant.[91]

In Exod 30, immediately after the establishment of the priesthood, Yhwh gives Moses the instructions for the construction of the incense altar (vv. 1-10), placed in front of the veil that is above the ark of the Covenant in close contact with the Holy of Holies "where Yhwh offers the presence of his word"[92] (v. 6). Aaron has to burn incense on it every morning and every evening (vv. 7-8); no sacrifice is to be made on this altar, except in the case of the Yom Kippur (v. 10). The use of the term דם in v. 10 is very significant: when the people have just concluded the covenant, Yhwh entrusts Moses with the task of consecrating Aaron and his sons for worship. In ch. 30, Yhwh reveals that they will also have to be even ministers of reconciliation between God and the people in the solemn rite performed once a year. The blood that made Israel God's people is the same instrument that allows the reconstitution of the bond broken by sin on the Day of Atonement (Exod 30:10), through the mediation of Aaron and the priests (Exod 29).

In conclusion, the passages that will be analyzed in the next chapters are unified by the fact that human blood (for Moses) and especially animal blood is shed and "circulates," indicating and actualising a relationship with God, that is in some way similar to the bond that unites "kinsmen."[93]

3. Excursus: The Nile Becomes Blood

The first occurrence of דם in Exodus is outside the series that moves our research, but it is a musical note that starts the symphony of voices tuned to the motif of blood and, in a way, marks them with its tone. For this reason, the reading of Exod 4:18-31 is preceeded by a quick examination of Exod 4:9, in which the term דם appears for the first time in Exodus, and of 7:14-25, in which what is anticipated in 4:9 really happens.

3.1 *The Sign of Blood (Exod 4:9)*

The narrative of Exod 3:1–4:17 is dedicated to Moses' encounter with Yhwh.[94] Exod 3 consists of the narration of Yhwh's apparition (vv. 1-6), the

---

[90] See W.K. GILDERS, *Blood Ritual*, 41.
[91] F. SERAFINI, *L'alleanza levitica*.
[92] Priotto, 554, my translation.
[93] See L. CAGNI, "Il sangue nella letteratura assiro-babilonese," 58; see also *COS*, II, 89.
[94] Regarding a bibliography on Exod 3–4, see J. JEON, *The Call of Moses*.

sending of Moses (vv. 7-10), a first series of two objections by Moses (vv. 11-12.13-15, with the revelation of Yhwh's name), and a discourse in which God anticipates some elements of his mission[95] (vv. 16-22).

The fourth chapter surprisingly begins with a new objection, "They will not believe me, nor listen to my voice, but will say, 'Yhwh did not appear to you'" (4:1). Since God has revealed to Moses that the elders of the people will welcome him (see 3:18), the envoy's reaction cannot be interpreted as a simple concern, but is a sign of disbelief.[96] Moses' own words portray a certain mistrust, because he does not merely express his doubts with a sentence, but exposes his thoughts with a double negation[97] and a statement, "they will not believe me, nor listen [...], but will say [...]." The sequence culminates with the quotation of words that Moses imagines the Israelites might utter; in this way the narrator reveals to the reader that Moses' point of view becomes dominant: the anticipatory quotation achieves an inversion in the montage of perspectives, making the point of view of the quoting forecaster prevail over that of the person that is quoted.[98] Without being certain of the outcome of his mission, then, Moses submits reality, which he does not know yet, to his vision.

God responds in three stages and orders Moses to take into his hand the instrument with which the shepherd defends the flock, the staff,[99] so that he may discover that the word of Yhwh can turn it into a snake (vv. 2-5), a mythical animal in Egyptian literature that is associated with the domain of life and death.[100] With this sign, Moses can recognise God's ability to dominate chaotic forces.[101] In addition to this, in the passage there is some irony: the magical arts practised in Egypt are mocked[102] (Is 19:1-15) and Moses himself, in fleeing from the serpent, demonstrates that the dominion over the mythological animal is not his personal prerogative, but a gift from God.[103]

---

[95] In regard to the literary and narrative structure of these chapters, see e.g. Nepi, I, 91.

[96] Priotto, 104; G.I. Davies, I, 317. See also *ExR* III, 12.

[97] The first two verbs used by Moses are in an inverted order ("to believe" and then "to listen"). This sequence shows even more explicitly what Moses' concern is: he does not think he will succeed in gaining the trust of the elders.

[98] Regarding the *pre-productive* quotations, see ch. II, § 3.2. See also M. STERNBERG, "Proteus," 138-139, "In terms of point of view ... the perspective of the ostensibly quoting forecaster ... predominates over that of the ostensibly quoted forecastee."

[99] Houtman, I, 29; Propp, I, 209; Fischer – Markl, 65; Priotto, 104; G.I. Davies, I, 317.

[100] See ch. II, § 3.1.3.

[101] Nepi, I, 107; Priotto, 105.

[102] Nepi, I, 108. See R. DAVID, *Religion and Magic in Ancient Egypt*.

[103] Priotto, 105.

Thereafter, God commands Moses to put his hand close to his chest and withdraw it to see it become "leprous like snow"[104] (מצרעת כשלג); then Yhwh tells him to repeat the same gesture and see his hand return to its usual vitality (4:6-7). The disease designated in Hebrew by the word צרעת was particularly feared, as a source of impurity and a cause of a death-like condition.[105] The sign, therefore, means that the activity of Moses, symbolically expressed with by reference to his hand,[106] will find its strength in Yhwh, who will grant Moses a power over death (see Num 12:13). After this, Yhwh gives him a final sign.

> [8] "And it will be, if they will not believe you nor listen to the voice of the first sign, they will listen to the voice[107] of the second. [9] If they will not believe even these two signs nor listen to your voice, you will take from the water of the Nile and pour it on the dry ground. And the water that you will take from the Nile will become blood on the dry ground."

The repetition of the verbs אמן and שמע and of the noun קול in v. 1 and in vv. 8-9 shows that the last two verses frame the passage with an explicit reference to Moses' objection (v. 1). The repetitious trend (the double use of the expression "if they will not believe you nor listen") prepares the statement of the third sign in v. 9, bringing the narrative intensity to its climax: the sign has the character of a portent whose degree is higher than that of the others. The ellipsis of the formula "and Yhwh said"[108] at the

---

[104] The scholars interpret צרעת and its derivate מצרעת as a reference to a skin disease in general; Houtman, I, 395-397; TH. HIEKE, "Leper, Leprosy I," 144-147, with bibliography.

[105] TH. HIEKE, *Levitikus 1–15*, 122; the sources of impurity were various: contacts with animals (Lev 11) and human corpses (Num 19:11-22); leaks of semen and menstruation (Lev 12; 15); skin diseases and other abnormalities that could affect the surfaces of buildings and clothing (Lev 13–14). The common denominator between these situations is the proximity to the sphere of death. See also Nepi, I, 107-108; Dohmen, 170.

[106] Priotto, 105. Regarding the symbolic interpretations, see Houtman, I, 398-399. See e.g. *ExR* III, 13, "Just as a leper defiles, so do the Egyptians defile you; and just as he becomes clean, so God will one day declare Israel clean." See also TERTULLIANUS, *De Resurrectione mortuorum*, XXVIII, 5 (CCLS 2,957): the signs of Exod 4:2-9, "sunt et quaedam ita pronuntiata, ut allegoriae quidem nubilo careant, nihilominus tamen ipsius simplicitatis suae sitiant interpretationem, quale est apud Esaiam: *Ego occidam et vivificabo* (Isa 26:20). Certe postea quam occiderit vivificabit. Ergo per mortem occidens per resurrectionem vivificabit."

[107] See G.I. Davies, I, 316. He notices that the term קול is unusual; it denotes "the meaning or content of the sign, an extension of the more natural use for the content of words." See Gen 45:16; Deut 5:28.

[108] See Houtman, I, 400; G.I. Davies, I, 316. Contrary to what Cassuto says (see *Exodus*, 47), the instructions of v. 7b are not a mere parenthesis describing Moses' actions during God's discourse (which continues from v. 7a directly to v. 8a); v. 8 begins with a conscious ellipsis.

beginning of v. 8, creates the impression of a fast pace and makes Yhwh's words more insistent.

The third sign is therefore directed at the Israelites, but indirectly involves the land of Egypt, because it affects the Nile, the great river celebrated as a divinity[109] that, thanks to the yearly flood, allowed the Delta area to receive a new deposit of mud and made all the surrounding regions fertile (Amos 8:8; 9:5; see Isa 19:5-8). If the Israelites will not believe, Moses will take some water from the Nile, pour it on the dry ground and witness its transformation into blood. The symbolic power of the colour red,[110] which often has a negative connotation,[111] creates a polar opposition between water and blood. Although the phenomenon could also be explainable from a material point of view,[112] the sign has a strong metaphorical charge: water, the source of life par excellence, is transformed into an element of death[113] (see Exod 7:21), the landscape becomes awful and gruesome.[114]

3.2 The Nile is Turned into Blood

The signs are performed by Moses before the people (4:30) and Israel believes (4:31). However, the first expedition to Pharaoh fails miserably, because he refuses to listen (5:1–7:6). Consequently, the Lord asks Moses and Aaron to repeat the sign of the staff in front of Pharaoh (7:9-10); but, since the court magicians manage to do the same thing (7:11-12), the king of Egypt refuses to listen to Moses and Aaron. At this point the narration of the plagues

---

[109] E. BRESCIANI, *Testi religiosi dell'Antico Egitto*, 177-180, "Inno al Nilo." See e.g., I, "Hail to you, o Nile! Who manifest yourself over this land and come to give life to Egypt!"; X, "Men offer the first-fruits of corn; all the gods adore you!" See *ExR* IX, 9, "Why were the *waters* first smitten, and with blood? Because Pharaoh and the Egyptians worshipped the Nile." See Cassuto, 48.

[110] M. GIRARD, *Symboles bibliques*. I, 963-970.

[111] In Egyptian literature, the colour red is linked to Seth, the god who kills Osiris, a deity linked to the desert ("red land"), "Seth was the 'Red One,' the ill-tempered god who personified anger, rage and violence, and who was often regarded as evil personified .... As a god of the desert or 'Red Land,' he opposed and threatened the vegetation upon which life itself depended" (R.H. WILKINSON, *The Complete Gods*, 197). See also Isa 1:18; Nah 2:4-5. In Rev 12:1-3 the dragon threatening the pregnant Woman is like the fire, and in 17:3-6 the woman Babylon is seated on a scarlet beast. See M. GIRARD, *Symboles bibliques*. I, 966-968.

[112] See Houtman, I, 403. At certain seasons of the year the Nile produces brown and red mud. Flavius Josephus believed that the sign of Exod 4:9 referred to the colour of the water, similar to blood, and not to an actual transformation; see *Ant. Jud.*, II, 273.

[113] A. RABINOWITZ, "Moses' Three Signs," 120-121; Houtman, I, 401.

[114] See L. RYKEN – J.C. WILHOIT – T. LONGMAN III, *Le immagini bibliche*, 1274. The sign of Exod 4:9 is interpreted as an omen; see 2 Kings 3:22 and Joel 3:3-4; see also M. SCANDROGLIO, *Gioele e Amos*, 224-225 (with bibliography).

unfolds as a story formed by a unified plot of several episodes. The narrative manifests the clear traces of a complex editing process[115] that forms a composition made of several layers.[116] From a structural point of view, the individual units create a regular ensemble which proceeds in steps and creates a progressive intensification; the majority recognises in Exod 7–10 a series of three cycles each formed by three plagues[117] and followed by the last plague (11:1-10) which stands out from the others and represents the climatic apex of the whole story.[118] Other authors have shown that, including 7:8-13 and considering 10:21-29 as the concluding episode, a concentric composition can be identified;[119] it has also recently been suggested that the narrative of Exod 7–11 was deliberately composed by bringing together different structural subdivisions, and placing the seventh plague in a priviledged position in comparison with the others.[120]

The text of Exod 7:7 has the function of a conclusion.[121] The narrative of the plagues, instead, begins with an introductory passage, 7:8-13, in which Moses and Aaron, having received instruction from Yhwh, perform before Pharaoh the sign of the staff turned into a serpent. A confirmation of the priviledged position of these verses is found in v. 13, where an essential

---

[115] In the first plague there is an obvious literary tension: while in Exod 7:17 Yhwh asks Moses to say to Pharaoh that it is he (Moses) who will strike the Nile with his staff, in 7:19 Moses is told to command Aaron to take his staff in his hand. See K.A. CHERNEY, JR., "The Plague Narrative (Ex 7:8 – 10:29)," 86-87. Ps 78 and Ps 105 also show that Exod 7–11 has undergone a complex redactional history, because the two psalms do not report the same number of plagues (Ps 78 relates seven, Ps 105, eight).

[116] Chapters 7–11 have been considered as the combination of two (Propp, I, 310-317; Dozeman, 48-51; R.E. FRIEDMAN, *The Bible with Sources*, 130-137) or three sources (See Childs, 131-151); scholars recognise the decisive role of P: see B. LEMMELIJN, "Setting and Function," 443-447; ID., "The So-called 'Priestly'", 481-511; his hypothesis is that P collects the previous material and edits it into a single whole (J. VAN SETERS, *The Life of Moses*, 100-112). See K.A. CHERNEY, JR., "The Plague Narrative (Ex 7:8 – 10:29)," 86-89.

[117] See Cassuto, 92; Childs, 160; Sarna, 28; Propp, I, 321; Houtman, II, 19; Nepi, I, 177-178; M. GREENBERG, "The Redaction of the Plague Narrative," 243-252.

[118] See J.-L. SKA, "*I nostri padri*", 50.

[119] The structure would be as follows: A (7,8-13: staff); B (7,14-25: water turned into blood); C (7,26 – 8,11: frogs); D (8,12-15: gnats); E (8,16-28: flies); E' (9,1-7: the plague); D' (9,8-12: boils); C' (9,13-35: hail); B' (10,1-20: locusts); A' (10,21-29: shadow). See Nepi, I, 176-177; K.A. CHERNEY, JR., "The Plague Narrative (Ex 7:8 – 10:29)," 83-92.

[120] The seventh plague is the most developed, and presents a long discourse of God and the first Pharaoh's repentence (9:27). See J. GROSSMAN, "The Structural Paradigm," 593. See also S.B. NOEGEL, "The Significance of the Seventh Plague," 532-539.

[121] See L. INVERNIZZI, "*Perché mi hai inviato?*", 418-419. The stereotypical formula that expresses age, found in Exod 7:7 ("Moses was *son of* eighty years"), is normally used at the end of a crucial experience (Cassuto, 90-91).

element of the plot of Exod 7–11 appears (to which the reader has already been prepared, from 4:21): the hardening of Pharaoh's heart.[122]

The first plague (Exod 7:14-25) has unique characteristics. The scene, in fact, begins with a discourse articulated in two parts (vv. 14-18 and v. 19), in which Yhwh reveals to Moses that Pharaoh's heart has become immovable (v. 14); thus, the hardening of the heart, recalled with other words becomes a preliminary motif that precedes all the episodes and remains valid for the continuation of the story.[123]

Thereupon, Moses receives precise instructions, as would be the case for a prophet:[124] he has to go unto Pharaoh at dawn,[125] taking advantage of the early hours of the morning, when the sovereign bathes in the Nile (which was a custom in the Egyptian royal family;[126] see Gen 41:17; Exod 2:5). He has to stand by the riverside to confront him,[127] with the staff in his hand (Exod 7:15). Thanks to some lexical echoes,[128] the scenario calls to mind Exod 2,1-10, and the reader, spectator of the scene on the Nile, knows that he can expect an extraordinary divine intervention.

The words of v. 15 are completed in the following verses by a quotation in which Yhwh anticipates to Moses what he will have to say (pre-productive quotation[129]). In this way, God's envoy is characterised as a

---

[122] Cassuto, 92-93; Childs, 151-153; Propp, I, 286; Priotto, 161. In regard to the hardening of the heart, see ch. II, § 3.2.1. The contact between Exod 7:8-13 with what follows, and in particular with 7:14-25, is confirmed by the strong connection between v. 13 and vv. 14, 22, as well as the similarity between the succession of signs that Moses has to perform before the people (4:1-9: first the staff, then the leprous hand, and then the waters of the Nile transformed into blood) and the sequence of 7:8-13 and 7:14-25. Other authors consider that the account of the plagues begins directly in 7:14 and that 7:8-13 is connected with what precedes; for a discussion of this argument, see Houtman, II, 11.

[123] Childs, 153; Priotto, 164.

[124] See Childs, 153, "God gives him exact instructions where is he to go, and what he is to say, just as he did to Elijah (1 Kings 17:3) and to Isaiah (Isa 7:3)."

[125] The first, fourth and seventh plagues begin with a reference to the morning (see 8:16; 9:13). This detail helps to associate Moses' mission to a prophetic calling, since even Samuel confronted the ruler Saul in the morning (1 Sam 15:12). See Propp, I, 323.

[126] The image of the deity bathing is well known in Egyptian literature, not least because the Egyptians are a civilisation born beside the waters of the Nile. See E. BRESCIANI, *Letteratura e poesia*, 193-194.448.

[127] The construction קרא + ל + נצב infinite construct, appears in Exod 5:20 and in Num 22:34, and in both cases it expresses an intention to hinder the path of a person in order to confront him. See Cassuto, 97.

[128] Some scholars note a connection between Exod 7:15 and 2:3-4: Moses is laid "on the banks of the Nile" (על־שפת היאר, 2:3), while his sister "stays" (יצב, 2:4; is a secondary root derived from נצב; see JM § 77b). See Cassuto, 97; Priotto, 164.

[129] See ch. II, § 3.2.

messenger who bears the divine word, obeying a prophetic mandate.[130] After recalling the request made to Pharaoh in Exod 5:1 ("let my people go") and his refusal to obey (7:16, "and behold you have not listened until now"), Moses must proclaim a divine message, formulated in 7:17-18:

> Thus says Yhwh (כה אמר יהוה), "By this[131] you shall know (בזאת תדע) that I [am] Yhwh (אני יהוה). Behold, I myself strike (אנכי מכה) with the staff that is in my hand the water that is in the Nile, and it shall turn into blood. The fish of the Nile shall die and the Nile shall stink and the Egyptians shall grow weary[132] to drink the water from the Nile."

The words that Moses has to utter exalt Yhwh through the repetition of the pronouns אני and אנכי, normally pleonastic, which highlight here the antithesis between Yhwh and Pharaoh.[133] Above all, the reader perceives a connection between 7:17 ("By this you shall know that I Yhwh") and the first meeting of Moses and Aaron with Pharaoh,[134] in 5:2, "Pharaoh said, 'Who is Yhwh, that I should listen to his voice and let Israel go? I do not know Yhwh (לא ידעתי את־יהוה)."' This echo confirms that the book of Exodus, in its entirety, is characterised by the plot of revelation:[135] if Pharaoh does not know who Yhwh is, he will recognise (ידע) his identity through the sign of blood.

The pronoun אנכי refers to God, but the action is performed by the person holding the staff, thus by Moses (7:15). The two actions overlap perfectly:

---

[130] See Childs, 153; Priotto, 164. In Exod 7:17 Moses uses the messenger formula "thus says Yhwh" (כה אמר יהוה), a stylistic feature characteristic of prophetic literature and very common in the book of Exodus (4:22; 5:1; 7:26; 8:16; 9:1.13; 10:3; 11:4). See L. INVERNIZZI, *"Perché mi hai inviato?"*, 147-148.

[131] The feminine (זאת) can be used as a substitute for the neuter (GK § 136*b*). In 7:17 it refers to that which follows (is proleptic), as is the case when the same term precedes the verb ידע (Gen 42:33; Num 16:28; Josh 3:10; Ps 41:12). See G.I. Davies, I, 500.

[132] The root לאה appears only 18x in the Hebrew Bible and can be translated "to grow weary" (regarding the *niphal*, see Ps 68:10; for the *hifil*, Jer 12:5). The same root can also mean "to be fed up," especially in the *nifal* (e.g. Isa 1:14; Jer 15:6), or, more broadly, "to be incapable, not able" (see Jer 20:9 where the verb is in parallel with לא יכל). See G.I. Davies, I, 501; Houtman, II, 34.

[133] See WO § 16.3.2*d*. See also Priotto, 165.

[134] Propp, I, 324; G.I. Davies, I, 504-505.

[135] The narrative of Exod 1–15 does not consist only of a plot of resolution: in various circumstances it is said that the purpose of the liberation is a "recognition" experienced by the Egyptians (ידע, 14:4, 18) and by the people (ראה, 14:30-31). Moreover, it is "an ignorance that triggers the process of oppression," when the text recalls that the first Pharaoh "had not known Joseph" (1:8, לא ידע את־יוסף), and when the new Pharaoh refuses to listen to Moses and Aaron because he ignores Yhwh (5:2; in Exod 7–12 the verb ידע is used in a similar way in 8:6, 18; 9:14, 29; 10:2; 11:7). See J.-L. SKA, *"I nostri padri"*, 40.

God is responsible for the turmoil thrown into Egypt (7:17), Moses is the executor of this intervention.[136]

The link between the verb נכה with the juridical domain and the context (7:14-25) allows the reader to interpret Yhwh's "blow" as a reference to a just sanction;[137] this interpretation is in line with the exegesis proposed in the *midrash* of Wis 11:6-7: the plague is a *contrapasso*, a punishment that strictly corresponds to the crime committed by the Egyptians, who are guilty of having stained the Nile with the blood of the Hebrew children[138] (Exod 1:22). The only difference is that this time the transformation takes place on a large scale,[139] the whole Nile will be turned into blood.

Normally, water is a source of survival;[140] in this case it will become a deadly instrument (and not only a symbol), as the first Christian interpretation of Exod 7:14-25 recalls (in Rev 8:8-9 and 16:3[141]) and as is reflected in its consequences: fish will die, the water of the Nile will no longer be drinkable. For a civilisation born on the shores of a large river, fishing was a very common activity, so much so that some species of fish were even considered sacred.[142] Consequently, the death of the fish is an ominous and fatal event that

---

[136] Priotto, 165; Durham, 97.

[137] The root נכה "to strike" can be used to indicate correction, especially when related to מוסר (Jer 2:30; 5:3; Prov 23:13; see also Prov 19:25); the same verb can express the afflictive measure that penalises the person guilty of a wrongdoing (Isa 11:4; see Gen 8:21; Exod 3:20; Lev 26:24; Deut 28:22). See P. BOVATI, *Re-establishing Justice*, 375, note 86.

[138] The *midrashic* style of Wis 10–19 is widely recognised by the authors; see L. MAZZINGHI, *Sapienza*, 21-23. For Wis 11:7 the waters of the Nile are turned into blood as a manifest reproof (εἰς ἔλεγχον) for the mass genocide of all infant male (Exod 1:22); the antithesis opposes the fate of the Egyptians to that of the Israelites, to whom God gave water in abundance even in the desert (Wis 11:7; *ibid.*, 455.457). The interpretation is suggestive, because it is not found in rabbinic exegesis and therefore seems to be an autonomous and original tradition. See also Rev 16:4-7 in which those who have shed the blood of the saints will drink the waters turned to blood, as punishment.

[139] Several attempts have been made to explain the plague as a natural phenomenon. The classic study dates from 1957: G. HORT, "The Plagues," 87-95; the author believes that the reddening of the waters of the Nile could occur during the period of the floods. See also K.A. CHERNEY, JR., "The Plague Narrative (Ex 7:8 – 10:29)," 89-92; G.I. Davies, I, 506.

[140] The essential role of water for Egyptian culture is clearly manifested in creation myths; the deity Nun is represented as a primordial ocean from which emerges a hill where the creative work begins; see E. BRESCIANI, *Testi religiosi dell'Antico Egitto*, 5-7.

[141] See U. VANNI, *Apocalisse*. II, 342.

[142] I. SHAW – P. NICHOLSON, *Dictionary of Ancient Egypt*, 100-101.248-249. In the Egyptian text *The prophecies of Neferti* the death of the fish is interpreted as a dramatic calamity, "the land is burdened with misfortune" (see *COS*, I, 108).

will have a widespread effect: the Nile will stink[143] (באש is linked to מות in Isa 50:2) and the Egyptians will no longer drink its waters.

Moses does not answer and in v. 19 Yhwh speaks again, this time to present an even more detailed and dramatic picture of the action that is about to take place. It will be Aaron who will raise up the staff and stretch out his hand over the waters of the Egyptians,[144] and the effects of his gesture will extend to a much broader scale. By virtue of a *crescendo*, which is expressed through the succession and accumulation of elements: rivers, canals, ponds, all the pools of water,[145] a distressing feeling arises in the reader. The result of Aaron's action is narrated with the repetition of the verb היה and of דם, "so that they may become blood;[146] and there shall be blood (ויהיו דם והיה) throughout all the land of Egypt, in woods and stones." The resemblance with Gen 1:3[147] creates a connection between the power of the Creator's word and the devastation produced in 7:19. The death will spread everywhere, even throughout the water found "in woods and stones."[148] In vv. 20-21 Yhwh's commands are carried out.

---

[143] When the subject of באש is food (the manna in Exod 16:20), water (7:21) or earth (8:10), the verb can be translated "to become fetid, to rot."

[144] Traditionally, v. 19 has been interpreted as part of the P layer, for the most part because of the tension with v. 15. See Childs, 131; J. VAN SETERS, *The Life of Moses*, 104; B. LEMMELIJN, "The So-called 'Priestly',", 494, n. 33; K.A. CHERNEY, JR., "The Plague Narrative (Ex 7:8 – 10:29)," 87-89. Rabbinic tradition notes the problem and solves it with originality: since the river protected Moses at the time of his birth (Exod 2:1-10), the task of striking it cannot be entrusted to him (*ExR* IX, 9; Rashi Exod, 49, לפי שהגן היאר על משה).

[145] The Hebrew expression כל־מקוה מימיהם can be translated as "all the pools of water"; קוה means "to gather" and when the noun is related to מים it indicates a cistern (Lev 11:36) or refers to the sea, as a whole collected "in one place" (Gen 1:9), unlike the dry land (Gen 1:10). Therefore, they are not spring waters (Rashi Exod, 49). Since the reference to all the pools of water is the fourth element in the sequence, it could be considered as a conclusive expression that has a comprehensive function: see Houtman, II, 35; G.I. Davies, I, 501.

[146] The sentence ויהיו דם manifests the result of the imperative which precedes it, "stretch out your hand ... *so that* they may become blood." See JM § 116*d*; Childs, 122.128. The phrase והיה דם בכל־ארץ מכרים, instead, introduces a subsequent future condition with a consecutive nuance, "and there shall be blood"; see JM § 119*c.e*; Houtman, II, 36.

[147] In Gen 1:3 we find a construction similar to that of Exod 7:17: the imperative of היה is followed by the object of the verb, the *wayyiqtol* of היה and the repetition of the same term: יהי אור ויהי אור. In Exod 7:17 the *weyiqtol* of היה introduces, maintaining a deontic mode, the consequences of the imperatives that precede it, and is followed by the term דם; the root היה is repeated immediately afterwards but in the *weqatal*, the verbal mode used to express a succession of events in the future, and is followed again by the term דם.

[148] The LXX translates literally: ἐν τε τοῖς ξύλοις καὶ ἐν τοῖς λίθοις, "in the woods and the stones"; see also the Peshitta and *TgN*. Nevertheless, the pair of terms עץ and אבן is used elsewhere in the Hebrew Bible for idols (Deut 4:28; 28:36; 2 Kings 19:18; Ezek 20:32) and *ExR* IX, 11 interprets it in this way; *TgO* and *TgPsJ* translate instead "in vessels of wood

Exod 7:17-19	Exod 7:20-21
[17] "Thus says Yhwh, "By this, you shall know that I Yhwh. Behold, I myself strike with the staff (מכה במטה) that is in my hand the water that is in the Nile, and it shall turn into blood (ונהפכו לדם). [18] The fish of the Nile shall die and the Nile shall stink and the Egyptians shall grow weary to drink the water from the Nile." [19] *Yhwh said to Moses, "Say unto Aaron, "Take your staff and stretch out your hand over the waters of Egypt, over their streams, over their ponds and over all their pools of water, so that they may become blood; and there shall be blood (והיה דם) in all the land of Egypt, and in wood and stones'."*	[20] Moses and Aaron did so, as Yhwh commanded and he lifted up (the hand) with the staff[149] and smote (וירם במטה ויך) the waters that were in the Nile, in the sight of Pharaoh, and in the sight of his servants; and all the waters that were in the Nile were turned into blood (ויהפכו...לדם). [21] The fish that were in the Nile died; and the Nile stank, and the Egyptians could not drink water from the Nile. *And there was blood (ויהי הדם) throughout all the land of Egypt.*

Verses 20-21 tells of the fulfilment of the divine word and together with vv. 17-19, have been considered as a combination of J and P (in the table, in italics).[150] Moses and Aaron obey God's command together, but the subject carrying out the action remains ambiguous, "and he lifted up (the hand) with the staff." Therefore, the question of the agent who "strikes" the Nile remains unresolved, leaving the field open to at least three possibilities: God, Moses or Aaron.[151] The correspondence and coincidence between Yhwh and his envoys cannot be stated more clearly.

The interplay of repetitions between the word spoken by God (vv. 17-19) and its fulfilment (vv. 20-21) creates a noteworthy effect: while in v. 17

---

and stone" as does the Vulgate (*tam in ligneis vasis quam in saxeis*); this translation has become the most widely followed; see Rashi Exod, 49; Childs, 122; G.I. Davies, I, 498. Other authors prefer to interpret wood and stone as a part referring to the whole, i.e. "buildings constructed with wood and stone": see Houtman, II, 37; Propp, I, 325 [he quotes Lev 14:45; 1 Kings 5:32; 2 Kings 12:13]; Priotto, 166.

[149] The sentence וירם במטה is elliptic, because the object of the verb רום is omitted, "raised (his hand) with the staff"; see G.I. Davies, 502. The LXX translates the phrase interpreting במטה as the object (with the Vulgate and the Peshitta), "raised the staff"; see GK § 119q; JM § 125m.

[150] See Childs, 131; J. VAN SETERS, *The Life of Moses*, 104; G.I. Davies, I, 506.

[151] Some authors believe that the subject of רום is Aaron; see Peshitta; Propp, I, 325; Houtman, II, 38. In effect, the sentence is ambiguous, it can also refer to Moses (Cassuto, 99; Childs, 128), and it cannot be excluded that the subject is God himself (Priotto, 146, n. 6). See the discussion in G.I. Davies, I, 507.

Yhwh says that he will simply strike the Nile with the staff, in v. 20 it is specified that the staff is "lifted up," creating an image that will be recognisable further ahead, in Exod 14:16. The fulfilment, moreover, is characterized with greater intensity; if in v. 17 it was only stated that "the waters" will change into blood, in v. 20 the object is described with more details, "all the waters that were in the Nile." The omissions are also significant. In the account of vv. 20-21, many details which were mentioned in v. 19 are omitted, especially those concerning the extension of the plague to all the water reserves of Egypt, and only the conclusion is repeated: v. 21 "and there was blood in all the land of Egypt." Thus, specifying in v. 20 that these prodigies took place "in the sight of Pharaoh and in the sight of his servants," the narrator relates the plague, focusing especially on the point of view of Pharaoh and his court.

In v. 22 the narrator recounts that the magicians of Egypt are able to imitate the work of Moses and Aaron. This connection is tangible, thanks to an evident repetition.

Exod 7:20	Exod 7:22
Moses and Aaron did so (ויעשו־כן), as Yhwh commanded (כאשר צוה יהוה)	The magicians of Egypt did so (ויעשו־כן), with their secret arts (חרטמי מצרים בלטיהם)

Without specifying the exact nature of the spell cast by the Egyptian magicians (this is a *blank*[152]), the text focuses on the opposition between Moses and Aaron, on the one side, and the magicians of Egypt on the other. The differences between the two verses are substantial: the performance of the enchanters is not based on obedience to a mandate ("as Yhwh had commanded"), but is achieved only "by their witchcraft," or "with their secret arts"[153] (בלטיהם), they are autonomous from any transcendent force.

---

[152] Regarding the ellipsis and *blanks* (i.e., insignificant gaps) in narrative texts, see J.-P. SONNET, "Analisi narrativa," 61; M. STERNBERG, *The Poetics*, 236. Investigating the nature of an intentional ellipsis can be misleading; in this case, it is not important to specify the characteristics of the prodigy performed by Egyptian magicians (Propp, I, 325), nor question which waters can turn into blood (Cassuto, 99).

[153] The noun לט is a rare term; it refers to the "spells" of the Egyptian magicians (Exod 8:3, 14), and designates elsewhere the word communicated privately (1 Sam 18:22), or the action done in secret (1 Sam 24:5; see Ruth 3:7). For this reason, some authors emphasise the occult character of the Egyptian incantations (Rashi Exod, 50: בלטיהם is "an incantation that they say in a whisper and in secrecy," בלט ובחשאי; Cassuto, 99, "secret arts"; Priotto, 166, "arti segrete"; Propp, I, 325, "mysteries").

The reaction of Pharaoh is described in vv. 22b-23 with a progressive sequence of verbs: he hardens his heart,[154] and does not listen, turns his back and directs his attention elsewhere,[155] he enters in his house, refusing to take the event he has witnessed to his heart. These actions are very useful for profiling Pharaoh's character. The trait that will mark his behaviour, until the end of the story, is the hardening of his heart and his refusal to listen to the word of the envoys. Nevertheless, the narrator adds other elements that he could have omitted: he turns his back to the people, and returns to the safety of his home. In front of the catastrophe, described in great detail in v. 19, Pharaoh does not spring into action, and this shows that he has no interest in what is happening, he has no concern for his subjects, who are forced to work hard digging around the Nile to find other sources of water[156] (7:24). Furthermore, the repetition of the term לב in vv. 22-23 reveals that Pharaoh's attitude originates in his inner life and is connected to his lack of understanding:[157] Pharaoh should "know," he should recognise the Lord (7:17), but this does not happen.

In conclusion, the close reading of 7:14-24 confirms the impression had in the analysis of 4:9: the water of the Nile transformed into blood becomes a deadly "blow" for the entire nation of Egypt. With this sanctioning intervention, Yhwh reveals his power over life and death and manifests his extraordinary sovereignty over creation (see 7:19, "so that they become blood, and there will be blood" and the connection with Gen 1:3) precisely through the fluid that will appear several times in the book of Exodus, marking its plot.

---

[154] Regarding the hardening of the heart, see ch. II, § 3.2.1.

[155] The verb פנה "to turn, to turn away" indicates the act of turning in a certain direction (Deut 1:7; 2:8); it designates also the action of turning away from a place (see Gen 18:22; Deut 9:15) or detaching oneself from a person (see Exod 10:6; 2 Kings 5:12). When a person turns away from God, normally has the intention of worshiping idols (Jer 2:27).

[156] The verb חפר indicates the act of digging to find a source of water: Gen 21:30; 26:15, 18-19, 21-22, 32. See Priotto, 167.

[157] The syntagma שית + לב (Exod 7:23) "to put one's heart" means "to take something to heart" (for men, 2 Sam 13:20; Ps 62:11; for God, Job 7:17) "to observe carefully" (Ps 48:14), "to consider," as an act of intelligence (see Prov 22:17; 24:32).

CHAPTER II

The Journey of Moses (Exod 4:18-31)

The motif of the shed blood reappears for the first time since 4:9 in 4:24-26, a mysterious and obscure text, embedded within the narrative of 4:18-31,[1] which tells of the journey of Moses to Egypt after Yhwh had entrusted him with the task of guiding the people. In contrast with the more common usage of the term, this occurrence of דם is unique, since the expression "bridegroom of blood" does not appear elsewhere in the Hebrew Bible and the practice of circumcision is never described with an explicit reference to the spilling of blood. Consequently, it seems fitting to begin with the analysis of Exod 4:18-31. In order to analyse the different episodes, the tensions found within the text will be considered and the solutions elaborated by the scholars who applied the historical-critical method reviewed, thus enabling the possibility of a close reading of the passage that will illustrate the coherence of the narrative in its final form and the decisive role of vv. 24-26 for the plot.

1. Literary Criticism

Exodus' third and fourth chapters occupy a prominent place in the discussion on the formation of the Pentateuch and they provide much food for thought regarding the history of Exodus' redaction. Therefore, it seems necessary to dedicate several paragraphs to the literary criticism of these verses. We shall begin by analysing the tensions which suggest that the text is made up of composite material; next, we shall present the solutions put forwards by different scholars; finally, the results of this research will be evaluated.

1.1 The Difficulties of the Text and the Hypotheses of Literary Criticism

1.1.1 The Relations Between v. 18 and v. 19

The text presents a first "fracture" in vv. 18-19: in v. 18 Moses reveals to Jethro his desire to return to Egypt, but in v. 19 it seems that he no longer wants to make the journey, and so Yhwh must send him again. Wellhausen noted that

---

[1] A. GOROSPE, *Narrative and Identity*, 93-96.

this difficulty can be resolved if the reader goes directly from 2:23a to 4:19[2] and therefore Noth already considered 3:1 – 4:18 to be a late addition inserted between 2:11-23a and 4:19-20[3]. In fact, the text of 4:19 gives the impression that Yhwh is speaking for the first time and, moreover, in 3:1 – 4:18 there is no reference to the events narrated in 2:11-23. Finally, the name of Moses' father-in-law is Jethro in 3:1 and 4:18, while in 2:18 he is called Reuel. Various explanations have been given by scholars regarding these difficulties.

Those who supported the *classical documentary hypothesis* held that v. 18 was part of the Jahvist[4] (J) document, while v. 19 belonged to the Elohist[5] (E). Recently, this explanation has been further developed, nevertheless the two verses continue to be considered as part of two separate sources[6].

Exegetes who believe that the *Jahvist editor* is from the Exilic period, propose two different solutions. Levin, presumes that 4:18 is part of the J redaction, while 4:19 would be a late non-J addition[7]. Van Seters, instead, conjectures that the late editor J is responsible for the composition of the majority of the non-P text, including the entire account of Moses' vocation (3:1 – 4:17) and 4:18. In his opinion, Exod 4:19 belongs to P[8].

Ultimately, even those who question the existence of source J (and E), consider the discontinuity between v. 18 and v. 19 as an obvious textual difficulty. Blum believes that Exod 3:1 – 4:18 is a late insertion by *KD*.[9] Berner identifies twelve different layers in Exod 3–4 and attributes vv. 18-19 to the post-P redaction[10]. K. Schmid is of a different opinion and considers Exod 3–4 as a composition that is entirely work of the post-P redactor[11]. According to Gertz, instead, vv. 18-20a, 24-26, 27-31* are parts of the non-P layer, to

---

[2] See J. WELLHAUSEN, *Die Composition*, 71.

[3] So, Noth, 22-23.

[4] Concerning the bibliography of recent research on the editorial history of the Pentateuch and regarding the acronyms used for the different layers (J; D; P; non-P; post-P, H; etc.), see B.A. ANDERSON, *An Introduction*, 42-59.

[5] J. WELLHAUSEN, *Die Composition*, 70-71.

[6] See e.g., P. WEIMAR, *Die Berufung des Mose*, 267-282, who believes that v. 18 belongs to J, while v. 19 is the result of the redactional work of the editor that created JE. See also Propp, I, 190-194; Childs, 93-96.

[7] See CH. LEVIN, *Der Jahwist*, 75-76.326-333.

[8] J. VAN SETERS, *The Life of Moses*, 64-67.

[9] E. BLUM, *Studien zur Komposition*, 20. He later revised his position and attributed Exod 4:1-17, 27-31 to the post-P editor; see ID., "Die Literarische Verbindung," 123-140.

[10] See C. BERNER, *Die Exoduserzählung*, 133-136.

[11] K. SCHMID, *Genesis and Moses Story*, 172-193. Exod 3–4 is a later text which (together with Gen 15, e.g.) allowed, at the time of the final redaction of the Pentateuch, the connection between the two narratives that originally developed independently: the Patriarchs story and the Exodus.

which the post-P redaction adds more verses[12] (4:20b, 21-23). Recently, Jeon has proposed another explanation: he believes that vv. 19-20a belong to the story of Moses' flight (2:10-23a; 4:19-20a), a layer which is older than the vocation narrative (3:1 – 4:18).[13]

1.1.2 Yhwh's Direct Speech in vv. 21-23

Moses decides to return to Egypt and Yhwh addresses him once more, revealing his intention to "harden the heart" of Pharaoh (v. 21) and telling Moses the words which he must say to the king of Egypt (vv. 22-23). Scholars have interpreted these verses as a late addition. As a matter of fact, vv. 21-23 seem redundant when compared with 3:18-22 and the threatening tone of v. 22 apparently contradicts the invitation to "negotiate" in 3:18; moreover, Aaron does not appear on the scene.[14] In addition, the "signs" (אתות) of 4:1-9 are to be performed for the Israelites, that they may recognise the authority of Moses, while the "wonders" (מופת) of 4:21 must be performed before Pharaoh.[15] Finally, the motif of the hardening of Pharaoh's heart (4:21), understood as an action performed directly by God, is considered to be a late theological development of P.[16]

According to the *classical documentary hypothesis*, vv. 21-23 are part of document E[17] or the result of the JE redaction.[18] Levin believes that vv. 21-23 are a later addition to the work of J.[19] Van Seters, instead, thinks that these verses are the work of P.[20] Lastly, the defenders of the model connected to the redactional criticism tend to recognise in vv. 21-23 the hand of the post-P editor.[21] The main argument in favour of this explanation is lexical: the verses mix both P and non-P terminology.[22]

---

[12] J.C. GERTZ, *Tradition und Redaktion*, 254-334.
[13] J. JEON, *The Call of Moses*, 147-148.154.
[14] J. JEON, *The Call of Moses*, 149.
[15] J.C. GERTZ, *Tradition und Redaktion*, 332-333.
[16] See e.g. J. VAN SETERS, *The Life of Moses*, 68.
[17] J. WELLHAUSEN, *Die Composition*, 70-71; Propp, I, 191-192; he considers the reference to the hardening of the heart to be the result of a late redaction.
[18] See P. WEIMAR, *Die Berufung des Mose*, 228-318.319-331.
[19] See CH. LEVIN, *Der Jahwist*, 333.
[20] J. VAN SETERS, *The Life of Moses*, 68.
[21] K. SCHMID, *Genesis and Moses Story*, 172-193; see also J.C. GERTZ, *Tradition und Redaktion*, 254-334; J. JEON, *The Call of Moses*, 158.
[22] In v. 21 there is a greater concentration of the P vocabulary: מפתים + לפני פרעה (see Exod 7:9) e חזק + לב + פרעה (e.g., Exod 7:13, 22; 14:4, 8). In vv. 22-23, on the contrary, expressions frequently used in non-P prevail: שלח את בני ויעבדני (e.g., Exod 5:1; 7:16, 26); מאן + שלח + ל (Exod 7:14, 27). See J. JEON, *The Call of Moses*, 149.

1.1.3 The Meeting with Yhwh (vv. 24-26) and the Return (vv. 27-31)

The episode narrated in vv. 24-26 raises many problems. Considering the unexpected and hostile manner in which Yhwh behaves towards Moses (v. 24, "he tried to kill him"), the passage does not really seem to fit the context. Many solutions have been proposed and in general the exegetes are deeply divided: for the defenders of the documentary hypothesis, this is an old text that belongs to the J document,[23] or represents the most recent result of the JE redaction.[24] Levin believes that these verses are a late non-J addition, posterior to the exilic Jahwist,[25] while Van Seters considers them to be a "midrashic" supplement, added after the elaboration of the P document, since they presuppose both J and P.[26] Blum thinks that vv. 24-26 are a pre-exilic text belonging to *KD*,[27] Schmid considers the entire Exod 4 to be post-P,[28] and then post-exilic. Gertz interprets vv. 24-26 as part of the most archaic layer of the narrative.[29] Finally, according to Jeon vv. 24-26 are the oldest passage in these chapters, linked to the story of Moses' flight from Egypt.[30]

With regard to vv. 27-31, on the other hand, we are presented with two scenarios. For some scholars, the passage is divided into several layers and Schmidt summarises the problem as follows: v. 29 and v. 31b are said to be part of the older source to which vv. 27-28 and 30-31a are added in order to include the figure of Aaron.[31] Other exegetes, however, consider vv. 27-31 as a unified text: for Propp it is part of the E document,[32] Levin believes that vv. 27-31 are a passage added by the late non-J editor,[33] Van Seters thinks that it is the work of the exilic redactor J,[34] and, ultimately, Gertz believes that this passage is a result of the post-P redaction[35].

---

[23] See J. WELLHAUSEN, *Die Composition*, 71; Propp, I, 191.

[24] P. WEIMAR, *Die Berufung des Mose*, 267-282.

[25] CH. LEVIN, *Jahwist*, 332.

[26] J. VAN SETERS, *The Life of Moses*, 68.

[27] E. BLUM, *Studien zur Komposition*, 12.

[28] K. SCHMID, *Genesis and Moses Story*, 172-193. See C. BERNER, *Die Exoduserzählung*, 133-136, who considers vv. 24-26 as the last of the twelve editorial layers of Exod 4.

[29] J.C. GERTZ, *Tradition und Redaktion*, 327-330.

[30] J. JEON, *The Call of Moses*, 151.

[31] Schmidt, 235-237. The character of Aaron is introduced in these verses without a clear transition and the change from a singular verb (הלך) to a plural (אסף) in v. 29 suggests that Aaron's name was added later. See also E. BLUM, *Studien zur Komposition*, 28; J. JEON, *The Call of Moses*, 152.

[32] Propp, I, 191.

[33] CH. LEVIN, *Jahwist*, 333.

[34] J. VAN SETERS, *The Life of Moses*, 68-70.

[35] J.C. GERTZ, *Tradition und Redaktion*, 334.

1.2 *An Evaluation of the Different Redactional Reconstructions*

The solutions listed present a complex scenario, in which, faced with the same textual phenomena, scholars reconstruct a completely different history of composition. The most convincing hypotheses are those that interpret 4:18-31 as a passage that belongs to the last stage of the composition of the Pentateuch. However, this brief examination of the different hypotheses shows that the "scientific" certainty of the editorial analysis is sometimes contradicted by its results: often the same critical arguments produce contradictory consequences and far from being "proofs," they instead become matter of further discussion.[36] Therefore, keeping in mind the positive indications derived from the consideration of the fractures in the text, 4:18-31 will be read from three perspectives.

Firstly, from the hermeneutical point of view, a *fragmentary reading* will be avoided. The risk of such an approach, which is excessively oriented to identifying the layers of the composition, is that of reducing exegesis to redactional history, limiting the interpretation to the operation of isolating the various textual "atoms" without considering the whole.[37]

In the second place, the literary coherence of the canonical form will be considered as a crucial element to understand the meaning of the text.[38] This approach proceeds from the conviction that the whole governs the fragments. The arrangement of the elements, in fact, and the narrative created by the juxtaposition of different parts (even if it occurred at different moments in the history of the text) produce a meaningful effect that cannot be forgotten. After noting the "tensions" of the text with the *source-oriented* exegesis, the literary coherence of the final form of the text will be analysed (according to a *discourse-oriented*[39] approach).

Finally, the reading of 4:18-31 will have a *theological interest* and will stem from the principle that the combination of different layers in the biblical canon creates a new result. Even if the final text maintains its ruptures, it has a communicative intention that is fulfilled in the act of reading.[40] The theology of the passage is expressed precisely through the combination of

---

[36] J.-P. SONNET, "La Bible et l'histoire, la Bible et son histoire," 7-10.
[37] See L. INVERNIZZI, *"Perché mi hai inviato?"*, 71-73, and especially the footnote 164.
[38] G. FISCHER, *Jahwe unser Gott*.
[39] The distinction between *source-oriented* and *discourse-oriented* approaches dates back to M. STERNBERG, *The Poetics*, 15. While the research on the sources and editorial history of the Scriptures (*source-oriented*) questions the historical genesis of texts and focuses on what is "behind" and before the text, "discourse-oriented analysis ... sets out to understand not the realities behind the text but the text itself as a pattern of meaning and effect."
[40] See S. RIMMON-KENNAN, *Narrative Fiction*, 44.

different materials into a single composition that harmonises the elements and expresses them together.

2. Exodus 4:18-31: Delimitation of the Text and Plot

In order to grasp the meaning of the text, it is crucial to specify the precise delimitation of the passage to be analysed in order to show how the whole of Exod 4:18-31 is determined by the harmony of an action expressed in a multiplicity of small episodes.

2.1 *The Narrative Units of the Text*

The delimitation of Exod 4:18-31 is a debated matter. According to Noth, the narrative arc, of which vv. 18-23 are a part, begins in 2:11.[41] Other authors think that the beginning of the narrative is found in 2:23.[42] Furthermore, several scholars extend the end of the narrative unit to ch. 7, and include in it the narrative of Moses' second calling presented in 6:2 – 7:7.[43]

The majority, however, believe that 3:1 is an important shift in the narration and that it marks the beginning of a new unit.[44] In fact, considering the dramatic criteria,[45] in 3:1 there is a noteworthy change of place: from Egypt (2:23) the scene moves to Midian. Moreover, from a syntactical point of view, the verse begins with a *w-x-qatal*,[46] a hebrew stylistic feature that often indicates the opening of a new unit.[47]

Identifying the conclusion of the narrative, instead, is rather more difficult.[48] The passages of Exod 4:17 and 4:18 are in fact strictly connected: in v. 18 Moses accepts the mission which has been entrusted to him by God in the previous chapters;[49] moreover, the use of verb הלך in v. 18 and in vv. 19, 21, 27, 29 creates a link between vv. 18-31 and ch. 3[50] (in which הלך appears in

---

[41] Noth, 22-23.
[42] See Cassuto, 16-20 (Exod 2:23 – 4:31); Houtman, I, 322-325 (Exod 2:23 – 4:19); see also P. WEIMAR, *Die Berufung des Mose*, 23-26.59-60 (Exod 2:23 – 5:5).
[43] Durham, 27-30; Dozeman, 94-98; CH. LEVIN, *Jahwist*, 322-333.
[44] See e.g., Childs, 51-52; G. FISCHER, *Jahwe unser Gott*, 22; E. BLUM, *Studien zur Komposition*, 17-20; J. VAN SETERS, *The Life of Moses*, 35-38.
[45] The main parameters for the delimitation of a narrative text are the dramatic criteria: space, time, action, characters. See J.L. SKA, *"I nostri padri"*, 15-16.
[46] See GK § 142*d*; JM § 155*k*.
[47] G. FISCHER, *Jahwe unser Gott*, 22.
[48] Childs, 52.
[49] See E. BLUM, *Studien zur Komposition*, 18. See also Propp, I, 180-182.
[50] Priotto, 86.

3:10, 16, 22). Nevertheless, 4:17 determines a break in the narrative:[51] in v. 18 Moses returns to Midian and the characters are different (he talks with Jethro and not with Yhwh); moreover, the actions described in v. 18 are difficult to isolate from that which is narrated later, especially the departure for Egypt. Thus, in 4:18 a course of action begins that, albeit related to 3:1 – 4:17, is marked by another unifying motif: Moses' return to Egypt.

The end of the story that begins in 4:18 is much more distinguishable. In fact, Exod 4, in concluding with the people of Israel believing in the words of Moses (4:31) presents a situation of extraordinary agreement. These last verses are similar to an epilogue[52] and in 5:1 the narrator explicitly indicates that he is about to introduce a different action, which takes place "afterwards" (ואחר, 5:1) and in another place, that is, at Pharaoh's court.[53]

Therefore, although the break between v. 17 and v. 18 is less obvious, the two extreme limits of the passage are precisely v. 18 and v. 31 of Exod 4. That being said, because of the fragmentation noted in the text, the question is now whether or not the passage should be divided into other parts.

2.2 *Exod 4:18-31: A Unified Plot*

The narrative of Exod 4:18-31 is structured as a unified plot,[54] yet is composed by different scenes, each one being functional to the others.[55] The table lists the division of episodes based on the change of characters and actions.

Verses	Characters	Action
v. 18:	Moses, Jethro	Moses asks for permission from Jethro to return to Egypt.
v. 19:	Moses, Yhwh	Yhwh says to Moses to return to Egypt.

---

[51] See J. VAN SETERS, *The Life of Moses*, 35-38; J. JEON, *The Call of Moses*, 75-76.

[52] J.-L. SKA, *"I nostri padri"*, 54: in the epilogue, suspense and tension disappear. See Propp, I, 258; L. INVERNIZZI, *"Perché mi hai inviato?"*, 135.

[53] See L. INVERNIZZI, *"Perché mi hai inviato?"*, 136-137.

[54] Regarding the distinction between "unified plot" and "episodic plot," see J.-L. SKA, *"I nostri padri"*, 38-39. Biblical narratives are normally made up of several episodes; Ska considers that there are some plots in which "all the episodes are relevant to the narrative and have a relevance for the outcome of the events recounted" while there are others in which "the order of the episodes can be changed, and the reader can skip an episode without harm." The distinction is maintained in this work, but the conclusions are diminished: although episodic plots present a series of scenes that seem fragmentary, in reality, it is the order of a text that creates the meaning; see M. PERRY, "Literary Dynamics," 311-361.

[55] Nepi, I, 123-138; Utzschneider – Oswald, 136-144; Priotto, 109.

v. 20:	Moses, wife, and sons	Moses returns to Egypt with his wife and sons.
vv. 21-23:	Moses, Yhwh	Yhwh speaks to Moses and discloses His "program."
vv. 24-26:	Moses, Yhwh, Zipporah, son	Yhwh seeks to kill Moses, Zipporah intervenes.
vv. 27-28:	Moses, Yhwh, Aaron	Yhwh speaks to Aaron, he leaves and encounters Moses.
vv. 29-31:	Moses, Aaron, Israelites	The brothers meet with the people.

The different scenes are unified by a central element: Moses who has already been told to go to Egypt (3:10.16), in 4:18 manifests his intention to return to Jethro,[56] leaves after the word of God (4:19-20) and encounters the elders and the people in 4:27-31. The narrative arc, therefore, begins with Moses' preparations for his departure and ends with his arrival in Egypt.

The narrative opens without any exposition or presentation of the characters, because it is strictly connected to what is told before (3:1 – 4:17). The first episode, therefore, goes from v. 18 to v. 23: Moses decides to leave (v. 18), he receives a word from Yhwh that spurs him on in his decision (v. 19), and he finally leaves (v. 20). Yhwh, for His part, continues to dialogue with Moses even after the departure (vv. 21-23).

The journey of Moses' return to Egypt is abruptly interrupted by a new intervention of Yhwh (vv. 24-26). The divine attack surprises[57] the reader, because Yhwh tries to kill Moses just as he accepts the mission and leaves. The story then becomes even more complicated, the tension reaches its climax and the intervention of Zippora (4:25) is the turning-point that brings the plot towards its resolution.[58] The final excerpt (v. 26) has the tone of a partial conclusion, because the narrator comments on what happened[59] and clarifies the contact between the enigmatic expression חתן דמים and circumcision.

The last part of the narrative is developed in two phases. God speaks to Aaron and urges him to go into the desert and meet Moses (v. 27); they meet again at the "mountain of God" and Moses reports all that God has

---

[56] Priotto, 109.

[57] Surprise (with suspense and curiosity) is one of the three "universals" of narrative literature; see M. STERNBERG, *The Poetics*, 309-320.

[58] The *climax* of a narrative is the point of highest suspense, the pivotal moment of a plot. See J.-L. SKA, *"I nostri padri"*, 51.

[59] The biblical narrator is normally "discreet," and his voice does not resonate except for necessary indications. See M. STERNBERG, *The Poetics*, 120-121. The interventions of the narrator are often concentrated in the conclusions (like Exod 4:26), see J.-L. SKA, *"I nostri padri"*, 55. Regarding narrative strategies of conclusion, see S. ZEELANDER, *Closure*.

said to him (v. 28). Moses and Aaron gather the elders of the Israelites, Aaron speaks to the people and they believe his words (vv. 29-31). The journey therefore has a successful conclusion: the Israelites manifest their trust in Yhwh, they bow down their heads and worship (v. 31).

3. The Departure of Moses (Exod 4:18-23)

The scene described in Exod 4:18-23 is marked by some significant repetitions, which are highlighted in the following translation; this complex of references reinforces the impression that the final form of the text is coherent.

¹⁸ And Moses |went| and **returned**[60] to Jethro, his father-in-law and said to him:
"|Let me go|, please, to **return**[61] to my brothers who are in Egypt, to SEE whether they are still alive."
And Jethro said to Moses:
"|Go| in peace."
¹⁹ Yhwh said to Moses in Midian:
"|Go|, **return** into Egypt, for they are dead, all the men that sought your life."
²⁰ And Moses <u>took</u> his wife and his children and made them ride upon an ass and he **returned** to the land of Egypt. And Moses <u>took</u> the rod of God *in his hand*.
²¹ Yhwh said to Moses:
"While you |go| and **return** into Egypt, SEE all the wonders which I have put *in your hand*, so that you may perform them before Pharaoh;[62] I will, instead, harden his heart and he will not let the people go.[63] ²² And you will say to Pharaoh, "Thus says Yhwh: 'Israel is My son, My firstborn. ²³ I said to you,

---

[60] The verbs הלך and שוב are an hendiadys (Gen 32:1; Num 24:25).

[61] The two cohortatives depend on a first cohortative and therefore manifest the purpose of what is said before, "let me go, *to* return... *to* see" (JM § 116*b*; GK § 108*d*); see also G.I. Davies, I, 352, although he translates only the second cohortative with a final nuance.

[62] See Houtman, I, 428-429. He translates v. 21, "remember to do before Pharaoh all the wonders I have empowered you to do" (see e.g., NRSV; NJPS), because he believes that what is said after ראה (the Hebrew sentence, כל המפתים אשר שמתי בידך) is not the verb's object and that the term המפתים is anticipated for the sake of emphasis. This construction would be formed by a *casus pendens*: GK § 112*mm*; 143*d*; JM § 156*j*; 176*j*. See Childs, 90; G.I. Davies, I, 353. The translation presented interprets המפתים as the object of the verb "to see," respecting the order of the MT; see Priotto, 84; Propp, I, 182-183; A.E. GOROSPE, *Narrative and Identity*, 97, note 22. Cassuto translates the sentence accentuating the consecutive connection, "see all the portents ... *so that* you may do them before Pharaoh" (see *Exodus*, 55).

[63] The *waw* of ואני is disjunctive, "while you... I, on the other hand...." See A.E. GOROSPE, *Narrative and Identity*, 97, note 23; see also Childs, 91; Priotto, 84. Houtman believes that ואני is emphatic: *Exodus*. I, 429 and JM § 146*a*.

let My son go, that he may serve Me, but you have refused to let him go. Behold, I am going to slay[64] your son, your firstborn.'"

v. 18. "Let me go, please." The cohortative אלכה has a deontic modal nuance[65] and expresses the intense desire of Moses; the particle נא shows that Moses is asking for permission.[66]

"Whether they are still alive." This is an indirect interrogative clause.[67] The expression is not a simple request for information (whether they are in good health[68]), but shows a genuine concern[69] (he wants to know if they survived).

"Go in peace." With the expression שלום + ל + הלך the speaker wishes the person he addresses a safe journey to his destination[70] (Judg 18:6; 1 Sam 1:17).

The Septuagint repeats part of Exod 2:23 at the end of v. 18: μετὰ δὲ τὰς ἡμέρας τὰς πολλὰς ἐκείνας ἐτελεύτησαν ὁ Βασιλεὺς Αἰγύπτου, "after those many days the king of Egypt died." It is probable that the Greek translation makes the transition between v. 18 and v. 19 less abrupt, by presenting the reason that caused Moses' delayed departure: his fear of Pharaoh.

v. 21. "While you go." The preposition ב can show the simultaneity of two actions: while Moses returns to Egypt, he must see the wonders that Yhwh puts in his hands.[71]

v. 23. "My son." The Septuagint has τὸν λαόν μου, "my people." This translation is to be considered a secondary one, inasmuch as it displays a harmonization with Exod 5:1; the MT, instead, is a *lectio difficilior*.[72]

3.1 *The Decision of Moses*

3.1.1 The Request to Jethro and his Blessing (v. 18)

The episode begins abruptly, Moses reacting immediately to Yhwh's last words, "Moses went and returned." The two *wayyiqtol* form a hendiadys

---

[64] The participle of the verb הרג introduces an action that is about to take place. JM § 116*p*; 121*e*; see A.E. GOROSPE, *Narrative and Identity*, 97, n. 28.

[65] See A. GIANTO, "Mood and Modality," 183-198.

[66] See GK § 110*c*; JM § 114*d*. Regarding the particle נא, see also JM § 105*c*; Exod 5:3; 2 Sam 15:7; Ruth 2:2; see A.E. GOROSPE, *Narrative and Identity*, 96, n. 13; Rashi Exod, 29: Moses "had sworn to him not to leave without his permission (כי אם ברשותו)."

[67] See GK § 150*i*; JM § 161*f*.

[68] Cassuto, 53; the author quotes Gen 45:3.

[69] See A.E. GOROSPE, *Narrative and Identity*, 97, n. 16. In the Joseph cycle, the same direct interrogative (ה + עוד + חיה) can be understood more clearly if interpreted literally: Gen 43:7, 27; see also 1 Kings 20:32. See also Childs, 90; Houtman, I, 420.

[70] See e.g., Houtman, I, 420; G.I. Davies, I, 353.

[71] A.E. GOROSPE, *Narrative and Identity*, 97, n. 19; see JM § 166*l* e GK § 164*g*.

[72] Propp, I, 189.

indicating some urgency regarding the main action,[73] that of "returning." With his decision "to go" (הלך), Moses obeys the word of Yhwh (לך: 3:10, 16; 4:12) and, after much resistance, acts without hesitation.

The second verb, שוב, is followed by the preposition אל and refers to a movement towards an individual and not towards a place:[74] after the encounter with God, Moses immediately decides to go to meet a particular person, Jethro. The relevance of Reuel/Jethro for Moses can be desumed thanks to several elements. Firstly, he is one of the few characters in the first chapters of Exodus whose name is explicitly mentioned.[75] The narrator, then, describes Reuel/Jethro as a priest of Midian and as a father of seven daughters[76] (2:18). Moreover, the narrator presents him as a very solicitous character since, as soon as he learns of Moses' compassionate intervention on behalf of his daughters (2:17), he invites him to share bread in his house. Finally, the words pronounced by Jethro in this occasion (2:20) characterise him as a very hospitable and open person.[77] Moses accepts his invitation[78] and takes one of his daughters as his wife. Hence, after years of stable

---

[73] The verb הלך, especially when it is found in the imperative, can be similar to an auxiliary verb (JM § 177e-f). Regarding the combination of the imperative of הלך and שוב, see 1 Kings 19:15.20; 2 Kings 1:6; see also Deut 20:5-8; 1 Sam 9:5. The same couple of verbs reveal in 2 Kings 19:36 the haste of Sennacherib's escape (see Josh 22:9).

[74] See L. INVERNIZZI, *"Perché mi hai inviato?"*, 292-293 especially footnote 3.

[75] The proper name is an important element in the characterisation of main and secondary characters; A. NEPI, *Dal fondale alla ribalta*, 23-24; A. REINHARTZ, *"Why Ask My Name?"*

[76] The first distinctive trait of Jethro's family is the great religiosity: before knowing the name of Moses' future father-in-law (2:18), the reader knows that he is a priest (2:16); moreover, the name Reuel can mean both "God is friend" and "friend of God." Jethro's family is certainly blessed: the reference to the seven daughters may have a symbolic meaning (see 1 Sam 2:5); in addition, the etymological meaning of the second name of Moses' father-in-law, Jethro, may derive from the Hebrew verb יתר "to give prosperity." See Nepi, I, 73; Priotto, 73-74.

[77] Reuel asks two questions to the daughters ("Where is he? Why have you left that man?") and adresses them with an imperative ("Call him that he may eat bread!"); his direct speech manifests his great concern for Moses: the benefactor must be helped! Regarding Jethro / Reuel, see A. LEVEEN, "Inside Out," 395-417; A. SAN MARTÍN JARA, "Moisés, Jetró," 265-288; D. LUCIANI, "Jéthro," 177-186.

[78] In v. 21 (ויאל משה לשבת את האיש) the verb יאל means literally, "to accept an invitation" (Judg 17:11). The *TgN* translates "to begin" (שרה), the Vulgate, instead, *iuravit ergo Moses*, probably assuming that MT uses the verb אלה, "to swear" (see also Simmacus and *ExR* I, 33). This last translation is noteworthy, because it presupposes a bond between Reuel and Moses; in fact, the oath was a formula that could also be used to seal the covenant between two parties (Gen 21:23-24; 26:28-31). These translations are coherent with the scenario of Exod 4:18 in which Moses, in order to be able to leave, feels the need to ask for permission (see Rashi Exod, 29; *bNed* 65a).

residence in Midian,[79] it is not surprising that first of all Moses wants to take leave of Jethro. In fact, for many years he has been the only father figure, so to speak, in Moses' life (see Exod 18).

"Let me go, please, to return." The syntax reveals that this sentence is a deferential request.[80] Nevertheless, the threefold use of the cohortative mode (ואראה; ואשובה; אלכה) shows that Moses does not simply ask for permission, but profoundly manifests his great desire to leave.[81] Scholars note the evident repetition of the two roots הלך and שוב, used first by the narrator and then by Moses.[82] This literary strategy reveals that there is an absolute correspondence between Moses' actions and his desire. The verb שוב is applied again to designate a movement towards people; in this case, Moses wants to reach out to the Israelites, whom he calls his "brothers."[83] The last time he "went out to his brothers" (Exod 2:11) he had not yet been sent by God, now the circumstances have deeply changed (3:10). Moses' "broken" identity (Hebrew, adopted by an Egyptian woman, married to a Midianite[84]) is being restored. Hence, in accordance with a canon very common to universal literature, Moses undertakes a journey which prefigures the route that Israel will take to return to the land.[85] Moses can begin this journey only because Yhwh has called him (3:1–4:17) and although he has made the decision to move (v. 18), a new divine intervention is still required before he can leave for good (v. 19).

Why does Moses not reveal to Jethro everything he has heard from God? Does he prefer to hide the real motive of his journey from Jethro?[86] It is

---

[79] According to Exod 7:7, Moses is eighty years old when he speaks with Pharaoh. Acts 7:23, 30, 36 describes Moses' life as a period of time divided into three parts of 40 years each: the first at Pharaoh's court, the second in Midian at Jethro's and the third after his call.

[80] A. GOROSPE, Narrative and Identity, 119-120. Before leaving, the man who lives with his wife in his father-in-law's house must ask for permission (Gen 31:1, 17-18, 26-30).

[81] Regarding the cohortative, see WO § 34.5.1a.

[82] Cassuto, 53; Houtman, I, 419.

[83] See Propp, I, 215; Priotto, 109; Fischer – Markl, 71.

[84] J. NOHRNBERG, Like Unto Moses, 135; D. LUCIANI, "Çippora (Ex 4,24-26)," 176-178; Luciani thinks that Moses' objections in Exod 3:11 and 4:1, 10 are clearly signs of his hesitation before the divine call, but they can also be an indication of his lack of self-confidence, due to the difficulty of his personal condition.

[85] See E. DI ROCCO, Raccontare il ritorno. The work is focused on the literary motif of the hero's return, common in biblical and extra-biblical literature (especially the Iliad and the Odyssey). Ibid., 7, "At the origins of literature and of the history of mankind there is the journey, therefore, an itinerary that begins as a wandering and soon ... turns into nostos, animated by the desire to return." The return of Moses to Egypt, "provides a model that will be replicated and extended, with variations, in the Hebrews' journey to Canaan" (Ibid., 79).

[86] Ibn Ezra, 29; see Propp, I, 215; Nepi, I, 124. Propp suggests a literary contact between Exod 4:18 and the literary genre of the "Sojourner's Tale," in which the guest deceives the

preferable to consider it as the caution of someone who has not yet fully grasped the enormity of the experience that he has lived.[87]

"Go in peace." The words of Jethro (לך לשלום) are astonishing. In fact, the verb הלך echoes again, and this repetition does not seem to be accidental: Moses' movement ("he went") and his request ("let me go!") are accepted. Moreover, Jethro's first word – "Go!" – fits in perfectly with Yhwh's (לך, 3:10, 16; 4:12). Furthermore, Jethro's reaction differs profoundly from Laban's answer to Jacob's request (Gen 31): regardless of the loss to his family, the father-in-law immediately grants Moses permission to depart.[88]

Hence, the expression לך לשלום does not seem to be a mere formality.[89] Biblical "peace" refers both to the harmony between two parties and to the condition of well-being of the person who possesses it.[90] The father-in-law's words, thus, can have a double meaning: certainly, Jethro wishes Moses' purpose to be fulfilled, but in doing so he uses a term that manifests his bond with his son-in-law. The exhortation "go in peace!", in fact, can be employed to manifest communion between two people (see 1 Sam 20:42[91]). Moreover, the noun שלום in Exodus is especially reserved to Jethro[92] (see Exod 18:7, 23), making of Moses' father-in-law a true "peacemaker." Finally, this farewell can be associated with a blessing,[93] for a blessing is often addressed by a person in a superior position[94] and the roots שלם, "to be completed, to keep peace" and ברך, "to bless," are sometimes connected (2 Sam 8:10; Ps 29:11). These considerations reveal a slight similarity between Jethro's gesture and a father's blessing.[95] Despite the departure, Moses' father-in-law greets him,

---

host in order to be able to free himself. This interpretation ignores the characterisation of Jethro/Reuel in 2:16-22 and especially in ch. 18.

[87] See Houtman, I, 421: Moses knows that he must return to Egypt but he has not yet reached the point where he can speak openly about his mission. See also Childs, 102.

[88] See Hamilton, 78.

[89] Priotto, 109-110; Fischer – Markl, 71. Against this view: Houtman, I, 420.

[90] In regards to the term שלום, see F.J. STENDBACH, "שלום," 13-46. The noun primarily indicates well-being, as can be noted when it is connected with טוב: Jer 33:9; Lam 3:17; see Ps 122:7. When שלום is opposed to the terminology of war, it clearly expresses the agreement between two parties: see 1 Kings 20:18; Mic 3:5; Zech 9:10; see also 1 Kings 5:26 where the peace between two rulers evolves into a ברית, a covenant.

[91] See D.T. TSUMURA, *The First Book of Samuel*, 524.

[92] Priotto, 110.

[93] See Fischer – Markl, 71, who refer to Jacob, 101, "A more logical reason for seeking Jethro's consent was the Torah's desire for Moses to begin this difficult mission with the *farewell blessing of a father-in-law*."

[94] In 1 Sam 1:17 the priest encourages Hannah by telling her to go in peace; see also 1 Sam 20:42; 2 Kings 5:19.

[95] F. FICCO, *"Mio figlio sei tu,"* 134-137.

guaranteeing the continuity of their relationship (they will meet again in Exod 18), and, by allowing the son-in-law to go his own way, he encourages the freedom and dignity of Moses without binding him to himself.[96]

3.1.2 The New Intervention of Yhwh (v. 19)

The great "tension" between v. 18 and v. 19 has already been noted:[97] if Moses has decided to set off, why does God send him again?[98] The most astonishing element, manifested by the Hebrew text of v. 19, is precisely that God speaks to Moses "in Midian," when the journey has not yet begun. What in fact emerges here is that the narrator is concerned to show that the real cause of Moses' departure is God,[99] and also intends to highlight discreetly Moses' hesitation. The Septuagint's insertion at the end of v. 18,[100] is a sign that the Greek translator intended to reduce the discrepancy between v. 18 and v. 19; repeating what has already been said in Exod 2:23, the reader is given a plausible reason for Moses' delay: he fears Pharaoh. The second part of the sentence uttered by Yhwh confirms this hypothesis, "for they are dead, all the men that sought your life."[101]

The word of encouragement is introduced by two imperatives which constitute a directive illocutionary act[102] and form a counterpoint with v. 18: לך שב, "Go, return!" The narrator uses the literary technique of repetition[103] to show that God's word is in harmony with what Moses has already decided to do (v. 18: אלכה נא ואשובה, "let me go, please, to return"); the choice of words is, however, significant because it is the first time that Yhwh uses the verb שוב, qualifying Moses' journey as a "return."[104]

We could say that the narrator does not provide the reader with all the information about the story, to stimulate his curiosity[105] and to make him ask,

---

[96] F. FICCO, "La relazione padre-figlio nel Pentateuco," 19-21.

[97] See ch. II, § 1.1.1.

[98] See Ibn Ezra, 29, who interprets ויאמר as a pluperfect, "and God had said." In this way, v. 19 would present a word that God addressed to Moses before he made the decision to leave.

[99] Priotto, 110; Nepi, I, 124.

[100] See ch. II, § 3.1.

[101] See Childs, 102; see also, Houtman, I, 422; Fischer – Markl, 72.

[102] Regarding the *Speech Act Theory*, see E.M. OBARA, "Le azioni linguistiche," 83-117; L. PÉREZ-HERNÁNDEZ, *Speech Acts in English*, 18-23.35-68.

[103] Regarding the technique of repetition in narrative texts, see M. STERNBERG, *The Poetics*, 365-440; A. WÉNIN, "Joseph interprète des rêves en prison (Genèse 40)," 259-273; P.L. DANOVE, "A Method for Analyzing," 55-84.

[104] See above, § 3.1.1.

[105] Curiosity derives from a lack of information, it is oriented to the past and does not generate any tension towards the future of the story; see M. STERNBERG, *The Poetics*, 283.

"why didn't Moses leave?" The reader in 4:19 does not have access to Moses' thought[106] and, thus, in the process of reading, has to fill in a gap.[107] He only knows that Moses has been characterised in v. 18 as a very determined person; for this reason, the sudden discovery that he is still in Midian creates a friction with the reader's perception of what has happened: although Moses was animated by a strong intent, he was unable to move.

In conclusion, Yhwh's words reveal the actual reason for this delay: Moses did not leave because he felt in danger, Pharaoh "sought his life"[108] (2:15). The construction נפש + בקש expresses a "perversion," since a verb normally used for a positive desire (בקש, "to seek") is employed for the firm intent to kill.[109] Therefore, since Moses is paralysed by his terror of death, he needs the authoritative word of Yhwh to depart.[110] God invigorates his desire and gives him the strength to carry it out.

3.1.3 Moses Sets Off Taking the Rod of God in his Hands (v. 20)

"And Moses took his wife and his children." Verse 20 does not contain any direct discourse and presents a long sequence of actions of which Moses is the subject. This dynamism, together with the repetition of the verb לקח, exalts the new impulse received through God's commission. Hence, the *tempo* of the narrative is accelerated[111] and the emphasis falls on the gestures that the narrator chooses to relate. Moses "takes" his wife and children. The stay in Midian has been "fruitful" and Moses has had another son besides Gershom;[112] the name of the second son, Eliezer, will be revealed to the reader in 18:4. The second son is probably a very young child, or even a newborn; in fact, if they were both grown up, it would be hard to imagine that only one mount would suffice for the mother and two children.[113]

---

[106] Regarding the reading positions, see M. STERNBERG, *The Poetics*, 163; J.-P. SONNET, "L'analisi narrativa," 54-55.

[107] See M. STERNBERG, *The Poetics*, 186.

[108] G.I. Davies, I, 357.

[109] When the object of the verb בקש is negative, the "harm" of a person (רעה, 1 Sam 24:10; Ps 71:13, 24), or the "life" (נפש, 1 Sam 20:1; 22:23; 25:29; 1 Kings 19:10; Jer 11:21) the meaning of the expression is deeply negative.

[110] B. COSTACURTA, *La vita minacciata*, 257: fear "can be overcome by the intervention of another person who presents himself as capable of helping and saving." (My translation)

[111] Regarding the difference between "narrative time" and "narration time" see J.L. SKA, *"I nostri padri"*, 23-24, who quotes S. CHATMAN, *Story and Discourse*.

[112] Some scholars prefer to read בנו instead of בניו precisely because according to Exod 2:22 and 4:25 Moses has only one son; Noth, 19; Propp, I, 216. Cassuto suggests to read בנה, a form that can be both singular and plural (*Exodus*, 54). See Houtman, I, 427-428.

[113] Nahmanides, 29.

"And made them ride upon a donkey." In an essential narrative like this, why insist on this detail? It could be a sign of Moses' new status,[114] or a common literary convention to narrate the journey of a family (see Gen 42:27). Possibly, the scene simply expresses Moses' concern for his wife and children,[115] since the donkey was considered a valuable possession (Gen 32:6; see Exod 20:17; 21:33; Deut 22:3); a humble beast of burden employed for work (Gen 42:26) and not for war, it was nevertheless a prestigious mount (Judg 5:10; 2 Sam 16:2). Thus, the journey of Moses' return to Egypt begins with a tone both humble and noble.[116]

"He returned .... And Moses took the rod of God." A new repetition of the verb שוב emphasizes again Moses' obedience. Nevertheless, scholars detect an imprecision in the logical order of events: the narrator should have first recalled that Moses took the staff and only then should he have referred to his return.[117] This alteration is coherent, because it can be an expedient[118] to emphasise the value of the "staff,"[119] which contained divine qualities.[120] This special instrument, in fact, has the ability to dominate the snake (4:9), a mythical animal and symbol of death,[121] an underground deity worshipped in

---

[114] See Propp, I, 216. The expression החמר has been considered as indefinite, even if the noun is preceded by the article (see GK § 126*q-r*; JM § 137*m*; Houtman, I, 428); it is more likely that the use of the article could be the sign that the narrator refers to a specific donkey.

[115] Jacob, 103; Fischer – Markl, 71.

[116] Rabbinic exegesis has interpreted החמר in a messianic sense: the use of the article indicates that Moses' donkey is a special donkey, the same animal that belonged to Abraham and that the Messiah will ride; Rashi Exod, p. 37, "it was the same donkey that Abraham saddled for the binding of Isaac (Gen 22:3), and that the King Messiah will one day appear on, 'humble, riding on a donkey' (Zech 9:9)." See also *PRE*, XXXI; Utzschneider – Oswald, 138, who points out that Christian art has interpreted Exod 4:20 as a prefiguration for the flight of the Family of Nazareth (Mat 2:13-15; Giotto, *Flight into Egypt*, Scrovegni Chapel).

[117] See Houtman, I, 428; Rashi, p. 37. The inverted order of the elements leads several authors to consider that Exod 4:20b is part of a different layer compared to 4:20a. See e.g. J. JEON, *The Call of Moses*, 135-136.152.

[118] M. STERNBERG, "Telling in Time (1)," 901-948; ID., "Time and Space," 81-145.

[119] Propp, I, 216. The author recalls that אלהים can be used as a superlative; see Ezek 1:1; 40:2; see also JM § 141*n* and 1 Sam 14:15.

[120] The Septuagint has: τὴν ῥάβδον τὴν παρὰ τοῦ θεοῦ. *TgO* and *TgN* say: the rod "with which miracles had been done before Yhwh." *TgPsJ* adds: the rod "that he had taken from the garden of his father-in-law, and it was from the sapphire of the throne of glory. Its weight was forty seahs, and on it was clearly inscribed the great and precious name, and with it, miracles had been done before Yhwh." See *MekhY*, 252-253. See also Sarna, 23; *ExR* 27,9.

[121] In the Epic of *Gilgamesh*, it is a snake that takes the plant of life away from the hero who seeks immortality. See *Gilgamesh*, XI, 303-306 (p. 723), "Gilgamesh found a pool

Egypt[122] and regarded as a powerful image for royalty.[123] The delay in mentioning the "staff," however, could be a sign of a new hesitation in Moses, who performs the most urgent action last.[124]

"In his hand." The position of בידו in the sentence is also uncommon:[125] the order is normally לקח + ב + יד + object (Gen 22:6; Exod 34:4; Deut 1:25; 1 Kings 14:3; see Gen 43:12), in this case בידו is placed at the end of the sentence, putting the hand of Moses in a special place, considering the syntax. In doing so, the narrator begins to create the overlap that, through the synergy of the hands of God, Moses and Aaron, will constitute the Exodus narrative as a celebration of Yhwh's choice to save the people with the collaboration of men.[126]

3.2 *The First-born Son and Pharaoh's Heart*

Moses has just left and Yhwh, as in 4:19, intervenes again with unexpected direct speech (4:21-23), which has been evaluated by critical exegesis as redundant in comparison with 3:16-22.[127] In this way, the narrator shows the reader that God does not merely want Moses to react, but accompanies him on the journey with His word.[128] Yhwh acts in

---

whose water was cool / he went down into it to bathe in the water. / A snake smelled the fragrance of the plant / it came up and took the plant away."

[122] In the Cairo Museum there is a statue of a healer discovered in Atribi, covered with inscriptions which celebrate the merits of Gedhor the healer, a magician. See E. JELINKOVA REYMOND, *Les inscriptions* and E. BRESCIANI, *Testi religiosi dell'Antico Egitto*, 339, "Ged-Hor-the-Healer ... for I have done good to all the inhabitants of Atribi and to anyone who passed by on the road, to free them from the venom of male and female snakes and all kinds of reptiles. I did the same for all the inhabitants of the domain of the necropolis, to keep the dead alive and to rid them of every biting snake." (My translation)

[123] See J.D. CURRID, *Ancient Egypt*, 93-95, especially the text quoted at p. 95, found in the temple of Horus at Edfu, where the king is depicted as a shepherd leading a flock with a snake-shaped rod.

[124] Exod 4:20 is similar to Gen 22:3, because the logical order of actions is not as expected: Abraham saddles the donkey, takes the servants, Isaac, and only then cuts the wood for the burnt offering. See J. JOOSTEN, "Une lecture du texte hébreu," 4.

[125] In 1 Sam 17:40, the word order is similar to the sequence found in Exod 4:20.

[126] See L. INVERNIZZI, *"Perché mi hai inviato?"*, 314-315, n. 60. The author notes the "interplay between different hands" in the book of Exodus, since God's statement that He will deliver the people from the "hand" of Egypt (3:8). The hand of Yhwh (3:20; 6:8; 7:4-5; 9:3, 15) interacts with that of Moses (7:15, 17; 9:22, 35; 10:12, 21-22) and of Aaron (7:19; 8:1-2, 13). This cooperation is reproduced in the narrative of the crossing of the Sea: the hand of Yhwh (14:31; 15:17), his right hand (15:6[*bis*], 12) acts together with the hand of Moses (14:16, 21, 26-27). See also Dohmen, I, 174,

[127] See above, ch. II, § 1.1.2.

[128] Priotto, 111; Nepi, I, 125; Houtman, I, 428-429.

advance, by revealing to Moses what he will do and by instructing him on what he must say; hence, God places Moses in a higher position than the other characters, and the narrator achieves the same effect for the reader, who is forewarned of what will happen.[129]

In vv. 22-23, a speech that Moses must communicate to Pharaoh is quoted. The literary phenomenon is called a "preproductive quotation,"[130] and it consists in a representation of the words that a character is going to say in the future. This occurrence falls within the broader framework of speech quotations. The quotation is *pre*productive because it refers to the future, and the quoter is free to reproduce the source or not; moreover, the source is subjected to the quotation, since the quoter could also decide to alter the phrase that he quotes.

Thus, by giving Moses a discourse to quote, Yhwh engages the freedom of his agent, generating *suspense* in the reader:[131] will Moses reproduce the words or betray them? Moreover, since Yhwh is clearly in a superior position, by offering the words to Moses, God imposes upon him a point of view. Thus, God trains Moses by teaching him what to say,[132] and by revealing him significant aspects of Yhwh's nature.

3.2.1 The Divine Wonders and Pharaoh's Heart (v. 21)

"While you go and return into Egypt." Once again, Yhwh uses הלך and שוב in sequence. The repetition is a conscious one and it allows the reader to

---

[129] See M. STERNBERG, *The Poetics*, 163-167 regarding reading positions; Exod 4:21-23 is a special case: the reader is in a higher position than the characters, but this effect is achieved through a speech of God that is addressed to Moses.

[130] See M. STERNBERG, "Proteus in Quotation-Land," 107-156; L. INVERNIZZI, "*Perché mi hai inviato?*, 109-134. In general, the quotations of a discourse are characterised by a number of "universals," namely recurrent and essential elements: (1) *mimetic relation*: a quotation imitates and may not precisely reproduce a word already said or to be said (which Sternberg calls "source"); (2) *frame*: the quoter's discourse incorporates the quotation in a new context; this shift may change the meaning of the source; (3) *communicative subordination*: the new context makes the source in some way subordinate to the quotation, which may use it to achieve different effects; (4) *perspectival montage*: the point of view of the person quoted is embedded in a new perspective, that of the quoter.

[131] M. STERNBERG, *The Poetics*, 264: the essential feature of any narrative is suspense, i.e. the "incomplete knowledge about a conflict … looming in the future."

[132] M. STERNBERG, "Proteus in Quotation-Land," 139, "the quoter is free to envisage and express the future speech from his own viewpoint and create the future speaker in his own image." Ibn Ezra explicitly interprets the words of God addressed to Moses as a teaching: "*And he said.* [God told Moses] this in Midian. And behold he taught him (הודיעו) that *Hashem* would harden his heart."

discern the narrator's intention: he wants to show that, after sending Moses (לך, 3:10, 16; 4:12), God again describes his journey as a "return" (4:19).

"See all the wonders which I have put in your hand, so that you may perform them before Pharaoh." The use of ראה, "to see," is awkward, and some authors interpret it as a simple interjection.[133] If the verb is taken in its more common meaning, how can Moses see the wonders before they are realised?[134] Should he perform the signs ahead of time?[135] Does it refer to the "signs" that Moses has already witnessed[136] (4:1-9)? If so, why is the Hebrew term מופת used, which has never been found before in Exodus and is explicitly employed for the account in Exod 7–11?[137]

The answer to these questions is found through the examination of the root ראה; this verb can also mean "to consider" and can refer to an experience in which perception and understanding intermingle.[138] This is especially the case when the verb ראה is in the imperative and is not immediately connected with the sense of sight[139] (see Exod 10:10; Deut 4:5; 11:26; 30:15; 1 Sam 25:35). The repetitions in the narrative confirm this assumption, because in vv. 18-23 the root ראה is used again in a similar way: Moses asks permission to return to Egypt "to see," namely to realise, whether his brothers are still alive (4:18); God, however, asks him to "see" the signs and wonders. Yhwh therefore wants to adjust Moses' intention: he should not only worry about the fate of his brothers, but he should concentrate on the signs he has received from God.

The reader would expect the term אות, used in 4:8, 28, 30 for the "signs"[140] which Moses must perform before the people of Israel and the elders. The choice of המפתים, instead, generates an effect of great "surprise" because it is

---

[133] G.I. Davies, I, 353.

[134] Nepi, I, 125.

[135] Propp, I, 216.

[136] Houtman, I, 428-429; Utzschneider – Oswald, 138.

[137] The use of the term מופת prompts some scholars to believe that vv. 21-23 belong to another editorial layer: see e.g., J.C. GERTZ, *Tradition und Redaktion*, 332-333.

[138] H.-F. FUHS, "ראה," 233. Visual perception and the act of understanding are explicitly linked in those texts in which ראה and ידע are used as synonyms, in contexts where ראה could mean other than "to see": Deut 11:2; 1 Sam 24:12; Isa 44:9; see Isa 41:20; Jer 5:1.

[139] Some scholars translate ראה with a verb which belongs to the semantic field of knowledge: see e.g., Propp, I, 216; Houtman, I, 428; Utzschneider – Oswald, 136.

[140] The "sign" (אות) is a material and external reality that refers to a content of a different level or quality; in Exodus, the noun is correlated to external acts that manifest and express the liberation: the Passover rite (13:9) and the consecration of firstborn sons (13:16). See Houtman, I, 363. The sign provides a deeper knowledge, in various circumstances אות is correlated to ידע, "to know": Exod 8:18-19; 10:2; 31:13; Jer 44:29; Ezek 14:8.

the first occurrence of מופת in the Hebrew Bible.[141] This very fact prompts the interpretation of this term as being a reference to the signs that Moses will perform in the future and not as an allusion to the wonders of 4:2-9.[142] Even the text itself specifies that these are "prodigies" that Moses will have to execute "before Pharaoh" and not before the Israelites (as the signs in 4:2-9). The noun מופת is often used, together with אות, as an allusion to the wonders of the Exodus[143] that God accomplishes by altering the natural order of things.[144] Such prodigies can reveal the legitimacy of the prophetic call, and they represent an important element for Moses' mission.[145] Therefore, Moses must "see," namely consider and keep in mind, the wonders he will do in the future, recognising "the purely theological character of his mission,"[146] for the wonders he will perform have been placed in his hand directly by God. Lastly, the repetition of the term יד in v. 20 and v. 21 establishes a connection between the "divine rod" and the wonders, and confirms the interplay between the hand of God and that of Moses (v. 20).

"I will, instead, harden his heart." In v. 21 the narrator presents the first recurrence of a complex motif.[147] Scholars have especially noted the apparent inconsistency between the occasions in which it is God who "hardens" Pharaoh's heart — mentioned in a direct speech by God (חזק *piel*, 4:21; 14:4, 17; קשה *hiphil*, 7:3; כבד *hiphil*, 10:1) or by the narrator (חזק *piel*, 9,12; 10,20, 27; 11,10; 14,8) — and those in which Pharaoh is the agent of this hardening,

---

[141] Regarding surprise, see M. STERNBERG, *The Poetics*, 309.

[142] Priotto, 110.

[143] The hendiadys "signs and wonders" appears in the narrative of the plagues (Exod 7:3) and in other texts: Deut 4:34; 6:22; 7:19; 26:8; 29:2; 34:11; Jer 32:20-21; Pss 78:43; 105:27; 135:9; Neh 9:10.

[144] The terms אות and מופת can qualify the "miracle," i.e. the event of divine origin that changes the natural order of things. With regard to מופת, see 2 Chron 32:24; as to the term אות, see 2 Kings 20:9.

[145] See J.-L. SKA, "La sortie d'Égypte (Ex 7–14)," 195-198. The prophet is accredited and confirmed by the performance of prodigious signs, and Moses is considered the greatest prophet of all (Deut 34:10-12). Some scholars recall the ambiguous nature of these prodigies, which can also be performed by false prophets, such as the magicians of Egypt (Exod 7:11-12; Deut 13:1-2); P. BOVATI – P. BASTA, *"Ci ha parlato,"* 87-88-118-120. The terms אות and מופת in Isa 8:18 and 20:3 allude to some "prophetic choices" made by Isaiah (having children and going barefoot and naked for three years). See Ezek 12:6, 11; 24:24, 27; Zech 3,8.

[146] Priotto, 110.

[147] See R. WILSON, "The Hardening of Pharaoh's Heart," 18-36; J.-L. SKA, *Le passage*, 47-60; P. GILBERT, "Human Free Will and Divine Determinism," 76-87; D.G. COX, "The Hardening of Pharaoh's Heart," 292-311; M. MCAFFEE, "The Heart of Pharaoh," 331-354. J. GROSSMAN, "The Structural Paradigm," 588-610.

both when the subject of the verb is Pharaoh's "heart" (חזק *qal*, 7:13, 22; 8:15; 9:35; כבד *qal*, 9:7, see 7,14 in which the adjective *kābēd* is used) and when it is the king of Egypt himself (כבד *hiphil*, 8:11, 28; 9:34). The historical-critical exegesis considers 4:21 to be part of a late layer[148] and the difference in vocabulary as a signal demonstrating the presence of different sources.[149]

This theme appears especially in the narrative of the plagues,[150] and it is expressed with verbs that have different nuances of meaning.[151] The root חזק *piel*, used in 4:21, has often been translated in a positive way, "to strengthen."[152] For some scholars, in fact, the verb belongs to the semantic field of *holy war* (see Josh 11:20) and would serve to indicate the strengthening of the antagonist: God gives strength to Pharaoh so that His victory over the king of Egypt may be celebrated even more. This explanation, however, does not seem to be the most adequate for 4:21.[153] With convincing arguments, in fact, Ska proposes to interpret חזק piel as an allusion to the *prophetic judgment*. As a matter of fact, the conclusion of the narrative arc of Exod 1–15 reveals that the hardening of Pharaoh's heart is a preparation for the manifestation of God's glory (14:8, 17-18). This literary motif, therefore, is intended to show how God's revelation takes place for those who refuse to acknowledge him;[154] Yhwh causes obstinacy, because His word and the plagues manifest and accelerate a process already active in Pharaoh's heart,

---

[148] See Propp, I, 196; he believes that Exod 4:21b is to be ascribed to the final editor of the Torah. Jeon, quoting other authors, thinks that 4:21-23 is part of the post-P redaction of the Pentateuch; see J. JEON, *The Call of Moses*, 149-150 (with bibliography).

[149] See J. JEON, *The Call of Moses*, 180.219. He considers חזק as a "priestly" (P) term, while the root כבד is a sign of the non-P layer (post-P).

[150] The references to God's hardening of Pharaoh's heart, apart from two initial direct speeches by Yhwh (Exod 4:21; 7:3), are concentrated in the last plagues and in the account of the Red Sea crossing (9:12; 10:20, 27; 11:10; 14:4, 8, 17). See especially, J. GROSSMAN, "The Structural Paradigm," 588-610.

[151] M. MCAFFEE, "The Heart of Pharaoh," 331-354.

[152] So R. WILSON, "The Hardening of Pharaoh's Heart," 18-36. Other authors come to similar conclusions: see P. GILBERT, "Human Free Will and Divine Determinism," 76-87; M. MCAFFEE, "The Heart of Pharaoh," 331-354.

[153] See J.-L. SKA, *Le passage*, 47-53. Regarding the Pᵍ vocabulary, it must be acknowledged that חזק *hiphil* means "to strenghten" in special circumstances: when used in parallel with אמץ *piel* (Deut 3:28; Ps 27:14); when related to the invitation "Do not be afraid!" (Deut 31:6; Josh 10:25) or connected to the assistance formula "I will be with you" (Deut 31:23; Josh 1:9). In addition, Josh 11:20 alone is not sufficient to interpret לב + חזק in a positive way; moreover, even the syntagma לב + אמץ can have a negative meaning (Deut 15:7; 2 Chron 36:13). Finally, there is a biblical recurrence in which חזק in adjectival form followed by לב indicates an attitude of rebellion (Ezek 2:4; see also 3:7).

[154] J.-L. SKA, *Le passage*, 57-60; ID., *Le livre*, 69-77.

which makes him incapable of listening to God.[155] Consequently, divine action and Pharaoh's stubbornness can coexist.[156]

God "hardens" Pharaoh's heart		Pharaoh "hardens" his heart	
Direct speech	Narrator	Narrator Subj.: the heart	Narrator Subj.: Pharaoh
4:21 (חזק *piel*) 7:3 (קשה *hif.*)		7:13,22 (חזק *qal*) 8:15 (חזק *qal*) 9:7 (כבד *qal*) 9:35 (חזק *qal*)	8:11,28 (כבד *hif.*)
10:1 (כבד *hif.*) 14:4,17(חזק *piel*)	9:12 (חזק *piel*) 10:20, 27 (חזק *piel*) 11:10 (חזק *piel*) 14:8 (חזק *piel*)		9:34 (כבד *hif.*)

The first two passages in which the motif of the "hardened heart" appears are 4:21 and 7:3; in those texts, it is Yhwh who, through direct speech, states that He will harden Pharaoh's heart. These passages report a private revelation of God to Moses, intended to prepare the divine messenger in advance. The first recurrence in which the narrator reports Pharaoh's obstinacy, however, is found in 7:13; this negative attitude is attributed there to the king of Egypt and not to God. Only in 9:12 the narrator reveals to the reader that it is God who has hardened Pharaoh's heart.[157] Thus, the distribution of the different reoccurrences maintains a certain balance between two apparently opposing elements: Pharaoh is responsible of his decisions, but the whole process is governed by God.

From a *theological* point of view, therefore, it must be acknowledged that the last editors of the Pentateuch preferred to maintain in the final text a tension between the passages in which it is God who hardens Pharaoh's heart and those in which the obstinacy is the effect of Pharaoh's choice. Without exaggerating speculative deductions (such as the opposition between human freedom and divine omnipotence), it can be admitted that the king of Egypt is

---

[155] See J.-L. SKA, *Le passage*, 59. God, compared to Pharaoh, "l'oblige à aller jusqu'au bout dans sa logique de refus, à jouer son va-tout dans sa contestation d'un pouvoir supérieur au sien. Israël, de son côte, saura par la suite qui est ce Dieu qui l'a pris à son service." See also, Priotto, 111.

[156] The connection between the hardening of the heart and the inability to listen is attested in Exod 7:13, 22; 8:15; 9:12; see also 11:1, 9; Ezek 2:4-5; 3:5-7.

[157] See P. GILBERT, "Human Free Will and Divine Determinism," 81, "It is important to note that the narrator attributes the responsibility of 'hardening' Pharaoh's heart to Yahweh only after the sixth plague (Exod 9:12)."

not a "flat" character,[158] merely functional to the plot.[159] He is a true antagonist,[160] a complex "hero," whose multiple and sometimes contradictory traits make him a "round" character.[161] Thus, even if the biblical narrative remains very sober with respect to the psychological dimension of characters,[162] it cannot be said that Pharaoh is portrayed as a person lacking in deliberation. These opposing indications, therefore, manage to hold together, thanks to the medium of storytelling, Pharaoh's "negative" freedom[163] and the divine omnipotence which makes the obstinacy of the king of Egypt part of His plan of salvation;[164] in this regard, the emphatic use in v. 21 of the pronoun אֲנִי, "I," before the verb חזק, is noteworthy. Yhwh prepares Moses by revealing the secret that will enliven the story which follows. The drama exposed in the narrative of the plagues and in the liberation will be governed by the presence of a "dual causality,"[165] Pharaoh's free will and Yhwh's sovereign authority.

---

[158] See P. GILBERT, "Human Free Will and Divine Determinism," 78. A character is "flat" when he has only one quality and is static, while he is "round" when his behaviour evolves in the story. See E.M. FORSTER, *Aspects of the Novel*.

[159] Commenting Exodus, Origen shows that the same contrast between Pharaoh's freedom and divine omnipotence is found in the Pauline texts: in Rom 2:5 the responsibility for sin is attributed to the human heart ("because of your hard and impenitent heart you are storing up wrath for yourself"), while in 9:18 God's work is exalted, almost independently of human freedom ("He 'hardens' whomever He wills"); ORIGENES, *HEx*, IV, 2 (Sources Chrétiennes 321,120-122).

[160] The antagonist is a co-primary character. In the case of Exod 1–15, Pharaoh is a "structural" character, i.e. he is present throughout the narrative and plays a crucial role in the plot. See A. NEPI, *Dal fondale alla ribalta*, 46-47.

[161] The king of Egypt has a significant presence in the scene. His characterisation is marked by a stubborn resistance to Yhwh and a pride capable of challenging God himself (see 5:2). Moreover, in the account of the plagues, the narrator reports many words that show the contradictions of his soul: he promises to let Israel go (8:4, 21, 24; 10:24), he asks Moses and Aaron to pray to Yhwh (8:4, 24) and acknowledges his own sin (9:27; 10:16), but until the very end he insists on not allowing Israel to leave (14:5).

[162] Ancient literature tended to be prudent in exploring the psychology of characters; see S. CHATMAN, *Story and Discourse*, 113, quoted in J.L. SKA, *"I nostri padri"*, 131.

[163] Nepi, I, 127.

[164] Augustine believes that the "hardening" of Pharaoh's heart is a fruit of his personal decision which God directs nonetheless to a higher good. AUGUSTINUS, *Questiones in Heptateuchum*, II, 18 (CCSL 33,76), "Utitur ergo Deus bene cordibus malis, ad id quod vult ostendere bonis, vel quod facturus est bonis. ... Ut ergo tale cor haberet Pharao, quod patientia Dei non moveretur ad pietatem, sed potius ad impietatem, vitii proprii fuit: quod vero ea facta sunt, quibus cor suo vitio tam malignum resisteret iussionibus Dei ... dispensationis fuit divinae."

[165] Regarding the "dual causality," see Y. AMIT, "The Dual Causality Principle," 385-400; ID., "Dual Causality – An Additional Aspect," 105-121.

3.2.2 The First-born Son of God (vv. 22-23)

The divine direct speech evolves into a quotation that anticipates what Moses will have to say to Pharaoh. The "messenger formula" used in v. 22 — "thus says Yhwh" — creates a connection with the prophetic style.[166] God trains Moses by anticipating words that he has to report to Pharaoh and shows His envoy that his mission will consist in quoting another source: Moses will be responsible of his words, but he will primarily have to make reference to the Word received from God.[167]

Moses' education begins with the revelation of Israel's special status,[168] "And you will say to Pharaoh: 'Thus says Yhwh: Israel is My son, My first-born.'"[169] The first-born was allocated with a "double portion" of the inheritance[170] and this privilege constituted him as a special heir with respect to the other sons. This condition was especially defined by the relationship with the father, because it was the *pater familias* who recognised the right of the heir (see the verb בכר, "to give the right of the firstborn," in Deut 21:16[171]).

Therefore, Yhwh acknowledges in Israel the condition of a first-born son and simultaneously presents Himself as the origin of a relationship that the Israelites receive as a gift without having any active role (as what happens to every son). This revelation establishes that the bond with God is not merely incidental, but defines Israel's identity so radically that it cannot be erased.[172] Moreover, this revelation highlights the gratuitousness of divine

---

[166] The prophet "judges history and directs it with the vertical force of His word" acting as a "man of the Word ... a spokesman for God." (My translation); P. BOVATI, "Alla ricerca del profeta (1)," 25. The so-called "messenger's formula," "thus saith the Lord," expresses precisely the event of the divine call and the inspiration of the prophet. In the Pentateuch, this formula appears only in Exodus and is used by Moses in the dialogue with Pharaoh: 4:22; 5:1; 7:17.26; 8:16; 9:1.13; 10:3; 11:4; the only exception is 32:27. L. INVERNIZZI, *"Perché mi hai inviato?"*, 147-148, n. 39.

[167] See L. INVERNIZZI, *"Perché mi hai inviato?"*, 354-356. Exod 4:22-23 is a preproductive quotation (see above, § 3.2) of the words that Moses will have to address to Pharaoh; the passage is similar to 3:16-17, since these verses present another preproductive quotation. In both cases the quotation that Moses has to say includes words said by Yhwh ("you shall say ..., 'Thus says Yhwh'"), i.e. a reproductive quotation (M. STERNBERG, "Proteus," 125-127).

[168] For Rashi Exod, 30, the specification "first-born" emphasises Israel's importance, "It is an expression of greatness (לשון גדלה), as in, 'I will appoint him firstborn' (Ps 89:28)."

[169] See Priotto, 111.

[170] F. FICCO, *"Mio figlio sei tu,"* 57-60.127-128.

[171] B. WELLS, "The Hated Wife," 131-146. The law of Deut 21:15-17 limits the arbitrariness of the father in giving the right of the first-born.

[172] F. FICCO, *"Mio figlio sei tu,"* 85-86.

love and the depth of His compassion.[173] Since the vocabulary of fatherhood and sonship is often used in Scripture referring to the covenant,[174] by using this terminology, Yhwh manifests in advance what will be fulfilled on Sinai (Exod 19–24), namely the covenant.[175] Yhwh binds the people to Himself through a structure of reciprocal rights and duties, and elevates the Israelites to the dignity of His free ally. Precisely by virtue of this bond, Yhwh has the duty to redeem His enslaved son.[176]

The next verse (v. 23) creates a complication in terms of time. The most natural sequence[177] would be, "²² Israel is My Son, my firstborn ²³ I am going to slain your son, your firstborn." Instead, Moses has to remind Pharaoh to let the people go so that they can serve God. The use of עבד, "to serve," makes the petition particularly strong, since the verb is crucial for the Exodus narrative,[178] since serving God means to be associated with a different ruler than Pharaoh.[179] The quotation is elaborated rhetorically:

Thus <u>says</u> Yhwh:	²² כה <u>אמר</u> יהוה
MY SON, my firstborn is Israel,	בני בכרי ישראל
I <u>said</u> to you:	²³<u>ואמר</u> אליך
"**Let** MY SON **go**, that he may serve Me,"	**שלח את־בני** ויעבדני
but you have refused to **let him go**.	ותמאן **לשלחו**
Behold, I am going to slay,	הנה אנכי הרג
YOUR SON, your firstborn	את־בנך בכרך

The first noteworthy aspect is the repetition of אמר: Moses must quote God's words ("thus says Yhwh") which, in turn, refer to what Yhwh said to Pharaoh

---

[173] Houtman, I, 430; Nepi, I, 128; Priotto, 111.

[174] See P. NISKANEN, "Yhwh as Father," 397-407; D.R. TASKER, *Ancient Near Eastern*, 176-178; S.W. HAHN, *Kinship by Covenant*, 31-32. 2 Kings 16:7 is normally quoted as support for the association between the semantic field of covenant and that of family.

[175] The bond between Yhwh and the people is already mentioned in Exod 3:7, where Israel is explicitly defined as "*my* people." A vast bibliography exists on the subject of the covenant; regarding the current state of research, see e.g. E.W. NICHOLSON, *God and His People*; S.W. HAHN, *Kinship by Covenant*, 1-22. See also, R.J. BAUTCH – G.N. KNOPPERS, *Covenant in the Persian Period*.

[176] See Propp, I, 217, who quotes Gen 14:12-16, which tells the story of Abraham's intervention on behalf of a relative (Lot, his brother's son) who had become a slave.

[177] L. INVERNIZZI, "*Perché mi hai inviato?*", 161-162.354-356.

[178] The wordplay between "slavery" and "service," based on the use of the same Hebrew root עבד to describe both forced labour in Egypt (1:13-14; 5:18; 6:5) and service to Yhwh (3:12; 7:16, 26; 8:16; 9:1; 10:3, 8, 11, 24; 12:31), is exploited by Auzou's commentary, entitled precisely *De la servitude au service*.

[179] The root עבד can indicate the covenant relationship between a ruler and his vassal (see 2 Kings 16:7). P. KALLUVEETTIL, *Declaration*, 66-79.

earlier ("I said to you"). In this way an opposition is created between the "messenger," Moses, who faithfully transmits the divine word, and Pharaoh who rejects it (ותמאן). The choice of the verb מאן, framed by the repetition of שלח, is quite significant because in the Exodus it is almost always used for the decision to prevent Israel's departure[180] and, in 10:3 is connected with Pharaoh's obstinate pride:[181] "how long will you refuse to humble yourself before me (מאנת לענת מפני)?" In these verses Yhwh anticipates the final dialogue with Pharaoh in ch. 11,[182] evoking the story of Exod 7–11 in which, with a progressive manifestation through signs of increasing intensity, God will reveal Pharaoh's persistence and punish his guilt.[183] Through the repetition of the terms בן and בכור, God demonstrates that He acts according to the *contrapasso*, מידה כנגד מידה, "measure for measure."[184]

At the end of the analysis conducted so far, one question still remains: why is there a further preproductive quotation in 4:22-23, after 3:18?[185]

Exod 3:18	Exod 4:22-23
"Yhwh, God of the Hebrews, has met with us and now, let us go, please, for a three days' journey into the desert, so that we may sacrifice to Yhwh, our God."	Thus says Yhwh, "My son, my firstborn is Israel. I **said** to you, "Let My son go, that he may serve Me," but you have refused to let him go. Behold, I am going to slay, your son, your firstborn"

In 3:18 Yhwh tells Moses that he will be accompanied by the elders, whereas according to 4:22-23 he has to go alone. In addition, the two quotations refer to different circumstances: while 3:18 recreates the situation of the first meeting with Pharaoh, in which Moses' language is more moderate,[186]

---

[180] The verb מאן indicates Pharaoh's refusal of the request to let the people go (מאן + ל + שלח: 7,14, 27; 9,2; 10:4; see Jer 50:33); in Exod 16:28, instead, מאן refers to the rebellious behaviour of the people (see Jer 3:3).

[181] The verb ענה *nifal* denotes the condition of the suffering servant: נגש והוא נענה, "oppressed and he was afflicted" (Isa 53:7); see K. BALTZER, *Deutero-Isaiah*, 414. Another possibility would be to translate, "oppressed and he let himself be humiliated," interpreting the *nifal* as a reflexive verb; see J. GOLDINGAY – D. PAYNE, *Isaiah 40–55*, II, 309, who propose precisely to compare Isa 53:7, with Exod 10:3 and Ps 119:107.

[182] The terms used in 4:23 reoccur several times in Exod 5–11 in the same contexts: שלח (5:1-2; 6:1, 11; 7:2, 14, 16, 26-27; 8:4, 16-17; etc.); עבד (7:16, 26; 8:16; 9:1, 13; 10:3, 7-8, 11). PSam introduces the passage of 4:22-23 again after 11:4, to make the correspondence between these moments even more obvious.

[183] See P. BOVATI, *Parole di libertà*, 73.

[184] Cassuto, 57-58; Sarna, 24; E.L. GREENSTEIN, "The Firstborn Plague," 556-557.

[185] See L. INVERNIZZI, *"Perché mi hai inviato?"*, 147-162.172-181.

[186] The expression נלכה־נא can be translated by emphasising that the cohortative has a precative deontic mode; see A. GIANTO, "Mood and Modality," 183-198. In this case, the

4:22-23 alludes to a moment in which the negotiations with the king of Egypt have already failed. The tone of the second quotation, therefore, is much more assertive than the first and this increased firmness does not seem to fit in at this point in the narrative. As previously shown, if one of the reasons for Moses' hesitation was precisely the fear of death (4:19), Yhwh reassures him by guaranteeing that He knows Pharaoh's reactions in advance. In conclusion, God constitutes Moses as a prophet and reveals to him that Israel has a son's status — a secret that he will keep[187] until the moments before his death.[188] Finally, he reassures him by announcing in advance His dominion over history.

4. The Night of Moses and Zipporah's Intervention (Exod 4:24-26)

Aristotle stated that the best plot combines resolution and revelation.[189] This very mixture is decisive for Exod 1–15, in which the narrative of liberation is accompanied by a progressive revelation of Yhwh's identity[190] (see 14:30-31). The same phenomenon is present in 4:18-31,[191] especially in vv. 18-23: the narrative could end with Moses' decision to leave (v. 20), but instead the narrator introduces divine words (vv. 21-23) addressed to him, which contain the extraordinary revelation of Israel's sonship. As with any plot of revelation, the reader is drawn into Moses' experience and learns with him the meaning of Yhwh's words.[192]

---

manifestation of desire (GK § 108*b*; JM § 114*f*) becomes a deferential appeal: the use of the polite particle נא is a signal that the tone is polite; M. GREENBERG, *Understanding Exodus*, 123.

[187] Moses will never actually say to Pharaoh, "Thus says Yhwh: My son, my firstborn is Israel;" this revelation will remain a private knowledge. Furthermore, the comparison between Exod 4:23 and 10:3 shows that Moses and Aaron, quoting verbatim 4:23, change the term בן "son," with עם "people": 4,23 שלח את־בני ויעבדני – 10,3 שלח עמי ויעבדני.

[188] Moses makes explicit reference to divine fatherhood in Deut 1:31; 8:3, 5; 32:6.

[189] See ARISTOTLE, *Poetics*, § 11, 1452a 33-37. The plot is the logical structure of events that sustains the narrated story and leads it to its resolution, whether this takes place through a *peripeteia* (περιπέτεια, a resolution through the reversal of the circumstances), or through a *recognition* (ἀναγνώρισις, a transition from ignorance to knowledge). In the paragraph quoted, Aristotle argues that the best form of recognition is that which is achieved through the mere telling of the story, i.e. when the action is revealing in itself. In this regard a contemporary scholar states, "if the best kind of recognition 'comes from the story itself' in the chain of events, it must reveal both to the characters involved and to the audience the reason for those events, the causes" (My translation; P. BOITANI, *Riconoscere è un dio*, 52).

[190] J.-P. SONNET, "L'analisi narrativa," 57.

[191] M. STERNBERG, *The Poetics*, 177. The plot of revelation is ubiquitous in biblical narrative. If the events narrated in the history of salvation can be defined as "merits," "merit itself consists less in innate virtue than in the capacity for acquiring *and* retaining knowledge of God's ways."

[192] On the subject of Moses' journey interpreted as a path of progressive acquisition of new knowledge, see E. DI ROCCO, *Raccontare il ritorno*, 78.

Nevertheless, after these verses, the story develops through a very obscure episode, which shifts from the plot of revelation to a test that surprises the traveler (Moses).[193] Moses' journey is interrupted by an incident that triggers the "universals" of narrative — suspense, curiosity and surprise — and shakes the reader.[194] The nocturnal aggression fits in very well with the literary model of *nostos*, mentioned above, since the return to the homeland can be accompanied by a strong experience that causes a transformation in the hero and matures him.[195]

The passage in 4:24-26 is enigmatic and almost impenetrable.[196] The connection of these verses with the context is extremely problematic[197] and their function in the narrative is veiled. The short episode, moreover, has all the essential elements of a true "rite of passage,"[198] because it takes place in a "liminal" area, an unspecified territory which become similar to a border that he has to cross. In addition, the test will cause a reunion between Moses and the Israelites. The close reading of the text will reveal more details, but will be prefaced with a paragraph devoted to the textual problems, and another dedicated to the current state of research.

4.1 *Translation and Textual Problems*[199]

²⁴ And so it happened, on the way, at a night encampment, that Yhwh encountered him and sought to kill him. ²⁵ Then Zipporah took a flint and cut

---

[193] Trial is an important element in the literary archetype of the hero's return journey: E. Di Rocco, *Raccontare il ritorno*, 20, "research and the pursuit of knowledge would not have the significance they normally have … if they did not also exploit Trial myth, the value of which derives from its connection with man's ethical essence." (My translation)

[194] D. Luciani, "Çippora (Ex 4,24-26)," 179, n. 86-87. Regarding *suspense*, curiosity and surprise, see R. Baroni, *La tension narrative*. Exod 4:24-26 would be similar to a "cold shower." See also Th. Römer, "De l'archaïque au subversif," 1-12.

[195] Concerning the *nostos*, see ch. II, § 3.1.1. Ulysses, like Moses, experiences a test on the border between life and death before returning to his homeland: he descends into Hades and receives a revelation about the final destiny that awaits every man (Homer, *Odyssey*, XI, 152-207). See E. Di Rocco, *Raccontare il ritorno*, 18, "Learning the ultimate truth of death is a fundamental step in the journey." (My translation)

[196] Flavius Josephus and Philo do not even mention this episode. The argument *ex silentio* desumes that for these interpreters the passage must have been obscure. Childs, 95.

[197] See J. Morgenstern, "The 'Bloody Husband,'" 43; According to the author, Exod 4:24-26 has no connection with the preceding and subsequent context. For a contrary view, see B.P. Robinson, "Zipporah to the Rescue," 450-453.

[198] A. Gorospe, *Narrative and Identity*, 163. The three phases of the rite in Exod 4:18-31 are as follows: separation (vv. 18-20), transition (vv. 24-26), incorporation (vv. 27-31); A. Van Gennep, *The Rites of Passage*, 11.

[199] G. Vermes, "Baptism and Jewish Exegesis," 307-319.

off her son's foreskin and touched his feet with it and said, "For you are a bridegroom of blood to me!" [26] And he withdrew from him. Then, in that time, she said, "bridegroom of blood," because of the circumcision.

v. 24. "On the way." The Peshitta specifies the subject of the verb, "Moses," before this indication, in order to immediately clarify one of the obscure features of the episode (see below).

"At a night encampment." The term מלון stems from the root לין and designates the place to stay overnight.[200] Noteworthy is the play on words with the root מול, "to circumcise."[201]

"Yhwh." The Septuagint, *TgO*, *TgN* have "the angel of Yhwh," presumably to safeguard divine transcendence.[202]

v. 25. "Touched his feet." The Septuagint translates προσέπεσεν πρὸς τοὺς πόδας, "fell at his feet." The Greek translation, then, interprets the suffix of רגליו as referring to the angel of Yhwh and imagines the woman prostrating herself before God in an attitude of supplication.[203]

"For you are a bridegroom of blood to me." The initial כי has a causal meaning, because of the context, since the direct speech of Zipporah serves as an explanation to the strange gesture.[204] The Targumim present explanatory paraphrases.[205] The Septuagint interprets as follows: ἔστη τὸ αἷμα τῆς περιτομῆς τοῦ παιδίου, "the blood from my baby's circumcision has stopped,"[206] namely the rite was accomplished in the best possible way and the blood stopped flowing. The Septuagint repeats the same sentence in v. 26, insisting on the success of the rite and on the obtained atonement.[207]

---

[200] Houtman, I, 433; the term מלון is also used for an animal shelter (Gen 42:27; 43:21) or for the encampment of an entire army (Josh 4:3, 8; Isa 10:29). The Septuagint translates the term with the Greek τὸ καταλύμα, "lodging place," the Vulgata has *diversorium*.

[201] W.H. PROPP, "That Bloody Bridegroom," 495, n. 2.

[202] Childs, 96; L.H. FINK, "The Incident at the Lodging House," 236-241; Priotto, 84. Simmachus and Theodotion keep κύριος, while Aquila has θεός. See G. VERMES, "Baptism and Jewish Exegesis," 310; A. LE BOULLUEC – P. SANDEVOIR, *L'Exode*, 103.

[203] Priotto, 84; A. LE BOULLUEC – P. SANDEVOIR, *L'Exode*, 103. *TgO* simplifies v. 25 translating "approached him," while the Peshitta has "he grabbed his feet."

[204] W.H. PROPP, "That Bloody Bridegroom," 496, n. 8.

[205] *TgO*, "With this blood of circumcision, a bridegroom has been given to us"; *TgN*, "The bridegroom sought to circumcise, but his father-in-law did not permit him. Now may the blood of his circumcision atone for this groom" (see *TgPsJ*).

[206] The difference between MT and LXX in v. 25 is already noted by Origen (*Contra Celsum*, 5, 48). Aquila and Theodotion try to translate the expression "bridegroom of blood" literally and render: νυμφίος αἱμάτων. G. VERMES, "Baptism and Jewish Exegesis," 307-319.

[207] A. LE BOULLUEC – P. SANDEVOIR, *L'Exode*, 104.

v. 26. "Then she said." The particle אז may indicate a logical consequence of what has been said above[208] ("this happened... and then [after]...") or may refer to a moment in the past[209] ("then, in that time"). In this context, the second nuance seems most appropriate, since in v. 26 many terms from v. 25 are repeated: the commentary in v. 26 refers to the sentence spoken in the previous verse and tries to clarify it.[210]

4.2 *An Obscure Text*[211]

The main difficulty for the interpretation of Exod 4:24-26 is the presence of personal pronouns whose reference is not specified.[212] The subject is Yhwh, but the other agents involved are unclear:[213]

- Yhwh attacks Moses, Zipporah circumcises Gershom and touches Moses' feet.[214] The circumcised son could be Eliezer[215] and not Gershom.
- Yhwh attacks Moses, Zipporah circumcises Gershom and lays his foreskin at the feet of Yhwh[216] (the angel of Yhwh). Again, some authors believe that Eliezer was the son circumcised during the night and not Gershom.[217]
- Yhwh attacks one of Moses' sons (Gershom or Eliezer), Zipporah circumcises one of them and touches the feet of Moses with the foreskin.[218]
- Yhwh attacks one of Moses' sons (Gershom or Eliezer), Zipporah circumcises one of them and lays his foreskin at the feet of Yhwh.[219]

---

[208] The term אז is translated διότι by the Septuagint. Regarding the term אז used to draw a logical conclusion, see WO § 39.3.4f.

[209] Childs, 99. The particle אז is used for a past event linked to another in a sequence of time (Josh 20:6; Judg 8:3); the Vulgate translation, *postquam*, is noteworthy in this regard. The term can also denote an indefinite time in the past (Gen 4:26), as well as a reference to a precise moment, without this indicating a temporal succession (Josh 14:11).

[210] Childs, 99-100.

[211] In Luciani's study there is an acute treatment of the current state of research on Exod 4:24-26: D. LUCIANI, "Çippora (Ex 4,24-26)," 162-165. See also J.T. WILLIS, *Yahweh and Moses in Conflict*.

[212] D. LUCIANI, "Çippora (Ex 4,24-26)," 161. In the three verses there are various ambiguous pronouns, "Yhwh encountered *him*, and sought to kill *him*" (v. 24); "touched *his* feet and said: For *you* are for me a bridegroom of blood" (v. 25); "and he withdrew from *him*" (v. 26).

[213] See, e.g., *bNed* 32a.

[214] B.P. ROBINSON, "Zipporah to the Rescue," 447-461; P.T. REIS, "The Bridegroom of Blood," 324-331.

[215] Ibn Ezra, 31.

[216] *TgN*, "brought it to the feet of the destroyer (וקרבת לרגלוי דמחבלה)." *TgPsJ*, "brought the foreskin of the circumcision to the feet of the angel destroyer (לריגלוי דמלאך חבלא)."

[217] R.B. ALLEN, "The 'Bloody Bridegroom,'" 259-269.

[218] L.H. FINK, "The Incident at the Lodging House," 236-241; J.M. COHEN, "*Hatan damim*," 120-126.

- Yhwh attacks one of Moses' sons (Gershom or Eliezer), Zipporah circumcises one of them and touches the feet of the circumcised son with the foreskin.[220]

Didier Luciani, after careful examination, points out that the majority continue to see Moses as the character attacked by Yhwh; as to the identity of the circumcised son, there is a slight preference for Gershom; the feet can be either those of Moses or those of Yhwh (present at the scene through His angel). Reading the text, these hypotheses will be considered and evaluated.

4.3 *The Nocturnal Attack (v. 24)*

The short episode of Exod 4:24-26 unfolds with a series of very quick actions. Yhwh's attack is sudden and fast, it strikes Moses in an instant. The narrator begins by identifying the precise space and time. Everything happens "on the way." As previously shown, the "hero" is on his way back to Egypt; the nocturnal encounter therefore interrupts the journey and jeopardises its success.[221] The attack of a person on the road is a quite well attested motif in the Old Testament. In effect, the condition of the traveler is that of a vulnerable person exposed to danger.[222] Hence, hospitality towards the pilgrim was sacred, while aggression was forbidden.[223]

"At a night encampment." This further clarification presents a clearer picture of both Moses' and his family's situation, and also suggests an essential time indication: the attack happens during the night. This aspect confirms the contact between 4:24-26 and the story of Jacob,[224] particularly the events

---

[219] F. LINDSTRÖM, *God and the Origin of Evil*, 41-55. See also, T.J. LEHANE, "Zipporah and the Passover," 46-50, who thinks that the circumcised son is Gershom.

[220] H. KOSMALA, "The Bloody Husband," 14-28. Zipporah circumcises Gershom, touches his feet and calls him חתן דמים, before Yhwh. See A.J. HOWELL, "The Firstborn Son of Moses," 63-76; see also, *bNed* 32a; see J. MORGENSTERN, "The 'Bloody Husband,'" 35-70; K.A. CHERNEY JR., "The Enigmatic Divine Encounter," 195-203.

[221] D. LUCIANI, "Çippora (Ex 4,24-26)," 170-171.

[222] A. GOROSPE, *Narrative and Identity*, 122-123. Regarding the traveler's exposure to danger, see Gen 28:20-21; 42:38; Exod 18:8; Ezra 8:21-22; see also Exod 23:20.

[223] See M. MALUL, "Some Aspects of Biblical Hospitality," 233-251. While a law explicitly forbidding attacks on travelers is not present in the OT, some norms forbid treacherous attacks (Exod 21:14) or ambushes (Num 35:20-21; Deut 19:11-12); see Hos 6:9; Mic 7:2 and A. GOROSPE, *Narrative and Identity*, 123.

[224] See J.W. ROGERSON – R.W.L. MOBERLY – W. JOHNSTONE, *Genesis and Exodus*, 118. There are several similarities between the story of Jacob and that of Moses, "[1] Both Jacob and Moses do something that is morally ambiguous (deception of father [Gen. 27], killing of Egyptian [Exod. 2.11-15]) and consequently have to flee. [2] Both have a meeting with God at a holy place, which prepares them for what lies ahead (Bethel, Gen. 28:10-22; Horeb, Exod. 3.1-4.17). [3] Both meet their wife-to-be at a well (Gen. 29.1-14; Exod. 2.15b-21). [4] Both are told by God to return to the place from which they are originally fled (Gen. 31.13;

narrated in Gen 32–33.[225] The night is loaded with symbolism, so much so, that the anthropologist Durand called "nocturnal regime" one of the two main "semantic basins" into which symbols common to different cultures can be gathered.[226] In the night the light is darkened,[227] eyesight fades, perception is more difficult and intellectual faculties can be deceived. The attacker, therefore, does not merely strike a vulnerable traveler, but does so at a time of poor visibility and under the cover of darkness.

Finally, v. 24 adds a third element that makes Moses even more defenseless. Since he stays in a place where he can spend the night (במלון), it can be assumed that he is asleep.[228] Sleep is a significant anthropological phenomenon (1 Kings 19:5) that manifests itself both as a suspension of consciousness, necessary for the recovery of energy, and as a refuge into which the person can retreat.[229] Subsequently, it can be a time of exposition to peril (see Ps 57:5) and people can be more easily attacked[230] (Lev 26:6; Job 11:19; Prov 3:24). It is not surprising, therefore, that Scripture associates sleep with death (Jer 51:57; Ps 13:4).

"Yhwh encountered him." The verb פגש normally has a neutral sense and can indicate the simple gathering of people (see 1 Sam 25:20; Jer 41:6); in

---

Exod. 4.18-20). [5] Both have mysterious and threatening night encounters with God (Gen. 32.22-32; Exod. 4.24-26)."

[225] B.P. ROBINSON, "Zipporah to the Rescue," 451-452. The author specifies some contacts between Exod 4:24-26 and Gen 32–33: (a) Both texts are initiatory tales that are decisive for the growth of the protagonist: the hero is wounded, but at the end of the fight he is transformed; (b) God has previously manifested himself to both (Gen 28; Exod 3); (c) The hero is attacked on his way back to his birthplace; (d) The attack takes place during the night (Gen 32:22); (e) The two narratives have a common terminology: נגע (Gen 32:26; Exod 4:25); (f) The brother of the protagonist meets him (Gen 33:4; Exod 4:27-28). See also, R.B. ALLEN, "The 'Bloody Bridegroom,'" 265 and Nepi, I, 132.135-137, who connects Exod 4:24-26 to what happened to the prophet Elijah (touched, נגע, by the angel, 1 Kings 19:5.7), to Abraham (Gen 22) and other biblical figures; he even makes reference to the NT and to Jesus' "agony" in Gethsemane (Luke 22:43).

[226] G. DURAND, Les structures anthropologiques, 217-433.

[227] Regarding the symbol of light, see O. PETTIGIANI, Dio verrà certamente, 119-127.

[228] The verb לין, from which there derives the noun מלון, "night encampment," can be used within the semantic field of sleep: Gen 28:11; Rt 3:13; see W. BEUKEN, "שכב," 1309.

[229] The phenomenon of sleep has an anthropological and a symbolic significance, whereas insomnia is, "vigilance sans refuge d'incoscience, sans possibilité de se retirer dans le sommeil comme dans un domaine privé" (E. LÉVINAS, Le temps et l'autre, 27). In another work Lévinas says the following, "il faut précisément se demander, si impensable comme limite ou négation, le 'néant' n'est pas possibile en tant qu'intervalle et interruption, si la conscience avec son pouvoir de sommeil, de suspension, d'époché, n'est pas le lieu de ce néant-intervalle" (E. LÉVINAS, De l'existence à l'existant, 105).

[230] W. BEUKEN, "שכב," 1309. In falling asleep, man expresses trust with the posture of his body; regarding the association of שכב and בטח, see Ezek 34:25; Hos 2:20; Ps 4:9; Job 11:18.

two occurrences it is employed for the fierce and brutal aggression of a bear (Hos 13:8; Prov 17:12), and thus it cannot be excluded that the root implies a certain hostility (2 Sam 2:13). This nuance is undoubtedly the most appropriate for Exod 4:24.[231]

The reader is puzzled by the choice of the name "Yhwh." The alternation between "Yhwh" and "Elohim" has been analysed in the historical-critical studies on the Pentateuch.[232] Recently, some attention has been directed to the literary and semantic implication of this alternative use of God's proper names in Exodus.[233] Linguistic science has indeed shown that the choice of one name over another can have a semantic value. "Yhwh" expresses the closeness of God and has a great pragmatic force.[234] This use is paradoxical because while in 4:24-26 everything seems dark and the features are blurred, the narrator describes God very clearly. Moses has just recognised Yhwh[235] and the reader already knows that it is the same God who revealed Himself in Exod 3–4, who attacks his messenger.

"And sought to kill him." The use of בקש, "to seek," followed by מות hiphil, "to kill" (2 Sam 20:19; 1 Kings 11:40; Jer 26:21; Ps 37:32), is astonishing, because in the Bible the subject of this action is never Yhwh. Why does He behave in this way? From ancient times, attempts have been made to answer this question by justifying the divine aggression:

+ Moses is attacked because he did not circumcise his son. Rabbinic tradition suggests two mitigating factors to justify his behaviour: his father-in-law explicitly forbade circumcision[236]; Moses chose not to circumcise Eliezer, his

---

[231] Priotto, 112.
[232] B.A. ANDERSON, *An Introduction*, 39-48.
[233] L. INVERNIZZI, *"Perché mi hai inviato?"*, 155-158, which refers to the semantic-cognitive studies; see T.R. WARDLAW, *Conceptualizing Words*; reading Exod 3:1-15 (*Ibid.*, 92-115), he concludes that the use of "Elohim" may be a linguistic sign which indicates that Moses still has an obscure knowledge of Yhwh.
[234] Polak believes that in the Pentateuch, the name "Yhwh" is used to emphasise God's closeness and relationship with men, while "Elohim" refers to God in a more general way, highlighting especially His universal authority and the distance separating Him from men. See F.H. POLAK, "Divine Names," 159-178.
[235] F.H. POLAK, "Divine Names," 171. The author states that while in Exod 3 the alternating use of the divine names "Elohim" and "Yhwh" can be explained as a symptom of an imperfect knowledge of God, Moses' explicit use of the name "Yhwh" in 4:1 shows that a progressive change of perspective has taken place in him.
[236] See *TgN*, "But Zipporah took a flint and cut off the foreskin of her son and brought it to the feet of the destroyer, and she said: 'For the bridegroom sought to circumcise, but his father-in-law did not permit him. Now may the blood of this circumcision atone for this groom'" (see also, *TgPsJ*). See *MekhY* on 18,3, p. 275; the rabbinic text imagines that after Moses has asked for Zipporah's hand, Jethro imposed a condition on the marriage: the first

newborn son, in order not to risk his life and, at the same time, to rapidly obey God's command[237] (4:19, "Go!").

+ Yhwh attacks Moses because of his previous blood crime[238] (2:11-12).

+ Yhwh assaults him because he married a foreign woman[239] or because of his hesitation in accepting the mission.[240]

Many, however, consider these solutions to be exaggeratedly anchored in the logic of retribution and, on the contrary, emphasise the mysterious and tremendous character of every encounter with God.[241] Indeed, the text gives no indication that this trial is a reaction to a previous fault. Instead, the reader easily finds connections with Jacob's story: both characters experience a "dark night" (Gen 32:24-32) in which they "fight with God," in a struggle in which their silhouettes are not perceivable,[242] and that finally leaves the character wounded.[243] The context also helps to clarify an important trait of Moses' experience:

Exod 4:19	Exod 4:24
For they are dead, all the men that **sought** (המבקשים) your life (את־נפשך).	Yhwh encountered him (ויפגשו) and **sought** (ויבקש) to kill him (המיתו).

The poetics of these repetitions[244] are subtle, since the verb בקש appears two times, with two different subjects. God prompts Moses to leave precisely by reassuring him about those who might threaten his life on his return to Egypt. Nonetheless, in v. 24 it is Yhwh himself who becomes a threat.[245] Thus, the echo created between the divine word in 4:19 and the narrator's voice in 4:24 has an astonishing effect: the very thing that Moses most fears befalls him, and it is Yhwh Himself who tries to kill him, not Pharaoh! Yhwh

---

son must belong to the idol, while the others may belong to God. Moses accepted and "It was for this that the angel at first wished to kill Moses."

[237] Some rabbinic texts justify Moses' behaviour interpreting it as a sign of concern for the health of his newborn Eliezer: Rashi Exod, 31; *ExR* V, 8; *bNed* 31b-32a.

[238] W.H. PROPP, "That Bloody Bridegroom," 501-505.

[239] O. KAISER, "Deus absconditus and Deus revelatus," 73-88.

[240] B.P. ROBINSON, "Zipporah to the Rescue," 447-461; See G.W. ASHBY, "The Bloody Bridegroom," 203-205; J.A. WALTERS, "Moses at the Lodging Place," 407-425.

[241] D. LUCIANI, "Çippora (Ex 4,24-26)," 169; A. DE PURY, "Le Dieu qui vient en adversaire," 49-50; TH. RÖMER, "De l'archaïque au subversif," 6-8; B.P. ROBINSON, "Zipporah to the Rescue," 456.

[242] So, R. FORNARA, *La visione*, 336-338.

[243] See R. FORNARA, *La visione*, 340-341.

[244] J.-P. SONNET, "L'analisi narrativa," 76-77.

[245] See above, ch. II, § 3.1.2. See also, Priotto, 112; B.P. ROBINSON, "Zipporah to the Rescue," 447-461; D. LUCIANI, "Çippora (Ex 4,24-26)," 171-172.

brings Moses face to face with a mortal danger, so that, by confronting the fears from which he has fled, he may have an experience of death and life that will generate in him a true transformation.[246]

4.4 *The Intervention of Zipporah (v. 25)*

"Then Zipporah took a flint." The reflexes of Moses' wife are quick and the few details narrated contribute to create an impression of extreme speed. Moses' life is once again saved by a woman,[247] as had already happened at the moment of his birth, when the harmonious action of his mother, his sister and Pharaoh's daughter enable the newborn child to "overcome" the waters of the Nile (2:1-10). The insistence on the name of Moses' wife is significant for the plot, because the introduction of the proper name in a narrative is crucial for the description of a secondary character. Zipporah is a foreign woman like Pharaoh's daughter, her attitude can be considered "subversive" and at the same time extraordinary.[248]

"She cut off her son's foreskin." As shown above[249], the verb כרת, taken from the semantic field of the covenant, is employed only here to indicate circumcision.[250] The contact between the ritual performed by Zipporah and the semantic field of the covenant is noted by some authors;[251] this association will be developed while commenting v. 26.

"She touched his feet with it." The scene is captured from an external perspective. The narrator does not specify to whom the personal pronouns refer and consequently the reader cannot clearly determine the identity of the actors. The only certain element is the person being circumcised, the son. Rabbinical tradition has suggested that the circumcised son was the newborn Eliezer (18:3), because Gershom must have been too old already[252] (2:21-22).

---

[246] Something similar to what had happened to Jacob occurs to Moses. See R. FORNARA, *La visione*, 344-345; the struggle with God (Gen 32:24-32) marks an important stage in the transformation of Jacob: he is defenseless and receives the blessing by God.

[247] A. NEPI, *Dal fondale alla ribalta*, 73.

[248] TH. RÖMER, "De l'archaïque au subversif," 8-10. See also M.M. TALBOT, «Tsipporah», 8; she considers the affective implications of the expression "*her* son" in v. 25, together with the consequences of the circumcision and of the threat of God, who attacks Moses and seeks to kill him.

[249] See ch. I, § 2.

[250] The expressions used for circumcision are, מול + בשר ערלה (Gen 17:11, 14, 23-25; Lev 12:3; see Deut 10:16) or just מול (Gen 17:26-27; 21:4).

[251] TH. RÖMER, "De l'archaïque au subversif," 10; D. LUCIANI, "Çippora (Ex 4,24-26)," 180. For the relationship between circumcision in Gen 17 and the covenant, see D.A. BERNAT, *Sign of the Covenant*; J.J. KRAUSE, "Circumcision and Covenant," 151-165.

[252] See *TgN* e *TgPsJ*.

However, the person who receives the action (the touching of the feet) must be Moses, since Zipporah moves precisely to save him from the nocturnal attack. Moreover, the gesture must concern the feet, because here the term רגל cannot be a euphemism for "genitals" as in other cases;[253] it would be difficult, in fact, to imagine Zipporah being able to touch Moses' genitals during the fight.

Together with rabbinic tradition,[254] some scholars have long recognised the connection between this mysterious act of Zipporah and the Passover narrative: as shown in the first chapter of this monograph,[255] the verb נגע is used for blood manipulation in both 4:25 and 12:22.[256] The rite, therefore, prefigures something that will happen later on, during the night of *Pesach*, when the blood will signal the members of the people, the "first-born son" of God (4:22), and save the lives of the Israelites (12:22). The interplay of references between the two texts helps us to interpret Moses' aggression as a "paschal" experience and to identify the blood as a decisive instrument for salvation.

"For you are a bridegroom of blood to me." The narrator quotes what Zipporah says and gives her direct speech a pragmatic effect: her words fulfill their meaning the very moment they are uttered,[257] as occurs in an illocutionary declarative act.[258] Moses is saved by means of blood and words. The meaning of the sentence, however, remains relatively enigmatic, so much so, that the Septuagint has tried to clarify its sense,[259] *TgO* has created a connection between the blood ritual and Moses' salvation,[260] and the *TgPsJ* and *TgN* have added the reason for the divine aggression.[261]

---

[253] G.I. Davies, I, 354.

[254] See *MekhY*, 24 on Exod 12:6, "Therefore the Holy One, may He be blessed, assigned them two duties, the duty of the paschal sacrifice and the duty of circumcision, which they should perform so as to be worthy of redemption." See also, *ExR* XIX, 7.

[255] See ch. I, § 2.

[256] So B.P. ROBINSON, "Zipporah to the Rescue," 453; W.H. PROPP, "That Bloody Bridegroom," 511; T.J. LEHANE, "Zipporah and the Passover," 46-50; D. LUCIANI, "Çippora (Ex 4,24-26)," 172. See Dohmen, 177.

[257] See A. GOROSPE, *Narrative and Identity*, 145-147.212-213.

[258] Regarding illocutionary declarative acts, see J.R. SEARLE, *Expression and Meaning*, 26.

[259] The Septuagint of 4:25 says as follows, "... and fell at his [of the angel] feet and said, 'The blood of the circumcision of my son is staunched.'" The Greek translation removes the term "bridegroom," perhaps in order to delete an element of complication. According to this version, Zipporah simply acknowledges that the flow of blood has stopped. In v. 26, the Septuagint adds a preposition that creates a causal link between the verses, "And He departed from him, *because* (διότι) she said, 'The blood of the circumcision of my son is staunched.'" G. VERMES, "Baptism and Jewish Exegesis," 310.

[260] *TgO* to Exod 4:25 translates as follows, "She *approached him* [Yhwh's angel] and she said, 'Because of the blood of this circumcision, a bridegroom may be given to us.'" In 4:26

In order to clarify the meaning of the expression "bridegroom of blood," we must firstly introduce some clarifications regarding the terms. The noun חתן is polysemic[262] and designates kinship acquired by marriage[263] rather than by consanguinity.[264] Generally, the term is translated "son-in-law," indicating the married man from the point of view of his father-in-law; by extension, it can also refer to the "bridegroom" or the "betrothed," the *fiancé*. The noun, therefore, alludes to the incorporation of an individual into a family group.[265] The plural דמים is unusual and unexpected, since it often refers to murder;[266] for this reason, some scholars interpret this expression as an allusion to the blood shed by Moses (in 2:12), and explain the ritual performed by Zipporah as an act of atonement.[267] Since there are other examples where the plural is used to indicate bleeding, or an effusion of blood (see Ezek 16:6, 9), its reference to a violent death does not seem necessary. Hence, Zipporah's words indicate that Moses acquired a new legal status precisely because of the blood that is spilled and manipulated.

The clarifications made so far are still insufficient to penetrate the mystery of Zipporah's words which remain enigmatic. Thereby, it is the narrator himself, in v. 26, who clarifies their meaning.

4.5 *The Bridegroom of Blood and Circumcision (v. 26)*

"And he withdrew from him." The verb רפה *qal* means "to release," the object is often the term יד "hand" (see 2 Sam 4:1; Isa 13:7; Jer 6:24) and the meaning of the expression is "to let one's hand drop," as a sign of discouragement. In our particular case, the preposition מן suggests that the object of the verb is implicit ("and [God] loosened [his grip] on him") or

---

this aspect is emphasized even more clearly, "She said, 'Had it not been for the blood of this circumcision, the bridegroom would have been declared guilty and killed.'"

[261] *TgPsJ* to Exod 4:25 says, "She cut off the foreskin of Gershom her son and brought the circumcision of the foreskin to the feet of the Destroying Angel and she said, 'The bridegroom tried to circumcise but his father-in-law prevented him. Now, may this blood of circumcision atone for my bridegroom.'" (see *TgN*) See G. VERMES, "Baptism and Jewish Exegesis," 312-313.

[262] See T.C. MITCHELL, "The Meaning of the Noun *ḥtn*," 93-112; A.J. HOWELL, "The Firstborn Son of Moses," 63-76; D. LUCIANI, "Çippora (Ex 4,24-26)," 165-166.

[263] T.C. MITCHELL, "The Meaning of the Noun *ḥtn*," 99.

[264] E. KUTSCH, "חתן," 295-297.

[265] See H. KOSMALA, "The Bloody Husband," 14-28.

[266] Si veda c. I, § 1.1.

[267] W.H. PROPP, "That Bloody Bridegroom," 495-510, and particularly the footnote 10; דמים could therefore be a plural *abstractionis* (GK § 124d-f; JM § 136g); see *TgO* Exod 4:25; S. FROLOV, "The Hero as Bloody Bridegroom," 520-523.

that the verb should be interpreted as a reflexive form[268] ("he withdrew from him"). The image is that of Moses being released from a hold.

"Then, in that time, she said, 'bridegroom of blood,' because of the circumcision."[269] The explicit intervention of the narrator is intended to clarify the expression "bridegroom of blood,"[270] with a gloss[271] introduced by אז. To help interpret Zipporah's words, the expression חתן דמים is linked to circumcision. How does this clarification explain the previous verse?

Common practice also in neighbouring cultures,[272] circumcision has a symbolic character.[273] For many scholars, this rite was originally a precondition for marriage[274] (see Gen 34:14-15[275]). From a narrative perspective, however, the reader of Exodus understands the circumcision mainly thanks to Gen 17.[276] Abram is ninety-nine years old (17:1), he has already had Ishmael, with Hagar, and thirteen years have passed since his birth (16:16). God appears once again, invites Abram to leave (in 17:1 the verb הלך is used, as in 12:1), and reveals His desire to "establish" a covenant with him (17:2). Abram falls to the ground and does not react; at this point Yhwh renews the promise of an extraordinary descendance (17:4-7), and gives Abram a new name, Abraham, "father of a multitude" (17:5). Abraham's silence (17:9) betrays the astonishment of a person surprised by

---

[268] K.A. CHERNEY JR., "The Enigmatic Divine Encounter," 198.

[269] In the expression למולת the preposition ל is commonly translated "with reference to"; see Childs, 100; GK § 119u. The term מולת, from the root מול, "to circumcise," is translated with a plural of abstraction, "the circumcision" (W.H. PROPP, "That Bloody Bridegroom," 497); Frolov, instead, prefers the plural "the circumcised people" (S. FROLOV, "The Hero as Bloody Bridegroom," 521); finally, Cassuto proposes "circumcisions" by emending למולת with למולים (Cassuto, 61). The most convincing explanation is the first: JM § 90f; Childs, 104; Dozeman, 146; see also, K.A. CHERNEY JR., "The Enigmatic Divine Encounter," 198.

[270] The biblical narrator is omniscient, but discreet, namely he does not often express his point of view; in Genette's language, he only rarely uses "zero focalization" (see G. GENETTE, Figures III, 183-224; J.-L. SKA, "I nostri padri", 107-121).

[271] D. LUCIANI, "Çippora (Ex 4,24-26)," 167.

[272] N. WYATT, "Circumcision and Circumstance," 405-431; A. FAUST, "The Bible, Archeology," 273-290.

[273] A. GOROSPE, Narrative and Identity, 124-125.130-140.

[274] W.H.C. PROPP, "The Origins of Infant Circumcision in Israel," 360.

[275] In Gen 34:14-15 Jacob's sons ask Hamor that Shechem circumcise himself before taking Dinah as his wife. See J. WELLHAUSEN, Prolegomena, 340; J. MORGENSTERN, "The 'Bloody Husband,'" 35-70; A.J. HOWELL, "The Firstborn Son of Moses," 73-74.

[276] Regarding the circumcision in Gen 17, see J. GOLDINGAY, "The Significance of Circumcision," 3-18; D.A. BERNAT, Sign of the Covenant; J.J. KRAUSE, "Circumcision and Covenant," 151-165.

an unexpected circumstance.[277] Yhwh, then, intervenes for a third time and specifies that the promise will have only one condition:[278] "let every male among you be circumcised" (17:10). Circumcision will thus become a "sign" of the covenant (17:11: אות ברית).

Why is circumcision a sign of the covenant?[279] The biblical covenant has a complex and articulated "structure"[280] which expresses at the same time God's free election and the laws that govern the bond with Him. Circumcision is precisely a condition by which the desire to establish a relationship with God is manifested. Whe just eight days old, the male son[281] passively suffers an operation that creates a permanent scar in his flesh.[282] The result is a mark that has a private character and is useful for the memory and the intelligence[283] of the circumcised person[284] (aspects that, at most, are shared with one's wife). In Gen 17, Yhwh promises Abraham that he will be fertile (vv. 2, 4-6) because of the bond that unites him to God, symbolically manifested through the sign of circumcision;

---

[277] A. WÉNIN, *Abramo e l'educazione divina*, 100; Abraham's silent reaction could be dictated by the surprise caused by Yhwh's unexpected words.

[278] Gen 17 is interpreted in two different ways. Some scholars believe that Gen 17 is a text which affirms the classical theology of P, marked by an exaltation of "pure grace." See CH. NIHAN, "The Priestly Covenant," 93. For many others, on the contrary, the obligatory nature of the circumcision of Gen 17 must be emphasised: see particularly D.A. BERNAT, *Sign of the Covenant*, 34-36. For this reason, it has been suggested that vv. 9-14 of Gen 17, which contain the command concerning circumcision, are in fact a late addition (connected to H); J. WÖHRLE, "Abraham amidst the Nations," 26; J. BLENKINSOPP, "The 'Covenant of Circumcision,'" 145-156. Krause believes instead that the final text is consistent and that there are no conditions for assuming that Gen 17:9-14, 23-27 are a late editorial insertion. He shows that Gen 17 certainly contains a "theology of grace" but asserts that there are also elements of normativity. "This promise to be God to his people cannot be realized except in a relationship" (J.J. KRAUSE, "Circumcision and Covenant," 161) and this relationship cannot be given without a free response and some conditions.

[279] D.A. BERNAT, *Sign of the Covenant*, 50-52.

[280] Some authors have noticed that the covenant structure is similar to that found in some Hittite and Neo-Assyrian vassalage treaties (see e.g., D.J. MCCARTHY, *Treaty and Covenant*; E. OTTO, *Das Deuteronomium*, with bibliography): preamble, historical prologue, recalling of the benefits granted by the sovereign that stipulates the pact, basic stipulation, covenant clauses (laws), invocation of witness, blessings and curses.

[281] Rabbinical tradition recalls that circumcision is practised "where is distinguishable a male from a female" (Rashi Gen, 165: שהוא נכר זכר לנקבה; see *bShab* 108a).

[282] A. WÉNIN, *Abramo e l'educazione divina*, 104.

[283] M. FOX, "Sign of the Covenant," 557-596.

[284] See e.g., J. GOLDINGAY, "The Significance of Circumcision," 5.16; D.A. BERNAT, *Sign of the Covenant*, 37, "Circumcision should be understood as a sign for the one who affixes it, the Israelite male, not YHWH."

therefore, circumcision expresses a "lack" through a wound in the flesh[285] (Lev 19:23), but, at the same time, procures a blessing[286] and fertility, through the relationship with Yhwh.[287]

So how does the reference to circumcision illuminate the episode of 4:24-26? Firstly, it must be acknowledged that v. 26 does not entirely remove the difficulties of the text. We can desume that the obscurity of this passage has to be intentional:[288] such an interplay of lights and shadows is intended to enhance the mystery,[289] because when Yhwh goes to meet His faithful, the latter cannot have a full understanding of what is happening, being suspended between confusion and trepidation before the absolute divine transcendence.[290] On the other hand, as previously shown,[291] an intentional play on words associates the rite of circumcision and the darkness of the night, thanks to the assonance between the noun מלות, "circumcision," and the term used for the night encampment (מלון). Darkness, then, is a useful scenario to show that the rite affects, and in some way transforms, the entire family of Moses: the difficulty found by the reader in identifying the various characters in the scene (see above, § 4.2) may actually indicate that everyone is involved in the encounter with Yhwh.

In a positive way, the reference to circumcision manifests the transition that, through the experience of 4:24-26, takes place in the life of Moses: from a condition of a person "hung on the balance" (Hebrew, Egyptian, Midianite), he becomes a Hebrew messenger of God.[292] In the calling of 3:1 – 4:17, Moses was constituted as a messenger by means of the word of

---

[285] A. WÉNIN, *Abramo e l'educazione divina*, 104: the scar is "a loss that affects the very organ that can lead man to believe that he is complete, and brings him back to his own limit." (My translation)

[286] See GenR 46,4, "Is said *'orlāh* for the tree, is said *'orlāh* for man. As the *'orlāh* of the tree (see Lev 19:23) is the place where it bears fruit (מקום שהוא עושת פרות), even the *'orlāh* of man is the place where he bears fruit (מקום שהוא עושת פרות)." Regarding circumcision as a necessary step to becoming fertile, see A. GOROSPE, *Narrative and Identity*, 133.

[287] So A. GOROSPE, *Narrative and Identity*, 133. See also, A. WÉNIN, *Abramo e l'educazione divina*, 104, "To accept circumcision ... means to accept to be wanting, not to have everything nor to be everything .... In this sense, circumcision is in itself a sign of covenant which, in order to be realised, supposes that a want is assumed." (My translation)

[288] D. LUCIANI, "Çippora (Ex 4,24-26)," 173-174.

[289] A. GOROSPE, *Narrative and Identity*, 165-168.194. The episode narrated in 4:24-26 reproduces a transitional, "liminal" experience: the encounter takes place in an unspecified place; the person being attacked remains passive and is not named throughout the scene; God's action is disconcerting. See also, N. STAHL, *Law and Liminality*, 13.

[290] B. COSTACURTA, *La vita minacciata*, 124-145; Exod 19:16-25; 20:18-21; 2 Sam 6:1-22.

[291] See above, § 4.1.

[292] D. LUCIANI, "Çippora (Ex 4,24-26)," 180.

Yhwh; in 4:24-26, instead, Moses is "physically" reached by God and, in a way, "reborn" thanks to a ritual that, on the borderline between life and death, creates a new bond between him and Yhwh.[293]

In conclusion, Genesis informs us that circumcision makes a person part of the people of Israel (Gen 17:10-14). Furthermore, Moses and the reader hold a secret revelation: the Israelites are considered by Yhwh as a "first-born son"[294] (4:22). Thus, both Moses and the reader will recognise that the blood of the circumcision of 4:24-26, while creating a closer bond between Moses and his family with the Israelites, at the same time constitutes them as part of the people that is God's "first-born son." Thus, it establishes a connection between Moses and the Israelites that is similar to that which unites members of the same family. Consequently, the blood renders Yhwh, Moses, and his family "kinsmen" not according to the flesh, but in virtue of the covenant (see above, § 4.4).

5. The Encounter with the Israelites (Exod 4:27-31)

After Moses' mysterious night, the narrator presents, at a faster pace, Aaron's call and the meeting with his brother (v. 27), the exchange between Moses and Aaron (v. 28) and the gathering with the elders and the people (vv. 29-31). These brief episodes conclude the narrative arc begun in 3:1. From this moment onwards, Moses will begin to take his first steps towards accomplishing his mission.

5.1 *Translation and Textual Problems*

²⁷ And Yhwh said to Aaron:

"Go to meet Moses, in the wilderness!"

And he went and encountered him at the mountain of God and kissed him. ²⁸ And Moses told Aaron all the *words* of Yhwh with which He had sent him and all the signs that He had commanded him[295]. ²⁹ And Moses and/with Aaron went and gather all the elders of the Israelites. ³⁰ Aaron *spoke*/uttered all the *words* that Yhwh said/*spoke* to Moses and performed all the signs in the sight of the people. ³¹ And the people believed. And they heard that Yhwh had

---

[293] A. GOROSPE, *Narrative and Identity*, 198-200.

[294] Gorospe sees a connection between 4:22, the revelation to Moses of Israel's privileged status as first-born, and the episode of 4:24-26, but explains it differently. See A. GOROSPE, *Narrative and Identity*, 197, "The threat of death to Moses' family reveals the extent to which God as a father will act for the sake of firstborn Israel."

[295] The Hebrew expression כל־דאתת אשר צוהו can presuppose לעשות, "to do," as an implicit element of the sentence (see Deut 6:24; 24:18). See *TgPsJ* Es 4,28; Houtman, I, 451.

visited the Israelites, and that He had seen their affliction; so they knelt down and prostrated themselves.

v. 29, "And Moses and/with Aaron went." The Peshitta, the Vulgata, and the PSam have a plural verb, וילכו "and they went," which agrees with the following plural *wayyiqtol*, ויאספו "and they gather." MT is to be preferred as a *lectio difficilior*, both because a verb in the singular can have a plural meaning,[296] and because the shift from singular to plural can be intentional (see below, § 5.3).

v. 31, "And the people believed. They heard." Again, the Peshitta and the PSam begin with a plural verb "they believed"; MT is a *lectio difficilior* because the term עם can be both singular and plural.[297] The Septuagint, in this case, has two peculiarities: it smoothes out the difficulties of the verse by translating all the verbs in the singular and renders וישמעו with the Greek ἐχάρη, probably reading וישמחו; the majority of interpreters maintain the MT[298], even if שמע appears in an unusual position in the sentence[299].

5.2 *The Encounter with Aaron (vv. 27-28)*

"And Yhwh said to Aaron." The narrator introduces Aaron more concisely, in a verse that has a theological focus,[300] because Aaron, like Moses, departs only after an encounter with God. The first part of v. 27 has a peculiar position on the temporal axis: since Yhwh has already revealed to Moses that Aaron is on his way (4:14, "behold he is going out to meet you"), v. 27 must be somewhat retrospective. Nevertheless, the use of the *wayyiqtol* form creates the impression of a certain temporal succession between v. 26 and v. 27[301]. Thus, v. 27 can be both a clarification of what has already been narrated (Aaron that decides to move in obedience to the

---

[296] Propp, I, 190; GK § 146*f-h*. Houtman (I, 452) assumes that the *waw* before Aaron's name functions as a *waw concomitantiae* (GK § 154*a*); in such cases, the verb preceding the pair of terms can be either singular or plural (JM § 150*p-q*).

[297] Propp, I, 190; Houtman, I, 453.

[298] See Cassuto, 63; Childs, 91; Dozeman, 148; Priotto, 84; Dohmen, I, 179; Utzschneider – Oswald, 135. A cautious advocate of the Septuagint is Propp (I, 190).

[299] Houtman, I, 453.

[300] Priotto, 114.

[301] Propp, I, 220. The retrospective character of v. 27 has prompted some authors to translate the verb אמר with the pluperfect; a translation with the past is preferred because, as Propp himself notes, if the author had wanted to achieve this effect he would have used a construction like the *w-X-qatal*; see also, Childs, 104. This tension between v. 26 and v. 27 has led to speculate that vv. 27-28 are editorial; see above, § 1.1.3.

divine will) and, respecting the sequence of events, a reference to an encounter with God that reinforces the decision already made by Aaron.

"Go to meet Moses, in the wilderness!" Yhwh's words to Aaron help clarify his character and are of great value, as they are the only direct speech in the episode of vv. 27-31. The imperative of הלך is employed for Moses both in the vocation narrative (3:10, 16; 4:12) and in 4:19. The repetition of the same root in 4:27, therefore, establishes a connection between the call of Moses and that of Aaron,[302] both wandering and embarking on a journey. Yhwh's words surprise the reader, for they do not specify the point of departure nor the place of the meeting; it is only said that Aaron has to go "into the wilderness." This description emphasises the journey all the more: it is not the place from which Aaron moves that is important, but only that he must leave. Moreover, Exodus has prepared the reader to recognise the desert as the "liminal" place where, despite the lack of secure points of reference, God can be encountered[303] (see 3:1); the desert is a "theological space." Aaron must therefore travel without being certain of a destination and relying only on the divine word.

"And he went and encountered him at the mountain of God." The new repetition of the verb הלך is intended to emphasise Aaron's perfect and immediate obedience.[304] Like Moses (הלך in 4:18, 21), Aaron too departs after a word from Yhwh, but, unlike the former, he does so without hesitation. The action is described rapidly, the meeting at the mountain of God, Horeb,[305] is brief yet its geography is of theological importance.[306] The speed of the representation is intended to show the favourable outcome of Aaron's momentum: since he trusted Yhwh and headed towards the wilderness, it is not long before he finds the same mountain that provided the setting for the story of Moses' vocation (3:1).

The usage of the verb פגש in 4:27 referring to the meeting between Moses and Aaron is significant, because the same root is employed in v.

---

[302] See Priotto, 113, "The narrator, by using the two terms 'wilderness' and 'mountain of God,' wants to lead even Aaron to the place of Moses' call."

[303] Dohmen, 178-179. Priotto, 113-114.

[304] Cassuto, 62, "He was told to go, *and forthwith went.*"

[305] See Ibn Ezra, 32, who explicitly presents the connection between the mountain and Horeb (as seen in Exod 3:1); see also the Peshitta.

[306] Houtman, I, 450; he reports that some scholars consider the indication of the mount of God to be an irregularity, because Moses had already left the mountain of God in Exod 4:18. The commentator notes that Flavius Josephus (see *Ant. Jud.*, II, 279-280) and Philo (*De vita Mosis*, I, 85) do not mention the mountain of God; Flavius Josephus, for example, suggests that the meeting took place on the border with Egypt. The reference to Horeb, then is theological rather than geographical.

24.³⁰⁷ While the verb normally has a neutral sense, in v. 24 it indicates the hostile aggression of Yhwh.³⁰⁸ This repetition can only be intentional, because the root is not so common; the narrator wants to discreetly indicate an association between the two events,³⁰⁹ the meeting of Aaron and Moses and the attack of God in the night. Since Aaron's movement has a "theological" quality, because he is sent by God, the root פגש could imply that while previously Yhwh met Moses in a threatening and tremendous way, now He manifests Himself through Aaron's visit.

"And kissed him." In such an essential story, the detail of the kiss could easily have been omitted. Instead, the narrator chose to emphasise the gratuitous outpouring of love³¹⁰ and Aaron's bond with Moses.³¹¹ In 4:14 God had answered to Moses' last objection by making reference to his brother ("is there not Aaron, your brother, the Levite?"). Since the reader is familiar with Genesis and the drama of brotherhood, he is astonished by the divine word; God reacts to Moses' resistance with the gift of a brother and with a singular promise of communion (4:14, "he will see you and rejoice in his heart") and collaboration (4:15, "you will speak to him and put words in his mouth"). This perspective is fulfilled by Aaron's greeting in 4:27.³¹²

"And Moses told Aaron all the words ... and all the signs." In v. 28, the contents of Moses' words to Aaron are omitted, but the act of telling is particularly highlighted. The repetition of כל emphasises the perfection of the communication: Moses holds nothing back and reports everything that has happened to him.³¹³ Moreover, Moses' account is balanced, because it is composed by words and signs in the precise order in which he received them from Yhwh. By virtue of this obedience, the reader can recognise

---

³⁰⁷ See Cassuto, 62: the repetition of פגש is a wordplay which highlights the opposition between the dangerous confrontation with God (Exod 4:24) and the happy conjuncture in which Moses and Aaron find themselves together (4:27). See also, Fischer – Markl, 78.

³⁰⁸ See above, § 4.3.

³⁰⁹ Priotto, 114, note 133.

³¹⁰ Fischer – Markl, 78.

³¹¹ The Hebrew root נשק is used to designate the greeting of close relatives (parents and children, Gen 31:28), especially when they meet after having been separated for a long time (brothers, Gen 33:4; 45:15; also, Exod 18:7). The kiss may be followed by intense emotional manifestations, such as weeping (Gen 33:4; 45:15; 50:1; 1 Sam 20:41) or embracing (Gen 48:10). See Cassuto, 62; Houtman, I, 451.

³¹² The midrash interprets Aaron's gesture as an expression of harmony between brothers in which rivalry is absent; *ExR* V, 10, "*And he kissed him.* Why? Each one rejoiced at the other's greatness (זה שמח בגדלתו של זה זה שמח בגדלתו של זה)." The Septuagint explicits the reciprocity of this kiss: κατεφίλησαν ἀλλήλους.

³¹³ Priotto, 114.

Moses as a true prophet, as the spokesman of God's word[314] (see 3:14-16; 4:22-23). In fact, the choice of words is deliberate: every word in v. 28 has already been used in Exod 3–4,[315] with the only exception of the verb צוה which, before 4:28, is found only in 1:22. Thanks to this new element, Moses' point of view is probably manifested:[316] the messenger interpreted the divine imperatives as a command and his response is an act of obedience, thus acknowledging in advance the divine sovereignty.[317]

5.3 *Moses and Aaron before the Israelites (vv. 29-31)*

"And Moses and/with Aaron went." The passage begins with a syntactical difficulty, for the singular וילך, "and went," does not fit in with the subsequent plural verb ("they gathered"). Even if Hebrew syntax allows this use of the singular, the verb might discreetly suggest that the first fruit of the brothers' meeting is an absolutely concordant action: Moses and Aaron obey the same command (לך, "go," 3:10.16; 4:12.19.27) and move as if they were one (וילך, "and went," 4:29).

"And gather all the elders of the Israelites." Many significant elements of the story are omitted:[318] what happened to Moses' family? Did Zipporah leave with him (this question will only be answered in Exod 18)? What happens during the journey? Where does the meeting take place? The narrator does not want these questions answered,[319] because the purpose of the verse is to show that v. 29 fulfils what was announced in 3:14-17: Moses can finally speak to the Israelites and to the elders.[320] From a geographical point of view, moreover, the lack of a

---

[314] See above, § 3.2.2.

[315] The reference to "words" and "signs" is clearly connected to Exod 3:1 – 4:17: the root דבר is used in 4:10, 12, 14-16, while the noun אות also appears in 3:12; 4:8-9, 17. In addition to this obvious contact, the repetition of the root שלח "to send" is noteworthy: it is employed to indicate Moses' sending in general (3:10, 12-15, 20; 4:13).

[316] Regarding the "point of view", see e.g., J.-L. SKA, *"I nostri padri"*, 107-125; J.-P. SONNET, "À la croisée des mondes," 75-100.

[317] L. INVERNIZZI, *"Perché mi hai inviato?"*, 200-201. The use of צוה in Exod 5:6 characterises Pharaoh as a king who opposes Yhwh (Exod 1:22). In fact, in Gen–Exod the verb is especially reserved for God (see Gen 2:16; 3:11, 17; 6:22).

[318] Childs, 104.

[319] Regarding "gaps" and "blanks" in narrative, see ch. I, § 3.2, note 152. The omissions of elements that have no particular meaning for the plot, allows the reader to better grasp the aspects that should instead attract his attention; M. STERNBERG, *The Poetics*, 236.

[320] Cassuto, 62. Rabbinic tradition accords great value to the elders, who represent the "wings" that enable Israel to fly; this idea explains the choice of gathering the elders first. See *ExR* V, 12, "R. Akiba said: Why is Israel compared to a bird? Just as a bird can only fly with its wings, so Israel can only survive with the help of its elders."

precise location creates a contrast with vv. 27-28, where two places are mentioned (the wilderness and the mountain of God); this is a further confirmation that the concept of space, in these verses, has a "theological" value and is noted only if it is related to Yhwh. Therefore, there is no mention of Egypt in v. 29 nor in the following verses. Instead, what is especially emphasised is the exchange between the two brothers and the people.

"And Aaron spoke all the words that Yhwh spoke to Moses and performed all the signs." The messengers obey Yhwh's instructions, but in v. 30 the narrator exalts the root דבר by repeating it three times.[321]

Exod 4:28	Exod 4:30
And Moses told Aaron all the **words** of Yhwh (כל־דברי יהוה) of Yhwh with which he had sent him and all the *signs* (כל־האתת) that He had commanded him	**Spoke** (וידבר) Aaron all the **words** (את כל־הדברים) that Yhwh **spoke** (דבר) to Moses and performed all the *signs* (האתת) in the sight of the people

While in v. 28 the report of Moses maintains the balance between words and signs (both דבר and אות are used once), in v. 30 the emphasis is on the speech act, rather than on the signs. With this shift of emphasis, the narrator probably wants to show that the heart of Moses' and Aaron's mission will be to report God's words and not just perform spectacular signs. Moreover, the text insists on Aaron's obedience to the divine order of 4:15-16: with a "turn of phrase" (v. 30, "he spoke all the words he had spoken") it is shown that Aaron does not deliver his own speech, but offers his voice to Yhwh and acknowledges Moses' role as the first carrier of this word. The communion of the messengers is full: Moses gives the word to Aaron, Aaron refers back to the words that God addressed to Moses and performs the signs of 4:1-9.[322]

"And the people believed. And they heard." The fears of Moses[323] (4:1) are contradicted: the people believe and listen. The use of שמע is a confirmation of the first impression regarding v. 30: the word has primacy; they believe because they listen[324] and not because they see the signs performed "in their sight." Furthermore, the verbs ויאמן and וישמעו raise two

---

[321] Priotto, 114; Cassuto, 62.

[322] Propp believes that in this verse the reference to Aaron as the author of the signs is unusual. However, in the light of Exod 4:15 ("I will teach you what you must do"), the reader can also think that the signs are part of what is taught by God to his messengers. Propp, I, 221; see Houtman, I, 452-453.

[323] Priotto, 114. In Exod 4:1 the sequence of the verbs "to believe" and "to listen" is identical to the succession found in 4:31.

[324] *ExR* V, 13, "Do not conclude that they believed only after they had seen the signs; no! *They heard that the Lord had visited*: they believed because they heard, not because they saw signs."

questions: why the alternation between singular and plural? Why are they presented in this order? Certainly, the narrator emphasises the promptness and unanimity of the response of faith: the singular is used because the people "believe" almost as if they were a single individual. The logical order of the verbs, instead, is reversed to show that trust precedes all other replies. The second verb, then, focuses on the reaction of the individuals and introduces the contents understood by the people.

"And they heard that Yhwh had visited the Israelites, and that He had seen their affliction." The narrator explains to the reader what has convinced the people; the choice to report only a few elements reveals that, as in every act of communication, the people have retained only some information. What did they understand of Aaron's discourse? By comparing 3:16 and 4:31, they certainly grasped that God wants to "visit" their affliction.

Exod 3:16	Exod 4:31
Go and gather the elders of Israel and say to them, "Yhwh, the God of your fathers has appeared to me, the God of Abraham, of Isaac, and of Jacob saying: 'Indeed, I have visited you (פקד פקדתי אתכם) and what has been done to you in Egypt'"	And the people believed. And they heard that Yhwh had visited (כי פקד יהוה) the Israelites, that He had seen (וכי ראה את ענים אתכם) their affliction.

The verb פקד, when used in a juridical context, falls within the paradigm of inquiry;[325] in 4:31 it refers back to what the Israelites heard from Joseph, "Indeed, God will visit you[326] (פקד יפקד אלהים אתכם)" (Gen 50:25). The people thus recognise that Yhwh is willing to take their plight to heart and intervene to restore justice against their oppressors. This interpretation is consistent with what has been narrated in previous chapters, as the judicial scenario is repeatedly employed to describe God's intervention against Egypt;[327] especially from 2:23 onwards, the oppressed people vent their grief in a cry which is both act of entrustment[328] and an act similar to that of an official complaint[329] (3:7, 9).

---

[325] P. BOVATI, *Re-establishing Justice*, 244-245. See Ps 17:2-3.

[326] Commenting on Exod 3:18 ("they will hear your voice"), Rashi interprets Exod 3:16-18 connecting the text with Gen 50:24-25; see Rashi Exod, 23.

[327] See P. BOVATI, *Parole di libertà*, 68-69. The verb גאל (Exod 6:6; 15:13; see Pss 74:2; 77:16; 78:35; 106:10), and the root פדה, used above all in Deuteronomy (see Deut 7:8; 9:26; 13:6), help to show that liberation is an act of due justice.

[328] See P.D. MILLER, *They Cried*, 44.55.

[329] See P. BOVATI, *Re-establishing Justice*, 314-317. The plaintiff is an aggrieved party who perceives the circumstances he experienced as an injustice and for this reason he

The use of the verb ראה is therefore not surprising, because, as פקד, it can designate an inquiry.[330] This choice, however, is unexpected, because ראה does not appear in 3:16, but only in 2:23 and in 3:7.

Exod 2:24-25	Exod 3:7, 9	Exod 4:31
[24] God heard their moaning, and God remembered his covenant with Abraham, Isaac, Jacob. [25] God saw (וירא אלהים) the Israelites, and God knew	[7] Yhwh said, "Indeed I have seen (ראה ראיתי) the misery of my people in Egypt […]. [9] And I have also seen (וגם ראיתי) the oppression with which Egypt oppresses them"	And the people believed. And they heard that Yhwh had visited (פקד) the Israelites and that He had seen (ראה) their affliction

The comparison of the three passages therefore shows that Moses told the people, through Aaron, an aspect of his personal revelation received in 3:7: Yhwh wants to intervene against Egypt because He has seen the affliction of Israel. The people, therefore, have grasped an essential aspect of God's description, because ראה is the only root that Yhwh repeats twice when He introduces Himself to Moses (3:7, 9). At the same time, v. 31 manifests to the reader, who possesses a fuller knowledge of Yhwh's thoughts and words (manifested in 2:24-25 and 3:7, 9), that the people lack several elements in order to fully understand the mystery of God: many actions are referred to God in 2:24-25 and 3:7,[331] but Israel only captures a part of them.

"So they knelt down and prostrated themselves." The roots קדד and חוה are always correlated[332] and indicate the act of the person kneeling and touching the ground with his face.[333] With this gesture, the individual normally manifests his submission to an authority, acknowledging its status.

---

appeals to the judge. The supplication is similar to formal complaint, because the oppressed person is marked by a state of weakness. Among the terms that designate this action of appeal the root צעק is common (2 Kings 8:3; Job 19:7; see also, Exod 3:7, 9).

[330] P. BOVATI, *Re-establishing Justice*, 244-245. Regarding the root ראה, see Exod 5:21; 1 Sam 24:16; Lam 3:59.

[331] In 2:24-25 and 3:7 Yhwh does not only say that He has "seen"; He turns towards the people with His senses and intelligence: in 2:24-25 God is the subject of four different actions, He hears the complaint, remembers the covenant, sees the Israelites and knows (ידע; ראה; זכר; שמע); in 3:7, instead, He uses only three verbs: ראה; שמע; ידע.

[332] In each of the fifteen recurrences of the verb קדד in the Hebrew Bible the action of kneeling is related to prostration. See Houtman, I, 454, who considers this pair as a hendiadys. Regarding וישתחוו as a *hishtafel* of חוה, "to worship, to prostrate," see JM § 59g.

[333] Houtman, I, 454. In 1 Sam 24:9; 28:14 and 1 Kings 1:31 kneeling is connected with the gesture of the face touching the ground: אפים ארצה.

The same two verbs are also employed to express reverence towards God.[334] The people, therefore, despite their partial knowledge of Yhwh, react to Aaron's words with a burst of faith. It is not explicitly stated in v. 31 that Israel trusts Moses as a messenger and a mediator[335], but Israel certainly recognises, through body language, the divine initiative behind the sending of Moses and Aaron,[336] anticipating what will be repeated in 12:27.[337]

Conclusion

Moses' journey to Egypt has significant shared features with a common motif of universal literature: the hero's return to his homeland. Nevertheless, our close reading has shown some variations on the theme. Egypt is not Moses' homeland and he must return to the people with the aim of bringing them out of the land of slavery. Furthermore, Moses is an exile in a double sense: despite his Hebrew origins, he was brought up in Egypt, far from his homeland, but after the killing of the Egyptian (2:11-14) he is forced to flee from Egypt as well. Finally, the "hero" is able to leave Midian only thanks to the impetus he receives from the divine word (§ 3.1.2), first with the calling of Exod 3–4 and then with the words that overcome his further resistance and provide him with new momentum (4:19).

The journey is not described in detail, and no precise geographical movements are reported; the reader only knows that Moses is sustained by the word of Yhwh (vv. 21-23). The special elaboration of God's words (Yhwh mentions the words that Moses will have to say in the future, § 3.2) shows that Yhwh does not merely reveal notions, but intends to train His messenger by giving him the words to say, revealing the extraordinary identity of Israel, God's first-born son, and thereby anticipating the results of his negotiations with Pharaoh: Yhwh will harden his heart and he will not let the people go. Therefore, Moses is warned from the very beginning: the events that are about to take place do not escape the divine rule upon reality (§ 3.2.2).

---

[334] Prostration was an act of homage that could be performed before a king (1 Sam 24:9; 1 Kings 1:16, 31) as well as in front of a prophet (1 Sam 28:14). In regard to the act of bowing down before God, see Exod 34:8; Num 22:31; Neh 8:6; 1 Chron 29:20; 2 Chron 20:18; 29:30; see also, Nepi, I, 133.

[335] L. INVERNIZZI, *"Perché mi hai inviato?"*, 135-136. The use of the verb אמן in 4:31 leaves open the question of the content of the people's trust. Moreover, v. 31 specifies that the people "heard" that Yhwh had visited the Israelites, and had seen their affliction, but there is no mention of the person of Moses.

[336] Houtman, I, 454, "from the context it is plain that the reverence is toward Yhwh."

[337] Priotto, 114.

The path of Moses is not without dangers![338] In the night, Yhwh attacks him and brings him to the limit between death and life (§ 4.3), foreshadowing what all the people will experience in the night (14:20) when they will think that God wants "to make them die"[339] (14:11). In this trial, the blood shed in the rite of circumcision, performed thanks to Zipporah's initiative, saves Moses from death. In this way, through the ritual sign of the wound inflicted to his son, which establishes a covenant bond with God, the blood makes Moses and his family be reborn as part of the people, "first-born" of Yhwh (4:22; see § 4.4 and § 4.5).

Once Moses has passed through this trial and crossed the "liminal" area, he can meet his brother again. Indeed, in vv. 27-31 the reader learns that even Aaron has been called by God and has immediately set out to meet Moses (§ 5.2). The reunion of Moses and Aaron is extraordinarily joyful. Since the book of Genesis is made up of several stories involving brothers who are divided by jealousy and hatred, the reader is certainly surprised by the great communion that follows the reunion of these brothers. Their mission is accomplished through the congregation of the people and the elders, to whom they announce the good news of deliverance. The story concludes with the immediate welcome of their words by the Israelites (§ 5.3).

The reference to the shed blood, therefore, is at the heart of this travel narrative, mentioned in an enigmatic episode marked by obscurity. Thanks to God's intervention, Moses, a fugitive and a wanderer, is once again directed towards his goal, his people. To reach it, however, he must pass through a trial from which he is saved precisely through blood (§ 4.5). Transformed by the confrontation with Yhwh, Moses can face Pharaoh for the first time (Exod 5:1 – 7:7); despite the failure of his first encounter, when sent to the king of Egypt a second time, he acts as an intermediary between Yhwh and Egypt in the long story of the plagues (Exod 7:8 – 10:29) until the final scourge which put an end to all negotiations. It is then that, once again, blood will mark out the border between life and death on the night of *Pesach*.

---

[338] E. Di Rocco, *Raccontare il ritorno*, 56. The journey of Ulysses is defined in the Odyssey as a "grievous return" (νόστον πολυκηδέ'; Homer, *Odyssey*, IX, 37).

[339] Nepi, I, 135. The author quotes Exod 16:2; 17:3 (see Num 20:4).

CHAPTER III

Blood and the Celebration of Passover (Exod 11:1–12:42)

The mission of Moses is marked by the nocturnal confrontation with Yhwh in which a "circulation" of blood transforms his entire family (4:24-26). After this episode, the term דם appears again in Exod 12, in Yhwh's instructions for the rite of the lamb (12:7, 13) and in Moses' transmission of these very words to the elders (12:22-23). These directions are framed by the announcement of the tenth plague (11:1-10) and its fulfillment (12:29-34). Moreover, the regulations for the paschal celebration (12:1-28) precede 13:17–14:31 and operate as a "liturgical memory" that prepares the way to the crossing of the sea.[1] The Exodus is thus anticipated in the rite that Yhwh entrusts to the people, so that they can preserve the remembrance of the salvific event for the generations to come.[2] This cultic act is fulfilled thanks to three crucial signs: the lamb (12:1-14), the unleavened bread (12:15-20) and the consecration of the firstborn sons (13:1-16).

In addition, while Moses is already prepared for the killing of the first-born sons (see 4:21-23[3]), the institution of Passover is an unexpected surprise[4] that sustains the hope of deliverance and prepare the Israelites so that they will be able to understand what they are about to experience.[5] The celebration thus anticipates the salvation that is to come and establishes a practice that will become crucial for the faith of Israel.[6] What role does blood play in this scenario? In the first chapter, lexical and thematic connections between the two sequences (4:18-31; 12:1-34) have been already highlighted.[7] In the close reading of 11:1–12:42, however, the analysis of these connections will be deepened, to show that before the crossing of the sea, the families of the

---

[1] Priotto, 213, who quotes Fretheim, 179-182.
[2] Fretheim, 182; see L.C. STAHLBERG, "Time, Memory, Ritual," 85, "Israel is enjoined to reenact ritually an event it has not yet enacted, to remember what will happen in the future." See also, R. GELIO, "Il rito del sangue," 467-468.
[3] See ch. II, § 3.2.1 e § 3.2.2.
[4] Fischer – Markl, 130.
[5] P. BOVATI, *Parole di libertà*, 85; R. DE ZAN, *Unius verbi*, 33-50.
[6] See P. BOVATI, *Parole di libertà*, 84-86.
[7] See ch. I, § 2.

Israelites live a "liminal experience," very similar to that of Moses[8] and overcome it through the signs of Passover and the blood of the lamb.[9] Therefore, Moses' return to Egypt represents a true model on which the Israelites' journey to the land is based.[10] Moreover, as is the case for Moses, the Passover ritual unites the families of Israelites in an intimate bond with Yhwh, similar to that shared by "blood relatives."[11]

The analysis, therefore, will firstly include a brief close reading of the announcement of the last plague (11:1-10) followed by an examination of the legal material in 12:1-28 and by the narrative of 12:29-42, in which the execution of the tenth plague and the departure of the Israelites are narrated. In each of these stages, a brief overview of the redactional hypotheses for the reconstruction of the texts' history will be provided.[12]

1. The Announcement of the Last Plague (Exod 11:1-10)

The passage of Exod 11:1-10 describes a "plague" that is different from the others, since it is presented as unique.[13] The chapter, in fact, does not begin with Yhwh's order to go to Pharaoh, like many of the previous episodes (7:14, 26; 8:16; 9:1, 13; 10:1), but with an instruction: the people must ask their neighbours for objects of gold and silver (see 3:21-22; 12,34-35); moreover, the scourge is called a "plague" for the first time (11:1, נגע) and does not require the staff; finally, the departure of Israel is now considered a certainty and not only a possibility.[14]

The passage has traditionally been interpreted as the result of a complex editing process. The verses that conclude the narrative of the ninth plague (10:28-29), in fact, contain words, according to which it seems that Moses will not meet Pharaoh again. Yet, the beginning of ch. 11 surprises the reader with an unexpected intervention of Yhwh (vv. 1-3), and vv. 4-8 introduce a new discourse of Moses before Pharaoh. A proposed solution is that vv. 4-8 are a direct continuation to 10:29.[15] Traditionally, exegetes who defend the classical

---

[8] See ch. II, § 4.3 – § 4.5.
[9] See ch. II, § 4, n. 201.
[10] See E. Di Rocco, *Raccontare il ritorno*, 79.
[11] C. Owiredu, *Blood*, 160-161.
[12] See É. Nodet, "Pâque, Azymes," 499-534. Exegetes have repeatedly turned to the comparative history of religions to explain the composition of Exod 12–13. The combination of Passover and Unleavened Bread would reflect the structure of a New Year festival of Canaanite origin. See Childs, 187-189.
[13] M. Greenberg, "The Redaction of the Plague Narrative," 243-252.
[14] Nepi, I, 178.
[15] Cassuto, 132; Houtman, II, 132; Fischer – Markl, 126.

documentary hypothesis have explained 11:1-10 as the result of three different sources: vv. 1-3 can be part of the document E, vv. 4-8 of J and vv. 9-10 of P[16]. This explanation has been revised and updated several times[17], without reaching unanimous agreement. The tension between the end of ch. 10 and 11:1 is significant, yet the episode is the outcome of conscious organization. In fact, the short narrative is composed by two discourses of Yhwh (vv. 1-2 and v. 9) that frame Moses' speech to Pharaoh (vv. 4-8a). In the three passages, the direct speech is followed by an intervention of the narrator (v. 3, v. 8b e v. 10). Therefore, the passage can be read as an organic ensemble, even if an undeniable fracture between 10:29 and 11:1 remains.

1.1 *The Instructions to the People (11:1-3)*

The narrative begins *ex abrupto* and quickly — "Yhwh said to Moses" — creating a rupture with 10:29: whereas at the end of ch. 10 Moses states that he will no longer see Pharaoh's face, in 11:4-8 the reader is in the presence of another official encounter between Moses and the king of Egypt. The narrator is certainly more interested in theology than in accuracy regarding Moses' place;[18] vv. 1-3 could therefore be a parenthesis in which past events are reported, and ויאמר could be translated with a pluperfect "he had said."[19] However, since in 10:29 Moses' departure from Pharaoh's presence (see 11:8) is not mentioned, vv. 1-2 may also report a "private" word from Yhwh to Moses while he is still before Pharaoh.[20]

---

[16] J. WELLHAUSEN, *Composition*, 68; Childs, 131.

[17] See G.I. Davies, II, 1-3. Exegetes agree that 11:9-10 is an expression of P, but they explain the composition of vv. 1-8 in different ways. Here are some proposals. L. Schmidt considers that v. 8b is to be read after 10:28-29 and that both are J, while he assumes that vv. 1.4-8a are part of the JE redaction and that vv. 2-3 are the expression of a late editorial layer (*Beobachtungen*, 50-57). Van Seters attributes vv. 9-10 to P and vv. 1-8 to the exilic source J, transposing 10:28-29 after 11:8a (*The Life of Moses*, 108.121). Blum thinks that vv. 4-8 belong to the pre-deuteronomistic account of the plagues while he considers vv. 1-3 to be the work of the editor who elaborated the Komposition-D (*Studien zur Komposition*, 13.35-36); of the same opinion is F. AHUIS, *Exodus 11,1 – 13,16*, 102-104. See J.C. GERTZ, *Tradition und Redaktion*, 180-182, for an even more complex hypothesis of reconstruction of the editorial history of vv. 1-8: the original layer consists only of part of vv. 4-8 to which vv. 1-3 and elements completing vv. 4-8 are added during the final editing.

[18] See Priotto, 197.

[19] See Ibn Ezra 72; Houtman, II, 129.

[20] See *ExR* XVIII, 1, "God rushed into Pharaoh's palace and spoke to Moses, as it says, *Yet one plague more will I bring upon Pharaoh* (Ex 11,1)"; see also, Cassuto, 131; Houtman, II, 130; Fischer – Markl, 124.

The setting of vv. 1-3 is completely different from that of the previous narratives, since the character of Pharaoh disappears and only Moses and Yhwh remain on the stage. The passage begins insisting on the divine initiative (v. 1), with the expression אביא, "I will let in." God is the main agent of Israelites' rescue that will take place in the night, when he will send "one more plague" (v. 1). The term נגע literally means "plague, disease" (Gen 12:17; Lev 13:2), and since the verb נגע is also translated "to strike,"[21] it can be rendered with a stronger nuance, "punishment" (2 Sam 7:14). In vv. 1-2, thereby, an irreversible and conclusive sentence is pronounced against Pharaoh and Egypt.[22] Yhwh says in v. 1 that, after the plague, it will be Egypt that will want to chase Israel away,

> Afterwards, he will let you go[23] (ישלח) from here. When[24] he will let you go (כשלחו), indeed, he will want to drive you out (גרש יגרש) completely[25] from here. (Exod 11:1)

The repetition of שלח creates a dynamism in *crescendo* that culminates in the verb used at the end of the series, since it is the only verb not preceded by the infinitive absolute, "he will want to drive you out."[26] The same verb is already used in 6:1, when, after the failure of Moses and Aaron's first meeting with Pharaoh (5:1-5), Yhwh says,

> Now you shall see what I will do to Pharaoh; for with a powerful hand (ביד חזקה) he will let them go (ישלחם), or rather with a powerful hand (ביד חזקה) will drive them out (יגרשם) from his land.

---

[21] Propp, I, 342. The term נגע designates violent crime (Deut 17:8).

[22] In the Book of Exodus, the substantive נגע is unique to 11:1 and indicates the divine intervention against Egypt. In the narrative of the plagues, in fact, different nouns are used: אות, "sign" (7:3; 10:1-2); מופת, "wonder" (7:3, 9; 11:9-10); מגפה, "scourge" (9:14); see Nepi, I, 175-176; Priotto, 198. Moreover, the place of נגע at the beginning of the sentence is unusual and is intended to catch the reader's attention; see JM § 155*o*; G.I. Davies, II, 8. Therefore, the expression עוד נגע אחד has the tone of a definitive statement; see Propp, I, 342. The numeral adjective אחד certainly means "*a single* plague"; see Gen 11:6; 34:16; 41:25; 42:11. *TgPsJ* of Exod 11,1 states that the blow "will be harder upon them than all of them."

[23] When the verb שלח has Pharaoh as subject, it can be translated "to let go"; see Exod 4:21.23; 5:1-2; 6:11; 7:2.14.

[24] The expression כשלחו has a temporal nuance, as the Septuagint translation emphasises (ὅταν δὲ ἐξαποστέλλῃ); see Houtman, II, 130.

[25] The term כלה is translated as an adverbial accusative, "completely," by the Septuagint (σὺν παντὶ) and the Peshitta (*klkwn*) and not as a derivative of the root כלה "to destroy"; Gen 18:21. See Houtman, II, 130-131, who thinks that this interpretation is coherent with the context: previously Pharaoh has never granted permission to leave without giving conditions.

[26] GK § 113*n*; JM § 123*e*. See also, S.N. CALLAHAM, *Modality*, 186. The infinitive absolute of 11:1 (גרש) has a "resultative" value: the verb is used in the apodosis of a conditional sentence that insists on the certainty of the result.

The duplication of יד deliberately creates a certain ambiguity in the qualification of the "hand": if in the first case, because of the context, the "powerful hand" is easily identifiable with that of God, in the second case it could also be the hand of Pharaoh.[27] The passage in 11:1 confirms this impression because the verb גרש *piel*, introduced by the infinitive absolute of the same root, increases the intensity of what is expressed by the simple שלח: Pharaoh will not only let Israel go, he will expel them harshly.[28]

After this, in v. 2 Yhwh asks Moses to speak "in the ears of the people (דבר נא באזני העם)." In this way, he insists, first and foremost, on the positive outcome of the verbal communication and on the accuracy with which the message must be conveyed, that it may reach everyone's ears.[29] The content of the message, on the other hand, refers to what the reader has already learned (3:21-22; see Ps 105:36-38): each member of the people will have to ask the Egyptians for gold and silver utensils. As in Exod 3, the appropriation of the Egyptians' goods is the result of a request (שאל) and the concession is a gift from people who have become "close"[30] (רעה). Nevertheless, unlike 3:22, the text of 11:2 does not specify that the handing over of precious objects will be a rightful plunder (3:22, "and you will spoil Egypt"), as a compensation for what was suffered during the cruel Egyptian slavery;[31] rather, what is emphasised is the voluntary nature of the dispensation.

Finally, v. 3 presents the narrator's point of view, who, from his privileged perspective, shows that the people were favoured "in the eyes of the Egyptians (בעיני מצרים)," thanks to an act of divine intervention[32] ("and Yhwh gave the people favour," ויתן יהוה את־חן את־העם). Yhwh, who hardens Pharaoh's heart

---

[27] See L. INVERNIZZI, *"Perché mi hai inviato?"*, 309-312. She notes that אשר can also introduce the object clause (JM § 157*b*) and believes that the verb עשה creates a contrast between the "work" that the Egyptians demand from the people (5:4, 9, 13) and God's intervention, his "work." Furthermore, she thinks that the double reference to the "powerful hand" in 6:1 refers in the first case to divine action and in the second to that of Pharaoh.

[28] The verb גרש *piel* indicates elsewhere the abrupt termination of a relationship: Gen 21:10; see Judg 9:41; 1 Kings 2:27.

[29] The syntagm דבר + ב + אזן expresses a successful communication (Gen 50:4; Deut 31:28).

[30] The Septuagint translates רעה with πλησίον "neighbour," the Vulgate with *amicus*. TgPsJ Exod 11:2 explicitly insists on the close relationship between Israelites and Egyptians, "each one from his Egyptian friend" using רחמיה instead of חבריה "close" (used by *TgO* and *TgN*).

[31] The handing over of precious objects is interpreted as an act of reparation for Israel's sufferings; see PHILO OF ALEXANDRIA, *De vita Mosis*, I, 141; in Wis 10:17 wisdom "rendered (ἀπέδωκεν) to the saints the salary (μισθόν) for their labours." Exod 3:22 can make reference to the plunder (Josh 7:21) or to the compensation after the *manumissio* (Deut 15:13-14; Cassuto, 44), while נצל means literally "to seize the spoils of war" (2 Chron 20:25).

[32] The construction נתן + חן + ב + עין always has God as its subject and indicates the favour found with superiors; Exod 3:21; 12:36; see Gen 39:21; Ps 84:12.

(see 4:21), also makes sure that the Egyptians treat the Israelites with benevolence.[33] Compared to 3:21, where this motif is already evoked, the text of 11:3 creates a *crescendo*, because the narrator does not limit himself to revealing the generosity of the Egyptians towards the people, but adds a consideration about Moses: "the man Moses[34] was much esteemed (lit. very great) in the land of Egypt, *in the sight* of Pharaoh's servants and *in the sight* of the people." The triple repetition of the term עין, "eye," and the double use of מצרים in the same verse achieve an effect of intensification: thanks to God's involvement, the the antagonists' point of view is radically transformed. However, even if the munificence is widespread, someone is missing: Moses is esteemed "great"[35] by Pharaoh's servants, by the Egyptians, but not by the King of Egypt.[36] His stubbornness causes his complete isolation.

1.2 *The Sentence (11:4-8)*

Pharaoh, indeed, ultimately disappears from the scene. In fact, although Moses' direct speech is delivered in his presence (see v. 8), the interlocutor is never mentioned; the interlude of vv. 1-3, therefore, and the unexpected passage of v. 4 gives the reader the impression that there is no longer any possibility of dialogue with the king of Egypt.[37] The "messenger formula,"[38] "thus says Yhwh" (v. 4), demonstrates again that God's messenger speaks because he obeys a command. The reader is unfamiliar with the message[39] which Yhwh delivered to Moses and therefore should not compare the quotation with the source;[40] the trustworthiness of Moses is taken for granted.

---

[33] Priotto, 198. See above, ch. II § 3.2.1, regarding the "dual causality."

[34] The expression האיש משה "the man Moses" must be interpreted as an epithet of respect; see Houtman, I, 7; Num 12:3. G.I. Davies, II, 13, believes instead that האיש is an anaphoric introduction (see 1 Kings 11:28).

[35] When the adjective גדול modifies the term איש, it can indicate a particularly rich (1 Sam 25:2) or honoured man (Esther 9:4).

[36] See Priotto, 199, note 21. In the narrative of the plagues, Pharaoh's servants are always associated with the king of Egypt (see 7:10; 8:7, 17, 20). The unity is already broken in 9:20 and 10:7 where Pharaoh's servants distance themselves from the decisions of their ruler. This fracture is completed in 11:3; see also Houtman, II, 131-132.

[37] Priotto, 199; G.I. Davies, II, 14.

[38] See ch. II, § 3.2.2.

[39] The PSam adds, after v. 3a, a repetition of vv. 4b-7 to make the text more harmonious; the achieved effect is clear: Moses comes before Pharaoh saying the words that he has received from Yhwh. See B. LEMMELIJN, *A Plague of Texts?* 205-207.

[40] Quotations of speech are usually reproductive, i.e. they are retrospective references to previously spoken words; the quotation mimetically re-presents the source, inserting it into a different discourse. See M. STERNBERG, "Proteus," 109-112.

"About[41] midnight, I myself am going out (אני יוצא) in the midsts of Egypt." (v. 4) The plague will take place during the night, a scenario that had symbolic meaning for Moses (4:24), but was also significant for the Egyptian people (10:21-23).[42] The reader therefore foresees that Pharaoh, in turn, will have to face a night-time attack by Yhwh. The personal pronoun אני, is itself pleonastic,[43] as it aims to exalt the divine initiative and His work. The verb יצא is one of the roots that expresses the liberation of the people,[44] although in this case it rather emphasises the movement by which Yhwh departs from His heavenly abode (Isa 26:21); it can also be correlated to the "military" domain[45] (Judg 4:14; 2 Sam 5:24; Isa 42:13). The expression is particularly significant because it is the first occurrence in the Hebrew Bible in which God is the subject of the verb "to go out." The reader's surprise, like that of Pharaoh, could not be greater: God Himself moves to aid his people. The choice to use the participle ("I am going out") confirms the imminence of God's intervention.[46] In v. 5, the certain fulfillment of what has been revealed to Moses in 4:23 is announced.[47]

Exod 4:23b	Exod 11:5
Behold, I'm going to slay (הרג) your son, your firstborn	And every firstborn in the land of Egypt shall die (ומת), from the firstborn of Pharaoh who sits on his throne, to the firstborn of the slave girl who is behind the millstones, and every firstborn of the cattle

The substitution of הרג with מות avoids specifying that "God kills,"[48] but the insistent recurrence of בכור, "first-born," shows that the "blow" inflicted by God is immense in scope:[49] not only the first-borns of Pharaoh will die (4:23), but all the first-born shall perish, from mankind to cattle, from the son of Pharaoh to the last slave girl. Pharaoh's first-born is identified as the son

---

[41] The expression כחצת הלילה is translated interpreting the preposition כ as having a temporal meaning and specifying that its use emphasises the approximate nature of the information "About midnight." See GK § 118u; Rashi Exod, 73. Houtman, II, 132.

[42] Regarding the symbolic meaning of the night, see ch. II, § 4.3, and Priotto, 199.

[43] See WO § 16.3.2d, "The referent of the pronoun may be involved in an *explicit antithesis* with another person or group of persons." See Gen 3:15; 2 Sam 17:15; 24:17.

[44] Fischer – Markl, 126; Priotto, 199.

[45] G.I. Davies, II, 14-15.

[46] JM § 121e, "The use of the participle to express the near future and the future in general is an extension of the use of the participle as present. A future action, mainly an imminent action, is represented as being already in progress."

[47] See ch. II, § 3.2.2.

[48] Priotto, 199.

[49] Nepi, I, 186.

of the one "who sits on the throne."⁵⁰ By striking him, Yhwh wounds the royal descendants in the heart. The "slave girl behind the millstones," instead, is a member of the lowest category in Egyptian society⁵¹. Together, these two elements form a merism, namely the combination of two contrasting components which indicate a plague that strikes without making any distinction between social classes⁵² and does not spare animals either.

God's intervention is unique and provokes an intense reaction, a cry never heard before and unrepeatable in the future⁵³ (v. 6). The root צעק is common in Exodus, but up to this point it has only been used for the Israelites⁵⁴ (צעקה, 3:7, 9; צעק, 5:8; also 8:8); in 11:6, it expresses a lament of higher intensity, because it is the first time in which צעקה is modified by the adjective גדלה. This term, then, suggests that the plague is a *contrapasso*: the Egyptians who harassed the Israelites will experience the same pain.⁵⁵ For the Israelites, on the contrary, Yhwh says, "a dog shall not sharpen his tongue" (v. 7). In contrast to what is awaiting the Egyptians, not even the snarling of dogs will be heard against the Israelites,⁵⁶ neither against humans nor against animals.⁵⁷

The consequence of the divine attack is explicitly stated in v. 7, "in order that you may know (למען תדעון) that Yhwh makes a distinction between Egypt and Israel (יפלה יהוה בין מצרים ובין ישראל)." Once again, Yhwh shows that the purpose of his action is the deeper revelation of his mystery: God acts so that even Egypt may "know"⁵⁸ that Yhwh "makes a distinction" (פלה);

---

⁵⁰ The MT הישב על־כסאו is polysemous and may therefore refer to either Pharaoh or his firstborn. *TgO* of 11:5 translates the text referring to his son, "who is preparing to sit on the throne of the kingdom." See Cassuto, 133; Houtman, II, 132-133. The translation "Pharaoh who sits in his own throne" is the most suitable, since it creates a striking contrast with the mention of the "slave girl behind the millstones." Durham, 149; Propp, I, 344; Priotto, 194.

⁵¹ Cassuto, 133. In the *Instruction of Ptah-hotep* there is an explicit reference to the slave woman who is behind the millstones (*ANET*, 412; Eccles 12:3). See Rashi Exod, 74, who justifies the harshness of the sanction as follows, "Because even they (the slave girls) made them (the Israelites) work and rejoiced at their distress."

⁵² Fischer – Markl, 126.

⁵³ Durham, 149; כמו + היה is already used in 9:18.24; 10:14.

⁵⁴ Propp, I, 344; G.I. Davies, II, 15.

⁵⁵ Nepi, I, 202.

⁵⁶ The MT has לא יחרץ־כלב לשנו, literally "a dog will not sharpen its tongue," (Josh 10:21). These words have been linked to the barking (or the snarling) of a dog: Cassuto, 133; Propp, I, 344; Houtman, II, 134; G.I. Davies, II, 11.

⁵⁷ Several authors believe that 11:7 refers to the calm of Israel's night: no alarm will be heard even from the dogs; see Nepi, I, 186; Priotto, 200. Propp, I, 344, instead, since "to sharpen the tongue" can refer to a hostile word (Gen 18:18; Ps 57:15), believes that this expression means here "to bark threateningly."

⁵⁸ Regarding the plot of revelation in the book of Exodus, see ch. I, § 3.2, note 135. G.I. Davies, II, 16, "The purpose of the final plague is still didactic."

He favours the Israelites[59] and judges with righteousness, separating the guilty from the innocent,[60] defending the victim and punishing the offender.

Moses' discourse concludes as follows, "all these servants of yours will come down[61] to me and will bow low to me" (v. 8). The repetition of עבד, "to serve," is intended to connect v. 3 with v. 8: the servants who will ask Moses to bring out the people are the same ones in whose eyes Yhwh's envoy "found favour" (v. 3, "a great man in the sight of Pharaoh's servants"). Furthermore, their "descent" has a symbolic value[62] because it emphasises the servants' shift of loyalty from Pharaoh to Moses. They will recognise him with a prostration, a gesture that is part of the ceremonial by which the subject pays homage to the king[63] (2 Sam 14:33; 1 Kings 1:53; Esther 3:2). Then, Moses reveals to Pharaoh that he will be respected as a ruler[64] and will receive honour precisely because of his mission, which associates him with Yhwh;[65] in Exodus, in fact, the root חוה, "to prostrate oneself," is employed to indicate the honour granted to God[66] (4:31).

The excerpt ends with a triple repetition of the root יצא, "to go out," which directs the reader towards the resolution of this first narrative arc. The servants will say, "*Get out*, you and all the people that is at your feet" and then Moses concludes his speech saying, "and after that I *could go out*." The verse ends with the third occurrence of יצא, "and he *went out* from Pharaoh." The repetition of the same verb highlights the movement that the people are about to begin;[67] the root יצא, however, is also used in 11:4 referring to Yhwh's intervention during the night. Through the technique of repetition, then, a link is made between Yhwh's action and people's movement: Moses, and the people, can depart thanks to the joint action of two causes: God will "go out" to strike

---

[59] The root פלה appears also in 8:18 e 9:4 and expresses the difference in the treatment of the Israelites and the Egyptians. See also 9:26; 10:23; 14:20; 33:16.

[60] The particle בין, "between," is often used in judicial contexts: בין + שפט (Gen 16:5; 31:53; Exod 18:16; Deut 1:16); בין + יכח (Gen 31:37; Job 9:33; see Isa 2:4).

[61] The demonstrative pronoun without article may follow the noun with suffix: Josh 2:20 "if you tell this matter (את־דברנו זה)"; see Judg 6:14; 1 Kings 10:8; GK § 126*y*. The value of this pronoun is deictic: Moses points to the servants present during his speech; Cassuto, 133.

[62] Priotto, 201. In the land of Canaan palaces were often located in higher spaces (2 Sam 11:9; 2 Kings 6:33); Propp, I, 344-345; Houtman, II, 135.

[63] See 1 Sam 25:23; 2 Sam 18:21; see also Exod 18:7 and ch. II, § 5.3, notes 335.

[64] Priotto, 201; G.I. Davies, II, 17.

[65] The Targumim do not render the text with the verb usually used for translating חוה: *TgO* and *TgPsJ* use בעי, "to beg, to ask"; *TgN* translates וישאלון בשלמי, "they will greet me." The embarassment of the translation is probably due to the fact that the verb "to adore" (סגד; see *TgO* Gen 18:2; 19:1; 22:5) is reserved to the worship of God.

[66] See also Deut 26:10; 2 Kings 17:36; Ezek 46:3; Pss 95:6; 99:5, 9; Neh 9:3.

[67] Fischer – Markl, 127.

Egypt, the servants will recognise this intervention and ask Moses to "go out." The verse ends with a harsh reaction from Moses, he leaves Pharaoh's "in a furious rage," which is not a mere emotional response, but an indignation towards Pharaoh's behaviour and towards the evil he has committed.[68]

1.3 *Epilogue (11:9-10)*

Pharaoh disappeared from the scene in vv. 1-8. Even if Moses addresses him (v. 8, "all these servants of *yours*"), there is no dialogue between them. The narrator thus demonstrates that Pharaoh's obstinacy is irreversible. It is therefore not surprising that in vv. 9-10 the king of Egypt is referred to using the third person: the character is no longer present in the story. The words uttered by Yhwh in v. 9 are pronounced after Moses' departure from Pharaoh[69] and, together with the narrator's commentary, they trigger a series of significant echoes with 4:21 and 7:3-4.

Exod 4:21	Exod 7:3-4	Exod 11:9-10
²¹ While you go and return to Egypt, see all the wonders (המפתים) which I have put in your hand; I will harden his heart (אחזק את־לב) and he will not let the people go (לא ישלח)	³ But I will stiffen (אקשה) the heart (את־לב) of Pharaoh and I will multiply (והרביתי) my signs (את־אתתי) and my wonders (ואת־מפתי) in the land of Egypt ⁴ Pharaoh will not listen (לא ישמע).	⁹ Yhwh said to Moses, "Pharaoh had not listened[70] (לא ישמע), in order that my wonders may be multiplied (רבות מופתי) in the land of Egypt." ¹⁰ Moses and Aaron had performed all these wonders (מופתים) before Pharaoh, but Yhwh hardened (ויחזק) the heart (את־לב) of Pharaoh, and he did not let the Israelites go (ולא שלח) from his land.

The noun מופת, "wonder," appears for the first time in 4:21, but is also found in 7:3-4 and 11:9-10; the only other recurrence of מופת in the account of the plagues is in 7:9. In 7:3 and 11:9 Yhwh specifies that the wonders will "be multiplied" (רבה). The motif of the hardening of the heart is attested to in all three passages (4:21; 7:3; 11:10), while the "failure to listen" (לא ישמע) appears

---

[68] Nepi, I, 186. Righteous anger for the evil committed is normally referred to God (Num 11:1, 33; 12:9; Deut 1:34); see Exod 16:20; 32:19, 22.

[69] Cassuto, 134-135; the author translates ויאמר as a pluperfect ("Yhwh had said") and considers the last verses of ch. 11 as a concluding recapitulation.

[70] The *yiqtol* ישמע must be translated as a paste tense, with a durative or frequentative aspect (GK § 107e): this form emphasises the repeated attempts to ask Pharaoh for the permission to leave. See L. SCHMIDT, *Beobachtungen*, 56; G.I. Davies, II, 11. Houtman translates ישמע with the future, "he will listen"; Houtman, II, 135; Childs, 127; Priotto, 201.

only in 7:4 and 11:9. The conclusion of chapter 11, therefore, through the word of Yhwh (v. 9) and a verse in which the narrator states his own point of view (v. 10), resumes the thread that has structured the narrative of the plagues and manifests again the "dual causality" that guides history with a complex synergy.[71] In 11:9 the "role" of human freedom is extremely exalted. Since Pharaoh repeatedly refuses to listen, his decision has a "providential" consequence, God multiplies the signs, מפתים, which are intended to favour a deeper knowledge of Him.[72] In v. 10, instead, the motif of the hardening of Pharaoh's heart is recalled to show that his persistence is part of God's plan and does not deny the absolute pre-eminence of divine causality.

2. The Passover

The rhythm of Exod 7–11 is persistent. With a succession of ten interventions, Yhwh tried to convince Pharaoh to let the Israelites go. Chapter 11 ended with the dramatic announcement of the last plague, the death of the firstborn (vv. 4-8). At this point, the flow of the narration is interrupted, at least until 12:29.[73]

Exegetes have long analysed the fruitful relationship between law and narrative,[74] and Bartor's monograph offers valuable tools for the narrative analysis of legal texts.[75] In the light of this work and some recent contributions,[76] Exod 12:1-28 is a body of cultic-legal texts embedded within

---

[71] Regarding the term מופת, the hardening of Pharaoh's heart and the principle of the "dual causality," see ch. II, § 3.2.1.

[72] The "wonders" are performed "in the sight of" (ל + עין, Deut 4:34; 6:22; 7:19; 29:2); the מופת can produce a better knowledge (ידע, see e.g. Ezek 24:24.27).

[73] The authors are practically unanimous in considering that 12:1 – 13:16 is a sequence distinct from what precedes and from the account of the liberation which follows (13:17 – 15:21). Compared to the previous section, the scenario changes and all the characters are different: Pharaoh is no longer on the scene and the action takes place among the Israelites. See Cassuto, 136; Childs, 178-182; Propp, I, 355; Fischer – Markl, 129-131; Priotto, 213-214; Dohmen, I, 286-287; G.I. Davies, II, 27. Some scholars, however, highlight the close relationship between ch. 12 and ch. 11; Houtman, II, 147; Durham, 88-89.

[74] S. CHAVEL, *Oracular Law*; M. RÖHRIG, "Gesetz und Erzählung," 407-421, with bibliography. See also H. NASUTI, "Identity, Identification," 9.

[75] A. BARTOR, *Reading Law as Narrative*.

[76] The relationship between law and narrative is based on three elements; see A. BARTOR, *Reading Law as Narrative*, 17, "(1) The Law collections (as well as individual laws) appear in the Pentateuch within a narrative frame ... (2) Several laws mention historical events that occurred in the past ... (3) A large proportion of all of the laws of the Pentateuch are constructed as 'miniature stories.'" Also M. STERNBERG, *Hebrews Between Cultures*, 520-537; ID., "If-Plots: Narrativity," 29-107; J.-P. SONNET, "If-Plots in Deuteronomy," 453-470;

a narrative. The instructions for the celebration of the Passover, in fact, are framed between the announcement of the tenth plague (11:1-10) and its execution (12:29-34). This phenomenon is attested to elsewhere in the Pentateuch and is very significant; it demonstrates that the law plays a central role in the development of the divine plan,[77] told in the narrative sections. In this case, it is the sacrifice of the lamb that has a prominent influence on the plot, because it is precisely the sign of the blood that causes Yhwh to "pass over" the houses of the Israelites striking only the Egyptians (12:23-24). The close reading, therefore, will allow to recognise the crucial role that Israelites' obedience (12,28) has for the progression of the plot.

2.1 Redactional Criticism of Exod 12

The coherent relationship between Exod 12:1-28 and the narratives that surround it (11:1-10 and 12:29-42), does not eliminate the textual problems of this literary unit; on the contrary, it is useful to carefully consider difficult transitions in order to reach a better understanding of the text. Therefore, the close reading is again preceded by the review of the most important redactional hypotheses regarding Exod 12:1-42.

2.1.1 The Passover Sacrifice and the Unleavened Bread (12:1-20)

The classical documentary hypothesis reconstructed the redactional history of Exod 12–13 in the following way: 12:1-20, 28 and 12:40–13:2 are considered as a product of the document P; 12:21-23, 27b, 29-39 were thought to be part of J (with the exception of 12:35-36, pertaining to E), 12:24-27a and 13:3-16 were counted among the Deuteronomistic passages.[78] This reconstruction has been questioned, yet continues to offer a point of departure. Exegetes agree that the main parts of Exod 12–13 are priestly in nature.[79]

---

the micro-plot of legal texts triggers similar dynamics to those found in narrative texts, because they too are characterised by suspense, curiosity and surprise.

[77] See A. BARTOR, *Reading Law as Narrative*, 20, "The laws play a central role in advancing the main story, as they constitute a necessary condition for the realization of the divine plan. The laws themselves [...] motivate the story plot, since the continued survival of the nation is dependent on receiving and observing them." The most characteristic example of this phenomenon is found again in the book of Exodus, in chapters 19–20, where the Decalogue (20:1-17) is sandwiched between Exod 19 and 20:18-21. See also J.-P. SONNET, "If-Plots in Deuteronomy," 461, "Deuteronomy is therefore built on the meeting and the coupling of two future oriented narrativities: the omnitemporality of the law and the particular, divinely bound, temporality of the promise."

[78] Childs, 184.

[79] Childs, 184; Propp, 373. Also see G.I. Davies, II, 30-31.

This conclusion, however, has been subject to numerous clarifications. First of all, in 12:1-11 there is an evident alternation between indications addressed to the people in the third person and those addressed in the second. Therefore, some scholars consider that vv. 1-11 are composed of a basic layer[80] (vv. 1.3.6b-8), and of an expansion[81] (vv. 2.4-6a.9-11), that is midrashic in nature[82]. According to this perspective, the writer P would have integrated an ancient document, written in the third person plural.[83]

The verses 12-13 have been regarded as a revision of v. 23, which was traditionally identified, together with vv. 21-22, as part of the oldest layer of ch. 12. Whereas in v. 23 the night attack is attributed to an "individual," the "destroyer" (המשחית, v. 23), in 12:13 this figure is explicitly identified with God.[84] Tucker demonstrates the link between vv. 12-13 and the P-narrative through the contact of these verses with the priestly lexicon.[85]

The following passage is redactional and links the Passover sacrifice to the ritual of unleavened bread[86] (vv. 15-17). In the next verses (vv. 18-20) Yhwh concludes His discourse with the dates of the feast.[87] This reconstruction has been recently deepened by Gesundheit, who attempts to demonstrate that the basic layer of vv. 1-20 is priestly in character and that phenomena of revision can be discerned in various verses, together with an inner-biblical exegesis similar in style to that of the post-biblical midrashim.[88]

2.1.2 The Word of Moses to the Elders (12:21-27)

Defenders of the classical documentary hypothesis suggested that vv. 21-23 are part of J, while vv. 24-27 are a portion of D.[89] Some recent interpreters, however, have reconsidered the link of vv. 21-27 with P, on two

---

[80] J.C. GERTZ, *Tradition und Redaktion*, 37.

[81] F. AHUIS, *Exodus 11,1 – 13,16*, 33-42 (with bibliography).

[82] The first expansion (vv. 4-6a) harmonises the opposition between what is said in v. 3 regarding the date of Passover ("the tenth of this month") and v. 6 ("you shall keep it until the fourteenth of the month"); F. AHUIS, *Exodus 11,1 – 13,16*, 35-36.

[83] R. RENDTORFF, *Die Gesetze*, 56-57. See J.C. GERTZ, *Tradition und Redaktion*, 32-35.

[84] See M.V. FOX, "Sign of the Covenant," 575, "In *Ex.* XII, 13 P is using an ancient concept, but for him it no longer belongs to the realm of magic, but to theology."

[85] See P.N. TUCKER, "The Priestly *Grundschrift*," 205-219; he recognises several lexical and thematic contacts between Exod 12:12-13 and two P texts, Gen 9:12, 16 and 17:9-13.

[86] P. WEIMAR, "Ex 12,1-14," 213, n. 80.

[87] I. KNOHL, *The Sanctuary of Silence*, 19.

[88] S. GESUNDHEIT, *Three Times*, 229.

[89] Childs, 184. See G.I. Davies, II, 83-84, who thinks that vv. 21-23, 25-27 belong to the "Plagues narrative" of J (7:15-17, 20*; 9:22-23a, 35; 10:12-15, 20; 10:21-23, 27), of pre-exilic times and prior to 12:1-20.

grounds: the priestly vocabulary[90] and the relation between 12:12-13 and 12:27.[91] Gesundheit, moreover notices the connection between vv. 22-27a, 28 and vv. 9-11.[92] This reconstruction has been a matter of debate. Berner, for instance, considers vv. 21-23 to be a post-P insertion;[93] Jeon argues that only vv. 24-27a are formed by an editorial post-P addition,[94] while, vv. 21-23 are part of an older layer. Tucker also advocates an early date for vv. 21-23.[95]

2.1.3 The Departure (Exod 12:28-51 and 13:1-16)

The narrative which was suspended after the announcement of the last plague (11:1-10) resumes in 12:29, in which the death of the firstborn of Egypt and the departure of the Israelites is narrated (vv. 29-42). The classical documentary theory had attributed vv. 29-34, 37a, 38-39 to the source J, vv. 35-36 to E and vv. 37b, 40-42 to P.[96]

Even recently, it is believed that vv. 29-39 are part of a source which predates P,[97] although there are several alternative hypotheses. Ahuis, for example, attributes vv. 29-33, 37-38 to J, while he considers vv. 34-36, 39 as an excerpt that must be ascribed to an editor D;[98] Gertz, on the contrary,

---

[90] The terms אזוב, "hyssop" (v. 22), טבל, "to dip" and משקוף "doorpost" (vv. 22-23), are frequent in P literature; J. VAN SETERS, "The Place of the Yahwist," 173. The same can be said of the expressions, "as a statute for you and your children forever" (v. 24) and "when you enter the land that Yhwh will give you" (v. 25); see Lev 23:10; 25:2. S. GESUNDHEIT, *Three Times*, 62.

[91] See J. WELLHAUSEN, *Die Composition*, 73.

[92] S. GESUNDHEIT, "Philology and Theory," 414-425; J. JEON, *The Call of Moses*, 160-162.

[93] C. BERNER, *Die Exoduserzählung*, 278-293.

[94] J. JEON, *The Call of Moses*, 160-164. Gesundheit, in a recent article, accepts the hypothesis that at least vv. 24-27 are an expression of the post-P redaction. Indeed, vv. 24-27 retain traces of the influence of the P-literature and the D-style, and therefore could be successive to both: דבר + שמר is common in D texts (Exod 12:24-25; see Deut 17:19; 29:8); the sentence of 12,24 לחק לך ולבניך עד עולם is attested in Lev 10:13, 15 (P); S. GESUNDHEIT, "Philology and Theory," 416-417.

[95] See P.N. TUCKER, "The Priestly *Grundschrift*," 210-211; ID., *The Holiness*, 77-93. The priority of vv. 21-23 over Exod 12:1-20 is demonstrated by various arguments: the narrative of divine instructions that Moses conveys to the elders may be Deuteronomistic in style (Utzschneider – Oswald, 271); 12:21-23 are connected with pre-P texts like 11:4-8 and 12:29-33 (Albertz, I, 188-192); the vocabulary is not exclusive of the P-literature (J.C. GERTZ, *Tradition und Redaktion*, 50). See also J. JEON, *The Call of Moses*, 164-165 who reconstructs the redactional history of 12:1-28 in three stages: the oldest text, from the monarchical period (12,21-23.27b); the institution of the Passover in the P style (12,1-14.15-20.28); the post-P expansion (12,24-27a).

[96] Childs, 184-185; Durham, 165. Propp attributes vv. 29-34 and vv. 37-39 to E, while considers that vv. 35-36 are part of J and that vv. 40-51 belong to P; see Propp, I, 375-376.

[97] Albertz, I, 212-219.

[98] F. AHUIS, *Exodus 11,1 – 13,16*, 67.70-72.

presents an even more fragmentary view: only v. 37a belongs to the earliest source, and vv. 35-36 derive from the final redaction together with vv. 33-34, 37b-39.[99]

The attribution of vv. 28, 40-41 to the P-literature is generally accepted,[100] while v. 42, since the reference to "night" creates a tension with the mention of the "day" in v. 41, can be seen as the result of the post-P redaction.[101] Even vv. 43-49, traditionally connected with the source P, have been explained as a post-P insertion,[102] framed by the resumption (*Wiederaufnahme*) of v. 41b in v. 51.[103] Exod 12:50-51 and 13:1-2 are usually attributed to P.[104]

Finally, since 13:3-16 is constituted by the motif of the "explanation of the Passover to the son," these verses are commonly associated to D[105] (see Deut 6:20-24). Some scholars have questioned this connection, suggesting that 13:3-16 is the work of the exilic Yahwist[106] or a post-P redaction.[107] The identification of 13:3-16 with the D-literature, however, is still the opinion of the majority.[108]

2.2 *Close Reading (Exod 12:1-20)*

Even if the "priestly" matrix remains quite dominant, scholars sustain that vv. 1-20 are the result of a complex editorial history. Nevertheless, the verses are dramatically unified: vv. 1-20 consist in a long discourse of Yhwh to Moses in which the essential instructions for the slaughtering of the Passover lamb and for the ritual of the unleavened bread are enumerated. Through these instructions, Yhwh establishes the paschal celebration and, at the same time, commits Himself (vv. 11-14, 17) announcing that the night will be

---

[99] J.C. GERTZ, *Tradition und Redaktion*, 59.176.183-186.
[100] G.I. Davies, II, 104.
[101] See e.g., J.C. GERTZ, *Tradition und Redaktion*, 58.
[102] G.I. Davies, II, 140-141.
[103] Regarding the *Wiederaufnahme*, see B.M. LEVINSON, *Deuteronomy*, 17-20.
[104] G.I. Davies, II, 159-161.
[105] For the supporters of the documentary hypothesis, Exod 12:24-27 and 13:3-16 are part of D (see e.g., Childs, 184) or in any case the result of a Deuteronomistic redaction (Durham, 161; Priotto, 227). Even Blum recognises in 12:21-27 and 13:3-16 the hand of D; see E. BLUM, *Studien zur Komposition*, 37-38.169-171 (with bibliography).
[106] J. VAN SETERS, "The Place of the Yahwist," 175.
[107] J.C. GERTZ, *Tradition und Redaktion*, 57-60.
[108] J. JEON, *The Call of Moses*, 169-173. In 13:3-16 the author identifies the terminology which can belong to the Deuteronomistic composition: בית עבדים (v. 3 and v. 14; see Deut 5:6; 6:12; etc.); יד חזקה (v. 9; see Deut 4:34; 5:15; etc.); שבע (v. 5 and v. 11; see Deut 6:11; 8:10, 12); the land flowing with milk and honey (v. 5; see Deut 6:3; 11:9; etc.); the education of children (vv. 14-15; see Deut 6:20-21). P. ALTMANN, *Festive Meals*, 186-190.

marked by His majestic deeds: God will pass through Egypt to save the Israelites and blood will play a significant role in this action.

2.2.1 The Passover Lamb (vv. 1-5)

> [1] Yhwh said to Moses and Aaron in the land of Egypt: [2] "This month is *for you* the beginning (ראש) of months, the first (ראשון) *for you* of the[109] months of the year."

The chapter begins with God addressing to Moses and Aaron. Aaron's involvement is not surprising, since in the future the keeping of the signs of worship will be entrusted to his descendants[110] (see Ezra 6:19-22; 2 Chron 30; 35). The specification "in the land of Egypt" is perhaps intended to distinguish the paschal prescriptions from the other laws promulgated at Sinai,[111] but it is more likely that it emphasises the point of departure whence Israel will start the journey towards the liberation.[112]

God's discourse begins in v. 2 with several repetitions. Yhwh constitutes a month different from the others, providing a solid foundation to the people of Israel,[113] "this month is for you the beginning," and creating a hierarchy between different moments in time.[114] The dual recurrence of the root ראש — "beginning of months" (ראש חדשים) and "the first for you" (ראשון לכם) — has the aim of unveiling that the month of the Exodus will become a focal point for the time to come, because it will be the "first" month in terms of quality; indeed, ראש creates an echo with Gen 1:1: Exodus, like creation, will stand in relation to the future epochs as the origin and foundation.[115] In addition, the

---

[109] The particle ל (לחדשי השנה) can be used as a genitive of relation (JM § 129*f*) with respect to ראשון. Houtman, II, 167; G.I. Davies, II, 37.

[110] Priotto, 215. See also *MekhY, Pisha*, § 1, pp. 1-3. See G.I. Davies, II, 45, who thinks that the remark of the proper name of Aaron is a sign of P style (see 6:13; 7:8; 9:8).

[111] Propp, I, 355. Cassuto believes that the addition "in the land of Egypt" is intended to affirm the temporary character of the directives found in vv. 2-13; see Cassuto, 136.

[112] Houtman, II, 166.

[113] The institution of the Sabbath will achieve a very significant hierarchization of time in some ways similar to what is attested in 12:1-2: the week will be ordered by the difference between the six days of work and the day "for Yhwh" (16:22-23; see 20:8-11), which will constitute the foundation for all the others (see 31:13-14).

[114] A.J. HESCHEL, *The Sabbath*, 8, "Judaism is a *religion of time* aiming at the *sanctification of time*.... Judaism teaches us to be attached to *holiness in time*, to be attached to sacred events, to learn how consacrate sanctuaries that emerge from the magnificent stream of the year. The Sabbaths are our great cathedrals." See Hamilton, 179.

[115] Utzschneider – Oswald, 235. The expression ראש חדשים can be translated "an outstanding month." When ראש is without article, can be used to exalt the quality of the term which it modifies (see Exod 30:23; Deut 33:15; Ezek 27:22; Song 4:14). Houtman, II, 167; Dozeman, 262; Priotto, 215-216; Nepi, 197. Even ראשון can have a similar nuance: Durham,

preposition with suffix לכם, "for you," used twice at close intervals, indicates that the new recapitulation of the calendar has a single direction: it *benefits* Moses, Aaron and the people.[116] Yhwh therefore defines the month of liberation as a crucial date for Israel memory and for their identity.[117]

3 Speak to the whole congregation of Israel, "On the tenth[118] of this month, they shall take[119] for themselves, every man[120] a lamb according to the house of the fathers/family, a lamb for a household. 4 But if the household is too small for a lamb,[121] then he[122] and the neighbour to his house shall take it according to the number of persons; you will calculate the lamb according to what each will eat.[123] 5 It shall be for you a lamb without blemish, male, born in the year, you may take it from the sheep or from the goats"

Moses and Aaron have to *speak* to Israel,[124] for the first time referred to as עדה, "assembly, congregation."[125] This term designates the gathering of a group that feels unified,[126] and indicates that Passover will be the essential element that will constitute the people as a "body."[127] Yhwh then, as

---

153; Propp, I, 384. See Rashi Gen, 2, "*In the beginning of.* R. Yitzhak said: the Torah should not have begun [here] but from: *This month shall be for you* (Exod 12:1)." See also Y. ZAKOVITCH, *"And You Shall Tell"*, 105.

[116] Utzschneider – Oswald, 235.

[117] J.T. THAMES JR., "Keeping the Paschal Lamb," 10.

[118] The components of a full date are two: the cardinal number and the unit of time, namely the "day" (Num 7:72), the "month" and the "year" (2 Kings 25:27). When the month is considered, as in Exod 12:3, the term יום can be omitted. GK § 134*n*; WO § 15.3.2*b*.

[119] The verb יקחו is preceded by a *waw* of apodosis introducing the command (WO § 31.5*c*) after the notification of time ("the tenth of this month"). See JM § 176*k*.

[120] The term איש can have a distributive meaning. See Gen 41:12; Num 5:10; 26:54; see anche JM § 147*d*.

[121]. The first מן can be interpreted as a comparative ("if the household were *too small*..."); see GK § 133*c*; WO § 14.4*f*; G.I. Davies, II, 37. The מן that precedes שה, "sheep/goat," could be a sign that the sentence is a combination of two phrases, "too small for being there a sheep/goat" (ימעט מהית שה) and "too small for a sheep/goat" (ימעט משה).

[122] The pronoun הוא does not refer to the "house," but to the head of the family, namely to the term איש employed in v. 3; see G.I. Davies, II, 38.

[123] The construction איש לפי אכלו, lit. "each for the mouth of his eating," is the object of the verb תסכו, "you will calculate." See G.I. Davies, II, 38.

[124] Rashi Exod, 76; Aaron must speak to the people on behalf of Moses (4:14-16); the plural of 12:3 is explained as a signal that there is full agreement between the two.

[125] The noun עדה derives from the root יעד, "to gather, to collect," and is used to describe the gathering of a group of men (Pss 1:5; 22:17); see Houtman, II, 168; G.I. Davies, II, 37.

[126] Utzschneider – Oswald, 239. See Priotto, 216.

[127] The term עדה is also correlated to another derived name of the root יעד, מועד "appointed time" connected with a feast (Exod 23:15; 34:18). This term alludes to the assembly of the community brought together for worship.

happened before (4:22-23), tells Moses the words he is going to say,[128] shaping His own messenger through His discourse (vv. 3-20).

The first important revelation that appears in Yhwh's words is the family dimension of the celebration. Although the orders are intended for the assembly of the people, the setting in which the ritual takes place is domestic and homely (בית is repeated twice in v. 3 and twice in v. 4) and the head of the family is the one who has the task of leading it.[129] On the tenth day of the month,[130] everyone has to get a sheep or a goat[131] "*for* [the] house of the fathers/family (לבית־אבת)," an animal "*for* each house (לבית)." The repetition of the preposition ל accentuates the link between the animal used and the relatives who celebrate Passover,[132] by suggesting that the victim will have a redeeming power *in favour of*[133] the members of the house (v. 5).

When the family unit is too small (v. 4), neighbours may join in, so that the celebration may be intimate and at the same time not limited to a restricted group.[134] In the latter part of v. 4 and in v. 5 Yhwh uses the second person plural, "you will calculate" (v. 4), "It shall be for you ... you may take it" (v. 5). Thereby, Yhwh does not speak to Moses and Aaron considering them as two persons who participate in the event as spectators, but He involves them in the same movement:[135] they too will have to live the same Paschal dynamism.

The purpose of v. 5 is to specify the qualities of the animal intended for the Passover ritual.[136] The head of small livestock, lamb or kid, must first of all be תמים, "complete, without blemish," which is common terminology in defining

---

[128] See ch. II, § 3.2.

[129] Priotto, 216.

[130] See Propp, I, 387-388. The "tenth day" held a crucial meaning in Israel (Lev 16:29; 23:27; 25:9; Josh 4:19; Ezek 40:1); Houtman, II, 170-171; G.I. Davies, II, 48.

[131] The noun שה indicates the head of small livestock and can refer to either the lamb or the kid (see 12:5).

[132] See JM § 133*d*; the particle ל means in general "for," expresses the idea of relationship and can replace the genitive (JM § 130*a*).

[133] The preposition ל can introduce the *dativus commodi*, "for the benefit of" (WO § 11.2.10*d*). See Jer 22:10. The reference to the "tenth day" could then create an allusion to the Day of Atonement rite; see Utzschneider – Oswald, 240.

[134] Rashi Exod, 77. Flavius Josephus attests that in the first century AD the minimum number of participants at Passover had to be ten, but he also states that the dinner could not have more than twenty participants; see *Bell.* VI,9,3, 423.

[135] A. BARTOR, *Reading Law as Narrative*, 24-27. A legal text consists of a legislator and an addressee who may participate to a greater or lesser extent in the formulation of the law. The direct reference to the addressee is intended to involve him personally.

[136] See S. GESUNDHEIT, *Three Times*, 46-48. The expansion to v. 3 provided by v. 5, made in the style of *halakhik midrashim*, accurately defines the characteristics of the animal required for the rite.

the animal suitable for sacrifice[137] (Exod 29:1; Lev 1:3, 10; 3:1, 6), without wounds or defects (מום, Lev 22:20-21; Deut 15:21[138]), male (Lev 1;3, 10) and "born in the year"[139] (Lev 9:3). Therefore, the Passover assumes some characteristics of the sacrifice,[140] even if the two actions remain distinct. In sacrificial practice, the animal or the offering are correlated with the offerer and in some way represent him: the sacrifice manifests the human aspiration for communication with God.[141] This nexus is also found in the choice of the Passover sacrifice, but with a difference: while in worship the offering is often intended to counteract the "evil" represented by impurity and sin, in the Passover the offerers are faced with another obstacle, Pharaoh and the Egyptians. The people are not impure or sinners, but they are prisoners; therefore, they can recognise themselves in the sacrificed animal in its being a "victim," which is a clear image of their situation of slavery and suffering.[142]

2.2.2 The Lamb's Blood (vv. 6-11)

> 6 You shall watch it until the fourteenth day[143] of this month, then the assembly of the congregation of Israel shall sacrifice him at twilight.[144] 7 They shall take

---

[137] Propp, I, 389; Priotto, 217; Utzschneider – Oswald, 240. The worship includes the possibility of sacrificing female animals (Lev 4:28, 32). For *ExR* XV, 12 Yhwh asks for a male animal because He will spare the firstborn sons of the Israelites.

[138] Houtman, II, 172.

[139] Propp, I, 389-390. The expression בן שנה means literally "son of a year." The reasons for this choice can be various: small cattle are usually more tender (Nepi, I, 197), and above all they are not made profane by the work (Num 19:2; Deut 21:3).

[140] J.T. THAMES JR., "Keeping the Paschal Lamb," 10-15; the ritual of Exod 12:1-13 is marked by a "syntax" (time, space, identity and ritual action) similar to that of a sacrifice, "There is no ritual purity requirement, but there are requirements for the physical state of participants (Exod. 12.11, 43-48); there is no sacred space, but there are well-defined spatial requirements (12.3-4,7); there is no dashing of blood on an altar, but there is manipulation of blood (12.7)" (*Ibid.*, 15).

[141] A. MARX, *Les systèmes sacrificiels*, 76, "le sacrifice a à faire avec la vie, non avec la richesse." *Ibid.* 221-222, "Le sacrifice a aussi une fonction proprement théologique… il témoigne d'une intense aspiration à établir une relation entre l'homme et Dieu."

[142] D. BARTHÉLEMY, *Dieu et son image*, 213, "s'il échappe à ce châtiment, ce n'est pas en tant qu'Israël, mais en tant que victime. Le sang de l'agneau pascal marquant ses portes manifeste sa situation actuelle de victime que motivera sa proche libération."

[143] See GK § 134*o-p*; WO § 15.3.2*b*.

[144] See Lev 23:5; Num 9:3, 5, 11. The expression בין הערבים has been a matter of discussion; the preposition בין is used with a local meaning (Gen 15:17), but in this case it could have a temporal aspect; WO § 11.2.6*b*. Since בין implies a reference to a time interval and the term ערבים is dual, בין הערבים can designate the period between the "two" evenings, the phase immediately following sunset: see Rashi Exod, 78; Ibn Ezra, 77; Childs, 182; Propp, I, 390-392; G.I. Davies, II, 39. The Septuagint translates πρὸς ἑσπέραν "in the

part¹⁴⁵ of the blood and put it on the two doorposts and the lintel, on the houses in which they eat it. ⁸ And they shall eat the flesh that night, roasted on the fire, with unleavened bread,¹⁴⁶ with bitter erbs they shall eat it.

The chosen animal must be kept with care until the fourteenth of the month, "you shall watch it."¹⁴⁷ At twilight, when night is falling, the animal will be sacrificed by the "assembly of the congregation of Israel." The repetition of עדה, already used in v. 3, is reinforced by the addition of the term קהל "assembly," which is actually pleonastic.¹⁴⁸ This expression accentuates the relationship between the family celebration and the simultaneous participation of the entire community of Israelites.¹⁴⁹ Although the Passover is a domestic feast, it has the strength to unite the people as an assembly. The expression "at twilight" offers a useful notification of time, simultaneously affirming that the sacrifice must be made at a moment of passage ("between the two evenings," בין הערבים), when twilight comes and begins to deepen. This expression creates the first connection with 4:24-26, in which a blood ritual saves Moses' life when the night is dark and obscure.¹⁵⁰

The first mention of "blood" is found in v. 7 which does not refer to human blood, as in the case of Moses (4:25), but it is related to the sacrificial victim;¹⁵¹ there are, however, several points of contact between the two stories. Blood must be scattered on the doorposts and on the lintel of the doors to mark the whole house,¹⁵² especially indicating the vulnerable boundary between the public and private spheres.¹⁵³ The act of smearing the doors with blood will indicate the house, in which one or

---

evening." Flavius Josephus places the Passover sacrifice between the ninth and eleventh hour (*Bell.* VI,9,3, 423); according *mPes* 5,1, instead, the sacrifice took place between half an hour after the eighth hour and half an hour after the ninth hour.

¹⁴⁵ The מן has a partitive meaning (JM § 133*e*; WO § 11.2.11*e*); see Exod 16:27; Lev 5:9.

¹⁴⁶ The Hebrew text simply has a *waw*: ומצות. See G.I. Davies, II, 40, who thinks that the *waw* could be translated "with" (JM §§ 150*p*; 151*a*). See also Houtman, II, 178.

¹⁴⁷ The term משמרת is used for the keeping of the Tabernacle (Num 3:38) and for the preservation of objects that keep the memory of the Exodus alive: the manna (Exod 16:23, 33-34) and Aaron's rod (Num 17:25). See J.T. THAMES, JR., "Keeping the Paschal Lamb," 9.

¹⁴⁸ S. GESUNDHEIT, *Three Times*, 51, note 14.

¹⁴⁹ See Cassuto, 138: the Israelites "become integrated into a single assembly by their united and simultaneous act of worship." See Houtman, II, 174-175; Propp, I, 390.

¹⁵⁰ See ch. II, § 4.3.

¹⁵¹ Exegetes recognise in the vocabulary of 12:7 a contact with P terminology (Exod 29:20; Lev 4:18; 8:15, 23; 9:9; 14:14; 16:18; Ezek 43:20; 45:19); see Utzschneider – Oswald, 241.

¹⁵² See Priotto, 218, note 54. The same preposition על is employed for the doorposts, the lintel and the house: the last noun refers to the preceding ones, which it recapitulates.

¹⁵³ See Gen 19:9; M. LURKER, *Dizionario delle immagini e dei simboli biblici*, 161-162.

more families have gathered for the celebration, thus distinguishing the organized space of the ritual action, from the chaotic exterior.[154]

Such a gesture, therefore, is principally apotropaic in character, inasmuch as it opposes the threat that is outside of the vulnerable threshold of the domestic space.[155] The blood has a "repellent" role as in Ezek 45:19, where the blood of the sin offering is applied to the doorposts (מזוזת, as in Exod 12:7), protecting the entrance of the temple and the courtyard doors.[156] The indication of Exod 12:7, moreover, finds an echo in Deuteronomy, when Moses orders the Israelites to write divine words on the doorposts (מזוזת) of their houses[157] (6:9; 11:20). Such an ideal exercise (few were able to write at the time) aims to encourage memory, involving every faithful in an act of reading everytime that they move from the private to the public sphere and vice versa,[158] and leaves a symbol on the door which will guard the area of the threshold.[159]

The blood also evokes another boundary, the one crossed when the child is born. This connection has been highlighted by Ilana Pardes:

> The blood that marks the Israelites is not only apotropaic. Its location on the two side posts of the door evokes natal imagery. The Israelites are delivered collectively out of the womb of Egypt.[160]

This suggestion is confirmed by the comparison of Exod 12:7 with 4:18-31: just as Moses joins Israel, God's "first-born son" (4:22), and is "born

---

[154] See W.K. GILDERS, *Blood Ritual*, 49, "The blood manipulation actions establish a boundary and construct a sphere for subsequent ritualized activity, the specialized eating of the lambs .... Finally, the precise and ordered placement of the blood indexes order, defining the marked homes as ordered space, which is marked off from chaotic space. Inside the blood-marked houses life is preserved."

[155] P.P. ZERAFA, "Il significato del sangue nella Pasqua biblica," 453-465; R. GELIO, "Il rito del sangue," 468-474; S.D. SPERLING, "Blood," 762; N. FÜGLISTER, "Blut," 533-534; C.A. EBERHART, "Blood. I. Ancient Near East and Hebrew Bible/Old Testament," 204.

[156] See J. MILGROM, "מזוזה," 803-804; W. ZIMMERLI, *Ezekiel 2*, 483.

[157] Houtman, II, 176; Priotto, 218. In the Ancient Near East, the symbol of a deity could be placed at the entrance of the temple in order to protect the building from evil forces threatening it from outside; see A. LANGE, "The Shema," 212.

[158] The act of placing the *Shema'* on doorposts has traditionally been interpreted as an action that favours memory; see A. LANGE, "The Shema," 212-213 in which some testimonies are reported: *Aristeas*, § 158, "on doors and gates he has prescribed that we set up the sayings to serve as a reminder of God"; see also PHILO OF ALEXANDRIA, *De specialibus legibus*, IV, 142; FLAVIUS JOSEPHUS, *Ant. Iud.* VIII, 213. See E.S. ALEXANDER, "Ritual on the Threshold," 100-130.

[159] A. LANGE, "The Shema," 214. Many of the *Mezuzot* found at Qumran were leather capsules containing pieces of skin with various biblical texts written on them. In the light of these discoveries, the author concludes, "The capsule which contained the Shema and other texts took on an apotropaic function." See also C. OWIREDU, *Blood*, 149-155.

[160] I. PARDES, *The Biography of Ancient Israel*, 26.

again" thanks to the blood of circumcision (4:24-26) which makes him a member of the people and thus a "family member" of God, likewise in 12:7 the new recurrence of the blood motif prepares the reader to expect a birth-like experience. The link between blood and birth is also found elsewhere in Scripture; for example, when in Ezek 16:6 Jerusalem is compared to an exposed child saved by Yhwh, the Scriptures say:

> I passed by you and saw you as you wallowed in your **blood**; and I said to you in your **blood**, "Live!" I said to you in your **blood**, "Live!"[161].

The triple repetition of the plural of דם suggests that the text alludes to a great amount of bloodshed[162] and highlights the paradoxical condition of birth. In fact, at the origin of every person death and life meet: the newborn child who has come into the world has managed to cross an essential threshold, but needs care in order to survive. The exposed child, thus, is in grave danger precisely because of abandonment; weak and defenceless, he is handed over to a destiny of death. Therefore, the passage of Ezek 16 exalts precisely the gratuitous solicitude of God who intervenes and gives life to the child with the power of His word.[163]

Yhwh, in v. 8, refers to the nighttime, and this motif once again associates 12:1-13 to 4:24-26.[164] In the same verse, Yhwh gives instructions on how to eat the animal: the meat must be roasted and cannot be eaten raw or boiled;[165] the dinner will be accompanied by unleavened bread[166] and bitter herbs,[167] evoking the cruel slavery: the reader in fact

---

[161] The v. 6 ends with the repetition of the two sentences, "I said to you in your blood, 'Live!'" (ואמר לך בדמיך חיי); this member creates a problem of textual criticism, because the Septuagint and the Peshitta report the sentence only once, and a difficulty of syntactic character: בדמיך can be linked with the preceding part of the sentence ("I said to you in your blood, 'Live!'") or with what follows ("I said to you, 'In your blood, live!'"). The Vulgate presents the two options in the same text: *et dixi tibi cum esses in sanguine tuo: Vive, dixi, inquam, tibi: in sanguine tuo vive*. O. PETTIGIANI, *"Ma io ricorderò"*, 125-126.

[162] Regarding the use of the plural form of דם, see ch. I, § 1.1.

[163] O. PETTIGIANI, *"Ma io ricorderò"*, 123-128.

[164] See ch. II, § 4.3.

[165] Propp, I, 395-396. The eating of boiled meat was common (Lev 6:21; 8:31); it is forbidden at Passover because the victim must be cooked preserving its integrity (see v. 9).

[166] The term מצה is of uncertain etymology and indicates a cake made of flour prepared in a hurry without having time to rise the dough (Gen 19:3; Judg 6,19). Some scholars argue that מצה is connected with μᾶζα, "barley cake," but the similarity seems to be fortuitous; Propp, I, 393-394.

[167] The noun מרור refers to bitter herbs that could serve as spices. The *Mishna* lists some possible types: lettuce, chicory, polygala, dandelion, watercress (see *mPes* 2,6).

instantly links מרור, "bitter," to the root מרר used in 1:14, "they bittered their lives (וימררו חייהם) with hard slavery."[168]

9 Do not eat of it raw or boiled in water,[169] but only roasted over the fire, the head, with[170] its legs and its entrails. *10 You shall not leave* any of it over until morning; anything that is left of it until the morning *you shall burn* over the fire. *11* In this manner[171] *you shall eat it*, with *your loins* girded, *your sandals* on *your feet, your rod* in *your hands*, you shall eat in haste (and with apprehension): this is Passover for Yhwh.

The second person plural (Italics in the translation) is again used in vv. 9-11, and this change gives to the discourse the compelling energy of a personal appeal.[172] What was said in v. 8 about cooking the animal is clarified and repeated in v. 9. The meat must not be raw, because otherwise the people would consume blood and fat[173] (Lev 3:17). The animal cannot even be boiled, since its integrity must be preserved.[174] Therefore, Yhwh says that the whole animal must be consumed during the night, with nothing left over[175] (v. 10), to manifest, through this ritual sign, the communion of all the family.[176]

The involvement of the participants reaches its climax in v. 11. Indeed, Yhwh's instructions do not merely provide a list of ritual prescriptions, but even recommend certain clothing, "with your loins girded, your sandals on your feet, your rod in your hands." In antiquity, one's dress had a relevant anthropological quality[177] because it could express an inner disposition (see Isa 59:17; Job 29:14). Loins are girded to facilitate work and to begin a journey;[178] the same applies to sandals[179] and to the staff (2 Kings 4:29). The garment

---

[168] See *mPes* 10,5; Rashi Exod, 78-79; Cassuto, 139; Propp, I, 394; Priotto, 219; Fischer – Markl, 133; G.I. Davies, II, 52. The expression "he has sated me with bitterness" (מרור) in Lam 3:15 alludes to an intense suffering.

[169] The adjective בשל means both "boiled" (1 Sam 2,13) and "raw" (2 Sam 13,8).

[170] The preposition על can mean also "in addition to, with," see G.I. Davies, II, 40.

[171] The adverb ככה, "thus," is proleptic and refers to what follows in the sentence; see GK § 110*i*; JM § 102*h*; Houtman, II, 182; see also Jer 13:9; 19:11.

[172] See above, ch. III, § 2.2.1.

[173] Propp, I, 394; G.I. Davies, II, 52.

[174] Cassuto, 139.

[175] See Propp, I, 395. The logic behind the requirement to consume the whole Passover animal is similar to the idea which governs other rituals: the consecration of the priests (Exod 29:34; Lev 8:32) and the peace offerings (Lev 7:15; 22:30).

[176] Houtman, II, 180-181.

[177] V.H. MATTHEWS, "The Anthropology of Clothing," 25-36.

[178] See Houtman, II, 182. The loins are girded when the person travels (2 Kings 9:1) or has to work (Prov 31:17). See also Isa 5:27; Jer 1:17.

[179] Propp, I, 200.395. Wearing sandals at home is a signal that someone intends to leave soon (see 12:11; 3:5).

imposed by Yhwh, thus, expresses the tension and readiness of the person who intends to move soon. The recommended clothing creates a connection with the dynamism already appreciated in Moses' return journey (4:18-31).

The Passover shall be eaten בחפזון, "in haste." The term חפזון is used only in Deut 16:13 and in Isa 52:12, where it is positioned parallel to מנוסה, "flight," and refers to the urgency and apprehension of one who flees from an imminent danger.[180] The evident wordplay between the two terms חפזון and פסח reveals once again that an active dynamism is essential to the Passover.[181]

The verse concludes with a crucial statement, "this is Passover for Yhwh." The noun פסח can define the rite in its entirety (Exod 12:43-44), as a synecdoche in which the whole is indicated by one part;[182] in v. 11 פסח designates the sacrificial animal (12:21; see Ezra 6:20). In the nominal clause, פסח is the predicate[183] and "Yhwh," preceded by the preposition, modifies the predicate, "for Yhwh" (ליהוה). This last expression appears for the first time in 12:11 and is repeated several times in Exod 12–13.

+ Feast (חג) for Yhwh (12:14; 13:6); Passover (פסח) for Yhwh (12:48); Passover sacrifice (זבח פסח) for Yhwh (12:27)
+ Night for Yhwh (12:42 [*bis*]); First-born for Yhwh (13:12[*bis*], 15)

The scheme demonstrates that several elements which constitute the Passover's feast refer explicitly to Yhwh ("for Yhwh"), to manifest that the "theological direction" is a crucial aspect of the whole ritual:[184] the relationship with God is the secret that sustains the whole Passover.

2.2.3 Yhwh Re-establishes Justice (vv. 12-13)

> [12] And I will pass through *the land of Egypt* in this night, and strike down every first-born in *the land of Egypt*, from man to beast; and on all the gods of *Egypt* I will execute judgements[185], I Yhwh. [13] The blood shall be a sign for you on the

---

[180] Priotto, 219-220; G.I. Davies, II, 41; the verb חפז expresses the haste of the fugitive (2 Kings 7:15), and can be a synonym of ירא, "to fear" (Deut 20:3; see Pss 31:23; 116:11).

[181] Utzschneider – Oswald, 243.

[182] E. OTTO, "פסח," 668. When the noun פסח is correlated with the verb עשה, "to do," it indicates the ritual (Exod 12:48; Num 9:2; Josh 5:10; 2 Kings 23:21-22; Ezra 6:19; 2 Chron 30:1-2). Normally, it is part of a non-verbal clause, as in Exod 12:11 (Lev 23:5; Num 28:16), or is correlated to the term חג, "feast" (חג הפסח, see Exod 34:25). Otherwise, the term פסח can refer to the animal, especially when it is preceded by שחט, "to slaughter" (Exod 12:21; 2 Chron 30:15) or זבח, "to sacrifice" (Deut 16:2; see Exod 12:27).

[183] G.I. Davies, II, 41. The noun פסח is the predicate of the non-verbal clause, because it is the element of the sentence that provides new information (WO § 4.5c).

[184] Propp, I, 399.

[185] The term שפטים is rendered "judgments" to preserve the plural in the translation.

houses where you are staying; I will see the blood and pass over you[186]; so that no plague will be among you for destruction[187], when I will strike *the land of Egypt*.

The passage is marked by a rapid shift to the first person singular: according to a procedure present in the legal texts of the Pentateuch and absent in the Ancient Near East, the divine lawgiver participates in the law He communicates, exposing Himself in the first person and taking a clear stance.[188] This phenomenon is undoubtedly conscious, because the verbs used for the divine actions form a sequence that revolves around a centre,[189] "I Yhwh."

Exod 12:12a	Exod 12:12	Exod 12:13
I will pass, strike, do אעשה ;והכיתי ;ועברתי	I Yhwh אני יהוה	I will see; pass over; strike בהכתי ;ופסחתי ;וראיתי

The repetition of the word "Egypt" (Italics in the translation) enhances the separation that Yhwh will create between the victim (Israel) and the persecutor (Egypt), re-establishing justice in the night of Passover. Moreover, for the first time in the Hebrew Bible, the verb עבר is used with a divine subject, communicating God's involvement in human vicissitudes;[190] in the night, God will pass through Egypt, making Himself extraordinarily near.[191] The reader cannot fail to notice a certain contiguity with the covenant ritual of Gen 15:17, in which God, in the form of a smoking brazier and burning torch, passes between the animals cut in two parts to forge a stronger bond with Abraham.[192]

---

[186] The sequence of inverted *qatal* with a future meaning could imply a logical relationship between the different elements of the succession; see GK § 124*g*. Davies believes that in this case it is a matter of purely temporal consequentiality, "*when* I will see the blood, I will pass over"; see JM § 167*b*; G.I. Davies, II, 41-42.

[187] While משחית in v. 23 designates a "destroyer" in v. 13 can be translated with a general meaning of "destruction" (Jer 51:25; Ezek 9:6; 21:36). See G.I. Davies, II, 42. In v. 13 the action is directly associated with God.

[188] A. BARTOR, *Reading Law as Narrative*, 24, "A lawgiver who employs emotionally charged language to describe an event is not merely presenting facts but is taking a position. By doing so, he exposes his persona."

[189] See Priotto, 220. In 12:12-13. In footnote 63, he quickly analyses the composition of the verse and shows that it is structured around the central statement "I Yhwh." See also Fischer – Markl, 134; Utzschneider – Oswald, 244.

[190] The verb עבר designates divine intervention, both when Yhwh manifests His glory (Exod 34:6; 1 Kings 19,11) and when punishes the evildoers (Amos 5:17). See Ezek 16:6, 8.

[191] *TgO* and *TgPsJ* of Exod 12:12 translate עבר with the *hitpaal* of גלה ("I will be revealed"), whereas *TgN* uses a form of circumlocution to respect the divine transcendence: "I will pass through my Memra (ואעבר במימרי)." These choices demonstrate that the targumim considered audacious the use of עבר with a divine subject.

[192] P. BOVATI – R. MEYNET, *Amos*, 198, note 52.

The term נכה and the construction שפטים + עשה in v. 12 refer to the execution of a sentence[193] with which Yhwh will smite the guilty in order to defend and save the victims. He exercises His authority and sovereign judgement[194] even against the deities of Egypt,[195] He has the strength to keep foreign gods under control.[196] The chain of actions concludes with a nominal clause, "I Yhwh," which has an exclamatory energy.[197] This formula is already used in Exod 6:2-8,[198] and does not have a merely notional meaning, but is rather un utterance with pragmatic meaning:[199] Yhwh offers His own person and His name[200] as the only guarantee for the Israelites.

The blood is mentioned for the second time in v. 13, specifying that "it will be for you (לכם) a sign on the houses where you are staying;" the preposition ל creates a correspondence between two aspects: the Passover, celebrated "for Yhwh" (ליהוה, v. 11), will be also in favour of the Israelites ("for you"). This clarification shows that the blood on the doorposts is not intended to "inform" Yhwh, but that it has the sole purpose of aiding the persons gathered in the house.[201] Moreover, v. 13 creates once again an

---

[193] The verb נכה "to strike" designates the sentence (Deut 25:2; Prov 17:26) or is used as a synonym of שפט "to judge" (Isa 11:4). The construction "to execute judgements" (עשה + שפטים) alludes to the final decision of the judicial authority and is normally used for an action accomplished by God (e.g., Ezek 5:10; 11:9). See Houtman, II, 184; Priotto, 220.

[194] The *MekhY* reports the opinion of a rabbi who associates the verb עבר with the movement of the king that inspects his territory, "Like a king who passes from one place to another" (*Pisḥa*, § 7, p. 38).

[195] After the first two verbs in the future tense (*w-qataltí*; JM § 119: ועברתי and והכיתי) and before two other similar forms (ופסחתי and וראיתי), the *yiqtol* אעשה, "I will do," is employed. The verb could express an assertive epistemic modality: the speaker commits himself to what he says, manifesting conviction "*Indeed*, I will make judgements." See A. GIANTO, "Mood and Modality," 183-198.

[196] See G.I. Davies, II, 54. He refers to Ps 82:1-2, 6-7 and Isa 24:21-23 in which Yhwh is depicted as a king ruling the heavenly court (see Pss 29:1-2; 97:7.9). The background could be Canaanite, in which *'Ilu* is king of an assembly of gods and is entitled to exercise his judgement over them; see Propp, I, 400. See also Cassuto, 140.

[197] See WO § 40.2.3*a*.

[198] L. INVERNIZZI, *"Perché mi hai inviato?"*, 323-325.

[199] See A.A. DIESEL, *"Ich bin Jahwe"*. In the light of the comparison with the Ancient Near Eastern literature, the clause אני יהוה can be a formula of self-presentation by which the absent person makes himself present through the act of speaking; see the Semitic royal inscriptions in which the king uses a similar formula, *KAI*, 10:1; 26:I:1; 181:1; 214:1; 216:1; 217:1. R. MÜLLER, "The Sanctifying Divine Voice," 70-84.

[200] See ch. II, § 4.3, note 235.

[201] The rabbinic tradition interprets לכם as a clarification that the "sign" of the blood is for Israel's benefit and not for Yhwh's. *MekhY*, *Pisḥa* § VII, p. 39, "A sign to you and not to Me, a sign to you and not to others." See Priotto, 221 and Propp, I, 401.

explicit link with 4:24-26, because here too the blood is a "sign," an external reality that refers to a deeper meaning. In 4:24-26, Moses is saved through circumcision, the "sign" *par excellence* that expresses the covenant with God;[202] in 12:13, the blood of an animal victim is a sign that a family is celebrating the paschal ritual.[203] In both cases blood indicates the union with Yhwh achieved through the ritual (circumcision and Passover) and manifests the bond that unites the "first-born son" (4:22-23) to Yhwh.

God's judgement against the Egyptians is performed after "taking vision" of the sign on the Israelites' doors (v. 13). When the verb ראה refers to God, it denotes an action that has redeeming consequences.[204] In the Book of Exodus, the root is very significant, because the story of liberation is triggered precisely when the narrator informs the reader that God "saw the Israelites"[205] (2:23; 3:7). In v. 13 the verb פסח is used for the first time, maintaining a certain polysemy; it can be translated both as "Yhwh passes over" and "Yhwh protects"[206] the people who are in the house to perform the paschal rites. The dramatic consequence of God's passage is enhanced by the repetition of the root נכה.

In conclusion, Yhwh gives the Israelites a ritual that allows them to face the "liminal" and terrible experience of the night without dying. Salvation is not achieved through a magical strategy, but rather it is the very presence of God ("I Yhwh") that liberates the Israelites and executes the Egyptians.[207]

2.2.4 The Memorial (v. 14)

> [14] And this day shall be a memorial for you, you shall celebrate it as a feast for Yhwh, throughout your generations, as an everlasting statute, you shall celebrate it.

---

[202] See ch. II, § 3.2.1 e § 4.2.

[203] See in this chapter the commentary on Exod 12:7 in § 2.2.2.

[204] See Deut 26:7; 1 Sam 1:11; Isa 38:5; Pss 9:14; 25:18.

[205] The verb ראה is one of the terms used for a formal inquiry (Gen 18:21; Job 10:6-7; 11:11). See P. BOVATI, *Re-establishing Justice*, 68-71.

[206] Regarding פסח, see Houtman, II, 183; Propp, I, 398-401. The verb פסח can be translated both "to pass over" and "to protect" (see Isa 31:5); the Septuagint of Exod 12:13 (and of 12:27) — "I will see the blood and pass over you (ראיתי את־הדם ופסחתי עליכם)" — renders פסח with the verb σκεπάζω, "to protect"; in 12:23, instead, it uses the verb παρέρχομαι, "to pass over"; see Rashi Exod, 80, "every [appearance of] פסח, means skipping and jumping (לשון דלוג וקפיצה)." See Priotto, 220, note 62. Rabbinic tradition noticed the ambiguity and maintained the two contrasting versions; see *MekhY, Pisḥa* § VII, p. 40, "Do not read ופסחתי (I will protect) but ופסעתי (I will step over). God skipped over the houses of His children in Egypt ... R. Jonathan says, 'I will pass over you.' This means, you alone will I protect but I will not protect the Egyptians."

[207] As in 8:18 and 10:23, Yhwh judges by separating the guilty from the innocent.

This passage marks the transition between the first part of the divine discourse, dedicated to the rite of the lamb, and the second, in which the ritual of unleavened bread is introduced.[208] Yhwh instructs the people to consider the day[209] of God's Passover as a "memorial." The term זכרון is derived from the root זכר and literally means "remembrance" (Eccles 1:11). The noun also refers to what promotes and strengthens the memory, whether it is a text (Exod 17:14) or an object (Exod 28:12.29; 30:16; Josh 4:7; Zech 6:14). The noun can also denote a feast (Lev 23:24) capable of making present the past event[210], through the celebration.[211]

The passage again recalls an important aspect of Passover: it must be celebrated as a feast "for Yhwh" (ליהוה, see v. 11), the rite expresses the relationship of the faithful with God. Moreover, the term חג, which normally indicates a feast of pilgrimage[212] (Zech 14:16, 18-19), in this case, qualifies the Passover as a cultic action characterised by joy and vitality.[213]

Finally, the ritual shall be transmitted "throughout your generations."[214] Sociologists have long shown that the term "generation" denotes a dual reality, which can be explained both by a vertical axis (from father to son) and by a "horizontal" dimension[215] (the generational group). This dual path is present in 12:14, since the formula לדרתכם "translates the role of each coeval group in the

---

[208] S. GESUNDHEIT, *Three Times*, 79-80.

[209] The expression "this day" extends the character of "memorial" both to the night of the Passover rite and to the day of deliverance (Exod 13:3), the first day of the feast of the Unleavened Bread. See Houtman, II, 186; Priotto, 221.

[210] J. ASSMANN, *Cultural Memory*, 37, the "figures to which memory attaches itself... are celebrated in festivals and are used to explain current situations." The symbolic elaboration of stories during a ritual contribute to establish the identity of the group that shares the same memory. See ID., "Memory, Narration, Identity," 3-18.

[211] Nepi, I, 196; Priotto, 221.

[212] The use of חג is explained by the authors in two different ways: the editor P intentionally means to create an anachronism by describing the feast with a term closer to him (see Ezek 45:21-24); the noun simply means "solemn feast." See G.I. Davies, II, 57.

[213] The "feast" is an action, חג is often the object of the verb עשה, "to do" (see Ezra 3:4; 6:22). Feasts are joyful celebrations of salvation (Hos 2:13; see Amos 8:10).

[214] See J.-P. SONNET, "'In quel giorno'", 18-20. The succession of generations "is what connects the two temporal moments of narrative communication ('on that day' / 'to this day')".

[215] See P. MOLLO, *The Motif of Generational Change*. The author analyses the various ways in which the generational change takes place. She notes that the term דור, "generation," is used for the generational group, while the Hebrew תולדת refers to the "genealogy," intended as a vertical designation (from father to son). The noun דור disappears after Judg 2:10 and 3:2, where it is used for the last time (with the only exception of 1 Chron 16:15) in the historical narration. From then on, responsibility is no longer attributed to men as a collective entity, but falls essentially on the kings.

inherited experience, lived throughout history, understood as tradition."[216] The generation of those who experienced the Exodus, therefore, has a considerable task in relation to their posterity: to bear witness.[217] This is why v. 14 speaks of an "everlasting statute," a command[218] addressed to the parents with respect to their children; they have to transmit the memory of liberation, through the Passover celebration. Therefore, the first "order" given by Yhwh to the people is to form the identity of their posterity through the memory of His wonders.[219]

2.2.5 The Unleavened Bread (vv. 15-20)

The ritual of unleavened bread, which probably originated in a context of sedentary farmers,[220] is introduced in v. 15. The text is organised according to a concentric composition constructed around v. 17.[221]

[15] Seven days[222] *you shall eat unleavened bread*. Indeed,[223] in the first (הָרִאשׁוֹן) day you shall remove[224] **leaven** (שְׂאֹר) from your houses, for whoever eats what is **leavened**,[225] that person shall be cut off[226] from Israel (מִיִּשְׂרָאֵל), from the first (הָרִאשׁוֹן) day to	[18] In the first month, on the fourteenth of the month, in the evening, *you shall eat unleavened bread*, until the twenty-first day of the month, in the evening. [19] For seven days, no **leaven** shall be found in your houses, because whoever eats from what is **leavened**, that

---

[216] J.-P. SONNET, "Generare, perché?" 152. The formula "from generation to generation" is used several times in the Psalms to express the involvement of each generational group with the transmission of the inheritance received from the fathers (Pss 33:11; 79:13; 49:12).

[217] See F. MARKL, "The Sociology." The term "generation" designates a social unit which shares a common destiny. The generation's consciousness and the cultural memory developed in Israel through the experience of trauma; J.J. AHN, "Diaspora Studies," 220-224.

[218] Nepi, I, 199.

[219] The distinction between "Passover in Egypt" and "Passover for subsequent generations" originates in rabbinic literature: see *MekhY, Pisha* § III, p. 18.

[220] J.A. WAGENAR, "Passover," 250-268; S. GESUNDHEIT, *Three Times*, 79-89.

[221] Utzschneider – Oswald, 247-248. The correspondences between v. 15 and vv. 19-20 can be the result of an editorial work (ch. III, § 2.1.1).

[222] See GK § 118k; JM § 138b; WO § 14.3.1d.

[223] The particle אַךְ in 12:15 could have an adversative sense, "but." See T. MURAOKA, *Emphatic Words*, 130; Houtman, II, 186. The context, however, prompts to interpret this particle rather with an asseverative meaning, since there is no contrast between the sentence which it introduces and the preceding phrase: see Childs, 182; G.I. Davies, II, 43.

[224] In Exod 12:15 the שבת *hifil* can be translated "to remove" (Ezek 23:27, 48).

[225] The two nouns שְׂאֹר, "leaven," and חָמֵץ, "leavened," namely the dough or bread (see Lev 2:11), are slightly different. The Septuagint translates both with the same word, ζύμη.

[226] The verb וְנִכְרְתָה is preceded by the *waw* of apodosis linked to the *casus pendens* "whoever shall eat." JM § 156e-f; G.I. Davies, II, 43.

the seventh. ¹⁶ On the first day (you shall have) a holy assembly, and on the seventh day (you shall have) a holy assembly. No work shall be done in them, but²²⁷ only what every person is to eat²²⁸ that will be prepared for you.	person shall be cut off from the congregation of Israel, whether he is a sojourner or a native of the land. ²⁰ Nothing **leavened** you shall eat, in all your dwellings you shall eat unleavened bread.
¹⁷ And you shall observe the Unleavened Bread because on this very day²²⁹ I brought your hosts out of the land of Egypt. And you shall observe this day throughout your generations, as an everlasting statute.	

Yhwh states in v. 15 that the feast will last seven days; in these days, only unleavened bread will be eaten. Once again, the material signs illuminate a deeper reality: on the first day one must "remove"²³⁰ leaven (שאר) and the unleavened bread must not retain any trace of what is old so as to explicit the expectation of "newness" that will be fulfilled during the Passover.²³¹ The consequence for offenders is severe: they are expelled from the community and uprooted from the people.²³² In v. 15 a metathesis is recognisable²³³ (highlighted in grey). This effect is obtained through the interchanging of ר and א, and the alternation between שׁ and שׂ. Indeed, the choice of שאר (instead of חמץ) seems to be intentional and the purpose could be to create a link with the word ראשון, "first."²³⁴ As in 12:2,²³⁵ also in 12:15, the root ראש could indicate what is "first" not only from a material point of view: searching for

---

²²⁷ In 12:16, אך has an adversative sense. T. MURAOKA, *Emphatic Words*, 130.

²²⁸ When the preposition ל follows a passive verb, can introduce the author of the action (JM § 132*f*).

²²⁹ The construction בעצם היום הזה is an idiomatic formula that literally means, "on the bone of this day." The term עצם, "bone, body," can be used to replace the reflexive pronoun "itself" and to emphasise the element to which it refers. JM § 147*a*; G.I. Davies, II, 45.

²³⁰ The verb שבת *hifil*, used to indicate the "removal" of all leaven, is interpreted as a subtle contact between the Feast of Unleavened Bread and Sabbath theology; see Priotto, 223, note 74. The first and seventh days of the feast, in fact, are days of rest (12:16; Propp, I, 402.

²³¹ Nepi, I, 199; Priotto, 223. In Leviticus it is stated that nothing leavened may be offered to God (Lev 2:11; see Exod 23:18; 34:25; Lev 6:10).

²³² Propp, I, 403-404; Houtman, II, 187; Priotto, 223. The expulsion from the community could be caused by transgressions against important cultic elements: the circumcision (Gen 17:14); the holy anointing oil (Exod 30:31-33); the Sabbath (Exod 31:14); the sacrifices (Lev 7:20-21); the Tabernacle (Lev 17:4); the Passover (Num 9:13); see also Lev 18:29; 20:6; 23:29-30; Num 15:30-31.

²³³ See I. KALIMI, *Metathesis*, 1: the metathesis is a "transposition, inversion or reversal of *contiguous letters* within a single word in comparison to another word within a text-unit, which tipycally comprises a sentence or a couple of sentences."

²³⁴ See I. KALIMI, *Metathesis*, 128-129.

²³⁵ Regarding the root ראש, see above, § 2.2.1.

leaven, with the transformation that it manifests, is an element that stands as a foundation for the Unleavened Bread ritual and expresses its main quality.

The first day and the seventh day of the feast there will be a "holy assembly" (v. 16, מקרא קדש) in which the Israelites will cease from all work[236]. Holiness, therefore, as with the Sabbath, will be manifested in the difference between these days and the others.[237]

In addition, the table shows that v. 17 is an essential part of vv. 15-20 and presents the theological sense of the ritual: "for on this very day I brought your hosts out of the land of Egypt." As in v. 11, Yhwh confirms His direct intervention to redeem the people, as the subject of the verb יצא *hifil*, "to bring out." However, this statement is surprising, since the Exodus is described as an event that has already taken place[238] ("I brought out"), creating a significant theological effect: liberation is announced as a certainty. The reference to the "hosts" of Israel, moreover, illustrates the Exodus as a military and cultic action at the same time: Israel will come out of Egypt as an army, but only after the ritual celebration of the Passover.[239] The v. 17 concludes in a similar way when compared to v. 14.

Exod 12:14	Exod 12:17
This day shall be a memorial for you, you shall celebrate (vb. חגג) it as a feast for Yhwh, throughout your generations, as an everlasting statute, you shall celebrate (vb. חגג) it.	You shall observe (vb. שמר) this day throughout your generations, as an everlasting statute.

The connections between v. 14 and v. 17 show that the Passover and the Unleavened Bread are feasts to be celebrated (חגג) and statutes to be "observed" (שמר); the use of the root שמר, in fact, means that the instructions about the rituals are interpreted both as an order to be obeyed and as a word to be treasured through an active participation of the subject.[240]

---

[236] In Lev 23:3 even the Sabbath is defined as "a holy assembly."

[237] The difference between holy and ordinary days is expressed through external signs: food, drink and clothing; see *MekhY, Pisḥa* § IX, p. 48. Regarding קדש, see ch. I, § 1.3.

[238] Houtman, II, 189.

[239] The noun צבא is normally used in a military context (Judg 4:7; 2 Sam 3:23; 10:7). Nevertheless, the verb צבא can even indicate the service in the sanctuary (Exod 38:8; Num 4:23; 8:25; 1 Sam 2:22); see Houtman, I, 520. The recurrences of צבאות in Exodus (6:26; 7:4) are usually interpreted in a military sense, since the exit from Egypt takes place through a clash between two hostile forces, Yhwh and Pharaoh (see L. INVERNIZZI, *"Perché mi hai inviato?"*, 407-408); furthermore, in Exod 13:18 it is explicitly stated that the people advance "in order of battle" (J.-L. SKA, *Le passage*, 14-17). See Propp, I, 405.

[240] The PSam of Exod 12:17 translates the MT ושמרתם adding עשה: ושמרתם ועשיתם "you shall observe and you shall do." The Septuagint, instead, simply renders ποιήσετε, "you will

Several elements mentioned in vv. 15-16 are developed in vv. 18-20.[241] In v. 18 Yhwh says that the Feast of Unleavened Bread begins exactly on the evening of the 14th of Nisan and ends on the 21st; the clarification serves to unify the ritual of the lamb with the celebration of the unleavened bread (Lev 23:32). Furthermore, in v. 19 it is specified that both the sojourner (גר) and the native are obliged to abstain from eating leaven. Finally, the people are defined more articulately, as the "congregation of Israel," and v. 20 repeats the command not to eat anything leavened, reinforcing what was already stated in v. 15.

2.3 Moses' Words to the Elders (vv. 21-28)

Once the divine discourse addressed to Moses is concluded, vv. 21-27 change register and narrate the words that Moses pronounces before the elders of Israel.[242] Normally, this factor has been underestimated, since many scholars considered this part as a repetition of some previous elements and explained this phenomenon only from a historical and redactional point of view.[243] On the contrary, the literary approach (discourse oriented) considers the order of the elements in the text as a significant feature for its meaning.[244] Thereupon, the reader must take the change of locutor seriously and ask himself whether Moses' words will faithfully report what he has heard from God; in this way, the reader is actively involved in the story, because he knows what Yhwh said to Moses, and he can recognise what Moses will choose to say.

2.3.1 The Blood on the Door (vv. 21-23)

²¹ Moses called all the elders of Israel and said to them, "Go,[245] pick a lamb for yourselves and for your families and slaughter the *Pesaḥ*.[246] ²² Take a bunch of

---

do." See Houtman, II, 189. This difference perhaps shows that the translators hesitate in translating שמר verbatim; the two verbs עשה and שמר express the careful observance of the law (Deut 4:6; 5:1, 32; 7:12; 16:12; see Lev 22:31; 25:18). The combination of these two roots presents the obedience to the law as an active commitment (עשה, "to do"), but also as an action that involves the person inwardly (שמר, "to observe"). See T. MURAOKA, "A Deuteronomistic Formula," 548-550; G. PAPOLA, *L'alleanza di Moab*, 92-93.

[241] See S. GESUNDHEIT, *Three Times*, 84-88.
[242] The elders are the representatives of the people to whom Moses is already sent in 3:16. They appear elsewhere in Exodus (17:5-6; 18:12; 19:7) and exercise various roles of responsibility (Deut 19:12; 21:2-4; 22:15; 25:7; 31:9). See Dozeman, 273.
[243] See in this chapter, § 2.1.2. See also W.K. GILDERS, *Blood Ritual*, 43.
[244] See M. PERRY, "Literary Dynamics," 35.
[245] The Hebrew root משך means "to draw, to pull." In this case it is associated to לקח and can be translated as "to go" (see the Septuagint; Judg 4:6). Propp, I, 407.
[246] In v. 21, as in vv. 11, 27, the term פסח refers to the slaughtered animal.

hyssops²⁴⁷, and dip it in the blood that is in the basin, and apply on the lintel and on the two doorposts²⁴⁸ some of the blood which is in the basin. Not a single one of you will leave the door of your house until morning. ²³ Yhwh will pass through to strike Egypt and will see the blood on the lintel and on the two doorposts and will pass over the door and will not allow²⁴⁹ the Destroyer to enter your houses to strike.

With his first words (v. 21), Moses alludes to the the lamb and to its slaughter²⁵⁰ (vv. 1-6) without insisting on the details that regard the sacrificial victim; this decision to omit some particulars and to translate the order received into practical and essential instructions may be a sign of the haste that characterises the whole rite (v. 11). The tone employed to address the elders is exhortatory and its formulation presupposes some urgency.²⁵¹ They have to take animals from the flock to slaughter them during the ritual. The Passover, on the contrary, is described by exalting the "benefit" it procures to the people who celebrate it (לכם, "for you"), as seen above;²⁵² the context is always domestic.²⁵³

In contrast to v. 21, instead, the vv. 22-23 recall with more details what the reader already knows from vv. 7, 12-13. In the text it is evident that there is a meaningful wordplay between the terms סף, "basin," פתח, "door" and the verb פסח "to pass over." Since סף is a polysemic noun, which can also be translated as "threshold" referring to the house,²⁵⁴ the wordplay

---

²⁴⁷ The traditional translation of אזוב is "hyssop," although the plant intended could rather be the "marjoram" (*Origanum majorana*); see Houtman, II, 192; Propp, I, 407; Dozeman, 273; G.I. Davies, II, 86.

²⁴⁸ The מן has a partitive value; WO § 4.4.1*b*; 9.6*b*; 11.2.11*e*.

²⁴⁹ When נתן is connected to another verb introduced by the preposition ל, it can be translated "to allow, to permit." See Gen 20:6; 31:7; Num 22:13; Pss 16:10; 66:9.

²⁵⁰ S. GESUNDHEIT, *Three Times*, 59. The author notices that vv. 21-27 cannot be an isolated fragment or a continuation of the non-P narrative on the tenth plague (11:4-8), because "when Moses … commands the people to slaughter the *Pesaḥ*, he refers to the *Pesaḥ* as something familiar ('slaughter *the Pesaḥ*')." See Houtman, II, 192.

²⁵¹ The sequence of two verbs, משך and לקח, is similar to the association of הלך and לקח, "Go, take!" (Exod 5:11; Hos 1:2), where הלך is used as an interjection (JM § 105*e*; Gen 37:20; 1 Sam 9:9). Rabbinic tradition tries to explain the use of משך, "to pull." *TgPsJ* Exod 12:21, "stretch out *their hands from the idols of the Egyptians*;" *MekhY, Pisha* § XI, p. 58.

²⁵² The preposition ל appears several times in Exod 12 and specifies who benefits from the action, the *dativus commodi* (JM § 133*d*; Houtman, II, 192). See vv. 2, 5, 13.

²⁵³ The noun משפחה normally implies a social nucleus smaller than the tribe, but broader than the family (see Exod 6:14-25), more often called "father's house" (Gen 24:7). The term, however, can also be translated simply as "family" (Gen 24:38).

²⁵⁴ The term סף means both "threshold" and "basin." The ancient versions reflect this ambiguity: the Septuagint has θύρα, "door," the Vulgate *limen*, "threshold." *TgO, TgN, TgPsJ*, renders מן, "container" (see Childs, 183). See *MehkY, Pisha* § XI, p. 59, "*Saf* here

between the three lexemes could be further evidence that the main setting in which liberation takes place is the door, the boundary area where the house is exposed to danger and is more fragile.

Exod 12:7, 12-13	Exod 12:22-23
⁷ They shall take part of the blood and put it (ונתנו) on the two doorposts and the lintel, on the houses in which they eat it.	²² Take a bunch of hyssops, and dip it in the blood that is in the basin (בסף), and apply (והגעתם) on the lintel and on the two doorposts some of the blood which is in the basin (בסף). Not a single one of you will leave the door (מפתח) of your house until morning.
¹² And I will pass through (ועברתי) the land of Egypt in this night, and strike down (והכיתי) every first-born in the land of Egypt, from man to beast; and on all the gods of Egypt I will execute judgements, I Yhwh.	²³ Yhwh will pass through (ועבר) to strike (לנגף) Egypt
¹³ The blood shall be a sign for you on the houses where you are staying; I will see the blood (וראיתי את־הדם) and pass over you (ופסחתי עלכם); so that no plague (נגף) will be among you for destruction (למשחית), when I will strike (בהכתי) *the land of Egypt.*	He will see the blood (וראה את־הדם) on the lintel and on the two doorposts, will pass over the door (ופסח על־הפתח) and will not allow to the Destroyer (המשחית) to enter your houses to strike (לנגף).

The comparison between vv. 7, 12-13 and vv. 22-23 is crucial for the interest of this research: Moses chooses not to devote much attention to the words received in vv. 1-6 and concentrates his discourse precisely on the rite of blood[255] (vv. 7, 12-13), giving it a special place.

In v. 22, Moses recalls v. 7, specifying that the blood's application will be carried out with a bundle of hyssop (or more precisely marjoram); thereby, he reveals that the gesture can be interpreted as an act of purification.[256] The comparison with v. 7, moreover, shows that Moses uses a different verb to indicate the act of smearing the blood on the Israelites' houses. In fact, he

---

means threshold, as in the passage, 'In their setting of their thresholds by My threshold' (Ezek 43:8). And it also says, 'And the posts of the thresholds were moved' (Isa 6:4). R. Akiba says: *Saf* here means vessel, as in the passage, 'And the cups, and the snuffers, and the basins' (1 Kings 7:50)." See Propp, II, 408; G.I. Davies, II, 87.

[255] Priotto, 226; Utzschneider – Oswald, 252.

[256] Marjoram is used in purification rituals to facilitate the application of blood (Lev 14:6; see Ps 51:9). See Propp, I, 407; regarding blood and purification, see ch. I, § 1.3.

employs נגע instead of the root נתן used in v. 7; the verb נגע appears earlier in 4:25, referring to the gesture of Zipporah who decides to touch Moses' feet with his son's foreskin[257] and it is rarely given this meaning.[258] In the words of Moses, therefore, his personal experience resounds and the vivid memory of the night attack of the Lord (4:24-26) becomes an interpretative key useful for understanding what the people are about to experience. When he is between life and death, Moses is saved thanks to the blood of circumcision that consolidates his (and his family's) relationship with God; in a similar way, the application of blood on the doors as a sign of the presence of Israelites gathered for the Passover[259] will be the cause of salvation, precisely because the Passover expresses and fulfils the bond between the Israelites and Yhwh. The repetition of נגע then establishes the extraordinary value of blood as an element capable of creating an authentic "circulation" of life between Yhwh and the Israelites through the Passover sacrifice.

Thereafter, Moses briefly recalls vv. 12-13; he again insists on the "passage" (עבר) of Yhwh and on His direct involvement against Egypt.[260] The choice of נגף to designate the blow inflicted by Yhwh, however, expresses in an even more explicit manner, the continuity between the nocturnal intervention of God and the action narrated in 7:27.[261] Nevertheless, the divine strike against Egypt is not described in all the features recounted in 12:12: nothing is said about the first-born and the harsh sanction; the formula "I Yhwh" is not repeated.

The references to v. 13, instead, are more detailed (as noted in the table): Yhwh "sees the blood," "passes over"[262] the door, without allowing the

---

[257] See ch. I, § 2; ch. II, §§ 4.3-4.5. Priotto, 226; G.I. Davies, II, 91.

[258] The root נגע designates blood manipulation only in Exod 4:25 and 12:22. See Propp, I, 407; Fischer – Markl, 136.

[259] Rabbinic tradition explains Moses' clarification by saying that the Angel of Destruction, passing through Egypt in the night, does not make any distinction between those who are outside the house, he strikes everyone; deliverance is offered to those who celebrate the Passover at home. See *MekhY*, *Pisha* § XI, p. 60, "This tells that the angel, once permission to harm is given him, does not discriminate between the righteous and the wicked, as it is said, 'Come, my people, enter thou into thy chambers,' etc. (Isa 26:20). And it also says, 'Behold, I am against thee, and will draw forth My sword out of its sheath, and will cut off from thee the righteous and the wicked' (Ezek 21:8)."

[260] See ch. III, § 2.2.3.

[261] Whereas the root נכה can have both divine and human subjects, the verb נגף is most often used for divine interventions (e.g., Josh 24:5; Judg 20:35; 1 Sam 4:3; 26:10); in Exodus נגף is employed in 7:27, and then used again in 32:35. G.I. Davies, II, 81.

[262] The verb פסח is polysemous (see above, notes 182 and 206); since v. 23, unlike v. 13, is not referred to the people, but deals with the Israelites' houses (פתח + על + פסח), it seems more correct to translate "to pass over" and not "to protect." See the Septuagint; Houtman,

"destroyer" to strike the Israelites. An obvious difference is that, while in v. 13 the term משחית had a generic meaning ("destruction"), in v. 23 it designates a personal entity, a "destroyer,"[263] often identified with an angel[264] (2 Sam 24:16; 1 Chron 21:15). Therefore, Moses' words faithfully report what he has received from Yhwh, while also retaining a "personal" character. He insists more precisely on the instruction regarding blood manipulation, and by repeating נגע, already used in 4:25, explicitly links his discourse to the experience of the nocturnal encounter with God (4:24-26). Finally, Moses exhorts the elders (vv. 21-22), and reassures them by announcing God's intervention (vv. 13, 23).

2.3.2 Passover: Word and Service (vv. 24-25)

The discourse of Moses has a sudden transition to the second person singular.[265] He unexpectedly addresses each Israelite as a father (v. 24, "for you and for your children"), revealing that the observance[266] (שמר, as in v. 17) of the instruction to celebrate Passover is an action that involves the relationship between parents and children.

Exod 12:24	Exod 12:25
(a) ²⁴ You shall **observe** this word (ושמרתם את־הדבר הזה) (b) as a statute for you and your children always.	(b') ²⁵ When you enter the land that Yhwh will give you, as He has said (כאשר דבר), (a') you shall **observe** (ושמרתם) this service (את־העבדה הזאת).

The reiteration of שמר in v. 24 and v. 25 is a remarkable literary phenomenon that creates a chiasm-like composition and a correspondence between the two objects of the same verb; the two verses are also joined by the repetition of the root דבר. The two statements repeated in the frame (a and a') introduce two definitions of the paschal ritual, understood as "word" and "service." The use of דבר in v. 25 induces the reader to translate the

---

II, 193; Priotto, 208-209. The *TgN* of 12:23 presents both the possibilities, "And He shall pass over, and the *Memra* of Yhwh shall protect the door (ויפסח ויגן מימריה דייי על תרע)."

[263] The Septuagint translates v. 23, emphasising the personal character of the executor of the divine sentence: τὸν ὀλεθρεύοντα.

[264] The noun משחית can refer to men: it is used both for the troops of the Philistines (1 Sam 13:17; 14:15) and for military deployments (Jer 4:7; 22:7). Regarding the association of the "destroyer" with an angelic figure, see *TgPsJ*; Cassuto, 143; Propp, I, 409; Priotto, 227; G.I. Davies, II, 91.

[265] In v. 24 the verb ושמרתם is conjugated in the second person plural ("and you shall observe"), while the suffix of the preposition ל is singular ("as a statute for you"). Hence, while reporting an instruction for all, Moses seeks to involve each person in these words. See G.I. Davies, II, 92.

[266] See in this chapter, § 2.2.5.

expression את־הדבר הזה in v. 24 literally, "this word."[267] The term דבר, then, contributes to describe Yhwh's instructions not just as notions or concepts; although they are different and varied, they also are unified as a single entity ("*this* word") and have the characteristics of a "word," accomplishing a personalistic communication between the speaker and the addressee.[268] Thereupon, the reader is not surprised when Moses in v. 25 says, "when you enter the land." In fact, God does not ask the people to observe a law without first announcing the gift he is about to make. The Passover is an instruction that comes second and is uttered after the first "word" that of the promise.[269]

The instructions to be observed, moreover, are defined in v. 25 as a "service," עבדה. The root עבד has a peculiar meaning in Exodus, because it refers to the forced labour to which the people are subjected (1:13-14) and also indicates Israel's worship of Yhwh[270] (3:12; 4:23; 7:16, 26). Therefore, Yhwh's utterance is both a word that the Israelites must keep, and a service that they must perform, especially when they will become a free people.[271]

The frame formed by the two orders includes both a temporal and a spatial aspect: God's instruction is transmitted from father to son ("as a statute for you and your children," v. 24) and will be put into practice after the liberation. The possibility of celebrating the Passover as a "service" is, in fact, a sign that the person no longer lives under the yoke of an oppressor and can freely enjoy the divine gift (v. 25, נתן) of the land.[272]

2.3.3 The Children's Question (vv. 26-28)

26 When your children will say to you, "What this service is for you?" 27 You shall say, "This[273] is the sacrifice of *Pesaḥ* to Yhwh, who passed over the

---

[267] The translation "this thing/practice" does not seem adequate for את־הדבר הזה. See Childs, 180; Propp, I, 356; G.I. Davies, II, 85; Utzschneider – Oswald, 249. In fact, when the noun דבר is preceded by שמר, the expression means "to observe the word" (Deut 12:28; 17:19; 29:8; Ps 119:9).

[268] P. GILBERT, *Le ragioni della sapienza*, 30, "The word has a personal origin and an equally personal intention." (My translation)

[269] The promise of the land is announced from the very first meeting with Moses, in 3:8, and is reiterated in 6:8 ("and I will bring you into the land"), after the unsuccessful experience before Pharaoh (5:1-20); see *MekhY*, *Pisḥa* § XII, p. 63.

[270] See ch. II, § 3.2.2, note 178.

[271] P. BOVATI, *Parole di libertà*, 89.

[272] The promise of the land is explicitly described as a "gift" in Exod 6, in v. 4, "and I also established my covenant with them to give (לתת) them the land of Canaan," and in v. 8, "in the land that I swore with a raised hand to give (לתת) to Abraham, Isaac and Jacob."

[273] The pronoun הוא is the subject of the nominal sentence, preceded by the predicate "the Passover sacrifice." See JM § 154g; G.I. Davies, II, 88.

houses of the Israelites in Egypt, when he struck Egypt and he saved our houses." The people knelt down and prostrated themselves. ²⁸ The Israelites went, did as Yhwh had commanded Moses and Aaron. So they did.

The formula "for your generations (לדרתכם)," employed in Exod 12:14, 17, consequently implies the involvement of children in the institution of the Paschal rite. In 12:26-27, Moses recreates a micro-narrative, staging the dialogue between children and parents on the occasion of Passover; hence, the involvement of successive generations is portrayed in dramatic form, simulating the child's question and the parent's answer.[274]

The question recalls v. 25, through the repetition of the noun עבדה, "service." The child, by introducing his words with מה, does not merely ask for information, but rather interrogates the parent about the meaning of the gestures made,[275] because the עבדה, "service," is easily confused with servitude. The issue, then, offers the parent the opportunity to educate the son through his words.[276]

Passover is defined as a "sacrifice for Yhwh (זבח־פסח ליהוה)," alluding especially to the rite of the lamb ("sacrifice"). Since the child's question arises immediately after the instructions related to the blood application (12:22-23), it is the context itself that prompts the reader to interpret the parent's words particularly as an explanation of the blood ritual. The answer reveals firstly the "theological" direction of these gestures, namely that they are made "for Yhwh" (vv. 11, 14), and as such, they manifest a relationship with Him. Then, the parent explains the feast (זבח־פסח, "sacrifice of *Pesaḥ*") with a wordplay that facilitates the child's understanding ("He passed over the houses of the Israelites," פסח על־בתי בני־ישראל)[277]. This interpretation is

---

[274] See J.-P. SONNET, *Generare è narrare*, 38. Fathers are invited to pass on their faith to their children not "in the manner of abstract erudition, but by the eminently concrete way of telling the story – in the form of the remote past, 'The Lord passed by.' The story told has the particularity of uniting fathers and sons in the same experience. The Lord, the fathers will say, 'saved our homes.'"

[275] The expression מה + noun + ל can be translated, "what X means for Y?" Josh 4:6; 2 Sam 16:2; Ezek 12:22; 37:18. See G.I. Davies, II, 88.

[276] Paternity requires an exercise that is not limited to material care. Raising a son demands for an educative activity carried on at diverse times that renders the parent similar to a teacher (Prov 2:1; 4:10, 20; 7:1). See F. FICCO, *"Mio figlio sei tu,"* 85-86.108-109; J.-P. SONNET, *Generare è narrare*, 39.

[277] The passage of vv. 26-27 is at the origin of the Passover Seder: the question-and-answer structure is attested already in *mPes* 10,4 ("why is this night different from all other nights?"). The *Mekhilta de r. Yshmael* collects together the questions of Exod 12:26; 13:8, 14; Deut 6:20 and in them represents four types of sons participating in the Passover; the question of Exod 12:26 is associated with the "wicked" son because he places himself outside the group that is celebrating the feast ("what does this service mean *to you*?"). See

communicated with a testimony[278] and the close repetition of the term "house" achieves an increasing participation in the fate of the Israelites, "(Yhwh) who passed over *the houses of the Israelites* when He struck Egypt and saved *our houses*." The "houses of the Israelites" become "our houses," the parent unites his own experience with that of the Israelites liberated from Egypt in order to demonstrate to the child that Yhwh's salvation achieved through blood is effective in their very celebration.[279]

Moses' instructions conclude with v. 27 in which, by means of various references, the narrator creates a correspondence with 4:29, 31.

Exod 4:29, 31	Exod 12:21, 27
[29] And Moses and/with Aaron went and gathered all the elders of the Israelites (את־כל־זקני בני־ישראל). [31] And the people believed. ... so they knelt down (ויקדו) and prostrated themselves (וישתחוו).	[21] Moses called all the elders of Israel (לכל־זקני ישראל) and said to them… [27] And the people knelt down (ויקד) and prostrated themselves (וישתחוו).

When Moses has to tell the people about the mandate he received from Yhwh, he first summons the elders (4,29), and yet after the words of Moses and Aaron, the elders are no longer mentioned. Nevertheless, it is still said that "the people believed" (4:31). The asymmetry may be intentional and its purpose could be to suggest that the elders faithfully conveyed to the people what they heard from God's envoys and that the people's reaction is prompt and firm.[280] In 12:21, 27 a similar correlation is found: it is not only the elders who manifest their desire to obey Moses' words, but all the people prostrate themselves.[281] In v. 28, a priestly phrase is added to what was said in the

---

MekhY, Pisha § XVIII, p. 113, "There are four types of sons: the wise, the simpleton, the wicked, and the one who does not know enough to ask. The wise—what does he say? 'What is the meaning of the testimonies and the statutes and the ordinances which the Lord our God hath commanded you?' (Deut 6:20). ... The simpleton—what does he say? 'What is this? And thou shalt say unto him: By strength of hand the Lord brought us out from Egypt, from the house of bondage.' (Exod 13:14) The wicked one—what does he say? 'What mean ye by this service?' (Exod 12:26). Because he excludes himself from the group."

[278] See M. RECALCATI, *Cosa resta del padre?*, 82. The good parent "is a father who embodies, in his own singular existence, the passion of desire …. But what does it mean to incarnate the passion of desire in one's own existence? In the first place, it means being able to bear witness to it." (My translation)

[279] J.-P. SONNET, *Generare è narrare*, 38; L.C. STAHLBERG, "Time, Memory, Ritual," 87.

[280] See Propp, I, 410.

[281] See ch. II, § 5.3.

previous verses.²⁸² The most significant element is the repetition of the verb "to do" (עשה), which expresses the people's obedience. The formulation is well known in the Pentateuch²⁸³ and emphasises the perfect consonance between divine order and human execution.²⁸⁴ The remarkable aspect of v. 28 is that the narrator does not merely say "they did as," but specifies that they "*went* and did."²⁸⁵ This addition, only apparently superfluous, manifests the impetus and the extreme dynamism that characterises the people's decision: they obey with enthusiasm to Yhwh's instructions.

3. Death of the First-born and Liberation (Exod 12:29-42)

The narrative thread, interrupted by the institution of Passover, is taken up again in v. 29 with the account of the plague announced in 11:1-10 and with the narration of Israel's departure from Egypt (vv. 35-42). The Israelites have obeyed the words of Moses and Aaron, and their liberation takes place precisely "after" this perfect obedience (vv. 27b-28). The shift to the narrative register, therefore, is intended to show the reader that what is still only foreshadowed and awaited, actually takes place on the night of the Israelites' deliverance thanks to their perfect observance.

3.1 *The Death of First-born Sons (vv. 29-34)*

The opening passage (vv. 29-30) explicitly recalls the announcement of the plague of Exod 11:1-10, with some minor variations. While 11:4 uses the verb "to go out," (יצא) in order to emphasise the divine movement in the direction of the people and the "military" nature of the intervention,²⁸⁶ in 12:29 the narrator is more succinct. He says that "Yhwh struck every first-born," and employs the divine name that expresses God's closeness to his people.²⁸⁷ The verb נכה, as seen in 12:12-13, denotes the blow dealt by the Lord to Egypt, and the words order exalts the work of God.²⁸⁸ In both cases, the plague affects

---

²⁸² Childs, 200; Priotto, 228; S. GESUNDHEIT, *Three Times*, 65; G.I. Davies, II, 102.

²⁸³ See Exod 7:6; 39:32.43; 40:16; Num 1:54; 8:20; 17:26. The formula can have a conclusive character (analeptic), but it can also introduce new content (proleptic).

²⁸⁴ See *MekhY*, *Pisha* § XII, p. 67, "This makes known the excellency of Israel, that they did exactly as Moses and Aaron told them ... What is the purport of the words, "so they did"? Simply, that Moses and Aaron also did so."

²⁸⁵ Rashi Exod, 86, "The verse considers it for them as they had [already] done [the commandments] (מעלה עליהם הכתוב כאלו עשו)."

²⁸⁶ Regarding 11:4-6, see c. III, § 1.2.

²⁸⁷ See c. II, § 4.3, note 234.

²⁸⁸ The clause ויהוה הכה reverses the natural verb-subject order common in Hebrew prose, probably to place emphasis on the subject, Yhwh: see Propp, I, 410; G.I. Davies, II, 113.

both Pharaoh and the person living in the poorest of conditions[289] (Wis 18:11); its execution is complete, not even the cattle are spared.

Exod 11:4-6	Exod 12:29-30
[4] About midnight, *I myself am going out in the midsts of Egypt* [5] And every firstborn in the land of Egypt shall die, from the firstborn of Pharaoh who sits on his throne, *to the firstborn of the slave girl who is behind the millstones*, and every firstborn of the cattle. [6] There shall be a great cry in all the land of Egypt, *which has never been and will not be repeated*.	[29] In the middle of the **night**, *Yhwh* struck[290] every firstborn in the land of Egypt, from the first-born of the Pharaoh who sits on his throne, *to the first-born of the prisoner who is in the dungeon*,[291] and every first-born of the cattle. [30] Pharaoh rose up in the **night**, he and all his servants and all of Egypt; and there was a great cry in Egypt, *for there was not a house where there was not someone dead*.

An unexpected detail is added in v. 30: "Pharaoh rose up in the night, he and all his servants and all of Egypt." The narrative, therefore, insists on the tragedy experienced in the night and the state of deep anguish.[292] At a time when the sovereign should rest,[293] the pain spreads and involves everyone (כל appears twice). Just as foretold in 11:6, the grief is expressed as a "great cry"[294] (3:7, 9). Eventually, in 12:30, it is stated, as a hyperbole,[295] that there was no house without someone dead. The

---

[289] In 12:29 the text mentions the prisoner in the dungeon and this category is even lower than that of the slave girl of 11:5. Rabbinic literature speculates on the reasons for not sparing even prisoners; see *MekhY, Pisḥa* § XIII, p. 68, "But how did the captives sin? It was only that the captives should not say, 'Our deity brought this visitation upon the Egyptians. Our deity is strong, for it stood up for itself. Our deity is strong, for the visitation did not prevail over us.' Another Interpretation: This is to teach you that the captives used to rejoice over every decree which Pharaoh decreed against Israel. For it is said, 'He that is glad at calamity shall not be unpunished' (Prov 17:5)."

[290] After the formula ויהי, the construction *waw* + name followed by a *qatal*, ויהוה הכה, expresses the simultaneity of events (JM § 166*d*; see Gen 15:12; Josh 2:5).

[291] The term בור literally means "cistern" (Gen 37:20; Lev 11:36), but together with בית refers to a building used as a prison (Jer 37:16). See Houtman, II, 196.

[292] The "night" is often the time when the sleep of the suffering person is shaken by anguish: Pss 6:7; 22:3; 77:3.

[293] See *MekhY, Pisḥa* § XIII, p. 69, "*And Pharaoh Rose Up*. I might understand this to mean after the third hour of the day [nine o'clock in the morning], for such is the custom of kings that they get up after the third hour. But Scripture says, 'in the night.'" See also Cassuto, 145.

[294] See ch. III, § 1.2.

[295] A hyperbolic style is frequent in the book of Exodus. See Propp, I, 347.

disappearance of the first-born, the "first fruits of all their strength" (Ps 105:36), is the end of Egypt's hope with regards to the continuation of life.

³¹ He called Moses and Aaron by night and said, "Get up, *go out* from among my people, both you and the Israelites (גַּם־אַתֶּם גַּם־בְּנֵי יִשְׂרָאֵל); *go,* serve Yhwh *as you have said*. ³² Take also your flocks (גַּם־צֹאנְכֶם) your herds (גַּם־בְּקַרְכֶם) *as you have said* and *go,* and bring your blessing upon me also (גַּם־אֹתִי)."

The nocturnal scenario is evoked again: the term לילה is used three times in vv. 29-31 so that darkness becomes a nightmare that envelops everything.²⁹⁶ The repetition shows that Pharaoh does not hesitate even for a moment to revoke the word pronounced in 10:28 ("beware of ever seeing my face again") and summons Moses and Aaron at an unusual time.²⁹⁷ His actions are frantic, his words characterise him as a terrified man,²⁹⁸ since in just two verses he uses a series of seven verbs.²⁹⁹ Since Pharaoh and the Egyptians arose in the night (קום, v. 30³⁰⁰), the Israelites must get up (קום, v. 31) at once and "go out from among my people." The reference is not geographical (he would have said "go out of my land"), but anthropological: Pharaoh demands that Israel definitively break every relation with the Egyptians. The imperative "go" (הלך), is then repeated twice to insist on the haste with which the Israelites must leave Egypt. At the centre of these verbs we have עבד, a key word for the Exodus.³⁰¹ Pharaoh, who had enslaved the people (1:13-14; 2:23), now admits defeat and urges the Israelites to "serve Yhwh." Moreover, although this has not previously been a guarantee for the truth of his words,³⁰² the king of Egypt chooses to employ the name of God that expresses a certain closeness, "Yhwh." He who did not want to acknowledge God ("who is Yhwh?" 5:2) now refers to him in the most correct way.

Finally, v. 31 concludes with an expression that is repeated in the exact same way in v. 32: כדברכם, "as you have said." In this manner, Pharaoh "materially" connects his own words to what he previously heard from Moses

---

²⁹⁶ Cassuto, 145.

²⁹⁷ Priotto, 230; G.I. Davies, II, 117.

²⁹⁸ Cassuto, 145.

²⁹⁹ The imperatives are associated with each other asymmetrically and simply juxtaposed. Even the last *w-qatalti* (וברכתם) is similar to an imperative (JM § 119*l*).

³⁰⁰ See Houtman, II, 198.

³⁰¹ See ch. II, § 3.2.2, note 178.

³⁰² See ch. II, § 4.3, note 234. In the "Plagues narrative" Pharaoh addresses Moses using the tetragrammaton and showing a certain respect for Yhwh (see 8:4, 24; 9:27-28; 10:10-11, 16-17, 24); in all these cases, however, the king of Egypt revokes the decisions in favour of the people of Israel. This fact leads the reader to interpret the words used by Pharaoh in 12:31 with caution.

and Aaron;³⁰³ moreover, the exact replication of כדברכם is intended to show that the king of Egypt does indeed intend to bow low before the word of God. This time, he recognises it as an order which he cannot but obey.³⁰⁴ The Israelites can leave³⁰⁵ with the flocks and the herds: the frequent repetition of גם, "also," highlighted in the text, has the role of showing that Israel departs without leaving anything behind, neither humans nor animals.³⁰⁶

The discourse of Pharaoh concludes with a request, "and bring your blessing upon me also." These words could be ambiguous, Pharaoh's previous behaviour.³⁰⁷ However, in many ways, the petition appears unique: the verb ברך is used for the first time by the king of Egypt and is the first recurrence of this root in the book of Exodus; moreover, although the repetition of the particle גם ("me too," גם־אתי) may simply be an effect of the ruler's frenetic nervousness, it just may be a sign that he really does want to be blessed together with the Israelites.³⁰⁸ Moses thus fulfils the vocation of Abraham, called by Yhwh to "become a blessing" for the nations.³⁰⁹ Furthermore, Pharaoh's blessing closes a cycle: Israel's first encounter with

---

[303] The words employed in 12:31-32 recur frequently in the account of the plagues, "to go" (הלך, 8:21, 24; 10:24); "to serve Yhwh" (יהוה + עבד; 10:7-8, 11); "flocks and herds" (צאן and בקר; 10:9, 24). See *MekhY*, *Pisḥa* § XIII, p. 71; Cassuto, 145, "*according to your word – as you, not I, have said. He now withdraws all his previous restrictions, one by one.*"

[304] See G.I. Davies, II, 117. The emphasis on Pharaoh's internal change, in comparison with what is narrated in Exod 5:1-4, is obtained also "by the repeated statement that it is the words of Moses and Aaron ('as you said' in vv. 31 and 32: cf. 7:16, etc.) that are to be determinative of the future, not Pharaoh's own."

[305] Houtman, II, 199; Propp, I, 411. In 12:31-32 Pharaoh may have intended to permit a celebration in the desert, and have expected the Israelites to return after three days (see 8:23); Ibn Ezra believes that in 14:5 he seems surprised that the Israelites fled (pp. 102-103). This interpretation, however, does not seem realistic; G.I. Davies, II, 117-118.

[306] Priotto, 230, note 105; Propp, I, 411; Houtman, II, 198; G.I. Davies, II, 117-118. The authors note the quadruple repetition of the particle גם, "also," in vv. 31-32 and state that this is a literary device to indicate Pharaoh's complete agreement to grant all the Israelites' requests.

[307] After some plagues, Pharaoh asks Moses and Aaron to pray for him and for Egypt (8:4, 24; 9:28; 10:17); in other cases, he is willing to allow them to go out and sacrifice to Yhwh (8:21; 9:28; 10:10-11, 24) or he acknowledges his sin before the Lord (9:27; 10:16). On all these occasions, however, the king of Egypt retracts.

[308] See Houtman, II, 199. Pharaoh asks Moses "to put in a good word for him with Yhwh since he let Israel go, so as to move Yhwh to reward him by sending blessing instead of calamity."

[309] In Gen 12:2 Yhwh said to Abraham, "so that you will be a blessing." Abraham's blessing is such that he becomes "the blessing's channel for others." (A. WÉNIN, *Da Adamo ad Abramo*, 171) See V.P. HAMILTON, *Genesis 1–17*, 373; F. GIUNTOLI, *Genesi 12–50*, 17; Isa 19:24; Zech 8:13.

Egypt was characterised by Jacob's blessing of the king, and now the last by Moses and Aaron's.[310] The blessing follows the greeting of two people who are separating (2 Sam 13:25), but it can also come as a consequence of the slaves' liberation (Deut 15:18; 24:13[311]). In addition, since the reader is prepared by Genesis to interpret the blessing as a reference to fruitfulness (Gen 1:22, 28; 9:1; 17:16, 20; 22:17), Pharaoh's request can be explained as the expression of the wish to receive the same blessing that made Israel fruitful (Exod 1:7). Be that as it may, Pharaoh's words do not cease to be problematic because the king of Egypt demands the blessing only for himself, uninterested in the destiny of the Egyptians' (7:23).

³³ Egypt urged[312] the people on, impatient[313] to send them away from the earth, because they said, "All of us are dead." ³⁴ The people brought the dough before it was leavened, the kneading bowls for the dough[314] wrapped in their cloaks upon their shoulders.

The text is elliptical, it does not relate any reaction of Moses and Aaron, it does not even say that a meeting ever took place.[315] This omission is certainly not a *gap*[316] but a *blank*:[317] the reader must not dwell on the precise course of events, but rather concentrate on Pharaoh's words, a sign of terror and trepidation.

The speech of the king of Egypt is followed by the reaction of the Egyptians, who prompt the Israelites to leave the country quickly. The verb חזק might be ironic,[318] since in Exodus it is used for Pharaoh's obstinacy

---

[310] Cassuto, 145-146; Priotto, 230; G.I. Davies, II, 118.

[311] The connection between the request for a blessing and the liberation of slaves is first made by Daube in 1963 in his work entitled *The Exodus Pattern in the Bible*, 51-52.

[312] The Hebrew ותחזק is unusual because, although proper names of a country (in this case "Egypt") may be followed by a feminine singular verb (GK § 122*i*; JM § 134*g*), in the book of Exodus the masculine plural is more frequent (e.g. Exod 1:13; 7:5).

[313] The infinitive construct of מהר preceded by ל has an adverbial value; Propp, I, 356; G.I. Davies, II, 114. The Septuagint tranlates שלח with ἐκβαλεῖν, "to chase away."

[314] The noun משארת appears only a few times in the Hebrew Bible (7:28; Deut 28:5.17), and the meaning is clarified in Exod 12:34. It denotes an object employed for the dough of unleavened flour, easily loaded on one's shoulders. The Septuagint reports instead τὰ φυράματα αὐτῶν, "their parts of dough" and the rabbinic tradition explains the noun as a derivative of the root שאר "to remain": the Israelites carry on their shoulders the remains of the bitter herbs and unleavened bread consumed in the night (Rashi Exod, 88; *TgO*).

[315] Utzschneider – Oswald, 256.

[316] The rabbinic tradition proposed a suggestive explanation of the ellipsis, *MekhY, Pisha* § XIII, p. 69, "*And He Called for Moses and Aaron*. This tells that Pharaoh went around in the whole land of Egypt, asking: Where does Moses dwell, where does Aaron dwell?"

[317] See ch. I, § 3.2, note 152.

[318] Propp, I, 411; Priotto, 231.

(4:21; 7:13, 22; 8:15; 9:12, 35; 10:20, 27; 11:10) and for the act of restraining the Israelites by force (9:2). The verb manifests also some insistence (see 2 Sam 24:4; 2 Kings 4:8), because the Israelites are perceived as a mortal and imminent danger[319] ("we all die").

The departure of the Israelites is described in v. 34, however, many facets are omitted given that there are no temporal or spatial indications, nor news of the journey (vv. 37-38). On the contrary, the narrator relates many details regarding the decision to bring the dough before the leavening,[320] thus showing that Israel's departure took place in great haste.[321]

And so, the growing intensification in the narrative of the plagues reaches its greatest degree. What was announced with the plague of water turned to blood, an ominous symbol foretelling death and famine,[322] becomes tragically real with the death of the first-born of Egypt.

3.2 *The Departure of the Israelites (vv. 35-42)*

> 35 And the Israelites had done according to the word of Moses, and had asked from Egypt for silver and gold, and for clothing. 36 And Yhwh had granted the people grace in the sight of Egypt, and they had granted their requests, and they had spoiled Egypt.

The passage interrupts the narrative thread with a small independent narration of events that cannot be subsequent to vv. 29-34 and therefore took place in the past.[323] The verses recall what is foreshadowed in 3:21-22 and what is said by Yhwh in 11:2-3:[324] God had told Moses to "speak" (11:2, דבר) to the people and the Israelites acted according to his word (12:35, דבר); the instruction was to ask (שאל, 3:22; 11:2; 12:35) for gold,

---

[319] The Egyptians use the *qatal*, "all of us are dead," כלנו מתים, expressing a past action that has been already performed and completed. Sometimes the *qatal* can denote actions which are fulfilled at the precise moment when the sentence is uttered (e.g., Gen 23:11; Jer 40:4; Ps 2:7) or to exalt the certain fulfilment (completed aspect) of what is said (Gen 15:18); see JM § 112g where even Exod 12:33 is quoted.

[320] The term בצק denotes a mixture of flour and kneaded water (or oil; 2 Sam 13:8; Jer 7:18; Hos 7:4).

[321] *MekhY, Pisha* § XIII, p. 71, "*And the People Took heir Dough before it Was Leavened.* This tells that they had kneaded their dough but had not sufficient time to let it leaven before they were redeemed." See Propp, I, 412; Priotto 231.

[322] See ch. I, § 3.2.

[323] See A. NICCACCI, *Sintassi*, § 27. Dealing with the construction *waw-x-qatal* he notices that sometimes, especially at the beginning of a new narrative, information referring to events that are in the background develops into a small independent narrative also constituted by some *wayyiqtol* indicating a succession in the past (1 Sam 28:3).

[324] See in this chapter, § 1.1.

silver, and clothing³²⁵ (3:22; 12:35). Yhwh, as expected, grants grace in the eyes of Egypt (3:21; 11:3; 12:36) and Egypt is "spoiled" (נצל, 3:22; 12:36). In this passage, therefore, God's sovereign dominion ("he had granted grace") and the fulfilment of the divine word are again declared. Yhwh, in fact, has not only promised to bring the people out of Egypt, but has repeatedly assured that their departure would be a victorious triumph; great dignity and beauty are accorded to the liberated Israelites, they depart with all their possessions (v. 38) clothed in gold and silver.³²⁶

> ³⁷ And the Israelites journeyed from Rameses to Succoth, about six hundred thousand men on foot, men without the children. ³⁸ A mixed multitude also went up with them, the flocks and herds, very much livestock. ³⁹ They baked the dough that they had brought out of Egypt, like unleavened cakes, because it had not risen, since they had been driven out of Egypt and had not been able to linger, nor did they prepare provisions for themselves.

The verb נסע, "to depart," is a key word for the section that opens in this verse and tells of Israel's journey to the promised land.³²⁷ In an allusive way, this verb helps to indicate the continuous dynamism that will characterise the expedition. The Israelites definitively leave Rameses, the city of Jacob's settlement (Gen 47:11) and symbol of the dramatic period of forced labour (Exod 1:11). The first stop is Succoth. Even if the site's identification is uncertain,³²⁸ the name creates an assonance with the Feast of Tabernacles, when the Israelites recall the period of wandering in the desert,³²⁹ which will also be celebrated after Passover (Deut 16:13).

---

³²⁵ See Houtman, I, 381; שמלה, "Robe" in general, or "cloak" used to hold or load objects (Judg 8:25; 1 Sam 21:10) and protect from the cold (Deut 22:17). The repetition in vv. 34-35 clarifies how it is possible to have a cloth that can carry the vessels for the dough; see Propp, I, 412; G.I. Davies, II, 119.

³²⁶ The spoliation of Egypt was an occasion of embarrassment for rabbinic exegesis; see *EstR* VII, 13. However, gold and silver objects were explained as the rightful booty taken from the vanquished; see Childs, 201; J. VAN SETERS, *The Life of Moses*, 98.

³²⁷ The root נסע appears several times in the book of Exodus (13:20; 14:10; 15:22 16:1; 17:1; 19:2; 40:36, 37), referring to the journey through the desert towards Mount Sinai; 89 of נסע 146 ocurrences in the Bible are found in the Book of Numbers.

³²⁸ The site of *Sukkōt* could correspond to today's city of *ṯkw / Tjeku*, identified with the ruins of Tell el-Maskhuta; E. BLEIBERG, "The Location of Pithom and Succoth," 21-27. The same name also denotes a location in the Jordan River (Gen 33:17; Josh 13:27; Judg 8:5).

³²⁹ The term סכה literally means "hut" (Isa 1:8) and "military booth" (1 Kings 20:12, 16). The name סכות can be an allusion to the Feast of Tabernacles (Lev 23:42-43). See Propp, I, 413; Priotto, 232; Fischer – Markl, 140; Utzschneider – Oswald, 257-258.

The number of six hundred thousand men seems excessively large[330] and perhaps has the purpose of exalting the extraordinary fruitfulness of the Israelites and the divine blessing announced from the very first lines of the Book of Exodus (1:12, 20).[331] On the other hand, the use of רגלי, "person going on foot, infantryman," suggests a military nuance[332] which would present the journey in the desert as a march of a victorious army[333] under the guidance of Moses.[334]

The people are escorted by "a mixed multitude"[335] (v. 38) of non-Israelites. Although this detail has generated different interpretations,[336] it is likely that the narrator wants to associate the nations' fate to Israel's calling,[337] as a way of showing the participation of some foreigners in the night of their liberation and in the celebration of Passover.[338] The people, then, do not leave the

---

[330] Commentators agree that the number of six hundred thousand is unrealistic, e.g., Propp, I, 414; Priotto, 232. Hence, it has been suggested that the meaning of אלף is not "thousand," but it should be translated as "clan" or "military unit." See E.W. DAVIES, "A Mathematical Conondrum," 449-469.

[331] Nepi, I, 204; Priotto, 233; Utzschneider – Oswald, 258.

[332] The term רגלי is practically always employed (except Jer 12:5) in a military context, as a collective noun (Num 11:21; 1 Sam 4:10; 15:4; 2 Sam 10:6; 1 Kings 20:29; 1 Kings 13:7) or in the expression איש רגלי (Judg 20:2; 2 Sam 8:4; 1 Chron 18:4; 19:18).

[333] G.I. Davies, II, 115.

[334] D. MATHEWS, "Moses as a Royal Figure," 73-78. The author suggests that the character of Moses can be interpreted as a royal figure. He proposes a comparison with the Cyrus Cylinder, in which the Persian ruler is praised for his great army.

[335] The noun ערב is derived from the root ערב I, "to associate with, mix with" (HALOT), which in Jer 25:20; 50:37 refers to non-Israelites.

[336] The "mixed moltitude" can designate a group of mercenaries united with the Israelites through mixed marriages; Bar recalls the use of ערב in the context of Nehemiah's reform against mixed marriages (Neh 13:3) and takes seriously the association of the Israelites' march with that of an army. S. BAR, "Who were the 'Mixed Moltitude,'" 27-39.

[337] See A. SHERWOOD, "The Mixed Moltitude in Exodus 12:38," 139-154. The exegete analyses the composition of vv. 34-39: [A] unleavened dough (v. 34); [B] Israel demands the booty from the Egyptians (vv. 35-36); [X] vv. 37-38; [A'], unleavened dough is baked (v. 39). The accordance between the extreme parts (A and A'), in which the unleavened dough is mentioned, creates a frame that sorrounds two parts: vv. 35-36 and vv. 37-38. Since in vv. 35-36 the Israelites' booty is mentioned, Sherwood believes that in vv. 37-38 the large group of foreigners associated to Israel are considered as Yhwh's "booty." Although suggestive, this interpretation seems to be an overreading.

[338] See Fischer – Markl, 141; Utzschneider – Oswald, 258. Rabbinic exegesis has given great relevance to this indication (MekhY, Pisḥa § XIV, p. 75). See ExR XVIII, 10, "the virtuous among the Egyptians came, celebrated the Passover with Israel, and went up with them, for it says: *and a mixed moltitude also went up with them* (Exod 12:38)."

country of Egypt in a state of poverty,[339] rather they depart with all their possessions and their livestock. Pharaoh has granted permission to leave the country with their flocks (12:32), but it is only in v. 38 that the narrator specifies that the quantity of animals is enormous.[340] And so, with the departure, the very same thing that Pharaoh wanted to avoid happens (10:9.24).

The Israelites baked the mass into unleavened cakes (v. 39). The insistence on these details is intended to affirm that, as predicted in 11:1,[341] the Israelites were "driven out" (גרש), chased out of the country without having the chance to delay.[342] Their liberation become in this sense a sudden irruption and, since the people leave without having the time to take provisions,[343] the journey begins with a gesture of strong reliance on God.

In the conclusion[344] (vv. 40-42), the narrator clarifies the length of time spent in Egypt, "the residence[345] of the Israelites who had lived in Egypt, was four hundred and thirty years"[346] (v. 40). By repeating the precise number of years spent in Egypt, in v. 41, as well as in v. 17, two additional elements become evident.

---

[339] See *MekhY, Pisha* § XIV, p. 75, "*And Flocks and Herds even Very Much Cattle*. All these were implied in what God said to our father Abraham, 'And afterward shall they come out with great substance' (Gen 15:14) – at their going out from Egypt I will fill them with silver and gold."

[340] The adjective כבד can mean "numerous" (Gen 50:9; Num 20:20; 1 Kings 3:9). The expression כבד מאד, referring to cattle, designates the large number of animals Abraham owned (Gen 13:2).

[341] See in this chapter, § 1.1.

[342] In the construction וגם צדה לא־עשו, "nor provisions have done," the classical verb-object order is inverted and the object is preceded by גם to emphasise precisely the haste: they had not even managed to gather provisions for the journey. See G.I. Davies, II, 115.

[343] *MekhY, Pisha* § XIV, p. 76, "*Neither Had They Prepared for Themselves Any Victual.* This proclaims the excellence of Israel, for they did not say to Moses: How can we go out into the desert, without having provisions for the journey? But they believed in Moses and followed him."

[344] The verses 43-50 are normally assigned to the document P. See Childs, 184; Propp, I, 374-375; J. VAN SETERS, *The Life of Moses*, 122. Knohl, instead, thinks that they are a secondary edition of P, see I. KNOHL, *The Sanctuary of Silence*, 21-22.

[345] The noun מושב denotes the place of "residence," as in Lev 25:29; Num 15:2.

[346] The number found in 12:40 is problematic because the Exodus narrative offers no other elements to reconstruct the exact length of the Israelites' stay in Egypt. The calculation should begin from Jacob's arrival in Egypt (Gen 46), but the text of the genealogy in Exod 6:14-25 does not seem to cover a sufficient time span to justify four hundred and thirty years. The number may recall the round figure of four hundred years mentioned in Gen 15:13; in 1 Kings 6:1, however, four hundred and eighty years are counted from the exit out of Egypt to the building of the Temple and this correspondence does not seem to be accidental. See Houtman, I, 175-179; Propp, 415-416.

Exod 12:17	Exod 12:41	Exod 12:51
17 And you shall observe the Unleavened Bread because on this very day (בעצם היום הזה) I brought your hosts out of the land of Egypt. And you shall observe this day throughout your generations, as an everlasting statute.	At the end of the four hundred and thirtieth year, on this/that very day (בעצם היום הזה), all the hosts of Yhwh went out of the land of Egypt.	On this/that very day, (בעצם היום הזה) Yhwh brought the Israelites out of the land of Egypt, according to their hosts.

Once again, the departure of the Israelites is interpreted as the march of an army according to "rank."[347] Moreover, in v. 41 there appears the expression "on this/that very day," which is employed three times in this chapter. In v. 17 it is used in the present tense when Yhwh communicates with Moses the day of departure from Egypt, whereas in verses 41 and 51 it assumes a different meaning. Typically, one would expect the narrator to say "on that day" (ביום ההוא, Gen 15:18; Exod 13:8; 14:30). Instead, the narrator says בעצם היום הזה, "on this/that very day." This construction probably has an anaphoric meaning and refers to v. 17, "on the very day in question, on that very day."[348] However, it cannot be ruled out that the choice of this particular formulation can have a deictic intention,[349] creating a metalepsis, a link between the world of the story and that of the reader-hearer,[350] prolonging the celebration of Passover into the present ("this day") of every generation.[351]

---

[347] See, in this chapter, notes 336-338.
[348] See J.-P. SONNET, "'Today,'" 502, "The deictic pointing, however, can also refer to elements mentioned in the ongoing speech, either earlier ("anaphora") or later ("cataphora"). The pointer הוא, in particular, can be used as anaphora: in many cases 'in that day,' means 'in the aforementioned day,' and so is it in Deuteronomy where the expression 'in that day (ביום ההוא)' used by the narrator in 27,11 and 31,22 refers to the (remote) day singled out in 1,3, the day of Moses' speeches in Moab. Analogously, the marker זה, 'this,' functions in some instances as an anaphora; this is the case in 32,48 when the narrator uses the phrase בעצם היום הזה, 'precisely in this day,' to refer anaphorically to the day established in 1,3."
[349] See G. BASILE – F. CASADEI – al., Linguistica generale, 377: deictic elements are personal pronouns ("I," "you"), demonstrative pronouns ("this") or adverbial expressions ("here," "today") which "relate utterances to spatio-temporal aspects, to the elements of the situation of enunciation and thus to the persons, objects, events, actions we intend to speak of. Deixis, in short, is the tangible manifestation of the way language and context relate to each other." (My translation)
[350] J.-P. SONNET, "'In quel giorno,'" 33. Narrative metalepsis occurs when certain aspects of the narrative cross the threshold of narrative and enter the reader's present; in the book of Deuteronomy, for example, the "today" in which Moses stands in front of the Jordan River merges with the "today" of the reader contemporary with the last editor of the book, who,

⁴² That was a night of vigil³⁵² *for Yhwh*, to bring them out from the land of Egypt: this same night is *for Yhwh*, a vigil for all the Israelites throughout their generations.

This verse is the epilogue which describes the night, especially through the repetition of שמרים, as a moment of obedience and expectation.³⁵³ The term in fact can be explained as a reference both to a vigilant attitude (see Ps 127:1) and to the observance of the Passover instructions³⁵⁴ (see שמר in 12:17, 24-25). The recurrence of ליהוה, "for Yhwh," then, calls to mind a refrain that has already appeared several times in the chapter³⁵⁵ (verse 11, 14, 27). Therefore, the actions performed in the night are once again qualified by their reference to God. Nevertheless, this repeated use of the preposition ל, "for," achieves an asymmetrical balance: what is done "for Yhwh" is at the same time "for" the Israelites ("to bring them out of the land of Egypt" and "for the Israelites and their generations"). Finally, v. 42 concludes with a new reference to subsequent generations, uniting the Israelites who were released on that night with those who will celebrate Yhwh's deeds in the future (12:14.17).

---

returning from exile, must adhere again to the covenant with God in order to enjoy the promise. Through him, every future reader is involved; see ID., "'Today,'" 498-518.

³⁵¹ See *MekhY*, *Pisḥa* § XIV, p. 79-80, "In that night were they redeemed and in that night will they be redeemed in the future."

³⁵² The term שמרים, "vigil," is a plural *abstractionis* (JM § 136g.i).

³⁵³ The *Targum Neofiti* reports a famous interpretative gloss on this verse, the so-called "Poem of the Four Nights." "The first night was when the Lord was revealed over the world to create it, the world was formless and void, and darkness was spreading over the face of the deep, and the Memra of the Lord was light and shone. So He called it the first night. The second night was when the Lord was revealed to Abram at one hundred years, and Sarah his wife was ninety years, to fulfill what the Scripture says, 'Behold, Abram at one hundred years old begat, and Sarah his wife at ninety years old gave birth.' And Isaac was thirty-seven years old when he was offered up on the altar. The heavens were lowered and came down, and Isaac saw their form. So his eyes became dim at their form, and He called it the second night. The third night was when the Lord was revealed to the Egyptians at midnight. His hand was slaying the firstborn of the Egyptians, while His right hand was protecting the firstborn of Israel to fulfill what the Scripture says, 'Israel is My firstborn son.' So He called it the third night. The fourth night will be when the world will be complete; at the time for it to be redeemed, the yokes of iron will be shattered, and the evil generations will be destroyed, and Moses will go up from the midst of the wilderness." See R. DI PAOLO, "Il poema delle Quattro notti," 83-105.

³⁵⁴ Propp, I, 416; G.I. Davies, II, 116.

³⁵⁵ See, in this chapter, § 2.2.2.

Conclusion

Several significant results have been obtained through the exegesis of Exod 11 and 12:1-42. The most relevant phenomenon of these chapters, from a literary and narrative point of view, is the insertion of 12:1-28 between the announcement of the last plague (11:1-10) and the account of its execution (12:29-34). The close reading has shown that where the reader would expect the words predicted by Yhwh to be fulfilled, he is surprised to find in its place a long section in which various instructions for the Passover are presented (12:1-28). This "intrusion" achieves the effect of connecting the legal section to the narrative section and constitutes the sequence "narrative – legal text – narrative" as a critical element for the understanding of these chapters. The obedience of the Israelites and the execution of what was indicated in the words of Yhwh and Moses (v. 28) is therefore the focal element that causes the resolution of the plot.

The essential instructions for the celebration of the rite of the lamb and of the rite of the unleavened bread are listed in vv. 1-20. The first reference to blood appears in v. 7 when Yhwh states that the household, after slaughtering the lamb (12:3-6), must smear some of the animal's blood on the doorposts and lintel of the house. After the instructions concerning the family dinner (12:8-11), Yhwh announces a majestic intervention. In the night he will "pass by" making himself incredibly close, striking Egypt and offering himself as a guarantee ("I Yhwh," see 12:12-13). At that moment, the blood on the gates will be a "sign" to the Israelites that Yhwh will "pass over" their homes. In vv. 15-20, while introducing the ritual of the Unleavened Bread, Yhwh presents the Exodus as an event that has already taken place ("I have brought out your hosts," v. 17) and as a "legacy" to be carried on to future generations (vv. 14.17).

The term "blood" appears a third time, in vv. 21-27, when Moses reports Yhwh's words to the elders. The close reading has shown that the change of locutor is crucial to the narrative. Moses does not simply reproduce Yhwh's discourse, he reduces the description of the lamb's slaughter (vv. 1-6) or the dinner (vv. 8-11) and devotes more attention to the rite of blood (vv. 22-23 recall vv. 7, 12-13). In this way, Moses makes an explicit connection between 12:22 and 4:25 through the repetition of the rare term נגע. And furthermore, he makes his own personal experience resound in his words, by alluding to his nocturnal confrontation with God (4:24-26).

The various recurrences of the term דם (vv. 7, 13, 22-23), therefore, create a connection between Exod 12 and the blood ritual of 4:24-26. The circumcision, "sign of the covenant" (Gen 17:11), allows Moses to overcome the danger. Attacked by Yhwh, he can survive thanks to the tangible sign that manifests his family bond with Yhwh. Similarly, the application of blood on the doors is a sign (Exod 12:12-13) of the presence of Israelites celebrating Passover "for Yhwh." The ritual expresses the relationship with Yhwh and it is precisely the relationship with Yhwh that enables the people to obtain salvation.

Moses' instructions conclude with a reference to another significant aspect of the paschal ritual. The celebration is defined as a "word" (v. 24) and a "service" (v. 25), an action ("service") which has a communicative intention ("word"). For this reason, the ritual action must be complemented by the word of the parent who, as a witness, will have the task of telling future generations about the experience of liberation (vv. 26-27).

After the narrator has informed the reader of the perfect obedience of the Israelites (v. 28), he picks up the narration again by recounting the execution of the plague (vv. 29-34) announced in 11:1-10 and with a description of Israel's departure from Egypt (vv. 35-42). In the middle of the night Yhwh strikes the Egyptians, causing tremendous anguish by occupying every space with a shadow of death (vv. 29-30.33). Pharaoh, having summoned Moses and Aaron, speaks frantically (vv. 31-32) and the people are quickly forced to leave, without any provisions other than the still unleavened bread dough (vv. 34.39). The Israelites however, do not leave in poverty, but together with large flocks (v. 38) and carrying precious objects given by the Egyptians (vv. 35-36). They are described as a large and triumphant army (vv. 40-42) whose departure manifests their new dignity.

Therefore, just as in the episode of 4:24-26 the blood "circulates" within Moses' family, in Exod 11:1 – 12:42 a similar experience is shared by all the Israelite families, thanks to a rite of passage in which all were involved. The trajectory generated by 4:24-26 and 11:1 –12:42 thus gives rise to an expectation: Will blood be a significant element in sanctioning the bond between God and the people as a whole?

CHAPTER IV

The Blood of the Covenant (Exod 24)

The plot of the book of Exodus is woven from many threads and thus forms a canvas in which the different colours create a harmonious ensemble; our attention has been caught by the scarlet of the blood in 4:24-26 and in 12:7, 13, 22-23 and many connections have been found between these two sections of the composition. In the night of liberation, blood plays a critical role and grants the slaves the possibility to leave Egypt and to enter the desert (15:22 – 17:16). After the visit of Jethro (18:1-27), Moses and the people stop at Sinai (19:1). On the mountain, Yhwh communicates his intention to seal a covenant with the people (19:3-6), Moses summons the elders and tells them of all that Yhwh intends to do; the people, then, enthusiastically accept (19:7-8). Afterwards, God manifests Himself in a wonderful theophany (19:16-19) and reserves for Moses a special revelation: the Decalogue (20:1-21) and the "Covenant Code" (20:22 – 23:33).

The subsequent sequence (ch. 24) marks the conclusion of the first part of their halt at Sinai, and recounts the stipulation of the covenant between God and the Israelites, using the same scarlet colour of blood (24:3-8) which appeared earlier in 4:24-26 and 12:7, 13, 22-23. The connections between 4:24-26 and 24:3-8 have already been shown; in both passages the root כרת is used (4:25; 24:8) and the motif of the covenant is alluded to.[1] The relationship between Exod 12 and Exod 24 appears to be established on a ritual level: unlike 4:25, the blood of 24:8 is obtained by means of the slaughtering of animals, as is the case in 12:7, 13, 22-23. Such a comparison manifests and supports the idea of a progressive dynamism with respect to the theme of blood in Exodus: the personal experience of Moses (4:24-26), which every single family of the Israelites shares on the night of Passover (12:7.13.22-23), now involves all the people gathered together on Mount Sinai (24:3-8). In fact, while the memory of the liberation in the night is still fresh, the Israelites relive a rite of passage that prolongs what began with the Passover, creating a deep bond between God and the people.

---

[1] See ch. I, § 2.

As elsewhere, the first stage of the analysis will be "genetic." The tensions and fractures of the chapter will be examined in order to discuss the main hypotheses for the reconstruction of its redactional history. This preliminary study will enable the reader to approach the *lectio cursiva* of Exod 24 with special attention to the most complex transitions.

1. Genesis of the Text

1.1 *Delimitations of the Scenes*

Delineating the various scenes which compose Exod 24 proves itself to be challenging. Firstly, v. 11 resembles a conclusion and vv. 12-18 seem to be part of a new section: in vv. 9-11 Moses fulfils what Yhwh had said in vv. 1-2, going up the mountain together with Nadab, Abiu and the seventy elders, where they "see" God and eat a meal (vv. 9-11). In v. 12, however, God addresses Moses directly, asks him once again to climb (עלה, «to go up») the mountain and to stay (היה), which is precisely what happens in v. 18 (where the same verbs are repeated). In v. 12, moreover, the text mentions the stone tablets on which God will write "the law and the command," anticipating what will be developed later (see 31:18). For such reasons, several authors consider vv. 1-11 and vv. 12-18 to be two different parts[2], while others divide the passage in a more articulated way.[3]

Many exegetes, however, consider the two parts of the chapter as a unity, interpreting vv. 12-18 as a transitional passage,[4] since it is common in the book of Exodus that the boundaries between episodes be marked by passages that can be related both to what precedes and to what follows.[5] According to this approach, vv. 12-18 are closely connected to the "Sinai pericope," but at the same time, they open up to what is about to be narrated in Exod 25–30. These authors normally emphasise the elements of continuity between vv. 1-11 and vv. 12-18: (a) the repetition of the verb עלה, "to go up" (in vv. 1.2.9 and vv. 12.13.15.18) and (b) the reference to Aaron and the elders (v. 1 and v. 14). These arguments are solid and the close reading that follows will offer further reasons in favour of the unity of the chapter.

---

[2] Nepi, II, 158-159; Priotto, 460; Fischer – Markl, 274-275.

[3] See Houtman, III, 285; he supposes that 24:1-2 is related to what precedes it, but explains vv. 3-11 and vv. 12-18 as being two different passages. Finally, he interprets vv. 12-18 as an introduction to the following section.

[4] See Childs, 497-502; Sarna, 150-155; Durham, 340-341; Dohmen, 198-199; see also P. ROCCA, *Gesù, messaggero*, 115, note 56.

[5] See the close reading of Exod 4:18 in ch. II, § 2.1.

1.2 Literarkritik

The chapter is characterised by some evident fractures: the instructions given by Yhwh to Moses in vv. 1-2 are executed only in vv. 9-11; moreover, from a syntactical point of view, the beginning of v. 1 is unusual; and for these reasons it seems that v. 3 is more easily associated with 23:33 rather than with v. 1.[6] Furthermore, vv. 3-8 maintain a certain independence from the context. Finally, the passage formed by vv. 12-18 is marked by several repetitions that seem redundant (for instance, the use of עלה, in vv. 12, 13, 15, 18).

The hypotheses concerning the redactional history of Exod 24 are various. Traditionally, vv. 1-15a, 18b have been interpreted as originating from source E.[7] There are various arguments in favour of this explanation: (i) in vv. 10, 11, 13 the divine name אלהים is used and (ii) in these verses there are several references to passages considered to be part of the E document (see 19:7-8; 20:21; see also Gen 22 and Num 11-12).

The remaining verses (vv. 15b-18a) have been associated with the document P,[8] especially because they employ the vocabulary of the priestly literature.[9] This conclusion is generally held by recent authors.[10] On the contrary, the assumptions concerning the origin of vv. 1-15a have been widely revised and ch. 24 is usually considered to be a work composed in more recent times.

Verses 3-8 form a unified and coherent passage[11] even though some have considered v. 3b and v. 7b to be doublets,[12] or have thought that vv. 3-8 were a combination of two different rituals, the first constituted by the reading of the book and the other by the blood manipulation.[13] Several reasons allow the interpreter to affirm that Exod 24:3-8 is an extremely late composition: (i) The complex ritual found in vv. 3-8 is unusual and unique; the fact that the covenant is based on a proclamation of the word implies a culture in which

---

[6] Childs, 498; the author notices that v. 1 does not begin with a *wayyiqtol*, but with a particular syntactic construction, "and to Moses he said (ועל־משה אמר)."

[7] See Durham, 340-341; R.E. FRIEDMAN, *The Bible with Sources Revealed*, 160.

[8] See W. BEYERLIN, *Sinaitic Traditions*, 14-18.30; Childs, 499-500; Propp, II, 147; Houtman, III, 298.

[9] The verb שכן and the nouns כבוד and סיני in vv. 16-17 belong to the vocabulary of P. Concerning שכן, see Exod 25:8; 29:45-46; 40:35; regarding כבוד, see Exod 40:34-35; Lev 9:6, 23. See Propp, II, 147.

[10] See E. BLUM, *Studien zur Komposition*, 89, note 194 e 312-313.

[11] E. BLUM, *Studien zur Komposition*, 51; L. SCHMIDT, "Israel und das Gesetz," 171.

[12] See J.L. SKA, "From History Writing," 161, note 70. The people's declarations in v. 3b and v. 7b cannot be considered as doublets, since they are slightly different.

[13] See W.M. SCHNIEDEWIND, *How the Bible Became a Book*, 124-125. Contrary to this approach, J.L. SKA, "From History Writing," 162-163.

the book is already present, and therefore a text dating to the Postexilic period. (ii) The consecration of Aaron and his sons (Exod 29; Lev 8), the most similar ritual to that of 24:3-8, is described in late P texts. (iii) In vv. 3-8 two elements which belong to distinct traditions are combined: the book, linked to Deuteronomistic literature, and ritual blood, a common theme in the P tradition.[14] Consequently, according to Ska, the text bears the "signature" of the scribes responsible for the last edition of the Torah and is "a clear example of the passage from orality to literacy."[15]

Several authors support the theory that even vv. 9-11 are Postexilic. (i) The presence of the category of "elders" does not give enough ground for an eventual hypothesis that affirms an early composition of this text, for this institution survives until the Postexilic period (see Isa 24:23). (ii) Similar vocabulary is found in vv. 9-11 and in late texts, such as Ezekiel (see Ezek 1:26; 10:1). (iii) The cultural background behind some of the statements is that of Mesopotamian temples, with which Israel came into contact in the exilic period.[16] From the same editorial layer we have vv. 1-2, which is a late passage linked to vv. 9-11 precisely because of the reference to the elders, and vv. 12-14. Thus, according to this reconstruction, chapter 24 would be an attempt to reconcile the two main groups of the Postexilic period, the priests and the elders.[17]

---

[14] Van Seters believes that Exod 24:3-8 is late, but related to the work of the post-exilic Yahwist; his grounds are less convincing than those offered by Ska (see next note). J. VAN SETERS, *The Life of Moses*, 284-286.

[15] J.L. SKA, "From History Writing," 165. The social *milieu* and occasion that made the birth of the Torah possible is not related to the so-called "Persian Imperial Authorization" (P. FREI, "Die Persische Reichsautorisation," 1-35; J.M. WATTS, ed., *Persia and Torah*). It does not seem plausible, in fact, to claim that the Persian authorities could have recognised in the Pentateuch, the instrument through which the Israelites could obtain their authorization. The Torah, indeed, is an extensive document written in Hebrew that concerns many "internal" issues and does seem suitable as a political instrument for obtaining rights from the Persian authorities. The writing of the Torah, on the contrary, owes much to two related circumstances: the emergence of a scribal culture dedicated to the writing of what is ancient and paradigmatic; the construction of a library in the Jerusalem temple in the post-exilic period. See S. NIDITCH, *Oral World and Written World*, 60-77; P.R. DAVIES, *Scribes and School*.

[16] See J.-L. SKA, "Le repas de Ex 24,11," 320-321, who quotes E. RUPRECHT, "Exodus 24,9-11," 139-173. The decisive ground for a late dating of Exod 24:11 is provided by the vocabulary used. The expression "like a work of sapphire plate" (Exod 24:10) is found in Ezek 1:26-27; 8:2; 10:1; moreover, sapphire is mentioned in late texts such as Exod 28:18; 39:11; Isa 54:11; Ezek 28:13; Job 28:6, 16; Lam 4:7. The term טהר "purity" (Exod 24:10) appears in recent texts (Lev 12:4,6; Ps 89:45) as do the verbs ראה and חזה (Exod 24:10-11; see Job 19:26-27; 23:9), and the noun עצם "self" (Exod 24:10; see Lev 23:28; Ezek 2:3; 24:2; 40:1).

[17] See J.-L. SKA, "Le repas de Ex 24,11," 322-327.

Accordingly, a Postexilic date for Exod 24 seems highly likely, given that these recent redactional analyses are based on solid grounds and present a coherent picture of this unique chapter, a chapter that has a pivotal function in the book of Exodus. Our close reading of the text, then, will prompt the reader to appreciate the order of such varied material and to clarify the role that blood plays in the narrative.

2. The Covenant Ritual

The episodes narrated in Exod 24 form a multifaceted entity, and reveal traces of a complex editorial elaboration. As a whole, however, the chapter maintains a logical structure and can be read highlithing the plot's unity.

An unexpected *waw-x-qatal* begins the narrative in 24:1 as a sharp *caesura*, introducing information not found in the main thread of the narrative.[18] After a long section in which Yhwh revealed to Moses the "Covenant Code" for the people (20:22 – 23:33), in 24:1-2 Yhwh addresses some private words to Moses only (vv. 1-2), and repeats what He had already anticipated in 19:24, "Go down, and come up, and Aaron with you."[19] The first part of the chapter (vv. 1-2) does not have an inciting moment of the plot, the narrator does not describe any action, but introduces Yhwh's discourse as a "program" that anticipates subsequent developments of the plot. Firstly, Moses will climb the mountain accompanied by Aaron, and his two sons, and seventy elders (v. 1; see vv. 9-11); later, he will approach Yhwh alone (v. 2; see vv. 12-18). Furthermore, in these first verses there begins a mechanism that will be repeated several times in the course of the chapter: the reader is led along certain "paths" through the plot which are then abruptly interrupted by the narrator spontaneously beginning a new and unexpected route. For instance, in this first passage what is said by Yhwh in v. 2 remains unresolved until v. 9.

The syntax of v. 3 marks a shift from the background (vv. 1-2) to the foreground of the action. The verses present a long series of verbs conjugated in the past form, the *wayyiqtol*, which creates the impression that Moses is fully active. In fact, the words uttered by the Israelites in v. 3 and v. 7, and by

---

[18] See A. NICCACCI, "Narrative Syntax of Exodus 19–24," 211.222.

[19] See J.-L. SKA, *"I nostri padri"*, 27-28. Through the analysis of narrative phenomena concerning the temporal order and sequence, the author notices that the Bible often makes use of the technique of overlapping (like tiles on a roof): the narrative returns to a previous narration to begin a new narrative: e.g., Exod 19:20 goes back to 19:16-19, just as Exod 24:1 recalls 19:13, 24 and Exod 33:1-6 resumes 32:34.

Moses in v. 8 are brief, while the actions are many and varied.[20] Moses comes down from the mountain and tells the people what he has heard from God, just as Yhwh said to him (see 20:22; 21:1). Suspense is increased through vv. 3-8, as they delay the execution of the orders pronounced in vv. 1-2 and provoke a strong effect of surprise. Moses writes Yhwh's words in a book (v. 4a) and also performs a unique ritual that Yhwh had not requested and which will never be repeated again in the Bible (vv. 4b-8).[21] This passage (vv. 3-8) will be of particular importance for us, as it involves the motif of blood manipulation. In these verses, as is the case in vv. 1-2 and in vv. 9-11, other "paths" are at first opened and then interrupted. In fact, what is recounted in v. 3 seems to reach a conclusion (Moses tells the people all that Yhwh has said and the people answer), whereas from v. 4 to v. 8 the narrator creates something similar to a "zigzag-like" itinerary that raises many questions in the reader: why does Moses decide to write Yhwh's words (v. 4a)? Why does the narrator immediately interrupt this narrative thread and give priority to the recounting of the sacrifices (vv. 4b-6)? Why does he cut off the plot again in verse 7? Finally, why does he conclude the narrative arc with the sprinkling of the blood and the words of Moses in v. 8? The *lectio cursiva* will attempt to formulate some answers to these questions.

The execution of the instructions given in v. 1 is found in vv. 9-11: without a word being spoken, Moses and the group which escorts him ascend the mountain and have a divine vision, at the end of which they eat and drink.

After this first passage, that seems concluded in v. 11, v. 12 marks the beginning of a new stage in the narrative: Yhwh tells Moses to "go up" (עלה) to Yhwh, recalling what is said in v. 2,[22] and to stay (lit. היה, "to be") on the

---

[20] The narrator in v. 3 recounts that Moses goes (בוא) and narrates (ספר) to the people what he has heard from Yhwh. The people answer and accept his words (ענה and אמר). Then, in v. 4 Moses writes (כתב) Yhwh's words, gets up early (שכם), and builds (בנה) an altar. Afterwards (v. 5), he sends (שלח) young men who offer (עלה) burnt offerings and sacrifice (זבח) peace offerings. Moses, then takes (לקח) half of the blood, and put it into basins (v. 6), takes (לקח) the Book of the Covenant (v. 7), reads it (קרא), after which the people again react to this proclamation (אמר). Finally, Moses takes (לקח) the blood, throws it (זרק) on the people and utters (אמר) some words in v. 8. All these verbs are conjugated in the *wayyiqtol*.

[21] See A. SCHENKER, "Les sacrifices d'alliance," 489, "La seule conclusion d'alliance accompagnée de sacrifices est finalement celle d'Ex XXIV." See also N. MACDONALD, "Scribalism and Ritual Innovation," 415-429; since there is no evidence to prove the existence of the covenant ritual and of the sprinkling with blood, he considers it to be a "scribal invention," namely a "textual speculation giving rise to ritual."

[22] See P. ROCCA, *Gesù, messaggero*, 115, note 56. Commenting on v. 2, Rashi considers the instruction given by Yhwh in this verse as a reference to the entry into the cloud (vv. 15-18); see Rashi Exod, 207.

mountain; the instruction is fulfilled in v. 18: "and he went up (ויעל) on the mountain and Moses stayed (ויהי) on the mountain for forty days and forty nights." In this second scene, the most important literary and narrative phenomenon is the repetition of עלה in vv. 12, 13, 15, 18: God asks Moses to go up and approach Him (v. 12), Moses goes up accompanied by Joshua (v. 13) resuming a movement announced in v. 1 and begun in v. 9. The narrator repeats the information concerning Moses' ascent in v. 15 and v. 18, where the reader is informed of Moses' solitary going up. This repetition of the verb עלה is intended to increase the *suspense*[23] because the arrival at the top of the mountain is continually delayed; thereby, the narrator dramatically reproduces the successive stages of a long climb.[24] This progressive advance also dramatises the gradual approach to God of the various characters who are involved in the story: the people on the slopes of the mountain; Aaron, his sons and the elders higher up; Joshua even closer to Moses; and then Moses alone in the climax of vv. 15-18.[25]

The chapter has its conclusion in vv. 15-18 which describe the meeting between the divine glory and Moses, in slow motion: Moses goes up and the cloud covers the mountain (v. 15); the glory settles on the mountain and the cloud stays there for six days; Moses can overcome the understandable reluctance to enter a devouring fire[26] (v. 17) because he is called by God (v. 16). Only on the seventh day can Moses enter the cloud (v. 18) and remain there.

Thus, the plot revolves around the divine word pronounced in vv. 1-2, that is slowly fulfilled in vv. 9-11 and in vv. 12-18. However, the course of events surprises the reader and, by means of the dramatic strategy of the interrupted "path," manifests the crucial character of vv. 3-8.

---

[23] A. NICCACCI, "Narrative Syntax of Exodus 19–24," 224.

[24] See P. ROCCA, *Gesù, messaggero*, 116, note 58; the triple repetition of the verb עלה gives the impression "of Moses' challenging climb to the summit of Sinai." (My translation)

[25] See Meyers, 205; Janzen, *Exodus*, 189. The episode of 24:12-18 depicts Moses as a person who has a greater intimacy with God in comparison to those who accompany him in vv. 9-11. He can, in fact, enter the devouring fire without being consumed (see 3:6); moreover, whereas the vision of the people's representatives in vv. 9-11 is limited, Moses' encounter with God is prolonged and closer (Moses "enters" the cloud, while the others see only from afar). Finally, during the encounter, Yhwh speaks at length with his prophet (see Exod 25–30); see P. ROCCA, *Gesù, messaggero*, 119-120.

[26] See P. ROCCA, *Gesù, messaggero*, 118; Exod 24:17 is an insertion between Moses' call (v. 16) and his entry into the cloud (v. 18) that heightens the suspense and generates some surprise in the reader.

2.1 God's "Program" (vv. 1-2)

¹ And He had said to Moses[27], "Come up towards Yhwh, you and Aaron, Nadab and Abihu, and seventy elders of Israel. You will bow low from afar[28] ² then Moses alone shall come near Yhwh, while[29] they shall not come near, nor the people shall come up with him."

Yhwh's speech to Moses begins with a quick transition: the syntax of v. 1 establishes a clear distinction between that which was said previously and that which is spoken by Yhwh in vv. 1-2, a discourse directed especially to Moses.[30] The first word addressed to Moses, "Come up to Yhwh," is unusual because God speaks of Himself in the third person.[31] The narrator introduces, from the very first line, the divine name by which God's familiarity and closeness are expressed more explicitly.[32] Nevertheless, the use of the third person is perhaps intended to put some distance between Yhwh and Moses (Yhwh does not say to Moses, "Come up to me," as in v. 12).

The repetition of what is said in Exod 19:24 – "Go, come down, and then you will come up with Aaron" – complements what was previously stated with some additional elements. While in 19:24 the elders and the sons of Aaron are not mentioned, in 24:1 it is specified that along with Moses also Nadab, Abihu and seventy elders will "come up". For someone reading the Pentateuch in Persian times, an indication such as this could have a crucial meaning, as it

---

[27] The form *waw-x-qatal* at the beginning of an episode introduces a prelude to the narrative that follows; thereupon the verb אמר can be translated as a pluperfect. See A. NICCACCI, *Sintassi*, § 27. Rashi Exod, 206-207, notes the problem and believes that the words of Exod 24:1-2 were uttered before Exod 20; la *MekhSh*, § LXXXII, 1, states that Exod 24:1 justifies the initiative taken by Moses in 19:3 ("Moses came up to God").

[28] The verb והשתחויתם is translated by the Septuagint with the third person plural; the MT is to be maintained because it is more attested in ancient translations and is a *lectio difficilior*: the Greek translation presupposes that only the companions will have to stop, while Moses will be allowed to climb higher up, adding τῷ κυρίῳ at the end of the verse, specifying to whom they must prostrate themselves. Childs, 498; Durham, 339; Propp, II, 138.

[29] The words of Yhwh in vv. 1-2 begin with an imperative (v. 1, עלה), two *w^eqatal* (והשתחויתם, v. 1 and ונגש, v. 2) and two *waw-x-yiqtol* (יגשו and יעלו, v. 2). The repetition of *w^eqatal* can express a succession of future actions (JM § 117d); the *waw-x-yiqtol*, on the contrary, express contemporaneity. The syntax indicates a clear difference between what Moses will do ("he will come near") and what the others will have to do ("while they shall not come near"). See A. NICCACCI, "Narrative Syntax of Exodus 19–24," 215.

[30] See above, note 27. See Propp, II, 292.

[31] Childs, 503-504; Propp, II, 292. The problem is noticed in rabbinic tradition, since *TgPsJ* of Exod 24:1 translates the passage as follows, "And Michael, the prince of wisdom, said to Moses at the seventh day of the month." This reading interprets the verse considering what is said in Exod 23:20-24 about the angel sent by Yhwh.

[32] See ch. II, § 4.3.

situates, gathered around Moses, the founding figures of the two main institutions of the Postexilic period, the priests and the elders.[33] On the other hand, the narrative has prepared the reader for the prominent role played by the elders: Yhwh asks Moses to go to them, before entering before Pharaoh (3:16, 18) and they meet in 4:29; then the elders are present when Moses transmits the instructions concerning the celebration of the Passover received from Yhwh (12:21), they witness the gift of water in 17:5-6, they participate in the meal of Moses and Jethro (18:12), and are mentioned in the Sinai pericope (19:7). The difference between 24:1 and these other occurrences is that the narrator refers to a precise number:[34] the group of the elders is now established.

The individuals gathered around Moses should fall prostrate (חוה), thereby manifesting full recognition of divine authority.[35] Their attitude will thus reflect what Yhwh proclaimed in the Decalogue, "You shall not make for yourself a carved image ..., you shall not bow down (תשתחוה) to them and serve them" (Exod 20:4-5). Worship, in fact, manifests a rejection of idolatry and communicates an exclusive reverence, directed towards Yhwh[36]. However, the group "will bow down from afar" to show their respect for God (see Gen 33:3) and above all to express their awareness of His transcendence.[37]

The divine discourse continues in v. 2 with a distinct trait in comparison to v. 1; whereas in the first part of His discourse Yhwh addresses Moses in the second person singular ("come up") and the group with the second person plural ("you will bow low"), in v. 2 He uses the third person: "then Moses alone shall come near Yhwh, while they shall not come near." Even

---

[33] E. BLUM, *Studien zur Komposition*, 339-345. The Priestly ministry is strictly connected to Moses because Aaron and his sons are consecrated precisely by him (Exod 29; Lev 8); even the elders are connected to Moses, since they help him to "bear the burden of the people" and receive part of the Spirit that is upon Moses (Num 11:16-17, 24-30; see Ezek 8:11). The priesthood and the elders are the two institutions that survived the catastrophe of the exile, whose authority is legitimized by Exod 24; J.-L. SKA, "Le repas de Ex 24,11," 317, note 51.

[34] The number seventy can have different meanings: the elders have a representative function for the whole of Israel, formed by seventy clans (Gen 46:8-27); in Exod 1:5 "seventy" is already referred symbolically to Israel; see Sarna, 150. Moreover, in the Ancient Near East, the ideal number of guests at a sumptuous banquet was seventy; see Propp, II, 293 and T.C. VRIEZEN, "The Exegesis of Exodus xxiv 9-11," 100-133.

[35] See ch. II, § 5.3, notes 333, 335; Sarna, 150, note 9, where he quotes Gen 27:29; 33:3; 49:8 and the oriental texts collected in *ANET*, 483-490. The prostration could be a symbolic expression of the willingness "to accept the yoke of God's kingdom" (Cassuto, 310).

[36] Fischer – Markl, 267-268.

[37] See Sarna, 151, where the author quotes *PRU*, IV, 221.226: the vassal bowed before the sovereign to show diplomatic courtesy.

though the unexpected change of person is possible in Hebrew,[38] the direct speech of v. 2 is intended to encourage the reader's involvement:[39] Yhwh addresses the reader considering him as "spectator" in order to define the positions of the characters in the next scenes.

The recurrence of the verb נגש, "to come near," used to indicate the personal encounter between individuals, is the main literary phenomenon of v. 2.[40] The repetition of נגש manifests a disproportion between Moses and the others: Moses will have a precise destination (נגש + אל + יהוה, "to come near Yhwh"), while for the other members of the people it is not specified ("they shall not come near"); there can be no greater emphasis regarding Moses' extraordinary intimacy with Yhwh, than is affirmed here[41] (see 20:21).

2.2 *From the Word to the Book (vv. 3-4a)*

> [3] And Moses went and recounted to the people *all the words of Yhwh* and all His judgements. And all the people answered with one voice[42] and said, "*All the words that Yhwh has uttered, we will do!*"[43] [4] And Moses wrote down *all the words of Yhwh*.

The narrator brings the reader back to the foreground of the story but creates an effect of suspense by interrupting the movement prepared by the previous verses. He postpones the account of the fulfillment of the

---

[38] The difficulty created by the third person plural used in v. 2 is noted in the comment of Ibn Ezra (p. 207). Propp believes that v. 2 quotes the words that Moses must utter before the people (Propp, II, 293).

[39] Concerning the narrative strategies employed to involve the reader, see e.g., J.-P. SONNET, "L'analisi narrativa," 54.

[40] The verb נגש has often a personal subject and object, and designates a meeting between individuals (Gen 27:22; 44:18; 45:4; Num 32:16; 1 Kings 20:13, 22).

[41] In Exod 19:22 נגש defines the special condition of priests, who are designated as "those who come near Yhwh" (הנגשים אל־יהוה). The same root is used (especially by P) to denote the priestly ministry: נגש + אל + God (see Ezek 44:13); נגש + אל + altar (Exod 28:43; 30:20; Lev 21:23; see Num 4:19; 8:19). See H. RINGGREN, "נגש," 233-234.

[42] The noun קול can have the function of an "adverbial accusative," i.e. an accusative indirectly subordinate to the verb to which it refers: גדול + קול + קרא, "to speak out loud" (Ezek 8:18; 11:13); גדול + קול + זעק, "to cry out loud" (2 Sam 19:5); see Deut 27:14. Joüon explicitly links this use of קול to Exod 24:3 (JM §§ 125*s*; 126*d*).

[43] At the end of v. 3, the Septuagint adds the words καὶ ἀκουσόμεθα, "and we will listen," relating v. 3 to v. 7 and repeating the same answer of the people in both cases. The MT is the *lectio difficilior* and must be maintained; see E. TOV, "Textual Problems," 13, who believes that the Septuagint's harmonisation of v. 3 and v. 7 has theological motives, for "some readers cannot imagine that Scripture would use different wordings under the same circumstances."

instructions given in vv. 1-2 and introduces an interlude.⁴⁴ The recurrence of the term עם, "people," in v. 2 and v. 3, nevertheless, may offer the reader the first clue that may explain the apparent suspension of the "program" announced in vv. 1-2: when Yhwh informs Moses that the people will not go up with him (v. 2, "nor shall the *people* come up with him"), Moses decides to involve the Israelites (v. 3, "he told the *people*") by telling them the words that he heard from God (as in 20:22 and 21:1). Only when the whole people can participate in Moses' journey will he climb the mountain, escorted by a delegation of the Israelites.

In these verses, the verb ספר *piel*, "to count, to relate,"⁴⁵ is the first surprise, as it has not previously been used in this context.⁴⁶ The choice of this root, in fact, creates a strong resonance with the term ספר, "book," employed in v. 7, and presents Moses as a model for future scribes.⁴⁷ When he received Jethro's visit, Moses had already formulated a "narrative" of God's works (ספר *piel*, 18:8); in Exod 24:4 Moses directly transmits His words.⁴⁸ This circumstance anticipates what will happen when Moses will once again recount God's deeds and words to the people (see Deut 1:1), just before they enter into the land.

The triple repetition of the expression "all the words" found within this passage, emphasises the integrity of Moses' narration. His communication is completely faithful and presents exactly what he received from God. Exegetes agree that the expression "all the words of Yhwh and all the judgments" refers to what is found in the previous chapters.⁴⁹ The syntax of this expression leaves the reader in suspense, so much so that the expression ואת כל־המשפטים has been considered as a late addition.⁵⁰ Furthermore, the

---

⁴⁴ B.P. ROBINSON, "The Theophany," 156.

⁴⁵ Usually, the root ספר *piel* has a personal subject (see Gen 29:13; 37:9-10) and is used for the narration of events (עשה, Gen 24:66; 2 Kings 8:4) and for the reporting of words (דבר, 1 Sam 11:5; 1 Kings 13:11); see also Num 13:27; Judg 6:13.

⁴⁶ Several times Yhwh states that Moses shall communicate to the people all that he has heard, but Yhwh never employs the verb ספר. He uses instead אמר (19:3; 20:22), נגד (19:3) and the construction פנים + ל + שים (19,7; 21,1).

⁴⁷ See J.L. SKA, "From History Writing," 165: the root ספר "is present in the noun *sōpēr*, 'scribe,' which does not appear in this context but is probably in the background of the scene. Moses, who writes Yhwh's words, is of course a 'scribe,' a 'writer.' ... Moses is the first *sōpēr* and can be seen as the model and ancestor of all the *sōpĕrîm* of Israel."

⁴⁸ Priotto, 462.

⁴⁹ Some scholars believe that דבר refer to the Decalogue (דברים, Exod 20:1) and משפט, on the other hand, to the Covenant Code (משפטים, Exod 21:1); see Childs, 505; Priotto, 462-463; Nepi, II, 145; anche Propp, II, 293. Others think that Exod 24:3 alludes to what is told between 20:22 and 23:33; Cassuto, 311; Sarna 159; Houtman, III, 289; Dohmen, II, 200-201.

⁵⁰ The specification ואת כל־המשפטים is considered to be an addition; see J.L. SKA, "From History Writing," 160-161. In fact, וכל־המשפטים seems out of place, since the phrase's order

repetition of the term דבר reveals that the covenant is first and foremost constituted by *words*.[51] In fact, before being a set of laws, the covenant establishes personal communication and a relationship between two parties. On a dramatic level, this impression is also confirmed through the answer of the Israelites, who acknowledge that the laws are mainly "words" (and do not refer to "judgements") by using the root דבר twice (v. 3 "all the words that uttered Yhwh," כל־הדברים אשר־דבר יהוה).

After hearing "all" the words and judgments, "all" the people react (v. 3). This repetition of כל, "all," rings in the ears of the reader, who understands that if Moses delivered what he received without leaving anything out, the people in turn respond with a reply that is full and complete.[52] *All* the people say "we will do *all* the words of Yhwh." By staging the direct speech of the people, the narrator also shows that the divine word has a performative energy: Yhwh's words produce a real effect in the listener (it is a directive illocutionary act), they challenge him and demand an assent (which they formulate through a commissive illocutionary act).[53]

The people's reply is unanimous, they answer "with one voice." The term קול creates a strong connection between 24:3 and chapter 19. Indeed, the theophany of Exod 19 proceeds from the visual to the auditory register precisely by virtue of the fourfold repetition of קול in v. 16 and v. 19.[54] While the people hear the loud sound of a horn, Moses listens directly to the voice of God (v. 19); after Moses' discourse (24:3) it is instead the voice of the people that resounds, manifesting that they consent to Yhwh through a "physical" gesture, which expresses the participation of their whole person.[55] In the short scene, then, by means of the interlocutor's consensus an act of communication is fulfilled.[56] It's pragmatic purpose is

---

"all the words of Yhwh and all his judgments" is unusual in comparison with the more normal structure, "all the words and all the judgments of Yhwh." Furthermore, the term משפט is not used in people's answer.

[51] See ch. III, § 2.3.2, note 272.

[52] Priotto, 463.

[53] See E.M. OBARA, "Le azioni linguistiche," 83-117. Concerning the classification of illocutionary acts, see L. PÉREZ-HERNÁNDEZ, *Speech Acts in English*.

[54] See P. ROCCA, *Gesù, messaggero*, 96-98. In Exod 19:16, 19 the reader's attention is caught by the auditory rather than the visual aspect. The two verses play on the polysemy of the term קול: v. 16bc "there was *thunder* and lightning (קלת ובקרים), and a dense cloud on the mountain, and a *sound* of shofar (וקל שפר) very loud," and v. 19 "and the *sound* of the shofar (קול השופר) grew louder and louder, while Moses spoke and God answered him with a *voice* (יעננו בקול)."

[55] See Fischer – Markl, 268: the people agree to adhere to the divine word "auch mit seiner körperlichen Sinnlichkeit und Ausdrucksfähigkeit."

[56] Analysing the deictic "today" in Deuteronomy, Sonnet shows the difference between the use of this term in the covenant narrative culminating in Exod 24:3, 7 and the use of

to generate the same response in the reader, for he has the opportunity to hear the words of Yhwh first hand together with Moses, and is now urged to unite his own personal reaction to that of the Israelites. Consequently, just as the people's speech act[57] consists in a promise of fidelity, that same reality is also offered to every believer engaged in the act of reading.[58]

The last words of v. 3 could conclude the narrative arc and open directly onto v. 9; however, the reader is surprised by v. 4, because, without God having said anything to Moses[59] (as is the case in 17:14), he decides to write down "all the words of Yhwh." The new occurrence of the costruction כל־דברי יהוה reveals the absolute correspondence between the oral report and the written text.[60] The excerpt shows, therefore, that although the initiative to write belongs to Moses, the book is an inspired text[61] because it communicates "all the words of *Yhwh*."[62] The writing of a text is a significant anthropological phenomenon: the word is fixed, so that it can be preserved even when the writer is far from the reader and so that it is made available for the memory of posterity.[63] The act of writing down corresponds to the desire to fix what Yhwh has said and anticipates the transition from orality to written culture, which will be decisive in the theology of Deuteronomy.[64] In comparison with

---

"today" in the book of Deuteronomy; while on Mount Sinai the narrator reproduces the people's answer, in Deut this does not happen because there is no need for it: the internal and external addressees are more clearly aligned; J.-P. SONNET, ID., "'Today,'" 498-518.

[57] The *Speech Act Theory* analyses those phrases which present information and, simultaneously, perform an action. See J.L. AUSTIN, *How to Do Things with Words*; J. SEARLE, *Speech Acts*.

[58] Reflecting on illocutionary acts, Searle first analyses the structure of the "promise." The sentence uttered by the Israelites belongs to the set of propositions that Searle assigns precisely into this category, "in expressing that p, S predicates a future act (A) of S" (J. SEARLE, *Speech Acts*, 57).

[59] Nepi, II, 145; Priotto, 463.

[60] J.-P. SONNET, "Lorsque Moïse," 509.

[61] See P. BASTA – P. BOVATI, *"Ci ha parlato"*, 169-173: writing bridges the distance between author and reader with a definitive formulation of the divine word that has been revealed. The text, however, has some weaknesses, since the written form is "silent," and not as effective as the oral word. To be "opened" the book needs an interpreter (Isa 29:11; Rev 5:1-4).

[62] See J.L. SKA, "From History Writing," 165. Concerning Exod 24:3-8 the author says the following, "This is the only case in the whole Bible that offers a complete description of all the different procedures connected with a *sēper*, a 'scroll' ... Moses *tells* (*waysappēr*) the people or reports to them Yhwh's words, which means he transmits to them the content of an oral, divine tradition, and the people *answer* .... Then, he *writes down* (*wayyiktōb*, v. 4) these same words, Yhwh's words, in a *sēper*."

[63] See P. RICOEUR, *Du texte à l'action*, 156, "l'écrit conserve le discours et en fait une archive disponible pour la mémoire individuelle et collective."

[64] See J.-P. SONNET, "The Fifth Book," 207-213. The phenomenon of writing appears in Deuteronomy at different stages. The first "writer" is God, who has written His words on

the oral announcement, the book will have the advantage of being duplicable and therefore ratifiable[65] (see Deut 17:18-19), it will be destined to future acts of reading (Deut 31:9-13), and it will prolong the mission of Moses by crossing the Jordan and by "entering" the land (Josh 8:31).

2.3 *The Sacrifices (vv. 4b-5)*

> [4] He woke up early in the morning and set up an altar at the foot of the mountain and twelve pillars[66] for the twelve tribes of Israel. [5] He sent young people among the Israelites and they offered burnt offerings (ויעלו עלת) and sacrificed oxen as offerings of well-being[67] to Yhwh.

After mentioning the book, the narrative creates suspense and develops in another direction: Moses gets up early in the morning, showing great resolve,[68] and prepares to offer sacrifices. This rapid change of initiative introduces the scenario in which blood manipulation will take place. The ritual that is going to be narrated lies at the heart of the covenant and is of decisive interest in our research. How can this sudden turn of events be explained? Firstly, from a more material point of view, the decision to build

---

two tablets (Deut 5:22) and kept them in the ark (Deut 10:1-11). The book also appears in the royal court, because the ruler will have to write a copy of the Torah for himself (Deut 17:18-19). Finally, Moses will write down the Torah and give it to the priests and to the elders (Deut 31:9) so that it may become the object of future re-reading for generations to come (31:11).

[65] See Sarna, 151. Something similar to a ratification process takes place with the two tablets mentioned in Exod 31:18; 32:15; Deut 4:13; 5:22. The authors explain this dual action in relation to the stipulation of ancient vassalage treaties, since it was common for such documents to be duplicated so that each contracting party could possess its own copy. The possibility of being copied gave the treaty a legal character. See e.g., D.L. BAKER, "Ten Commandments, Two Tablets," 8-9; D. CHARPIN, *"Tu es de mon sang,"* 126, "Au fil des siècles, l'écrit fut en quelque sorte sacralisé."

[66] The Septuagint and PSam translate "stones" instead of "pillars." This change is a harmonisation with Josh 4,20; see Childs, 498; Propp, II, 139. The MT is therefore to be maintained. After numerals (in this case "twelve"), the noun is often conjugated in the singular (1 Kings 20:1; Isa 7:23); see GK § 134*f*; WO § 7.2.2*b*.

[67] The two terms זבחים and שלמים are separated and not connected in the more common construct form זבחי שלמים, "sacrifices of well-being" (see PSam). The second term is in apposition to the first; see Propp, II, 139.

[68] In two other passages the narrator reports Moses' decision to wake up early in the morning (8:16; 9:13), and in both cases it is Yhwh who asks him to do this. The verb שכם, "to get up early," can be also translated "to be eager for" (Zeph 3:7) and, when used in the absolute infinitive and paired with another verb, it can have adverbial value, "persistently" (see Jer 7:13.25; 25:4; 44:4). When followed by the specification בבקר, "in the morning," and the subject is personal, the verb indicates an action carried out with urgency. See Gen 19:27; 21:14; 22:3. A.L.B. PEELER, "Desiring God," 194.

an altar and twelve pillars at the foot of the mountain must be genuinely considered. Moses seems to be acting contrary to what is said in vv. 1-2: Yhwh asks Moses to come up (v. 1, עלה) but instead he builds an altar "at the foot of the mountain," literally "under" it (v. 4, תחת ההר). However, since the Israelites gather precisely at the slopes of the mountain (19:17), Moses presumably decides to remain at the bottom in order to perform an action that involves the people in his ascent. Indeed, he makes the Israelites participate in his movement, because they too can "ascend" through the burnt offerings, which exalt precisely the vertical axis[69] (v. 5, ויעלו עלת, "they offer burnt offerings").

Finally, to understand this decision even better, the reader cannot but recall the instructions received by Moses in 20:22-26.[70] In this text, the verb עשה, "to make," is used twice in 20:23[71] (*"you shall not make* gods of silver near to me, nor *shall you make* for yourselves gods of gold," 20:23), and once in 20:24: "an altar of earth *you shall make* for me and sacrifice on it your burnt offerings and your peace offerings." Therefore, the building of the altar, with the subsequent practice of sacrifices, is evidently contrasted with idolatry. Thus, when Moses builds an altar in Exod 24:4, he remembers Yhwh's command of 20:24. Through the symbols of the ritual, therefore, the sacrificial action aims at promoting exclusive devotion to Yhwh.

Primarily, Moses defines the ritual space and identifies the two covenantal partners:[72] Yhwh is symbolically manifested by the altar; the twelve tribes of the Israelites are represented by twelve pillars, objects that are no longer linked to the Canaanite divinities[73] (see 1 Kings 18,31-32) and serve as monumental witnesses for the covenant that is about to be stipulated.[74] This distribution of elements allows the recognition of a similarity between Exod 24:3-8 and the texts previously examined, since

---

[69] L.M. MORALES, *The Tabernacle Pre-Figured*, 240-241.

[70] S. CHAVEL, "A Kingdom of Priests," 169-222.

[71] Propp, II, 183; Fischer – Markl, 243, "Selbst die wertvollsten Metalle ermöglichen nicht die Darstellung des Göttlichen. Zweimal, und damit betont, ist gesagt „nicht sollt ihr machen," und zwar „für euch" …. Stattdessen erklären V.24f, was „du machen sollst," und zwar „mir / für mich." Israel soll auf narzisstische *Götter Marke Eigenbau* verzichten und stattdessen die Du-Beziehung zu seinem Gott liturgisch symbolisieren (zu diesem Zusammenhang vgl. auch 23,13-33)."

[72] Cassuto, 311; Childs, 505; W.K. GILDERS, *Blood Ritual*, 37-38.

[73] See Exod 23:24; 34:13; Lev 26:1; Deut 7:5; 12:3.

[74] The term מצבה is used when different parties are stipulating a covenant and the pillars have the function of legal witnesses (Gen 31:44-45; Josh 22 and 24:26-27). Durham, 343; Propp, II, 294; Nepi, II, 146.

they are all rites of passage.[75] Indeed, the gathering of different entities (the twelve *stelae*) in a single position is a symbol of the unity that will be achieved when the ritual action will have taken place.[76] What is described in 24:3-8, therefore, is part of the "liminal" phase that began with the departure out of Egypt when the people were separated from their condition of slavery. On Mount Sinai they have had an experience on the edge between life and death (see 20:19b, "don't talk to us God, otherwise we will die"), and they will be regrouped as a new entity when they will enter the promised land.[77]

The sacrifices are entrusted to young men (v. 5), because the priests have not yet been consecrated (ch. 29). Rabbinic tradition explains this choice supposing that they are among the first-born and, therefore, they are somehow "consecrated"[78] (see 13:2; 22:28). Some members of the "second generation" probably accompany Moses for the sacrifice, because they will be those elected to enter the promised land, after the death of the first generation of Israelites who came out of Egypt (Josh 5:4-6); thereby, the rite of the covenant, while recalling the past (see 20:1, "I am Yhwh, who brought you out of Egypt"), opens up onto the future by involving those who will be protagonists at the moment of the entry into Canaan.

The sacrifices are the same as those mentioned in the "law of the altar" (20:22-26), burnt offerings and offerings of well-being.[79] Of these two kinds of sacrifices the latter are the most developed because the narrator specifies that the young men had to offer oxen, a particularly regarded animal[80] (Lev 1:3-17). This assertion does not surprise the reader, because

---

[75] See, ch. II, § 4 and A. VAN GENNEP, *The Rites of Passage*.

[76] See R. HENDEL, "Sacrifice," 376.

[77] See R.L. COHN, *The Shape of Sacred Space*, 12-13; W. VOGELS, "D'Égypte à Canaan," 21-35. See also R. HENDEL, "Sacrifice," 375, "The ceremony in Ex 24,3-8 takes place within the tripartite scheme of a rite of passage: separation, limen, and reaggregation. The separation is signified by the escape of the Israelite slaves from Egypt and their subsequent journey. The liminal stage is represented by the encounter with Yahweh at the holy mountain, Sinai/Horeb. At this holy place, far from human habitation, the ceremony in Ex 24,3-8 occurs.... The third and final stage of reaggregation is represented by the journey homeward; in this case a return to the ancestral home, the promised land. The entire account of the escape from Egypt, the covenant at Sinai, and the entry into the promised land can be viewed as an elaborate rite of passage."

[78] The *TgO* and *TgPsJ* translate 24:5 as follows, "and sent first-born sons of the Israelites" (וישלח ית בחורי בני ישראל). See Rashi Exod, 208; Cassuto, 311. Propp believes that young people are chosen because the elders will have to go up the mountain with Moses; see Propp, II, 294. The text does not offer any evidence to support this hypothesis.

[79] See ch. I, § 1.3.

[80] Priotto, 464.

the offering of well-being more clearly expresses the communion between Yhwh and the people,[81] and is thus better suited for covenantal contexts. The same sacrifices will be repeated when Aaron will build an altar for the golden calf[82] (32:5-6); this lexical echo has a dramatic effect, because the making of the golden calf is represented as an act of radical opposition to the pact between God and the people.

2.4 *The Rite of the Covenant (vv. 6-8)*

a   [6] Moses **took** half of the blood and put it in basins, while[83] he *poured* (זרק) half of the blood on the altar.

b   [7] He **took** the book of the COVENANT and read it in people's ears and they said:
"All that Yhwh has uttered, we will do and we will listen[84]!"

         [8] Moses **took** the blood and *poured* (ויזרק) on the people and said:

a'   "Behold the blood of the COVENANT that Yhwh stipulated with you upon all these words."

The description of vv. 6-8 is characterised by the anaphoric repetition of the same verb לקח, "to take," and by other repetitions (זרק and דם, in v. 6 and v. 8; ברית, in vv. 7-8) which create a concentric composition according to the scheme a-b-a'. The organisation of vv. 6-8 therefore, is extremely careful and describes the ritual as an intertwined combination of the blood manipulation and the proclamation of the book.

2.4.1 The Blood on the Altar (v. 6)

Firstly, Moses divides the blood into two parts, putting half of the blood into basins and pouring the other half upon the altar[85] (v. 6). The difference

---

[81] See Nepi, II, 147. The peace offering "expressed the 'fullness' (*shalom*) of communion and the confirmation of the covenant bond between the offerer and God." (My translation)

[82] Fischer – Markl, 269.

[83] The passage from the *wayyiqtol* (ויקח, "he took"; וישם, "he put") to the *waw-x-qatal* (וחצי הדם זרק, "while poured half of the blood") is used to describe two simultaneous actions; see A. NICCACCI, "Narrative Syntax of Exodus 19–24," 213.

[84] The Ambrosian Code of the Septuagint, the PSam, the Peshitta, and 4QpaleoExod[m] invert the order of the verbs "to do" and "to listen," and translate, "we will listen and we will do!" The MT is to be maintained because the inversion of the most correct order of the elements may be conscious; therefore, the MT is a *lectio difficilior*. See Propp, II, 139.

[85] See W.K. GILDERS, *Blood Ritual*, 40-41, "The altar – the locus at which, in this cultural system, a sacrifice is offered to Yahweh – receives the first blood manipulation. Then Moses reads Yahweh's words to the people. The privileging of Yahweh by the

between the various "ministers" of the cult employed in this complex ritual is very clear: while the young men are in charge of the sacrifices, the blood manipulation is reserved for Moses.[86] It is this aspect in particular that associates the rite of 24:6-8 with some Ancient Near Eastern texts, in which the bond between the two partners is formed through blood.[87]

The division of the blood into two parts could be a sign alluding to the fate of the person who betrays the covenant (Gen 15:17).[88] By shedding the blood on the altar, Moses signals Yhwh as the prime contractor in the covenant to be sealed. Indeed, in the ritually reorganised setting,[89] the altar is the symbol of Yhwh's presence[90] (Gen 33:20; 35:7; Exod 17:15-16). Thereupon, the blood circulating on the altar shows that the origin of the new bond between the two parties lies in Yhwh. When the reader is waiting to learn what happens with the second part of the blood, the ritual is suddenly interrupted.

2.4.2 The Proclamation of Yhwh's Words (v. 7)

Moses takes the "Book of the Covenant" (v. 7), namely the written text that manifests in its contents the covenant between God and the people,[91] and reads it to the people. The repetition of לקח as the initial term signals the profound correlation between the actions that Moses performs in v. 6 and v. 7. The new interruption consequently shows that after sprinkling the altar, Moses makes present, through the solemn proclamation of the words written in the book, Yhwh who is signified in it.

---

application of blood to the altar is subsequently reflected in the fact that Yahweh speaks first." The author quotes C. BELL, *Ritual Theory*, 98-100.

[86] See W.K. GILDERS, *Blood Ritual*, 38.

[87] Concerning Exod 24:3-8, Charpin says the following, "les contractants mettaient leur vie en jeu, acceptant en cas d'infidelité de subir la même sort que l'animal dont le sang était répandu sur eux" (D. CHARPIN, *"Tu es de mon sang,"* 261). A reference to kinship cannot be excluded, because, as seen in A.2730 (see ch. I, § 2, note 89), when exchange of blood takes place between two sovereigns "il s'agit bien de *consanguineité* et désormais, les rois sont considérés comme étant *réellement* 'frères." (*Ibid.*, 77)

[88] See Propp, II, 308-309: the subdivision of the blood "represents the sanguinary fate that awaits the traitor to the pact."

[89] C. BELL, *Ritual Theory*, 98.

[90] Cassuto, 312; Durham, 344; Nepi, II, 147; Priotto, 465. The identification between the altar and God can become evident in the name: אל אלהי ישראל, "El, God of Israel" (Gen 33:20) and יהוה נסי, "Yhwh is my flag" (Exod 17:15); see Exod 20:24; Josh 22:26-27; Judg 6:24.

[91] The expression "Book of the Covenant" consists in an objective genitive: "the book bearing the words of the covenant." See WO § 9.5.2.

Then, in v. 7, the narrator introduces a scene in which, thanks to the repetition of דבר, "to speak,"[92] the perfect equivalence between what Moses *says* (v. 3), *writes* (v. 4) and *reads* (v. 7) is manifested.[93] This ritual has an extraordinary pragmatic effect: the reader and the people on stage could not be any closer,[94] because they share one and the same experience. In this passage, in fact, the narrator employs the technique of the *mise en abyme*,[95] creating a strict connection between the book that is read in the story and the text that the reader has in his hands.

The blood on the altar and the reading of the book ritually manifest the priority of the divine initiative that, by stipulating a covenant,[96] "gives His blood" and "calls" the people through His word. This element fits in very well with the first words of the Decalogue in which, before the list of "laws," Yhwh offers first and foremost His own person, "I am Yhwh your God who brought you out of the land of Egypt"[97] (20:2). The foundation of the relationship lies in the divine gift.

The purpose of the proclamation, however, is to reach "the ears" of the people (v. 7), in order to fulfil the act of communication. The interlocutor is not only informed, but above all he is called to consent to the word that is proclaimed.[98] The Israelites' reply is quick and is reported in v. 7. As in v. 3, the people react with an illocutionary "commissive" act of speech. They commit themselves to what they say and they again express a promise of loyalty.[99] This second reply strengthens the association between the

---

[92] To be more specific, in v. 3 and v. 4 the narrator employs the construction כל־דברי יהוה to indicate both Moses' discourse and written text; in v. 8, however, Moses refers to blood manipulation, saying that it took place "upon all these words" (על כל־הדברים האלה).

[93] J.-L. SKA, "From History Writing," 165-166.

[94] See J.-P. SONNET, "'Lorsque Moïse,'" 511, "Grâce à cet effet de mise en abyme [Exod 24:7] le destinataire intra-diégétique et le destinataire extra-diégétique ... sont pour ainsi dire contigus (sans se confondre), et l'engagement éthique de l'un ne peut manquer d'affecter l'éthique de la réception de l'autre."

[95] See J.-P. SONNET, "Le Sinaï dans l'événement de sa lecture," 323-344; J.-L. SKA, "*I nostri padri*", 82-83. The expression *mise en abyme* refers to the literary phenomenon of the narrative within a narrative. Regarding Exod 24:3-8, Ska says that it is as if the act of reading were directly addressed to the reader, inviting him to an ethical reaction.

[96] Regarding the theme of the covenant, see ch. II, notes 281, 283.

[97] R. MEYNET, "I due decaloghi," 663; F. MARKL, *Der Dekalog*, 98.

[98] Beauchamp notices the great leap that takes place in Gen 1:28: for the first time God speaks to someone, "He said *to them*." See P. BEAUCHAMP, *Testamento biblico*, 19, "The man who speaks is always already called by God.... Creation therefore lasts to the extent that man can listen to the divine injunction and respond to it." (My translation)

[99] See above, ch. IV, notes 57-58; see also J. SEARLE, *Speech Acts*, 56-57; M. KISSINE, *From Utterances to Speech Acts*, 148-165.

characters and the readers/listeners: while the reader no longer has the privilege of listening to Moses (v. 3), he, like the Israelites, can have access to the book (v. 7). The connection between v. 7 and v. 3 also helps to establish the identity between the words heard by Moses and their written form.[100] However, the difference between the two formulations is significant and our interpretation of the text must take it to close scrutiny.

Exod 24:3	Exod 24:7
All the words (כל־הדברים) that has uttered Yhwh (אשר־דבר יהוה), we will do (נעשה)	All (כל) that has uttered Yhwh (אשר־דבר יהוה), we will do and we will listen (נעשה ונשמע)

The Israelites' reply to God is even more complete and absolute in v. 7 than in v. 3, because, instead of the expression "all the words," the simple כל, "all," is used.[101] On the other hand, as had happened with the words of Moses (v. 3), the people recognise the "inspired" character of the written text; they do not say "all the things that have been proclaimed," but "all that Yhwh has uttered." The book is thus accepted as the word of God and not as a mere "letter" produced by men. Finally, the Israelites express their promise through an expression never used before[102]: נעשה ונשמע, "we will do and we will listen." This conclusion is very surprising, when compared to v. 3, because the elements' order seems to be reversed; normally a person "hears" first and then "does."[103] This reversal has been interpreted in many ways.

Primarily, this peculiar sequence could be a further accentuation of the book's value: while the people commit themselves to the fulfilment of what is stated in the written text, at the same time they promise that they will

---

[100] Priotto, 465.

[101] In both v. 3 and v. 7, the construction of the sentence especially emphasises the object of the main verbs; the normal order of the Hebrew sentence is verb – subject – object, whereas in these two verses there is an inverted sequence: object – verb, with the subject being implied. See JM § 155*k*.

[102] The two *yiqtol* (נעשה ונשמע) have a deontic modality and express a willingness to undertake a commitment. See A. GIANTO, "Mood and Modality," 183-198; A. NICCACCI, "Narrative Syntax od Exodus 19–24," 216.

[103] See above, note 84. Propp believes that, in v. 7, the narrator consciously employs the rhetorical figure of the *hysteron proteron*, i.e. the enunciation of a sequence in the opposite order in comparison with the one expected. See Propp, II, 295-296. See anche *MekhSh*, LXXXII:II, 1*d*, "Because they put doing first, Moses said to them, 'Is it possible to do something without hearing? Hearing allows one to do!' They repeated, saying, 'We will do and we will hear' (Exod 24:7), we will do what we will hear! This teaches that they accepted upon themselves both doing and hearing before the giving of the Torah."

continue to listen to its proclamation in the future.[104] Moreover, from an anthropological point of view, it should not be forgotten that the perfect acceptance of a word pronounced by others presupposes the assent of the will: to listen means to "do" something,[105] for this reason the verb עשה precedes שמע, "hearing a voice speaking to you is tantamount to accept, *ipso facto*, the obligation to the person who speaks."[106] Finally, if it is true that v. 7 could mean that listening and understanding grow with practice,[107] the sequence formed by the verbs "to do" and "to listen" outlines a progressive movement whose aim is the listening and not the practice: the consensus to and the acceptance of what Yhwh has said is not brought to completion in what is full understanding ("we will listen and *we will know*") or in full compliance ("we will listen and *we will do*"), but rather in a growing docility and in continuous openness to the word[108] ("we will do and *we will listen*").

2.4.3 The Sprinkling of Blood (v. 8)

The ritual ends with a second sprinkling of blood (v. 8), directed towards the people; the sacrificial pattern reappears and the "path" interrupted by v. 7 is concluded. Through this blood manipulation, Moses performs another rite

---

[104] Priotto, 466. The interpretation of שמע, "to listen," as a verb which designates "obedience" (Cassuto, 313; Propp, II, 295-296), does not seem convincing, because this docile attitude is already presupposed by the root עשה, "to do."

[105] P. BOVATI, "La dottrina," 17-64. The act of listening is a complex phenomenon that includes various aspects. To listen, the individual must "stretch out the ear," paying attention (Prov 2:2; 4:20); listening is fostered by silence (Job 4:16), consolidated by "custody" (שמר, Prov 2:1; 4:4), fulfilled through understanding (see Deut 28:49). The act of listening, therefore, presupposes a decision of the listener.

[106] E. LÉVINAS, *Quattro letture talmudiche*, 93. The author comments on *bShab* 88a, "r. Eleazar said, 'When the Jewish people accorded precedence to the declaration 'We will do' over 'We will listen,' a Divine Voice emerged and said to them: Who revealed to my children this secret that the ministering angels use? As it is written: *Bless the Lord, you angels of His, you mighty in strength, that fulfill His word, hearkening unto the voice of His word* (Ps 103:20). At first, the angels fulfill his Word, and then afterward they hearken.'" Regarding this passage Lévinas says the following, "Secret of angels and not child consciousness .... *We will do and we will listen*, what seemed contrary to the logical order is the order proper to angelic existence." (My translation; *ibid.*, 89) Afterwards, he adds, "knowledge without faith is logically tortuous; the examination that precedes adhesion – that excludes adhesion, that rejoices in temptation – is, first of all, a degenerative process of reason and, only as a consequence of this, is it a corruption of morality." (My translation, *ibid.*, 93)

[107] Fischer – Markl, 271. The speech act "impliziert sie eine umso größere Bereitschaft zur tatkräftigen Umsetzung der göttlichen Worte. Manchmal wachsen das Hören und Verstehen im Tun."

[108] The act of listening presupposes an inner disposition to welcome and trust the listener. See P. BOVATI, "La dottrina," 26-28.

that is unique in the Bible and that symbolically unites the altar and the Israelites' assembly, God and the people.[109] The continuity with the previous verses is achieved thanks to several repetitions: as Moses "took" the book, so he "took" the blood; moreover, the writing is called "book of the covenant," and, in the same way, the blood is called "blood of the covenant." Therefore, the book and the blood become symbols of the covenant between God and the people (even in v. 8 what we have is an objective genitive[110]). The connection between the two components of the rite could not be stated more clearly.

This ritual has been interpreted as a kind of "priestly consecration" of Israel. The arguments in favour of this explanation are various: there is a literary relationship between 24:6-8 and the consecration of the priests (29:19-21; Lev 8:14-15, 22-24, 30); moreover, 24:3-8 is closely related to 19:3-8 in which God promises to make of Israel a "priestly kingdom."[111] This explanation is certainly correct, and will be developed in the next chapter. However, this "priestly" interpretation is questionable, inasmuch as it does not sufficiently consider the order of Exodus narrative. While it is true that from a historical and redactional point of view, an association between Exod 24 and Exod 29 can be established, from a purely literary perspective it is rather Exod 29 that depends upon Exod 24 and not *vice versa*, since the consecration of the priests (Exod 29) is narrated successively.

The reader of Exodus, however, is not surprised by the fact that Moses decides to seal the covenant with blood, because he already read 4:18-31 and 12:7, 13, 22-23.[112] Moses had a "liminal" experience in 4:24-26: attacked by God, he was saved through the blood of circumcision, which bound him to Yhwh[113] and turned him into a "bridegroom of blood" for Zipporah. In Exod 12, Moses gave the people the instructions about the Passover as he had received them from God;[114] in the ritual, the animal

---

[109] See W.K. GILDERS, *Blood Ritual*, 41, "Blood from the animals is divided into two parts and applied in an identical manner to the altar and to the people. Clearly, an existential relationship is established in this way between the altar and the people. As I have noted, many interpreters understand the altar to represent Yahweh. Thus, we may speak of a relationship between Yahweh and the people being indexed by the blood manipulations."

[110] See above, note 91.

[111] J.-L. SKA, "From History Writing," 164, note 83. In the footnote he quotes, E. RUPRECHT, "Exodus 24,9-11," 167; E. BLUM, *Studien zur Komposition*, 50-52; A. SCHENKER, "Les sacrifices d'alliance," 481-494. See also T.B. DOZEMAN, *God and the Mountain*, 112-118; Nepi, II, 148.

[112] A.L.B. PEELER, "Desiring God," 187-205.

[113] See ch. II, § 4.

[114] See Propp, II, 309, "Exodus 24 may in fact be read as the mirror image of the *Pesaḥ*. The blood ritual in Exodus 12 initiates Israel's freedom; the blood ritual of Exodus 24

blood, smeared on the doorposts and on the lintel, becomes both a sign that the family gathered in the house "belongs" to Yhwh, and a reason for the salvation of the people.[115] Therefore, Moses knows from experience that the ritual blood signifies and realises a bond (a family relationship) through its manipulation and circulation,[116] expressing a mystery of life and death and causing what can be described as "rebirth."[117]

In all the Exodus passages that refer to ritual blood, the family scenario is discreetly evoked.[118] Through blood (4:24-26), Moses becomes part of God's "first-born" people (4:22-23); the Israelites, who are already "sons" are "reborn" on the night of the Passover[119] (12:1-43), after the killing of the first-born sons of the Egyptians (11:1-10). Therefore, even Exod 24:3-8 can be interpreted following the same trajectory: through the blood rite, the Israelites "reborn" are promoted to a more mature stage of their relationship with God (24:3-8). At this moment, the word becomes the privileged *medium* for communication, and the "son" is now able to listen and to interact with Yhwh. This progress is certainly no less true for Moses,[120] than it is for the people. It is in this ch. 24 that the Israelites are, for the first time, the object of an articulate revelation and they reply to

---

terminates it. Released from involuntary servitude to Pharaoh, Israel voluntarily enters Yahweh's servitude."

[115] See ch. III, § 2.2.2 e § 2.2.3.

[116] See T.J. LEWIS, "Covenant and Blood Rituals," 343. He refers to Robertson Smith studies – mostly *The Religion of the Semites* of the year 1889 – to explain Exod 24 and he quotes Abusch, who affirms that the sacrifice of blood "can be an artificial means of creating relationships," establishing a bond "that supercede the natural blood-ties produced through women's childbirth." See Z. ABUSCH, "Blood in Israel and Mesopotamia," 679; N. JAY, *Throughout Your Generations*.

[117] Propp recognises a link between the covenant ritual and the rite of passage in which the candidate symbolically dies and is reborn; see Propp, II, 308, "Exodus 24 describes a rite of passage whereby Israel enters into vassalage under Yahweh. Initiation rituals generally feature an inflicted trauma, real or symbolic, that symbolizes the candidate's death. Having survived his ordeal, he is symbolically reborn to a new status."

[118] See ch. I, § 2; c. II, § 4.4 and § 4.5.

[119] See ch. III, § 2.2.2.

[120] In *The life of Moses*, Gregory of Nyssa establishes Sinai as the place where Moses' faith reaches its most perfect peak. He describes the revelation on the mountain as an event that involves, in a singular way, the sense of sight; see GREGORIUS NYSSENUS, *De vita Moysis*. II, 163, "for in this is the true knowledge of what we seek, to see through (the experience of) not seeing" (ἐν τούτῳ γὰρ ἡ ἀληθής ἐστιν εἴδησις τοῦ ζητουμένου καὶ ἐν τούτῳ τὸ ἰδεῖν ἐν τῷ μὴ ἰδεῖν). On the other hand, Gregory recalls that, precisely because of the impossibility of perfectly grasping God by sight alone, Sinai is above all an interlocutory event; see *Ibid.*, II, 165, "Having arrived there (on the mountain), he (Moses) learns again from the divine word what he had previously learned through the darkness" (ὁ δὲ ἐκεῖ γεγονώς, ἃ προεπαιδεύθη διὰ τοῦ γνόφου πάλιν διὰ τοῦ λόγου διδάσκεται).

Yhwh's words with two statements reported by the narrator (24:3, 7), thus creating the first real exchange of words between Yhwh and the people.

After the sprinkling rite, Moses addresses the people and says, "Behold the blood of the covenant which Yhwh has stipulated with you, upon all these words." The Hebrew construction עַל כָּל־הַדְּבָרִים הָאֵלֶּה, is introduced by the preposition עַל, a particle that indicates the ground on which an act is based, or the reason for it.[121] Thereupon, Moses recognises the blood as a symbol that manifests and signifies the covenant between God and the people ("blood of the covenant"), but recalls that the bond between the two parties is founded especially on the act of speech ("upon all these words"). In this way, he sanctions the pact with a performative declaration similar to that pronounced by Zipporah on the night of the divine aggression.[122]

Exod 4:25	Exod 24:8
Then Zipporah **took** a flint and cut off (כרת) her son's foreskin and touched his feet with it and said, "For you are a bridegroom of blood to me!"	Moses **took** the blood and poured on the people and said: "Behold the blood of the covenant that Yhwh stipulated (כרת) with you upon all these words."

The blood manipulation is followed in both cases by two statements that produce an effective result. Thus, in v. 8 the blood circulates, pulsating from the altar to the people, thanks to the mediation of Moses. Yet, in the same verse it is also the words of the two parties that circulate, creating a close link between the two acts. Through the writing and reading of the "book of the covenant" (v. 7), Yhwh's word is definitively given to the people; with the sprinkling of the "blood of the covenant" (v. 8) the bond is definitively established. Therefore, even when he has to recognise the efficacy of the rite of blood ("behold the blood of the covenant," v. 8), Moses cannot but return to the act of speech ("which Yhwh has made with us on the basis of all these words," v. 8). The transformation of the Israelites is complete, their "flesh" is touched by the blood, their heart by the word.

2.4.4 Concluding Remarks: the Blood, the Word and the Oath

Moses' words pronounced in v. 8 conclude the sequence of interrupted "paths" which constitute the main pattern of vv. 3-8, creating in the reader the impression of a profound interconnection between linguistic act and

---

[121] See WO § 11.2.13e; Lev 5:22; Isa 60:7; Jer 6:14; Ps 31:24.

[122] In both cases, the sentences pronounced by Zippora and Moses belong to the group of the so-called "primary performative propositions," like "I promise," "I bet," "I declare." See J.L. AUSTIN, *How to Do Things with Words*, 69.

ritual. This articulation between word and gesture is pertinent to the context of the covenant, since the stipulation of a pact could be expressed through an oath (made up of gestures and words) in which the contracting parties committed themselves to mutual faithfulness.

An oath is an act of speech that reinforces what it refers to,[123] namely, a "linguistic act intended to confirm a meaningful proposition (a *dictum*), whose truth or effectiveness it guarantees"[124] and a statement that is fulfilled by facts.[125] In the ancient world, this solemn commitment was a necessary form of contract used in various fields; in Greece, for example, oaths were regarded as a valuable instrument of social and political cohesion.[126] In the Ancient Near East, on the other hand, treaties of vassalage consisted mainly of an oath of loyalty to the sovereign.[127] In Israel, the connection between oath and covenant is widely recognised by the exegetes, and several authors have studied the biblical attestations of this relationship.[128] For instance, Williamson states firmly, "an oath was indeed an indispensable aspect in the ratification of a covenant."[129]

The oath was usually followed by a gesture which symbolically expresses – "physically" – the commitment of the parties to the agreement.[130] Evidence from the Ancient Near East, dating back to the 2nd millennium BC, shows that the covenant between two kings, who concluded the agreement in person, could take the form of a solemn statement accompanied by the ritual

---

[123] See É. BENVENISTE, "L'expression du serment," 81-82. Regarding the praxis of oaths in Greece he says the following: "c'est une modalité particulière d'assertion, qui appuie, garantit, démontre, mais ne fonde rien.... le serment n'est que par ce qu'il renforce et sollennise: pacte, engagement, déclaration."

[124] G. AGAMBEN, *The Sacrament of Language*, 5.

[125] G. AGAMBEN, *The Sacrament of Language*, 21.

[126] See I. BERTI, "Now let Earth be my Witness," 181-209. Oaths were commitments made in trials or in treaties stipulated between cities (e.g. between Athens and Argo, see THUCYDIDES, *The Peloponnesian War*, V, 47, 8). They were used for the foundation of a new colony (*IGT*, 49) and for contracts (*IG* XII, 9, 191).

[127] See M. ZEHNDER, "Building on a Stone?" 511-535, who analyses the parallels between Hittite treaties, the treaty of vassalage of Esarhaddon (VTE) and the Deuteronomy.

[128] The root שׁבע, "to swear," is often correlated with ברית, "covenant." A covenant is an act which the person "swears" (ברית can be the object of שׁבע in Deut 4:31; 7:12; 8:18; also 31:20; Judg 2:1) or a contract sealed through an oath (Josh 9:15; 2 Kings 11:4; Ezek 16:8; Ps 89:4).

[129] P.R. WILLIAMSON, *Sealed with an Oath*, 39.

[130] G. AGAMBEN, *The Sacrament of Language*, 5-6, "on the essentially verbal nature of the oath (even if it can be accompanied by gestures, like raising one's right hand) there is agreement among the majority of scholars."

gesture of slaughtering a sacrificial victim.[131] Even when the covenant was concluded from a distance, external signs confirmed the commitment, such as washing one's hands or touching one's throat. Especially this second sign expressed the willingness to show that the partners' faithfulness reaches the point of putting their very lives on the line.[132]

Among the gestures that were performed in the oath, some included bloodshed. The connection between the act of speech and blood, for instance, is quite well attested to in ancient Greece: the oath ceremony could be sealed with the blood manipulation of a slaughtered animal[133] and a curse pronounced aloud, which foreshadowed the fate of perjury. These actions symbolically manifested the irreversibility of what was promised:[134] just as the sacrificial victim suffers a fate that cannot be changed in any way, so the pact cannot be revoked. In the Ancient Near East, on the other hand, such gestures are used on the occasion of treaties signed from a distance: through blood[135] the two parties communicated the strength of the new bond, which was similar to the relation that united blood relatives.[136]

---

[131] A letter written to Zimri-Lim (king of Mari from 1775 to 1761 a.C.) from Yasim-El, representative of the kingdom, describes a covenant sealed by the king of Andarig, Atamrum and the king of Karana, Asqur-Addu; see D. CHARPIN, *Archives Royales de Mari*. XXVI/2, 404, especially l. 62-64, "ils conclurent l'alliance et l'ânon fut immolé; ils se firent jurer l'un à l'autre un serment par le dieu."

[132] See D. CHARPIN, *"Tu es de mon sang,"* 70-72.

[133] AESCHYLUS, *Seven Against Thebes*, 43-53. Seven warriors swear to defeat Thebes or to die in the attempt thereof; while pronouncing the oath, they slaughter a bull on a shield and touch the blood, performing a sign that expresses their commitment to the cause, "they splashed their hands in bull blood, they swore by the trinity of battle, Ares, god of strife, Enyo, goddess of frenzy, and Phobos god of fear." See also XENOPHON, *Anabasis*, II, II, 8-9, "the two parties ... made oath that they would not betray each other and that they would be allies .... These oaths they sealed by sacrificing a bull, a boar and a ram over a shield, the Greeks dipping a sword in the blood and the barbarians a lance."

[134] I. BERTI, "Now let Earth be my Witness," 204, "The founding element of an oath was based on the irreversible nature of the act, which was made clear by the complete destruction of the sacrificial victim (either through combustion or dismemberment), of an object ... or the shedding of liquid on to the ground (wine or blood)."

[135] In Tell Leilan's archive (2nd millenium B.C.) a king refers to the covenant with Till-Abnû linking blood to the oath: "I undertook a journey and brought blood of Till-Abnû. Before we start on the campaign let us touch his blood, and let us swear an oath" (PIHANS 117, § 185; see J. EIDEM, ed., *The Royal Archives from Tell Leilan*, 257-258).

[136] A letter from among those of Tell Leilan exalts the role of blood in the covenant between the king of Eluhut and the ruler of Apum: "and you and I meet (and) swear an oath to each other and blood bonds are established between us" (PIHANS 117, § 89; see J. EIDEM, ed., *The Royal Archives from Tell Leilan*, 161).

Therefore, the passage analysed (Exod 24:3-8) can be illuminated more precisely by these considerations about the oath. Even though the root שבע, "to swear," is not explicitly employed in the text, the ritual is described essentially as a solemn exchange of words whose articulation suggests that this ritual is somehow connected to an oath.[137]

The text in fact insists on the speech acts performed by the two parties and on the reciprocity of the pledge. Moses reports the divine words and the narrator, through the repetition of the root דבר, manifests Yhwh's desire to create a relationship with the people.[138] Thereby, His word turns into a real action,[139] since Moses not merely transmits notions but the divine call to stipulate a covenant. This announcement achieves its effect when the people react with a "commissive" act of speech,[140] ("we will do!") which can be compared to a solemn oath (v. 3). Yhwh's word is then fixed in a written text that establishes the act of speech as permanent[141] (v. 4a). The proclamation (v. 7) completes the communication and is followed by an even stronger promise of fidelity uttered by the Israelites – "we will do and we will listen" – which precedes and favours the conclusion of the covenant rite.

---

[137] P.R. WILLIAMSON, *Sealed with an Oath*, 41, "Interestingly, there is no explicit mention of a divine oath in the context of the ratification of the covenant at Sinai, although a few texts appear to allude to such an oath (e.g. Num. 14:16; Deut. 28:9; 31:23; Ezek. 16:8, 59–60; 20:5–6; Mic. 7:20; see also Exod. 13:11), and such is certainly mentioned explicitly in relation to the renewal or remaking of this covenant on the plains of Moab (see Deut. 29:12, 14, 19)."

[138] Great reflection has been given by recent philosophy to the difference between the notion of "discourse," abstract and universal, and the concept of "word." See P. GILBERT, *Pazienza d'essere*, 131, "Considering the "word" as such, and not only as a term or a discourse, implies the consideration of the person who takes the floor and who addresses his interlocutor, establishing an interpersonal relationship with him." (My translation)

[139] P. GILBERT, *Le ragioni della sapienza*, 32, "Language is constituted by a communicative act, in which the subject who speaks addresses another subject recognised as such, in that the word is heard, accepted, shared in the social construction of a common sense." (My translation)

[140] See J.R. SEARLE, *Speech Acts*, 58, "'I promise' and 'I hereby promise' are among the strongest illocutionary force indicating devices for commitment provided by the English language. For that reason, we often use these expressions in the performance of speech acts which are not strictly speaking promises, but in which we wish to emphasize the degree of our commitment."

[141] See P. RICOEUR, *Du texte à l'action*, 154, "si on entend par parole … la réalisation de la langue dans un événement de discours … alors chaque texte est par rapport à la langue dans la même position d'effectuation que la parole. En outre, l'écriture est, en tant qu'institution, postérieure à la parole dont elle paraît destinée à fixer par un graphisme linéaire toutes les articulations qui ont déjà paru dans l'oralité."

The ritual, like an oath, is then composed by some gestures. Every human act is symbolic,[142] especially when it is performed during a ritual. Hence, the sacrificial actions, mentioned in vv. 5-6 and in v. 8, are extremely significant and must be deciphered to fully understand the meaning of the stipulation. The emphasis placed on peace offerings (v. 5), noted above, certainly demonstrates that the people confirm their assent (v. 3) with a sign that manifests the offering of themselves to God.[143] However, the use of blood, divided into two basins and poured first on the altar (v. 6) and then sprinkled on the people (v. 8), completes the meaning of the sacrifices with other aspects. First of all, it expresses the *irreversibility* of the bond sanctioned by the covenant: the blood, the "life" of the sacrificial victim, is completely poured out so that the commitment of the contracting parties may be full and firm. Furthermore, the reference to the "cutting" of the covenant (in v. 8 כרת literally means "to cut"), that alludes to the "cut" of Moses' foreskin (כרת, 4:25) and to the circumcision, suggests that the covenant is similar to a "cut" that is final and stable in nature.[144]

On the other hand, while the blood manifests that the oath is no longer revocable, its circulation between the altar and the people, obtained through the ritual, exhibits another aspect of the covenant: as in 4:24-26 and 12:7, 13, 22-23, the blood creates a sort of "kinship" between Yhwh, Moses and all the Israelites.[145] The use of blood in 24:3-8, therefore, generates an interaction between the impression of an irreversible pact, produced by the sign of a victim whose life is interrupted, and the idea of a bond between two partners that become similar to blood relatives after sealing the covenant: the circulation of blood has the purpose of generating a *"communication of life"*.

---

[142] See P. RICOEUR, *Du texte à l'action*, 214, "Une action offre la structure d'un acte locutionnaire. Elle a un contenu *propositionnel* susceptible d'être indentifié et réidentifié comme étant le même."

[143] See D. CHARPIN, *"Tu es de mon sang,"* 51-55. A sacrifice concluding the covenant has the symbolic purpose of showing the fate of those who violate the pacts. See e.g., the treatise between Aššur-nerari V (754-745 a.C.) and Mati'-ilu, king of Arpad; after the slaughtering of the lamb, he states, "this head is not the head of a spring lamb, it is the head of Mati'-ilu …. If Mati'-ilu [should sin] against this treaty, so may, just as the head of this spring lamb is c[ut] off …, the head of Mati'-ilu be cut off" (SAA 2, 2: I, 21-28).

[144] T.J. LEWIS, "Covenant and Blood Rituals," 344; in the convenant's ritual of the Ancient Near East "curses and associated blood rituals were used to express commitment and to instill fear among potential transgressors." He even states that "the biblical idiom 'to cut a covenant' (*krt bryt*) has traditionally been associated with the cutting of sacrificial animals [see Gen 15; …]. The cutting and bloodletting associated with covenant rites could also be understood as acts of (self-)imprecation." See Ger 34,18.

[145] T.J. LEWIS, "Covenant and Blood Rituals," 343.

The reflection on the connection between Exod 24 and the practice of oaths leads to a concluding consideration: the difference between the testimony of comparative literature and similar occurrences in the Scriptures is *theological*. In 24:3-8 a bilateral bond is indeed established, and the people are called to assume precise obligations. Nevertheless, the Sinai covenant, prior to being a simple pact between two contracting parties, is a revelation of divine loving-kindness,[146] namely, it shows that divine action precedes man's reply, and that the gift of God precedes the law. This aspect is explicitly revealed in 24:3-8 precisely because of the interconnection between the word and blood. Yhwh addresses the people first of all in order to communicate and to offer the possibility of a relationship with Him and only afterwards asks for the Israelites' reply. The blood, in turn, circulates first on the altar and only subsequently is poured out on the people, to indicate that the irreversible decision is first taken by God and then requested from the people. In both cases, the blood and the word attest that the foundation and origin of the new bond resides in Yhwh. Prior to being a human commitment, the covenant is a divine oath, a word that is fulfilled in the very utterance[147] and a "gift of life" that is obtained through the sprinkling of blood.

2.5 *The Vision of God (vv. 9-11)*

⁹ Then Moses, and Aaron, Nadab and Abihu, and seventy from the elders of Israel went up. ¹⁰ And they saw the God of Israel[148], under his feet there was[149] like a lapis lazuli floor, like the very heaven[150] in terms of purity[151].

---

[146] P.R. WILLIAMSON, *Sealed with an Oath*, 98, "It would seem, however, that these laws [Exod. 20–23] had a revelatory purpose; just as ancient law-codes generally made a statement about the king who had promulgated them, so the covenant obligations revealed at Sinai disclosed something of the nature and character of Yahweh."

[147] G. AGAMBEN, *The Sacrament of Language*, 21. Regarding a text of Philo (PHILO, *Legum allegoriae*, III, 204-208) the author presents some reflections on the connection between the concept of oath and the divine word, "1. The oath is defined by the verification of words in facts .... 2. The words of God are oaths. 3. The oath is the *logos* of God, and only God swears truly.... 5. Since we know nothing of God, the only certain definition that we can give of him is that he is the being whose *logoi* are *horkoi*."

[148] The Septuagint translates ויראו את אלהי ישראל with the Greek phrase, καὶ εἶδον τὸν τόπον, οὗ εἱστήκει ἐκεῖ ὁ θεὸς Ισραηλ, "and they saw the place where the God of Israel was." The translator has the theological intention to nuance the statement of v. 10, since the vision of God was considered a delicate matter; see E. TOV, "Textual Problems," 16-17. Even the targumim avoid saying directly that the group saw God, "they saw the glory of the God of Israel." (*TgPsJ*); "they saw the glory of Yhwh's Sheknah" (*TgO*; *TgN*). See also E. WYCKOFF, "When Does Translation Become Exegesis?" 675-693.

¹¹ And on the noble men¹⁵² of the Israelites He did not stretch His hand¹⁵³: they contemplated God¹⁵⁴, and they ate and drank.

The rite of vv. 3-8 takes place at the foot of Sinai. In this passage, however, by climbing the mountain with Aaron, his sons, and the seventy elders, Moses fulfills what Yhwh had said in vv. 1-2 marking the beginning of a new "narrative path" that will gradually lead him to be alone in the presence of Yhwh (vv. 12-18). Why then insert the interlude of vv. 3-8? The sequence of events shows that the vision (v. 10), the meal (v. 11), and the ascent (vv. 12-18) occur only when the gift of the covenant reaches the goal for which it was intended, its reception by the Israelites (v. 3 and v. 7); only when the bond with Yhwh is established can the representatives of the people ascend with Moses and come closer to God.

Furthermore, the ascent movement of this episode represents a reversal of the motif of the "return journey," identified above.¹⁵⁵ After the liberation (Exod 12–15), the first stages in the desert (15:22 – 18:27) and the stay at Mount Sinai (19–23), the reader would expect that the people are ready to leave. Surprisingly however, the journey from Egypt to the land is interrupted precisely in Exod 19:1, and the entry into this land is not narrated in the book of Exodus. The narrative continues until the end of the

---

¹⁴⁹ The *wayyiqtol* ("they saw") followed by a nominal clause ("under his feet ...") expresses a simultaneity between the reference to the general vision of God, and the description of this spectacle. A. NICCACCI, "Narrative Syntax of Exodus 19–24," 214.

¹⁵⁰ The Septuagint translates the expression כעצם השמים with a circumlocution to preserve divine transcendence: ὥσπερ εἶδος στερεώματος τοῦ οὐρανοῦ, "like the shape of the sky firmament." See Priotto, 460, note 8. Regarding the translation of עצם, see GK § 139g; JM § 147a.

¹⁵¹ The preposition ל introduces a specification, "as to, in terms of." See WO § 11.2.10d; Gen 17,19; 41,19. The term טהר is rare, is connected to the root טהר, "to be pure" (Lev 12:4), and could be translated even "splendour, clarity." See Childs, 498.

¹⁵² The noun אציל is a *hapax*. For the *HALOT* it comes from an Arabic word meaning "of noble descent." Clines believes that it derives from a root אצל, "to set aside" (see Num 11:17, 25). See Childs, 499; Houtman, III, 294-295.

¹⁵³ The first part of Exod 24:11 introduces a piece of information that lies in the background, since the sentence is constituted by a *waw-x-qatal*. See A. NICCACCI, "Narrative Syntax of Exodus 19–24," 209. The Septuagint paraphrases and interprets as follows, καὶ τῶν ἐπιλέκτων τοῦ Ισραηλ οὐ διεφώνησεν οὐδὲ εἷς, "and of the elect of Israel not a single one was lost." The aim of this translation, perhaps, is to blur the anthropomorphism (God "stretching his hand"). See Propp, II, 140.

¹⁵⁴ The version of the Septuagint, καὶ ὤφθησαν ἐν τῷ τόπῳ τοῦ θεοῦ ("and appeared in the place of God"), expresses the same concern of God's transcendence as in 24:10; the targumim give the same reading of 24:10 (see above, note 148). See E. TOV, "Textual Problems," 17.

¹⁵⁵ See ch. II, § 3.1.1.

book eventually building up a "vertical" axis, by reporting Yhwh's words to Moses on the mountain (25–31), the crisis of the golden calf in the valley, Moses' repeated descents and ascents (32–34), and the making of the Tabernacle (35–40). This structural consideration is very significant for the interpretation of the Book of Exodus as will be shown.

The passage begins with a daring expression (as shown by the Septuagint's translation[156]) – "and they saw the God of Israel." The impossibility of "seeing" God, in fact, is often mentioned in the Bible (see Exod 33:20). Despite the audacity of these words however, the text has various elements that mitigate its content. In reality, the group does not see "Yhwh," but "Elohim." The diverse naming of God safeguards divine transcendence, since "Elohim" is the appellation which refers to God in a generic way.[157] Furthermore, the content of the vision is not specified. Instead, the text simply states that they see what is beneath God's feet without providing specifications regarding any of His features. In this way the narrator draws on two similes:[158] Yhwh is visible yet His mystery remains unreachable.

The people who are with Moses on the mountain see "like a lapis lazuli floor"[159] (v. 10). In this verse, the narrator shares the point of view of the observers, and the preposition כ shows that the vision is described using a symbolic representation. The colour blue evokes the divine world,[160] and the bright and magnificent vision illustrates that the presence of God is seen in an

---

[156] See above, ch. IV, note 148.

[157] See ch. II, notes 234, 235; Cassuto, 314.

[158] See R. FORNARA, *La visione*, 386-387. Not of this opinion, Propp, II, 296, "Although the elders see Yahweh, the author conceals the divine appearance from the reader." Divine appearances are not simply hidden from the reader, but it is precisely those who see who cannot comprehend God with their eyes.

[159] The Hebrew construction כמעשה לבנת הספיר means, literally, "like a lapis lazuli brick work." For the Israelites, bricks are a memory from the past (see לבנה in 1:14; 5:7, 8, 16-19). For the *TgPsJ* this memory is still alive, "like a lapis lazuli stone, bringing to remembrance the slavery to which the Egyptians subjected the Israelites"; see Rashi Exod, 209. The ספיר corresponds to the "lapis lazuli"; see J.E. HARRELL – J.K. HOFFMEIER – K.F. WILLIAMS, "Hebrew Gemstones in the Old Testament," 19, "Lapis lazuli is a rock consisting of dark blue lazurite with minute golden specks of pyrite and white patches or veins of calcite. This understanding of lapis lazuli might stand behind the description of the barrier looking like the sky, the gold-like particles resembling stars in a dark-blue celestial canopy and the white patches looking like clouds."

[160] The colour blue is expressed through the Hebrew term תכלת; this pigment is used for different elements of Israel's cult: the Tabernacle (26:1, 4), the veil behind which the ark of the testimony is kept (26:31, 36), the curtain at the gate of the court (27:16), the priestly garments (28:5-6, 8, 15, 31, 33). See R. FORNARA, *La visione*, 389 and Propp, II, 296, which recalls that Ba'al's palace is made of gold, silver and lapis lazuli (*KTU* 1.4, V, 18-19, 33-35).

unconcealed way. The indication that there is a glimpse of the "feet" of God suggests that Moses and the people with him see a throne (Isa 6:1; Ezek 1:26), the seat of a sovereign, metaphor for God's royal power. This interpretation is consistent with the context because the idea of divine kingship recurs often in Exodus and even determines its plot on several occasions.[161]

The second metaphor used to describe the vision of God is that of the bright sky. The term טהר, "purity," comes from טהר, "to be pure," and in this context can be also translated as "splendour" (Job 37:21). Heaven is the region above which God dwells (Gen 28:17; 1 Kings 8:27; see 1 Sam 2:10; Ezra 1:2) and that is linked to the symbol of light which often appears in theophanic contexts.[162]

The "noble men" of Israel are granted to see God, who in turn does not "stretch out His hand" against the, to strike them[163] (v. 11). Therefore, their experience is extraordinary, because seeing God is normally forbidden and can cause imminent death[164] (Exod 19:21; Num 4:20). However, after the rite of the blood covenant, Israel's "noble men" can approach God, now that they have become closely related to Yhwh. Just like what happened to Moses (4:24-26), the circulation of blood prevents the notables from succumbing to the threatening divine presence. Furthermore, the reference to visual perception is enhanced thanks to the use, in v. 10, of a synonym of ראה, the verb חזה, and thanks to the repetition of the object seen in v. 11 ("and contemplated God," ויחזו את־האלהים). This verse, then, describes a gradual deepening of the representatives' perception. In fact, when the root חזה appears in a theophanic context, it can have an intensive nuance.[165] Moreover, as a technical verb for the prophetic vision,[166] חזה reveals the great intimacy into which the "noble men" are admitted.

---

[161] J.-L. SKA, *Le passage*, 58, "Dieu affirme son droit celui d'être reconnu comme supérieur à Pharaon en son propre pays. Le roi d'Egypte se trouve face à un pouvoir d'un autre ordre que le sien, pouvoir qu'on pressent universel." See Nepi, II, 150.

[162] The symbol of light is associated with God and alludes to transcendence. Especially in Ezek 1 and 10:4 the "light" is an essential component of the theophany (see Ezek 43:2; Hab 3:4, 11). See O. PETTIGIANI, *Dio verrà certamente*, 123-126.

[163] The expression "to stretch the hand" very often has a negative connotation both when the subject is human (Gen 22:12; 37:22; 1 Sam 24:7, 11; 26:9, 11, 23; Ps 55:21), and when it is divine (Exod 3:20; 9:15; Job 2:5).

[164] See R. FORNARA, *La visione*, 93-94.

[165] See R. FORNARA, *La visione*, 48-56.387. He quotes Exod 24:11; Isa 33:17; Pss 11:7; 17:15; 63:3 and Job 19:26-27. See also Childs, 507.

[166] See J.-L. SKA, "Le repas d'Ex 24,11," 312-314. The vision is a gift that God grants to prophets to legitimise their mission and to sustain their vocation (see Exod 3:1-6; Isa 6:1-11; 1 Kings 22:19-22). See Childs, 507; See R. FORNARA, *La visione*, 51; Priotto 468.

The vision ends with a disconcerting detail[167] – they "ate and drank." This experience of familiarity is unique in the Bible[168] and has been interpreted by some, because of its context, as a covenant meal.[169] Critics, however, have long specified that this interpretation seems unlikely because God does not participate in the banquet with the other contracting parties.[170] The text, then, may allude to a cultic meal, according to a ritual already mentioned in Deuteronomy (12:7, 18; 14:26; 27:7; see 1 Chron 29:22). Some scholars suggest that the meaning of the expression may find its true perspective if interpreted as a way of indicating a joyful reaction[171] (1 Kings 4:20; Jer 22:15; Eccles 5:17). All things considered, the most plausible interpretation of the meal is that it be considered as a royal banquet:[172] the divine king admits the representatives of a people newly "consecrated" by means of the covenant, in order to share the intimacy of the table,[173] and thereby manifest his full communion with them.

2.6 Moses climbs the Mountain with Joshua (vv. 12-14)

¹² And Yhwh said to Moses, "Come up to Me on the mountain, and stay there. And I will give you the tablets of stone, the Torah[174] and the commandment which I have written to instruct them." ¹³ And Moses arose with Joshua his

---

[167] See Ibn Ezra, 211-212, who gives two interpretations of Exod 24:11. He believes that the group of nobles comes down from the mountain and consumes the meat that had been slaughtered for the sacrifice by the young men (24:3-8). Another interpretation is that the text mentions the meal to distinguish the elders, Aaron and the sons from Moses: the former need to eat, while Moses can fast for forty days and forty nights (34:28). See Cassuto, 315.

[168] See B.P. ROBINSON, "The Theophany," 168.

[169] See Childs, 507. The v. 11 "places the whole account into the context of a covenant meal (Gen 31,46.54; Es 18,12)." See Noth, 159; D.J. McCARTHY, *Treaty and Covenant*, 162.265-166; Durham, 345; Propp, II, 297.

[170] See J.-L. SKA, "Le repas d'Ex 24,11," 306-307.

[171] See E.W. NICHOLSON, "The Origin," 149-150; ID., *God and His People*, 131-132.

[172] See J.-L. SKA, "Le repas d'Ex 24,11," 316-320. Regarding the subject of "eating at the king's table," he quotes, 1 Sam 20:29; 2 Sam 9:11; 1 Kings 2:7; Esther 1:3; see 2 Sam 19:34. See also Priotto, 469.

[173] See B.P. ROBINSON, "The Theophany," 155-173. See anche R. FORNARA, *La visione*, 391-392. Remembering the connection of Exod 24 with the consecration of the priests in Lev 8–9, the author notes that in both cases a meal is eaten before God (Exod 24:9-11; Lev 8:31-32; see Exod 29:32-34). He also shows a relationship between Exod 24:9-11 and Ezek 1:26, where, after the vision, God asks the prophet to "eat" the scroll (Exod 2:8; 3:1-2).

[174] See Childs, 499. The *waw* that introduces התורה can be a simple conjunction ("the stone tablets *and* the Torah") or an explanatory *waw* ("the stone tablets, i.e. the Torah," GK § 154*a*); several authors favour the second solution: Houtman, III, 300; Propp, II, 140; Priotto, 471. See Childs, 499 for a more detailed discussion.

attendant, and Moses went up[175] into the mountain of God. [14] Meanwhile, he said[176] to the elders, "Stay here for us[177] until we return to you; and behold, Aaron and Hur are with you, let whoever has legal matters,[178] approach to them."

This excerpt picks up again from what Yhwh had said in v. 2: after the ascent of the group, Yhwh asks Moses once again to ascend the mountain, but this time, he has to do it alone. The ellipsis in the narrative between v. 11 and v. 12 encourages the active participation of the reader, who must reconstruct what happened and if need be, like the people's representatives, "participate" in their contemplation.[179] In this new episode, however, the auditive dimension becomes dominant, as opposed to the visual dimension that was previously developed, and the special role of Moses is evident.

Exod 24:1	Exod 24:12
"Come up to Yhwh (עלה אל־יהוה)"	"Come up to me (עלה אלי)"

Whereas God had previously addressed Moses in the third person, in v. 12, with the expression "to me,"[180] Yhwh addresses His interlocutor directly and invites the latter to a personal dialogue. In addition, Moses' experience here, unlike that which was described in vv. 9-11, is not occasional, since it will last several days ("and stay there").

The direct speech of v. 12 is quite articulate and composed of many elements. First of all, Yhwh specifies the purpose of Moses' ascent which is to receive the written tablets directly from God. The term לוח denotes the

---

[175] The Septuagint translates the singular Hebrew verb into the plural ἀνέβησαν εἰς τὸ ὄρος, "they went up the mountain." This translation results in a smoother logical succession. The MT, in fact, mentions Moses and Joshua, but then it says that Moses goes up the mountain alone. See Propp, II, 140; E. Tov, "Textual Problems," 9-10.

[176] The Septuagint renders the singular Hebrew verb with the plural, "they said" (εἶπαν); the MT is the *lectio difficilior*. See Propp, II, 140; E. Tov, "Textual Problems," 10.

[177] The MT, שבו לנו, literally "stay *for us*," is translated differently by the ancient versions: *TgN* has, "stay *for me*"; the Septuagint has ἡσυχάζετε αὐτοῦ, "rest here," while the Vulgate has *expectate hic*, "wait here." The MT is the *lectio difficilior*.

[178] The MT has an idiomatic expression, מי־בעל דברים, "whoever masters/has words/legal matters" (see Gen 37:19; Isa 50:8). The particle מי is similar to an indefinite pronoun and can therefore be translated as "whoever," see WO § 18.2e.

[179] The narrator consciously omits some information between v. 11 and v. 12. This literary strategy is perhaps intended to encourage the reader's participation in the act of reading. J.-L. SKA, *"I nostri padri"*, 33, "the text does not mention what happened between the two verses. Only the reader can provide the missing information. The ellipsis allows the reader to participate actively." (My translation)

[180] See Priotto, 474; the author notices the relationship between 24:12 and 19:4.

scriptural tablet carved in stone.[181] The precise description of the material is intended to show the importance of the written text[182] and to express the desire for long term preservation (see Deut 27:8; Josh 8:32). The content of the tablets is designated through the hendiadys: והתורה והמצוה, "the Torah and the commandment." The first *waw* (והתורה) has explanatory value ("I will give you the tablets of stone, namely the Torah and the commandment"), otherwise the reader would have to assume that God had written several documents[183] ("I will give you the tablets of stone, and the Torah and the commandment"). In this sense, the expression "the Torah and the commandment" clarifies the meaning of what is written in the tablets,[184] revealing it to be at the same time both an instruction and a commandment, namely a collection of words that has an "educational" purpose while possessing the force of an imperative.

After a sequential reading of vv. 1-11 and vv. 12-18, the reader may notice some tension within the texts. On the one hand, Moses puts into writing the words received from Yhwh (v. 4), on the other, he learns that he will be given the tablets, as a gift (נתן), which will be written directly by Yhwh and which he will have to keep as an eternal testimony.[185] In this way, the written text is the result of both human (v. 4) and divine enterprise (v. 12); it is the Word of God in human words.

---

[181] See A. BAUMANN, "לוח," 496-499. The Decalogue is written on two stone tablets (Exod 31:18; 32:15; Deut 4:13; 5:22; 9:9-11, 15, 17; 1 Kings 8:9; 2 Chron 5:10). After the crisis of the golden calf, the tablets of the Torah are shattered to be replaced by two others, identical to the former (Exod 34:1, 4, 19, 28-29; Deut 10:1-5).

[182] In some Assyrian and Babylonian documents, reference is made to legal acts of particular value written on two tablets and reproduced in duplicate. In Hebrew as in Akkadian, then, the mention of "stone" indicates the stable character of the document. See Dohmen, II, 213-214.

[183] See Childs, 499; Houtman, II, 300; Priotto, 474-475.

[184] See Houtman, II, 300. Deuteronomy states several times that the stone tablets contain the Decalogue (4:13; 5:22; 10:4); however, rabbinic tradition has suggested that the content of the tablets was a combination of the Decalogue and the precepts. See *TgPsJ* Exod 24:12, "I will give you the tablets of stone in which the rest of the words of the Torah are intimated, as well as the six hundred thirteen commandments." See also Rashi Exod, 211, "All 613 commandments are included in the Ten Commandments."

[185] Rabbinic tradition assumes that there is a very close connection between v. 12 and vv. 9-11: the stone of which the tablets are made is lapis lazuli, just like the floor under the divine throne. See *Midrash, Tanchuma*, Eikev 9, "'From where did he carve them?' One answered, 'From under [God's] throne of glory.' And the other said, 'God created a quarry in the midst of his tent, and he quarried two tablets of stone from there. And he took the remnants from there and he became wealthy from there, since they were from lapis lazuli (ספירים).'" See R. HINCKLEY, "Sapphire *Lukhoth*," 28.

Lastly, God's word to Moses in v. 12 concludes with an important repetition. Immediately after the content of the tablets is defined as "the Torah and the commandment (והתורה והמצוה)," Yhwh specifies, through the repetition of the root ירה, that the purpose of the written tablets is the education of the people: "Which I have written to instruct them (להורתם)." This conclusion thereby shows that the didactic character of Yhwh's words prevails[186] over the imperative one.

In the next verse, the first ascent up the mountain is staged by Moses (v. 13) who, on the basis of a new impulse, "rises"[187] and starts to climb in the company of Joshua, his assistant[188] (Exod 33:11; Num 11:28; Josh 1:1). During the time of the ascent, Moses asks the elders to wait for them until they return ("stay *for us*," v. 14) and to govern in their (Moses' and Joshua's) place until then (the expression can be even translated, "stay *in our place*").[189] Thus, Moses establishes both Aaron and Hur, who have been with him since the confrontation between the Israelites and the Amalekites[190] (Exod 17:10, 12), as leaders and as judges declaring, "whoever has legal matters, get close to them"[191] (v. 14). This detail is intended to prepare the reader for what is about to happen in Exod 32, in particular, the sin of idolatry with the golden calf.[192] Moses trusts in Aaron and so he transfers his judicial authority to him. And not only that, rather than leave him alone, he gives him Hur as an assistant. Despite of all this however, in the end Aaron will betray the trust he had been given and will lead the people into idolatry (32:1-2).

---

[186] In the Book of Exodus, the verb ירה appears in strategic passages: in 4:12, 15 Yhwh reassures Moses by saying that He will instruct him and his brother, teaching them what they shall speak; in 15:25, however, Yhwh responds to the murmurings of the people by "teaching" (ויורהו) a wood capable of making the waters sweet. The noun תורה, on the other hand, denotes both the instructions concerning the Passover (12:49; 13:9), and the directives concerning the Manna (16:4, 28). Eventually, Moses himself is described as an intermediary between God and the people who will make them know the Torah (18:16, 20).

[187] Earlier, in the book of Exodus, קום manifests a broader meaning than simply "to stand up." In 2:17, for instance, the verb designates the impetus that drives Moses to defend the daughters of Reuel from the shepherds' bullying at the well in Midian.

[188] See Rashi Exod, 212; when Moses comes down from the mountain, the Scriptures says, "Joshua heard the voice of the people as they cried out." (Exod 32:17) This excerpt suggests that Joshua is not with the Israelites when they cast and shape the golden calf.

[189] Propp, II, 299.

[190] Propp, I, 617-618. The name Hur is of Egyptian origin, although the character is identified with a Jewish notable (31:2).

[191] The verb נגש has a judicial nuance and can denote the movement towards a court of law: Deut 25:1; Isa 41:1; see Gen 18:23. P. BOVATI, *Re-establishing Justice*, 218-221.

[192] Propp, II, 299; Priotto, 475.

2.7 *Moses enters the Cloud (vv. 15-18)*

¹⁵ Then[193] Moses[194] went up on *the mountain*, and the **cloud** covered the *mountain*. ¹⁶ And YHWH's glory dwelt on *mount* Sinai, and the **cloud** covered it six days. He called to Moses from the midst of the **cloud**. ¹⁷ Now the appearance of YHWH's glory was like a devouring fire on the top of the *mountain* in the sight of the Israelites[195]. ¹⁸ And Moses entered inside the **cloud** and went up on the *mountain*. And Moses remained on the *mountain* forty days and forty nights.

This passage presents the last stage of Moses' ascent and concludes with his entry into the cloud (v. 18). The text is structured on the basis of some significant repetitions (highlighted in the translation). The various members create a drawn-out series of actions: the movement of Moses (15a, "he went up") corresponds to that of the cloud and of Yhwh's glory (15b-16b); Yhwh calls Moses (16c) and he enters the cloud and remains on the mountain for forty days and forty nights (v. 18); then, in v. 17 the series of actions is interrupted by a symbolic description of Yhwh's glory, which generates some suspense before the plot's resolution is described in v. 18.[196] The "intrigue" of this small scene, therefore, is centered around the movements of two key characters — Moses and Yhwh — towards one another which culminates with Moses' entry into the cloud after being invited and admitted to an encounter with God[197] (v. 18).

The glory of God is described through a typical "form" of divine revelation, the cloud, a natural element that is extremely common in Yhwh's theophanies[198] (see 16,10). Here, it is presented as covering the mountain

---

[193] In order to understand the plot of 24:12-18, it is important to note the change of time in vv. 13-15: Moses goes up the mountain (עלה, *wayyiqtol*, v. 13), and this action is in the foreground; at the same time, he tells the elders to wait for him (ועל־הזקנים אמר, *waw-x-qatal*, v. 14). The repetition of the verb עלה *wayyiqtol* in v. 15, therefore, is intended to resume the thread of the discourse after the pause in v. 14, and can be translated as "*then,* Moses went up." See A. NICCACCI, "Narrative Syntax od Exodus 19–24," 213.

[194] The Septuagint adds the name of Joshua in v. 15, specifying that he remains with Moses during the ascent. Here again, the variant attempts to clarify a literary problem of the passage. See Propp, II, 141; E. TOV, "Textual Problems," 10.

[195] The nominal clause of v. 17 introduces information that is in the background, namely a description of the divine glory as perceived from the Israelites' point of view. See A. NICCACCI, "Narrative Syntax od Exodus 19–24," 214; Houtman, III, 304.

[196] See P. ROCCA, *Gesù, messaggero*, 118.

[197] See P. ROCCA, *Gesù, messaggero*, 116-121.

[198] See R. FORNARA, *La visione*, 183-184. There are several texts in which a cloud is a sign of the divine presence: Exod 13:21-22; 14:19-20; 19:9, 16; 33:10; 40:38; Num 9:15-16, 20-21; 14:14; Ezek 1:4.

where the glory of Yhwh dwells (v. 16a) for six days (v. 16b). However, although the cloud veils and hides what is inside, it visibly reveals a presence, so much so that even the people from afar will be able to catch a glimpse of the divine glory (v. 17). On the other hand, the term כבוד literally designates the "weight" (derived from כבד, "to burden," see Exod 5:9) and figuratively the "prestige" that is recognisable through appearances.[199] When כבוד is used for God[200], it refers to His visible "manifestation" which both inspires terror (Deut 5:24) and shines with light.[201] The combination of luminous and obscure symbols is common in Exodus; the glory of Yhwh is often manifested in the cloud (Exod 16:10; 40:34-35; see Num 17:7), thus creating a dialectic between visibility and non-visibility of the divine presence.[202]

Then the text says that Yhwh "dwells" in the cloud. The use of שכן establishes an extremely close link between Mount Sinai and the Tabernacle (משכן). In fact, the Sanctuary, which is built after receiving instructions from God (Exod 25–27), will later be considered to be like "Sinai in motion."[203] However, the new reference to the cloud in v. 16, "and the cloud covered *him* for six days," gives rise to several interpretations. The pronominal suffix, in fact, can refer to both the mountain and the glory of Yhwh.[204] This ambiguity is perhaps intentional and its purpose may be that of preserving the distance between Moses and Yhwh. Even though the mountain is "covered" by the cloud, it is not the only inaccessible thing, because the same is also true for the glory of Yhwh. The verse focuses on the separation between the two subjects and on the six days that Moses has to wait. This duration has been explained in various ways. For instance, Yhwh took six days to create the world[205] and

---

[199] The "glory" is connected with brightness (זרח, see Isa 60:1-2). The term כבוד can also be translated as "wealth" (Gen 31:1; see 1 Kings 3:13; Prov 3:16).

[200] Redactional critics have commonly associated the term כבוד with the terminology belonging to priestly literature (P); see M. WEINFELD, "כבוד," 32-35.

[201] In a theophany, "Yhwh's glory" is very often correlated to the verb ראה, "to see" (Exod 16:7, 10; 33:18; Lev 9:6, 23; 14:10, 22; 16:19; 17:7; 20:6; see Isa 35:2; Ezek 1:28; 3:23). See R. FORNARA, *La visione*, 186-187.

[202] See R. FORNARA, *La visione*, 395-396.

[203] See Cassuto, 316; Propp, II, 299; Priotto, 476; Fischer – Markl, 276, "Gottes Vergegenwärtigung hier am Sinai ist Vorausbild dafür, wie er ab der feierlichen Erfüllung des Heiligtums mit seiner Herrlichkeit in 40,34f das Volk weiter mit seiner Gegenwart begleiten wird." See *TgN* e *TgPsJ*, "and the glory of Yhwh's Shekinah remained on Mount Sinai."

[204] See R. FORNARA, *La visione contraddetta*, 393. The Hebrew noun כבוד, "glory," is masculine and therefore the pronominal suffix of ויכסהו can refer also to "Yhwh's glory."

[205] Fischer – Markl, 276. See anche Propp, II, 675-676; J.L. SKA, "Il libro dell'Esodo," 132-135. In the creation narratives of the Ancient Near East, the myth normally ends with the construction of the temple dedicated to the creator god (as Marduk in the *Enuma Eliš*

also takes six days to prepare the model of the Tabernacle[206] (Exod 25:9, 40). The length of time could also refer to the period that Moses needed for his purification[207] (see 29:30, 35, 37). A subtle allusion to the "Sabbath" cannot be excluded; in fact, Moses had already been instructed by God on the difference between the six ordinary working days and the "Sabbath," the latter being holy because it is "consecrated" and dedicated to the celebration of the relationship with Yhwh[208] (16:23, שבת־קדש ליהוה). In fact, Moses himself will encounter the divine glory on the seventh day. This expectation shows that Moses is unable to contemplate Yhwh's glory immediately: entry into the cloud is a concession and a gift, not a right.

This consideration is also confirmed in the last segment of the verse: Moses does not enter the cloud simply because he decides to do so, but because he is called (קרא) by God. The use of the verb קרא highlights the difference between the experience described in vv. 9-11 and Moses' encounter with Yhwh. Contrary to the former case, in which there is no divine discourse, in v. 16 Moses is directly invited by the voice of God[209] to a meeting in which there will be an exchange of words.[210] Moreover, this excerpt (v. 16) is an echo of what Moses has already experienced in his first meeting with God who visibly manifested Himself in the burning bush. Even on that occasion, in fact, Moses heard a voice speaking his name (3:4) and had a visual experience which led to a dialogue with Yhwh[211] (3,6 – 4,18).

In v. 17 there is an interlude which slows the narrative down and allows the narrator to share with the reader the Israelites point of view ("in the sight"). While still safeguarding the special privilege of Moses, v. 17 informs the reader that even now the Israelites continue to take part in the same experience

---

and Ba'al in the texts found in Ugarit). Therefore, Ska believes that the references to Gen 1:1 – 2:3 found in the last chapters of Exodus (including Exod 24:16; see nota 56), indicate that the construction of the Tabernacle in Exod 40 is the fulfillment of a process that was started in Gen 1:1 – 2:3.

[206] See Nepi, II, 163; Sarna, 154.
[207] Cassuto, 316.
[208] Priotto, 476.
[209] See Houtman, I, 45. He notices that קרא designates "the seeking of contact with someone by the use of the voice," because it is often followed by *verba dicendi* (אמר, Exod 1:18; 8:4, 21; דבר, 34:31; דברים, 19:17).
[210] P. ROCCA, *Gesù, messaggero*, 125, "*Words* are the overwhelming novelty of Sinai."
[211] See R. FORNARA, *La visione contraddetta*, 363-364; from Exod 3:4 onwards, the narrative of Moses' vocation accords to the sense of hearing a much more significant role, "putting aside the visual phenomenon, the account concentrates on the long dialogue between God and man." (My translation) See also P. ROCCA, *Gesù, messaggero*, 136. There is an important lexical contact between ch. 3 and ch. 24: God calls Moses "from the midst of the cloud" (מתוך הענן, 24:16), as he did "from the midst of the bush" (מתוך הסנה, 3:4).

of their representatives, previously mentioned (24:9-11). Despite the limitations, given the distance of the Israelites, the vision they have of God is still real. As a matter of fact, the narrator highlights that they grasp the "appearance" (מראה) of God's glory, even if they cannot see it directly.[212] Yhwh's glory is described through the visible metaphor of "a devouring fire." The symbol of fire is very common in theophanies,[213] because it is an element in which both destructive and vital forces are contemporarily present.[214] The description of the glory calls to mind the scene of the burning bush (3:2) even though the two episodes are evidently different. In the latter, Moses sees a fire that does not "devour" the bush, whereas in the former the people see a "devouring fire." Furthermore, contrary to the bush theophany which attracts Moses and brings him closer, the theophany witnessed by the Israelites emphasises the distance between them and God.[215]

The narrative concludes with Moses entering the cloud (v. 18). The use of עלה ("and he went up"), seems unusual because it was repeatedly mentioned that Moses "went up" on the mountain (vv. 12, 13, 15). Indeed, the repetition unifies ch. 24 by showing that the divine order given in v. 2, is finally fulfilled, even though Moses' ascent is carried out in distinct stages.[216] And so, this chapter, characterised by a plot full of intricacies, comes to its resolution: Moses has a special encounter with God, consisting in a protracted exchange of words.[217] The remarkable quality of the circumstance is confirmed by its duration: Moses stays (היה, v. 18 and v. 2) in the cloud forty days and forty nights. Moses had waited six days before he was summoned by Yhwh, but his stay on the mountain is far longer, to the point of exceeding all expectations. Obviously, the number "forty" has a symbolic meaning; however, in this case, it not only suggests that a great amount of time is required for Moses' formation,[218]

---

[212] Propp, II, 299; Sarna, 154.

[213] There are several expressions belonging to the semantic field of "fire" which appear in theophanic contexts: אש (Exod 19:18; 40:38; Lev 9:24; Num 9:15-16; Deut 1:33; 4:36; 5:5; 18:16; 1 Kings 18:38; Ezek 8:2); עמוד אש (Exod 13:21-22; Num 14:14); לפיד אש (Gen 15:17; see Exod 20:18).

[214] See M. GIRARD, *Symboles bibliques*. I, 144-181.200-207. See also T.J. LEWIS, "Divine Fire," 796-798; D.E. GRANT, "Fire and the Body of Yahweh," 143-146.

[215] See Propp, II, 299-300; R. FORNARA, *La visione contraddetta*, 397.

[216] See P. ROCCA, *Gesù, messaggero*, 118.

[217] See P. ROCCA, *Gesù, messaggero*, 119-120. Aaron, Nadab and Abiu, togheter with the seventy elders, see God, Moses "enters" the cloud; moreover, while the vision of the group does not evolve in a dialogue, the encounter of Moses is a long experience of listening.

[218] See Sarna, 155. Forty days defines the duration of the flood (Gen 7:4, 12, 17; 8:6); it represents the period of Elijah's rest on Mount Horeb (1 Kings 19:8); Israel wandered in the

but also that his training through divine instructions will be perfect and complete.[219]

Conclusion

Recent criticism tends to interpret Exod 24 as the fruit of a rather late editorial work and this hypothesis seemed reasonable, given the peculiarity and uniqueness of the rite described in the chapter. Nevertheless, the tensions of the text do not prevent the interpreter from proposing a close reading of the chapter. Exod 24 consists of two distinct but deeply connected parts, vv. 1-11 and vv. 12-18. The motif of blood appears in the first part of the chapter, in v. 6 and v. 8, within an episodic plot consisting of a remarkable series of courses of action that begin and then are left in suspense, only to be resumed again and linked together later.

Yhwh addresses Moses directly (Exod 24:1-2), laying out His "program" and foretelling what will happen next, distinguishing Moses from the others for his special condition and, at the same time, legitimising the ancestors of the two most significant post-exilic institutions, the priesthood (Aaron) and the elders. The divine instructions are not carried out immediately, because the plot is interrupted by an interlude (vv. 3-8). Moses remains at the foot of Sinai and performs a series of actions on behalf of the people. Why does he stay? The repetition of the term עם in vv. 2 and v. 3 allows the reader to give an answer to this question. Moses understands that he cannot proceed without the people's involvement in the process.

Acting as a model for the scribes, Moses recounts (ספר, *piel*) "all the words of Yhwh" (v. 3). The frequent repetition of the root דבר confirms the idea that Yhwh's covenant is constituted above all on the basis of words aimed at establishing a relationship with the people. In fact, the communication of Yhwh's words fulfils its purpose when the people answer and, as such, perform a "commissive" linguistic act: "all the words that Yhwh has uttered, we will do!" (v. 3). The episode could well have ended with the people's reaction; instead, it continues with Moses who decides to write down what Yhwh has said (v. 4a).

The next day, however, this initiative to write is interrupted and focus is now given to the "sacrificial plot." Moses delimits the sacred space,

---

desert for forty years (Exod 16:35; see Amos 5:25; Ps 95:10). The number indicates a long and complete period of time (2 Sam 5:4; 1 Kings 2:11; 11:42; 2 Chron 9:30).

[219] H. MAHFOUZ, "Appearing to them for Forty Days (Ac 1,3)," 363-384. When Luke says in Acts 1:3 that Jesus appears to the apostles for forty days, he is alluding to Exod 24:18.

positioning the symbolic elements which represent God and the Israelites, namely the altar and the twelve pillars.

The blood of the slaughtered animals is collected into two basins and sprinkled on the altar (v. 6). The rite is interrupted and the focus shifts to Moses who proclaims all the words of Yhwh written in the book (v. 7). The proclamation of Yhwh's words achieves its purpose once again when the Israelites answer saying, "all that Yhwh has uttered, we will do and we will listen." At the end, Moses then sprinkles the people with the remaining blood (v. 8). This interweaving of interrupted and recontinued actions shows the profound interconnection between ritual gestures and spoken words, especially when considering the repetition of the term ברית, in v. 7 and v. 8. This consideration allows the reader to notice a certain similarity between the covenant in 24:3-8 and the practice of the oath-taking through blood rituals. The blood manipulation expresses the irreversible character of the covenant and represents the bond which the blood "circulation" between the altar and the assembly establishes between the two parties.

After these events, Moses "comes up" (v. 9) on the mountain together with Aaron, his two sons and the seventy elders, the representatives of the people who can contemplate God (vv. 10-11), and eat and drink before Him without being harmed by His threatening presence (v. 11). The encounter between Moses and God, on the other hand, is described much more gradually. The cloud (v. 15) prevents Moses from having immediate access to the divine glory and compels him to wait for six days (v. 16). And while the people contemplate from afar the divine glory (v. 17), Moses experiences something similar to what he had felt the first time he was called by God (3:1-6): he approaches a flame, but is not "devoured" by it (v. 18). In the next chapter, special attention will be given to the instructions surrounding the consecration of Aaron and his sons (29:1-35).

Thus, the appearance of blood is crucial for the plot of Exod 24 and its function is of great importance. The representatives of the people can go up the mountain, approach God, and live an experience of extraordinary communion with Him precisely because they, together with all the Israelites, have been transformed by the oath taken in vv. 3-8.

Chapter V

The Consecration of Aaron and his Sons (Exod 29:1–30:10)

Moses' exclusive encounter with Yhwh (25,1–31,18) constitutes a very significant part of the book of Exodus; the fulfilment of God's instructions together with the construction of the Tabernacle, marks its conclusion and point of arrival (35,1–40,38). The plot formed by the structure "instruction-realisation" is interrupted by chs. 32–34 in which the episode of the golden calf is narrated, a breach of the covenant that happens while Moses is still on mount Sinai (see 32:19). The sequence of events thus indicates that the people were able to build the Tabernacle only after the unexpected gift of divine mercy (ch. 34). The received forgiveness, in fact, makes the Israelites capable of carrying out the required work in full accordance with the word of Yhwh, "according to all that Yhwh had commanded Moses, so the Israelites did all the work. Moses saw all the work, and behold, they had done it as Yhwh had commanded, so they did it. And Moses blessed them" (39:42-43). The lexical relationship between this section and Gen 1:1–2:3[1] demonstrate that, after sinning, the Israelites obey Moses' and God's instructions in a perfect way.

The instructions addressed to Moses thus constitute a continuation of the revelation of the Decalogue (Exod 20) and of the "Covenant Code" (Exod 21–23), and mark a crucial moment in the book of Exodus. First, Yhwh orders Moses to collect a contribution for the construction of the Tabernacle (25:2-7) that is going to be shown, in the form of a model[2] to Moses (25:8-9). This initial indication is followed by the description of the essential

---

[1] See E. BLUM, *Studien zur Komposition*, 306-307. The connections between Exod 39–40 and Gen 1:1 – 2:3 are unequivocal: (i) "God saw everything (וירא אלהים את־כל) that he had made and behold (והנה) it was very good" (Gen 1:31a) – "Moses saw all the work (וירא משה את־כל־המלאכה), and behold (והנה) they had made it as Yhwh had commanded, so they made it" (Exod 39:43a); (ii) "Thus were finished (ויכלו) the heaven and the earth and all their host" (Gen 2:1) – "And it was finished (ותכל) all the work of the Tabernacle of the Tent of Meeting" (Exod 39:32); (iii) "God finished (ויכל אלהים) ... the work (מלאכתו) that He had been doing" (Gen 2:2a) – "And Moses finished (ויכל משה) the work (את־המלאכה)" (Exod 40:33b); (iv) "And God blessed (ויברך אלהים) the seventh day" (Gen 2:3) – "Moses blessed them (ויברך אתם משה)" (Exod 39:43b).

[2] See J. ROUX, "Modèle en haut ou modèle en bas?" 37-56.

components of the Tabernacle (chs. 26–27), the priestly garments of Aaron (28:1-43) and the instructions for the priestly consecration (29:1-35).

The main subject of this chapter is represented by the sequence of Exod 29:1 – 30:10. In chapter 29, the appearance of blood (vv. 12, 16, 20-21) calls to mind what has previously happened in the Exodus[3] (above all in Exod 24). The purpose of this part is to show that the ritual signifies and expresses the new bond between God and the priests, associating the act of consecration with the stipulation of the covenant.[4] The complex ritual is made up of a three-stage movement: the sin offering, the burnt offering,[5] and the last sacrifice, an offering of well-being.

The chapter concludes with the reading of Yhwh's indications for the consecration of the altar of burnt offerings (29:36-37), the institution of the daily sacrifice (29:38-42), and the construction and use of the altar of incense (30:1-10). The articulation of the subsequent paragraphs follows the form of the previous sections. Firstly, the difficulties and tensions of the text will be highlighted, entering into dialogue with the studies dedicated to redactional analysis. Then the second unit will be dedicated to the *lectio cursiva* of the text, paying particular attention to the literary coherence of the final whole. The specific interest of this work, the motif of the shed blood, will lead the close reading as a guiding concept. What is the function of blood in the complex ritual of consecration? How does the blood used in Exod 29 relate to what has been seen in the other chapters?

1. Genetic-historical Approach

In general, even if Exod 25–31 are attributed to the document P,[6] the scholars tend to recognise the traces of different editorial layers in them. The subject has been addressed several times, yet the agreement among exegetes is not unanimous, especially with regard to ch. 26.[7] Nevertheless,

---

[3] See ch. I, § 2.
[4] See F. SERAFINI, *L'alleanza levitica*.
[5] Concerning the sacrificial terminology, see ch. I, § 1.3.
[6] See Propp, II, 365-366, for a detailed analysis of the vocabulary, style and theology of Exod 25–31. According to the author, in this sequence a significant accumulation of terms and stylistic features characteristic of P literature are found.
[7] Traditionally, Noth already considered ch. 26 to be made up of two distinct traditions concerning the sanctuary, understood as a mobile tent (26:7-14) or as a building (26:1-6, 15-29); see Noth, 170-174; see also T. POLA, *Die ursprüngliche Priesterschrift*, 237-240. Some, however, continue to think that ch. 26 is homogeneous, especially on the basis of the references to the materials used for construction, gold (v. 6) and bronze (v. 11); see: P.P. JENSON, *Graded Holiness*, 101-106. In fact, gold is reserved for the objects inside the

the articulation of chs. 26–27 can be considered as coherent and well organized.[8] These two chapters, together with ch. 25, are part of a unique and ancient *stratum* of P,[9] while some authors believe that chs. 28-29 may be the result of a later addition.[10] This position has recently been revised and clarified by Christoph Nihan in a monograph devoted to the priestly document where he provides solid evidence to support the thesis that Exod 28–29 belong to the older P narrative.

The chapters cannot be considered separately, because it would not make sense to introduce instructions for the making of priestly garments (Exod 28), without then referring to the consecration of the priests (Exod 29).[11] Moreover, these chapters are essentially consistent with the P literature, especially considering their terminology and theology.[12] Nihan, then, presents a close comparison between Exod 29 and Lev 8–9, in which the ritual of the consecration of the priests is celebrated according to Yhwh's instructions received from Moses.[13] In Exod 29, as well as in the previous chapter,[14] there are perhaps only a few interpolations:

Tabernacle, while bronze is relegated to the constructions that are placed outside. Therefore, since bronze is mentioned in v. 11 as the material for construction, it may be assumed that in vv. 7-14 the divine instructions refer to the outer covering of the Tabernacle. CH. NIHAN, *From Priestly Torah*, 40-41.

[8] The hypothesis that ch. 27 was added later can also be refuted; see P.P. JENSON, *Graded Holiness*, 89-114. The specification that bronze is the material used for what is placed outside the Tabernacle (27:2-3, 10, 19) only makes sense in comparison with the golden interior (26:6). CH. NIHAN, *From Priestly Torah*, 41-42.

[9] Some authors tend to consider 25:10-40 to be a secondary and more recent passage because of the similarity between this part and 1 Kings 6–8. See discussion in S. OWCZAREK, *Die Vorstellung*, 59-63.

[10] See Noth, 186-191. The scholar considered that Exod 29 was later than chs. 25–28. On the basis of this position Exod 28–29 was subsequently identified as an editorial layer more recent than Exod 25–27. An even more "minimalist" hypothesis is defended by Pola, who believes that the P^g layer is simply formed by 25:1, 8-9 together with 29:45-46 and 40:16, 17, 33); see T. POLA, *Die ursprüngliche Priesterschrift*, 292-298.

[11] *Pace* Noth, 188. He suggests that Exod 28–29 are part of two different layers. This separation does not explain the double reference to the consecration of the priests at the beginning (v. 1) and at the end (v. 41) of chapter 28. CH. NIHAN, *From Priestly Torah*, 52.

[12] See above the footnote 6.

[13] See CH. NIHAN, *From Priestly Torah*, 124-147. The differences between Lev 8 and Exod 29 are not sufficient to consider these chapters as the result of two distinct editorial layers. The repetition of the formula כאשר צוה יהוה, "as Yhwh has commanded" (Lev 8:4, 9, 13, 17, 21, 29, 36) clearly shows that Lev 8 is considered to be the faithful execution of what is said in Exod 29. See also G. DEIANA, *Levitico*, 105.

[14] Regarding Exod 28, vv. 3-5 are normally interpreted as an addition, because whereas in v. 2 God directly asks Moses to prepare the priests' garments, in vv. 3-5 the task is

+ Exod 29:21. The instruction to take the blood from the altar and sprinkle Aaron's robes has created several problems for scholars. In P we find no other reference to a similar use of blood. Furthermore, the sprinkling of blood around the altar (29:20) usually marks the conclusion of the rite.[15] The interpolation connects the consecration of Aaron with that of his sons and of their garments.[16]

+ Exod 29:27-28. These verses have no counterpoint in Lev 8 and seem to be an addition. The consecration of the breast and leg of the ram as a portion for Aaron and his sons (vv. 27-28) is a doublet of what is already said immediately before and contradicts it: while the leg is burnt with the rest of the animal (vv. 22-25), only the breast is given to the priests (v. 26). A similar specification is found in very late texts[17] (Lev 7:28-36).

+ Exod 29:29-30. The transmission of the priestly garments from Aaron to his sons interrupts the thread of the discourse; the continuation of v. 26 may be found in v. 31,[18] because the logic order formed by the succession of v. 26 and v. 31 corresponds to the sequence found in Lev 8:29, 31.

+ Exod 29:36-37. The instruction concerning the seven days in which the atoning rite must be performed in order to consecrate the altar is not found in Lev 8 and could therefore be an addition.[19] In fact, the motif of the "anointing of the Altar" is attested in more recent passages[20] (see 30:28-29).

The passage formed by vv. 38-42 is normally regarded as a recent supplement,[21] because it seems out of place in comparison with the conclusion of the instructions reserved for Aaron (vv. 36-37). Moreover, the prescription of a double sacrifice, in the morning and in the evening, could belong to a later period.[22] The last verse (v. 42b) is closely related to vv. 43-46, because of a certain homogeneity of the themes evoked (especially the motif of the "meeting, encounter," יעד) and because of the repetition of the term שמה, "there." However, a slight fracture between vv. 38-42 and what follows can be recognized in the text, for in v. 42 it is said that God will

---

entrusted to "wise" craftsmen. This indication in fact corresponds to the instructions given in 31:1-11, which are commonly interpreted as a late addition; see Noth, 179.

[15] See Lev 1:5, 11; 3:2, 8, 13; 7:2; 8:19; 9:12, 18. The Septuagint takes the Hebrew phrase וזרקת את־הדם על־המזבח סביב, "and sprinkled the blood on the altar, round about" (Exod 29:20), and puts it at the end of v. 21. This change suggests that the translator interpreted the transition between v. 20 and v. 21 as a non-linear progression.

[16] CH. NIHAN, *From Priestly Torah*, 129.

[17] CH. NIHAN, *From Priestly Torah*, 130-131.

[18] See Noth, 190-191; CH. NIHAN, *From Priestly Torah*, 131-132.

[19] See Noth, 191.

[20] CH. NIHAN, *From Priestly Torah*, 132-133.

[21] See Noth, 191.

[22] See CH. NIHAN, *From Priestly Torah*, 36, note 83.

gather the people to speak to Moses (לדבר אליך), in contrast to the statement in v. 43, "I will meet with the Israelites." Perhaps, v. 42 intends to harmonize vv. 43-46 with what is said in 33:7-11 with respect to Moses' privilege, namely his exclusive access to the tent.[23]

The passage of 29:43-46 is instead considered as part of the older priestly layer (P^g).[24] Even those who explain Exod 29 as a late composition believe that vv. 43-46 are an expression of P literature in its early stages,[25] and therefore this hypothesis does not create any problems.

With regard to chapters 30–31, on the other hand, the exegetes generally agree that they are late additions.[26] The description of the altar of incense (30:1-10), in particular, seems out of place, for one would assume that it should be described together with the other elements used for the service in the tent (25:10-40; 26:31-37); moreover, this altar appears after a divine discourse that has the characteristics of a conclusion[27] (29:43-46). This impression is confirmed by the textual tradition of Exod 30:1-10, which tends to place the passage in a different place.[28] The altar of incense is also unknown to the author of Lev 16, since it is not mentioned during the *Yom Kippur* ritual[29] (in contrast to 30:10). Thereby, 30:1-10 may be the expression of a preoccupation connected with the situation of the authors of Pentateuch's last redaction.[30]

Thus, chapter 29 is part of the oldest layer of the P-document, in which Yhwh gives Moses precise instructions for the consecration of Aaron and

---

[23] See CH. NIHAN, *From Priestly Torah*, 37.

[24] A significant exception to this consensus is Knohl, who attributes 29:43-46 to H, especially for Yhwh's use of the first person and for the formula אני יהוה אלהיהם. See I. KNOHL, *The Sanctuary of Silence*, 18.81.102. This hypothesis is criticised with various arguments by CH. NIHAN, *From Priestly Torah*, 34, note 72. In Knohl's hypothesis, it is particularly difficult to consider that Exod 6:2-8 is part of H (יהוה אני appears in 6:2, 6-8). Without this excerpt, the P-narrative would be deprived of an essential junction.

[25] See e.g. CH. FREVEL, *Mit Blick auf das Land*, 98-103, who presents a critical reading of T. POLA, *Die ursprüngliche Priesterschrift*, 235. According to this author, whereas v. 42b seems to be united with vv. 43-44 by the triple repetition of שמה, "there," v. 43 is rather part of a distinct unit; the section of vv. 43-46 is in fact syntactically unified by the use of the *we-qatal*. V. 42, on the other hand, is a typical transitional verse which completes the discourse of the preceding verses and, at the same time, creates a transition to the following ones; therefore, it is reasonable to think that it has been composed later.

[26] G. VON RAD, *Die Priesterschrift*, 61; Childs, 529-530; E. BLUM, *Studien zur Komposition*, 308-309; H. SHAPIRA, «Making Sense», 23-42.

[27] For discussion of Exod 30:1-10, see CH. NIHAN, *From Priestly Torah*, 31-33.

[28] For more details regarding the textual criticism of Exod 30:1-10, see, ch. V, note 195.

[29] J. WELLHAUSEN, *Composition*, 139-140.

[30] Propp, II, 369.

his sons. The final result creates an organic composition in which the blood appears during the consecration in three different moments (in 29:12, 16, 20-21) and, after the institution of priesthood, becomes an instrument of expiation and reconciliation (30:10).

2. A Close Reading of Exod 29

2.1 *The Context (Exod 25–28)*

The complex section of Exod 25–29 is primarily organized and structured on the basis of the names used for the Sanctuary. In the first part, the attention is focused around the "Tabernacle" (משכן, 25,1 – 27,19) while, in the second, the expression "Tent of Meeting" is employed[31] (אהל מועד, 27:20–29:46). The two terms refer to the same reality but present it from different perspectives. The noun משכן, a derivative of שכן, "to dwell," does not surprise the reader, for he recognises in it an extension of what Moses witnessed in 24:16, "and dwelt Yhwh's glory on Mount Sinai." Thus, if משכן insists on Yhwh's residence among the people,[32] אהל מועד, on the contrary, emphasises that the Tabernacle[33] will be a place of "meeting." In fact, it is through the worship in the sanctuary, that a true encounter with God is possibile.[34] The first words addressed to Moses (25:1-9) introduce and prepare the instructions that will be announced later, insisting especially on the divine

---

[31] The chapters 25–27 are characterized by a great number of repetitions of the term משכן, "Tabernacle" (25:9; and 26:1, 6-7, 12-13, 15, 17-18, 20, 22-23, 26-27, 30, 35; 27:9, 19). The expression אהל מועד, "Tent of Meeting," on the contrary, appears several times in subsequent chapters (27:21; 28:43; 29:4, 10-11, 30, 32, 42, 44; 30:16, 18, 20, 26, 36; 31:7).

[32] See ch. IV, § 2.5. See also G. PAXIMADI, *E io dimorerò*, 19.

[33] The scholars have noticed that Exod 26 combines two different conceptions of the sanctuary, the Tabernacle (26:1-6) and the Tent (26:7-14, although it is never referred to as the "Tent of Meeting"); see G. PAXIMADI, *E io dimorerò*, 104-107, in which the rhetorical art with which the final editors combine the two passages into a unified whole is studied. See also the *excursus* of CH. NIHAN, *From Priestly Torah*, 39-41, in which a history of the research regarding Exod 26:1-14 is presented. The idea of a combination of two layers is discarded by the author, because 26:1-6 and 26:7-14 are not true "doublets" and chapter 26 is coherent and well-organised.

[34] See R.E. HENDRIX, "*Miskan* and *'ohel mo'ed*," 213-223; ID., "The Use of *Miskan*," 13, "Thus, in all contexts within Exod 25-40 the biblical writer has masterfully controlled the use of *miskan* and *'ohel mo'ed* in order to clarify the dual nature of YHWH's habitation. That habitation was to be understood as a transient dwelling place, such as was consistent with the dwelling places of nomadic peoples; hence the choice of *miskan*. However, that habitation also had the continuing function of fostering the cultic relationship, and this aspect was best expressed by the choice of *'ohel mo'ed*."

foundation of worship and on the theological orientation of what God asks of the Israelites.[35]

Exod 25:2	Exod 25:8
Speak to the people of Israel, and they gather (ויקחו) **for me** (לי) a contribution (תרומה). From every man whose heart offers it, you shall take **my** contribution (תרומתי).	And let them make (ועשו) **for me** (לי) a sanctuary so that I may dwell in their midst.

The covenant newly sealed (Exod 19–24) influences these first words in an indirect yet effective way. Worship, in fact, will be an expression of the relationship with God ("they gather *for me*," v. 2; "let them make *for me*," v. 8) and, on the other hand, it will question the freedom of the Israelites, prompting them to a generous and deliberate participation[36] (v. 2). After the list of valuable and refined[37] materials needed for worship,[38] the passage concludes by insisting that the purpose of the sanctuary will be the great proximity of God: "I will dwell in their midst" (שכנתי בתוכם, v. 8). Yhwh achieves unprece-dented closeness through a form of presence and "residence" among the people.[39]

The chapter continues with the description of the Tabernacle, starting from the central core and culminating in the outer parts. The Ark of the Covenant (25:10-16), made of acacia wood and covered with gold, contains the Testimony, the stone tablets of the Law[40] (24:12). The seat cover (propitiatory, 25:17-21, כפרת) is a golden plaque adorned with two cherubim on both ends which, together with the Ark, manifests the presence of God.[41] In the central verse of the chapter[42] (v. 22) it is stated

---

[35] Priotto, 503.
[36] The Hebrew expression מאת כל־איש אשר ידבנו לבו, literally means "from every man whose heart prompts it" (25:2); the verb נדב, in fact, means "to move, to incite" and designates the act of volunteering (Judg 5:2.9) or the spontaneous offering (Ezra 1:6; 2:68; 3:5). See Rashi Exod, 215; Propp, II, 372.
[37] See Houtman, III, 337-338. The materials used for the construction of the inner parts are much more valuable than those employed for the outer parts, according to a gradation very common in the P literature. See P.P. JENSON, *Graded Holiness*.
[38] See G. PAXIMADI, *E io dimorerò*, 52-53.
[39] The subject of שכן is often divine. See Num 5:3; 35:34; 1 Kings 6:13; 8:12; Isa 8:18; 33:5; 57:15; Ezek 43:7.9. Regarding שכן, see also ch. V, note 181.
[40] The connection between the "testimony" of Exod 25:16 and the stone tablets written by the hand of God is guaranteed by the use of verb נתן, "to give," both in 25:16 ("the ark of the testimony which I will give you") and in 24:12 ("I will give you the stone tablets"). See Propp, II, 383; G. PAXIMADI, *E io dimorerò*, 82; Priotto 509.
[41] See Propp, II, 466-468; G. PAXIMADI, *E io dimorerò*, 82-86. The propitiatory and the two cherubim represent God's empty throne (R. DE VAUX, *Ancient Israel*, 299-300). He continues to maintain that the Ark and the propitiatory have a "royal function" (*Ibid.*, 84),

that Yhwh will speak to Moses on behalf of the Israelites from the space left empty between the two cherubim: at the heart of the Tabernacle there is nothing but the Word.[43]

The description of the Tabernacle then continues with the mentioning of the table where the loaves of the presence are placed (25:23-30), a sign that Yhwh wants to offer the Israelites the possibility of sitting at his royal table,[44] thus prolonging what happened in 24:11.[45] The chapter culminates with the description of the golden lampstand (25:31-40), the *m^{e}nôrāh*; this object, which is placed inside the Holy (25:35, 37), is commonly associated with cosmic symbolism[46] and with the image of the tree.[47] Light sources were quite common in local shrines (1 Sam 3:3) because shining light is a powerful symbol of God[48] (2 Sam 22:29).

The instructions concerning the objects kept in the Holy and in the Holy of Holics are complemented by the orders which describe the construction of the Tabernacle itself[49] (Exod 26). The sanctuary is covered with ten cloths woven out of precious materials (26:1); the whole building is roofed with an outer curtain of goat's hair (26:7-11) and two further covers made

---

also because representations of royal thrones in the Ancient Near East were often supported by images of winged sphinxes (see O. KEEL, *The Symbolism*, 169-171). Regarding cherubs, see A. WOOD, *Of Wings and Wheels*, 23-34.

[42] See G. PAXIMADI, *E io dimorerò*, 77.

[43] See ch. IV, §§ 2.2 – 2.4. Priotto, 511. Referring to God's presence, Priotto says that "it will not be a material presence, but a dialogical presence." (My translation)

[44] The "breads of the presence" are mentioned in 25:30 and in Lev 25:7-9, where they are granted a detailed analysis. In the Ancient Near East, the practice of preparing food for the deity is well attested (W.G. LAMBERT, "Donations of Food," 191-201), but various authors believe that the acacia table and the breads placed in the Holy have rather a symbolic meaning: see Cassuto, 337; Durham, 362; Houtman, III, 393-394. The presence of the loaves reveals that the people's encounter with Yhwh will reproduce the same communion that is created between those who share of the one table. See Priotto, 513.

[45] See ch. IV, § 2.5.

[46] Propp, II, 510. The seven-armed lampstand could evoke the seven major celestial bodies recognised by ancient astronomy (see Zech 4:10).

[47] The contact between the lampstand and the symbol of the tree is attested to in Mesopotamian iconography. L. YARDEN, *The Tree of Light*, figg. 6.213.214; C. MEYERS, *The Tabernacle Menorah*, 98-105; G. PAXIMADI, *E io dimorerò*, 86-92.

[48] Propp, II, 512, "Light betokens Yahweh's presence and his favor (e.g., Num 6:25; Isa 10:17; 60:1; Micah 7:8; Ps 4:7; 27:1 …)." See also Priotto, 515; Nepi, II, 169.

[49] See G. PAXIMADI, *E io dimorerò*, 123-127. It should be noted that the description of the Tabernacle agrees with pavilions made of a wooden structure and covered with curtains, dating from the Bronze Age. See J.K. HOFFMEIER, "The Possible Origin," 167-177. In addition, connections with Ugaritic literature are mentioned (R.J. CLIFFORD, "The Tent of El," 221-227) together with the Tabernacle's similarity to a sanctuary found in Timna (N. AMZALLAG, "Moses' Tent of Meeting," 298-317).

of ram skins and goatskins (26:12-14). The internal structure is formed by an acacia wood frame (26:15-30) and the spaces are delimited by the veil that separates the "Holy of Holies" from the "Holy" (26:31-35) and the curtain which signals the entrance of the Tent (26:36-37).

The external elements of the Tabernacle, the altar (27:1-8) and the enclosure (27:9-19), are described in chapter 27. The altar, made of acacia wood with a bronze covering,[50] is the most widely used instrument for sacrificial worship. In it, the Israelites express their relationship with God through the sacrificial offering.[51]

The lamp lighting service[52] is mentioned even before the priests are introduced (27:20-21); the order of the elements – the lamp (27:20-21), the priestly garments (ch. 28) and the priestly consecration (ch. 29) – shows that the presence of God in the Tabernacle, with its light,[53] precedes and supports the worship carried out by the priests. The oil used is made of the highest quality,[54] the rite of lighting the lamp is part of the set of ritual gestures that must be performed in the sanctuary on a daily basis, a practice that will be called *tāmîd*[55] (27:20).

Finally, the description of the priestly garments (Exod 28) offers the opportunity to analyse the meaning of priesthood, starting from the "symbols" that they will wear.[56] They are "clothes of holiness" (בגדי קדש, 28:2), which therefore suggest a special union with God.[57] The names of the Israelites are inscribed on the garments, so that the people can access the presence of Yhwh, through the mediation of the High Priest.

---

[50] Concerning recent archaeological discoveries regarding the altar with four horns at the four corners (27:2), see S. GITIN, "The Four-Horned Altar," 95-123.

[51] See Priotto, 524-528; G. PAXIMADI, *E io dimorerò*, 130-134. The sacrificial worship, far from being an expression of the desire to "feed" the divinity, is a symbolic sign for the act of entrusting to God. See G.J. WENHAM, "The Theology of Old Testament Sacrifice," 75-87; M.J. SELMAN, "Sacrifice in the Ancient Near East," 88-104.

[52] Verse 20 is distinct from what precedes it because of a ואתה (see also 28:1, 3) with which Yhwh addresses Moses directly and urges him with a greater energy; Propp, II, 427.

[53] Priotto, 530.

[54] The oil obtained directly from crushing the olives by hand was considered to be more refined than the oil obtained from the press. See Propp, II, 428, who refers to *mMen* 8,4-5; see also Cassuto, 369-370; Durham, 379.

[55] The noun תמיד, "regularly, continually," will be used as a technical term for the daily liturgy in the Temple. See J. MILGROM, *Leviticus 1–16*, 456-457. See below, ch. V, § 2.7.

[56] Priotto, 530-531. Clothing, in general, has not only an external value, but can be a symbol which expresses interior qualities (2 Sam 13:18; Dan 5:16; Nepi, II, 171).

[57] Regarding the root קדש, see ch. I, § 1.3.

The *ephod* is woven of the same material as the veil (28:5) and is a mysterious object decorated in gold[58] made up of shoulder-pieces decorated with two lazuli stones with the names of the twelve tribes of Israel engraved in them (28:9-14); these gems "make present"[59] (זכרון) the people when the High Priest enters the Tabernacle. The *breastpiece*[60] (28:15-21) is related to the *ephod* (28:22-28) and adorned with three rows of four precious stones representing the people, as a "remembrance" (זכרון, v. 29). The breastplate contains the *'ûrîm* and the *tummîm* (v. 30), objects of obscure origin that had an oracular function,[61] and are placed on the breast of the High Priest so that he may carry "the right of the Israelites on his heart" (v. 30); that is, he may embody some sort of "judicial" function.[62]

The priestly garments are then completed by the robe (מעיל, Exod 28:31), worn under the *ephod* (29:5; Lev 8:6-7), whose hems are adorned with pomegranates[63] made of blue, purple and scarlet yarns, and by golden bells

---

[58] See G. PAXIMADI, *E io dimorerò*, 163-166. For some scholars the *ephod* is nothing more than a reference to priestly garments in general; for others it is a garment that was meant to cover the High Priest's genitals and was reminiscent of the ancient loincloth mentioned in 2 Sam 6:20 (see also 1 Sam 2:18; 22:18; 2 Sam 6:14).

[59] The term זכרון, "remembrance, memorial," is used twice in 28:12, "and you shall put the two stones to the shoulder-pieces of the *ephod*, as stones of *remembrance* (אבני זכרון) for the Israelites, and Aaron shall carry their names before Yhwh upon his two shoulder-pieces *for remembrance* (לזכרון)." The High Priest "carries" the Israelites with him, so that Yhwh may remember them. See Rashi Exod, 245-246; Priotto 534; Propp, II, 438, who says that in P literature "the root *zkr* 'remember' is thematic: God 'remembers' his Covenant with Israel (Gen 9:15–16; Exod 2:24; 6:5; Lev 26:42, 45; also Jer 14:21; Ezek 16:60; Amos 1:9; Ps 106:45). The very names of Israel's sons thus may be a kind of covenant witness—legal signatures, so to speak—reminding Yahweh and Israel of their mutual obligations."

[60] See G. PAXIMADI, *E io dimorerò*, 166-173.

[61] Several topics are discussed in relation to the *'ûrîm* and *tummîm* (see Lev 8:8; Num 27:21; Deut 33:8; 1 Sam 28:6; Ezra 2:63; Neh 7:65). Concerning the etymology, the terms are interpreted as derivatives of the roots אור, "to illuminate," or ארר, "to curse," and of the root תום, "to be perfect." A parallel with an Assyrian text presenting a divination ritual is also a matter of discussion (A.-M. KITZ, "The Plural Form of *'ûrîm* e *tummîm*," 401-410).

[62] P. BOVATI, *Ristabilire la giustizia*, 158. See Num 27:21; Deut 17:9.12; 19:17; Hos 5:1.

[63] The pomegranate tree is a symbol of fertility and life, because of its ability to produce flowers and fruit all year round. An Egyptian Love Song represents a pomegranate which states, "[I am the most beautiful tree] in the garden, because I remain in every season" (E. BRESCIANI, *Letteratura e poesia*, 443). Pomegranates are among the fruits found by the explorers in the promised land (Num 13:23); the bronze capitals that were placed in front of the entrance to Solomon's Temple were decorated with two pomegranates (1 Kings 7:18); they are mentioned in the Song of Songs to describe the beloved man (4:13; 6:11) or the beloved woman (7:13; 8:2). See also N. AVIGAD, "The Inscribed Pomegranate," 157-166. The use of the pomegranate as a symbol of the divine word does not seem to be excluded (Prov 25:11); see Houtman, III, 511-512.

(28:33). The sound of these chimes prevents Aaron from dying when he enters the Holy (28:35) and demonstrates that the admission into the sphere of holiness is a concession and not a right.[64] On the other hand, the noise of these ornaments involves the people in the High Priest's rituals, because they are alerted to his movements precisely because of these little bells.[65] The diadem is embellished with a gold frontlet (28:36) which is literally called "flower"[66] (ציץ; see Lev 8:9, ציץ הזהב נזר הקדש); the most remarkable aspect of this lamina is the inscription "Holy for Yhwh,"[67] a sign of Aaron's exclusive relationship with Yhwh (see Wisd 18:24[68]).

2.2 *The Garments and the Anointing Ritual (vv. 1-9)*

The consecration of Aaron and his sons is relevant to this study precisely because the ritual has blood as an essential component. Chapter 29 is divided into three parts: the clothing of the future priests and the sacrifice of the bull (vv. 1-14), the sacrifice of the two rams (vv. 15-25) and the distribution of parts of the sacrificed animals to the priests (vv. 26-35). The vv. 36-37 are the concluding elements and the sacrifice of the two rams is the central part of the chapter.[69]

> ¹ This is what you **will do** for them, to consecrate them so that they may serve as priests to me. Take one bull of the herd, and two rams without blemish, ² and unleavened bread and unleavened cakes,[70] with oil mixed in and unleavened

---

[64] Critics have suggested that the bells had an apotropaic value, see J. MILGROM, *Leviticus 1–16*, 508; Noth, 183; Nepi, II, 173. The explanation is problematic because the bells are only part of the High Priest's robes, yet the other priests too can enter the Holy. Maybe they had the function of announcing the entry of the High Priest in the Tabernacle (Cassuto, 383). Their sound could have the purpose of prompting the High Priest to absolute obedience with respect to the instructions (Priotto, 538). Alternatively, the chimes could be the signal for the beginning of a "rite of passage" between the sacred and profane spheres (G. PAXIMADI, *E io dimorerò*, 177-178; Propp, II, 446).

[65] See Propp, II, 446.

[66] Various archaeological discoveries show that the royal diadem could have been a floral crown or could be decorated with flowers (G. STEINS, "ציץ," 1031). Flower decorations are not uncommon in cultic environments, as in Solomon's Temple (1 Kings 6:18, 29, 32, 35).

[67] See ch. I, § 1.3.

[68] See L. MAZZINGHI, *Sapienza*, 705-706.

[69] See the structure proposed by Paximadi in the book dedicated to the rhetorical analysis of Exod 25–31 and in particular the chapter consecrated to Exod 29; G. PAXIMADI, *E io dimorerò*, 184-208.

[70] The Septuagint reports only the expression ἄρτους ἀζύμους, "unleavened bread," translating in a simpler way the apparently redundant Hebrew, "unleavened breads and unleavened cakes." The TM's choice seems deliberate; see Propp, II, 349.

wafers smeared with oil.[71] Of choice wheat flour **you shall make them**. 3 And you shall put them in one basket and present them in the basket, along with the bull and with the two rams.

The consecration ritual has a coherent character, so much so that Yhwh refers to it with a singular, "this is the thing"[72] (וזה הדבר). The repetition of the verb עשה, "to do," referred to Moses, helps to establish the unique position of God's messenger.[73] The emphasis in v. 1 is placed on the destination and purpose of the ritual action, "you will do *for them* (להם), *to* consecrate them (לקדש אתם), *so that* they may serve as priests (לכהן) *to* me (לי)." Moses will act with Aaron and his sons and for their benefit,[74] in order to enable them to access the sphere of divine transcendence, its holiness.[75] The significant aspect of this verse (see 28:1, 3-4, 41) is exposed in the last expression, "to serve as priests *for me*." Whereas the priesthood could seem to be simply a cultic function for the people's advantage, it expresses in fact a special form of relationship with God.[76] This detail prepares in advance the consideration that priesthood establishes a deep bond with God, precisely thanks to a rite consisting of three different blood manipulations.

The ritual act will include the sacrifice of a bull, the sacrifice of two rams, and the offering of loaves and cakes. The animals must be without blemish, intact (v. 1, תמימים) and gifts must be offered[77] (v. 3, והקרבת) to God together in one basket.

---

[71] The Hebrew construction משחים בשמן, "anointed with oil," is not reported by PSam, 4QPaleoExod^m. The testimony of the other ancient translations makes this shorter text less likely; see Propp, II, 349.

[72] The Septuagint translates the first words of v. 1 with an unexpected plural: καὶ ταῦτά ἐστιν ἃ ποιήσεις, "and *these* is what you will do." See Propp, II, 349.

[73] See Propp, II, 454. The verb עשה indicates always an action performed by Moses (29:2, 35-36, 38-39) in connection with the consecration of the priests. On the contrary, even though in most cases the verbal subject of the root קדש is Moses (29:1, 27, 36), the end of the chapter specifies that Yhwh is the author of the consecration (29:44).

[74] The expression להם of v. 1 introduces the indirect object, "to them," but can also be used for the *dativus commodi* (JM § 133*d*) which indicates to whose advantage the action is performed ("for them"). See Houtman, III, 529.

[75] See ch. I, § 1.3.

[76] The proposition ל cannot be translated here as a *dativus commodi*, because the priesthood is certainly not a function that obtains an advantage to Yhwh. With this particle, then, a relation is expressed; see WO § 11.2.10*d*. The construction כהן + ל + Yhwh is frequent in the Hebrew Bible: Exod 29:44; 40:13; Lev 7:35; Ezek 44:13; 2 Chron 11:14.

[77] The verb קרב *hifil* holds a cultic nuance, both when it is referred to the animal sacrifice (Lev 4:3.14; 8:18; 16:6; Num 6:14) and when used for cakes (Lev 7:12) and breads (Lev 21:8, 17, 21; Num 28:2).

⁴ And you shall bring near Aaron and his sons up to the entrance of the Tent of Meeting and wash them with water. ⁵ You shall take the garments and clothe Aaron with the tunic, with the robe of the *ephod*, with the *ephod*,⁷⁸ with the breastplate and gird it with the decorated band of the *ephod*. ⁶ And you will put the turban on his head and place the holy diadem⁷⁹ on the turban.

The recurrence of the verb קרב in v. 4 creates a correspondence between the offering (v. 3) and the choice of Aaron and his sons. Such a close resonance shows that, while Moses puts Aaron and his sons in a position to approach the holiness of God (קרב *hifil*, "you shall bring near"), at the same time, he "offers" them,⁸⁰ through ablution rites which have a purifying character.⁸¹ Aaron is the first consecrated to be clothed with the "symbolic" vestments⁸² (vv. 5-6), and especially with the *ephod*.⁸³

⁷ And you shall take the anointing oil and pour it on his head and anoint him. ⁸ And you shall bring near his sons and clothe them with tunics. ⁹ And gird them with the sash,⁸⁴ Aaron and his sons⁸⁵, and shall bind to them turbans. And the priesthood shall be for them an everlasting statute.⁸⁶ And you shall fill the hand of Aaron and the hand of his sons.⁸⁷

---

⁷⁸ Th PSam is longer than MT, "... with the tunic *you shall gird him with a sash and put a mantle on him, and you shall put on him* the ephod." This text may be the result of a harmonisation with Lev 8:7. The shorter version of the Septuagint and of the Vulgate, on the contrary, may be due to haplography. See Propp, II, 350.

⁷⁹ The Hebrew term employed in 29:6 is נזר, "diadem," unlike the noun ציץ, "flower," used in 28:36. Two nouns are used as synonyms, because the Septuagint renders both terms with the Greek πέταλον which means, literally, "leaf."

⁸⁰ The verb קרב *hifil* designates the act by which a person is destined to the service of the priesthood (Exod 28:1; 40:12; Lev 8:24; Num 3:6). See Propp, II, 456.

⁸¹ See J. AUNEAU, "Le bain," 103-111; see also Houtman, III, 532.

⁸² Regarding the symbolic meaning of Aaron's garments, see above ch. V, § 2.1.

⁸³ G. PAXIMADI, *E io dimorerò*, 159-163. See ch. V, § 2.1.

⁸⁴ The Septuagint, the Peshitta, the Targumim and PSam have the plural, "sashes"; this is a harmonisation with the other plural terms in vv. 8-9 (tunics, turbans). The MT is the *lectio difficilior* and therefore must be maintained. See Propp, II, 350.

⁸⁵ The mention of Aaron seems unusual, because Yhwh has already begun to give instructions concerning the consecration of his sons; therefore, the Hebrew אהרן ובניו, "Aaron and his sons," is omitted in the Septuagint and the Vulgate. The MT can be maintained, because there is no mention of the girdle on v. 5 and this object appears only in v. 9; see Propp, II, 350; G. PAXIMADI, *E io dimorerò*, 185.

⁸⁶ The Hebrew text has been translated, "and the priesthood shall be theirs by eternal decree." See e.g., NJPS; Childs, 518; Durham, 391. Regarding the construction ל + היה pronominal suffix + ל, "X will be for Y, as Z," see Gen 6:21, "will be as nourishment for you and for them" (והיה לך ולהם לאכלה); also Gen 17:8; 20:12; Exod 2:10; 4:16; 13:9; 15:2.

⁸⁷ The Septuagint and *TgN* translate יד in both cases with the plural.

Oil, a staple food (see 1 Kings 17:12), is among the main products of agriculture (see Deut 7:13), valued for its medicinal qualities (Isa 1:6; see Luke 10:34), and used in the preparation of perfumes (see Ruth 3:3; Eccles 9:8). In Israel, as in other nations of the Ancient Near East, oil was used for the royal investiture ceremony.[88] A great quantity of oil is poured on Aaron's head in a ritual action that is reserved for him[89] (Lev 21:10, 12; see Ps 133:2). The priest's anointing may have a purifying value,[90] because it followed immediately the ablution (v. 4); however, it certainly had the purpose of exalting the dignity of the anointed person bound to God by a kind of special covenant.[91] No wonder then that in v. 21 the oil, mixed with sacrificial blood, will be used again in the consecration ritual.

The mention of Aaron in v. 9 reunites the father with his sons, "and the priesthood shall be for them an everlasting statute (לחקת עולם)." This expression has already been analysed in the commentary on Exod 12:14, 17,[92] and refers both to the everlasting lamp (27:21) and to the priestly garments (28:43). With these words it is essentially stated that the priesthood is both a divine decision that will remain stable in the succession of generations to come,[93] and a command that determines the condition of the priest with a bond similar to a covenant.[94] The verse concludes with a stereotypical expression which alludes to the consecration, "and you shall fill the hand of Aaron and the hand of his sons."[95]

---

[88] The Hittite kings were anointed with oil, as were the Israelites (1 Sam 16:1-13; 1 Kings 9:1-3). See H.A. HOFFNER, "Oil in Hittite Texts," 111. Paximadi recalls that the oil was also used in anointing ceremonies for Egyptian high officials, but not for Pharaoh. See G. PAXIMADI, *E io dimorerò*, 204, who quotes S.E. THOMPSON, "The Anointing," 15-25.

[89] The title "the anointed priest" (הכהן המשיח) is especially reserved for the High Priest; see Sarna, 187; see also Lev 4:3; 6:15 and *TgPsJ* which specifies כהנא רבא, "Great Priest".

[90] With regard to the Ancient Near Eastern background, in the rite of installation of the priestess of Ba'al at Emar, in the 13th century BC, the priestess receives a double anointing on her head in order to be purified. See D.E. FLEMING, *The Installation*, 176-182.

[91] See G. PAXIMADI, *E io dimorerò*, 205-207; D.J. MCCARTHY, "Hosea XII 2," 14-20.

[92] See ch. III, § 2.2.4.

[93] The instructions regarding the Passover (Exod 12:14) and the lamp of the Tabernacle (27:21) state that the feast (12:14) will be an eternal statute for the generations to come (לדרתיכם), just as the lamp will be lit before Yhwh in every generation (27:21, לדרתיכם).

[94] The term חקה, "statute," belongs to the semantic field of the "law," since it appears in correlation with תורה (Jer 44:10; see Num 19:2; 31:21), מצוה (Gen 26:5; Deut 30:10; 2 Kings 17:13), משפט (1 Kings 2:3; 2 Kings 17:34) and עדות (Jer 44:23). The connection with ברית (1 Kings 11:11; see 1 Kings 23:3) also shows that חקה is part of the covenant domain.

[95] The construction יד + מלא, "to fill the hand," can denote the priestly consecration (Exod 28:41; 29:29, 33, 35; see Judg 17:5.12; 1 Kings 13:33; 2 Chron 13:9). See Sarna 185.

2.3 *The Sin Offering (vv. 10-14)*

(a) [10] You shall bring near the <u>bull</u> before[96] the *Tent of Meeting*
(b)   and Aaron and his sons shall lay their hands on the head of the <u>bull</u>.
(b') [11] And you shall slaughter the <u>bull</u>
(a')   before YHWH, at the entrance of the *Tent of Meeting*.

The recurrence of אהל מועד, "Tent of Meeting," in v. 10 and in v. 11 creates a chiastic construction: the two mentions of the Tent frame the two ritual actions, the laying on of the hands on the bull and the slaughtering of the animal. The gesture of the "laying on of hands" is a sign that expresses the symbolic transfer of sin onto the animal,[97] but it can also simply express the offerer's willingness to entrust himself to God.[98]. The act, in any case, signals a link between the person offering the animal and the victim.[99]

The sequence of actions culminates in the insistent specification that the ritual will take place in the presence of God[100] (v. 11, "before Yhwh, at the entrance of the Tent of Meeting"). Moreover, v. 11 is the first passage in chapter 29 in which Yhwh is explicitly named. This lexical choice, together with the double repetition of אהל מועד, generate a movement that terminates with the name "Yhwh" and indicates the theological orientation which characterises the entire ritual.

[12] And *you shall take* from the **blood** of the bull and "give" it on the horns of the altar with your finger and all the **blood**[101] you shall pour at the base of the altar.[102] [13] *You shall take* all the fat that covers the entrails and the lobe of the

---

[96] The Hebrew לפני אהל מועד, "before the Tent of Meeting," is translated by the Septuagint as follows, ἐπὶ τὰς θύρας τῆς σκηνῆς τοῦ μαρτυρίου, "at the door of the Tent of Meeting." This translation is a harmonisation with 29:11; the MT is a *lectio difficilior*.

[97] See Propp, II, 457; Sarna, 188; Durham, 395; Priotto 544; Nepi, II, 174. See also Lev 8:10-12; 16:21. Contrary to this approach, Milgrom believes that the laying of hands indicates ownership of the sacrificed animal (*Leviticus 1–16*, 151-153); see Houtman, III, 535. The two hypotheses are not mutually exclusive.

[98] See A. MARX, *Les systèmes sacrificiels*, 108-110. The sin offering presents a sequence of actions which signal the gift of the animal: the presentation of the victim, the willingness to offer it for Yhwh, and the immolation of the victim.

[99] W. GILDERS, *Blood Ritual*, 81, "The hand-pressing indexes a relationship between offerer and animal. It is the one who offers the animal who presses his hand on its head."

[100] Priotto, 543.

[101] The Hebrew את־כל הדם, "all the blood," is rendered by the Septuagint as follows, τὸ δὲ λοιπὸν πᾶν αἷμα, "all the *remaining* blood" (see Vulgate: *reliquum autem sanguinem*). The translation specifies that the second mention of the blood in v. 12 refers to the remaining part of the blood already smeared on the horns of the altar.

[102] The term יסוד, "foundations," could simply refer to the lower part of the altar (27:4-5) or to its base. Propp notices that 1 Kings 18:32, and perhaps Ezek 43:13-17, describe the altar as a structure surrounded by moats. See Propp, II, 458.

liver, the the two kidneys and the fat that is above them and you shall burn them on the altar. ¹⁴ And the flesh of the bull, and its skin, and its dung, you shall burn in the fire outside the camp. It (the bull) is a sacrifice for sin.

The consecration is therefore preceded by a sacrifice in which there is an initial blood manipulation. The gesture performed at this moment by Moses (v. 12) has a priestly character[103] and is intended to signal the altar as the emblem that manifests God's ritual presence. However, in the light of this research, it should be noted that the blood is used differently from what has been seen before: in the first passage analysed in fact the blood touches Moses' feet (4:25); in the Passover it is applied to the doors of the Israelites' houses (12:7, 13, 22-23); during the covenant's ceremony, it is poured on the altar and sprinkled on the people (24;8). In this case just the horns of the altar are sprinkled and the remaining blood is poured on its base, but it is not sprinkled or poured on others. The sacrifice of these verses is in fact defined as a "sin offering" (חטאת, v. 14), a rite in which the blood manipulation involves only the altar.[104]

The sacrifice continues with a holocaust of the animal's fat and organs[105] which are its most delicious parts (see Num 18:12); they must "be turned into smoke" (קטר *hifil*) on the altar (v. 13). Flesh, skin and dung must be burnt outside the camp so that sin could be symbolically banished.[106] This clarification is added to describe the ritual as a sin offering out of the ordinary because nobody will eat the leftover meat.[107] The rite, so to speak, combines the elements of the sin offering with those of the burnt offering.[108]

Therefore, the first act that Moses will have to perform in the newly built Tabernacle is a חטאת (v. 14). This indication aligns this excerpt with a very common element of P's theology. The Tent is not defiled by some external

---

[103] See W. GILDERS, *Blood Ritual*, 121, "In P, only priests apply blood to the altar. Thus, we should conclude that Moses is functioning as a priest on this occasion …. It is the first action that is clearly a priestly prerogative."

[104] A. MARX, *Les systèmes sacrificiels*, 119; see also, ch. I, § 1.3.

[105] Firstly, Moses must take את־כל־החלב המכסה את־הקרב, "all the fat that covers the entrails." See J. MILGROM, *Leviticus 1–16*, 205, who specifies that the חלב is the layer of fat that surrounds the organs. Moreover, the Hebrew expression ואת היתרת על־הכבד denotes the caudate lobe of the liver (Propp, II, 458). The allusion to the liver may have a polemical tone towards neighbouring cultures and the practice of divination (Ezek 21:26); Sarna, 188.

[106] According to Milgrom, the area outside the camp is not immediately an unclean area, but rather a neutral zone (Lev 14:40-41, 45). See J. MILGROM, *Leviticus 1–16*, 261-264.

[107] Propp, II, 459. In the sacrifice for ordinary sin the priest could eat leftover meat (Lev 6:18-22), but in the most significant violations (those committed by the people or by the priests themselves) the meat had to be destroyed (Lev 6:23). See Houtman, III, 536.

[108] Concerning the sacrifices, see ch. I, § 1.3.

power, but it is the sin of the people that make it impure.[109] In this circumstance sin is not treated as a subjective reality, but as an "objective" energy.[110] The Torah in fact "indissolubly unites sacred and profane law,"[111] worship and conduct of life, as it can be deduced from the structure of Exodus: the people can receive a priestly dignity ("priestly kingdom" and "holy nation," Exod 19:6) primarily thanks to a covenant that dictates the conditions for a righteous behaviour (the Decalogue, Exod 20, and the Covenant Code, Exod 21-23) and only later introduces the instructions concerning worship (see Exod 25–30). Therefore, before consecrating Aaron, Yhwh asks Moses to manifest an implicit acknowledgement of guilt through a sin offering in order to achieve, through blood, an actual reconciliation.

The function of the blood used in the first rite is expiatory[112] and its manipulation is intended to purify the altar contaminated by the sins of the people.[113] In fact, sin is the truly chaotic force that can undermine the sanctity of the Tabernacle and that threatens the relationship between the Israelites and God. Therefore, blood appears also here in a "liminal" situation[114] and has the repulsive power to expel the deadly menace of sin.

2.4 *The Sacrifice of the First Ram (vv. 15-18)*

The second ritual act of Aaron's consecration is the sacrifice of the first ram; the procedure resembles the sin offering (vv. 10-14), but the differences are relevant.

---

[109] See J. MILGROM, *Leviticus 1–16*, 260; G. PAXIMADI, *E io dimorerò*, 197. According to the classical gradual progression towards the most holy place, typical of the P literature, the sin of the individual defiles the altar of burnt offerings (Lev 4:34), the fault of the people or that of Aaron involves the altar of incense (Lev 4:6-7.17-18), the offenses of an entire year penetrate into the Holy of Holies and requires the special purification of the *Yom Kippur* (Lv 16).

[110] See J.-L. SKA, "'Misericordia voglio,'" 133, "The notion of guilt in the Bible is ... a matter of 'state' rather than of intention." (My translation)

[111] J.-L. SKA, "'Misericordia voglio,'" 138.

[112] See ch. I, § 1.3. See also A. MARX, *Les systèmes sacrificiels*, 123-124. The blood employed in the חטאת is explicitly correlated to the root כפר, "to atone," in Exod 30:10 (see Lev 6:23; 16:16-18; Ezek 43:20; 2 Chron 29:24). Blood has the capability to repel sin and remove it from the sanctuary.

[113] The excerpt of Lev 16:19 clarifies the meaning of blood manipulation during the sin offering, "so he will purify (וטהרו) it and sanctify (וקדשו) it (the altar) from the impurities of the Israelites." See W. GILDERS, *Blood Ritual*, 131, "The Priestly tradent who composed Lev 16:19b attributes an instrumental effect to blood manipulation actions. The daubing of blood onto the horns of the altar achieves the instrumental effect of eliminating impurity. The act cleanses (וטהרו) the altar. The sprinkling of blood is the medium for communicating holiness to the altar (וקדשו)." See also J. MILGROM, *Leviticus 1–16*, 1037-1039.

[114] See ch. II, § 4.5, note 290.

Exod 29:15	Exod 29:10
The first ram you shall take, and Aaron and his sons shall lay their hands on the head of the ram.	*You shall bring near* the bull *before the Tent of Meeting* and Aaron and his sons shall lay their hands on the head of the bull.

In comparison with the ram's sacrifice, the sin offering (v. 10) specifies the location of the ritual: "before the Tent of Meeting." This detail shows that the rite can create the conditions for entering the Tent and thus for approaching God. In both cases, the text insists on the gesture of the laying of the hands to establish a link between the offerer and the victim itself.[115]

Exod 29:12-13	Exod 29:16-17
12 And *you shall take* from the **blood** of the bull and "give" it (ונתתה) on the horns of the altar with your finger and all the **blood** you shall pour (תשפך) at the base of the altar. 13 *You shall take* all the fat that covers the entrails and the lobe of the liver, the two kidneys and the fat that is above them and you shall burn them on the altar. 14 And the flesh of the bull, and its skin, and its dung, you shall burn in the fire outside the camp. It is a sacrifice for sin.	16 You shall slaughter the ram[116] and *shall take* its **blood** and throw it (וזרקת) upon the altar, all around. 17 And you shall cut the ram into pieces, and wash its entrails and its legs, and put over the pieces and its head. 18 And you shall burn the whole ram on the altar. (a) It is a burnt offering for Yhwh, (b) pleasing savour[117], (a') an offering made by fire for Yhwh.

The blood manipulation (v. 16) offers some points of resemblance with the first sacrifice; however, the ritual described is much less elaborate, Moses will only throw the blood on the altar (וזרקת על־המזבח סביב). Nonetheless, the root זרק, "to throw, to sprinkle," creates a closer connection with Exod 24 than with 29:12 (where נתן and שפך are employed), since the narrator uses the verb זרק in both texts (see 24:6, 8). Thereupon, the symbolic significance of

---

[115] W. GILDERS, *Blood Ritual*, 81.

[116] The Septuagint (and Vg) has καὶ σφάξεις αὐτόν, "and you shall slaughter *it*," instead of Hebrew ושחטת את־האיל, "and you shall slaughter *the ram*." The Septuagint may be the oldest version; MT adds את־האיל to clarify the object of the verb. See Propp, II, 351.

[117] The Septuagint (εἰς ὀσμὴν εὐωδίας), the Peshitta (*lryh nyḥ'*) and *TgN* (לריח דרעוא) translate "*as* pleasing savour," but this translation presupposes a different Hebrew text from the MT, לריח ניחוח in spite of ריח ניחוח. This rendering is a harmonisation with other passages in the same chapter (29:25, 41; see Lev 8:21). See Propp, II, 352.

this manipulation signals the relationship between the animal blood and the altar, and thus the "covenant" between the offerer and God, represented by the altar.[118]

The table above highlights the sober style of v. 14 in comparison with the developed description of the ram's sacrifice in v. 18. The text states that the ram is a "burnt offering for Yhwh" and at the same time, "an offering made by fire for Yhwh." The reiteration of ליהוה evidently manifests that the sacrifice is an act of entrustment to Yhwh.[119] Therefore, in the heart of the passage, the sacrifice is designated as a "pleasing savour." The alliteration ריח ניחוח often appears in sacrificial language[120] and is even attested in Ancient Near Eastern literature.[121] It indicates the divine appreciation of the offering[122] and should not be interpreted as a food anthropomorphism.[123] This stereotypical formulation, alluding to "sensitive" pleasure,[124] shows that the integral gift of the victim is particularly "attractive" to Yhwh because it manifests the human desire to be reunited with God.

Therefore, in the ram's offering Moses must perform a burnt offering in which the entire victim is consumed by fire. If the first חטאת demonstrates that the ritual has the power to repel sin, the sacrifice of the first ram, on the other hand, is an expression of the offerer's donation to God. In this case, the blood signals the altar as the place of God's presence and shows that the ritual is addressed precisely to Yhwh as the recipient of the offering.

---

[118] W. GILDERS, *Blood Ritual*, 81-82, "The altar is a locus of the divine presence. The application of blood to it has indexed a relationship between the offerer and the altar. This relationship between offerer and altar must involve a relationship between the deity whose altar it is and the offerer.... The offerer, whose relationship with the animal was indexed by the hand-pressing gesture, is linked to the altar, and to Yahweh, whose presence is manifested at the altar, by the transfer of the animal's blood to the altar."

[119] Priotto, 545; A. MARX, *Les systèmes sacrificiels*, 112, "La seconde phase du rituel, qui culmine dans la combustion de la matière sacrificielle dans son intégralité, fait de l'holocauste l'expression la plus accomplie du don sacrificiel."

[120] The expression "pleasing savour" denotes the burnt offering (29:41; Lev 1:9, 13, 17) and the sacrifice of well-being (Lev 3:5, 16; 17:6); see also Lev 2:2, 9, 12 and Num 15:7. Propp, II, 462. It is important to notice that Exod 29:18 is the first occurrence of the construction ריח ניחוח in the Book of Exodus (29:25, 41), where it only appears in ch. 29.

[121] See *Gilgamesh*, XI, 161-163, "The gods smelled the savour / the gods smelled the sweet savour / the gods gathered like flies around the sacrificer."

[122] In *TgO* and in *TgPsJ* the MT is translated specifying that, "this is a holocaust before Yhwh to be accepted with favour (לאתקבלא ברעוא)." See Sarna, 189; Priotto, 545.

[123] See G. PAXIMADI, *E io dimorerò*, 131-133; R. ABBA, "The Origin," 123-138. Paximadi notes that while in the Canaanite context the meat offered to the deity had to be cooked, in the P ritual only the parts of the animal that were to be eaten, were boiled (see 29:31).

[124] The term ריח is often used in the Song of Songs (1:3, 12; 2:13; 4:10-11; 7:9, 14) as a metaphor for love; see G. BARBIERO, *Song of Songs*, 55.

2.5 The Sacrifice of the Second Ram (vv. 19-25)

¹⁹ An *you shall take* the second ram and Aaron and his sons shall lay their hands upon the head of the ram. ²⁰ You shall slaughter the ram and you *shall take* of its **blood** and you will "give" (ונתתה) it upon the lobe of Aaron's ear[125] and upon the lobe of the right ear of his sons, and upon the thumb of their right hands, and upon the great toe of their right foot and you shall smear (וזרקת) the **blood** upon the altar, round about.[126] ²¹ And you *shall take* of the **blood** which is on the altar and of the anointing oil and you shall sprinkle (והזית) upon Aaron and his garments, and upon his sons and upon their garments with him. He shall be sanctified, and his garments, and his sons, and his sons' garments with him.[127]

The sacrifice of the second ram begins again with the laying of hands on the animal's head (v. 19). The most significant difference from the other sacrifices, however, appears in the blood manipulation. Whereas in the first offerings, only the altar is "touched" by blood, in this ritual the same fluid is "given" (v. 20) and sprinkled (v. 21) upon Aaron and his sons.[128]

The act of spreading the earlobe, the thumb of the right hand and the big toe of the right foot with blood (v. 20) has given rise to various interpretations. First of all, the gesture is explained as the manifestation of a total consecration (from top to bottom) to Yhwh.[129] In addition, the ritual action alludes to the symbolic function of the parts of the body to which the blood is applied. In the biblical world in fact the right side is usually preferred (Gen 48:13-14, 17), and therefore the application can mean that the touched limbs will be adequate for their task: the ear would thus be perfectly capable of hearing God's words, the hand and the foot, of accomplishing them and "walking" in them.[130] The rite of blood, moreover, has been

---

[125] The Septuagint is longer than the MT: ἐπὶ τὸν λοβὸν τοῦ ὠτὸς Ααρων τοῦ δεξιοῦ καὶ ἐπὶ τὸ ἄκρον τῆς χειρὸς τῆς δεξιᾶς καὶ ἐπὶ τὸ ἄκρον τοῦ ποδὸς τοῦ δεξιοῦ, "on Aaron's *right* earlobe, *on the thumb of his right hand and on the big toe of his right foot*" [in italics the parts that do not appear in the MT]. Together with the PSam, this longer lesson seems to be a harmonisation with Lev 8:23; see Propp, II, 352.

[126] The Septuagint reports the sentence "you shall smear the blood upon the altar, round about" at the end of v. 21.

[127] PSam and 4QPaleoExod^m introduce the v. 21 after the v. 28. This choice is perhaps deliberate and is intended to harmonise these verses with Lev 8:24-25, 30 (Propp, II, 353).

[128] The ritual variation of the sacrifice of the second ram is so drastic that Gilders speaks of a real "plot twist." W. GILDERS, *Blood Ritual*, 101.

[129] Houtman, III, 541, "The cleansed future priests are now made completely holy. Sanctified from top to bottom, they are now in a new state of life."

[130] See PHILO OF ALEXANDRIA, *De vita Mosis*, II, § 150, "… the earlobe, the extremity of the hand and of the foot, all on the right, to signify that what is perfect must be both in the word and in the work and in all life; for the ear is the judge of the word, the hand the symbol

interpreted as an expression connected with life; this conclusion is usually deduced from the surprising contact between the manipulation performed during Aaron's consecration and the ritual found in the cleansing of the leper[131] (Lev 14:14, 25). This consideration creates a link with what has already been seen in Exodus and especially in 4:24-26: ritual blood is an element that possesses an extraordinary vital force, since it is capable of saving Moses from mortal danger.[132] Finally, these interpretations do not exclude the most obvious element: the ram of consecration is intended to bring about a permanent union between God (signified by the altar) and the priests,[133] continuing a trajectory already traced in 24:3-8, in which the covenant between Yhwh and the people is sealed through a ritual of blood.[134] The blood thus makes explicit its special ritual power to bind the offerer and God.[135] The last gesture (v. 21) creates some friction with the context.[136] The meaning of the sprinkling can be derived from the repetitive cadence of the verse:

| And you shall sprinkle (והזית) upon Aaron and his garments, and upon his sons and upon their garments with him. | He shall be sanctified (וקדש הוא), and his garments and his sons and his sons' garments with him. |

The sprinkling (חזה) of blood mixed with oil can "sanctify" (קדש), namely separate Aaron and his sons from the mundane sphere so that they might be more closely united to God.[137] The combination of these two substances manifests the extraordinary character of the ritual.[138] Thus, the gesture expresses again an action of Yhwh[139] with which He binds the priests to the altar and

---

of the work, the foot the symbol of the path with respect to life." See Sarna, 189. The action, "may well symbolize the idea that the priest is to attune himself to the divine word and be responsive to it in deed and direction in life"; see also Nepi, II, 174; Priotto, 546.

[131] See G. PAXIMADI, *E io dimorerò*, 200-201. Both the rite of priestly consecration and the ritual performed for the cleansing of the leper are "rites of passages" in which a separation, a "liminal" experience and a new aggregation take place.

[132] Milgrom interprets all the priestly consecration as a "rite of passage." See J. MILGROM, *Leviticus 1–16*, 566-569, and particularly p. 569.

[133] See W. GILDERS, *Blood Ritual*, 101.

[134] See G. PAXIMADI, *E io dimorerò*, 202. Regarding Exod 24 and Exod 29, the author recognises a relationship, since "the fundamental gesture is the same: the distribution of the blood between the consecrated persons and the altar." (My translation)

[135] See A. MARX, *Les systèmes sacrificiels*, 122-123.

[136] See above, ch. V, § 1.

[137] Regarding "holiness," see ch. I, § 1.3. Concerning the oil, see above ch. V, § 2.1.

[138] Houtman, III, 541, "the combination of two cleansing substances enhances the purifying effect."

[139] See J. MILGROM, *Leviticus 1–16*, 532.

thus to Himself.[140] The passage, therefore, intensifies what appeared already in v. 20, manifesting even more clearly the "unitive" character of blood.

²² You shall take the fat of the ram and the tail[141], and the fat that covers the entrails, and the lobe of the liver and the two kidneys, and the fat that is upon them, and the right thigh; for it is a ram of consecration[142]. ²³ And one loaf of bread, one cake of oiled bread, and one wafer, from the basket of unleavened bread that is before **YHWH**, ²⁴ and you shall put the whole upon the palms of Aaron and upon the palms of his sons and elevate them as a *tĕnûpâ* before **YHWH**.

Just like the bull sacrifice, the best parts of the meat, the fatty parts, are offered to the Lord (29:13), together with bread (the loaf, the flatbread, the crushed bread). Everything is placed on the palms of Aaron's hands and those of his sons, to perform the תנופה (v. 24), the wave-offering that emphasises the act of the offerer, namely the donation that the worshipper had brought to Yhwh.[143] The ritual of consecration, therefore, ends with the burnt offering of what is in the hands of the priests. Verse 25 recalls v. 18 very closely, through a series of recognisable repetitions.

Exod 29:18	Exod 29:25
You shall make the whole ram burn upon the altar. It is a burnt offering *for Yhwh*, pleasing savour, an offering made by fire *for Yhwh*.	You shall take them from their hands, and make them burn upon the altar as a burnt offering for a pleasing savour *before Yhwh*. This is an offering made by fire *for Yhwh*.

---

[140] See G. PAXIMADI, *E io dimorerò*, 208, "the rite of blood consecrates Aaron and his sons, clothing them with a sacredness that has God (represented by the altar) as its source and that unites the consecrated priests to the altar (and therefore to God)." (My translation)

[141] The Septuagint does not contain the Greek equivalent of the term אליה, "tail," perhaps because of a haplography (Propp, II, 352).

[142] The Hebrew construction איל מלאים הוא can be translated literally, "this is a ram for the filling." The verb מלא is used in the syntagm מלא + יד, "to fill the hand," as a technical expression for the priestly consecration. The noun מלאים is often employed in similar contexts (Lev 7,37; 8:22, 28-29, 31, 33) and can refer to the gemstones to be set (Exod 25:7; 35:9; 1 Chron 29:2). See Ibn Ezra, 259; Propp, II, 463. The Septuagint translates as follows, ἔστιν γὰρ τελείωσις αὕτη, "this is a perfection." This translation refers the phrase to the whole ritual, considered as "perfect." See also Rashi Exod, 259; Priotto, 491.

[143] See J. MILGROM, *Leviticus 1–16*, 461-473. Through the gesture of elevation there is a reference to a "passage" of ownership (Num 8,13-15). Concerning Exod 29, he writes, "is similar to the well-being offering.... One might expect that the portions designated for the altar or given to the officiating priest would require *tĕnûpâ*, unlike the rest of the meat, which is eaten by the offerers and therefore does not require *tĕnûpâ*. This is in fact the case. The suet, the bread, and the thigh, which are offered on the altar, as well as the breast, which is delivered to the officiating priest, undergo *tĕnûpâ*, whereas the rest of the meat and bread are eaten by the offerers without *tĕnûpâ*. The difference between them is clear: *tĕnûpâ* transfers the offering from the domain of the offerers to the domain of the Lord" (*Ibid.*, 465).

From this comparison it can be seen that the two ram's sacrifices conclude with a strong insistence on the theological meaning of the rites. The difference between the two offerings consists of the consciousness that the first burnt offering is exclusively "for Yhwh," (ליהוה, is repeated twice) while the second is "before Yhwh" (לפני יהוה) and "for Yhwh" (ליהוה). Thus, as will also be seen in the subsequent verses, in the sacrifice of the second ram the emphasis on sharing ("before") is stronger than the mere insistence on offering ("for").

2.6 *The Conclusion of the Consecration (vv. 26-37)*

The burnt offering is not the last gesture of the ritual, which instead concludes with a list of the animal's parts assigned to the priests (vv. 26-28). The sacrifice of the second ram in fact is an offering of well-being[144] in which the offerers receive part of the sacrificed animal and can eat of it. The ritual act thereby reveals more explicitly the unity and the relationship between Yhwh and the newly consecrated priests. The animal's breast and the thigh are handed over to God by means of the תנופה, "rite of elevation," and through a תרומה, "rite of raising, giving."[145] The ram's breast, however, is assigned to Moses, who acts as a priest.[146] Aaron and his sons are then entitled, through an "eternal decree," to receive a portion of the offered animals.

The instructions concerning the priestly garments (vv. 29-30) interrupt the thread of the discourse,[147] showing that they are not the exclusive property of Aaron, but his vestments are to be transferred to the person who, as his successor, will have the right to use them.[148] In addition, the text recalls the fact that the holy garments are to be worn by Aaron's heir

---

[144] See ch. I, § 1.3. See also A. MARX, *Les systèmes sacrificiels*, 112-117.

[145] See J. MILGROM, *Leviticus 1–16*, 473-481. The term תרומה literally means "raising," as it is derived from the root רום, "to raise." The lexicographic survey conducted by Milgrom actually shows that the basic meaning of the expression is rather "gift" for Yhwh or for the priests. The main difference between the תרומה and the תנופה consists in the fact that while the latter is realised לפני יהוה, "before Yhwh," or, sometimes ליהוה, "for Yhwh," the תרומה is always performed ליהוה, "for Yhwh." Therefore, the תנופה is a ceremony that takes place in the sanctuary, while the תרומה takes place outside the Tabernacle. The author explains the apparent interchangeability of the two terms in 29:26-27 as follows, "*tĕrûmâ*, then, is a necessary step preceding *tĕnûpâ*. An offering requiring *tĕnûpâ* must undergo a previous stage of *tĕrûmâ* ... its separation from the profane to the sacred" (*Ibid.*, 476).

[146] Houtman, III, 544; Propp, II, 464.

[147] See above, ch. V, § 1.

[148] See Propp, II, 465; since the expression בגדי הקדש, "holy garments," refers normally to Aaron's vestments (40:13; Lev 16:32), it must be assumed that Exod 29:29 is related to the succession of the first-born heirs, destined to hold the office of High Priest.

for seven days (v. 30) and so the reader discovers for the first time that the rite of consecration lasts for seven days[149] (v. 35; see Ezek 44:26-27).

The last verses (vv. 31-34) supplement the instructions enumerated in vv. 26-28 concerning the consummation of the sacrifice of well-being. The ram's flesh must be boiled in a holy place[150] (v. 31) and eaten together with the loaves that are in the basket at the entrance of the Tent of Meeting (v. 32). Thereby, the relationship produced by the rite, and especially through the blood manipulation, becomes more "familiar" thanks to the "shared meal." A similar progressive intimacy appeared already in Exod 24, when Moses and the representatives of Israel had the opportunity of eating and drinking on the mountain[151] (24:11).

The purpose of the rite is clarified in v. 33, "and they shall eat those things with which atonement has been made for to them (אשר כפר בהם) to fill their hand and to sanctify them (לקדש אתם)" (v. 33). Thereby, Yhwh closely relates the sacrifice of the ram to the act of atonement[152] and thus to reconciliation, manifested most explicitly in the sacrifice of the bull, the חטאת (vv. 10-14). The interpreter cannot fail to note that the verb כפר occurs in the passive conjugation *pual*. Therefore, the atonement is described as a divine action which in no way can be attributed to the simple initiative of the priest or, in this case, of Moses.[153] However, v. 33 combines expiation and sanctification in a single sentence,[154] thus revealing that the atonement is not an end in itself, but its purpose is to produce the priestly participation in the divine holiness. The leftovers of the ritual must be burnt in the fire[155] (v. 34).

---

[149] See Propp, II, 465.

[150] The meat must be boiled (Exod 12:9; Deut 16:7; 1 Sam 2:13-15; Ezek 24:5) at the entrance of the Tent of Meeting (Lev 8:31).

[151] Concerning Exod 24:9-11, see ch. IV, § 2.3. See also A. MARX, *Les systèmes sacrificiels*, 123. The sprinkling of blood in 24:6-8 and 29:20-21 "n'établit, dans tous ces cas, qu'une relation médiate et prépare une relation plus intime: en Ex. xxiv, le repas des représentants d'Israël en présence de Dieu (Ex. xxiv 9–11); en Ex. xxix // Lev. viii, le partage des prêtres avec Yhwh de la viande et des pains."

[152] Regarding the root כפר, see ch. I, § 1.3.

[153] See Propp, II, 467. In other occasions, the forgiveness of sins is alluded through circumlocutions which show that only God can forgive. See e.g., ונסלח לו, "and it will be forgiven to him" (Lev 4:26.31.35; 5:10.13.16.18.26; 19:22).

[154] See discussion of the root קדש in ch. I, § 1.3.

[155] See Propp, II, 468. The limitation concerning the consummation of the sacrifice also appears in the istructions for the Passover (12:10) and for other offerings (Lev 7:15; 22:30). This directive assumes that the sacrifice must be consumed within a limited time frame and in a space close to the altar, because holiness does not survive long and cannot be far from the presence of God. See also J. MILGROM, *Leviticus 1–16*, 219-220.

Yhwh's instructions for the consecration of Aaron conclude with the clarification that the rite of consecration continues for seven days, "You shall do to Aaron and his sons according to all that I have commanded you (אשר צויתי). During seven days[156] shall you fill their hand" (v. 35). The consecration of the priests is then related with that of the altar (vv. 36-37), since Moses has to sacrifice a sin offering on each of the seven days of consecration[157] (v. 36). The bull's offering in fact is intended to "remove sin" from the altar through the rite of atonement. The consequence of these rites will be the special link of the altar with the divine holiness.

Exod 26:33b	Exod 29:37b
The veil shall separate for you the Holy and the Holy of Holies (קדש הקדשים).	The altar will be a most holy reality[158] (קדש קדשים), everything that touches the altar will be holy.

The expression used in 29:37 closely resembles the phrase that defines the innermost part of the Tabernacle, where the Ark of the Covenant is kept. There is only one difference between the two statements. In the description of the Tabernacle, the construction קדש הקדשים is determined, "*the* Holy of Holies," while in 29:37 קדש קדשים does not have the article, "a most holy reality."[159] The assonance, however, creates a link between the altar, the Holy of Holies – the holiest space in the Tent – and God Himself, so much so that the divine holiness bestowed on the altar can be transmitted through simple contact.[160]

2.7 *The Daily Burnt Offering (vv. 38-46)*

The last verses[161] of chapter 29 introduce the reference to a crucial component of the Israelite cultic system, the daily and perennial sacrifice.

---

[156] J. MILGROM, *Leviticus 1–16*, 537-540. The consecration of the altar is also a ritual that must be repeated for seven days (29:36-37).

[157] See Propp, II, 469. The sacrifice of the bull, presumably, is the same that is mentioned in 29:10-14, repeated every day. Milgrom notices, however, that in 11QTemple XV,14-18 the rite of consecration of the High Priest is described, specifying that two bulls were offered each day as a sin offering. See J. MILGROM, *Leviticus 1–16*, 562.

[158] The construction קדש קדשים is a "superlative"; see Sarna, 191; Priotto, 548.

[159] See WO § 9.5.3*j*. The authors translate the superlative of Exod 29:37, קדש קדשים, "most holy," but render the expression שיר השירים of Song 1:1, "*The* Song of Songs."

[160] See Houtman, III, 547-549; Priotto, 548; the *TgPsJ* of v. 37 presupposes that no other person than Aaron and his sons may touch the altar, "All from the sons of Aaron who come near to the altar shall be sanctified, but it is impossible for the rest of the people to come near, lest they be burned in the flaming fire that would come forth from the holy things."

[161] See above, ch. V, § 1, for discussion of vv. 38-42.

In v. 38 in fact it mentions as follows, "this is what you shall do on the altar: two lambs of a year day by day, always (תמיד)." In addition to the acts performed regularly inside the Tabernacle (lighting the lamps, 27:20; offering incense, 30:7-8; presenting the loaves, 25:30), the daily burnt offering on the external altar will be a decisive component of Israel's sacrificial worship.[162] The lamb must be one year old[163] (29:38), like the one used for the Passover (12:5). This similarity creates an evident connection between the daily sacrifice and the liberation from Egypt. This sacrifice is accompanied by the offering of fine flour mixed with oil and a libation of wine (v. 40), still defined as a "pleasing savour" for Yhwh:

Exod 29:18b	Exod 29:25b	Exod 29:41b
Burnt offering for *Yhwh*, pleasing savour, offering made by fire for *Yhwh*.	Burnt offering for pleasing savour, before *Yhwh*. This is an offering made by fire for *Yhwh*.	For pleasing savour, an offering made by fire for *Yhwh*.

Although the motif of the "pleasing savour" and the offering of bread mixed with oil and wine may evoke the meal and the table, the rites have a clearly symbolic value.[164] These rituals are performed outside the Tabernacle[165] and not within it, as if to dismiss the belief that the offering could really be able to "nourish" God. Moreover, the "theological" emphasis, achieved through the repetition of the term "Yhwh," together with the use of the term מנחה, show that the sacrifice and the handing over of particular goods are a sign of an individual offering to God made in order to recognise His sovereignty.[166]

The successive passage, v. 42, provides the transition between the description of the daily sacrifice and the final instructions of vv. 43-46.

---

[162] Regarding the daily offering, see Num 28:3-8.

[163] See Lev 9:3; 12:6; 23:12; Num 6:12; see also Ezek 46:13; Mic 6:6.

[164] Regarding the expression "pleasing savour," see ch. V, § 2.4, and P.P. JENSON, *Graded Holiness*, 162-163, "The turning of some or all of the sacrifice into smoke is an appropriate symbol of an immaterial God who dwells in heaven."

[165] G. PAXIMADI, *E io dimorerò*, 224.

[166] See G. PAXIMADI, *E io dimorerò*, 133-134. The author recognises that the term מנחה, "offering" (29:41), can be a manifestation of submission. The noun מנחה, in effect, denotes both the voluntary "gift" (Gen 43:11, 15, 25; 2 Kings 8:9; 20:12; see Ps 45:13) and the compulsory "tax" (2 Kings 3:4; 17:3; 2 Chron 17:11; see 2 Sam 8:2, 6; 1 Kings 5:1; 10:25). Paximadi, then, says as follows, "the sacrificial meaning of the term מנחה derives from this 'profane' meaning, and emphasises the value of sacrificial worship as a recognition of God's sovereignty." (My translation) See Propp, II, 472."

⁴² It shall be the regular burnt offering (עלת תמיד) throughout your generations (לדרתיכם) at the entrance of the Tent of Meeting (פתח אהל־מועד), before Yhwh, where I will meet (אשר אועד) with you there (שמה), to speak (לדבר) unto you, there (שם).

The daily burnt offering (עלת תמיד) is described here as a ritual capable of transmitting to future generations (לדרתיכם), from father to son, what is communicated in the signs of the ritual.[167] This particular act of worship, therefore, is part of a broader movement that begins with the Passover (12:14, 17, 42) and results in the preservation of the memory of God's saving actions through the regular repetition of gestures that can affect the day-to-day life of each and every Israelite.

Moreover, though Yhwh indicates that the sacrifice is to be performed at the entrance of the Tabernacle, in His presence[168] ("before Yhwh"), He also emphasises that the initiative of the meeting lies with Him, "I will meet with you."[169] The reiteration of the root יעד (four times in vv. 42-44[170]) shows that Yhwh insists precisely on His desire to reveal Himself in order to create a relationship with the people.[171] Furthermore, the recurrence of the adverb שם (and שמה), "there," establishes the unique quality of the Tabernacle as being the concrete space where the transcendent God is made near.[172] The use of the root יעד, then, generates a distant echo with 25:22 where, at the end of the description of the ark, Yhwh states that it is precisely from the ark that He will make Himself known. In both cases, communication will take place through an act of speech[173] (לדבר). This transitional excerpt (v. 42), then, consists of two diametrically opposed movements: the people's movement towards God through an offering, and Yhwh's movement towards the people. This meeting will happen thanks to Moses, who will be the first to receive God's words[174] (לדבר אליך) and transmit it.

---

[167] Concerning the reference to future generations, see ch. III, § 2.2.4.

[168] See P.P. JENSON, *Graded Holiness*, 113.

[169] The root יעד *nifal* means both "to meet" (with אל, Num 10:3-4) and, when it refers to God, "to reveal oneself" (with ל, Exod 25:22; 29:43; 30:6.36; Num 17:19). In the latter occurrences, the main agent is God and the aspect of the *yiqtol* is iterative (אועד); יעד creates an echo with the denomination אהל־מועד, "Tent of Meeting." See Ibn Ezra, 263.

[170] The expression אהל־מועד appears in v. 42 and v. 44, while יעד is found in vv. 42-43.

[171] Cassuto, 388.

[172] See P.P. JENSON, *Graded Holiness*, 113, "The purpose of the sanctuary is that God might meet (נועד, Exod. 29.42, 43) with his people, and the frequent use of 'Tent of Meeting' (אהל מועד, e.g. Exod. 29.44) is a transparent reminder of this theme."

[173] Priotto, 550.

[174] Il PSam, the Septuagint and Vg translate the MT לכם, "for you," with the singular, assuming that the verb ויועד refers to Moses. The variant attempts to smooth out a text which is complex in itself, "I will meet *with you* (people), there, to speak *unto you* (Moses), there." It

Exod 29:42b	Exod 29:43
(a) Where I will meet (אשר אועד) with you, there (לכם שמה),	(a) I will meet there (ונעדתי שמה) with the Israelites (לבני ישראל),
(b) to speak unto you, there (שם).	(c) and it shall be sanctified by My glory.

The repetitions especially accentuate the divine decision to meet the Israelites in a specific place. Unlike v. 42, however, v. 43 focuses particularly on the divine glory, and the reader already knows that Moses can contemplate it in a close and intimate way (24:16).[175] It is Yhwh's glory that has the power to make holy the Tent of Meeting and all that is found therein.[176]

Yhwh reveals the extent of His sanctifying deeds in v. 44. His active intervention will extend to the Tabernacle, to the altar, as well as to Aaron and his sons. Yhwh also alludes to the underlying idea behind all that has been developing since Exod 25 until the priestly consecration: the Tabernacle and the priests are "for Yhwh."[177]

Exod 29:45	Exod 29:46
I will dwell[178] amongst (ושכנתי בתוך) the Israelites, and I will be (והייתי) to them (להם) God (לאלהים).	And they will know (וידעו) that I Yhwh (אני יהוה), their God (אלהיהם) that I brought them out of the land of Egypt in order to **dwell**[179] amongst them (לשכני בתוכם), I Yhwh (אני יהוה), their God (אלהיהם).

The verses are structured according to a series of repetitions, and is governed by a gradual progression: v. 45 is shorter, whereas v. 46 is more developed. The recurrence of שכן, "to dwell," is critical for the Book of

---

can be assumed that, while Moses can approach Yhwh (25:22), the people must remain outside the Tabernacle: Yhwh will be met with the people *through Moses*; see Propp, II, 355.

[175] See ch. IV, § 2.5.

[176] The v. 43, actually, is ambiguous, because the subject of the verb ונקדש, "will be sanctified," is not determined. The Septuagint, the Peshitta, *TgO* and *TgPsJ* translate with the first person singular, referring the verb to God, "*I* will be sanctified." The Vulgate translates, *et sanctificabitur altare in gloria mea* ("the *altar* will be sanctified in my glory"), introducing a subject not found in the MT. The verb can refer to "everything" (the people, the Tent, etc.), but especially to the Tent; Sarna, 192; Houtman, III, 552; Priotto, 550.

[177] Regarding the construction לכהן לי, see above, ch. V, § 2.2.

[178] The Septuagint translates the MT ושכנתי with the Greek expression ἐπικληθήσομαι, "and I shall be invoked," softening the uncomfortable idea of God dwelling among men. See J.W. WEVERS, *Notes*, 487. The MT is confirmed by the attestation of the other translations.

[179] The Septuagint translates v. 46 euphemistically: ἐπικληθῆναι αὐτοῖς καὶ θεὸς εἶναι αὐτῶν, "to be *invoked* by them and to become *their God*." The Greek text is not supported by the other ancient translations and is therefore a secondary lesson.

Exodus, since the divine glory "dwells" on Mount Sinai[180] (24:16) and admits Moses to a meeting with Yhwh. There, God reveals to Moses that the sanctuary (מקדש) will be built so that Yhwh may dwell among the people (ושכנתי בתוכם, 25:8). Then, at the end of the Exodus' narrative, the narrator states that "the cloud dwells" in the newly built Tent (שכן עליו הענן, 40:35), and that the "glory of God fills the Tabernacle (משכן)." Therefore, the priestly consecration is part of an arc of events which begins with the "descent" of the divine glory on Sinai and ends with the construction of the Tabernacle as God's "habitation" in the midst of the people. The journey inaugurated through the liberation (Exod 12) can be resumed with new vigour precisely because it is now the divine glory that will guide the people during their campaign.

The purpose of the double occurrence of שכן is meant to assert the permanent residence of God[181] and above all, His proximity (בתוך, "among") to the Israelites.[182] This intention is verbalised in a significant phrase, "I will be, to them, God" (v. 45), which recalls 6:7a which says as follows, "I will take you to be my people and I will be your God (והייתי לכם לאלהים)." This formula expresses the mutual relationship of two individuals juridically bonded.[183] For this reason, the same formula is used in covenant texts.[184] In

---

[180] See ch. IV, § 2.5 and Cassuto, 316.

[181] The root שכן qal is frequently associated with ישב, "to dwell" (Judg 5:17; Isa 13:20; 18:3; 32:16; Ger 17:6) and correlated to the term אהל (Gen 9:27; Isa 13:20; Ps 78:55).

[182] See Propp, II, 376. Regarding Exod 25:8, Propp notices that while Yhwh could have concluded "and I will dwell in it," referring to the Tabernacle, He instead says "and I will dwell among them," thus manifesting his intention to dwell with the Israelites (Lev 15:31; 16:16; 22:32; 26:11-12; Num 16:3; 18:20; 35:34).

[183] See L. INVERNIZZI, "Perché mi hai inviato?", 345, note 144. The formulas of Exod 6:7 are used in several areas: לקח + ל + ל introduces a marriage ("to take for oneself as a wife," e.g., Gen 12:19; 24:67; 25:20; 28:9; 34:4, 21; Exod 6:20, 23, 25; Deut 21:11; 24:3-4; 25:5; Judg 3:6; 14:2; 1 Sam 25:39, 40; 2 Sam 12:9-10; 1 Kings 4:15; Ezek 44:22; Ruth 4:13), an adoption (Esther 2,7.15) the acquisition of slaves (2 Kings 4:1; see Gen 43,18; Job 40,28). However, the construction ל + ל + היה can be used for enslavement ("to become a slave to," Gen 44:9; 1 Sam 8:17; 17:9; 27:12; 2 Sam 8:2, 6; Jer 34:16; 2 Chron 12:8; 36:20; see Isa 56:6), for marriage (Ruth 1:11; see also איש + ל + היה, Lev 21:3; 22:12), for adoption ("I will be a father to," normally referred to God, 2 Sam 7:14; 1 Chron 17:13; 28:6; see also, Judg 17:10), for adoptive children ("to become a son for," Exod 2:10; theologically in 2 Sam 7:14; 1 Chron 17:13; 22:10), for the acquisition of a property (Lev 25:45; Num 32:22; Ezek 45:8).

[184] The construction ל + ל + היה can express the covenant: "to be God for the people" + "to be people for God" (Lev 26:12; Jer 7:23; 11:4; 24:7; 30:22; 31:1, 33; 32:38; Ezek 11:20; 14:11; 37:23, 27; Zech 8:8), just "to be God for the people" (Gen 17:7, 8; 28:21; Lev 11:45; 22:33; 25:38; 26:45; Num 15:41; Deut 26:17; 29:12; 2 Sam 7:24; Ezek 34:24; 1 Chron 17:22) or "to be people for God" (Deut 4:20; 7:6; 14:2; 26:18; 27:9; 29:12; 1 Kings 11:17; Zech 2:15). Therefore, it is not surprising that rabbinic exegesis has interpreted God's

Exod 29:45, this construction communicates Yhwh's intention behind the complex instructions for the construction of Tabernacle and the institution of the sacrificial worship, namely, the divine desire "to be for" the Israelites.

The next verse reiterates what is stated in v. 45 with an intensification. Above all, the Israelites "will know (וידעו) that I (am) Yhwh, their God." The narrative arc of the Exodus, in fact, is triggered by a major complication: first there is a Pharaoh who "had not known" Joseph (1:8) that oppresses the Israelites; then there is God who, on the contrary, "knows"[185] the people's suffering (2:25; 3:7) and then finally there is a new Pharaoh who declares that he does not know Yhwh (5:2). These repetitions create an escalating complication of the plot and increase the suspense. The plot of revelation comes to a resolution when the people (6:7; 10:2, 26), Moses (7:17) and the Egyptians themselves (7:5; 11:7; 14:4, 18; see 8:6, 18; 9:14, 29) know Yhwh through His works.[186] Although the crossing of the sea (Exod 14) is an indisputable revelation of God's power, it does not make up a definitive knowledge of God. This is because the people will continue to "know" God through the gift of "manna."[187] The text of 29:46, therefore, fits into this trajectory and prolongs what has been recounted in this plot of revelation. In fact, the conclusion of all the instructions of Exod 25–29 indicates that worship (and therefore blood manipulation) eventually permits the people to

---

theophany on Sinai and the gift of the covenant as an act of a spousal nature. See *MekhY, Baḥodesh*, § III, p. 306, "And Moses Brought Forth the People Out of the Camp to Meet God. Said R. Jose: Judah used to expound, 'The Lord came from Sinai' (Deut. 33.2). Do not read it thus, but read, 'The Lord came to Sinai,' to give the Torah to Israel. I, however, do not interpret it thus, but, 'The Lord came from Sinai,' to receive Israel as a bridegroom comes forth to meet the bride."

[185] In the Book of Exodus, there are several occurrences in which the verb ידע has God as its subject: God knows in advance that the king of Egypt will let the people go (3:19), even though he is aware that they do not fear him (9:30); God also knows in advance Aaron's qualities (4:14) and reveals them to Moses to encourage him.

[186] Concerning the plot of revelation, see ch. II, § 4. See L. INVERNIZZI, *"Perché mi hai inviato?"*, 171-172. Exod 5:2 introduces a much stronger complication in the Book of Exodus than that of 1:8. In fact, whereas in the case of the first chapter, Pharaoh did not know Joseph, a human character, in 5:2 it is explicitly said that the new king of Egypt does not know Yhwh. If the first lack of knowledge caused the oppression of the Israelites, Pharaoh's ignorance of Yhwh will create much more problems for the people. See also Sarna, 5; Propp, I, 37.272.630.

[187] The root ידע appears three times in Exod 16: Moses and Aaron reassure the people, as follows, "this evening *you will know* that Yhwh has brought you out of the land of Egypt" (v. 6); God himself, announcing the gift of food in the desert, declares that just after eating *"you will know* that I am Yhwh, your God" (v. 12); the last recurrence is v. 16, "The Israelites saw it and said to each other, 'What is it?' for *they did not know* what it was."

progress in the knowledge of Yhwh[188] ("they will know that I Yhwh") and especially of His disposition to enter into a relationship with them[189] (*"their God"*). The excerpt of 29:46 thus reproduces a similar situation to that of Exod 24:3-8.[190] The "actual" union with God, obtained through worship and through the use of blood in the priestly consecration, is accompanied by an "epistemic" consideration ("they will know that I Yhwh," v. 46): the aim of worship is not only an external transformation, but also an interior change of perception and awareness.

Finally, the concluding passage offers further clarifications regarding the content of this new knowledge obtained through worship, "They will know ... that I brought them out of the land of Egypt *in order to dwell* among them." (v. 46). Again, this is not a mere repetition of what has already been said in v. 45, because here, Yhwh specifies an important component of the liberation: the *purpose*. God did not liberate the people simply to take them to another place, but rather He liberated them in order to dwell among them through His presence in the Tabernacle.[191] The verse ends the long divine discourse with a seemingly superfluous phrase, אני יהוה אלהיהם. Yhwh's last words are a self-presentation[192] and a consideration that accentuates once again His "relational orientation"[193] (*"their* God"). After the instructions for the priestly consecration, Yhwh offers Himself to the Israelites and reveals that the purpose of the Exodus was not merely their deliverance from the Egyptians because true freedom can only be acquired through the covenant.[194] The priesthood, established through sacrifice and blood, will be the instrument that will help to seal and deepen this relationship.

---

[188] See Cassuto, 388, "*And they shall know*, through seeing the tabernacle that symbolizes My presence in their midst." See Nepi, II, 175. Durham, with respect to the phrase "I am Yhwh who brought you out of the land of Egypt," states that it is a "autokerygmatic statement" (see 6:2; 20:2); Durham, 397.

[189] The root ידע belongs to the terminology of the covenant and expresses the mutual recognition (Isa 1:2; see Jer 2:8; 4:22); see also H.B. HUFFMON, "The Treaty Background," 31-37; P. BOVATI, "La 'nuova alleanza,'" 201-202.

[190] See the concluding reflections of ch. IV, § 2.2.

[191] See Ibn Ezra, 264; Houtman, III, 553; Priotto, 550.

[192] Regarding the phrase אני יהוה, see ch. III, § 2.2.3.

[193] The pronominal suffix can emphasise the mutual belonging of two parties, see M.S. SMITH, "'Your People,'" 256.

[194] Cassuto, 388-389, "*that I might dwell among them*, in order to make a covenant with them so that they shall be to Me a people, and I shall be to them a God and cause My presence to dwell in their midst." See also Fischer – Markl, 309, "Wie die Begegnung am Sinai das unmittelbare Ziel des Exodus gewesen war ('zu mir,' 19,4), so ist *Gottes bleibende Gegenwart* in Israel das *Ziel des ganzen Geschichte*."

3. The Altar of Incense (Exod 30:1-10)

The place of this passage has generated much discussion, also because of an uncertainty registered at a textual level.[195] The description of the altar of incense, in fact, would have been more appropriate at the end of Exod 26, where the presentation of the Tabernacle is concluded. The present position of this passage in the text, however, is perfectly coherent because the instructions concerning the altar of incense are not only descriptive in character, as is the case with other objects within the Tabernacle (Exod 25–26), but give indications as well for the ritual that may be performed on it.[196] This last detail presupposes the presence of priests (Exod 29). The passage is divided into three parts: the construction of the altar (vv. 1-5), the specification of its location (v. 6) and the description of its usage[197] (vv. 7-10).

> [1] *You shall make* an altar to burn the incense,[198] of acacia wood *shall you make it*, [2] a cubit shall be its length, a cubit its breadth, square it shall be, and two cubits its height, its horns shall be part of it. [3] You shall overlay it with pure gold, its top, its sides around, and its horns; you shall make a gold border around it. [4] *You shall make* two golden rings for it, below its border, on two opposite sides *you shall make* them, and they shall be[199] as places for poles, to lift it with them. [5] *You shall make* the poles of acacia-wood and overlay them with gold.

The altar of incense, also called the "golden altar" (39:38; 40:5, 26; Num 4:11), has long been regarded as a late element of a post-exilic cultic

---

[195] On the assumptions of editorial criticism concerning 30:1-10, see above, ch. V, § 1. The PSam and the manuscript 4QpaleoGen-Exod^m place 30:1-10 in a different position in comparison with the MT, immediately after 26:35, where the description of the Tabernacle ends. See P.W. SKEHAN – E. ULRICH – J.E. SANDERSON, "4QpaleoGen-Exod^m," 53-132.

[196] Cassuto, 389-390; Sarna 193; Dohmen, II, 275; Priotto, 550; see Propp, II, 334.716-717, who considers the position of 30:1-10 in the MT to be a *lectio difficilior* in comparison with that of PSam. See also G. PAXIMADI, *E io dimorerò*, 216-221; especially on page 221 where the author summarises the results of the research by stating that the altar of incense "could not be presented together with the candlestick and the table, because its meaning is completely different: it does not symbolise the presence of God, but man's worship of that presence." (My translation) See also C. MEYERS, "Realms of Sanctity," 33-46.

[197] Regarding the rhetoric of 30:1-10, see G. PAXIMADI, *E io dimorerò*, 212-213.

[198] The hebrew construction מזבח מקטר קטרת means literally, "altar, holder for burning incense." The noun מקטר is a derivative of the verb קטר, "to burn" and modifies the term מזבח, "altar," which has a more general value. The Septuagint simplifies the MT, harmonising the text with 37:25, θυσιαστήριον θυμιάματος, "an altar of incense."

[199] The PSam, the Septuagint, the Peshitta, the Vulgate trantlate the MT with the plural, "and they shall be." The singular can be retained if it is taken as a reference to the whole apparatus described in the verse; see G. PAXIMADI, *E io dimorerò*, 213.

practice,[200] even though archaeology has shown that this type of altar was used also in the pre-exilic period.[201] In accordance with P's classic process of gradual approach to holiness, the materials used for the construction of the altar are of the same kind as those used to build the Holy.[202] In fact, it is made of acacia wood (v. 1; see 25:10, 23, 28; 27:1), its shape is squared and, just as the outer altar, it has four horns[203] (v. 2). Moreover, it is also completely overlaid in pure gold [204] (v. 3; see 25:11, 24). At the centre of this passage, v. 6 creates an explicit reference to 29:42.

Exod 29:42	Exod 30:6
The regular burnt offering throughout your generations at the entrance of the Tent of Meeting, *before* Yhwh, where I will meet (אשר אועד) with you there (לכם שמה) to speak unto you, there.	You shall put it *before* the veil which is above the ark of the testimony, *before* of the propitiatory which is above the testimony[205], where I will meet with you there (אשר אועד לך שמה).

This connection[206] shows that Yhwh compares and relates the two altars. They are in fact located in two places of passage: at the Tent's entrance (29:42) and in front of the veil that separates the Holy from the Holy of Holies (30:6). Through worship, the two altars will allow the people access into the presence of God[207] ("where I will meet with you").

---

[200] Wellhausen was the first exegete to think that the use of burning incense could not be ancient. A plausible justification for this hypothesis was found in Lev 10:1-10 where incense is regarded as a forbidden cultic element. See J. WELLHAUSEN, *Prolegomena*, 67; G. PAXIMADI, *E io dimorerò*, 218-221.

[201] The altar described in 30:1-10 has close similarities to other altars which were common in Palestine, in both the Canaanite and the Israelite periods (see 1 Kings 7:48); see S. GITIN, "Incense Altars," 52-68; ID., "New Incense Altars," 43-49.

[202] See P.P. JENSON, *Graded Holiness*, 101.

[203] See M.D. FOWLER, "Incense Altars," 409-410.

[204] See Propp, II, 380. See also P.P. JENSON, *Graded Holiness*, 103, "Gold is rare, desirable, and very costly, and fittingly represents the dignity and power."

[205] The Septuagint translates v. 6a simplifying the MT, καὶ θήσεις αὐτὸ ἀπέναντι τοῦ καταπετάσματος τοῦ ὄντος ἐπὶ τῆς κιβωτοῦ τῶν μαρτυρίων, "And you shall set it before the veil that is on the ark of the testimony." The MT, however, introduces an apparently superfluous repetition, "before the propitiatory above the testimony" (as in the Peshitta, Vulgate and Targumim). The multiple attestation of the ancient translations prompts the interpreter to maintain the MT, considering also the possibility that the Septuagint can be the result of a haplography; see J.W. WEVERS, *Notes*, 491.

[206] See G. PAXIMADI, *E io dimorerò*, 214-215. Exod 29:42 and 30:6 are at the heart of the rhetorical structure found by the author.

[207] See G. PAXIMADI, *E io dimorerò*, 223.

The golden altar is described in great detail and with overloaded wording, while the allitteration[208] between פרכת ("veil") and כפרת ("propitiatory"), creates a progressive intensification. Yhwh says that the altar will be "before the veil," and then, even more precisely, He specifies that it will be "before the propitiatory," as an essential component of the ark where Yhwh reveals Himself. The reader reminds that it is precisely "from above the propitiatory" (25:22, מעל כפרת) that Yhwh "will meet with" Moses (25:22, ונועדתי) and speak with him[209] (25:22, ודברתי אתך). Thereby, through some gradual clarifications, the gift of the divine presence is again emphasizsed. Yhwh manifests Himself not only through ritual actions, but also through the words addressed to Moses and to all the people.

> [7] And Aaron shall burn fragrant incense[210] on it. Every morning, when he dresses the lamps, he shall make it burn. [8] When Aaron will stare up[211] the lamps at sunset (between the two evenings), he shall burn it, an everlasting (תמיד) incense before Yhwh *throughout your generations*. [9] You shall not offer on it any foreign incense, or burnt offering, or grain offering and libation shall not pour on it. [10] Aaron shall make atonement on its horns once a year, with the blood of the sacrifice for the sin offering for atonement; once a year he shall make atonement on it *throughout your generations*. A most holy reality for Yhwh.

The meaning of incense burning has been explained in different ways. For some authors, the incense screens and protects the High Priest from the divine presence and its equally threatening "otherness."[212] Furthermore, a connection with the cloud that leads the people in the Exodus and manifests the divine presence, cannot be excluded.[213] More likely, when considered in the context of the Ancient Near East,[214] incense was intended to symbolically express

---

[208] The play on words between פרכת and כפרת is noticed by Houtman (III, 560) but not explained.

[209] See Priotto, 551-552.

[210] The expression קטרת סמים, "aromatic, fragrant incense," designates the incense allowed in worship (Exod 40:27; Lev 4:7; 16:12). Dohmen, 399.

[211] When the verb עלה *hifil* alludes to the lamps of the Tabernacle, may indicate the setting of these objects in an elevated position (Exod 25:37; 27:20); see Dohmen, 399. Rashi, however, translates עלה *hifil* "to burn" and refers the enigmatic phrase to the simple lighting of the lamps; Rashi Exod, 265, "*When he lights the lamp*. Literally, when he 'raises' the lamps – when he kindles the lights in order to raise their flame."

[212] See K. NIELSEN, "Incense," 406. Regarding Lev 16:12-13 the author asserts, "This incense cloud provides the high priest with cover against the divine wrath or the divine 'radiation.' The incense smoke gives protection." See W.K. GILDERS, *Blood Ritual*, 123.

[213] Propp, II, 514.

[214] See O. KEEL, *The Symbolism*, 125, which presents a representation of the Temple of Amun at Karnak in Egypt, during the time of Pharaoh Ay, illustrating a priest who offers incense in the sanctuary. This fragrant essence was also used in funerary practices (see *Ibid.*, 64) and was considered as a sign of homage to Pharaoh (*Ibid.*, 75). In Mesopotamia, too,

prayer[215] (Ps 141:2). The gestures performed on the inner altar, then, demonstrate that Israelite worship is not just a "material" ritual, but has an "interior dimension,"[216] because it expresses a desire for communion with God.[217]

However, the reference to the lighting of the lamps made "before Yhwh" (v. 8), recalls what has already been suggested previously (27:20-21), namely, that the perennial light is a sign of the divine presence.[218] Furthermore, the abrupt transition to the second person plural ("throughout *your* generations," v. 8) has a pragmatic strength[219] because it means that the gestures performed by Aaron every morning and every evening have an actual relevance in the life of the Israelites. At dawn and dusk, in fact, the Israelites will remember the wonders of the liberation through the daily sacrifice, while, on the other hand, they will recall the proximity of God through the lighting of the lamps. Furthermore, the reference to future generations (v. 8) enables the reader to identify in the daily service a powerful instrument for the transmission of faith from parent to child.[220]

No "foreign incense" (v. 9), namely, an essence that does not conform to the proper formula, should be offered on the altar[221] (see 30:34-37). As a result, this verse disqualifies all improper usages of the golden altar, such

---

perfumed smoke was part of the daily service in various temples: Nielsen reminds that a golden altar for incense was erected in the temple of Marduk Esagila during the reign of Assurbanipal; incense, in this case, had an expiatory value (*LAR*, II, 385-386; see also *LAR*, II, 38). See K. NIELSEN, "Incense," 406.

[215] See G. PAXIMADI, *E io dimorerò*, 224; see also Rev 8:3-4; J. MILGROM, *Leviticus 1–16*, 238. The burning of fragrant incense to the "Queen of heaven" (Jer 44:17) is a symbolic action performed to manifest the request for fertility and well-being.

[216] See G. PAXIMADI, *E io dimorerò*, 224.

[217] The bride of the Song of Songs appears in the desert clothed in a divine halo and the choir asks "who is she who rises from the desert like columns of smoke, perfumed with myrrh and frankincense (מקטרת מור ולבונה), of every merchant's aroma?" (3:6). Rashi, notices the link between this passage and the practice of the daily sacrifice in the inner altar of incense, "because of the cloud of incense which would rise straight up from the inner altar (על שם ענן הקטרת שהיה מתמר מעל מזבח הפנימי)."

[218] See above, ch. V, § 2.1.

[219] See ch. III, § 2.2.1.

[220] See ch. III, § 2.2.4. The occurrences of לדרתכם create a common thread in the book of Exodus. The instructions for the Passover celebration (12,14, 17, 42) should be transmitted to the following generations, and some of the "manna" should be preserved for posterity (16:32-33). Moreover, the sons will be educated through ritual gestures: the rite of the Menorah (27:21), the daily burnt offering and incense offering (29:42; 30:8, 10), the washing of the priests' hands (30:21), the anointing oil (and thus the priesthood, 30:31; see 40:15), the Sabbath (31:13, 16).

[221] See P.P. JENSON, *Graded Holiness*, 109-110.

as the unsuitable frankincense[222] or its use for cultic actions reserved for the external altar (burnt offering or libation). The only exception to this rule is the extraordinary sacrifice on the Day of Atonement, the *Yom Kippur*[223] (v. 10). During this solemn celebration, blood will be applied to the four horns of the altar and a special sin offering will be sacrificed in order to atone[224] (כפר) for the sins of all the people (see Lev 16:18-19). For this reason, the golden altar is defined in v. 10 as a "most holy reality" (קדש קדשים) just like the external altar[225] (29:37), emphasising once again the connection between the two altars.

Aaron, therefore, will act as an intermediary of reconciliation[226] between Yhwh and the people through this special rite that is performed once a year. The ritual will be described in more detail in Lev 16 and a comparison between Lev 16 and Exod 30:10 shows that in Exod the instructions refer exclusively to the blood manipulation upon the altar[227] (Lev 16:18-19) and omit the action performed in the Holy of Holies before the propitiatory (Lev 16:14-15)[228]. In both cases, the application of blood has a precise ritualistic meaning: it makes possible the re-establishment of the relationship between Yhwh and the people.[229] In Exod 30:10, the altar of Lev 16:18-19 is clearly identified with the

---

[222] The religious use of incense, very popular in the Ancient Near East, is questioned in v. 9. K. NIELSEN, "Incense," 405-406; Sarna, 194-195; Propp, II, 475. See also Lev 10:1-7.

[223] Regarding the *Yom Kippur*, see TH. HIEKE – T. NICKLAS, ed., *The Day of Atonement*.

[224] Concerning the root כפר, see ch. I, § 1.3.

[225] See above, ch. V, § 2.6.

[226] See ch. I, § 1.3, and the discussion on the construction כפר + על + altar / horns.

[227] After sprinkling the propitiatory with the blood of the bull and the blood of the goat (16:14-15), Aaron must sprinkle the horns of the incense altar with the blood of the bull and the blood of the goat seven times, to purify the altar and sanctify it from the impurities of the Israelites (16:18-19).

[228] The High Priest sprinkles seven times the blood of the bull with his finger upon the propitiatory (Lev 16:14), then, after having slaughtered the goat, he brings its blood beyond the veil of the Holy of Holies and sprinkles it over the propitiatory (16:15). The text then does not specify if the ritual is performed by the High Priest in the Tabernacle, but simply says, "and he shall make atonement for the Holy Place from the impurities of the Israelites and from their rebellions for all their sins. And so, he shall do for the Tent of Meeting..." (16:16). Moreover, the passage does not reveal precisely what the blood manipulation that Aaron will have to do involves, and normally scholars think that it is the action described in Lev 4:5-7, 16-18; the rabbinic tradition believes that he should do one sprinkling of blood upwards and seven downwards (*mYoma* 5,4). See W.K. GILDERS, *Blood Ritual*, 125.

[229] See W.K. GILDERS, *Blood Ritual*, 124. He explains the first sprinkling of blood in Lev 16:14-15 as follows, "every blood manipulation in the cult, whatever else it may be said to do, indexes a relationship between Yahweh, the priestly mediator, and the lay individual or the community of Israel."

altar of incense inside the Holy[230] and, after describing the movement that proceeds from the innermost part of the sanctuary (Holy of Holies, 16:14-15) to the Holy (16:18-19), the attention is directed to the final gesture. The special importance given in Exod 30:10 to "atonement" indicates that the sacrifice made is a חטאת, and as such, intends to deal with the threat that sin represents for worship and seeks to recuperate the union with God.[231]

Finally, the passage ends with a significant repetition of the expression לדרתכם, "throughout your generations," turning the Day of Atonement in one of those decisive gestures which will have the task of transmitting the memory of God's deeds to future generations. In fact, in the rite of *Yom Kippur*, the Israelites will have to remember each year that their relationship with God is based on His forgiveness and on His extraordinary willingness to reconcile. Similarly, both Exod 29:12 and 30:10 agree that blood is the ritual instrument that can restore the communion with God. The sequence of instructions and the use of the term דם in 29:12, 16, 20-21 and in 30:10, then, bring about an important development: what happens on the Day of Atonement becomes an extension of Aaron's consecration. In fact, thanks to the mediation of the priests, the possibility of being in relation with God is extended to the people (see Lv 16:16) through a ritual that, precisely through a blood manipulation, achieves the atonement for sin. This mention is particularly significant for the Book of Exodus, precisely because the reader, recognising in this passage the divine readiness for mercy, is prepared beforehand for what will be recounted in chapters 32–34 regarding the sin of the golden calf. Before the people's transgression (see 20:3-5; 32:1-4), Yhwh reveals that worship will offer the possibility of atonement and reconciliation through blood manipulation.

Conclusion

In the first paragraph, it was acknowledged that, apart from a few brief additions, Exod 29 belongs to the earliest layer of the P-document. The passage dedicated to the golden altar (30:1-10), on the contrary, is probably late. The *lectio cursiva* showed that the sequence of the elements

---

[230] The identification of the altar described in Lev 16:18-19 as the altar of incense, claimed without hesitation by rabbinic tradition, is considered by many as the result of a late reinterpretation. See W.K. GILDERS, *Blood Ritual*, 126.

[231] A. MARX, *Les systèmes sacrificiels*, 123-124. The first part of the *Yom Kippur* has a "negative" character since it eliminates sin and impurity in two phases: the atonement in the sanctuary (vv. 14-19) and the expulsion of sin to an uninhabited place (vv. 20-22). The second moment, on the contrary, is "positive," it recreates the relationship with God.

mentioned in 29:1 – 30:10 is logical: the institution of priesthood (29,1-35) must precede the consecration of the altar (29:36-37), the description of the daily sacrifice in the outer altar (29:38-42) and the explanation of the rites performed in the inner altar (30:1-10).

The blood manipulations described in Exod 29 have two fundamental purposes, expiation and consecration, and both are aimed at establishing a special relationship between Aaron, his sons and Yhwh. The consecration is a rite that Moses will do for the benefit of the priests so that they will exercise the priesthood in communion with God (v. 1, לכהן לי). They will be "presented" to Yhwh (קרב *hifil*, v. 4 and v. 8) in order to live their priesthood as a true "offering" of themselves to Him (see קרב in v. 3).

The consecration of Aaron and his sons, therefore, takes place with a ritual that consists of a dynamism in three stages, each characterised by a different blood manipulation. In the bull sacrifice (29:10-14), Moses must recognise that sin is an objective threat to the relationship with God through the sin offering. In the sacrifice of the first ram (vv. 15-18), the ritual manifests the desire of offering to God a perfect and complete gift, a "burnt offering for Yhwh" (v. 18). Only at this point, can he celebrate the actual consecration with the blood previously shed on the altar. With the immolation of the second ram (vv. 19-25), the blood shed on the altar is used to anoint the lobe of the right ear, the thumb of the right hand and the big toe of the right foot of Aaron and his sons (v. 20), bringing the touched limbs to a special proficiency. The sprinkling of blood mixed with oil, then, performs a "sanctification" (v. 21) that separates the priests from the "worldly sphere," to unite them more closely to God.

The sanctification of the altar (vv. 36-37) and the institution of the daily sacrifice (vv. 38-42) describe some of the essential elements of the future ministry of the priests in relation to the altar which will have the task of perpetuating the memory of the divine deliverance. Worship, in this way, will enable the people to know Yhwh more deeply (v. 46) and thus discover that Yhwh freed the Israelites not only to move them to another residence, but mainly to dwell among them, to manifest his closeness to them.

The analysis ends with 30:1-10, where the instructions for the construction, the consecration and the use of the golden altar are found. The most decisive aspect for this research, however, is precisely v. 10 in which the blood manipulation of the annual feast of the *Yom Kippur* is quickly described. The blood that was used in a complex series of rites to seal the special union between Yhwh, Aaron and his sons, will be used once a year, precisely through the mediation of the priests, to atone for guilt and thus rebuild the relationship between Yhwh and the people.

GENERAL CONCLUSION

At the end of this study, the main contributions of this monograph will be briefly summarised. First of all, the whole study will be re-read, focusing on two narrative and literary aspects: the meaning of the episodes' order and their relationship with the literary *topos* of the "return of the hero." Then, a reflection on two theological consequences of this study will be presented.

1. Narratological Considerations

1.1 *The Sequential Order of the Episodes*

The close reading of the episodes in which a blood ritual is mentioned has shown that the succession of scenes is significant for the meaning of the narration as a whole. The first episode analysed is Exod 4:18-31, a passage which recounts Moses' return to Egypt after the calling received in 3:1 – 4:17. During the night, Yhwh assaults Moses in a moment of particular fragility (4:24; ch. II, § 4.3). Moses finds himself alone, confronted with a mortal danger, but manages to survive thanks to Zipporah's rapid and solicitous intervention, which saves his life through a ritual of circumcision (4:26; ch. II, § 4.4). As a result, the blood ritual transforms Moses' condition so that he can become Zipporah's "bridegroom of blood" (4:25-26).

Blood, then, is also a determining element for the Israelites' celebration of Passover (12:7, 13, 22-23). Moses makes explicit the association between what he experienced (4:24-26) and the divine indications (12:1-20), in the speech he addresses to the elders and to the people (vv. 22-23; ch. III, § 2.3.1). As he reports Yhwh's words, he makes a significant lexical choice: whereas Yhwh said to "give" (נתן, 12:7) the blood on the doorposts and on the lintel, he says to "apply it" (נגע, 12:22), employing the same verb as in 4:25. Moses, therefore, interprets Yhwh's instructions (12:7, 13) in the light of his own personal experience and, then, the passage of 4:24-26 becomes a real key offered to the people to interpret the paschal blood manipulation. The experience of salvation lived by one individual is thus shared by all the families of the Israelites.

The blood plays a central role in the stipulation of Yhwh's covenant with the people on Mount Sinai (24:3-8; ch. IV, § 2.3 and § 2.4). For the second time, all the Israelites become witnesses of its special virtue. In this case, however, the Israelites are not engaged exclusively on a personal level, but they are involved as a people, gathered before Yhwh. Even in this circumstance, as happens during the "night of Moses," the new bond is sealed through a "cut" (כרת, "to cut" is employed in 4:25 and in 24:8), namely a ritual which, through blood, creates a close and irrevocable bond.

Finally, the blood is decisive during the consecration of Aaron and his sons (29:12, 20-21; see ch. V, § 2.3 and § 2.5). The future priests are "separated" from the people in order to take care of the daily sacrifice at the altar of burnt offerings (29:36-42; ch. V, § 2.7) and of the rituals that take place at the altar of incense, especially the annual *Yom Kippur* (30:10; ch. V, § 3). This special destination may have an explanation, which will become clearer in the next paragraph: since the proximity with blood exposes the person to a dangerous threshold between life and death, its manipulation is limited to those who have enjoyed a special proximity with it by means of the consecration.

Therefore, the blood manipulation, which marked the life of Moses and that of the people, after Exod 29 is reserved especially for Aaron and for his sons. The narrative sequence of the episodes examined, then, shows that Yhwh's action is often conveyed by blood and that, at the end of the Book, He multiplies and, at the same time, limits the number of mediators.

1.2 *Blood and the "Return of the Hero"*

Comparison with contemporary scholars,[1] allowed the reader also to recognise an allusion to a common *topos* of universal literature which recurs, with original features, in some of the texts analysed: the journey and in particular the *nostos*, namely, the hero's return home.[2]

Moses' condition is that of a person exiled "twice." Far from the land of his fathers, brought up in Egypt (Exod 2:1-10), when he is reached by God's call, he finds himself in Midian; his identity is "broken,"[3] son of the people of Israel, adopted by an Egyptian woman and married to a Midianite. Moses, therefore, embodies the archetype of the wandering man

---

[1] See ch. II, § 3.1.1, note 85.

[2] E. Di Rocco, *Raccontare il ritorno*, 8.

[3] The similarity between Moses and Odysseus is striking. Moses has a "broken" identity, Odysseus is a character with various facets, "Sings to me, O Muse, the multiform (πολύτροπον) man, who indeed wandered much after destroying the sacred city of Troy" (Homer, *Odyssey*, I, 1-2). See P. Boitani, *Riconoscere è un dio*, 63-73.

and his journey is explicitly defined as a *"return"* (the verb שוב is frequent in 4:18-21; ch. II, § 3.1.1 and § 3.1.2).

In literature, the wanderer's journey is often characterised by the surprise of a *trial*.[4] Moses, in turn, is attacked at night, when he is most vulnerable (4:24; ch. II, § 4.3). God's envoy, who could leave because those who sought his life were dead (בקש, 4:19), is confronted by the same God who encouraged him to go, and it is Yhwh Himself who "tries" (בקש, 4,24) to kill him. The location is "liminal," between life and death, the context resembles the scenario of a rite of passage.[5] Zipporah intervenes with a "circulation" of blood (4:26; ch. II, § 4.5) which turns Moses into a "bridegroom of blood." Since the term חתן indicates a kinship acquired by legal means (through marriage) and not a carnal bond, Zipporah's words express Moses' new incorporation into the family group (4:25; ch. II, § 4.4) and therefore the conclusion of the rite of passage: the blood ritual facilitate the crossing of the "liminal" area.

The narrator then reveals that Aaron, following a very similar itinerary to that of Moses (לך, 4:27; see ch. II, § 5.2), is led into the desert and meets his brother in this "liminal" space. The root פגש (4:27), already employed to designate Yhwh's aggression (4:24), confirms the "theological" character of the reunion: in the past Yhwh confronted Moses during the nocturnal attack, now He meets him through Aaron. The reunited brothers, therefore, arrive together to the Israelites, concluding the expedition (וילך, 4:29; ch. II, § 5.3).

The trial of Moses and his journey continue in the solemn march of the Israelites narrated in the text of Exod 11:1 – 12:42. In effect, the setting in both cases is nocturnal (11:4; see 4:24; ch. III, § 1.2), since the lamb must be slaughtered "between the two evenings" (12:6; see ch. III, § 2.2.2) and the term לילא is repeated three times (in 12:29-31; ch. III, § 3.1). Each family, moreover, must perform, just as Zipporah, a blood manipulation (12:7), spreading it on the vulnerable threshold of the house in order to face the test of Yhwh's threatening presence (ch. III, § 2.2). This narrative is thus similar to the rite of passage experienced by Moses because, through blood, the people can escape the mortal danger (see 12:33; ch. III, § 3.1) and cross the border between life and death.

---

[4] See H. FISCH, *Remembered Future*, 3. The myth of the wanderer is a "universal" type. During the quest, the hero often must face a test, "The hero's struggle, his disappointments, his death, and his final success or recovery are all comprised in this archetype of archetypes."

[5] See ch. II, § 4, note 290.

So far, the lexical and thematic connections with the archetype of the "hero's return home" are various: the journey, the trial (and the rite of passage), the itinerary which leads towards new incorporation. Exodus, however, proposes this literary *topos* from an original perspective. In fact, even though the Book does not end with people's entry into the land, in several passages the destination seemed to have been reached, as happens for example in 19:4, "You yourselves have seen what I did to Egypt, I lifted you up on eagle wings and brought you *to me*." The Israelites do not yet enter the land, but, somehow, they reach their destination, Yhwh Himself.[6] This particular angle of Exodus has been considered in the fourth chapter.

The dialectic between the "movement" and the arrival, also appears in Exod 24:1-18. The close reading has shown that 24:3-8 is connected with 4:24-26 and 12:7, 13, 22-23. The presence of God on Sinai constitutes a similar danger to that experienced by Moses (20:19; see ch. IV, § 2.5), and the covenant is established by means of a new rite of passage (ch. IV, § 2.3) whose aim is the relationship with God, sealed by the circulation of blood (ch. IV, § 2.4a and § 2.4c-d). Nonetheless, after the covenant with God is stipulated, the Israelite's expedition is not resumed, since the narrative describes Moses' gradual ascent to the top of the mountain (ch. IV, § 2.6 and § 2.7).

In the last part of Exodus, Moses and the people do not move, but God's envoy receives the instructions for the construction of the sanctuary and the people complete the work (Exod 25–40). However, the same chapter 29 ends with words that summarise and relaunch the opposition between the "return" and the end of the flight, "they will know … that I brought them out of the land of Egypt to dwell among them" (v. 46). God did not free the people only to take them elsewhere, but to dwell with them, to be "their God" and to create the possibility for a deeper "knowledge" of Him.[7] In Exod 29, moreover, elements proper to the rite of passage have been recognised. This consideration enables the reader to grasp once again the "dynamic" character of this section. The priestly consecration, in fact, is a ritual which clearly

---

[6] See e.g., the allegorical interpretation of Gregory of Nyssa, who considers Moses' experience on Sinai as a progressive approach to the knowledge of God. GREGORIUS NYSSENUS, *De vita Moysis*, II, 167, "First he (Moses) left the roots of the mountain …; then he welcomes in the ear the voices of the trumpets higher up (εἶτα δέχεται τῇ ἀκοῇ τὰς τῶν σαλπίγγων φωνὰς τῷ ὕψει τῆς ἀνόδου συνεπαιρόμενος). Then he penetrates the invisible sanctuary of God's knowledge (ἐπὶ τούτοις εἰς τὸ ἀόρατον τῆς θεογνωσίας ἄδυτον παραδύεται), but he does not remain here either and passes into the tent not built by human hands (ἀχειροποίητον μεταβαίνει σκηνήν): in fact, it is only at this point that those who have elevated themselves through such ascents reach the *frontier* (τῷ ὄντι γὰρ ἐπὶ τοῦτο καταντᾷ τὸ πέρας ὁ διὰ τῶν τοιούτων ἀνόδων ὑψούμενος)."

[7] See E. DI ROCCO, *Raccontare il ritorno*, 80; P. BOITANI, *Riconoscere è un dio*, 221-222.

marks a detachment from a previous condition, through the sin offering (חטאת, 29:10-14; ch. V, § 2.3) and the burnt offering (29:15-18; ch. V, § 2.4), and then, with the last sacrifice, fulfills the incorporation of Aaron and his sons in their new condition (29:19-21; ch. V, § 2.5).

Therefore, during the return journey, when Moses and the people approach the threshold of death, the blood enables them to cross the dangerous border. The ritual blood is then entrusted to the priests which will have the ministry to perpetrate this "passage" throughout the future generations. Concluding, the "path" of Exodus does not end in a "material" goal, but is accomplished with the revelation of Sinai and the construction of the Tabernacle. This consideration prepares the subject of the next paragraph, the relationship between blood and "word."

2. Anthropological and Theological Consequences

2.1 *The Use of Blood and the Speech Acts*

The previous considerations facilitate some remarks on an element that has emerged several times in the close reading of the passages studied. Philosophical speculation has long highlighted the circular tension between action and words. Action has a symbolic structure that contributes to turn it into an object of interpretation; for this reason, just as a word, every deed has a meaning which must be deciphered.[8] On the other hand, the words exchanged between people do not only convey content, but can have a performative and active character[9] creating an act of communication in which the locutor offers himself to the other person and demands from him attention and an open ear.[10]

---

[8] Regarding "action" Ricoeur said as follows, "La thèse est qu'il n'est pas d'action humaine qui ne soit déjà articulée, médiatisée, interprétée par des symboles .... L'action n'est pas à proprement parler symbolique puisqu'elle est réelle, mais elle est structurée par des symboles" (P. RICOEUR, "La structure," 31). As every symbol, each singular deed necessarily requires interpretation, "une interprétation qui respecte l'énigme originelle des symboles, qui se laisse enseigner par eux, mais qui, à partir de là, promeuve le sens, forme le sens" (ID., *Philosophie de la volonté*. II, 325-326).

[9] J.L. AUSTIN, *How to Do Things with Words*; J. SEARLE, *Speech Acts*. See E. LÉVINAS, *Totalité et Infini*, 189, "Le langage *effectue* l'entrée des choses dans un éther nouveau où elles reçoivent un nom et deviennent concepts, première action au-dessus du travail, action sans action, même si la parole comporte l'effort du travail, si pensée incarnée, il nous insère dans le monde, dans les risques et les aléas de toute action."

[10] Referring to Lévinas, Gilbert presents the distinction between the word used as a vehicle for the transfer of notions (thematisation) and the discourse which introduces personal exchange (communication), "Thematisation orders and objectifies our knowledge

This structural correspondence between "word" and "action" is recognisable even in the Book of Exodus. The liberation, which will be fulfilled with the entry into the land of Canaan, has an external and "active" character. According to this aspect, God redeems the people from the condition of slavery and oppression, and guides them towards the promised land. On the other hand, the very structure of the biblical book, highlighted in the previous paragraph, shows that a crucial objective of the journey of Exodus is precisely Mount Sinai. Thus, while Yhwh frees His people from slavery (plot of resolution), at the same time He promotes their freedom with the gift of His words (plot of revelation). Therefore, the Israelites will not only "see," but they will "know" that Yhwh brought them out of Egypt (Exod 6:7), namely the act of liberation will be revelatory and Yhwh's words uttered on Mount Sinai will become an actual principle of Israelites' freedom.[11]

In the texts analysed a similar interaction appears on several occasions: the ritual which includes the blood manipulation has an "active" efficiency and creates a new "ontological" condition, but this effectiveness is only achieved by involving the mediation of language.

The circumcision done by Zipporah averts the threat, Yhwh withdraws (4:26; ch. II, § 4.5) and transforms Moses "physically" in a "bridegroom of blood," but this ritual action is not extraneous to the "domain of the word." Primarily, Zipporah pronounces a sentence that performs a declarative illocutionary act and confirms her husband's new status ("bridegroom of blood you are to me," ch. II, § 4.4). Moreover, although Moses has already decided to leave (4,18; ch. II, § 3.1.1), Yhwh's act of speech is required for this resolution to be fulfilled (4:19; see ch. II, § 3.1.2).

The concatenation between the narration of the plague (proclamation, 11:1-10, and execution, 12:29-34) and the instructions for the Passover (12:1-28) shows once again how much the law, and therefore the word, plays a crucial role in the development of the divine plan. Obedience to the divine instructions (12:28) in fact is critical for the resolution of the complication. God asks Moses and the families of the Israelites to perform an apotropaic gesture of blood manipulation (12:7; see ch. III, § 2.2.2) in order to expel chaos and evil (see Ezek 45:19), so much so that in Moses' discourse (12:21-27), he reports above all the instructions concerning the blood (ch. III, § 2.3.1). Therefore, Moses himself defines the ritual

---

according to the canons of the contemporary science. Communication lies at the origin of thematisation. There is no word or discourse that does not arise from an exchange .... In communication, objectification is experienced as a desirable good, desired for the exchange with the other." (My translation of P. GILBERT, *Pazienza d'essere*, 143).

[11] P. BOVATI, *Parole di libertà*, 98.

precisely on the basis of the tension recognised so far (12:24-25; see ch. III, § 2.3.2).

> ²⁴ You shall observe *this word* (את־הדבר הזה) as a statute for you and your children always. ²⁵When you enter the land that Yhwh will give you, as He has said (כאשר דבר), you shall observe *this service* (את־העבדה הזאת).

Passover is both a "service," a ritual action, and a "word," namely, an act of communication. This particular dimension becomes concrete immediately afterwards thanks to the excerpt of 12:26-27 in which there is a potential dialogue between parent and child. The young one risks to interpret the rite as empty gestures imposed by others and thus remain as a distracted spectator. Questioning the parent about the meaning of the rite (12:26; ch. III, § 2.3.3), the child would ask him to exercise his care, and the father would educate his son through his own testimony and his words.

2.2 *The Blood Oath*

The connection between "blood" and "speech act" appears most visibly in ch. 24. The covenant rite begins with the transmission of what God has said to Moses (24:3; ch. IV, § 2.2). Challenged by this announcement, the people answer Moses, and the narrator reports the Israelites' words ("all the words that Yhwh has spoken, we will do," 24:3), presenting their promise as an illocutionary commissive act (ch. IV, § 2.2). The communication is fulfilled in the act of writing (24:4a) which fixes what has been said into a document.

At this moment, the plot is interrupted, Moses does not read the book, but prepares sacrifices (24:4b-5; ch. IV, § 2.3), expressing "materially" the intention to involve the people, represented through the *stelae* which symbolise the twelve tribes. Afterwards, the ritual continues with a complex blood manipulation (24:6; ch. IV, § 2.4a) which is interrupted by the proclamation of the words that Moses had written in the book (24:7; ch. IV, § 2.4b). Yhwh makes Himself "physically" present in the ritual through the blood, but leads the ritual even with His written and proclaimed word.

The exchange achieves its purpose and the people formulate another illocutionary commissive speech act (24:7; ch. IV, § 2.4b), an even more radical promise: they commit themselves with a statement which inverts the usual order of elements. By saying "we will do and we will listen," they ensure the reading of the book in future generations and, at the same time, they guarantee that the bond with God will be deepened through a continuous listening ("we will listen") and not only by means of a "practical" activity ("we will do").

The connection between the "material level" (blood) and the "speech act" is evident even in vv. 7-8. The repetition of the term ברית generates an interesting dialectic: the blood fulfills the relationship "materially" ("blood of the covenant," 24:8), but it cannot accomplish this purpose without the "word" ("on the basis of all these words," 24:8; ch. IV, § 2.4c). In contrast, the word proclaimed ("book of the covenant," 24:7) does not simply contain a list of notions, but performs an action and challenges the listeners. Therefore, a relationship between the covenant of Exod 24:3-8 and the institution of the oath is possible (ch. IV, § 2.4d).

An oath is in fact an act of speech which guarantees that the word spoken is truthful and, at the same time, that the speaker will not violate the promise uttered through his words.[12] Comparisons with studies devoted to this form of commitment in the Greek world and the Ancient Near East have shown that the oath was often complemented with imprecatory gestures which had the purpose to curse the perjury.[13] In this context, sacrifices that included blood manipulation were certainly performed (ch. IV, § 2.4d). Such a background can shed some light on Exod 24:3-8, since this covenant actually resembles an oath.

The blood shed after the promise, then, suggests the irrevocability of the covenant and insists on the irreversible character of the ritual actions. Through the circulation of the blood from the altar (24:6; ch. IV, § 2.4a) to the people (24:8; ch. IV, § 2.4c), the manipulation fulfills a communication of life between Yhwh and the Israelites that creates a bond comparable to that which unites "blood relatives."

Finally, at the end of Exod 29, Yhwh states precisely as follows, "I will meet with you there, to speak to you there" (29:42). Thereupon, the same tension identified in the previous texts is found, since Yhwh shows that worship will not consist only of external and mechanical actions (the sacrifice), but will be based on a word communicated to Moses and intended for the people (ch. V, § 2.7). Therefore, the priestly consecration, thanks to three blood rituals, has some aspects in common with a solemn promise, since every oath is somehow a *sacratio*, a consecration of the person to the word uttered.[14]

---

[12] G. AGAMBEN, *The Sacrament of Language*, 7.18.

[13] In the Greek world, oaths could end with a blessing for the faithful person and a curse for perjury (C.A. FARAONE, "Curses and Blessings," 140-158). See J.S. ANDERSON, *The Blessing and the Curse*.

[14] See G. AGAMBEN, *The Sacrament of Language*, 29. The oath is "a form of *sacratio* .... In both cases a man was rendered *sacer*, that is, consecrated to the gods and excluded from the world of men." Agamben refers to Benveniste who, considering the latin term *sacramentum*, believes that it "implies the notion of making '*sacer*.' One associates with the

3. Blood and the Covenant

3.1 *The Terminology of the Covenant*[15]

In this first paragraph the references (explicit or indirect) to the covenant are recalled. In Exod 4:18-31, the reader cannot forget that Moses is saved by circumcision (4:26) and that this practice is explicitly defined as the "sign of the covenant" (Gen 17:11-14; ch. II, § 4.4 and § 4.5). In addition, Moses is also spared from death thanks to a "cutting" of the foreskin, which is described with a characteristic term that belongs to the semantic field of the covenant (כרת, 4:25; ch. II, § 4.4). Similarly to what happens through circumcision, in fact, the covenant is ritually achieved through a "cut" (כרת, 24:8; ch. IV, § 2.4c), a "wound" that limits the contracting party, creating a permanent obligation, but that, at the same time, is vital and fecund.

In the Passover narrative, moreover, the covenant is indirectly recalled several times, mostly when the Passover is qualified as an action that people will perform "for Yhwh." The expression ליהוה in fact is correlated to חג, "feast," (12:14; 13:6), פסח, "Passover" (12:11, 27, 48), and is employed twice in 12:42, "That was a night of vigil *for Yhwh*, to bring them out from the land of Egypt: that is this night *for Yhwh*, vigil for all the Israelites throughout their generations" (ch. III, § 3.2). Thereby, the Passover is characterized by a relational orientation (ch. III, § 2.2.4) and by a reference to God (ch. III, § 2.2.2). During the night, when Yhwh "goes out" with his threatening presence, it is precisely the sign of blood on the door (12:13; ch. III, § 2.2.3) that indicates the special relationship between God and the family who is celebrating Passover and that prompts the exterminator to spare the lives of the people in the house.

In the passage of Exod 24:3-8, the covenant is mentioned explicitly and the term ברית is used twice (24:7-8) creating a close correlation between the proclamation of the written text, the "book of the covenant" (ספר הברית, 24:7; ch. IV, § 2.4b) and the sprinkling of the people with the "blood of the covenant" (דם הברית, 24:8; ch. IV, § 2.4c).

Conclusively, even though the term ברית does not appear in Exod 29, the priestly ministry alludes to the semantic field of the covenant (see ch. V, § 2.1). In effect, the use of the verb זרק (29:16) to indicate the blood manipulation creates a close lexical connection with the covenant ritual

---

oath the quality of the sacred, the most formidable thing which can affect a man: here the 'oath' appears as an operation designed to make oneself *sacer* on certain conditions" (É. BENVENISTE, *Dictionary*, 444).

[15] See c. II, § 3.2.2.

described in 24:6, 8, where the same verb is employed. In this way, the two ritual actions are ideally associated. The consecration of Aaron and of his sons, then, resembles the irreversible and vital covenant stipulated by God with the people. Finally, the third sacrifice of this ritual (29:19-21) is based on a circulation of blood that is firstly spread onto the altar and then sprinkled on Aaron and on his sons to consecrate them and unite them to God, just as is seen in 24:6-8 (ch. V, § 2.5).

3.2 *The Covenant's Structure: Origin, Law and Blessing*

Scholars have long recalled how the biblical covenant is not simply a concept, but first and foremost a structure, in which the different elements can be mentioned separately. The covenant is not reduced only to the notion of obligation,[16] but discloses an actual history that includes different elements: an origin in which the gratuitousness of God's gift is manifested, the laws which establish the rights and the duties of the contracting parties, and, at the end, the promise of blessing or the threat of curse for those who infringe the bond.[17]

The texts analysed have repeatedly shown that it is precisely through the ritual sign of the shed blood that the dialectic between the priority of the divine gift and the consequent participation of the Israelites is expressed. Yhwh's discourse to Moses (4:21-23) during the journey of his return to Egypt, shows that the salvation of the people has a remote origin. Certainly, Moses' call (3,1 – 4,17) is the expression of God's reaction to the people's cry (2:23-25; 3:7). Nonetheless, Yhwh Himself specifies that his intervention is based on a primordial relationship. In fact, Yhwh reveals to Moses that the people are His first-born son (ch. II, § 3.2.2), employing a lexicon used elsewhere to express the covenant between two parties (see 1 Kings 16,7). The divine gift, then, precedes and constitutes the covenant with Israel as a (blood) bond that can never be canceled.

On the other hand, the circulation of blood between Moses and his son alludes to the conditional structure of the covenant (ch. II, § 4.5). In Genesis, in fact, it is clear that circumcision is an imperative and therefore an essential requirement (Gen 17:10) to obtain blessing and fruitfulness (Gen 17:2, 4, 6; see ברך in 17:16, 20). The night of Moses (4:24-26)

---

[16] See P. BEAUCHAMP, "Propositions sur l'Alliance de l'Ancien Testament," 60-61, "Le texte de Alliance commence par le passé du prologue historique: comme il y a un passé de bienfaits, il peut y avoir un avenir de bénédiction, mais non par distribution garantie, car tout dépend du maintenant, comme réactualisation de l'origine".

[17] See P.R. WILLIAMSON, *Sealed with an Oath*, 19-34.

therefore proposes again the typical covenant tension between "gift" and "law." The haste and the mystery that cloaks the narrative suggest that the action has in God's grace its preponderant component. Nonetheless, Moses is saved only after the circumcision of his son, therefore after the obedience to a condition necessary to become a member of the people.

The liberation of Passover is repeatedly defined as the exclusive fruit of God's initiative, and it is precisely this salvific event that will represent the founding moment of the covenant on Sinai (Exod 20:2, "I am Yhwh, your God, who brought you out of the land of Egypt"). The divine intervention is original and surprising, and this aspect is accentuated in 11:4 through the root יצא — "I myself am going out in the midst of Egypt" (11:4; ch. III, § 1.2) — and in 12:12-13 with the root עבד: Yhwh will "pass over" Egypt to execute His sentence and to strike the first-born sons of the Egyptians (ch. III, § 2.2.3; see § 2.3.1). The deliverance is so certain that in 12:17 it is already given as a past event, "this very day I brought your hosts out of the land of Egypt" (ch. III, § 2.2.5).

However, Yhwh's instructions in Exod 12 are manifold and detailed (see ch. III, § 2.2.1 and 2.2.2) and the divine gift of deliverance is not without relation with the law. The preparation of the Passover, in fact, commits the Israelites to an obedience that makes the attention to detail an essential cornerstone of the celebration (12:7, 13, 22-23; see ch. III, § 2.3.1). The Israelites must express their tension for the departure even wearing a specific clothing (12:11 girded hips, sandals on their feet, staff in their hands; ch. III, § 2.2.2). The blood manipulation, then, is presented as the most significant action of the ritual, because only the instructions related to the blood are repeated by Moses with various specifics (12:22-23). Finally, the close reading has demonstrated how it is precisely the obedience to Yhwh's indications that allows the Israelites to overcome the danger represented by the exterminator and to leave the land of Egypt.

The text of 24:3-8 emphasises the divine origin of the covenant stipulated, especially through the ritual actions. Indeed, the blood that marks the altar and circulates on it shows precisely that the source of the new bond and the life offered to the people has its origin in God (24:6; ch. IV, § 2.4a). Similarly, this priority of God's grace is communicated through the repetition (24:3) and the proclamation of Yhwh's words, which manifest, even more explicitly, the priority of the divine initiative: it is God who "calls" the people and promotes them to a relationship with Himself (ch. IV, § 2.4b). Even in this narrative, however, the participation of the people is necessary, since it is precisely their double assent (24:3, 7; ch. IV, § 2.4b) that brings the rite to its conclusion (24:8; ch. IV, § 2.4c-d).

Finally, the dialectic between divine grace and human obedience appears also in Exod 29. Worship, in fact, is established on the ground of a divine action, as is explicitly stated in Exod 29:42, "It shall be the regular burnt offering throughout your generations ..., before Yhwh, where I will meet with you there, to speak unto you, there" (ch. V, § 2,7). The possibility to meet with God through worship is a gift and a concession, not a privilege. The decision to "separate" some members from the people to dedicate them especially to the ministry of the Tabernacle originates in Yhwh. On the other hand, through the sin offering, the burnt offering (ch. V, § 2.3 and § 2.4) and the sacrifice of the second ram (ch. V, § 2.5), the blood constitutes the priesthood as an irreversible condition, an "eternal statute" (חכת עולם, 29:9), a statutory decree that turns Aaron and his sons into ministers bound by duties and obligations (ch. V, § 2.2).

This research therefore concludes with a close reading of Exod 30:1-10, a passage in which the blood shed appears for the last time in the book of Exodus (ch. V, § 3), to indicate the *Yom Kippur* rites (30:10). The priests, united with God by a special relationship, will witness the virtue of ritual blood which, through its life-giving power, will not only be capable of creating the bond with the people, but will also be able to regenerate it through the forgiveness of sins. Thereupon, the motive of the covenant remains as an essential element for the Book of Exodus, but the *Yom Kippur* hinted at in ch. 30 shows that the covenant will be betrayed several times and that Yhwh will offer the people a blood ritual capable of reconstituting it. In this way, this conclusion remains open to further insights and can create a constructive dialogue with the New Testament: Jesus Christ, the "New Moses" circumcised in His first days (Luke 2:21, 27), will offer His very blood in order to re-establish the covenant with the people precisely through the forgiveness of sins (Matt 26:28).

BIBLIOGRAPHY

1. Commentaries on Exodus (Chronological Order)

ORIGENES, *Homiliae in Exodum – Homélies sur l'Exode* (ed. M. BORRET) (Sources Chrétiennes 321; Paris: Cerf, 1985).

Midrash Rabbah. Exodus (ed. H. FREEDMAN) (New York, NY: Soncino, 1983).

CARASIK, M., ed., *The Commentators' Bible. The Rubin JPS Miqra'ot Gedolot. Exodus* (Philadelphia, PA: JPS, 2005).

CASSUTO, U., *A Commentary on the Book of Exodus* (Jerusalem: Varda Books, 1947; ²1964).

NOTH, M., *Das zweite Buch Mose. Exodus* (Das Alte Testament Deutsch 5; Göttingen: Vandenhoeck & Ruprecht, 1959, ⁷1984).

AUZOU, G., *De la servitude au service. Le livre de l'Exode* (Paris: Éditions de l'Orante, 1961).

CHILDS, B.S., *Exodus: A Commentary* (OTL; London: SCM – Philadelphia, PA: Westminster, 1974).

DURHAM, J., *Exodus* (WBC 3; Waco, TX: Word Books, 1987).

SCHMIDT, W.H., *Exodus 1,1–6,30* (BKAT 2.1; Neukirchen–Vluyn: Neukirchener Verlag, 1988).

SARNA, N., *Exodus* (JPS.TC; Philadelphia, PA: JPS, 1991).

HOUTMAN, C., *Exodus*. I-IV (Historical Commentary on the Old Testament; Kampen: Kok 1993, 1996 – Leuven: Peeters, 2000, 2002).

PROPP, W.H.C., *Exodus 1–18. Exodus 19–40. A New Translation with Introduction and Commentary* (Anchor Bible 2a-b; New York, NY: Doubleday, 1999, 2006).

ROGERSON, J.W., – MOBERLY, R.W.L., – JOHNSTONE, W., *Genesis and Exodus* (Old Testament Guides 1-2; Sheffield: Bloomsbury, 2001).

NEPI, A., *Esodo (Capitoli 1–15). Esodo (Capitoli 16–40)* (DLP Lectio divina popolare, AT; Padova: Messaggero, 2002, 2004).

MEYERS, C., *Exodus* (New Cambridge Bible Commentary; Cambridge: Cambridge University Press, 2005).

DOZEMAN, T.B., *Exodus* (Eerdmans Critical Commentary – Old Testament; Grand Rapids, MI: Eerdmans, 2009).

FISCHER, G., – MARKL, D., *Das Buch Exodus* (Neuer Stuttgarter Kommentar Altes Testament 2; Stuttgart: Katholisches Bibelwerk, 2009).

HAMILTON, V.P., *Exodus: An Exegetical Commentary* (Grand Rapids, MI: Baker Academics, 2011).

UTZSCHNEIDER, H. – OSWALD, W., *Exodus 1–15* (IECOT; Stuttgart: Kolhammer, 2015).

PRIOTTO, M., *Esodo*. Nuova versione, introduzione e commento (I libri biblici. Primo Testamento 2; Cinisello Balsamo: Paoline, 2014).

ALBERTZ, R., *Exodus 1–18. Exodus 19–40* (Zürcher Bibelkommentar AT 2/1-2; Zürich: Theologischer Verlag, 2012, 2015).

DOHMEN, Ch., *Exodus 1–18. Exodus 19–40* (HThKAT, Freiburg im Breisgau: Herder, 2015, 2004).

DAVIES, G.I., *Exodus 1–18*. A Critical and Exegetical Commentary. 1. Chapters 1–10. 2. Chapters 11–18 (ICC; London: Bloomsbury – New York, NY: T&T Clark, 2020).

2. Ancient Works

2.1 Ancient Near East

CHARPIN, D. – *al.*, *Archives Royales de Mari*. XXVI, Archives épistolaires de Mari (ARM 26/2; Paris: Éditions Recherche sur les Civilisations, 1988).

EIDEM, J., ed., *The Royal Archives from Tell Leilan*. Old Babylonian Letters and Treaties from the Lower Town Palace East (PIHANS 117; Leiden: Peeters, 2011).

GEORGE, A.R., ed., *The Babylonian Gilgamesh Epic*. Introduction, Critical Edition and Cuneiform Texts. I (Oxford: University Press, 2003).

HEIMPEL, W., *Letters to the King of Mari*. A New Translation with Historical Introduction, Notes, and Commentary (Mesopotamian Civilizations 12; Winona Lake, IN: De Gruyter, 2003).

LAMBERT, W.G., – MILLARD, A.R., ed., *Atra-ḫasīs*. The Babylonian Story of the Flood (Oxford: Clarendon, 1969).

TALON, P., ed., *Enūma eliš*. The Standard Babylonian Creation Myth. Introduction, Cuneiform Text, Transliteration and Sign List with a Transilation and Glossary in French (SAA IV; Helsinki: The Neo-Assyrian Text Corpus Project, 2005).

WISEMAN, D.J., "Vassal Treaties of Esarhaddon," *Iraq* 20 (1958) i-ii.1-99.

WRIGHT III, B.G., *The Letter of Aristeas*. "Aristeas to Philocrates" or "On the Translation of the Law of the Jews" (CEJL; Berlin: De Gruyter, 2015).

2.2 Greek Literature and Church Fathers

AESCHYLUS, *Seven Against Thebes* (ed. A. HECHT – H.H. BACON) (The Greek Tragedy in New Translations; Oxford: University Press, 1973).

ARISTOTLE, *Poetics* (ed. S.H. BUTCHER) (London: Lynch, 1898).

AUGUSTINUS, *Questiones in Heptateuchum Libri VII* (CCSL 33; Turnhout: Brepols, 1958).

FLAVIUS JOSEPHUS, *Jewish Antiquities*. Book I-IV (The Loeb Classical Library. Josephus. in Nine Volumes, IV; Cambridge, MA: Harvard University Press, 1956).

GREGORIUS NYSSENUS, *De vita Moysis* (ed. M. SIMONETTI) (Scrittori Greci e Latini; Milano: Mondadori, 1984).

HOMER, *The Odyssey*. I (ed. A.T. MURRAY) (Cambridge, MA: Harvard University Press, 1945).

JOHANNES CHRYSOSTOMUS, *Ad illuminandos catecheses – Huit catéchèses baptismales* (ed. A. WENGER) (Sources Chrétiennes 50; Paris: Cerf, 1957).

PHILO OF ALEXANDRIA, *Legum allegoriae I–III* (Les Œuvres de Philon d'Alexandrie 2; Paris: Cerf, 1962).

——— , *De vita Mosis I–II* (Les Œuvres de Philon d'Alexandrie 22; Paris: Cerf, 1967).

——— , *De specialibus legibus* (Les Œuvres de Philon d'Alexandrie 25; Paris: Cerf, 1970).

XENOPHON, *Anabasis*. Books I-VII (The Loeb Classical Library; Cambridge, MA: Harvard University Press, 1980).

TERTULLIANUS, "De resurrectione mortuorum," *Quinti Septimi Florentis Tertulliani Opera*. II. *Opera montanistica* (CCSL 2; Turnhout: Brepols, 1954) 919-1012.

THUCYDIDES, *The Peloponnesian War* (ed. M. HAMMOND – P.J. RHODES) (Oxford: University Press, 2009).

3. Monographs and Studies

ABBA, R., "The Origin and Significance of Hebrew Sacrifice," *BTB* 7 (1977) 123-138.

ABUSCH, Z., "Blood in Israel and Mesopotamia," *Emanuel. Studies in Hebrew Bible, Septuagint and Dead Sea Scrolls in Honor of Emanuel Tov* (ed. S.M. PAUL – *al.*) (Leiden: Brill, 2003) 675-684.

AGAMBEN, G., *The Sacrament of Language. An Archaeology of the Oath* (Homo Sacer, II, 3) (Stanford, CA: University Press, 2011).

AHN, J.J., "Diaspora Studies," *The Oxford Encyclopedia of Biblical Interpretation*, I (ed. S.L. MCKENZIE) (Oxford: University Press, 2013) 217-225.

AHUIS, F., *Exodus 11,1 – 13,16 und die Bedeutung der Trägegruppen für das Verständnis des Passa* (FRLANT 168; Göttingen: Vandenhoeck & Ruprecht 1996).

ALEXANDER, E.S., "Ritual on the Threshold: Mezuzah and the Crafting of Domestic and Civic Space," *Jewish Social Studies* 20 (2014) 100-130.

ALLEN, R.B., "The 'Bloody Bridegroom' in Exodus 4:24-26," *Bibliotheca Sacra* 153 (1996) 259-269.

ALTER, R., *The Art of Biblical Narrative* (New York, NY: Basic Books, 1981).

ALTMANN, P., *Festive Meals in Ancient Israel.* Deuteronomy's Identity Politics in Their Ancient Near Eastern Context (BZAW 424; Göttingen: De Gruyter, 2011).

AMIT, Y., "The Dual Causality Principle and Its Effects on Biblical Literature," *VT* 37 (1987) 385-400.

─────── , "Dual Causality – An Additional Aspect," *In Praise of Editing in the Hebrew Bible: Collected Essays in Retrospect* (HBM 39; Sheffield: Phoenix Press, 2012) 105-121.

AMZALLAG, N., "Moses' Tent of Meeting – A Theological Interface between Qenite Yahwism and the Israelite Religion," *SJOT* 33 (2019) 298-317.

ANDERSON, B.A., *An Introduction to the Study of Pentateuch* (T&T Clark Approaches to Biblical Studies; New York, NY: T&T Clark, 2017).

ANDERSON, J.S., *The Blessing and the Curse.* Trajectories in the Theology of the Old Testament (Eugene, OR: Cascade Books, 2014).

ASHBY, G.W., "The Bloody Bridegroom: The Interpretation of Exodus 4:24-26," *ExpTim* 106 (1995) 203-205.

ASSMANN, J., *Cultural Memory and Early Civilization.* Writing, Remembrance, and Political Imagination (Cambridge: University Press, 2011).

─────── , "Memory, Narration, Identity: Exodus as a Political Myth," *Literary Construction of Identity in the Ancient World* (ed. H. LISS – M. OEMING) (Winona Lake, IN: Eisenbrauns, 2010) 3-18.

AUNEAU, J., "Le bain de purification des Prêtres; Lv 8,6 et parallèles," Κεχαριτωμένη. *Mélanges René Laurentin* (Paris: Desclée, 1990) 103-111.

AUSTIN, J.L., *How to Do Things with Words* (Cambridge, MA: Harvard University Press, 1962).

AVIGAD, N., "The Inscribed Pomegranate from the 'House of the Lord'", *BA* (1990) 157-166.

BAKER, D.L., "Ten Commandments, Two Tablets. The Shape of the Decalogue," *Themelios* 30 (2005) 6-22.

BALTZER, K., *Deutero-Isaiah.* A Commentary on Isaiah 40–55 (Hermeneia; Minneapolis, MN: Fortress Press, 2001).

BAR, S., "Who were the 'Mixed Moltitude'?," *HebStud* 49 (2008) 27-39.

BARBIERO, G., *Cantico dei Cantici.* Nuova versione, introduzione e commento (I libri biblici. Primo Testamento 24; Milano: Paoline, 2004).

BARONI, R., *La tension narrative: suspense, curiosité et surprise* (Paris: Seuil, 2007).

BARR, J., *The Semantics of Biblical Language* (London: Oxford University Press, 1961).

BARTHÉLEMY, D., *Dieu et son image.* Ébauche d'une théologie biblique (Trésors du christianisme; Paris: Cerf, 1963, 2013).

BARTON, J., *The Nature of Biblical Criticism* (Louisville, KY: Westminster John Knox, 2007).

BARTOR, A., *Reading Law as Narrative. A Study in the Casuistic Laws of the Pentateuch* (SBL.AIL 5; Atlanta, GA: SBL, 2010).
BASILE, G., – *al.*, *Linguistica generale* (Manuali universitari 83; Roma: Carocci, 2010).
BAUMANN, A., "לוּחַ," *ThWAT*, IV, 496-499.
BAUTCH, R.J., – KNOPPERS, G.N., ed., *Covenant in the Persian Period. From Genesis to Chronicles* (Winona Lake, IN: Eisenbrauns, 2015).
BEAUCHAMP, P., *L'un et l'autre Testament*. I. Essai de lecture (Paris: Seuil, 1976).
——— , "Propositions sur l'Alliance de l'Ancien Testament comme structure centrale," *Pages exégétiques* (LeDiv 202; Paris: Cerf, 2005) 55-86; orig. *Revue de Sciences Religieuses* 58 (1970) 161-193.
BELL, C., *Ritual Theory, Ritual Practice* (New York, NY: Oxford University Press, 1992).
BENVENISTE, É., "L'expression du serment dans la Grèce ancienne," *Revue de l'histoire des religions* 134 (1948) 81-94.
——— , *Dictionary of Indo-European Concepts and Society* (Chicago, MA: University of Chicago Press, 2005, orig. 1969).
BERNAT, D.A., *Sign of the Covenant. Circumcision in the Priestly Tradition* (SBL.AIL 3; Atlanta, GA: SBL, 2009).
BERNER, C., *Die Exoduserzählung: Das literarische Werden einer Ursprungslegende Israels* (FAT 73; Tübingen: Mohr Siebeck, 2010).
BERTI, I., "'Now let Earth be my Witness and the Broad Heaven above, and the Down Flowing Water of the Styx…' (Homer, Mas XV, 36-37): Greek Oath-Rituals," *Ritual and Communication in the Graeco-Roman World* (ed. E. STRAVRIANOPOULOU) (Kernos supplément 16; Liége: Presses universitaires de Liège, 2006) 181-209.
BEUKEN, W., "שׁכב," *ThWAT*, VII, 1306-1318.
BEYERLIN, W., *Origins and History of the Oldest Sinaitic Traditions* (Oxford: Basil Blackwell, 1965).
BIBB, B.D., "'Blood, Death, and the Holy' in the Leviticus Narrative," *The Oxford Handbook of Biblical Narrative* (ed. D.N. FEWELL) (New York, NY: Oxford University Press, 2016) 137-146.
BLEIBERG, E., "The Location of Pithom and Succoth," *Ancient World* 6 (1983) 21-27.
BLENKINSOPP, J., "The 'Covenant of Circumcision' (Gen 17) in the Context of the Abraham Cycle (Gen 11:27–25:11). Preliminary Considerations," *The Post-Priestly Pentateuch. New Perspectives on Its Redactional Development and Theological Profiles* (ed. F. GIUNTOLI – K. SCHMID) (FAT 101; Tübingen: Mohr Siebeck, 2015) 145-156.
BLUM, E., *Studien zur Komposition des Pentateuch* (BZAW 189; Berlin: De Gruyter, 1990).

BLUM, E., "Israël à la montagne de Dieu. Remarques sur Ex 19–24; 32–34 et sur le context littéraire et historique de sa composition," *Le Pentateuque en question. Les origines et la composition des cinq premiers livres de la Bible à la lumière des recherches récentes* (ed. A. DE PURY) (Genève: Labor et Fides, 1991) 271-295.

————, "Die Literarische Verbindung von Erzvätern und Exodus: Ein Gespräch mit neueren Endredaktionshypotesen," *Abschied Vom Jahwisten: Die Komposition des Hexateuch in der jüngsten Diskussion* (ed. K. SCHMID) (Berlin – New York, NY: De Gruyter, 2002) 123-140.

BOITANI, P., *Riconoscere è un dio. Scene e temi del riconoscimento nella letteratura* (Saggi 944; Torino: Einaudi, 2014).

BOVATI, P., *Re-establishing Justice. Legal Terms, Concepts and Procedures in the Hebrew Bible* (JSOT.S 105; Sheffield: Academic Press, 1994).

————, "La dottrina dell'ascolto nell'Antico Testamento," in *Ascolto, Docilità, Supplica* (DSBP 5; Roma: Borla, 1993) 17-64.

————, "Alla ricerca del profeta (1). Una presenza singolare nel cammino," in *"Così parla il Signore." Studi sul profetismo biblico* (Biblica; Bologna: EDB, 2008) 17-35.

————, "'Non so parlare' (Ger 1,6). La parola come atto profetico," in *"Così parla il Signore." Studi sul profetismo biblico* (Biblica; Bologna: EDB, 2008) 53-76.

————, "La 'nuova alleanza' (Ger 31,31-34)," in *"Così parla il Signore." Studi sul profetismo biblico* (Biblica; Bologna: EDB, 2008) 183-210.

————, *Parole di libertà. Il messaggio biblico della salvezza* (Biblica; Bologna: EDB, 2012).

————, "Teologia Biblica e ispirazione. Problemi e aperture," in J.-P. SONNET, ed., *Ogni Scrittura è ispirata. Nuove prospettive sull'ispirazione biblica* (Lectio 5; Roma: G&B Press – Cinisello Balsamo: San Paolo, 2013) 283-303.

BOVATI, P. – MEYNET, R., *Il libro del profeta Amos*. Seconda edizione rivista (Rhetorica Biblica et Semitica 21; Leuven – Paris – Bristol, CT: Peeters, 2019).

BRESCIANI, E., *Letteratura e poesia dell'Antico Egitto* (I millenni; Torino: Einaudi, 1969).

————, *Testi religiosi dell'Antico Egitto* (I Meridiani. Classici dello Spirito; Milano: Mondadori, 2001).

CAGNI, L., "Il sangue nella letteratura assiro-babilonese," *Sangue e Antropologia Biblica*. I (ed. F. VATTIONI) (CSSC 1; Roma: Pia Unione Preziosissimo Sangue, 1981) 47-85.

CALLAHAM, S.N., *Modality and the Biblical Hebrew Infinitive Absolute* (AKM 71; Wiesbaden: Harrassowitz, 2014).

CANOBBIO, G., "La morte di Gesù: sacrificio o fine dei sacrifici?," *Il sacrificio* (Quaderni teologici del Seminario di Brescia 29; Brescia: Morcelliana 2019) 141-164.

CASSUTO, U., *A Commentary on the Book of Genesis*. II (ed. I. ABRAHAMS) (Jerusalem: Magnes Press, 1964).

CHARPIN, D., *"Tu es de mon sang."* Les alliances dans le Proche-Orient ancient (Docet Omnia 4; Paris: Les Belles Lettres, 2019).

CHATMAN, S., *Story and Discourse. Narrative Structure in Fiction and Film* (Ithaca, NY: Cornell University Press, 1978).

CHAVEL, S., *Oracular Law and Priestly Historiography in the Torah* (FAT.2 71; Tübingen: Mohr Siebeck, 2014).

———, "A Kingdom of Priests and its Earthen Altars in Exodus 19–24," *VT* 65 (2015) 169-222.

CHERNEY JR., K.A., "The Enigmatic Divine Encounter in Exodus 4:24-26," *Wisconsin Lutheran Quarterly* 113 (2016) 195-203.

———, "The Plague Narrative (Ex 7:8 – 10:29). Structure, Source Criticism and Naturalistic Explanations," *Wisconsin Lutheran Quarterly* 116 (2019) 83-92.

CLIFFORD, R.J., "The Tent of El and the Israelite Tent of Meeting," *CBQ* 33 (1971) 221-227.

COCCO, F., *The Torah as a Place of Refuge*. Biblical Criminal Law and the Book of Numbers (FAT.2 84; Tübingen: Mohr Siebeck, 2016).

COHEN, J.M., "*Hatan damim*: The Bridegroom of Blood," *JBQ* 33 (2005) 120-126.

COHN, R.L., *The Shape of Sacred Space*. Four Biblical Studies (AAR.SR 23; Chico, CA: Scholars Press, 1981).

COSTACURTA, B., *La vita minacciata*. Il tema della paura nella Bibbia ebraica (AnBib 119; Roma: PIB, 1997).

COX, D.G., "The Hardening of Pharaoh's Heart in its Literary and Cultural Contexts," *Bibliotheca Sacra* 162 (2006) 292-311.

CURRID, J.D., *Ancient Egypt and the Old Testament* (Grand Rapids, MI: Baker Book, 1997).

DANOVE, P.L., "A Method for Analyzing the Semantic and Narrative Rhetoric of Repetition and Their Contribution to Characterization," *EstBíb* 76 (2018) 55-84.

DAUBE, D., *The Exodus Pattern in the Bible* (London: Faber and Faber, 1963).

DAVID, R., *Religion and Magic in Ancient Egypt* (London: Penguin, 2002).

DAVIES, E.W., "A Mathematical Conondrum: The Problem of the Large Numbers in Number i and xxvi," *VT* 45 (1995) 449-469.

DAVIES, P.R., *Scribes and School: The Canonization of the Hebrew Scriptures* (Library of Ancient Israel; Louisville, KY: Westminster John Knox, 1998).

DEIANA, G., *Levitico*. Nuova versione, introduzione e commento (I libri biblici. Primo Testamento 3; Milano: Paoline, 2005).

DE PURY, A., "Le Dieu qui vient en adversaire. De quelques différences à propos de la perception de Dieu dans l'Ancien Testament," *Ce Dieu qui vient: études sur l'Ancien et le Nouveau Testament offertes au Professeur B. Renaud à l'occasion de son soixante-cinquième anniversaire* (ed. R. KUNTZMANN) (LeDiv 159; Paris: Cerf, 1995) 45-68.

DE VAUX, R., *Ancient Israel*. Its Life and Institutions (New York, NY – Toronto – London: McGraw-Hill Book Company, 1961).

DE ZAN, R., *Unius verbi Dei multiplices thesauri*. Lettura liturgica della Bibbia: appunti per un metodo (Bibliotheca "Ephemerides Liturgicae". Subsidia 196; Roma: CLV, 2021).

DIESEL, A.A., *"Ich bin Yahwe."* Der Aufstieg der Ich-bin-Aussage zum Schlüsselvort des alttestamentlichen Monotheismus (WMANT 110; Neukirchen-Vluyn: Vandenhoeck & Ruprecht, 2006).

DI GIOVANBATTISTA, F., *Il sistema sacrificale israelitico*. Alla luce della Pasqua e nella Tradizione Rabbinica (Ecclesia Mater. Studi 8; Città del Vaticano: Lateran University Press, 2016).

DI PAOLO, R., "Il poema delle Quattro notti," *Liber Annuus* 60 (2010) 83-105.

DI ROCCO, E., *Raccontare il ritorno*. Temi e trame della letteratura (Studi e ricerche. Critica letteraria 726; Bologna: Il Mulino, 2017).

DOZEMAN, T.B., *God and the Mountain*. A Study in Redaction, Theology and Canon in Exodus 19–24 (SBL.MS 37; Atlanta, GA: Scholars Press, 1989).

DOZEMAN, T.B. – EVANS, C.A. – LOHR, J.N., ed., *The Book of Exodus: Composition, Reception, and Interpretation* (VT.S 164; Leiden: Brill, 2014).

DURAND, G., *Les structures anthropologiques de l'imaginaire*. Introduction à l'archétypologie générale (Collection Études Supérieures 14; Poitiers: Bordas, 1969).

EBERHART, C.A., "Blood. I. Ancient Near East and Hebrew Bible/Old Testament," *EBR*, IV, 201-212.

FARAONE, C.A., "Curses and Blessings in Ancient Greek Oaths," *JANER* 5 (2006) 140-158

FAUST, A., "The Bible, Archeology, and the Practice of Circumcision in Israelite and Philistine Societies," *JBL* 134 (2015) 273-290.

FEDER, Y., "On *kupurru, kipper* and Etymological Sins that Cannot be Wiped Away," *VT* 60 (2010) 535-545.

———, *Blood Expiation in Hittite and Biblical Ritual*. Origins, Context, and Meaning (SBL.WAWSup 2; Leiden – Boston, MA: Brill, 2011).

FICCO, F., *"Mio figlio sei tu" (Sal 2,7)*. La relazione Padre-figlio e il salterio (TG.T 192; Roma: PUG, 2012).

———, "La relazione padre-figlio nel Pentateuco," *Parole di Vita* 65/3 (2020) 19-24.

FINK, L.H., "The Incident at the Lodging House," *JBQ* 21 (1993) 236-241.

FISCH, H., *Remembered Future*. A Study in Literary Mythology (Bloomington, IN: Indiana University Press, 1984).

FISCHER, G., *Jahwe unser Gott*. Sprache, Aufbau und Erzähltechnik in der Berufung des Moses (Ex 3-4) (OBO 90; Freiburg: Universitätsverlag, 1989).

FISHBANE, M., *Biblical Interpretation in Ancient Israel* (Oxford: Clarendon, 1985).

FLEMING, D.E., *The Installation of Baal's High Priestess at Emar*. A Widow on Ancient Syrian Religion (HSS 42; Atlanta, GA: Scholars Press, 1992).

FOKKELMAN, J.P., *Narrative Art in Genesis*. Specimens of Stylistic and Structural Analysis (Studia Semitica Neerlandica 17; Assen – Amsterdam: Van Gorcum, 1975).

FORNARA, R., *La visione contraddetta*. La dialettica fra visibilità e non-visibilità divina nella Bibbia ebraica (AnBib 155; Roma: PUG, 2004).

FORSTER, E.M., *Aspects of the Novel* (Cambridge: Penguin, 1927).

FOWLER, M.D., "Incense Altars," *ABD*, III, 409-410.

FOX, M.V., "The Sign of the Covenant: Circumcision in Light of Priestly *'ôt* Etiologies," *Revue Biblique* 81 (1974) 557-596.

FREI, P., "Die Persische Reichsautorisation: Ein Überblick," *ZABR* 1 (1995) 1-35.

FREVEL, CH., *Mit Blick auf das Land die Schöpfung erinnern*. Zum Ende der Priestergrundschrift (HBS 23; Freiburg im Breisgau: Herder, 2000).

FRIEDMAN, R.E., *The Bible with Sources Revealed*. A New View into the Five Books of Moses (San Francisco, CA: Harper, 2003).

FROLOV, S., "The Hero as Bloody Bridegroom: On the Meaning and Origin of Exodus 4,26," *Biblica* 77 (1996) 520-523.

FÜGLISTER, N., "Blut," *LThK*, II, 531-534.

FUHS, H.-F., "ראה," *ThWAT*, VII, 233-267.

GELIO, R., "Il rito del sangue e l'identificazione del *negeph lemašḥît*," *Sangue e Antropologia Biblica*. II (ed. F. VATTIONI) (CSSC 1; Roma: Pia Unione Preziosissimo Sangue, 1981) 467-476.

———, "Sangue e vendetta," *Sangue e Antropologia Biblica*. II (ed. F. VATTIONI) (CSSC 1; Roma: Pia Unione Preziosissimo Sangue, 1981) 515-528.

GENETTE, G., *Figures III* (Poétique; Parigi: Seuil, 1972).

GERTZ, J.C., *Tradition und Redaktion in der Exoduserzählung* (FRLANT 189; Göttingen: Vandenhoeck & Ruprecht, 2000).

GESUNDHEIT, S., *Three Times a Year* (FAT 82; Tübingen: Mohr Siebeck, 2012).

———, "Philology and Theory: Exodus 12:21-27 as a Case Study," *VT* 70 (2020) 414-425.

GIANTO, A., "Mood and Modality in Classical Hebrew," in S. ISRAEL – al., ed., *Past Links*. Studies in the Languages and Cultures of the Ancient Near East (Winona Lake, IN: Eisenbrauns, 1998) 183-198.

GILBERT, P., "Human Free Will and Divine Determinism. Pharaoh, a Case Study," *Direction* 30 (2000) 76-87.

GILBERT, P., *Corso di Metafisica*. La pazienza d'essere (Casale Monferrato: Piemme, 1997).

————, *Le ragioni della sapienza* (Philosophia 2; Roma: G&B Press, 2010).

GILDERS, W.K., *Blood Ritual in the Hebrew Bible: Meaning and Power* (Baltimore, ML: John Hopkins University Press, 2004).

GIRARD, M., *Symboles bibliques. Langage universel. Pour une théologie des deux Testaments ancrée dans les sciences humaines*. I-II (Montréal: Médiaspaul, 2016).

GIRARD, R., *Des choses cachés depuis la foundation du monde*. Recherches avec J.-M. Oughourlian et Guy Lefort (Paris: Grasset, 1978).

GITIN, S., "Incense Altars from Ekron Israel and Judah: Context and Typology," *Erez Israel* 20 (1989) 52-68.

————, "New Incense Altars from Ekron: Context, Typology and Function," *Erez Israel* 23 (1992) 43-49.

————, "The Four-Horned Altar and Sacred Space: An Archaeological Perspective," *Sacred Time, Sacred Place. Archaeology and the Religion of Israel* (ed. B.M. GITTLEN) (Winona Lake, IN: Penn State University Press, 2002) 95-123.

GIUNTOLI, F., *Genesi 12–50*. Introduzione, traduzione e commento (NVBTA 1^2; Cinisello Balsamo: San Paolo, 2013).

GOLDINGAY, J., "The Significance of Circumcision," *JSOT* 88 (2000) 3-18.

GOLDINGAY, J., – PAYNE, D., *A Critical and Exegetical Commentary on Isaiah 40–55*. Volume II (ICC; London – New York, NY: T&T Clark, 2006).

GOROSPE, A.E., *Narrative and Identity. An Ethical Reading of Exodus 4* (Biblical Interpretation Series 86; Leiden – Boston, MA: Brill, 2007).

GRANT, D.E., "Fire and the Body of Yahweh," *JSOT* 40 (2015) 139-161.

GREENBERG, M., *Understanding Exodus* (HBI 2; New York, NY: Behrman House, 1969).

————, "The Redaction of the Plague Narrative in Exodus," *Near Eastern Studies in Honour of W.F. Albright* (ed. H. GEODICKE) (Baltimore, MD – London: John Hopkins University Press, 1971) 243-252.

GREENSTEIN, E.L., "The Firstborn Plague and the Reading Process," *Pomegranates and Golden Bells. Studies in Biblical, Jewish and Near Eastern Ritual, Law, and Literature in Honor of Jacob Milgrom* (ed. D.P. WRIGHT – *al.*) (Winona Lake, IN: Eisenbrauns, 1995) 555-568.

GRENET, É., *Unité du "Je" psalmique* (LeDiv 273; Paris: Cerf, 2019).

GROSSMAN, J., "The Structural Paradigm of the Ten Plagues Narrative and the Hardening of Pharaoh's Heart," *VT* 64 (2014) 588-610.

HAHN, S.W., *Kinship by Covenant.* A Canonical Approach to the Fulfillment of God's Saving Promises (New Haven, CT: Yale University Press, 2009).

HAMILTON, V.P., *The Book of Genesis.* Chapters 1–17 (NICOT; Grand Rapids, MI: Eerdmans, 1990).

HARRELL, J.E., – HOFFMEIER, J.K. – WILLIAMS, K.F., "Hebrew Gemstones in the Old Testament: A Lexical, Geological, and Archaeological Analysis," *BBR* 27 (2017) 1-52.

HENDEL, R., "Sacrifice as a Cultural System: The Ritual Symbolism of Exodus 24,3-8," *ZAW* 101 (1989) 366-390.

HENDRIX, R.E., "*Miskan* and *'ohel moʻed*: Etymology, Lexical Definitions, and Extra-Biblical Usage," *AUSS* 29 (1991) 213-223.

——— , "The Use of *Miskan* and *'ohel moʻed* in Exodus 25–40," *AUSS* 30 (1992) 3-13.

HESCHEL, A.J., *The Sabbath.* Its Meaning for the Modern Man (New York, NY: Noonday Press, 1951, ²¹1994).

HIEKE, TH., *Levitikus 1–15* (HThKAT; Freiburg im Breisgau: Herder, 2014).

——— , "Leper, Leprosy I. Ancient Near East and Hebrew Bible/Old Testament," *Encyclopedia of the Bible and Its Reception* 16 (2018) 144-147.

HIEKE, TH., – NICKLAS, T., ed., *The Day of Atonement.* Its Interpretations in early Jewish and Christian Traditions (Themes in Biblical Narrative. Jewish and Christian Traditions 15; Leiden – Boston, MA: Brill, 2012).

HINCKLEY, R., "Sapphire *Lukhoth* and Blood-Sprinkled *Kapporeth.* The Ten Commandments in Context," *Logia* 21 (2012) 27-33.

HOFFMEIER, J.K., "The Possible Origin of the Tent of Purification in the Egyptian Funerary Cult," *Studien zur altägyptischen Kultur* 9 (1981) 167-177.

HOFFNER, H.A., "Oil in Hittite Texts," *BA* (1995) 108-114.

HORT, G., "The Plagues of Egypt," *ZAW* 69 (1957) 87-95.

HOWELL, A.J., "The Firstborn Son of Moses as the "Relative of Blood" in Exodus 4.24-26," *JSOT* 35 (2010) 63-76.

HUFFMON, H.B., "The Treaty Background of Hebrew ידע," *BASOR* 181 (1966) 31-37.

INVERNIZZI, L., *"Perché mi hai inviato?".* Dalla diacronia redazionale alla dinamica narrativa in Es 5,1–7,7 (AnBib 216; Roma: G&B Press, 2016).

JANOWSKI, B., *Sühne als Heilsgeschehen.* Studien zur Sühnetheologie der Priesterschrift und zur Wurzel KPR im Alten Orient und im Alten Testament (Neukirchen-Vluyn: Neukirchener Verlag, 1982).

JAY, N., *Throughout Your Generations Forever.* Sacrifice, Religion and Paternity (Chicago, IL: University of Chicago Press, 1992).

JELINKOVA REYMOND, E., *Les inscriptions de la statue guérisseuse de Djed-Her-le-saveur* (Le Caire: Institut Français d'Archéologie Orientale, 1956).

JENSON, P.P., *Graded Holiness. A Key to the Priestly Conception of the World* (JSOT.S 106; Sheffield: Academic Press, 1992).

JEON, J., *The Call of Moses and the Exodus Story* (FAT.2 60; Tübingen: Mohr Siebeck, 2013).

JONES, E., "Direct Reflexivity in Biblical Hebrew: A Note on נפש," *ZAW* 129 (2017) 411-426.

JOOSTEN, J., "Une lecture du texte hébreu," *L'épreuve d'Abraham ou la ligature d'Isaac (Genèse 22)* (ed. M. ARNOLD) (CEv. Supplément 173; Parigi: Cerf, 2014) 3-11.

KAISER, O., "Deus absconditus and Deus revelatus. Three Difficult Narratives in the Pentateuch," *Shall Not the Judge of All the Earth Do What Is Right? Studies on the Nature of God in Tribute to James L. Crenshaw* (ed. D. PENCHANSKY – P.L. REDDITT) (Winona Lake, IN: Penn State University Press, 2000) 73-88.

KALIMI, I., *Metathesis in the Hebrew Bible*. Wordplay as a Literary and Exegetical Device (Peabody, MC: Hendrickson Academic, 2018).

KALLUVEETTIL, P., *Declaration and Covenant*. A Comprehensive Review of Covenant Formulae from the Old Testament and the Near East (AnBib 88; Roma: PIB, 1982).

KEDAR-KOPFSTEIN, B., "דם," *ThWAT*, II, 253-267.

KEEL, O., *The Symbolism of the Biblical World* (New York, NY: The Seabury Press, 1978).

KISSINE, M., *From Utterances to Speech Acts* (New York, NY: Cambridge University Press, 2013).

KITZ, A.-M., "The Plural Form of *'ûrîm* e *tummîm*," *JBL* 116 (1997) 401-410.

KNOHL, I., *The Sanctuary of Silence*. The Priestly Torah and the Holiness School (Minneapolis, MN: Fortress Press, 1997).

KORNFELD, W., – RINGGREN, H., "קדש," *ThWAT*, VI, 1179-1204.

KOSMALA, H., "The Bloody Husband," *VT* 12 (1962) 14-28.

KRAUSE, J.J., "Circumcision and Covenant in Genesis 17," *Biblica* 99 (2018) 151-165.

KUTSCH, E., "חתן," *ThWAT*, III, 289-297.

LAMBERT, W.G., "Donations of Food and Drink to the Gods of Ancient Mesopotamia," *Ritual and Sacrifice in Ancient Near East* (ed. J. QUAGEBEUR) (Orientalia lovaniensia analecta 55; Leuven: Peeters, 1993) 191-201.

LANGE, A., "The Shema Israel in Second Temple Judaism," *Journal of Ancient Judaism* 1 (2010) 207-214.

LAWRENCE, J.D., *Washing in Water*. Trajectories of Ritual Bathing in the Hebrew Bible and Second Temple Literature (SBL.AB 23; Leiden – Boston, MA: Brill, 2006).

LE BOULLUEC, A. – SANDEVOIR, P., *L'Exode* (La Bible d'Alexandrie 2; Paris: Cerf, 1989).

LEHANE, T.J., "Zipporah and the Passover," *JBQ* 24 (1996) 46-50.

LEMMELIJN, B., "Setting and Function of Exod 11,1-10 in the Exodus Narrative," *Studies in the Book of Exodus*. Redaction – Reception – Interpretation (BEThL 126; Leuven: Peeters, 1996) 443-460.

——— , "The So-called 'Priestly' Layer in Exod 7,14–11,10, 'Source' and/or/nor 'Redaction,'" *Revue Biblique* 109 (2002) 481-511.

——— , *A Plague of Texts?* A Text-Critical Study of the So-Called "Plagues Narrative" in Exodus 7:14 – 11:10 (OTS 57; Leiden – Boston, MA: Brill, 2009).

LEVEEN, A., "Inside Out: Jethro, the Midianites and a Biblical Construction of the Outsider," *JSOT* 34 (2010) 395-417.

LEVIN, CH., *Der Jahwist* (FRLANT 157; Göttingen: Vandenhoeck & Ruprecht, 1993).

LÉVINAS, E., *Le temps et l'autre* (Bibliothèque des textes philosophiques; Paris: Presses universitaires de France, 2007, orig. 1948).

——— , *De l'existence à l'existant* (Bibliothèque des textes philosophiques; Paris: Presses universitaires de France, 2004, orig. 1963).

——— , *Quattro letture talmudiche* (Genova: Il Melangolo, 2008; orig. fr. 1968).

——— , *Totalité et infini*. Essai sur l'extériorité (La Haye: Nijhoff, 1961).

LEVINE, B., *In the Presence of the Lord*. A Study of Cult and Some Cultic Terms in Ancient Israel (SJLA 5; Leiden: Brill, 1974).

LEVINSON, B., *Deuteronomy and the Hermeneutics of Legal Innovation* (New York, NY: Oxford University Press, 1997).

——— , *Legal Revision and Religious Renewal in Ancient Israel* (Cambridge: University Press, 2008).

LEVY, T.E. – al., ed., *Israel's Exodus in Transdisciplinary Perspective: Text, Archaeology, Culture, and Geoscience* (Quantitative Methods in the Humanities and Social Sciences; Heidelberg – New York, NY – Dordrecht – London: Springer, 2015).

LEWIS, T.J., "Covenant and Blood Rituals. Understanding Exodus 24:3-8 in Its Ancient Near Eastern Context," *Confronting the Past*. Archaeological and Historical Essays on Ancient Israel in Honor of William G. Dever (ed. S. GITIN – J.E. WRIGHT – J.P. DESSEL) (Winona Lake, IN: Eisenbrauns, 2006) 341-350.

——— , "Divine Fire in Deuteronomy 33:2," *JBL* 132 (2013) 791-803.

LINDSTRÖM, F., *God and the Origin of Evil* (CBOTS 21; Lund: CWK Gleerup, 1983).

LOSS, N.M., "Carne, anima e sangue," *Sangue e Antropologia Biblica*. II (ed. F. VATTIONI) (CSSC 1; Roma: Pia Unione Preziosissimo Sangue, 1981) 403-412.

LUCIANI, D., *Sainteté et pardon*. I. Structure littéraire du Lévitique (BEThL 185A; Leuven – Paris – Dudley, MA: Peeters, 2005).

——— , "Çippora (Ex 4,24-26): le petit oiseau va sortir... et il échappe encore aux interprètes," *EstB* 70 (2012) 159-184.

LUCIANI, D., "Jéthro: L'objection subsidiaire de Moïse en Madian," *A Pillar of Cloud to Guide*. Text-Critical, Redactional and Linguistic Perspectives on the Old Testament in Honour of Marc Vervenne (ed. H. AUSLOOS – B. LEMMELIJN) (BEThL 269; Leuven: Peeters, 2014) 177-186.

LURKER, M., *Dizionario delle immagini e dei simboli biblici* (Cinisello Balsamo: Paoline, 1990).

MACDONALD, N., "Scribalism and Ritual Innovation," *HeBAI* 7 (2018) 415-429.

MAHFOUZ, H., "Appearing to them for Forty Days (Ac 1,3)," *EstBíb* 76 (2018) 361-384.

MALUL, M., "Some Aspects of Biblical Hospitality and their Significance," *Teshûrôt La Avishur*. Studies in the Bible and the Ancient Near East, in Hebrew and Semitic Languages. Festschrift Presented to Prof. Yitzhak Avishur on the Occasion of his 65th Birthday (ed. M. HELTSER – M. MALUL) (Tel Aviv: Archaeological Center, 2004) 233-251.

MANZI, F., "Sangue," *TTB*, 1232-1237.

MARKL, D., *Der Dekalog als Verfassung des Gottesvolkes*. Die Brennpunkte einer Rechtsermeneutik des Pentateuch in Ex 19–24 und Dtn 5 (HBS 49; Freiburg im Breisgau: Herder, 2007).

———, "The Sociology of the Babylonian Exile and Divine Retribution 'to the Third and Fourth Generation'", *The Dynamics of Early Judean Law*. Studies in the Diversity of Ancient Social and Communal Legislation (ed. S. JACOBS) (BZAW; Berlin: De Gruyter, 2020), Not yet published.

MARX, A., *Les systems sacrificiels de l'Ancien Testament*. Formes et fonctions du culte sacrificial à Yhwh (VT.S 105; Leiden: Brill, 2005).

MATHEWS, D., "Moses as a Royal Figure in the Pentateuch with the Edict of Cyrus as a Test Case," *Restoration Quarterly* 59 (2017) 65-78.

MATTHEWS, V.H., "The Anthropology of Clothing in the Joseph Narrative," *JSOT* 65 (1995) 25-36.

MAZZINGHI, L., *Il libro della Sapienza*. Introduzione – Traduzione – Commento (AnBibSt 13; Roma: G&B Press, 2020).

MCAFFEE, M., "The Heart of Pharaoh in Exodus 4–15," *Bulletin for Biblical Research* 20 (2010) 331-354.

MCCARTHY, D.J., *Treaty and Covenant*. A Study in Form in the Ancient Oriental Documents and in the Old Testament (AnBib 21; Roma: PIB, 1963).

———, "The Symbolism of Blood and Sacrifice," *JBL* 88 (1969) 166-176.

———, "Further Notes on the Symbolism of Blood and Sacrifice," *JBL* 92 (1973) 205-210.

———, "Il simbolismo del sangue (timore reverenziale, vita, morte)," *Sangue e Antropologia Biblica*. I (ed. F. VATTIONI) (CSSC 1; Roma: Pia Unione Preziosissimo Sangue, 1981) 19-35.

McCarthy, D.J., "Hosea XII 2: Covenant by Oil," *Institution and Narrative.* Collected Essays (AnBib 108; Roma: PIB, 1985) 14-20.

Meyers, C., *The Tabernacle Menorah.* A Syntetic Study of a Symbol from the Biblical Cult (Missoula, MT: Gorgias, 1976).

———, "Realms of Sanctity: the Case of the 'Misplaced' Incense Altar in the Tabernacle Text of Exodus," *Texts, Temples and Traditions.* A Tribute to Menahem Haran (Winona Lake, IN: Eisenbrauns, 1996) 33-46.

Meynet, R., "I due decaloghi, legge di libertà (Es 20,2-17 & Dt 5,6-21)," *Gregorianum* 81 (2000) 659-692.

Milgrom, J., "מזוזה," *ThWAT*, IV, 802-804.

———, *Leviticus 1–16. Leviticus 17–22.* A New Translation with Introduction and Commentary (AncB 3-3A; New York, NY: Doubleday, 1991, 2000).

Mitchell, T.C., "The Meaning of the Noun *ḥtn* in the Old Testament," *VT* 19 (1969) 93-112.

Mollo, P., *The Motif of Generational Change.* A Literary and Lexicological Study (Lewiston, NY: Edwin Mellen Press, 2016).

Morales, L.M., *The Tabernacle Pre-Figured.* Cosmic Mountain Ideology in Genesis and Exodus (Biblical Tool and Studies 15; Leuven: Peeters, 2012).

Morgenstern, J., "The 'Bloody Husband' (?) Exod. 4:24-26 Once Again," *HUCA* 34 (1963) 35-70.

Müller, R., "The Sanctifying Divine Voice: Observations on the אני יהוה-formula in the Holiness Code," *Text, Time and Temple.* Literary, Historical and Ritual Studies in Leviticus (ed. F. Landy – al.) (HBM 64; Sheffield: Phoenix Press, 2015) 70-84.

Muraoka, T., *Emphatic Words and Structures in Biblical Hebrew* (Jerusalem: The Magnes Press, 1985).

———, "A Deuteronomistic Formula, 'שמר + עשה,'" *VT* 52 (2002) 548-550.

Nasuti, H., "Identity, Identification and Imitation: The Narrative Hermeneutics of Biblical Law," *Journal of Law and Religion* 4 (1986) 9-23.

Nepi, A., *Dal fondale alla ribalta.* I personaggi secondari nella Bibbia ebraica (Epifania della parola. Nuova serie; Bologna: EDB, 2015).

Niccacci, A., "Narrative Syntax of Exodus 19–24," *Narrative Syntax and Hebrew Bible.* Papers of the Tilburg Conference 1996 (ed. E. Van Wolde) (BiblInterpS 29; Leiden – New York, NY – Köln: Brill, 1997) 203-228.

———, *Sintassi del verbo ebraico nella prosa biblica classica.* Seconda edizione riveduta e ampliata, a cura di G. Geiger (Studii Biblici Franciscani Analecta 88; Milano: Terra Santa, ²2020).

Nicholson, E.W., "The Origin of the Tradition in Exodus xxiv 9-11," *VT* 26 (1976) 148-160.

———, *God and His People.* Covenant and Theology in the Old Testament (Oxford: Clarendon, 1986).

NIDITCH, S., *Oral World and Written World: Ancient Israelite Literature* (Library of Ancient Israel; Louisville, KY: Westminster John Knox, 1996).

NIELSEN, K., "Incense," *ABD*, III, 404-409.

NIHAN, CH., *From Priestly Torah to Pentateuch. A Study in the Composition of the Book of Leviticus* (FAT.2 25; Tübingen: Mohr Siebeck, 2007).

———, "L'analisi redazionale," *Manuale di esegesi dell'Antico Testamento* (ed. M. BAUKS – CH. NIHAN) (Bologna: EDB, 2010) 121-166.

———, "The Priestly Covenant, Its Reinterpretations, and the Composition of 'P,'" *The Strata of the Priestly Writings. Contemporary Debate and Future Directions* (ed. S. SCHECTMAN – J.S. BADEN) (AThANT 95; Zürich: Theologischer Verlag, 2009) 87-134.

———, "The Templization of Israel in Leviticus: Some Remarks on Blood Disposal and *Kipper* in Leviticus 4," *Text, Time and Temple. Literary, Historical and Ritual Studies in Leviticus* (ed. F. LANDY – *al.*) (HBM 64; Sheffield: Phoenix Press, 2015) 94-130.

NISKANEN, P., "Yhwh as Father, Redeemer, and Potter in Isaiah 63:7," *CBQ* 68 (2006) 397-407.

NODET, É., "Pâque, Azymes et théorie documentaire," *Revue Biblique* 114 (2007) 499-534.

NOEGEL, S.B., "The Significance of the Seventh Plague," *Biblica* 76 (1995) 532-539.

NOHRNBERG, J., *Like Unto Moses. The Constituting of an Interruption* (ISBL; Bloomington, IN: Indiana University Press, 1995).

OBARA, E.M., "Le azioni linguistiche: l'influsso del testo sul contesto," *Comunicazione e pragmatica nell'esegesi biblica* (ed. M. GRILLI – M. GUIDI – E.M. OBARA) (Lectio 10; Roma: G&B Press – Cinisello Balsamo: San Paolo, 2016) 83-117.

OTTO, E., "פסח," *ThWAT*, VI, 659-683.

———, *Das Deuteronomium. Politische Theologie und Rechtsreform in Juda und Assyrien* (BZAW 284; Berlin – New York, NY: De Gruyter, 1999).

OWCZAREK, S., *Die Vorstellung vom "Wohnen Gottes inmitten seines Volkes" in der Priesterschrift. Zur Heiligtumstheologie der priesterschriftlichen Grundschrift* (Frankfurt am Main: Peter Lang, 1998).

OWIREDU, C., *Blood and Life in the Old Testament* (Durham Thesis; Durham 2004), http://etheses.dur.ac.uk/3096/1/3096_1121.pdf?UkUDh:CyT.

PAPOLA, G., *L'alleanza di Moab. Studio esegetico teologico di Dt 28,69 – 30,20* (AnBib 174; Roma: PIB, 2008).

———, *Deuteronomio* (NVBTA 5; Cinisello Balsamo: San Paolo, 2011).

PARDES, I., *The Biography of Ancient Israel. National Narratives in the Bible* (Berkeley, CA: University of California, 2000).

PASSERI, S., "Rinunciare al sacrificio? Un tentativo di rilettura etico-teologica," *Il sacrificio* (Quaderni teologici del Seminario di Brescia 29; Brescia: Morcelliana, 2019) 165-188.

PAXIMADI, G., *E io dimorerò in mezzo a loro*. Composizione e interpretazione di Es 25–31 (Retorica biblica 8; Bologna: EDB, 2004).

PEELER, A.L.B., "Desiring God: The Blood of the Covenant in Exodus 24," *BBR* 23 (2013) 187-205.

PENNA, A., "Il sangue nell'Antico Testamento," *Sangue e Antropologia Biblica*. II (ed. F. VATTIONI) (CSSC 1; Roma: Pia Unione Preziosissimo Sangue, 1981) 379-402.

PÉREZ-HERNÁNDEZ, L., *Speech Acts in English*. From Research to Instruction and Textbook Development (Studies in English Language; Cambridge: University Press, 2021).

PERRY, M., "Literary Dynamics: How the Order of a Text Creates Its Meaning [With an Analysis of Faulkner's 'A Rose for Emily']," *Poetics Today* 1 (1979) 35-64.311-361.

PETTIGIANI, O., *"Ma io ricorderò la mia alleanza con te."* La procedura del *rîb* come chiave interpretativa di Ez 16 (AnBib 207; Roma: G&B Press, 2015).

—————, *Dio verrà certamente*. La preghiera perversa alla luce di Os 6,3 (Studi e ricerche. Sezione biblica; Assisi: Cittadella, 2017).

PETTINATO, G., "Il sangue nella letteratura sumerica," *Sangue e Antropologia Biblica*. I (ed. F. VATTIONI) (CSSC 1; Roma: Pia Unione Preziosissimo Sangue, 1981) 127-134.

POLA, T., *Die ursprüngliche Priesterschrift*. Beobachtungen zur Literarkritik und Traditionsgeschichte von P^g (WMANT 70; Neukirchen-Vluyn: Neukirchener Verlag, 1995).

POLAK, F.H., "Divine Names, Sociolinguistics and the Pragmatics of Pentateuchal Narrative," *Words, Ideas, Worlds*. Biblical Essays in Honour of Yairah Amit (ed. A. BRENNER – *al.*) (HBM 40; Sheffield: Phoenix Press, 2012) 159-178.

PROPP, W.H.C., "The Origins of Infant Circumcision in Israel," *HAR* 11 (1987) 355-370.

—————, "That Bloody Bridegroom (Exodus IV 24-6)," *VT* 43 (1993) 495-518.

RABINOWITZ, A., "Moses' Three Signs: Symbol and Augury," *JBQ* 20 (1992) 115-121.

VON RAD, G., *Die Priesterschrift im Hexateuch literarisch untersucht und teologisch gewertet* (BWANT 65; Stuttgart – Berlin: Kohlhammer, 1934).

RECALCATI, M., *Cosa resta del padre?* La paternità nell'epoca ipermoderna (Temi; Milano: Cortina, 2011).

—————, *Contro il sacrificio*. Al di là del fantasma sacrificale (Temi; Milano: Cortina, 2017).

REINHARTZ, A., *"Why Ask My Name?"* Anonymity and Identity in Biblical Narrative (New York, NY – Oxford: University Press, 1998).

REIS, P.T., "The Bridegroom of Blood: A New Reading," *Judaism* 40 (1991) 324-331.

RENDTORFF, R., *Die Gesetze in der Priesterschrift: eine gattungsgeschichte Untersuchung* (FRLANT 44; Göttingen: Vandenhoeck & Ruprecht, 1954).

—————, "Leviticus 16 als Mitte der Tora," *BiblInterp* 11 (2003) 252-258.

RICOEUR, P., *Philosophie de la volonté*. II. Finitude et culpabilité, vol. II: La symbolique du mal (Points, Essays 623; Paris: Cerf, 1960).

―――, "La structure symbolique de l'action," *Actes 14ème Conférence Internationale de sociologie des religions (Strasbourg 1977)* (Lille: CISR, 1977), 29-50.

―――, *Du texte à l'action*. Essai d'hermeneutique. II (Points, Essays 377; Paris: Cerf, 1986).

RIMMON-KENNAN, S., *Narrative Fiction. Contemporary Poetics* (London: Methuen, 1983).

RINGGREN, H., "נגש," *ThWAT*, V, 233-237.

ROBINSON, B.P., "Zipporah to the Rescue: A Contextual Study of Exodus 4,24-26," *VT* 36 (1986) 447-461.

―――, "The Theophany and Meal of Exodus 24," *SJOT* 25 (2011) 155-173.

ROCCA, P., *Gesù, messaggero del Signore*. Il cammino di Dio dall'esodo al vangelo di Marco (AnBib 213; Roma: G&B Press, 2016).

DE LA ROCHETERIE, J., *La symbologie des rêves*. Le corps humain (Paris: Imago, 1984).

RÖHRIG, M., "Gesetz und Erzählung in Num 15: Redaktionsgeschichtliche Überlegungen zum Sabbatsünder (Nm 15,32-36)," *ZAW* 131 (2019) 407-421.

RÖMER, TH., "De l'archaïque au subversif: le cas d'Exode 4/24-26," *ETR* 69 (1994) 1-12.

ROUX, J., "Modèle en haut ou modèle en bas? Lecture d'Exode 24–40," *Quand Dieu montre le modèle*. Interprétations et déclinaisons d'un motif biblique (ed. M.-L. CHAIEB – J. ROUX) (Paris: Honoré Champion, 2016) 37-70.

RUPRECHT, E., "Exodus 24,9-11 als Beispiel lebendiger Erzähltradition aus der Zeit des babylonischen Exils," *Werden und Wirken des Alten Testaments. Festschrift für Claus Westermann* (Göttingen: Vandenhoeck & Ruprecht, 1980) 139-173.

RYKEN, L., – WILHOIT, J.C., – LONGMAN III, T., *Le immagini bibliche*. Simboli, figure retoriche e temi letterari della Bibbia (I dizionari San Paolo; Cinisello Balsamo: San Paolo, 2006).

SAN MARTÍN JARA, A., "Moisés, Jetró y el narrador. Análisis narrativo de Ex 18," *EstBíb* 69 (2011) 265-288.

SARNA, N.M., *Genesis* (JPS.TC; Philadelphia, PA: JPS, 1989).

SCANDROGLIO, M., *Gioele e Amos in dialogo*. Inserzioni redazionali di collegamento e aperture interpretative (AnBib 193; Roma: G&B Press, 2011).

SCHENKER, A., "Les sacrifices d'alliance, Ex XXIV, 3-8, dans leur portée narrative et religieuse: Contribution à l'étude de la *berît* dans l'Ancien Testament," *Revue Biblique* 101 (1994) 481-494.

SCHMID, K., *Genesis and Moses Story*. Israel's Dual Origins in the Hebrew Bible (Winona Lake, IN: Eisenbrauns, 2010).

SCHMIDT, L., *Beobachtungen zu der Plagenerzählung in Exodus VII 14 – XI 10*, (Studia Biblica 4; Leiden: Brill, 1990).

————, "Israel un das Gesetz: Ex 19,3-8b und 24,3-8 als literarischer und theologischer Rahmen für das Bundesbuch," *ZAW* 113 (2001) 167-185.

SCHNIEDEWIND, W.M., *How the Bible Became a Book. The Textualization of Ancient Israel* (Cambridge: University Press, 2004).

SCHWARTZ, B.J., "The Prohibitions Concerning the "Eating" of Blood in Levitucus 17," *Priesthood and Cult in Ancient Israel* (ed. G.A. ANDERSON – S.M. OLYAN) (JSOT.S 124; Sheffield: Academic Press, 1991) 34-66.

SEARLE, J., *Speech Acts. An Essay in the Philosophy of Language* (Cambridge: University Press, 1969).

————, *Expressions and Meaning. Studies in the Theory of Speech Acts* (Cambridge: University Press, 1979).

SEEBASS, H., "נפש," *ThWAT*, V, 532-555.

SELMAN, M.J., "Sacrifice in the Ancient Near East," *Sacrifice in the Bible* (ed. R.T. BECKWITH – M.J. SELMAN) (Grand Rapids, MI: Paternoster, 1995), 88-104.

SERAFINI, F., *L'alleanza levitica. Studio della b^{e}rît di Dio con i sacerdoti leviti nell'Antico Testamento* (Studi e ricerche. Sezione biblica; Assisi: Cittadella, 2006).

SHAPIRA, H., «Making Sense of the Incense Altar: Location in Sacred Space and Text», *JBL* 142 (2023) 23-42.

SHAW, I. – NICHOLSON, P., *The British Museum Dictionary of Ancient Egypt* (Cairo: British Museum Press, 1995).

SHERWOOD, A., "The Mixed Moltitude in Exodus 12:38: Glorification, Creation, and Yhwh's Plunder of Israel and the Nations," *HBT* 34 (2012) 139-154.

SKA, J.L., "La sortie d'Égypte (Ex 7–14) dans le récit sacerdotal (P^g) et la tradition prophétique," *Biblica* 60 (1979) 191-215.

————, *Le passage de la mer: étude de la construction, du syle et de la symbolique d'Ex 14,1-31* (AnBib 109; Roma: PIB, 1986).

————, "Le repas de Ex 24,11," *Biblica* 74 (1993) 305-327.

————, "From History Writing to Library Building: The End of History and the Birth of the Book," *The Pentateuch as Torah* (ed. G.N. KNOPPERS – B.M. LEVINSON) (Winona Lake, IN: Penn State University Press, 2007) 145-169.

————, *"I nostri padri ci hanno raccontato." Introduzione all'analisi dei racconti dell'Antico Testamento* (Biblica; Bologna: EDB, 2012).

————, "Il libro dell'Esodo. Questioni fondamentali e questioni aperte," ID., *Il cantiere del Pentateuco. 1. Problemi di composizione e di interpretazione* (Biblica; Bologna: EDB, 2013) 113-135.

————, "'Misericordia voglio e non sacrifici' (Os 6,6). I sacrifici nell'Antico Testamento e la critica profetica," ID., *Il cantiere del Pentateuco. 2. Aspetti letterari e teologici* (Biblica; Bologna: EDB, 2013) 121-142.

————, *Le livre de l'Exode* (Mon ABC de la Bible; Paris: Cerf, 2021).

SKEHAN, P.W. – ULRICH, E. – SANDERSON, J.E., "4QpaleoGen-Exod^m," *Qumran Cave 4, IV, Paleo-Hebrew and Greek Biblical Manuscripts* (DJD 9; Oxford: Clarendon Press, 1992) 53-132.

SMITH, M.S., "'Your People Shall Be My People.' Family and Covenant in Ruth 1:16-17," *CBQ* 69 (2007) 242-258.

SOGGIN, J.A., "Il sangue nel racconto biblico delle origini," *Sangue e Antropologia Biblica.* II (ed. F. VATTIONI) (CSSC 1; Roma: Pia Unione Preziosissimo Sangue, 1981) 413-423.

SONNET, J.-P., "Le Sinaï dans l'événement de sa lecture: la dimension pragmatique de Exode 19–24," *NRTh* 111 (1989) 323-344.

———, "'Lorsque Moïse eut achevé d'écrire' (Dt 31,24). Une 'théorie narrative' de l'écriture dans le Pentateuque," *Revue de Sciences Religieuses* 90 (2002) 509-524.

———, "À la croisée des mondes. Aspects narratifs et théologiques du point de vue dans la Bible hébraïque," *Regards croisés sur la Bible.* Études sur le point de vue (LeDiv; Paris: Cerf, 2007) 75-100.

———, "L'analisi narrativa dei racconti biblici," *Manuale di esegesi dell'Antico Testamento* (ed. M. BAUKS – CH. NIHAN) (Testi e commenti; Bologna: EDB, 2010) 45-85.

———, "The Fifth Book of the Pentateuch. Deuteronomy in Its Narrative Dynamic," *Journal of Ancient Judaism* 3 (2012) 197-234.

———, "La Bible et l'histoire, la Bible et son histoire: une responsabilité critique," *Gregorianum* 94 (2013) 1-23.

———, "If-Plots in Deuteronomy," *VT* 63 (2013) 453-470.

———, *Generare è narrare* (Vita e pensiero. Sestante 33; Milano: Vita e pensiero, 2015).

———, "'In quel giorno,' 'fino a questo giorno.' L'arco della comunicazione narrativa nella Bibbia ebraica," *Rivista liturgica* 105 (2018) 13-41.

———, "Generare, perché? Una prospettiva biblica," *Anthropotes* 36 (2020) 139-190.

———, "'Today' in Deuteronomy: A Narrative Metalepsis," *Biblica* 101 (2020) 498-518.

SPERLING, S.D., "Blood," *ABD*, I, 761-763.

STAHL, N., *Law and Liminality in the Bible* (JSOT.S 202; Sheffield: Academic Press, 1995).

STAHLBERG, L.C., "Time, memory, Ritual, and Recital: Religion and Literature in Exodus 12," *Religion and Literature* 46 (2014) 75-94.

STEINS, G., "ציץ," *ThWAT*, VI, 1028-1034.

STENDBACH, F.J., "שלום," *ThWAT*, VIII, 13-46.

STERNBERG, M., "Proteus in Quotation-Land: Mimesis and the Forms of Reported Discourse," *Poetics Today* 3 (1982) 107-156.

STERNBERG, M., *The Poetics of Biblical Narrative*. Ideological Literature and the Drama of Reading (Bloomington, IN: Indiana University Press, 1987).

―――― , "Telling in Time (1): Chronology and Narrative Theory," *Poetics Today* 11 (1990) 901-948.

―――― , "Time and Space in Biblical (Hi)story Telling: The Grand Chronology," *The Book and the Text: The Bible Literary Theory* (ed. R. SCHWARZ) (Oxford: Basil Blackwell, 1990) 81-145.

―――― , *Hebrews Between Cultures*. Group Portraits and National Literature (Indiana Studies in Biblical Literature, Bloomington, IN: Indiana University Press, 1998).

―――― , "If-Plots Narrativity and the Law-Code," *Theorizing Narrativity* (ed. J. PIER – J.A. GARCÍA LANDA) (Berlin: De Gruyter, 2008) 29-107.

TAGLIAFERRO, E., "Sangue: area lessicale nell'epica greca arcaica," *Sangue e Antropologia Biblica*, I (ed. F. VATTIONI) (CSSC 1; Roma: Pia Unione Preziosissimo Sangue, 1981) 173-221.

TALBOT, M.M., «Tsipporah, Her Son, and the Bridegroom of Blood: Attending to the Bodies in Ex 4:24-26», *Religions* 8 (2017) 1-15.

TASKER, D.R., *Ancient Near Eastern Literature and the Hebrew Scriptures about the Fatherhood of God* (StBL 69; New York, NY: Peter Lang, 2004).

THAMES JR., J.T., "Keeping the Paschal Lamb: Exodus 12.6 and the Question of Sacrifice in the Passover-of-Egypt," *JSOT* 44 (2019) 3-18.

THOMPSON, S.E., "The Anointing of Officials in Ancient Egypt," *JNES* 53 (1994) 15-25.

TOV, E., "Textual Problems in the Description of Moses's Ascent to Mt. Sinai in Exodus 19, 24, 32 and 34," *Gottesschau – Gotteserkenntnis*. Studien zur Theologie der Septuaginta. I (ed. E.G. DAFNI) (WUNT 387; Tübingen: Mohr Siebeck, 2017) 3-18.

TSUMURA, D.T., *The First Book of Samuel* (NICOT; Grand Rapids, MI: Eerdmans, 2006).

TÜCK, J.-H., ed., *Die Beschneidung Jesu*. Was sie Juden und Christen heute bedeutet (Freiburg im Breisgau: Herder 2020).

TUCKER, P.N., "The Priestly *Grundschrift*: Source or Redaction? The Case of Exodus 12:12-13," *ZAW* 129 (2017) 205-219.

―――― , *The Holiness Composition in the Book of Exodus* (FAT.2 98; Tübingen: Mohr Siebeck, 2017).

VAN GENNEP, A., *The Rites of Passage* (London: Routledge and Kegan Paul, 1960).

VANNI, U., *Apocalisse di Giovanni*. Secondo volume (Commenti e studi biblici; Assisi: Cittadella, 2018).

VAN SETERS, J., "The Place of the Yahwist in the History of Passover and Massot," *ZAW* 95 (1983) 167-182.

VAN SETERS, J., *The Life of Moses*. The Yahwist as Historian in Exodus-Numbers (Kampen: Kok Pharos, 1994).

VATTIONI, F., "Sangue: vita o morte nella Bibbia," *Sangue e Antropologia Biblica*. II (ed. F. VATTIONI) (CSSC 1; Roma: Pia Unione Preziosissimo Sangue, 1981) 367-378.

———, "'Sposo di sangue' (Es 4,24-26)," *Sangue e Antropologia Biblica*. II (ed. F. VATTIONI) (CSSC 1; Roma: Pia Unione Preziosissimo Sangue, 1981) 477-496.

———, "Il sangue dell'alleanza (Es 24,8)," *Sangue e Antropologia Biblica*. II (ed. F. VATTIONI) (CSSC 1; Roma: Pia Unione Preziosissimo Sangue, 1981) 497-513.

VERMES, G., "Baptism and Jewish Exegesis. New Light from Ancient Sources," *NTS* 4 (1958) 307-319.

VERVENNE, M., "'The Blood is the Life and the Life is the Blood': Blood as Symbol of Life and Death in Biblical Tradition (Gen 9:4)," *Ritual and Sacrifice in the Ancient Near East* (ed. J. QUAEGEBEUR) (Orientalia lovaniensia analecta 55; Leuven: Peeters, 1993) 451-467.

VOGELS, W., "D'Égypte à Canaan: Un rite de passage," *Science et Esprit* 52 (2000) 21-35.

VRIEZEN, T.C., "The Exegesis of Exodus xxiv 9-11," *OTS* 17 (1972) 100-133.

WAGENAR, J.A., "Passover and the First Day of the Festival of Unleavened Bread in the Priestly Festival Calendar," *VT* 54 (2004) 250-268.

WALTERS, J.A., "Moses at the Lodging Place: The Devil in the Ambiguities," *Encounter* 63 (2002) 407-425.

WARDLAW, T.R., *Conceptualizing Words for "God" within the Pentateuch. A Cognitive-Semantic Investigation in Literary Context* (LHB.OTS 495; New York, NY: T&T Clark, 2008).

WATTS, J.M., ed., *Persia and Torah: The Theory of Imperial Authorization of the Pentateuch* (SBLSymS 17; Atlanta, GA: SBL, 2001).

WEIMAR, P., *Die Berufung des Mose: Literaturwissenschaftliche Analyse von Exodus 2,23–5,5* (OBO 32; Göttingen: Vandenhoeck & Ruprecht, 1980).

———, "Ex 12,1-14 un die priesterschriftliche Geschichtsdarstellung," *ZAW* 107 (1995) 196-214.

WEINFELD, M., "כבוד," *ThWAT*, IV, 24-40.

WEISSENRIEDER, A. – DOLLE, K., ed., *Körper und Verkörperung. Biblische Anthropologie im Kontext antiker Medizin und Philosophie. Ein Quellenbuch für die Septuaginta und das Neuen Testament* (FoSub 8; Berlin – Boston, MA: De Gruyter, 2019).

WELLHAUSEN, J., *Prolegomena to the History of Israel* (Atlanta, GA: Scholars Press, 1994).

———, *Die Composition des Hexateuchs und der historischen Bücher des Alten Testaments* (Berlin: Georg Reimer, 1899).

WELLS, B., "The Hated Wife in Deuteronomic Law," *VT* 60 (2010) 131-146.

WENHAM, G.J., *Genesis 1–15* (WBC 1; Grand Rapids, MI: Eerdmans, 1987).

———, "The Theology of Old Testament Sacrifice," *Sacrifice in the Bible* (ed. R.T. BECKWITH – M.J. SELMAN) (Grand Rapids, MI: Paternoster, 1995), 75-87.

WÉNIN, A., "Joseph interprète des rêves en prison (Genèse 40). Quelques fonctions de la repetition dans le récit biblique," *Mémoires d'écriture: hommage à Pierre Gibert s.j. offert par la Faculté de Théologie de Lyon* (Le livre et le rouleau 25; Bruxelles: Lessius, 2006) 259-273.

———, *Da Adamo ad Abramo o l'errare dell'uomo. Lettura narrativa e antropologica della Genesi. I. Gen 1,1–12,4* (Testi e commenti; Bologna: EDB, 2008).

———, *Abramo e l'educazione divina. Lettura narrativa e antropologica della Genesi. II. Gen 11,27–25,18* (Testi e commenti; Bologna: EDB, 2017).

WESTERMANN, C., "נפש," *THAT*, II, 71-96.

WEVERS, J.W., *Notes on the Greek Text of Exodus* (SBL.SCS 30; Atlanta, GA: Scholars Press, 1990).

WILKINSON, R.H., *The Complete Gods and Goddesses of Ancient Egypt* (London: Thames and Hudson, 2003).

WILLIAMSON, P.R., *Sealed with an Oath. Covenant in God's Unfolding Purpose* (NSBT 23; Downers Grove, IL: Apollos, 2007).

WILLIS, J.T., *Yahweh and Moses in Conflict. The Role of Exodus 4:24-26 in the Book of Exodus* (Bible in History; Bern: Lang, 2010).

WILSON, R., "The Hardening of Pharaoh's Heart," *CBQ* 41 (1979) 18-36.

WÖHRLE, J., "Abraham amidst the Nations. The Priestly Concept of Covenant and the Persian Imperial Ideology," *Covenant in the Persian Period. From Genesis to Chronicles* (ed. R.J. BAUTCH – G.N. KNOPPERS) (Winona Lake, IN: Eisenbrauns, 2015) 23-39.

WOOD, A., *Of Wings and Wheels. A Synthetic of the Biblical Cherubim* (BZAW 385; Berlin: De Gruyter, 2008).

WYATT, N., "Circumcision and Circumstance: Male Genital Mutilation in Ancient Israel and Ugarit," *JSOT* 33 (2009) 405-431.

WYCKOFF, E., "When Does Translation Become Exegesis? Exodus 24:9-11 in the Masoretic Text and the Septuagint," *CBQ* 74 (2012) 675-693.

XELLA, P., "Il sangue nel sistema mitologico e sacrificale siropalestinese durante il Tardo Bronzo," *Sangue e Antropologia Biblica*. I (ed. F. VATTIONI) (CSSC 1; Roma: Pia Unione Preziosissimo Sangue, 1981) 105-126.

YARDEN, L., *The Tree of Light. A Study of the Menorah* (London: NCROL, 1971).

ZAKOVITCH, Y., *"And You Shall Tell Your Son...." The Concept of Exodus in the Bible* (Jerusalem: The Magnes Press, 1991).

ZEELANDER, S., *Closure in Biblical Narrative* (Biblical Interpretation Series 111; Leiden – Boston, MA: Brill, 2012).

ZEHNDER, M., "Building on a Stone? Deuteronomy and Esarhaddon's Loyalty Oaths (Part 2): Some Additional Observations," *BBR* 19 (2009) 511-535.

ZERAFA, P.P., "Il significato del sangue nella Pasqua biblica," *Sangue e Antropologia Biblica.* II (ed. F. VATTIONI) (CSSC 1; Roma: Pia Unione Preziosissimo Sangue, 1981) 453-465.

ZIMMERLI, W., *Ezekiel 2*. A Commentary on the Book of the Prophet Ezekiel, Chapters 25–48 (Hermeneia; Philadelphia, PA: Fortress Press, 1983).

INDEX OF AUTHORS

Abba, R.: 209.
Abusch, Z.: 171.
Aeschylus: 174.
Agamben, G.: 173; 177; 236.
Ahn, J.J.: 125.
Ahuis, F.: 99; 109-110.
Albertz, R.: 110.
Alexander, E.S.: 117.
Allen, R.B.: 76; 78.
Alter, R.: 20.
Altmann, P.: 111.
Amit, Y.: 69.
Amzallag, N.: 198.
Anderson, B.A.: 48; 79.
Anderson, J.S.: 236.
Aristotle: 73.
Ashby, G.W.: 80.
Assmann, J.: 124.
Augustinus: 69.
Auneau, J.: 203.
Austin, J.L.: 161; 172; 233
Auzou, G.: 71.
Avigad, N.: 200.
Baker, D.L.: 162.
Baltzer, K.: 72.
Bar, S.: 143.
Barbiero, G.: 209.
Baroni, R.: 74.
Barr, J.: 30.
Barthélemy, D.: 115.
Barton, J.: 20.
Bartor, A.: 107-108; 114; 121.
Basile, G.: 145.
Baumann, A.: 183.
Bautch, R.J.: 71.

Beauchamp, P.: 20; 25; 167; 238;
Bell, C.: 166.
Benveniste, É.: 173; 236; 237.
Bernat, D.A.: 81; 84; 85.
Berner, C.: 48; 50; 110.
Berti, I.: 173; 174.
Beuken, W.: 78.
Beyerlin, W.: 151.
Bibb, B.D.: 29.
Bleiberg, E.: 142.
Blenkinsopp, J.: 85.
Blum, E.: 48; 50; 52; 99; 111; 151; 157; 170; 191; 195.
Boitani, P.: 73; 230; 232.
Bovati, P.: 20; 23; 31; 42; 66; 70; 72; 93; 94; 97; 121; 123; 133; 161; 169; 184; 200; 221; 234.
Bresciani, E.: 38; 40; 42; 63; 200.
Cagni, L.: 21; 35.
Callaham, S.N.: 100.
Canobbio, G.: 15.
Cassuto, U.: 25; 32; 37-40; 44; 45; 52; 55; 56; 58; 61; 72; 84; 88; 89; 90-92; 98-99; 101; 104-107; 112; 116; 119; 122; 132; 137-140; 157; 159; 163; 164; 166; 169; 179; 181; 186; 187; 198; 199; 201; 217; 219; 221; 222.
Charpin, D.: 34; 162; 166; 174; 176.
Chatman, S.: 61; 69.
Chavel, S.: 107; 163.
Cherney Jr., K.A.: 39; 42; 43; 77; 84.
Childs, B.S.: 32; 39-41; 43; 44; 48; 52; 55; 56; 59; 60; 74-76; 84; 88; 91; 98; 99; 106-111; 115; 125; 129; 133; 136; 142; 144; 150;

151; 156; 159; 162; 163; 178; 180; 181; 183; 195; 203.
Clifford, R.J.: 198.
Cocco, F.: 24.
Cohen, J.M.: 76.
Cohn, R.L.: 164.
Costacurta, B.: 61; 86.
Cox, D.G.: 66.
Currid, J.D.: 63.
Danove, P.L.: 60.
Daube, D.: 140.
David, R.: 36.
Davies, E.W.: 143.
Davies, G.I.: 36; 37; 41-44; 55; 56; 61; 65; 82; 99; 100; 102-109; 111-116; 119-122; 124-126; 129-134; 136; 138-140; 142-144; 146.
Davies, P.R.: 152.
Deiana, G.: 29; 193.
De Pury, A.: 80.
De Vaux, R.: 197.
De Zan, R.: 97.
Diesel, A.A.: 122.
Di Giovanbattista, F.: 27; 28.
Di Paolo, R.: 146.
Di Rocco, E.: 58; 73; 74; 96; 98; 230; 232.
Dohmen, Ch.: 33; 37; 63; 82; 88; 89; 107; 150; 159; 183; 222; 224.
Dolle, K.: 16; 21.
Dozeman, T.B.: 19; 39; 52; 84; 88; 112; 128; 129; 170.
Durand, G.: 78.
Durham, J.: 42; 52; 104; 107; 110; 111; 112; 150; 151; 156; 163; 166; 181; 198; 199; 203; 205; 221.
Eberhart, C.A.: 16; 22; 24; 117.
Eidem, J.: 174.
Evans, C.A.: 19.
Faraone, C.A.: 236.

Faust, A.: 84.
Feder, Y.: 18; 22; 24; 26-30.
Ficco, F.: 34; 59; 60; 70; 134.
Fink, L.H.: 75; 76.
Fisch, H.: 231.
Fischer, G.: 36; 51; 52; 58-60; 62; 90; 97-99; 103-105; 107; 119; 121; 131; 142; 143; 150; 157; 160; 163; 165; 169; 186; 221.
Fishbane, M.: 20.
Flavius Josephus: 38; 74; 89; 114; 116; 117.
Fleming, D.E.: 204.
Fokkelman, J.P.: 23.
Fornara, R.: 80; 81; 179; 180; 181; 185-188.
Forster, E.M.: 69.
Fowler, M.D.: 223.
Fox, M.V.: 85; 109.
Frei, P.: 152.
Frevel, Ch.: 195.
Friedman, R.E.: 39; 151.
Frolov, S.: 83; 84.
Füglister, N.: 16; 117.
Fuhs, H.-F.: 65.
Gelio, R.: 16; 22; 97; 117.
Genette, G.: 84.
Gertz, J.C.: 48-50; 65; 99; 109-111.
Gesundheit, S.: 109; 110; 114; 116; 124; 125; 128; 129; 136.
Gianto, A.: 56; 72; 122; 168.
Gilbert, P.: 66-69.
Gilbert, P.: 133; 175; 234.
Gilders, W.K.: 17; 24; 26-28; 30; 32; 35; 117; 128; 163; 165; 166; 170; 205-211; 224; 226; 227.
Girard, M.: 16; 21; 22; 25; 38; 188.
Girard, R.: 15.
Gitin, S.: 199; 223.
Giuntoli, F.: 139.
Goldingay, J.: 72; 84; 85.

Gorospe, A.E.: 47; 55; 56; 58; 74; 77; 82; 84; 86; 87.
Grant, D.E.: 188.
Greenberg, M.: 39; 73; 98.
Greenstein, E.L.: 72.
Gregorius Nyssenus: 171; 232.
Grenet, É.: 25.
Grossman, J.: 39; 66; 67.
Hahn, S.W.: 71.
Hamilton, V.P.: 23-25; 59; 112; 139.
Harrell, J.E.: 179.
Heimpel, W.: 34.
Hendel, R.: 164.
Hendrix, R.E.: 196.
Heschel, A.J.: 112.
Hieke, Th.: 37; 226.
Hinckley, R.: 183.
Hoffmeier, J.K.: 179; 198.
Hoffner, H.A.: 204.
Homer: 74; 96; 230.
Hort, G.: 42.
Houtman, C.: 36-41; 43; 44; 52; 55; 56; 58-63; 65; 71; 75; 87-90; 92; 94; 95; 98-100; 102-107; 112-117; 119; 122-129; 131; 137-139; 142; 144; 150; 151; 159; 178; 181; 183; 185; 187; 197; 198; 200; 202; 203; 205; 206; 210; 211; 213; 215; 218; 221; 224.
Howell, A.J.: 77; 83; 84.
Huffmon, H.B.: 221.
Invernizzi, L.: 39; 41; 51; 53; 57; 63; 64; 70-72; 79; 91; 95; 101; 122; 127; 219; 220.
Janowski, B.: 29.
Jay, N.: 17; 171.
Jelinkova Reymond, E.: 63.
Jenson, P.P.: 192; 193; 197; 216; 217; 223; 225.
Jeon, J.: 35; 49; 50; 53; 62; 67; 110; 111.
Johannes Chrysostomus: 15.
Johnstone, W.: 77.
Jones, E.: 25.
Joosten, J.: 63.
Kaiser, O.: 80.
Kalimi, I.: 126.
Kalluveettil, P.: 71.
Kedar-Kopfstein, B.: 16; 22.
Keel, O.: 198; 224.
Kissine, M.: 167.
Kitz, A.-M.: 200.
Knohl, I.: 17; 27; 109; 144; 195.
Knoppers, G.N.: 71.
Kornfeld, W.: 31.
Kosmala, H.: 77; 83.
Krause, J.J.: 34; 81; 84; 85.
Kutsch, E.: 83.
Lambert, W.G.: 21; 198.
Lange, A.: 117.
Lawrence, J.D.: 29.
Le Boulluec, A.: 75.
Lehane, T.J.: 77; 82.
Lemmelijn, B.: 39; 43; 102.
Leveen, A.: 57.
Levin, Ch.: 48-50; 52.
Lévinas, E.: 78; 169; 233.
Levine, B.: 30.
Levinson, B.: 20; 111.
Levy, T.E.: 19.
Lewis, T.J.: 171; 176; 188.
Lindström, F.: 77.
Lohr, J.N.: 19.
Longman III, T.: 38.
Loss, N.M.: 16.
Luciani, D.: 29; 57; 58; 74; 76; 77; 80-84; 86.
Lurker, M.: 116.
MacDonald, N.: 154.
Mahfouz, H.: 189.
Malul, M.: 77.

Manzi, F.: 16; 25.
Markl, D.: 36; 58-60; 62; 90; 97; 99; 103-105; 107; 119; 121; 125; 131; 142; 143; 150; 157; 160; 163; 165; 167; 169; 186; 221.
Marx, A.: 16; 27; 28; 115; 205-207; 209; 211; 213; 214; 227.
Mathews, D.: 143.
Matthews, V.H.: 119.
Mazzinghi, L.: 42; 201.
McAffee, M.: 66; 67.
McCarthy, D.J.: 24; 27; 85; 181; 204.
Meyers, C.: 155; 198; 222.
Meynet, R.: 121; 167.
Milgrom, J.: 16; 25-27; 117; 199; 201; 205-207; 211-215; 225.
Millard, A.R.: 21.
Mitchell, T.C.: 83.
Moberly, R.W.L.: 77.
Mollo, P.: 124.
Morales, L.M.: 163.
Morgenstern, J.: 74; 77; 84.
Müller, R.: 122.
Muraoka, T.: 125; 126; 128.
Nasuti, H.: 107.
Nepi, A.: 33; 36; 37; 39; 53; 57; 58; 60; 63; 65; 69; 71; 78; 81; 95; 96; 98; 100; 103; 104; 106; 112; 115; 124-126; 143; 150; 159; 161; 163; 164-166; 170; 180; 187; 198; 199; 201; 205; 211; 221.
Niccacci, A.: 141; 153; 155; 156; 165; 168; 178; 185.
Nicholson, E.W.: 71; 181.
Nicholson, P.: 42.
Nicklas, T.: 226.
Niditch, S.: 152.
Nielsen, K.: 224; 225.
Nihan, Ch.: 17; 20; 28; 31; 85; 193-196.

Niskanen, P.: 71.
Nodet, É.: 98.
Noegel, S.B.: 39.
Nohrnberg, J.: 58.
Noth, M.: 48; 52; 61; 181; 192-194; 201.
Obara, E.M.: 60; 160.
Origenes: 69.
Oswald, W.: 53; 62; 65; 88; 110; 112-116; 120; 121; 125; 130; 133; 140; 142; 143.
Otto, E.: 26; 85; 120.
Owczarek, S.: 193.
Owiredu, C.: 17; 18; 25; 29; 98; 117.
Papola, G.: 26; 128.
Pardes, I.: 34; 117.
Passeri, S.: 15.
Paximadi, G.: 196-201; 203; 204; 207; 209; 211; 212; 216; 222-223; 225.
Payne, D.: 72.
Peeler, A.L.B.: 162; 170.
Penna, A.: 16; 23; 29; 31.
Pérez-Hernández, L.: 60; 160.
Perry, M.: 53; 128.
Pettigiani, O.: 78; 118; 180.
Pettinato, G.: 21.
Philo of Alexandria: 74; 89; 101; 117; 177; 210.
Pola, T.: 192; 193; 195.
Polak, F.H.: 79.
Priotto, M.: 32; 33; 35-37; 40-42; 44-46; 52-55; 57-60; 63; 66; 68; 70; 71; 75; 79; 80; 88-90; 92; 95; 97; 99; 100; 102-107; 111-117; 119-124; 126; 130-132; 136; 138-140; 142; 143; 150; 159-161; 164; 166; 168; 169; 178; 181-184; 186; 187; 197-199; 201; 205; 209; 211; 212; 215; 217; 218; 221; 222; 224.

Propp, W.H.C.: 19; 32; 36; 39; 40; 41; 44; 45; 48-50; 52; 53; 55; 56; 58; 61; 62; 65; 67; 71; 75; 80; 82-84; 88; 92; 100; 104; 105; 107; 108; 110; 112-116; 118-120; 122; 123; 126-133; 135-137; 139-144; 146; 151; 156-159; 162-166; 168-171; 178; 179; 181; 182; 184-186; 188; 192; 195; 197-203; 205; 206; 208-210; 212-216; 219; 220; 222-224; 226.
Rabinowitz, A.: 38.
von Rad, G.: 195.
Recalcati, M.: 15; 135.
Reinhartz, A.: 57.
Reis, P.T.: 76.
Rendtorff, R.: 29; 109.
Ricoeur, P.: 161; 175; 176; 233.
Rimmon-Kennan, S.: 51.
Ringgren, H.: 31; 158.
Robinson, B.P.: 33; 74; 76; 78; 80; 82; 159; 181.
Rocca, P.: 150; 154; 155; 160; 185; 187; 188.
de la Rocheterie, J.: 21.
Rogerson, J.W.: 77.
Röhrig, M.: 107.
Römer, Th.: 74; 80; 81.
Roux, J.: 191.
Ruprecht, E.: 152; 170.
Ryken, L.: 38.
San Martín Jara, A.: 57.
Sanderson, J.E.: 222.
Sandevoir, P.: 75.
Sarna, N.M.: 25; 32; 33; 39; 62: 72; 150; 157; 159; 162; 187; 188; 204-206; 209; 211; 215; 218; 220; 222; 226.
Scandroglio, M.: 38.
Schenker, A.: 154; 170.
Schmid, K.: 48-50.

Schmidt, L.: 99; 106; 151.
Schmidt, W.H.: 50.
Schniedewind, W.M.: 151.
Schwartz, B.J.: 26; 27.
Searle, J.: 82; 161; 167; 175; 233.
Seebass, H.: 25.
Selman, M.J.: 199.
Serafini, F.: 35; 192.
Shapira, H.: 195.
Shaw, I.: 42.
Sherwood, A.: 143.
Ska, J.L.: 20; 39; 41; 52-54; 61; 66-69; 84; 91; 127; 151-153; 157; 159; 161; 167; 170; 180-182; 186; 207.
Skehan, P.W.: 222.
Smith, M.S.: 221.
Soggin, J.A.: 16; 25.
Sonnet, J.-P.: 18; 20; 45; 51; 61; 73; 80; 91; 107; 108; 124; 125; 134; 135; 145; 158; 160; 161; 167.
Sperling, S.D.: 16; 117.
Stahl, N.: 86.
Stahlberg, L.C.: 97; 135.
Steins, G.: 201.
Stendbach, F.J.: 59.
Sternberg, M.: 20; 36; 45; 51; 54; 60-62; 64; 66; 70; 73; 91; 102; 107.
Tagliaferro, E.: 21.
Talbot, M.M.: 81.
Talon, P.: 21.
Tasker, D.R.: 71.
Tertullianus: 37.
Thames Jr., J.T.: 113; 115; 116.
Thompson, S.E.: 204.
Thucydides: 173.
Tov, E.: 158; 177; 178; 182; 185.
Tsumura, D.T.: 59.
Tück, J.-H.: 15.
Tucker, P.N.: 109; 110.

Ulrich, E.: 222.
Utzschneider, H.: 53; 62; 65; 88; 110; 112-116; 120; 121; 125; 130; 133; 140; 142; 143.
Van Gennep, A.: 74; 164.
Vanni, U.: 42.
Van Seters, J.: 39; 43; 44, 48-50; 52; 53; 99; 110; 111; 142; 144; 152.
Vattioni, F.: 16.
Vermes, G.: 32; 74; 75; 82; 83.
Vervenne, M.: 25.
Vogels, W.: 164.
Vriezen, T.C.: 157.
Wagenar, J.A.: 125.
Walters, J.A.: 80.
Wardlaw, T.R.: 79.
Watts, J.M.: 152.
Weimar, P.: 48-50; 52; 109.
Weinfeld, M.: 186.
Weissenrieder, A.: 16; 21.
Wellhausen, J.: 48-50; 84; 99; 110; 195; 223.
Wells, B.: 70.
Wenham, G.J.: 22; 23; 199.
Wénin, A.: 25; 60; 85; 86; 139.
Westermann, C.: 25.
Wevers, J.W.: 218; 223.
Wilhoit, J.C.: 38.
Wilkinson, R.H.: 38.
Williams, K.F.: 179.
Williamson, P.R.: 173; 175; 177; 238.
Willis, J.T.: 76.
Wilson, R.: 66; 67.
Wöhrle, J.: 85.
Wood, A.: 198.
Wyatt, N.: 84.
Wyckoff, E.: 177.
Xella, P.: 27.
Xenophon: 174.
Yarden, L.: 198.

Zakovitch, Y.: 113.
Zeelander, S.: 54.
Zehnder, M.: 173.
Zerafa, P.P.: 16; 117.
Zimmerli, W.: 117.

Finito di stampare presso Printbee Noventa Padovana - PD nel mese di giugno 2023